Orogenic Processes
in the Alpine Collision Zone

Guest Editors:
Nikolaus Froitzheim
Stefan M. Schmid

Swiss Academy of Sciences
Akademie der Naturwissenschaften
Accademia di scienze naturali
Académie des sciences naturelles

Fonds National Suisse
Schweizerischer Nationalfonds
Fondo Nazionale Svizzero
Swiss National Science Foundation

Swiss Journal of Geosciences
formerly Eclogae Geologicae Helvetiae

Vol. 101
Supplement 1
2008
Pages S1–S310
ISSN 1661-8726

Journal of the
Swiss Geological Society
Swiss Society of Mineralogy and Petrology
Swiss Palaeontological Society
and affiliated organisations and specialist groups:

Platform Geosciences of the Swiss Academy of Sciences
Swiss Geological Survey
Swiss Tectonic Studies Group
SwissSed
Swiss Group of Geophysicists

Contents

1661-8726/08/01S001-3
DOI 10.1007/s00015-008-1296-7
Birkhäuser Verlag, Basel, 2008

Swiss J. Geosci. 101 (2008) Supplement 1, S1–S3

Orogenic processes in the Alpine collision zone

Preface to the Special Issue of the Swiss Journal of Geosciences devoted to "Orogenic processes in the Alpine collision zone"

NIKOLAUS FROITZHEIM[1] & STEFAN M. SCHMID[2]
INVITED EDITORS

The papers in this issue further elaborate themes that were presented at the 8th Workshop on Alpine Geological Studies held in Davos on 10–12 October 2007. Ever since a first meeting in Grenoble that took place in 1993, a series of successor Workshops on Alpine Geological Studies has continued in a two-year rhythm: 1995 Basel, 1997 Biella – Oropa, 1999 Tübingen, 2001 Obergurgl, 2003 Sopron, 2005 Opatija. Gradually, the study area encompassed by these meetings was enlarged beyond the Alps in the strict sense. More and more contributions touched on other parts of the Alpine Collision Zone such as the Carpathians, Dinarides or Apennines. Thereby an increasing multitude of various processes leading to collisional mountain building in general could be discussed. Consequently, this present volume captures a part of that same diversity. The 8th workshop in 2007 took place with 117 participants from 14 countries.

Two excursions to the high Alpine area around Davos were carried out before and after the workshop, favoured by clear, warm autumn weather prevailing during the entire conference. The pre-workshop field trip on October 9, guided by Daniel Bernoulli (Basel) and Othmar Müntener (Lausanne), was devoted to the ophiolites at Totalp, where lithologies and structures testify for the exhumation of subcontinental mantle rocks during the opening of the Piemont-Ligurian Ocean in Jurassic times. The second field trip of October 13 was a hike from St. Antönien near Klosters to the Tilisuna area, across the Alpine nappe stack at the Penninic-Austroalpine boundary. It was guided by Thorsten Nagel and Niko Froitzheim (Bonn).

During the three lecture days, 50 oral and 68 poster contributions were presented, organized into the following topical sessions: (1) Alpine oceans: Rifting, break-up, spreading, and paleogeography, (2) Deep structure, lithospheric strength, and mantle dynamics, (3) From subduction to collision, (4) Tectonic and metamorphic processes and the role of HP/UHP metamorphic rocks, (5) Orogenic curvature and kinematics of the Alps-Carpathians-Dinarides, (6) Foreland and hinterland basins: What controls their evolution?, (7) From Neogene to present-day Alps: Neotectonics, brittle tectonics, big tunnels, and finally, (8) Coupling of climate, uplift, erosion, and topography.

Keynote lectures opening the sessions were given by Thorsten Nagel (Bonn), Edi Kissling (Zürich), Onno Oncken (Potsdam), Alfons Berger (Bern), Liviu Matenco (Amsterdam), Francois Roure (Paris), and Sean Willett (Zürich). Ben Reinhardt (Dornach) offered an introduction to the geologic results of and problems encountered along the construction of the Lötschberg and Gotthard base tunnels in a public lecture entitled "Lange Tunnels durch die Alpen: NEAT aus der Sicht des Geologen".

The presentations and discussions during the workshop reflected the recent development of Alpine geology. An increasing part of Alpine geological research deals with the Neogene to recent evolution of the earth's surface and with the interplay of climate and tectonics. For this field of research, the Alps are ideally suited as a natural laboratory because of their limited size, well-constrained boundary conditions, and high density of data. Several presentations dealt with the tectonic continuations of the Alps into the Carpathians and Dinarides and as far as the Balkan Peninsula. The intense, border-crossing research in these areas is proceeding towards a stage where a synoptic picture of the tectonic evolution emerges. Important new findings were also presented by several groups working on bio- and lithostratigraphy in basin sediments around the Alpine orogen. It was remarkable that this particular Alpine Workshop finally managed to bring together geologists working on the tectono-metamorphic history of the Alps with those working on the stratigraphy of adjacent foreland basins.

This volume captures some highlights of the 2007 Davos Workshop. 16 articles cover a multitude of Alpine-type working areas and processes active in collisional mountain building. We wish to acknowledge the financial contributions of the Schweizerische Nationalfonds and the Swiss Academy of

[1] Steinmann-Institut für Geologie Mineralogie und Paläontologie, Universität Bonn, Meckenheimer Allee 171, 53115 Bonn, Germany.
E-mail: niko.froitzheimuni-bonn.de
[2] Geologisch-Paläontologisches Institut, Universität Basel, Bernoullistr. 32, 4056 Basel, Switzerland. E-mail: stefan.schmid@unibas.ch

Sciences towards the organization of the 8th Workshop on Alpine Geological Studies in Davos and the publication of this series of articles. We are also grateful to the editors of the Swiss Journal of Geosciences who agreed to host this Special Issue, particularly Stefan Bucher who assisted the technical aspects of editing, and the three supporting Swiss Earth Science Societies (Swiss Geological Society, Swiss Palaeontological Society and Swiss Society of Mineralogy and Petrology) who, via their budgets, also substantially contributed towards the production costs.

A first series of papers focuses on the role of **sedimentary processes during orogeny**. The evolution of foreland and internal basins of orogens in general is discussed in a review paper by **Roure**, mostly by discussing industrial seismic data from basins in the Alpine-Mediterranean area and in America. This illustrates the influence of inherited structures such as platform-basin transitions and former rifts which are inverted during collisional deformation. In the case of intramontane basins, the paper illustrates the influence of tectonic escape, strain partitioning during oblique convergence, and post-orogenic collapse. Another and more specific study focussing on synorogenic basins is that of **Mikes et al.** who provide a provenance study for the Bosnian Flysch in the Dinarides, forming an intensely folded stack of Upper Jurassic to Cretaceous mixed carbonate and siliciclastic sediments sandwiched between the Adriatic Carbonate Platform and the Dinaride Ophiolite Zone. The authors conclude that Middle Jurassic intraoceanic subduction of the Neotethys was shortly followed by exhumation of the overriding oceanic plate. Following mid-Cretaceous deformation and thermal overprint, the depocentre of the Bosnian Flysch is reported to have migrated further towards SW, receiving increasing amounts of redeposited carbonate detritus from the Adriatic Carbonate Platform margin. The third contribution focussing on the importance of understanding sedimentary processes during orogeny is that of **Ortner et al.** who analyse and discuss Late Jurassic sediments and structures in the western part of the Northern Calcareous Alps. By doing so the authors discuss two competing processes that occurred along the same continental margin: (1) Chaotic breccia deposition near a major normal fault scarp that is part of a pull-apart basin associated with strike slip movements in their working area and (2) synchronous gravitational emplacement of exotic slides and breccias (Hallstatt mélange), triggered by Late Jurassic orogeny reported for the eastern part of the Northern Calcareous Alps. **Veselá & Lammerer** show that the geometry and sediment type of Latest Carboniferous to Triassic rift basins in the Western Tauern Window can be reconstructed in spite of strong Alpine deformation and metamorphism. This interdisciplinary study introduces formation names, presents U-Pb zircon data from meta-volcanic rocks that help constraining the age of the sediments, and shows how the sedimentary basins predetermined the geometry of Alpine thrusts.

The discussion on the mechanism of the **formation of the arc of the Western Alps** is enriched by a tectonic synthesis by **Dumont et al.** who present and review data on the multi-stage orogeny in the French external Alps (Dauphiné Zone). The authors show that Eocene or older deformation was followed by N to NW-oriented basement thrusting so far only reported for more internal parts of the Western Alps. The classical main WNW-directed compression represents a third event, indicating a rapid transition from northward-directed Alpine collision to the onset of westward thrusting that formed the Western Alpine arc. The fourth event is coeval with final uplift of the external massifs, producing strike-slip faulting and local rotations and significantly redeforming earlier structures.

A series of papers focuses on the **metamorphic evolution of the Alps**. Starting with the Palaeozoic evolution of migmatites from the Ötztal nappe, **Thöny et al.** using textural relations and microprobe analysis were able to separate pre-Variscan, Variscan, and Eo-Alpine parageneses and determined P-T conditions for these stages. U-Pb dating of monazites from the leucosomes on the microprobe yielded results in favour of an Ordovician-Silurian (441 ± 18 Ma) age of migmatization. The article of **Wiederkehr et al**. represents a combined structural and metamorphic study of Bündnerschiefer series at the front of the Adula Nappe in the eastern part of the Lepontine dome of the Central Alps. The authors demonstrate that these rocks experienced a Tertiary-age pressure-dominated metamorphism, characterized by the occurrence of Fe-Mg carpholite. This was followed, after isothermal decompression, by isobaric heating, leading to the temperature-dominated Lepontine metamorphic event. The authors discuss the heat source of Barrow-type metamorphism, arguing that such heating was caused by radioactive decay of accreted continental material. **Rütti et al.** report on the structural and metamorphic evolution of the Leventina Nappe that represents one of the lowermost exposed structural units of the Alpine nappe edifice and is also part of the Lepontine dome. However, maximum metamorphic pressure conditions did not exceed 8 and 10 kbar for the northern and southern parts of the nappe, respectively. These pressures, and temperatures between 550 °C and 650 °C, are interpreted to be related to the under thrusting of the thinned European margin into the crustal accretionary prism that initiated during late Eocene to early Oligocene times.

Another series of papers significantly contributes towards elucidating **the timing of orogeny based on radiometric methods and the manifold geodynamical consequences**. In the Western Alps, and using Lu-Hf geochronology, **Herwartz et al.** determined ages of ca. 42 and ca. 45 Ma for prograde garnet growth in eclogites from the Balma ophiolite unit on the southern side of the Monte Rosa massif, i.e. from a much debated piece of the internal Western Alps, derived from lithologies that record a former continent-ocean transition. The authors present isotope and trace element data in favour of a MORB character of the protoliths and they discuss the far-reaching paleotectonic implications of these data. **Kurz et al.** combine $^{40}Ar/^{39}Ar$ dating with micro-structural analyses for dating different stages of the eclogite-facies evolution in a part of the basement exposed in the Tauern Window of the Eastern Alps. They discuss isotopic signature and micro-tectonic processes that took place after peak pressure conditions were reached at 39 Ma and during exhumation until some 31 Ma ago. By showing how deformation during exhumation results in the resetting of the Ar isotopic system, they contribute towards a better understanding of the methods that may lead to unravelling the timing of subduction processes in the Alps.

Since some time, it is widely recognized that **thermochronological data on exhumation processes** enrich our thinking on driving forces and dating of exhumation. The article by **Luth & Willingshofer** presents a set of maps displaying the cooling history of the Eastern Alps, based on the existing thermochronological database. These maps highlight the diachronous exhumation of deep structural levels in a framework of combined east-west extension and north-south shortening in the eastern Alps. **Danišík et al.** show, using low-temperature thermochronological methods, that a Cretaceous-age granite in the Western Carpathians records a distinct thermal event during the Middle to early Late Miocene, likely related to mantle up welling, magmatic activity, and increased heat flow in the Carpathian-Pannonian region. Thereby the authors show that the Miocene thermal event had a regional character and affected large parts of the basement outcrops in the Western Carpathians north of the Pannonian basin.

Two articles address processes related to **orogen-parallel extension in the Eastern Alps and the formation of the Pannonian basin.** A fault slip analysis on the Neogene evolution of the Austroalpine basement units east of the Tauern Window in the context of orogen-parallel lateral extrusion and focussing on the Koralm basement is presented by **Pischinger et al.** Together with the stratigraphic evolution of the Styrian and Lavanttal Basins and related subsidence histories the authors reconstruct the late tectonic evolution and final exhumation of a part of the Eastern Alps that is adjacent to the Pannonian basin. **Fodor et al.** provide new up-to-date U-Pb radiometric data that imply an Early Miocene crystallization age for the Pohorje pluton located at the southeastern margin of the Eastern Alps, confirming interpretations based on K-Ar geochronology. The new data imply a temporal coincidence with magmatism in the Pannonian Basin system. K-Ar ages and zircon fission track data combined with structural investigations indicate rapid cooling of the pluton, interpreted as related to lateral extrusion of the Eastern Alps and/or back-arc rifting in the Pannonian Basin

Finally, two articles are devoted to the **tectonic evolution of the Carpathians, Dinarides and Balkanides,** i.e. orogens in Southeastern Europe that bridge an important gap between the Alps and the Hellenides and their extension into Turkey. **Ustaszewski et al.** present a restoration of the major tectonic units of the Alpine-Carpathian-Dinaridic system for Early Miocene times. They show how severely the late-stage tectonic evolution has modified the configuration that existed at the end of collision. The mid-Miocene to recent evolution is dominated by block rotations that resulted from the combined effect of ongoing indentation of Adria and subduction retreat in the East Carpathians. The authors further present vertical and horizontal seismic tomography sections of the mantle under the area, and discuss the relations between lithospheric slabs imaged under Alps and Dinarides. **Tückmantel et al.** provide new structural data on a yet badly understood part of the Balkanides in the northwestern part of the Rhodope metamorphic province. Alpine, amphibolite-facies gneisses in the area of Rila valley in western Bulgaria were exhumed by several distinct phases of extensional deformation. The most pronounced faulting occurred in the Eocene to Early Oligocene when a major normal fault exhumed rocks from the ductile middle crust to the surface, as evidenced by syn-rift deposits in the hanging wall.

1661-8726/08/01S005-25
DOI 10.1007/s00015-008-1285-x
Birkhäuser Verlag, Basel 2008

Swiss J. Geosci. 101 (2008) Supplement 1, S5–S29

Foreland and Hinterland basins: what controls their evolution?

FRANÇOIS ROURE

Key words: Foreland, hinterland, intramontane basins, inversion tectonics

ABSTRACT

Compressional systems are usually characterized by a positive topography above the sea level, which is continuously modified by the conjugate effects of tectonic contraction or post-orogenic collapse, thermo-mechanical processes in the deep lithosphere and asthenosphere, but also by climate and other surface processes influencing erosion rates.

Different types of sedimentary basins can develop in close association with orogens, either in the foreland or in the hinterland. Being progressively filled by erosional products of adjacent uplifted domains, these basins provide a continuous sedimentary record of surficial, crustal and lithospheric deformation at and near plate boundaries.

Selected integrated basin-scale studies in the Circum-Mediterranean thrust belts and basins, in Pakistan and the Americas, are used here to document the effects of structures inherited from former orogens, rifts and passive margins, active tectonics and mantle dynamics on the development and long term evolution of synorogenic basins.

Introduction

Flexure of the oceanic lithosphere as a response to the tectonic loading by accretionary wedges and slab pull has been well described in the vicinity of active subduction zones (Karig 1974; Karig & Sharman 1975; Leggett 1982; Watts et al. 1982; von Huene 1986; von Huene & Sholl 1991). Intra-oceanic flexural moats developing as a response to the load of intraplate volcanoes have been carefully studied in Hawai (Watts et al. 1980). An extensive literature deals with the significance of foreland flexural basins, which are known to develop on continental lithosphere as a response to the load of both collisional and Cordillera-type orogens. Thermo-mechanical controls, associated with the thermal state and layered composition of the lithosphere and accounting for spatial and temporal changes observed in the width and depth of foreland basins, have also been widely studied (Beaumont 1981; Royden & Karner 1984; Kusznir & Park 1984; Kusznir & Karner 1985; Kruse & Royden 1987; 1994). Although most erosional products sourced by the orogens are likely to be trapped in adjacent foreland basins, recording successively marine and continental sedimentation, differential uplift and subsidence associated either with a negative inversion of former thrusts (post-orogenic collapse) or with the development of back-thrusts can also account for

dominantly isolated, discontinuous depocenters in the hinterland. Ultimately, a part of synorogenic/synkinematic sediments does not reach the autochthonous foreland, being trapped in thrust-top or piggyback basins (Ori & Friend 1984; DeCelles & Giles 1996).

This study is focused on the control exerted on foreland basin evolution by pre-existing structures such as low-angle faults inherited from former orogens and high-angle faults inherited from the former rift architecture, as well as by lateral thickness and facies variations which are likely to occur in the post-rift sequences of former passive margins.

We will describe how active tectonics can induce the development of thrust-top and hinterland basins, and how post-orogenic mantle dynamics can impact the uplift and erosional history of the orogen itself, but also of adjacent foreland basins.

1 Lithological controls of passive margin series on the localization of decollement levels

Whereas the North American Cordillera and especially the Canadian Rocky Mountains show little evidence of major lateral thickness and facies variations in the pre-orogenic series, the current architecture of Circum-Mediterranean and Alpine foothills is dominantly controlled by the Tethyan rifting which

Institut Français du Pétrole, 1–4 Avenue de Bois-Préau, F-92852 Rueil-Malmaison Cedex & Department of Tectonics, VU-Amsterdam, the Netherlands.
E-mail: Francois.Roure@ifp.fr

operated in Triassic and Liassic times (Bernoulli & Lemoine 1980; Bernoulli 1981; Lemoine et al. 1981, 2000).

For instance, the occurrence or lack of Triassic salt have a strong influence on the development of transfer zones in the Jura Mountains and Sub-Alpine Chains (Guéllec et al. 1990; Philippe 1994; Philippe et al. 1996).

The marked contrasts in structural style among Mediterranean and Alpine thrust belts derived from the deformation of former passive margins of the Tethys are clearly related to the distribution of Cretaceous platform to basin transitions, as well as passive margins versus continental series. Seemingly, Mexican cordilleras such as the Zongolica and Sierra Madre thrust belts are also derived from the reactivation of Jurassic rift margins with wide Cretaceous prograding platforms, and share many similarities with Tethyan thrust belts from the other side of the Central Atlantic (Ortuño et al. 2003).

1.1 Architecture of platform to basin transitions in Albania

The Ionian Basin in Southern Albania is made up of dominantly Mesozoic thin-skinned tectonic units which have been detached from the infra-Triassic substratum along the basal Triassic salt. These thrust units involve relatively thin (about 1 km-thick) Mesozoic series of basinal affinities. Each unit is made up of Toarcian blackshales, Middle Jurassic cherts, Late Cretaceous carbonate turbidites and Eocene Scaglia-type fine-grained pelagic limestones, which are overlain by Oligocene siliciclastic synflexural series (Roure et al. 1995; 2004; Carminati et al. 2004). Farther north, these Mesozoic basinal series still belong to the autochthon in the Peri-Adriatic Depression, the main décollement of the northern Albanian foothills being located within Cenozoic series.

Up to 2 to 3 km-thick prograding Cretaceous platforms were built on both sides of the Mesozoic Ionian-Adriatic basin, accounting for the shallow water carbonate facies of the Sazani-pre-Apulian Platform domain in the west, and of the Kruja zone in the east (Roure et al. 1995; 2004).

Due to rheology contrasts between the massive platform carbonates and finely layered basinal series, but also between Mesozoic carbonates and siliciclastic Oligocene flysch, triangle zones have developed along these paleogeographic boundaries, accounting in both cases for the development of a regional backthrust and deeply buried duplexes (Fig. 1a, b).

In the northern transect (Fig. 1a), the Kruja units, made up of Cretaceous platform carbonates and Oligocene flysch, have been thrust over the siliciclastic series of the Peri-Adriatic Depression during a pre-Messinian thrusting episode. Subsequent deformation during the Pliocene involved the tectonic accretion of deeper platform duplexes, deformation propagating forelandward along a blind thrust, antithetic from a shallower east-verging backthrust.

In the southern transects (Fig. 1b), the foreland propagation of the frontal thrust is only visible in the northwestern side of the Sazani promontory (section 1), whereas farther south, it accounts for a west-verging blind thrust, propagating in the opposite direction from a shallower east-verging conjugate backthrust.

Both areas are yet underexplored, although they are likely to host hydrocarbon reserves in slope breccias near the transition between the Kruja and Sazani platform domains (known for their good reservoirs) and the Ionian and Peri-Adriatic basins (likely to have a good source rock potential; Roure et al. 1995, 2004).

1.2 The architecture of platforms to basin transitions in the French Alpine foreland

In southeastern France, triangle zones have also developed along the northern border of the Provençal Platform, accounting for the large backthrusts of the La Lance and Ventoux-Lure carbonate platforms, which are made up of Urgonian reefal facies and are widely thrust over coeval basinal facies of the Vocontian Trough (Roure et al. 1992, 1994a; Roure & Colletta 1996; Fig. 2).

La Lance structure

The deep architecture of the La Lance structure is related to the reactivation of a former Liassic basement-involving high-angle fault. A basal décollement is located in the Triassic salt series in the Vocontian Basin in the north, but in Jurassic blackshales in the south. A basement short-cut is evidenced at depth, with a south-verging reverse fault transporting passively the crest of the former Jurassic tilted-block (Fig. 2a). A blind antithetic north-verging backthrust has detached the Urgonian (Aptian) platform series, connecting the intra-Jurassic décollement in the south with a shallower décollement in the north, which propagated within the Lower Cretaceous basinal series of the Vocontian domain as far north as the Saou syncline.

Lateral thickness and facies variations of the Barremian-Aptian series can be clearly recognised on the seismic profiles, where the transition between thick prograding Urgonian series and thinner, isopachous basinal sequences can be picked very accurately.

Actually, regional-scale basinal inversion is also evidenced by the current position of the top Jurassic horizon, which is higher within the currently inverted basinal domain in the north, than in the ajdacent paleo-horst where Urgonian carbonates have been deposited in the south.

Although the seismic profile crossing the La Lance structure is of average quality at depth, the overall architecture of this structure fits quite well with the geometry expected for such localization of thin-skinned tectonics and wedging, associated with the reactivation of a deeper basement-involving fault, as predicted by analogue models (Fig. 2b).

The Ventoux-Lure structure

The Ventoux-Lure is a west-trending platformal unit which constitutes the eastern prolongation of the La Lance thrust sheet. As the latter, it is thrust northward over coeval basinal facies of the Vocontian Basin (Fig. 2). Although the surface ar-

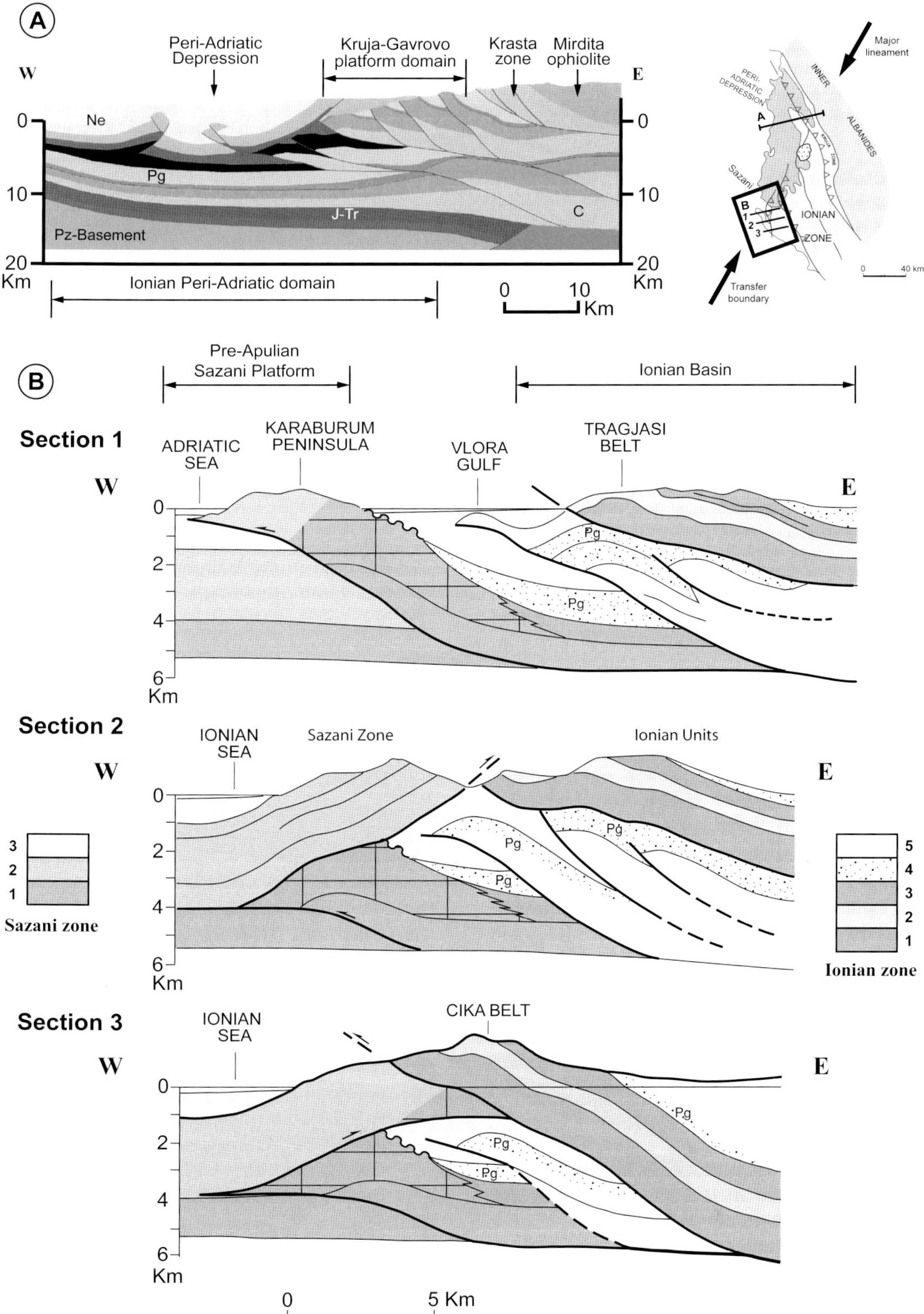

Fig. 1. Thin-skinned deformations associated with Mesozoic platform to basin transitions in Albania: a) Kruja duplexes and associated backthrust developing at the transition between the Kruja-Gavrovo Platform and the Ionian-Peri-Adriatic Basin; b) Serial sections in the Vlora area, outlining the lateral changes in thrust architecture at the transition between the Sazani-Pre-Apulian Platform and the Ionian Basin, with a progressive stacking of Ionian duplexes and development of a triangle zone. The Sazani units are made up of Mezozoic platform carbonates (2) and Neogene siliciclastic series (2). The Ionian units are detached along the Triassic salt (1), and comprise Jurassic (2) and Cretaceous (3) to Eocene basinal series, overlain by Oligocene turbidites (4) and Neogene clastics (5).

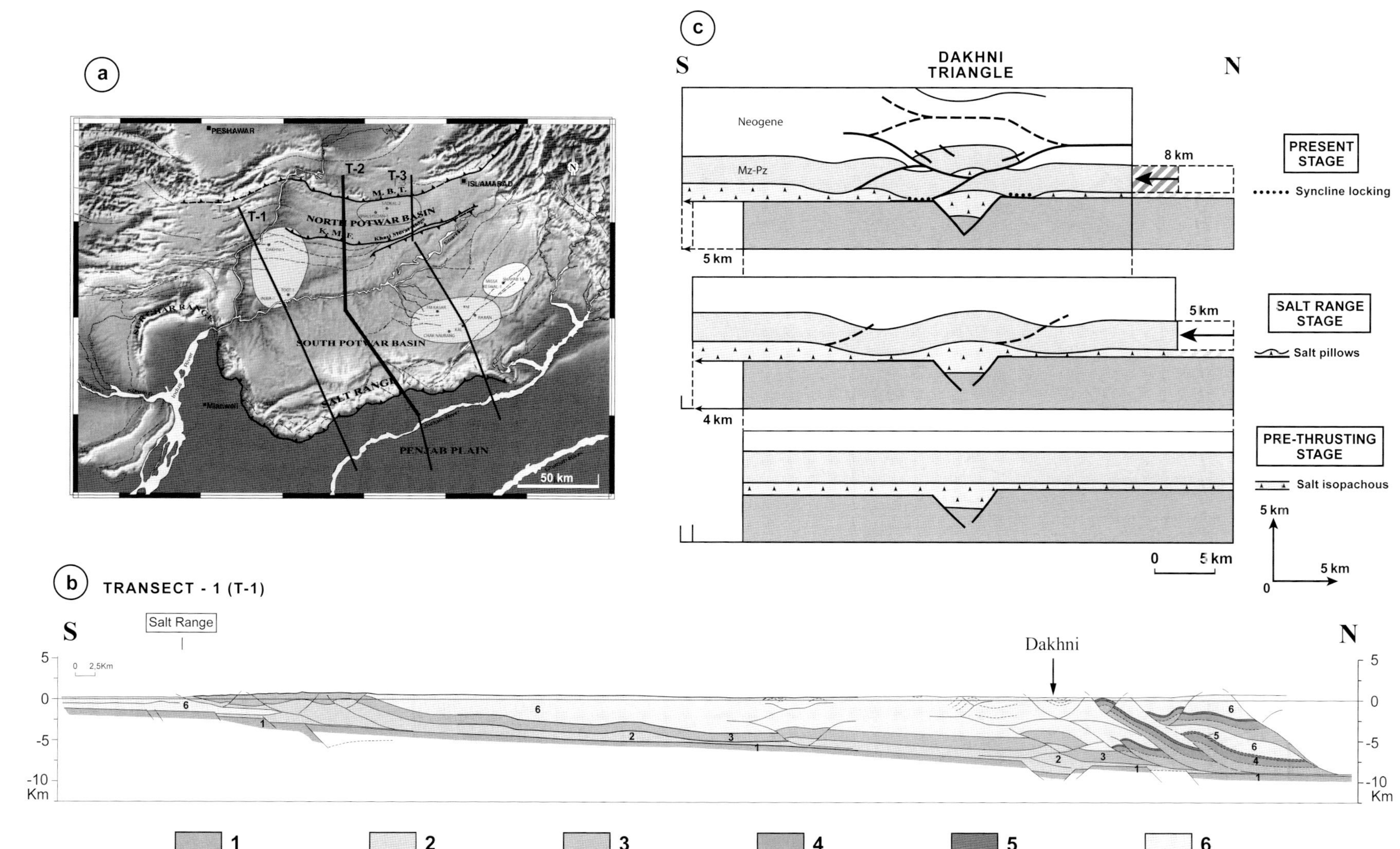

Fig. 5. Infra-salt Precambrian basins and localisation of Neogene deformation in the Potwar Basin (Pakistan): a) Location map; b) Regional section across the western part of the Potwar Basin from the Salt Range in the south to the Dakhni field in the north (modified after Grelaud et al. 2002); c) Evolutionary diagrams of the Dakhni structure, accounting for a first episode of thin-skinned tectonics, coeval with the thrust emplacement of the Salt Range frontal structure, and subsequent transpressional reactivation of basement structures (modified after Jaswal et al. 2004). Mz-Pz: Mesozoic and Paleozoic. The lithostratigraphy of the Salt Range-Potwar Basin comprises an infra-salt Precambrian substratum (1), Eo-Cambrian salt series (2), a Paleozoic sandstone and carbonate platform (3), Mesozoic to Eocene series (4 and 5), and Neogene Siwalik siliciclastic deposits (6).

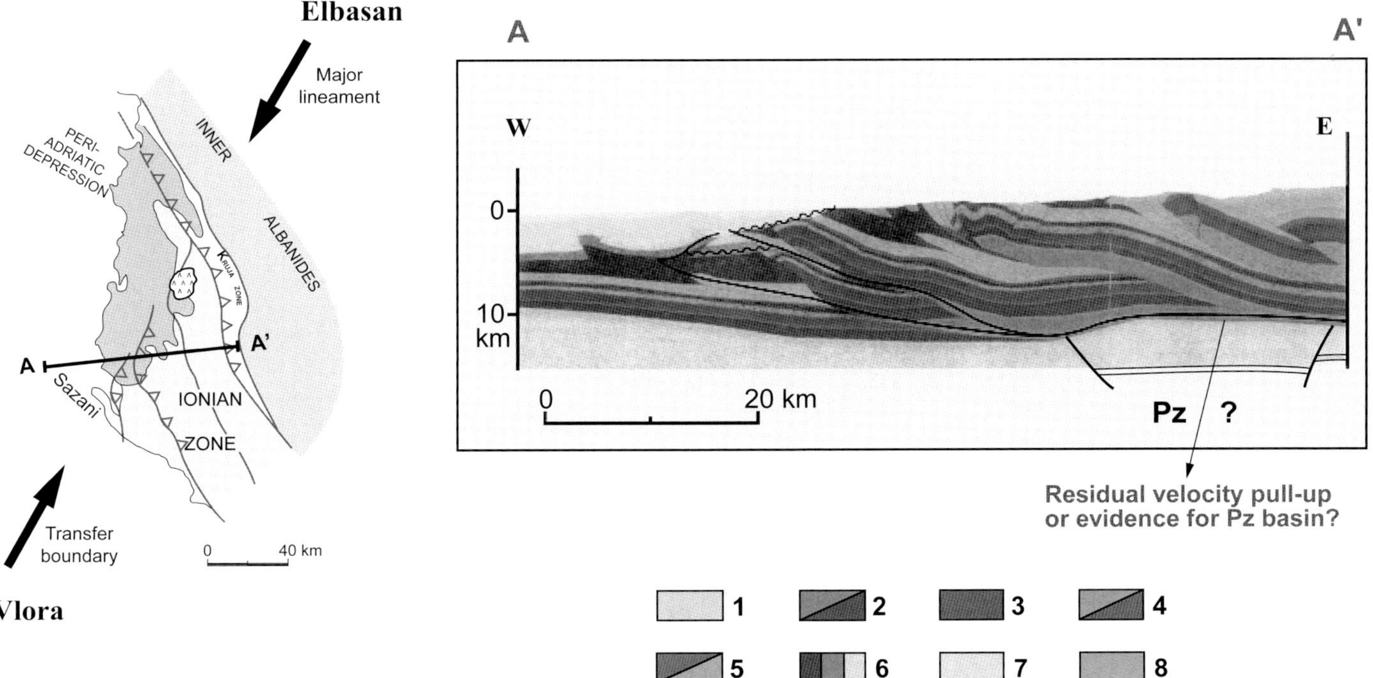

Fig. 6. The Vlora-Elbasan lineament in Albania: a lateral ramp connecting intra-Triassic and Cenozoic décollement levels.

almost horizontal, thus attesting for the late-stage, dominantly strike-slip motion along these former south-verging thrust contacts.

2.2 What is controlling the development of lateral ramps?

Scaled analogue models of thrust deformation have documented the influence of brittle-ductile coupling and thickness variations of décollement layers on the location of the active thrusts (Smit et al. 2003). Seismic profiles across lateral ramps and transfer zones do not differ too much from profiles crossing the frontal structures, although they accommodate a lot of "out-of-the-plane" motion. In Albania and in Eastern Venezuela, they provide a key for better understanding the deep controls accounting for the localization of the deformation along two well known transfer zones, namely the Vlora-Elbasan lineament and the Urica Fault:

Vlora-Elbasan lateral ramp (Albania)

The Vlora-Elbasan transfer fault constitutes a southwest-trending tectonic feature which separates the inverted Ionian Basin in the south from the Peri-Adriatic Depression in the north. It is related to a major lateral shift in the depth of the basal décollement, which is localized within the Triassic salt and evaporites in the south beneath the Ionian Basin, but ramps upward into the Oligocene and Neogene clastics of the flexural sequence further north beneath the Peri-Adriatic Depression (Roure et al. 1995, 2004).

Two different hypotheses have been proposed to account for this localization of the deformation (Fig. 6):

1) either the Vlora-Elbasan structure is located along a major paleogeographic facies boundary, accounting for the lack of Triassic salt in the north, the base of the Triassic remaining flat beneath the ramp;
2) or the main control is exerted by a high-angle fault in the basement, accounting for a vertical offset of the base of the Triassic series.

The latter explanation involving a southwest-trending fault may eventually be validated by depth migrating the time sections crossing the transfer zone. At this stage, an apparent antiformal deformation can be noticed below the basal intra-Triassic décollement, but there is not enough control on seismic velocities at depth yet to perform a confident depth migration of the lines. If still preserved after depth migration, this infra-salt doming would rather account for the reactivation of basement structures or inversion of a Paleozoic basin. Unfortunately, the resolution of potential data such as gravimetry is not sufficient to discriminate among the various hypotheses, due to the high density of shallow carbonates, and no deep seismic is yet available to document the presence or absence of an infra-salt basin.

Seismic profiles across the lineament account for a major change in the structural style, with a basal decollement located in the Triassic salt in the southeast, and in the Oligo-Miocene siliciclastics in the northwest. At intermediate depth (i.e. be-

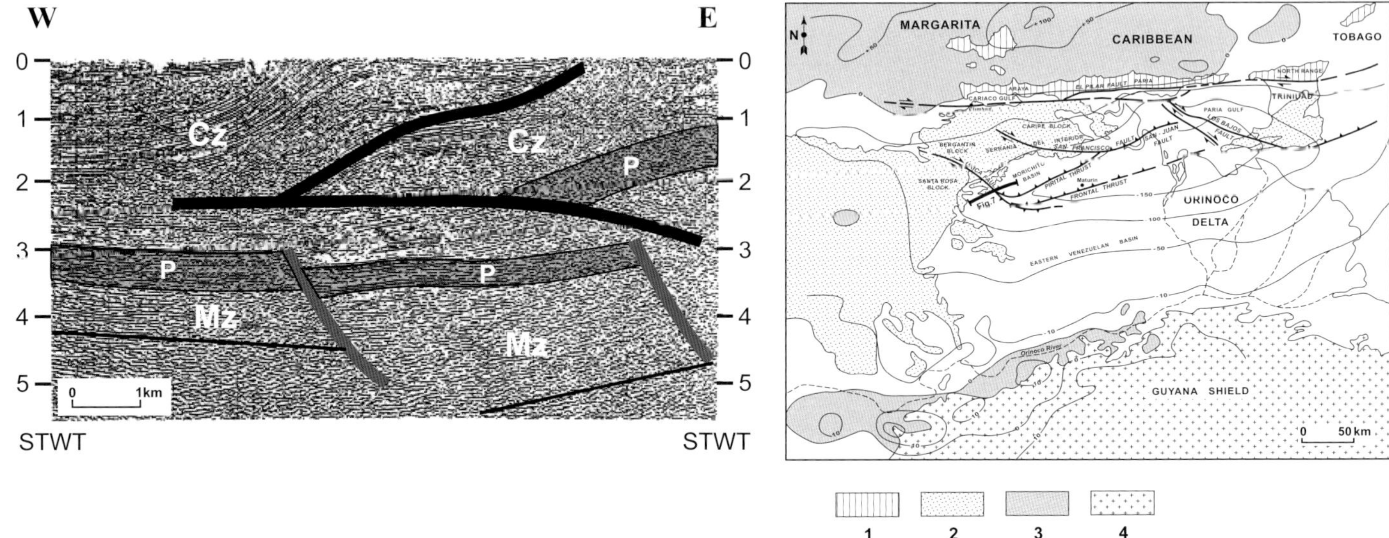

Fig. 7. The Urica transfer zone in Eastern Venezuela: a lateral ramp connecting Lower Cretaceous and intra-Miocene décollement levels (modified after Roure et al. 1994).

tween 4 and 8 km), the Vlora-Elbasan structure is best described as a lateral ramp). Deeper controls are still conjectural, being either related to a lateral change in the Triassic facies, or to a pre-existing Mesozoic or Paleozoic high-angle fault. The slight deformation observed at the base of Triassic series in the eastern part of the section could either be related to a velocity pull-up (underestimation of the seismic velocities during time-to-depth conversion of the section), or indicate inversion of a Paleozoic graben.

Urica lateral ramp (Eastern Venezuela)

The Urica Fault is a southeast-trending tectonic feature which constitutes the western border of the Serrania. At the surface, it is connected laterally with the regional north-verging back-thrust of the main Eastern Venezuelan tectonic front. East-trending seismic profiles across the Urica zone help constraining its architecture at depth (Roure et al. 1994b; Fig. 7):

–   To the east, the basal décollement beneath the Serrania is located in the Mesozoic series of the former passive margin, i.e., in Lower Cretaceous coal measures of the Barranquin Formation or in even deeper synrift Jurassic (?) series;
–   To the west, the basal décollement is shallower, being located in the synflexural siliciclastic series of the Carapita Formation;
–   The surface trace of the Urica Fault is related to an east-verging thin-skinned backthrust which roots within the intra-Carapita décollement;
–   The deep control of the Urica trend consists in a south-southeast-trending high-angle normal fault which crosses the Mesozoic series and the basement. Although it guides the Late Miocene to Pliocene tectonic inversion of the Serrania, this fault still preserves its normal offset at basement

level. This deep Urica fault was inherited from the Mesozoic rifting and accounts for an abrupt thickening of the Mesozoic series toward the northeast.

Figure 7 shows the rapid thickening of the Mesozoic rift sequence in the footwall of the thin-skinned detachment. At shallower level, the surface expression of this structure consists in a regional backthrust, whereas at deeper level, it is related to the reactivation and inversion of a Mesozoic high-angle fault system. The main Mesozoic depocenter is now inverted and dissected into numerous thrust sheets which account for a number of productive east-trending ramp anticlines at and near the main deformational front (i.e., the El Furial and Orocual trends), which are still deeply buried beneath the Neogene synorogenic series, and for the Serrania topography.

2.3 Inversion processes in Western Venezuela: From intra-plate basement short-cuts to foreland basement uplifts

Western Venezuela and Colombia are characterized by the occurrence of a Jurassic rifting episode which accounts for the development of north- and northeast-trending normal faults associated with Jurassic grabens. Outcrops in the Merida Andes and Sierra de Perija in Venezuela, and in the Eastern Cordillera of Colombia, help to study the Jurassic synrift sequences, which are dominantly made up of continental red beds and volcanics of the La Quinta Formation. The same series were also identified from subsurface drilling in the Maracaibo Basin, where industry seismic profiles helped to better understand the successive steps of basin inversion, from almost undeformed grabens still located at 2 or 3 km below the sea level along the western side of the Maracaibo Lake (Colletta et al. 1997; Roure et al. 1997; Fig. 8), up to the area of major foreland basement uplifts such as the Merida Andes and Eastern Cordillera, where

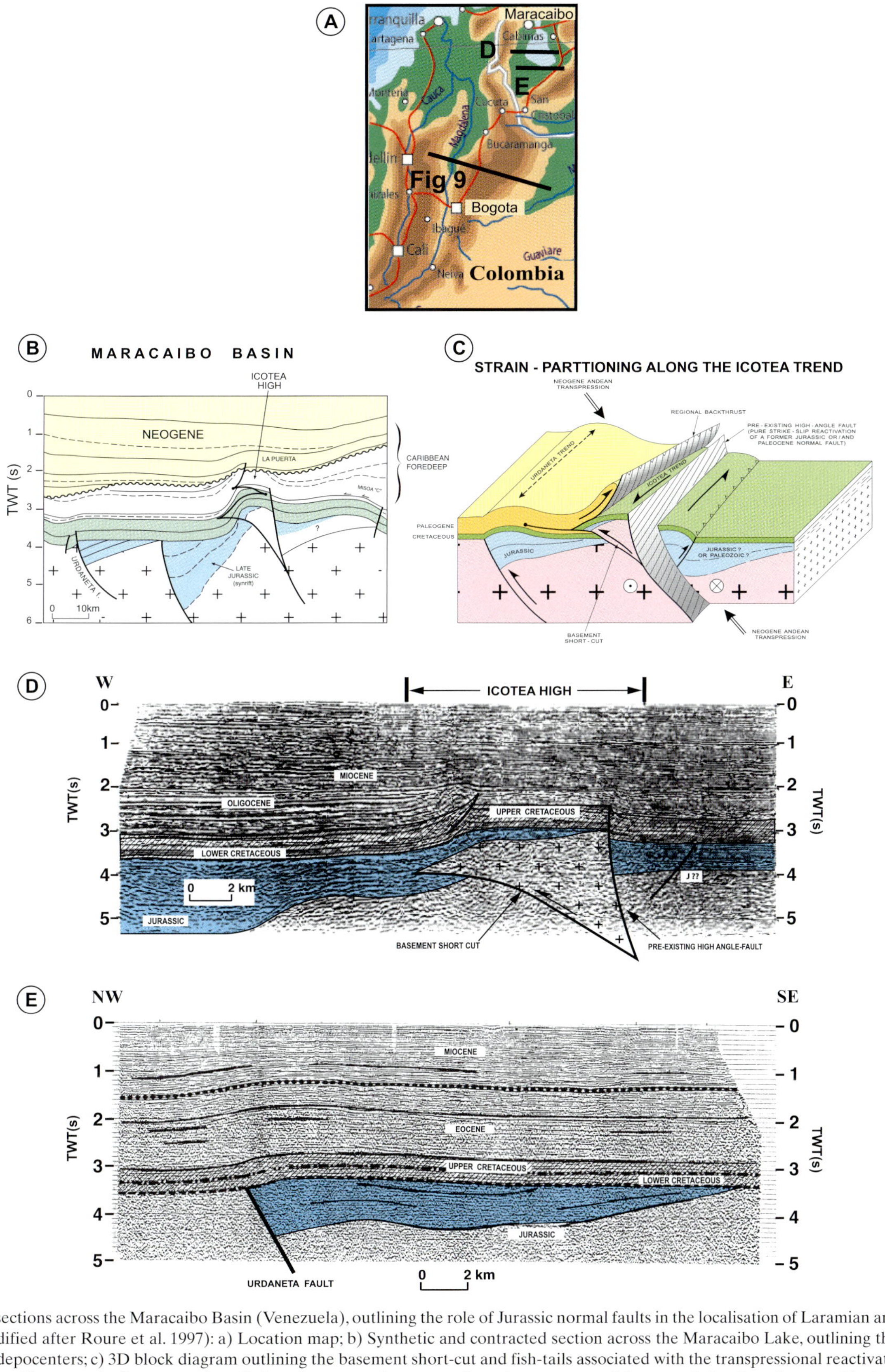

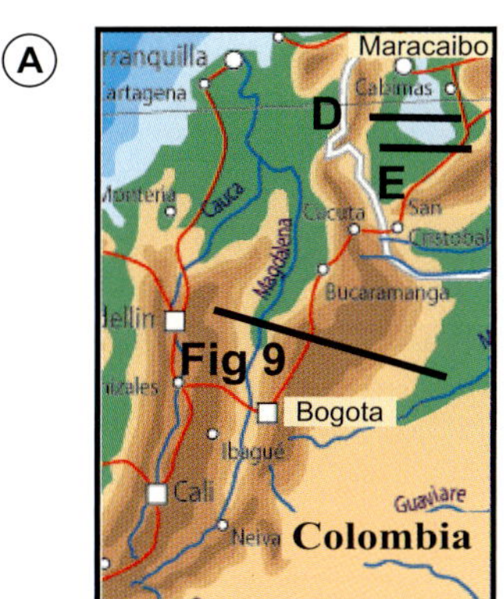

Fig. 8. Structural sections across the Maracaibo Basin (Venezuela), outlining the role of Jurassic normal faults in the localisation of Laramian and Andean inversion features (modified after Roure et al. 1997): a) Location map; b) Synthetic and contracted section across the Maracaibo Lake, outlining the distribution of the main Jurassic depocenters; c) 3D block diagram outlining the basement short-cut and fish-tails associated with the transpressional reactivation of the Icotea trend; d) Seismic profile across the Icotea trend, outlining a basement short-cut and passive transport of the pre-existing Jurassic normal fault. e) Profile across the Urdaneta Jurassic half-graben, outlining a slighter inversion.

the Jurassic series are now exposed at more than 2 or 3 km of elevation (Fig. 9).

Because paleostress directions changed with time (Freymüller & Kellogg 1993; Freymüller et al. 1993), from a dominantly north-south maximum principal stress during the Caribbean/Laramian deformation episodes (Late Cretaceous to Eocene), to a rather northwest-southeast attitude of the main horizontal stress during the Late Miocene-Pliocene Andean deformation, Jurassic normal faults of the Maracaibo Lake have been reactivated successively as right-lateral or left-lateral transpressional features, with also a few episodes of transtension.

Limited inversion occurred along the Urdaneta trend in the south of the Lake, where Jurassic grabens are still overlain by flat Albian and younger post-rift and synflexural series.

Farther north, incipient inversion accounts for the folding of the Albian unconformity, with zero displacement at the tip of the underlying Jurassic border fault (Fig. 8c).

In contrast, oblique inversion becomes the dominant structural style along the Icotea trend, in the north-central part of the Lake, where it accounts for localized basement highs. Careful analysis of seismic profiles shows that the main Jurassic normal fault has been passively uplifted but still preserves its normal offset along the eastern border of the Icotea High, whereas the western border of this anomalous topography is related to a west-verging late-stage reverse fault accounting for a basement short-cut (Fig. 8d).

East-west horizontal shortening is very limited in the area of Maracaibo Lake. Most thin-skinned tectonic structures are localized in the vicinity of the basement-involving inversion features and are related to transpression, with the occurrence of numerous fish tails and other local accommodation features induced by a mechanical decoupling between the rigid basement and more plastic sedimentary cover (Roure et al. 1997). Larger shortening accounts for the major foreland uplifts of the Merida Andes and Eastern Cordillera, which will be further discussed in Chapter 4.

3 Thrust-top basins as a mirror of sub-thrust tectonic accretion

Piggyback or thrust-top basins developing on top of the mobile allochthonous edifice have been identified first in the Apennines a long time ago (Ori & Friend 1984; Casero et al. 1991). They are also well documented in Sicily (Caltanissetta Basin; Roure et al. 1990b), as well as in Eastern Venezuela (Morichito Basin; Roure et al. 1994b) and in many other thrust belts where depocenters have developed at the rear of frontal anticlines, being either isolated or still in direct connection with coeval sediments infilling the adjacent foreland basin.

Although they commonly display contrasting lithofacies, usually shallow marine or continental, making direct chronostratigraphic correlations with the deeper-water foredeep sediments a bit challenging, their basal and successive internal unconformities usually provide unique constraints to document the timing of tectonic accretion. Progressive tilting of these imbricated unconformities and coeval lateral shifts of piggyback depocenters can be used also as additional templates to guide the geologist when addressing forward kinematic modelling and editing intermediate geometries between the present and pre-orogenic configurations:

- In the southern Apennines, subthrust accretion of deeply buried Mesozoic platformal duplexes beneath the basinal Lago-Negro nappes and Neogene clastics of the Bradano Trough accounts for the development of nappe anticlines, tectonic windows and klippen, which result from the refolding of former thrusts and coeval erosion (Fig. 10; Roure et al. 1990a, 1991). Piggyback basins can also develop above flat segments of the sole thrust, in the core of overlying nappe synclines, and help to decipher whether tectonic accretion operates farther east at the thrust front, or farther west, by underplating of deeply buried duplexes (Fig. 10, bottom; Hippolyte et al. 1991, 1994; Roure et al. 1991).
- Pleistocene piggyback depocenters observed along a famous seismic transect published by Pieri and Bally in the Northern Apennines (Pieri 1983; Fig. 11a) provide also evidence for post-Pliocene and still ongoing deformation along the basal décollement, which is located at more than 10 km depth in this portion of the Apennines (Scrocca et al. 2007). Along this transect, Pliocene and older outcrops are located in wide regional antiforms that developed above the ramps of the deeper, still active décollement, whereas Pleistocene depocenters are found above its flat segments (Fig. 11b). In this case, the lithospheric flexure also had a direct control on the subsidence pattern of the foothills, as evidenced by forward kinematic simulations (Zoetemeijer et al. 1992, 1993):

Tectonic accretion above a flat décollement surface would rather generate uplift and erosion in the hinterland, in an area where the main Pleistocene depocenter is located, thus implying that thrusting operated synchronously with ongoing flexural subsidence of the underthrust foreland lithosphere.

4 The development of hinterland basins: a combination of strain partitioning, strike-slip faulting and thrust reactivation

Collisional orogens like the Alps and the Pyrenees and American cordilleras like the Andes and Rocky Mountains do share a number of surficial similarities, although their driving mechanisms are quite distinct at a deeper level, with the juxtaposition of two continental lithospheres in the Alps and the Pyrenees, vs. complex interplays between the American continents and the subduction of the Pacific Ocean in the Andes and the North American Cordillera:

- In these two contrasting types of orogens, oblique convergence accounts for strain partitioning, most of the oblique component being frequently absorbed along active strike-slip faults which run parallel to the plate boundary (Chemenda et al. 2000; Martinez et al. 2002; Lingrey 2007). This is the case for instance with the Periadriactic Line in the Alps

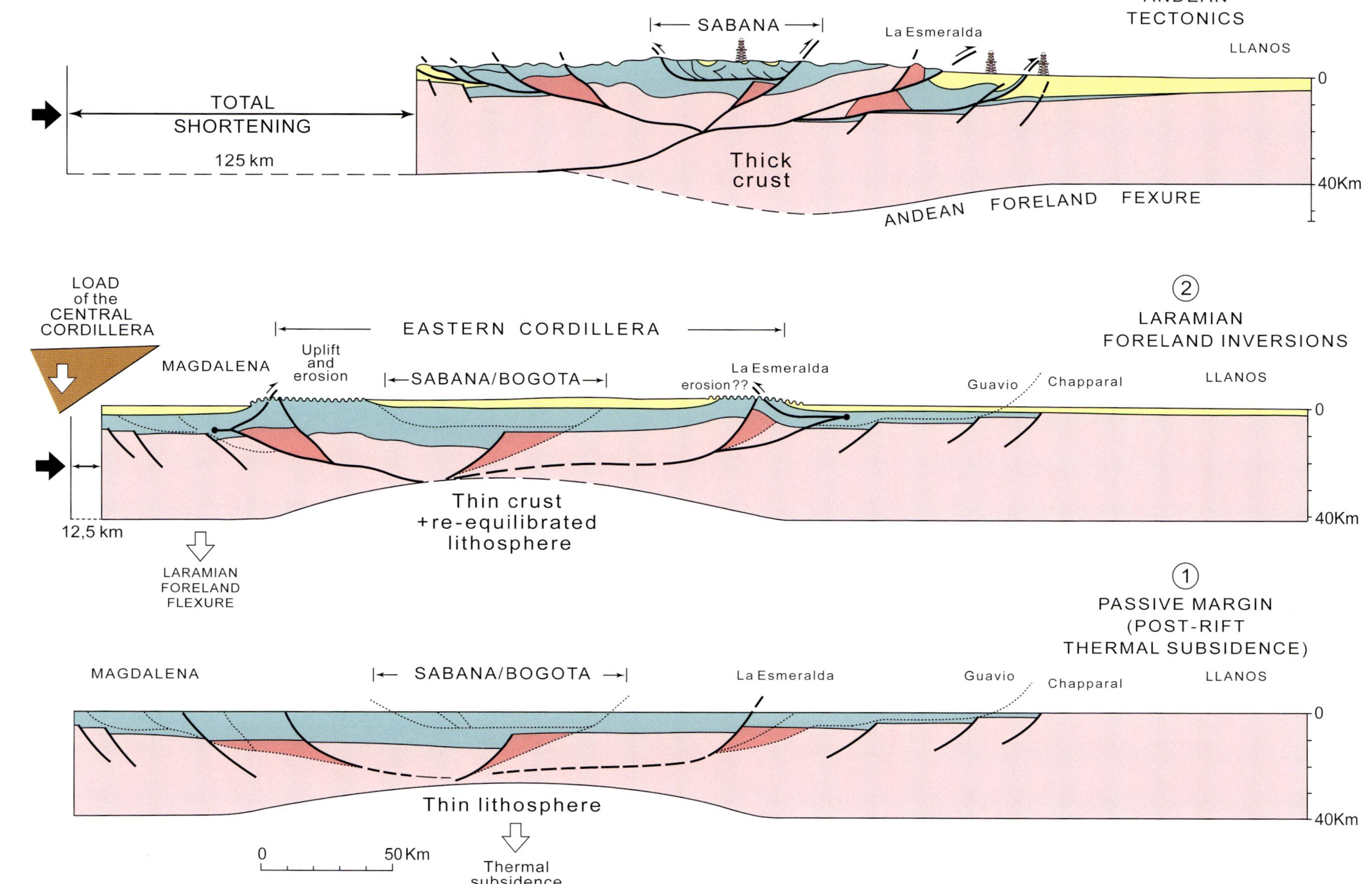

Fig. 9. Structural section across the Eastern Cordillera of Colombia, outlining the control of Jurassic normal faults and basement short-cuts in large Andean foreland basement uplifts (see location in Figure 8a): Top section: Present stage accounting for the overall inversion of the former Jurassic depocenter of the Sabana de Bogota. Notice also the geometry of the Esmeralda Fault, a former Jurassic fault which has been passively transported piggy-back of the main east-verging basement up-thrust of the Cordillera; Central section: reconstruction at the end of the Laramian orogeny (Eocene). Notice localized erosions in the Laramian foreland, due to incipient inversion features similar to the present-day Icotea trend of Western Venezuela (compare with Figure 8); Bottom: Pre-orogenic architecture of the transect.

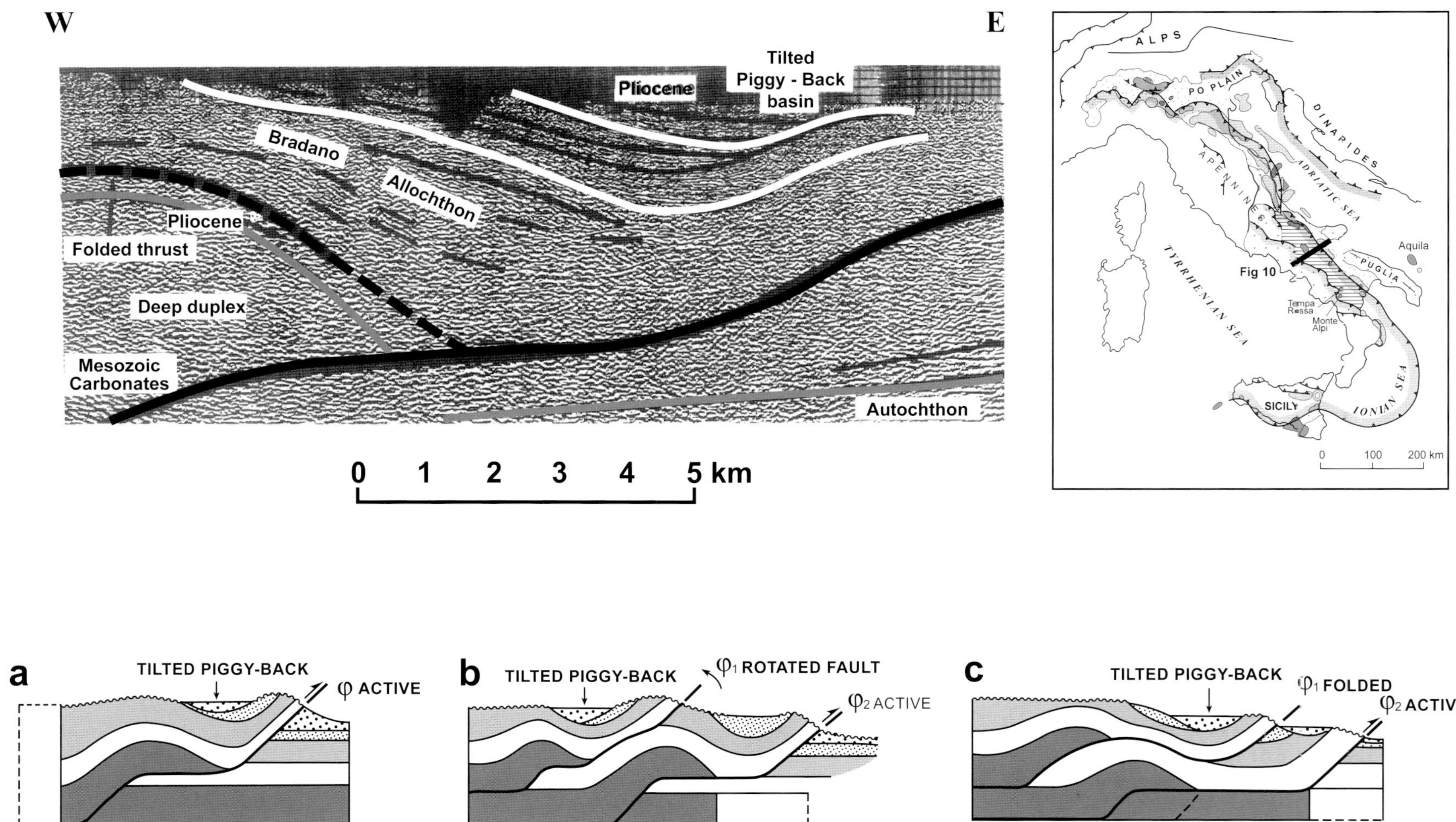

Fig. 10. Structural sections across the Southern Apennines, outlining the role of deep buried duplexes in the overall architecture and tilting of thrust top basins (modified after Roure et al. 1991).

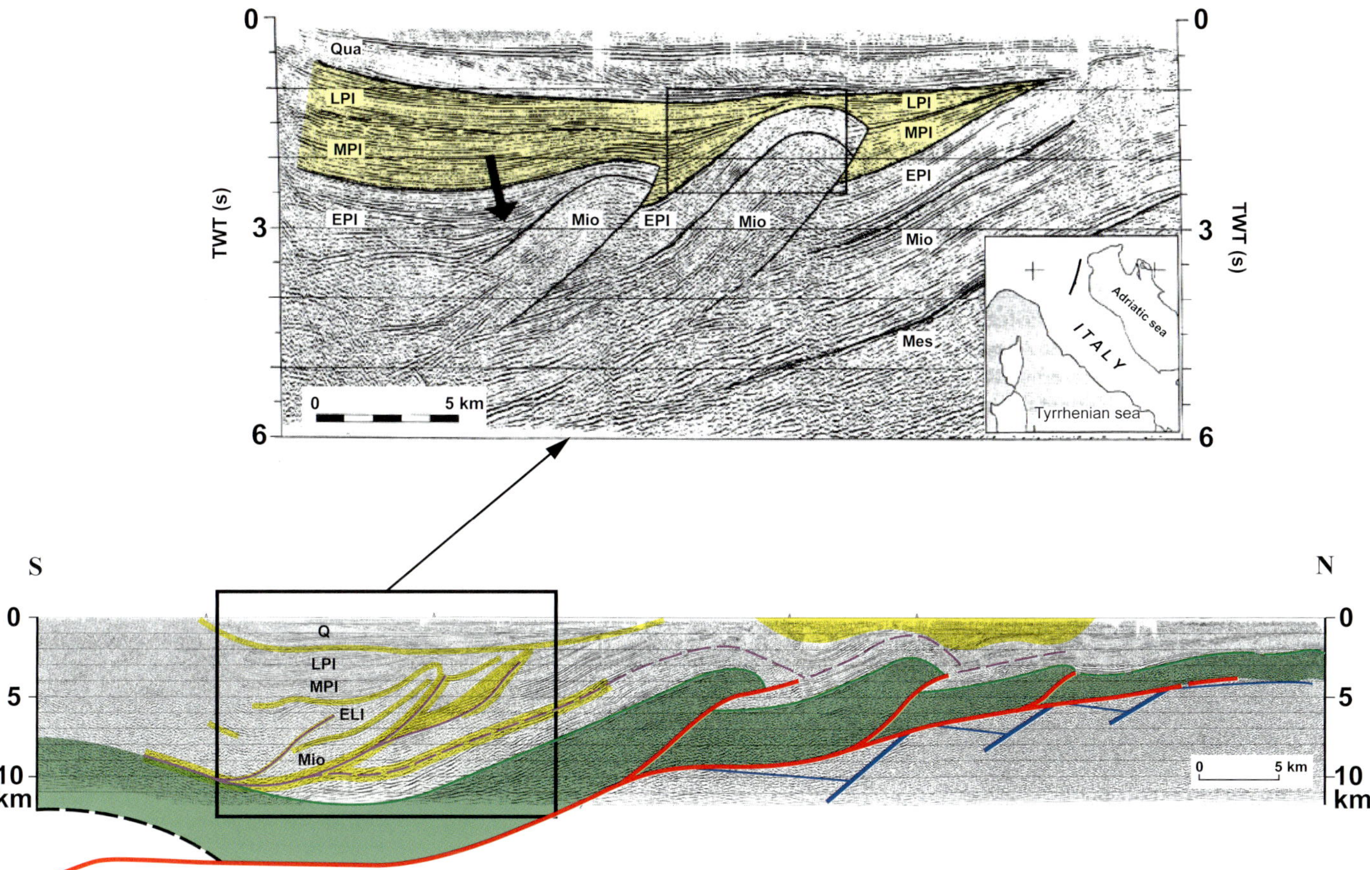

Fig. 11. Structural section across the Northern Apennines, outlining synkinematic Pliocene and Quaternary deposits, and ongoing displacement along the basal décollement (seismic profile from Pieri 1983): Top: Stratigraphic calibration of shallow horizons made by Pieri and Bally (in Pieri 1983). Black arrow and frame show progressive onlaps of growth stata on Pliocene anticlines. Bottom: Deep interpretation outlining the diachronous activation of an early intra-Miocene décollement level (pink, mainly active during the Lower and Middle Pliocene), and a deeper, younger intra-Triassic basal detachment (red, mainly Upper Pliocene but still active during the Quaternary).

(Schmid et al. 2004) and the North Pyrenean Fault in the Pyrenees. The indentation and eastward escape of the intra-Carpathian blocks account also for the post-collisional Miocene strike-slip dismembering of the former Pieniny Klippen Belt (Sauer et al. 1992). Strain-partitioning in areas of oblique convergence accounts also for the northward escape of the Salinian block west of the San Andreas Fault in California, for the southward motion of the Maracaibo indenter west of the Bocono Fault in the Merida (Venezuelan) Andes, and the eastward escape of the Carribean plate north of the El Pilar fault in Eastern Venezuela (Freymüller & Kellogg 1993; Freymüller et al. 1993). Due to partitioning, transport direction remains dominantly perpendicular to the thrust anticlines in the foothills.

– Overthickened crust of the Alps, other Tethyan/Mediterranean orogens and the American Cordilleras was affected by a ductile flow of the lower crust, associated with well-documented post-orogenic collapse and orogen parallel extension in the Basin and Range (Wernicke 1981), as well as in the Betic and Rif orogens and the intervening Alboran Sea

(Dewey 1988). Although aternative hypotheses involving a roll-back of the subduction and coeval back-arc opening have been proposed for the Pannonian Basin, the Aegean and Tyrrhenian domains, where no former high mountain plateau could account for a post-orogenic gravitational collapse, these areas display also evidence of reactivation of former thrust faults as low-angle normal faults. Negative inversion is effectively obvious at various scales within these three intra-arc systems, i.e. in Hungary (Horvath 1993; Peresson & Decker 1997; Tari et al. 1999; Horvath et al. 2006), in the Cycladic Islands and the Apennines (Bally et al. 1988; Ghisetti et al. 1993; Brun et al. 1994; Jolivet et al. 1994, 1998; Ghisetti & Vezzani 2002).

Paleomagnetic and microtectonic studies performed in the Southern Apennines and adjacent Bradadano and Puglia foreland (Hippolyte et al. 1991, 1994) could identify periods of strong coupling between the allochthon and the foreland, i.e., paleostress directions being then similar on both sides of the thrust front, separated by time intervals when a complete decoupling between the thrust belt and the autochthon prevailed, with very distinct paleostress directions.

Various parameters such as pore-fluid pressure in potential décollement levels and thermomechanical behaviour of the lower crust and sub-continental mantle probably control the coupling or decoupling between the orogen and its foreland, foreland inversions developing when all the tectonic stress propagates forelandward from the plate boundary during periods of strong coupling (Ziegler et al. 1998, 2002).

These successive changes in coupling and decoupling between the hinterland and the foreland, associated with deeper controls exerted by the structural grain of the crust (i.e., occurrence of pre-existing weakness and inherited structures in the crust), or with the negative inversion of former thrusts, are the main processes accounting for the localisation and development of hinterland basins:

4.1 Post-orogenic collapse and negative inversion of former thrusts

The negative, extensional inversion of former reverse faults is a common phenomenon in the hinterland of most orogens, where it accounts for the development of syn-extensional depocenters, i.e. in the Basin and Range province of the USA (Wernicke 1981), in the hinterland of the Canadian Rocky Mountains (Price 1986), in the Betic Cordillera and Alboran Sea (Dewey 1988), and the Tyrrhenian side of the Apennines (Jolivet et al. 1994, 1998). Additionally, localisation of the deformation along former orogenic structures has been envisioned or even demonstrated in many rift systems and passive margins, i.e. for the Jurassic basins of northern Colombia and western Venezuela, for various segments of the East African Rift, but also for the northwestern margins of the Atlantic Ocean and Gulf of Mexico, which were prone to reactivate former thrusts of the Appalachians and Ouachita Paleozoic orogens (Ando et al. 1983; Hatcher et al. 1989), as well as for the Caledonides in Scandinavia and off England (Séguret et al. 1989; Séranne et al. 1989, 1995; Séguret & Benedicto 1999; Séranne 1999).

In France, negative hinterland inversion associated with syn-extensional basin development has been well documented locally:

– In the Aquitaine Basin, the North Pyrenean deep seismic Ecors profile has evidenced the development of Permian grabens above reactivated Hercynian thrusts (Choukroune et al. 1990; Roure et al. 1996). Although there is no seismic profile yet available, the same process could probably account for many other post-Hercynian European basins, such as the Permian Lodève Basin in the vicinity of the Montagne Noire, where post-orogenic collapse has been well documented on the basis of microtectonic and petrofabric data (Faure & Becq-Giraudon 1993; Becq-Giraudon & van den Driessche 1994; Burg et al. 1994).

– In Languedoc, Oligocene extension associated with the opening of the Gulf of Lion and Western Mediterranean is known to have locally reactivated Pyrenean thrusts in the St-Chinian Arc and Montpellier fold (Benedicto 1996; Benedicto et al. 1996; Séguret & Benedicto 1999), thus accounting for the development of the Quarante Basin and adjacent roll-over regionally know as the La-Clappe anticline (Roure et al. 1988; Fig. 12).

4.2 Thrust-top pull-apart basins

The Vienna Basin is probably the most famous and archetype of thrust-top pull-apart basins, developing above the Alpine allochthon after its thrust emplacement, in connection with lateral eastward block escape along the Carpathian arc (Royden 1985; Sauer et al. 1992; Seifert 1996; Decker & Peresson 1996).

Other pull-apart basins have developed in the hinterland of Circum-Mediterranean thrust belts, i.e. in the Apennines and in North Algeria, as a result of local and temporal changes in the paleostress regimes, and in relation to strain partitioning.

The physiography and lozenge shape of the Chelif Basin in North Algeria is well identified on geological maps and landsat imagery (Fig. 13). This basin is located north of the Tellian thrust front, which reached its current position during the Langhian (Frizon de Lamotte et al. 2000; Roca et al. 2004; Benaouali et al. 2006). It is adjacent to a major east-trending lineament, known as the "Dorsale Calcaire", which separates the Kabylides crystalline basement in the north from the Tellian nappes in the south, and most likely behaved as a major strike-slip fault during the development of the Chelif Basin.

The Neogene sedimentary infill of the Chelif Basin comprises Burdigalian to Langhian synkinematic series, which were deposited in a piggyback position at the same time as the main southward thrust emplacement of the Tellian nappes, at a time when oblique convergence, transpression and strain-partitioning affected the plate boundary. These basal deposits were overlain by post-nappe Tortonian to Pliocene depocenters, which are spatially limited and controlled by active normal faults. These normal faults, locally exposed at the surface, can be also traced down to the deepest part of the basin on seismic profiles and are indicative of an Upper Miocene-Pliocene episode of transtension along the North African plate boundary. These faults are oblique (en échelon) with respect to the Dorsale Calcaire lineament. In a similar way as the El Pilar Fault in northern South America, the Dorsale Calcaire lineament accommodated the lateral shift of the Kabylides with respect to the Tell allochthon and underlying underthrust African foreland during a Tortonian to Pliocene post-nappes episode of transtension.

Plio-Quaternary inversion of these depocenters accounts for renewed transpression along the plate boundary, with folding and erosion of Pliocene series in the vicinity of the major border faults of the Chélif Basin.

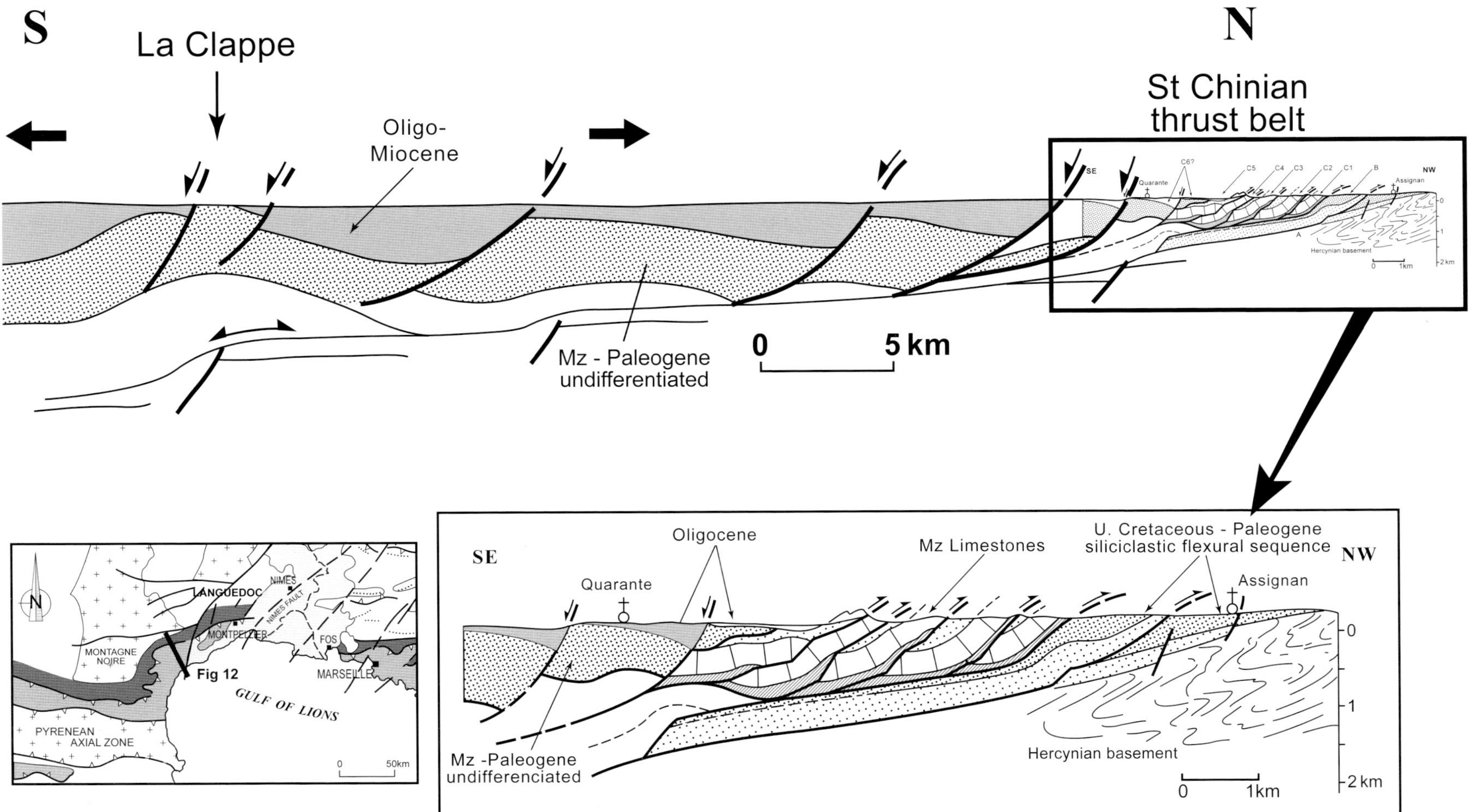

Fig. 12. Structural section across the Quarante Fault and La Clappe anticline in Languedoc (northern Pyrenean foreland), showing the Oligocene negative inversion of the former Late Cretaceous to Eocene north-verging thrust system.

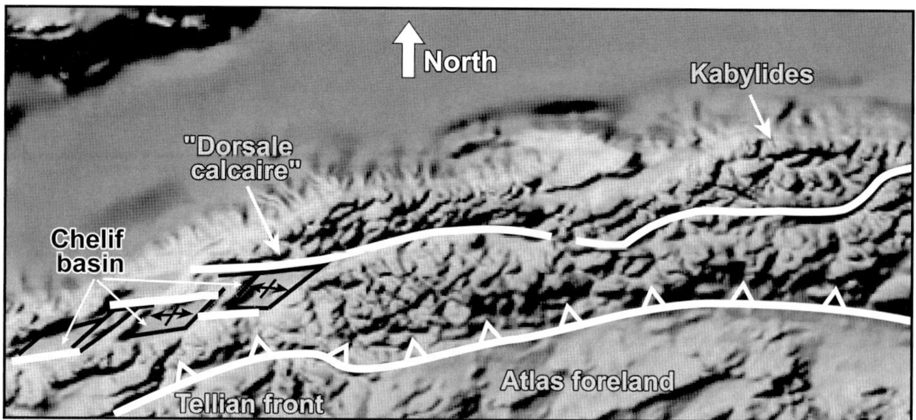

Fig. 13. Landsat image of Northern Algeria, outlining the distribution of thrust-top pull-apart depocenters of the Chelif Basin associated with a major east-trending lineament (Dorsale Calcaire), between the Tellian thrust front in the south and the Kabylides-Western Mediterranean plate in the north.

4.3 Intra-crustal backthrusts and development of intramontane basins

Analogue models accounting for the flow of the ductile lower part of an overthickened continental crust have been proposed to account for the development of pop-down intramontane basins such as the Magdalena Basin in Colombia (Davy & Cobbold 1991), where thick and dominantly continental Neogene deposits have been trapped between the growing topographies of the Central and Eastern Cordilleras.

Industry seismic profiles across the Llanos foothills, Garzon Massif and Middle Magdalena Basin help to constrain regional balanced cross-sections and to propose new interpretations for the crustal structure of this transect, whereby the regional west-verging backthrust of the Garzon Massif connects at depth with former Paleozoic east-verging thrusts (Fig. 14; Roure et al. 2003, 2005a; Toro et al. 2004; Sassi et al. 2007).

Occurrence of Paleozoic thrusts in the Llanos foreland has also been recognized farther north in the Barinas Basin in Venezuela (Fig. 14b). It is likely that this inherited structural grain of the South American foreland accounted for both the localisation of the Jurassic rifting (Fig. 9), and subsequent Andean foreland basement uplifts.

As such, the present day location of the Maracaibo Basin is very similar to the one of the Magdalena Basin. Although the Maracaibo area is mostly interpreted as a distinct microplate, it could also be adequately considered as an intramontane basin, which became isolated from the main Llanos foreland basin in the east due to the intervening Neogene basement uplift of the Merida Andes.

5 Mantle dynamics and post-orogenic uplift of foreland basins

5.1 Post-orogenic uplift and erosion of foreland basins

Many foreland basins are no longer close to the sea level, but have experienced uplift and erosion since the end of the main compressional/tectonic loading episodes (Fig. 15):

– In North Algeria, Langhian deep-water turbidites deposited near Tiaret in the foreland autochthon, immediately south of the Tellian thrust front, are presently located at an elevation of 1 km above sea level (Roca et al. 2004).

– In the Alberta Basin in Canada, up to 3 km of synflexural sediments were removed by erosion since the end of the Laramian/Cordilleran deformation, i.e. from Eocene onward (Faure et al. 2004; Hardebol et al. 2007). Worth to mention, the city of Calgary itself, which is located in the foreland autochthon, about 100 km east of the thrust front, currently displays an average elevation of 1 km above sea level, which is quite surprising for an ancient foredeep basin (Price & Fermor 1985; Price 1994; Fig. 15a).

– The same type of post-orogenic uplift and erosion of former flexural sequences occurred also along the western margin of the Gulf of Mexico, i.e. in the foothills of the Sierra Madre Oriental and adjacent coastal plain, which is actually superimposed on the former Cordilleran foreland basin. Up to 4 km of post-Laramian erosion is thus recorded in the Burgos Basin in the north, and about 2 to 3 km farther south in the Chicontepec Basin and in the Cordoba Platform in the Veracruz State (Fig. 15b) (Gray et al. 2001; Roure et al. 2008).

In Mexico, these post-orogenic uplift and unroofing processes have completely changed the former attitude of the basement, which is currently dipping toward the east beneath and in front of the Cordoba Platform, whereas it was dipping westward at the time of foreland basin development. Late Cretaceous to Paleocene turbidites and gravity slides infilling the former Chicontepec flexural basin currently display apparent downlaps toward the Faja de Oro or Golden Lane, whereas they were initially deposited as onlapping sequences, prior to post-orogenic tilting and unflexing of the foreland basement (Alzaga et al. 2007a, b).

Erosional products derived from the Sierra Madre itself, but also from post-Laramian uplift and unroofing of the adjacent foreland, account for a huge Oligocene to Neogene siliciclas-

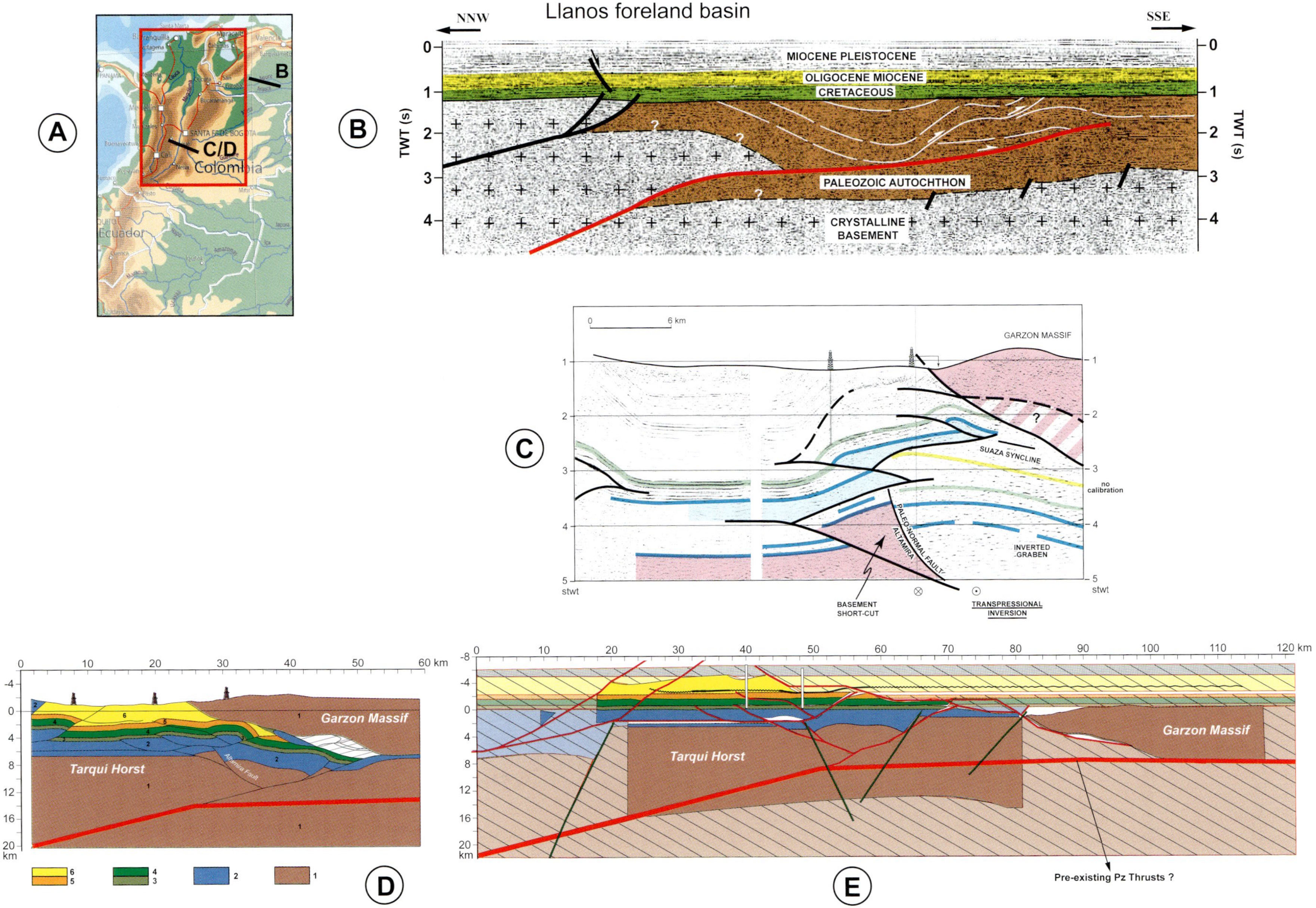

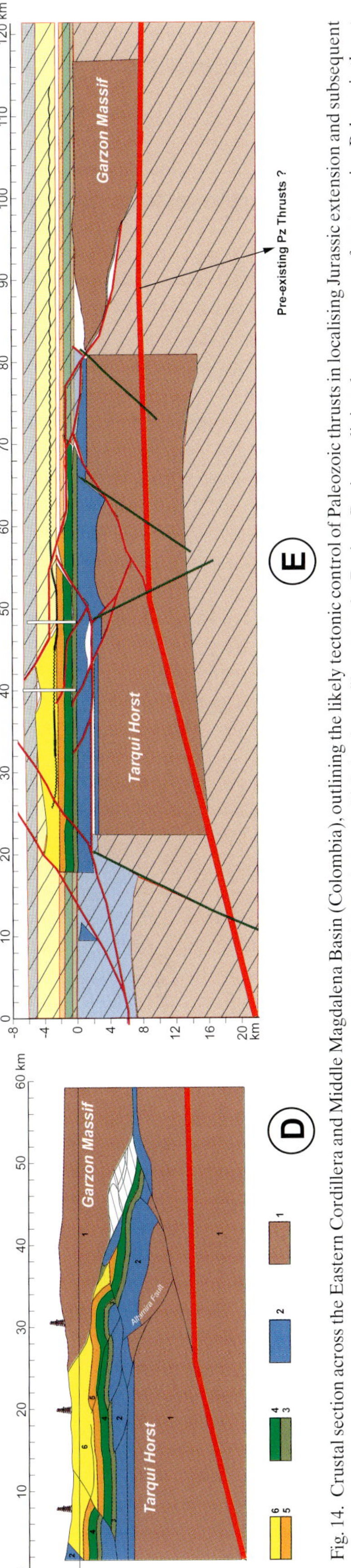

Fig. 14. Crustal section across the Eastern Cordillera and Middle Magdalena Basin (Colombia), outlining the likely tectonic control of Paleozoic thrusts in localising Jurassic extension and subsequent thrust development west of the Llanos foreland basin (after Roure et al. 2005): a) Location map; b) Seismic profile across the Barinas Basin, outlining the occurrence of east-verging Paleozoic thrust sealed by the Albian unconformity (modified after Colletta et al. 1997); c) Seismic profile across the Altamira High, outlining inversion features similar to the Icotea trend of western Venezuela (compare with figure 8). d and e) Balanced cross-section across the Middle Magdalena Basin and Garzon Massif, showing crustal wedging, development of a foreland-dipping monocline associated with the forelandward propagation of a blind thrust (thick red line) beneath the Llanos Basin and antithetic west-verging backthrust of the Garzon Massif. We assume that this east-verging blind thrust is directly controlled by the pre-existing Paleozoic structural grain, Paleozoic thrusts being successively inverted as normal detachments during the Jurassic rifting, and as thrust faults during both Laramian and Andean foreland inversions.

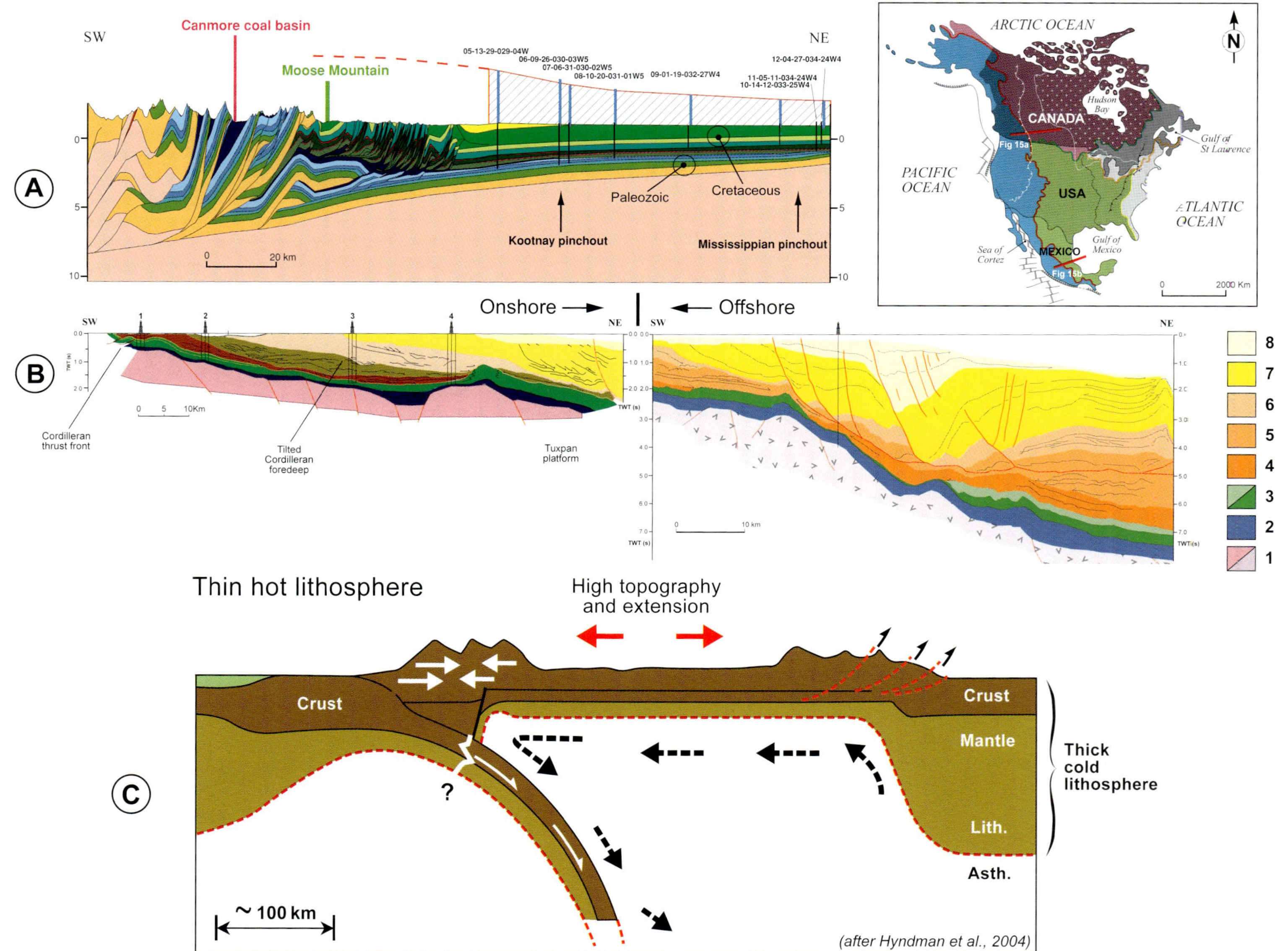

Fig. 15. Lithospheric sections in North America, outlining the role of asthenospheric rise in post-orogenic uplift and erosion of former flexural basins. Location of the sections a and b is shown on the map (blue, green, violet and grey patterns relate to the Pacific, Mississippi-Gulf of Mexico, Arctic and Arctic drainage areas, respectively). a) Structural section across the Canadian Rocky Mountains, outlining the amount of post-Laramian uplift and erosion (modified after Faure et al. 2004). Amounts of erosion have been derived from 1D thermal modelling on wells (foreland) and outcrop data (foothills). Thick Paleozoic (i.e. Cambrian, Devonian and Mississippian passive margin carbonates) and Cretaceous synflexural series are indicated by circles. Intervening Permian to Jurassic series are very thin along this section. b) Structural section across the Sierra Madre Occidental and Gulf of Mexico, outlining the post-orogenic tilting of the former Laramian foredeep basin and foreland basement, coeval with more than 4 km of erosion and denudation in the Sierra Madre and Cordoba Platform (modified after Alzaga et al. 2008). Colour code: (1) basement; (2) Jurassic; (3) Cretaceous; (4, 5 & 6) Paleogene; (7) Miocene); (8) Plio-Quaternary. c) Lithospheric section across the East-Pacific subduction and North American Cordillera, outlining the asthenospheric rise above the retreating subducted slab (modified after Hyndman et al. 2005). Deep mantle processes are advocated here to account for high elevations, rapid denudation and extension in the Basin and Range province, from Canada to Mexico.

tic sedimentary influx into the Gulf of Mexico, resulting in the building of overpressures in underlying Eocene shales and to the gravitational collapse of the margin (Alzaga et al. 2007a, b). Post-orogenic erosional products derived from the uplift of the Alberta foreland basin, which is devoid of any post-Cretaceous series, have also been certainly transferred either to the north into the Arctic, or to the south into the Gulf of Mexico, depending on the actual position of the continental divide between the Mississippi and Arctic basins during the Eocene and younger periods.

Apart from this Cordilleran example, where vertical motion is controlled by an astenospheric rise, post-orogenic uplift and erosion are also common processes in other orogens such as the Alps, the Carpathians, the Apennines-Maghrebides-Betics system, as well as in the Brooks Ranges, among others. Unlike in the Cordillera, where the subduction of the Pacific Ocean lithosphere beneath the orogen never stopped, alternative hypotheses involving a slab detachment, as described below, have been proposed to account for the recent vertical motion recorded in most Circum-Mediterranean and Alpine orogens (Wortel & Spakman 1992, 2000; van der Meulen et al. 1998; Frizon de Lamotte et al. 2000; Roca et al. 2004)

5.2 Mantle dynamics and coupling with surface processes

Mantle dynamics constitute the engine accounting for the post-Laramian uplift and erosion of the Canadian and Mexican forelands. Due to a corner effect of the Pacific subduction, hot mantle is progressively thinning and uplifting the North American lithosphere over an extremely wide surface, accounting for the post-Laramian collapse of the Cordilleran orogen coeval with the development of metamorphic core complexes and basin and range-type extension, for recent volcanic activity, but also for the wide doming and unroofing observed in the foreland, from Canada to southern Mexico (Price 1986; Hyndman et al. 2005; Fig. 15c).

In the Central Apennines, rapid changes observed during the Upper Pliocene and Pleistocene in the subsidence history of the Adriatic foredeep and coeval increase in the uplift rates of adjacent foothills have been interpreted as an evidence for slab detachment, the slab pull no longer contributing to the down-flexing of the Adriatic foreland lithosphere (van der Meulen et al. 1998; Wortel & Spakman 1992, 2000; Spakman & Wortel 2004). Although such process is still debated, it could actually be proposed also to account for the flexural rebound observed in the North Algerian foreland, south of the Tellian front.

Alternatively, asthenospheric rise and advection of hot mantle in the Western Mediterranean and Tyrrhenian back arc basins could easily explain such late stage vertical motion of the foreland lithosphere (Wortel & Spakman 1992, 2000; Spakman & Wortel 2004).

Conclusions

Strong coupling between the thrust belt and its foreland can occur at different times in both subduction-related (i.e. Cordilleran-type) or collision-related (i.e. Alpine-type) orogens, thus accounting for both early and late foreland inversion processes (Ziegler et al. 1998, 2002).

Since the mid 80's, deep crustal seismic imaging across many orogens such as the Alps, the Pyrenees and the North American Cordillera has provided direct controls on the deep architecture of the thrust systems, and a better understanding of the coupling between thin-skinned and thick-skinned tectonics, whereas since the 90's, mantle tomography is progressively documenting the occurrence or absence of lithospheric slabs beneath recent orogens. In many thrust belts where neither deep seismics nor mantle tomography is yet available, the pending question is to know whether slab detachment may account for rapid uplift and post-orogenic erosion of former foreland basins, as described in the Central Apennines by van der Meulen et al. (1998), or if mantle convection and asthenospheric rise alone can account for post-orogenic uplift, as evidenced in the Alberta and Veracruz basins.

Source to sink studies are also necessary to define the spatial and temporal coupling between erosion, sedimentary transfer and deposition. Until recently, most efforts were devoted to high resolution seismostratigraphic studies coupled with core and outcrop descriptions of the synflexural/synkinematic sedimentary infill of the foreland basins. Today, however, GPS measurements and thermo-chronometers such as Apatite Fission Tracks and U-Th, can provide direct control on the uplift and unroofing history of the hinterland. Ultimately, new techniques must still be developed to provide information on paleo-elevations, which are essential for discriminating between different tectonic models, e.g. orogenic collapse and rollback, and which are also likely to control the boundary conditions (hydraulic heads) required for computing the pore-fluid pressure evolution in adjacent low lands (Schneider 2003; Schneider et al. 2004; Roure et al. 2005b).

Further understanding of the coupling between deep (mantle) and surface (climate) processes in orogens and adjacent foreland basins constitutes one of the main current challenges for Earth scientists, which will require access to well documented data bases to feed numerical models, involving a lot of integration and multi-disciplinary team work. International networks such as the Transmed (Cavazza et al. 2004a, b) and ILP task forces and related workshops may help to initiate these new collaborations. Pioneer work is currently done in Europe (Topo-Europe programme), where continental topography has been indeed widely impacted by the Alpine orogen and recent mantle upwelling in the Western Mediterranean and West European rift system.

Acknowledgements

Bernard Colletta and William Sassi provided helpful comments on the initial draft. The manuscript benefited from the careful review from Liviu Matenco, Nikolaus Froitzheim, and an anonymous reviewer.

region: plate tectonics, basin formation and hydrocarbon habitats. American Association of Petroleum Geologists, Memoir 79, Ch. 34.

Roure, F., Brun, J.P., Colletta, B. & Van Den Driessche, J. 1992: Geometry and kinematics of extensional structures in the French Alpine foreland (basin of SE France). Journal of Structural Geology.

Roure, F., Brun, J.P., Colletta, B. & Vially, R. 1994a: Multiphase extensional structures, fault reactivation, and petroleum plays in the Alpine Foreland Basin of Southeastern France. In: Mascle, A. (Ed.): Hydrocarbon and petroleum geology of France. European Association of Petroleum Geoscientists. Special publication n°4. Paris, Springer-Verlag, 245–268.

Roure, F., Carnevali, J.O., Gou, Y. & Subieta, T. 1994b: Geometry and kinematics of the North Monagas thrust belt (Venezuela). Marine and Petroleum Geologists, 11, 347–361.

Roure, F., Casero, P. & Vially, R. 1991: Growth processes and mélange formation in the southern Apennine accretionary wedge. Earth and Planetary Sciences Letters, 102, 395–412.

Roure, F., Choukroune, P. & Polino, R. 1996: Deep seismic reflection data and new insights on the bulk geometry of mountain ranges. Comptes-Rendus de l'Académie des Sciences, Paris, IIa, 322, 345–359.

Roure, F. & Colletta, B. 1996: Cenozoic inversion structures in the foreland of the Pyrenees and Alps. In: Ziegler, P. & Horvath, F. (Eds.): PeriTethys Memoir 2. Museum national d'Histoire Naturelle, Paris, 173–210.

Roure, F., Colletta, B., De Toni, B., Loureiro, D., Passalacqua, H. & Gou, Y. 1997: Within-plate deformations in the Maracaibo and East Zulia basins, Western Venezuela. Marine and Petroleum Geology, 14, 139–163.

Roure, F., Faure, J.L., Colletta, B., Macellari, C. & Osorio, M. 2005a: Structural evolution and coupled kinematic-thermal modeling of the Uper Magdalena Basin in the vicinity of the Garzon Massif, Colombia. American Association of Petroleum Geologists, International Conference, Calgary, Abstract.

Roure, F., Howell, D.G., Guellec, S. & Casero, P. 1990a: Shallow structures induced by deep-seated thrusting. In: Letouzey, J. (Ed.): Petroleum tectonics in Mobile Belts. Technip, Paris, 15–30.

Roure, F., Howell, D.G., Moretti, I. & Müller, C. 1990b: Neogene subduction complex of Sicily. Journal of Structural Geology, 12, 259–266.

Roure, F., Nazaj, S., Mushka, K., Fili, I., Cadet, J.P. & Bonneau, M. 2004: Kinematic evolution and petroleum systems: an appraisal of the Outer Albanides. In: McKlay, K. (Ed.): Thrust Tectonics and Hydrocarbon Systems. American Association of Petroleum Geologists, Memoir 82, Ch. 24, 474–493.

Roure, F., Sadiku, U. & Valbona, U. 1995: Petroleum geology of the Albanian foothills. Americal Association of Petroleum Geologists, International Conference, Nice, Guide-Book, 100 pp.

Roure, F. & Sassi, W. 1995: Kinematics of deformation and petroleum system appraisal in Neogene foreland fold-and-thrust belts. Petroleum Geoscience, 1, 253–269.

Roure, F., Séguret, M. & Villien, A. 1988: Structural styles of the Pyrénées: a view from seismic reflexion to surface studies. Guide-Book American Association of Petroleum Geologists, Mediterranean basins conference, Nice, Field-Trip 3, 140 pp.

Roure, F., Swennen, R., Schneider, F., Faure, J.L., Ferket, H., Guilhaumou, N., Osadetz, K., Robion, Ph. & Vendeginste, V. 2005b: Incidence and importance of Tectonics and natural fluid migration on reservoir evolution in foreland fold-and-thrust belts. In: Brosse, E. et al. (Eds.): Oil and Gas Science and Technology, Revue de l'Institut Français du Pétrole, 60, 67–106.

Royden L.H., 1985: The Vienna Basin: a thin-skinned pull-apart basin. In: Biddle K.T., and Christie-Blick N. (Eds.): Strike slip deformation, basin formation and sedimentation. SEPM Society for Sedimentary Geology, Special Publication., 37, 313–338.

Royden, L.H. & Karner, G.D. 1984: Flexure of the lithosphere beneath Apennine and Carpathian foredeep basins: evidences for an insufficient topographic load. American Association of Petroleum Geologists Bulletin, 68, 704–712.

Sassi, W., Graham, R., Gillcrist, R., Adams, M. & Gomez, R. 2007: The impact of fedormation timing on the prospectivity of the Middle Magdalena sub-thrust, Colombia. In: Ries, A.C., Butler, R.W.H. and Graham, R.H. (Eds.): Deformation of the continental crust: The legacy of Mike Coward. Geological Society, London, Special Publication, 272, 473–498.

Sauer, R., Seifert, P. & Wessely, G. 1992: Guidebook to excursion in the Vienna Basin and the adjacent Alpine-Carpathian thrustbelt in Austria. Mitteilungen der Österreichischen Geologischen Gesellschaft, 85.

Schmid S.M., Fügenschuh B., Kissling E. & Schuster R., 2004: TRANSMED transects IV, V and VI: Three lithospheric transects across the Alps and their forelands. In: Cavazza W., Roure F., Spakman W., Stampfli G.M. & Ziegler P.A. (Eds.): The TRANSMED Atlas: The Mediterranean region from crust to mantle, Springer Verlag.

Schneider, F. 2003: Basin modeling in complex area: examples from Eastern Venezuelan and Canadian foothills. Oil and Gas Science and Tehnology, Revue de l'Institut Français du Pétrole, 58, 2, 313–324.

Schneider, F., Pagel, M. & Hernandez, E. 2004: Bsin-modeling in complex areas: example from the Eastern Venezuelan foothills. In: Swennen, R., Roure, F. & Granath, J. (Eds.): Deformation, fluid flow and reservoir appraisal in foreland fold-and-thrust belts. American Association of Petroleum Geologists, Hedberg Memoir, 1.

Scrocca, D., Carminati, E., Doglioni, C. & Marcantoni, D. 2007: Slab retreat and active shortening along the Central-Northern Apennines. In: Lacombe, O., Lavé, J., Roure, F. and Vergés, J. (Eds.): Thrust belts and foreland basins. Springer, Ch. 25, 471–487.

Séguret, M. & Benedicto, A. 1999: Le duplex à plis de propagation de rampes de Cazedarnes (Arc de St-Chinian, avant-pays nord-pyrénéen, France). Bulletin de la Société Géologique de France, 170, 1, 31–44.

Séguret, M., Séranne, M., Chauvet, A. & Brunel, M. 1989: Collapse-basin: a new type of extensional sedimentary basin from the Devonian of Norway. Geology, 17, 127–130.

Seifert, P. 1996: Sedimentary-tectonic development and Austrian hydrocarbon potential of the Vienna Basin. In: Wessely, G. & Liebl, W. (Eds.): Oil and gas in Alpidic thrustbelts and basins of Central and Eastern Europe. European Association of Geoscientists and Engineers, Special Publication, 5, 331–342.

Séranne, M., Chauvet, A., Séguret, M. & Brunel, M. 1989: Tectonics of the Devonian collapse basin of western Norway. Bulletin de la Société Géologique de France, 8, V, 489–499.

Séranne M., 1999: The Gulf of Lion continental margin (NW Mediterranean basins: Tertiary extension within the Alpine orogen. The Geological Society Special Publication, 156, 15–36.

Séranne M., Benedicto A., Truffert C., Pascal G. & Labaume P., 1995: Structural style and evolution of the Gulf of Lion Oligo-Miocene rifting: Role of the Pyrenean orogeny. Marine and Petroleum Geology, 12, 809–820.

Smit J.H.W., Brun J.P. & Sokoutis D., 2003: Deformation of brittle-ductile thrust wedges in experiment and nature. Journal of Geophysical Research, 108, B10, 2480.

Spakman, W. & Wortel, R. 2004: A tomographic view on Western Mediterranean geodynamics. In: Cavazza, W., Roure, F., Spakman, W., Stampfli G.M. & Ziegler, P.A. (Eds.): The Transmed Atlas. Springer, Berlin, Heidelberg, 31–52.

Swennen, R., Roure, F. & Granath, J. (Eds.) 2004: Deformation, fluid flow and reservoir appraisal in foreland fold-and-thrust belts. American Association of Petroleum Geologists, Hedberg Memoir, 1.

Tari G., Dövenyi P., Dunkl I., Horvath F., Lenkey L., Stefanescu M., Szafian P. & Toth T., 1999: Lithospheric structure of the Pannonian Basin derived from seismic, gravity and geothermal data. The Geological Society, London, Special Publication, 156, 215–250.

Toro, J., Roure, F., Bordas-Lefloch, N., Le Cornec-Lance, S. & Sassi, W. 2004: Thermal and kinematic evolution of the Eastern Cordillera fold and thrust belt, Colombia. In: Swennen, R., Roure, F. & Granath, J.W. (Eds.): Deformation, fluid flow and reservoir appraisal in fold and thrust belts. American Association of Petroleum Geologists, Hedberg Series, 1, 79–115.

van der Meulen, M.J., Meulenkamp, J.E. & Wortel, M.J.R. 1998: Lateral shifts of Apenninic fordeep depocenters reflecting detachment of subducted lithosphere. Eart and Planetary Science Letters, 154, 203–219.

von Huene, R. 1986: Seismic images of modern convergent margin tectonic structure. American Association of Petroleum Geology, Studies, 26, 1–60.

von Huene, R. & Scholl, D.W., 1991: Observations at convergent margins concerning sediment subduction, subduction erosion and the growth of continental crust. Reviews of Geophysics, 29, 3, 279–316.

Watts, A.B., Bodine, J.H. & Ribe, N.R. 1980: Observations of flexure and the geological evolution of the Pacific ocean basin. Nature, 238, 532–537.

Watts, A., Karner, G.D. & Steckler, M.S. 1982: Lithospheric flexure and the evolution of sedimentary basins. Philosophical Transactions of the Royal Society of London, Ser. A, 305, 249–281.

Wernicke, B. 1981: Low-angle normal faults in the Basin and Range Province: nappe tectonics in an extending orogen. Nature, 291, 645–647.

Wessely G., 1988: Structure and development of the Vienna Basin in Austria. In Royden L.H. and Horvath F., eds., American Association of Petroleum Geologists, Memoir, 45, 333–346.

Wortel, M.J.R. & Spakman, W. 1992: Structure and dynamics of subducted lithosphere in the Mediterranean region. Proceedings of the Koninklijke Nederlandse Akademie van Wetenschappen, 95, 325–347.

Wortel M.J.R. & Spakman W., 2000: Subduction and slab detachment in the Mediterranean-Carpathian region. Science, 290, 1910–1917.

Ziegler P.A., Bertottti G. & Cloetingh S., 2002: Dynamic processes controlling foreland development: the role of mechanical de(coupling) of orogenic wedges and forelands. In: Bertotti G., Schulmann K. & Cloetingh S. (Eds.): Continental collision and tectonic-sedimentary evolution of forelands, European Geophysical Society, 1, Stephan Müller Special Publication, 29–91.

Ziegler, P. & Roure, F. 1996: Architecture and petroleum systems of the Alpine orogen and associated basins. In: Ziegler, P. & Horvath, F. (Eds.): PeriTethys Memoir 2. Museum national d'Histoire Naturelle, Paris, 15–46.

Ziegler, P. & Roure, F. 1999: Petroleum systems of Alpine-Mediterranean foldbelts and basins. In: Durand, B., Jolivet, L., Horvath, F. and Séranne, M. (Eds.): Geological Society, London, Special Publication, 156, 517–540.

Ziegler P.A., van Wees J.D. & Cloetingh S., 1998. Mechanical controls in collision-related compressional intraplate deformation. Tectonophysics, 300, 103–129.

Zoetemeijer, R., Cloetingh, S., Sassi, W. & Roure, F. 1992: Stratigraphic sequences in piggyback basins: records of tectonic evolution. Tectonophysics, 226, 253–269.

Zoetemeijer, R., Cloetingh, S., Sassi, W. & Roure, F. 1993: Stratigraphic sequences in piggyback basins: records of tectonic evolution. Tectonophysics, 226, 253–269.

Manuscript received Dezember 19, 2007
Revision accepted June 27, 2008
Published Online first November 13, 2008
Editorial Handling: Stefan Schmid & Stefan Bucher

1661-8726/08/01S031-24
DOI 10.1007/s00015-008-1291-z
Birkhäuser Verlag, Basel, 2008

Swiss J. Geosci. 101 (2008) Supplement 1, S31–S54

Provenance of the Bosnian Flysch

Tamás Mikes [1,7,*], Dominik Christ [1,8], Rüdiger Petri [1], István Dunkl [1], Dirk Frei [2], Mária Báldi-Beke [3], Joachim Reitner [4], Klaus Wemmer [5], Hazim Hrvatović [6] & Hilmar von Eynatten [1]

Key words: Dinarides, Adriatic plate, ophiolite, flysch, Cretaceous, provenance, geochronology, biostratigraphy, mineral chemistry

ABSTRACT

Sandwiched between the Adriatic Carbonate Platform and the Dinaride Ophiolite Zone, the Bosnian Flysch forms a *c.* 3000 m thick, intensely folded stack of Upper Jurassic to Cretaceous mixed carbonate and siliciclastic sediments in the Dinarides. New petrographic, heavy mineral, zircon U/Pb and fission-track data as well as biostratigraphic evidence allow us to reconstruct the palaeogeology of the source areas of the Bosnian Flysch basin in late Mesozoic times. Middle Jurassic intraoceanic subduction of the Neotethys was shortly followed by exhumation of the overriding oceanic plate. Trench sedimentation was controlled by a dual sediment supply from the sub-ophiolitic high-grade metamorphic soles and from the distal continental margin of the Adriatic plate. Following obduction onto Adria, from the Jurassic–Cretaceous transition onwards a vast clastic wedge (Vranduk Formation) was developed in front of the leading edge, fed by continental basement units of Adria that experienced Early Cretaceous synsedimentary cooling, by the overlying ophi-

olitic thrust sheets and by redeposited elements of coeval Urgonian facies reefs grown on the thrust wedge complex. Following mid-Cretaceous deformation and thermal overprint of the Vranduk Formation, the depozone migrated further towards SW and received increasing amounts of redeposited carbonate detritus released from the Adriatic Carbonate Platform margin (Ugar Formation). Subordinate siliciclastic source components indicate changing source rocks on the upper plate, with ophiolites becoming subordinate. The zone of the continental basement previously affected by the Late Jurassic–Early Cretaceous thermal imprint has been removed; instead, the basement mostly supplied detritus with a wide range of pre-Jurassic cooling ages. However, a *c.* 80 Ma, largely synsedimentary cooling event is also recorded by the Ugar Formation, that contrasts the predominantly Early Cretaceous cooling of the Adriatic basement and suggests, at least locally, a fast exhumation.

Introduction

One of the most peculiar tectonostratigraphic units of the Dinaride orogen within the SE European Alpine system is the 'Zone Bosniaque', defined by Aubouin et al. (1970). It is tectonically sandwiched between the most external belt of Dinaride ophiolites in the NE that are floored by continental basement nappes (Aubouin 1973) derived from the Adriatic plate (Schmid et al. 2008), and units of the vast Mesozoic carbonate platform of the Adriatic plate in the SW. The Bosnian Zone mainly comprises thick Late Jurassic to Cretaceous flysch successions and other gravity flow deposits. We will collectively refer to them as Bosnian Flysch hereafter.

The presence of ophiolitic detritus in the Bosnian Flysch had been recognized early on (Blanchet 1966; Blanchet et al.

1969; Charvet 1970; Olujić et al. 1978), yet subsequent tectonic models assessed its importance diversely. Aubouin (1973) suggested that, as a result of Late Jurassic obduction, a foredeep was formed in front of the ophiolite nappes and demonstrated that the lower part of the Bosnian Flysch represents a synorogenic sequence. This view was largely contended by Lawrence et al. (1995), Tari & Pamić (1998), Tari (2002) and Schmid et al. (2008). As a marked contrast, Pamić (1993) and Pamić et al. (1998) interpreted the Bosnian Flysch in terms of a sequence that was deposited on the NE passive margin of Adria; a genetic relationship of the sediments to the Dinaride ophiolites was not part of their models.

In this study a wide range of sedimentary provenance information was acquired to decipher the Late Mesozoic evolution of the source area geology of the Bosnian Flysch. The data were

[1] Abteilung Sedimentologie/Umweltgeologie, Geowissenschaftliches Zentrum Göttingen, Goldschmidtstr. 3, D-37077 Göttingen, Germany.
[2] De Nationale Geologiske Undersøgelser for Danmark og Grønland – GEUS, Øster Voldgade 10, DK-1350 København K, Denmark.
[3] Rákóczi utca 42, H-2096 Üröm, Hungary.
[4] Abteilung Geobiologie, Geowissenschaftliches Zentrum Göttingen, Goldschmidtstr. 3, D-37077 Göttingen, Germany.
[5] Abteilung Isotopengeologie, Geowissenschaftliches Zentrum Göttingen, Goldschmidtstr. 3, D-37077 Göttingen, Germany.
[6] Geološki zavod Bosne i Hercegovine, Ustanička 11, BH-71210 Ilidža-Sarajevo, Bosnia and Herzegovina.
[7] Current address: Institut für Geologie, Universität Hannover, Callinstr. 30, D-30167 Hannover, Germany.
[8] Current address: E.ON Ruhrgas AG, Huttropstr. 60, D-45138 Essen, Germany.
*Corresponding author. E-mail: tamas.mikes@geo.uni-goettingen.de

used to assess existing models addressing the geodynamic setting of the flysch. Arenitic samples were analysed for a detailed description of the lithology and age of the source rocks, using heavy mineral signatures and zircon chronology. Clay mineralogical and whole-rock geochemical methods were employed to characterize the provenance of the fine-grained sediments. Our investigations were completed by calcareous nannofossil and carbonate microfacies data, which put additional constraints on the biostratigraphic range.

Tectonic framework

The Bosnian Flysch is a 500 km long belt of Late Jurassic to Cretaceous, mixed siliciclastic-carbonate sequences incorporated into the Dinaride nappe pile (Fig. 1). In the NE, the flysch is tectonically overlain by the East Bosnian–Durmitor thrust sheet, which passively carries Dinaride ophiolite units previously thrusted onto it (Schmid et al. 2008; cf. 'Zone Serbe' of Aubouin 1973). The East Bosnian–Durmitor unit wedges out towards the NW, where the Dinaride Ophiolite Zone appears to directly overlie the Bosnian Flysch.

To the SW, the Bosnian Flysch is structurally underlain by the Pre-Karst Subzone and in turn by the Main Karst Zone (Aubouin et al. 1970). The latter largely corresponds to the Adriatic Carbonate Platform (AdCP; defined by Vlahović et al. 2005), the former representing the heteropic, distal slope and basin facies bordering the platform. Overall facies distribution suggests that the AdCP, the largest Mesozoic platform of the Adriatic plate, acted as a palaeotopographic entity during the evolution of the Bosnian Flysch basin (Charvet 1980; Vlahović et al. 2005). It was converted into the tectonic footwall of the Outer Dinaride thrust pile during Tertiary compression (Aubouin et al. 1970; Chorowicz 1977; Charvet 1980; Tari 2002; Mikes et al. 2008; Schmid et al. 2008).

With respect to the tectonic setting of the ophiolites and other Inner Dinaride structural elements (Fig. 1), we will largely follow the interpretation of a recent kinematic reconstruction made by Schmid et al. (2008). These authors suggest that both belts of Dinaride Triassic-Jurassic oceanic units represent displaced fragments of an initially single zone of Neotethyan ophiolites, which were obducted westwards onto the Adria passive margin in the Late Jurassic. Cretaceous and Tertiary thrusting gave rise to far-travelled thrust sheets composed of continental and ophiolitic series. The orogen-parallel, allochthonous Palaeozoic to Triassic units represent continental basement nappes derived from the distal Adriatic plate, exposed mostly in elongated tectonic windows. This interpretation of Schmid et al. (2008), which is in line with ideas already put forward by, for example, Bernoulli & Laubscher (1972), Charvet (1980), and Gawlick et al. (2008), fundamentally differs from most other models which either propose two Mesozoic oceanic branches separated by a continental microplate (e.g. Dimitrijević &

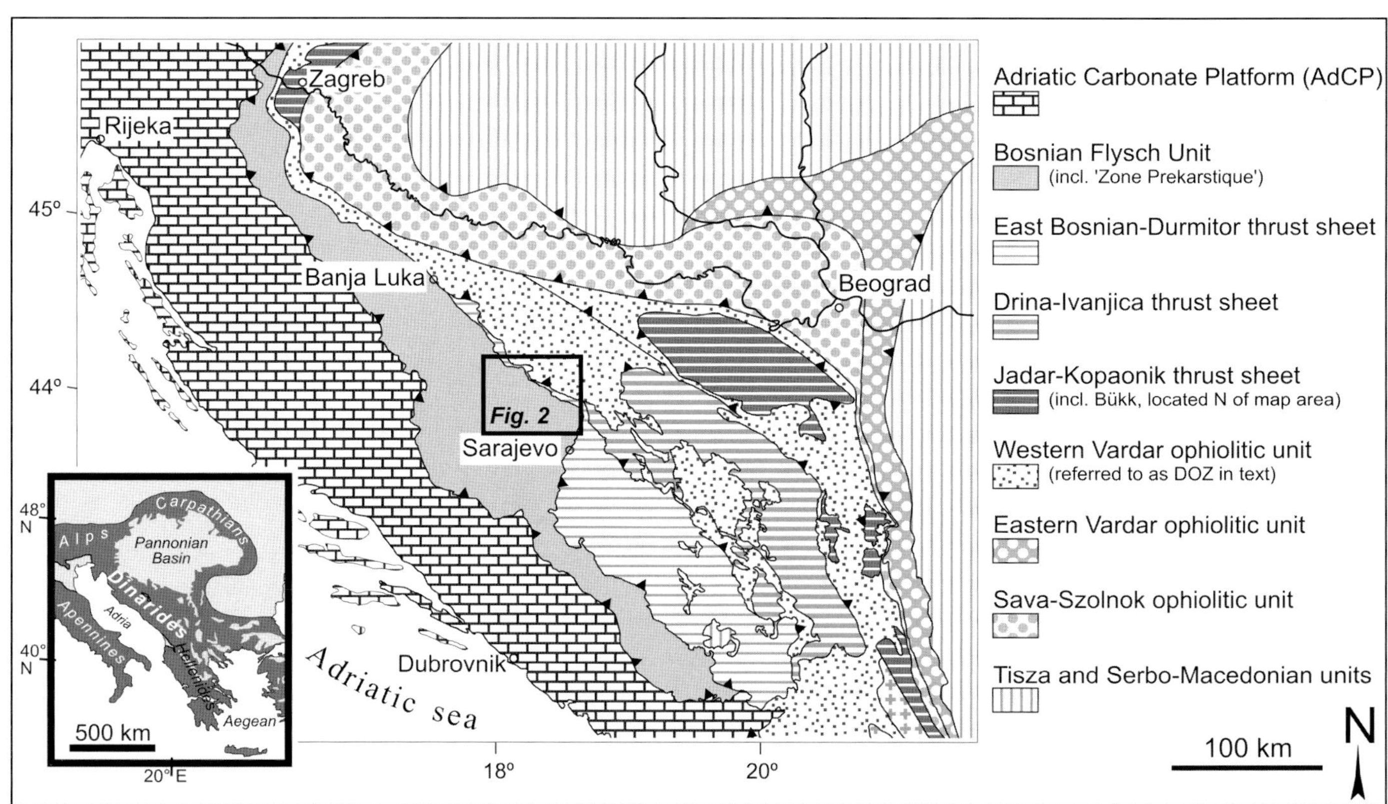

Fig. 1. Map showing the major structural units of the Dinarides (after Schmid et al. 2008, slightly modified).

Dimitrijević 1973; Robertson & Karamata 1994; Karamata 2006), or assume that the basement nappes were derived from the European margin by out-of-sequence thrusting (Pamić et al. 1998; Hrvatović & Pamić 2005).

The Jurassic ophiolites structurally above the Bosnian Flysch were described as the Dinaride Ophiolite Zone (DOZ; Dimitrijević & Dimitrijević 1973; Pamić et al. 2002), Central Dinaride Ophiolite Belt (CDOB; Lugović et al. 1991; Babić et al. 2002) or as the zone constituting the most externally transported thrust sheets of the Western Vardar ophiolites (Schmid et al. 2008), or as forming part of the Serb, Golija and Drinjača Zones according to Charvet (1978, 1980). These ophiolites, from now on referred to as Dinaride Ophiolite Zone (DOZ) in this paper (Fig. 1), were recently argued to have been formed in an intraoceanic supra-subduction zone setting (Bazylev et al. 2006; Smith 2006; Lugović et al. 2006, 2007). K/Ar, Ar/Ar and Sm/Nd age data of sub-ophiolitic metamorphic soles from Dinaride and Albanian ophiolites range between 178 and 161 Ma, indicating Middle Jurassic intraoceanic thrusting (e.g. Okrusch et al. 1978; Parlak & Delaloye 1999; Dimo-Lahitte et al. 2001; Olker et al. 2001; Smith 2006 and references therein). Petrology of these soles indicates both basaltic (e.g. Pamić et al. 1973; Majer et al. 2003; Operta et al. 2003; Schuster et al. 2007) and sedimentary protoliths (e.g. Karamata et al. 1970; Schreyer & Abraham 1977; Okrusch et al. 1978; Carosi et al. 1996).

Ophiolite obduction onto the Adriatic margin was completed by Late Jurassic. Deep crustal levels of both the oceanic plate and the Adriatic margin suffered largely coeval (Okrusch et al. 1978; Majer & Lugović 1991; Milovanović et al. 1995; Most 2003) HP/LT metamorphic overprint as is indicated by blueschist facies rocks occasionally found among the metapelitic to metabasic rocks on both the upper and lower plates (Majer 1956; Charvet 1978; Djoković 1985; Majer & Lugović 1991; Mutić & Dmitrović 1991; Milovanović et al. 1995; Belak & Tibljaš 1998; Most 2003). Stratigraphic constraints on the timing of obduction are provided by the Tithonian-Berriasian age of the oldest sedimentary deposits sealing the ophiolites, consisting of alluvial coarse-grained siliciclastic strata and shallow marine carbonates interfingering with them (Blanchet et al. 1970; Charvet & Termier 1971; Charvet 1973, 1978; Neubauer et al. 2003). This unconformity is, however, a diachronous surface; the overlying sediments become progressively younger towards the more internal domains of the DOZ, attaining Cenomanian age at the NE border of the DOZ. These observations indicate an overall transgressive trend from the Tithonian/Berriasian to the Cenomanian, perpendicular to the strike of the orogen (Charvet 1978, 1980).

The continental units in the Dinarides are considered to represent major, Adria-derived thrust sheets (see discussion in Schmid et al. 2008). The East Bosnian–Durmitor Unit, together with the Drina–Ivanjica, Jadar, Kopaonik, Medvednica and the displaced Bükk units, are dominated by Palaeozoic to Triassic (meta-)sediments (Podubsky 1970; Rampnoux 1970; Djoković 1985; Dimitrijević 1997; Pamić & Jurković 2002), which underwent regional thermal overprint ranging up to anchi- to epizonal conditions. Early Cretaceous cooling is widely demonstrated by K/Ar age data yielding 135 ± 11 Ma in the Drina–Ivanjica Unit (Milovanović 1984), 118 ± 4 Ma (Belak et al. 1995) as well as 107 ± 8 Ma (Judik et al. 2006) in the metapelites of the Medvednica Unit, as well as various ages from 133 to 98 Ma in the Bükk Unit, in agreement with zircon FT age data in this latter tectonic unit (see details in Árkai et al. 1995). Ar/Ar age spectra of detrital white mica from the Ljig Flysch (in the External Vardar Subzone *sensu* Dimitrijević 1997; covering the Jadar-Kopaonik thrust sheet according to Schmid et al. 2008) contain a 110 Ma age component (Ilić et al. 2005). Finally, the Palaeozoic of the Bosnian Schist Mts., considered to represent the basement of the Pre-Karst Subzone (Aubouin et al. 1970; Schmid et al. 2008) also records this thermal event (K/Ar ages from 121 to 92 Ma) prior to its main phase of Paleogene cooling (see details in Pamić et al. 2004).

Geological setting of the Bosnian Flysch

The Bosnian Flysch forms a rather uniform belt (Figs. 1 & 2): formations comparable to those investigated by this study crop out to the NW of Central Bosnia, in the Zrinska Gora and in the Slovenian Trough (Aubouin et al. 1970; Cousin 1972; Babić & Zupanič 1976; Bušer 1987; Hrvatović 1999; Rožič 2005), and in form of the so-called Durmitor Flysch in the SE (Dimitrijević & Dimitrijević 1968; Blanchet et al. 1969; Rampnoux 1969; Aubouin et al. 1970).

Two distinct lithostratigraphic units characterize the Bosnian Flysch (Olujić 1978; Hrvatović 1999). The lower, turbiditic to monotonous pelagic series is more than 1000 m thick, and is dominantly composed of siliciclastic sandstones, marls, shales, cherty micritic limestones, and grey radiolarites (Vranduk Formation). It stretches along the NE side of the flysch zone (Fig. 2), i.e. nearer to the overriding nappes and hence more proximal to the source. The upper succession (Ugar Formation; Fig. 2) occurs SW of the Vranduk Formation and is carbonate-dominated, comprising thin-bedded marly to micritic limestones and red or grey shales, intercalated with calcareous turbidites, and finally, coarse catastrophic carbonate mass flow deposits, several tens of metres thick, in the upper part of the sequence. These carbonate debrites contain large Scaglia Rossa clasts up to several metres across, exhibiting internal slump folds. Stratigraphic thickness of the Ugar Formation exceeds 2000 m (Hrvatović 1999). Palaeotransport indicators suggest S- to SE-directed shedding for the Vranduk Formation and SE to NE-directed transport for the Ugar Formation (Dimitrijević & Dimitrijević 1968; Hrvatović 1999). The described formations largely correspond to the sediments of the 'flysch bosniaque interne' of Charvet (1978).

Calpionellids and foraminifera suggest that the age of the Vranduk Formation in Central Bosnia ranges mostly from the Tithonian to the Berriasian-Valanginian, whereas the Ugar Formation ranges in age from Late Albian to Maastrichtian, and locally, into the Paleocene (Cadet 1968; Dimitrijević & Dimitrijević 1968; Charvet 1978; Olujić et al. 1978). To the

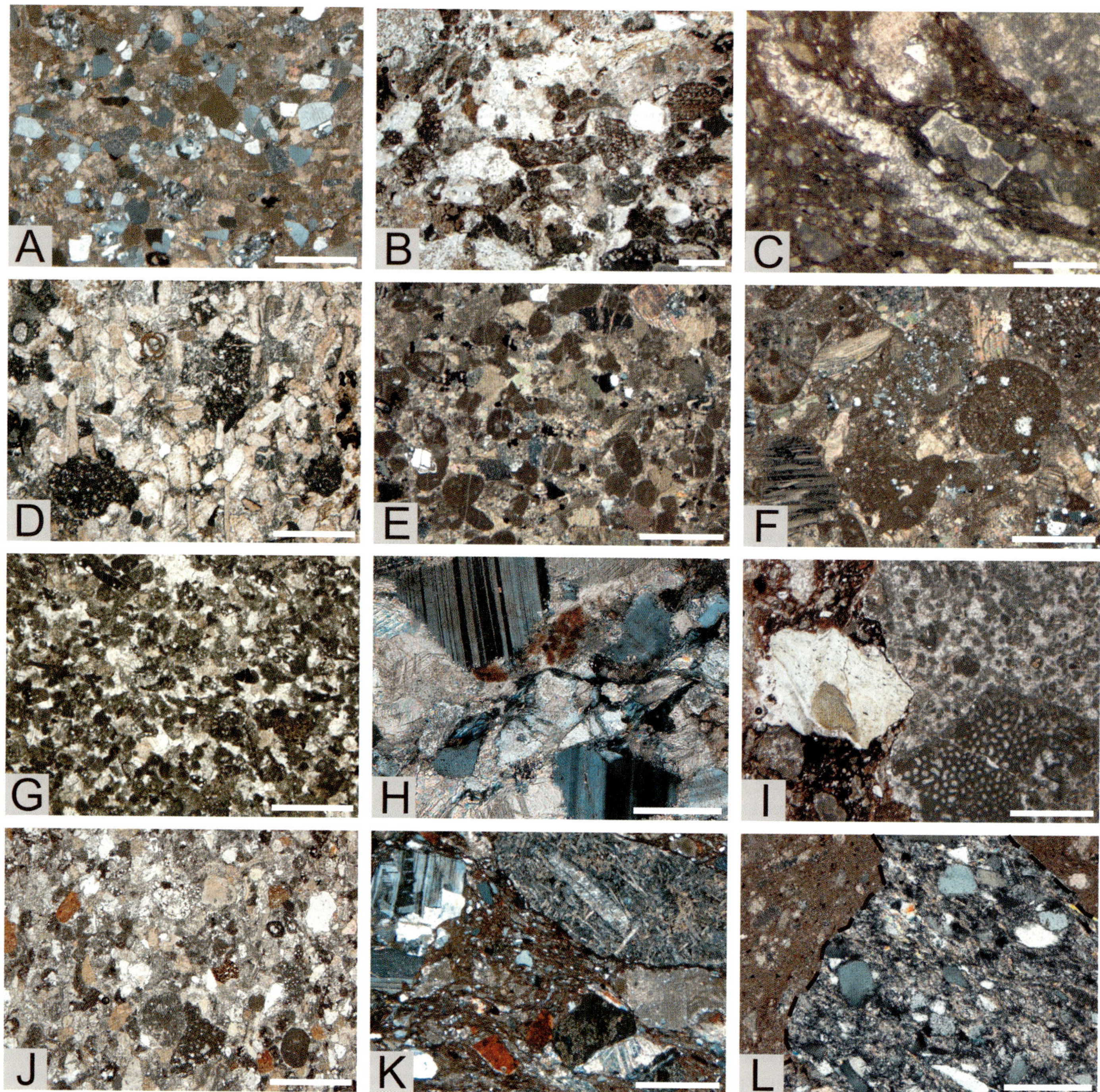

Fig. 4. Microphotographs of characteristic sandstone and breccia varieties in the Ugar Formation (a–f) and Vranduk Formation (g–l) of the Bosnian Flysch. +N: crossed polarizers. Scale bar is usually 1 mm; for (c), (h) and (l) it is 250 μm. (a) Calcareous litharenite with mono- and polycrystalline quartz, chert, micritic carbonate rock fragments and bioclasts, +N, sample BO-31, (b) calcarenite with orbitolinids, BO-18, (c) bicarinate *Globotruncana* sp. indicative of a Senonian age, in the red pelitic matrix of a matrix-supported carbonate breccia, BO-36, (d) calcarenite with bioclasts and red biomicrite rock fragments, BO-129, (e) calcarenite with peloids, bioclasts, chert and monocrystalline quartz, +N, BO-5, (f) calcarenite with orbitolinid and rudistid bioclasts, +N, BO-51, (g) calcarenite (pelsparite) with chert and Cr-spinel grains, BO-85/b, (h) grains of plagioclase feldspar and quartz in calcareous litharenite, +N, BO-75, (i) carbonate rock fragment (pelbiosparite) and chlorite flake in polymict breccia, BO-82/2, (j) calcareous litharenite with quartz, chert, radiolarite, carbonate rock fragments, orbitolinid and rudistid bioclasts, TD-153, (k) plagioclase, mafic volcanic lithoclast of intersertal texture, serpentinite and radiolarite in polymict breccia +N, BO-82/6, (l), sublitharenite rock fragment in red finer-grained matrix, +N, TD-145.

Many samples in the Vranduk and Ugar formations show characteristic diagenetic features. Replacement of chert and radiolarite lithoclasts and plagioclase by calcite patches or rhombohedra is common. Calcite often replaces clayey sandstone matrix, in addition to abundant calcite veinlets crosscutting the texture. Small (80–160 µm), euhedral albite crystals may occur in carbonate-dominated arenites. There, they replace micritic carbonate, especially peloids and ooids.

Whole-rock geochemistry

The concentrations of major and several trace elements of pelites and arenites from both flysch formations are summarized in Table T4 (electronic supplement). In the sample set a strong negative correlation appears between SiO_2 and CaO, with SiO_2 concentrations varying between 1.6 and 77.0 wt% and CaO ranging from 0.5 to 54.1 wt% (Fig. 5a). This trend is accompanied by a likewise strong positive correlation between CaO and LOI, indicating that CaO is almost entirely carbonate-bound. Therefore, a continuous mixing trend can be established from the Vranduk to the Ugar formations, where the siliciclastic proportions are diluted by additional carbonate to various degrees. A slight deviation from this trend is mostly related to pelitic samples having higher Al-proportions in the silicate fraction. Our data show a conspicuous stratigraphic control on this mixing trend, with the Vranduk Formation being richer in siliciclastic material, whereas in the Ugar Formation the carbonate proportion increases up to almost those of pure calcarenites. However, as is evident from Fig. 5a, the transition is continuous with minor overlap, and shows no abrupt jump in carbonate content between the Vranduk and Ugar formations.

Fig. 5b shows the results of the major element discrimination procedure based on Herron (1988). Most pelites fall into the field of shale, with no pelite sample reaching the Fe-shale field. By contrast, both the coarse and the fine- to medium-grained sandstones have mostly Fe-sand compositions. The reason for this behaviour is probably not only related to Fe, as most variation in analyses of carbonate-corrected Fe_2O_3 occurs in a limited range of 4 to 8 wt%; this is only slightly above the normal range for lithic arenites and greywackes from active continental margin settings (Pettijohn et al. 1987) and seems largely unrelated to grain-size. Rather, the tendency of K_2O to be enriched in pelites (2 to 3.5 wt%) relative to the sandstones, which show low K_2O concentrations (0.2 to 1.5 wt%) when compared to average lithic arenites and greywackes (Fig. 5b), could account for the high $Fe_2O_3^t/K_2O$ ratios of the sandstones.

Selected provenance-sensitive trace elements of the pelitic samples are displayed in the Cr/V vs. Y/Ni plot (Fig. 6a; McLennan et al. 1993) in order to reveal the nature of the source lithologies. Cr/V ratios range from ~0.5 to 3.4 and thus largely exceed average upper continental crust (UCC; according to McLennan 2001), whereas Y/Ni ratios are lower compared to UCC. For the sake of comparison, a mixing curve was also constructed, using the composition of the UCC and that of ultramafic rocks as end-members (Fig. 6a). The majority of the sam-

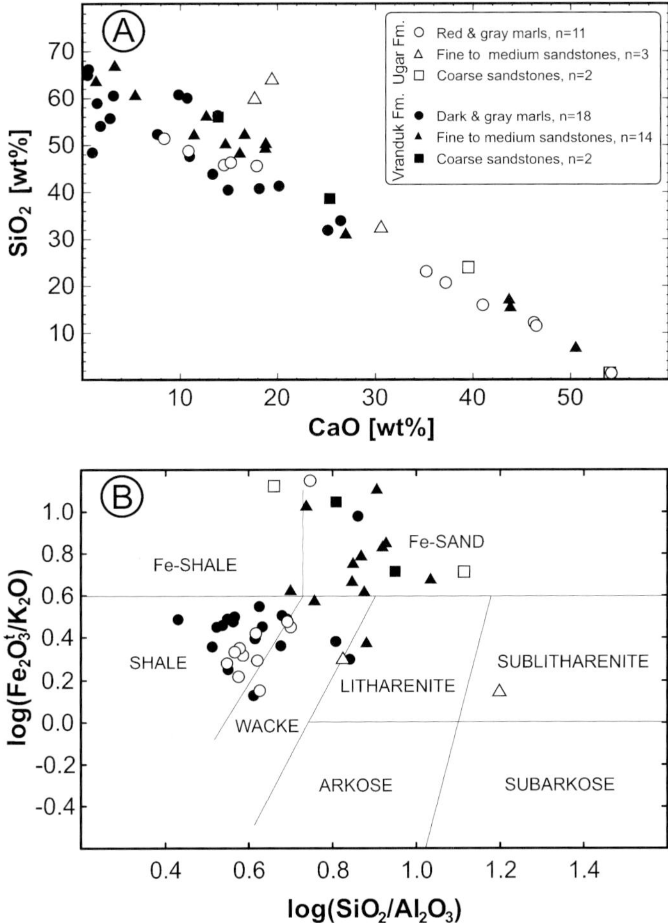

Fig. 5. Major element geochemistry of the Bosnian Flysch sediments. (a) CaO–SiO_2 plot showing the generally higher carbonate contents of the Ugar Formation samples in comparison to the Vranduk Formation (b) Classification of the flysch sandstones based on the ratios of SiO_2/Al_2O_3 vs. $Fe_2O_3^t/K_2O$ (Herron 1988).

ples plots close to this mixing curve, which provides evidence for the simultaneous erosion of ultramafic lithologies and felsic rocks. The proportion of mafic and/or ultramafic source rocks in most Vranduk samples is distinctly higher compared to the Ugar Formation. UCC-normalized, carbonate-corrected concentrations of Cr and Ni in shales and marls are in agreement with this observation (Fig. 6b). Ni, for example, is up to 10 times enriched in the Vranduk Formation (two more samples have even higher enrichment factors of 12 and 19). Contrary to this, the Ugar samples scatter around the UCC values, with the enrichment factors of Ni ranging between 0.4 and 3 only.

Cr/Ni ratios of the arenitic samples do not exceed (with one exception in Ugar Formation) the Cr/Ni values for ultramafic (2.0) and mafic rocks (5.2) of the Dinarides (Lugović et al. 1991; Robertson & Karamata 1994; Pamić et al. 2002). This indicates that sediment recycling and, consequently, Cr-spinel concentration is not a significant process in forming these sandstones (von Eynatten 2003). Similarly, Zr/Sc ratios (McLennan et al. 1993) give no evidence for significant zircon concentration.

Table 1. Semi-quantitative heavy mineral composition of arenites in the DOZ mélange, and the Vranduk and Ugar formations. Symbols refer to species abundances: triangles: predominant; x's: common; circles: subordinate; dots: very rare.

profile	unit	sample	spl	zrn	rt	grt	tur	ap	others
Stavnja Valley	Ugar Fm.	BO-17	X	○	○				mnz
		BO-117	○	○	·	○	○		ep, mnz
		BO-16	X	○	X	○	○	·	ttn
	Vranduk Fm.	BO-61	▲	X	○		X		ttn
		BO-59	▲	○	X		X		
		BO-115/b	▲	○	○		·		
		BO-52	▲	○	○		○	·	ttn, mnz
	DOZ mél.	BO-25	○	○	X	X		·	ep, ttn, zo, ky
		BO-23	○	X	○	X		·	cld, ttn, ep, zo, ky
		BO-22	○	X	○	X		·	ttn, ep, zo, czo, ky
Bosna Valley	Vranduk Fm.	BO-12	X	○	○	X		·	mnz
		BO-87	▲	○	X	X	X		
		BO-95	X	○	○	○	·	·	mnz, ttn
		BO-72	▲	X	○	○	·	·	mnz, xnt, ttn, aug
		BO-73	X	X	○			·	ttn
		BO-75	▲	X	○	·	·	·	ttn
		BO-92	X	○	X				
		BO-4	▲	X	X	○	·		ep

Clay mineralogy

Chlorite and illite are the dominant clay-sized minerals in pelites from all units examined. In addition, kaolinite in the Ugar Formation, and serpentine in the DOZ mélange matrix, amount to c. 10% of the entire clay mineral assemblages. Kaolinite and serpentine were not detected in the Vranduk Formation pelites. Smectite was detected in several Vranduk samples (refer to Petri 2007 for details); the proportion of smectite in the illite/smectite mixed layer structures is usually below 10% in both the <2 µm and <0.2 µm size fractions.

In the Vranduk Formation, the Kübler Index (KI) of the <0.2 µm size fractions ranges between 0.40 $\Delta°2\Theta$ and 0.98 $\Delta°2\Theta$. In the <2 µm size fractions of the same samples the KI values are consistently lower, varying between 0.24 $\Delta°2\Theta$ and 0.63 $\Delta°2\Theta$. These inconsistencies probably indicate the disturbing effects of detrital micas in the <2 µm fraction. The KI data, along with the presence of kaolinite, suggest that both the Vranduk and the Ugar formations experienced only diagenetic overprint. No obvious trend was revealed by the regional distribution of these data.

Heavy mineral analysis

The heavy mineral spectra of the Bosnian Flysch are overall dominated by Cr-spinel, especially in the Vranduk Formation. Further species include zircon, rutile, garnet, tourmaline, and lesser amounts of apatite, titanite and monazite (Table 1). In the DOZ mélange, however, Cr-spinel is not predominant, tourmaline and monazite are even absent. The main constituents of the heavy mineral spectra in the DOZ mélange are garnet and zircon, chloritoid, kyanite, clinozoisite, and epidote. Epidote is locally also found in the Vranduk Formation (Table 1).

The Ugar Formation differs from the Vranduk Formation by its subordinate garnet content, lower average Cr-spinel concentrations, and the absence of epidote (Table 1). Whereas a significant proportion of zircon in the Vranduk Formation is euhedral with sharp crystal edges, zircons of the Ugar Formation are typically subhedral to rounded, indicating a metapelitic or a mature sedimentary source. Chemical compositions of selected heavy mineral species were determined by electron microprobe to characterize their source in more detail.

Tourmaline: Nearly all crystals are derived from metapelitic sources and have largely similar compositions in the Vranduk and Ugar formations (Fig. 7). The proportion of magmatic-derived tourmaline crystals is rather small. Well-preserved growth zoning and compositional polarity, as revealed by back-scattered electron images, correspond mainly to a Barrovian type, lower-grade metamorphic source (cf. Henry & Dutrow 1996). If outer and inner metamorphic zone compositions are plotted separately (Fig. 7), a general enrichment in Fe is observed towards the rims. The predominance of tourmaline of metamorphic origin is also supported by their high Ti and low Zn contents (up to 0.2 and 0.015 atoms per formula unit, respectively; see Viator 2003).

Garnet chemistry indicates a variety of source lithologies, the most dominant being greenschist facies metamorphic rocks. Additional garnet populations are also observed: one is derived, according to the classification scheme of Morton et al. (2003), from amphibolite to granulite facies metapelites (along the almandine-pyrope join) and another one from amphibolite to eclogite facies metabasic rocks (pyrope- and grossular-rich almandines; Fig. 8). The share of the contribution from these high-grade sources is the highest in the DOZ mélange (in total ~45%). It then decreases significantly in the Vranduk Formation (~21%) and probably disappears in the Ugar Fm (only a few grains that plot close to the boundaries; Fig. 8). A small proportion of garnets (c. 4%) derived from skarns or low-grade metabasic rocks also occurs in the DOZ mélange.

Rutile crystals are mostly (60 to 80% throughout the samples) of metapelitic origin, as deduced from their Cr/Nb ratio (Zack et al. 2004a, Triebold et al. 2007), with the remaining portion derived from metamafic lithologies (Table T6, electronic supplement). Zr-in-rutile thermometry was performed on the metapelitic crystals in order to assess source rock metamorphic conditions following Zack et al. (2004b) and Watson et

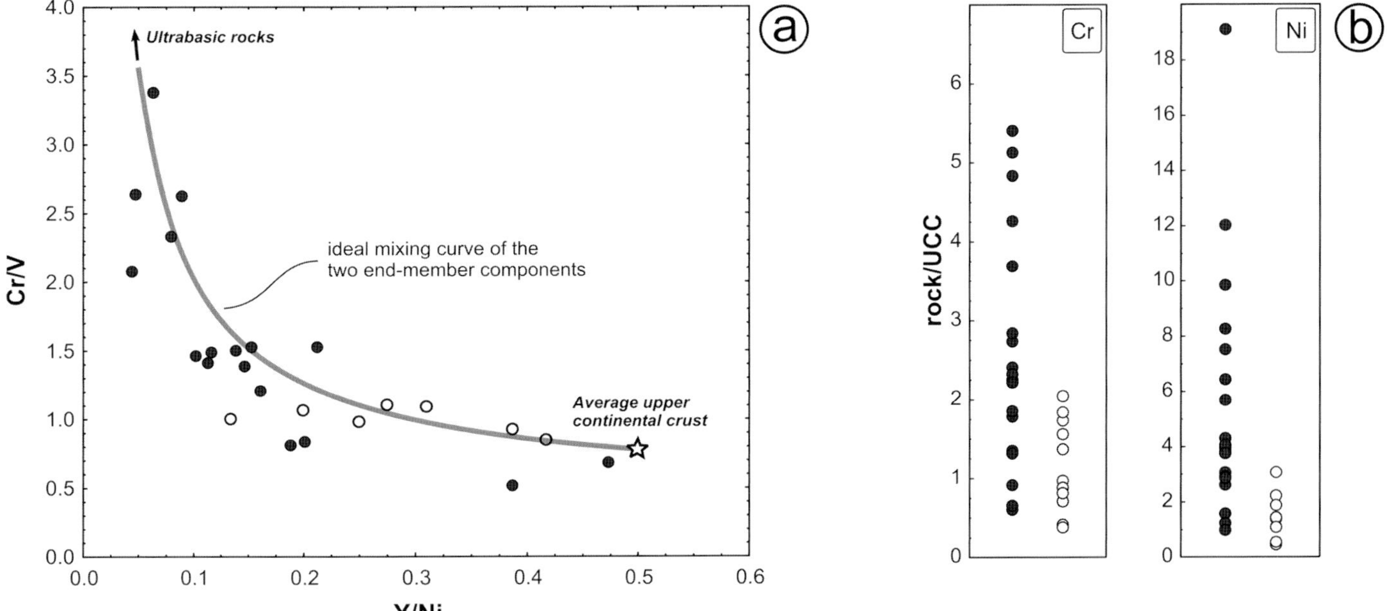

Fig. 6. Trace element geochemistry of the Vranduk and Ugar formations. Only pelitic samples are plotted. Filled and open circles indicate Vranduk and Ugar samples, respectively. (a) Cr/V vs. Y/Ni ratio-ratio plot (McLennan et al., 1993). Calculated mixing line uses ultrabasic (Cr/V = 40; Y/Ni ~ 0.0003 – Turekian & Wedepohl 1961) and average upper continental crust (UCC) compositions (Cr/V = 0.775; Y/Ni = 0.5 – McLennan 2001) as end-members. Star refers to UCC composition. Four Ugar Formation samples had Y and/or Ni concentrations below instrumental detection limit and are excluded from the plot. (b) Carbonate-corrected, UCC-normalized values of Cr and Ni.

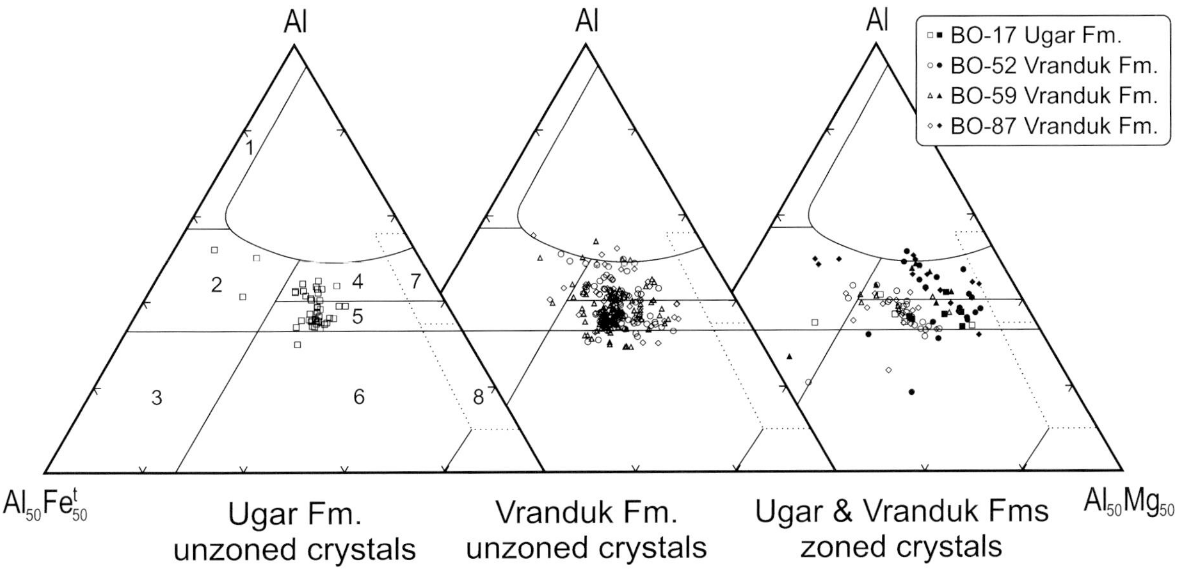

Fig. 7. Ternary plots of tourmaline chemistry in the Fe-Mg-Al system and assignment to probable source lithologies according to Henry & Guidotti (1985). Numbered fields correspond to tourmaline compositions dominantly in (1) Li-rich granitoids, (2) Li-poor granitoids, (3) hydrothermally altered granites; (4–5): metapelites (4) coexisting or (5) not coexisting with an Al-saturating phase, (6) skarns, (7) metamorphosed ultramafic rocks, (8) metacarbonates. Filled symbols: cores or inner zones, open symbols: composition of outer zones or of unzoned crystals.

al. (2006). The results yield a broad distribution of calculated metamorphic temperatures between *c.* 500 and 900 °C (Fig. 9). The distribution of inferred temperatures in the detritus of the DOZ mélange most likely consists of two components: one ranging from 650–700 to ~850 °C, and one between 500 and 650–700 °C. This is different from the both flysch formations which appear to be dominated by a single population with a mode between 550 and 700 °C (Fig. 9).

Cr-spinel chemical data from the DOZ mélange and from the Vranduk and Ugar formations indicate abundant (65 to

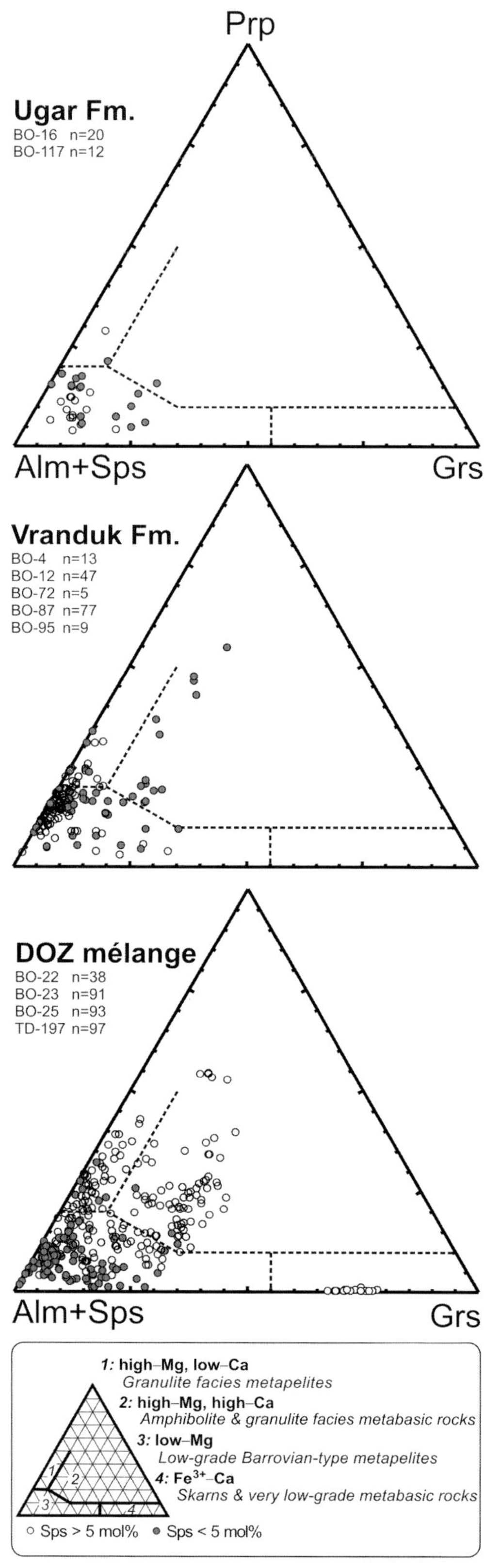

Ugar Fm.
BO-16 n=20
BO-117 n=12

Vranduk Fm.
BO-4   n=13
BO-12 n=47
BO-72 n=5
BO-87 n=77
BO-95 n=9

DOZ mélange
BO-22 n=38
BO-23 n=91
BO-25 n=93
TD-197 n=97

Fig. 8. Comparison of detrital garnet compositions from the DOZ mélange and the Vranduk and Ugar formations, according to the classification scheme of Morton et al. (2003). Alm: almandine, Sps: spessartite, Prp: pyrope, Grs: grossular.

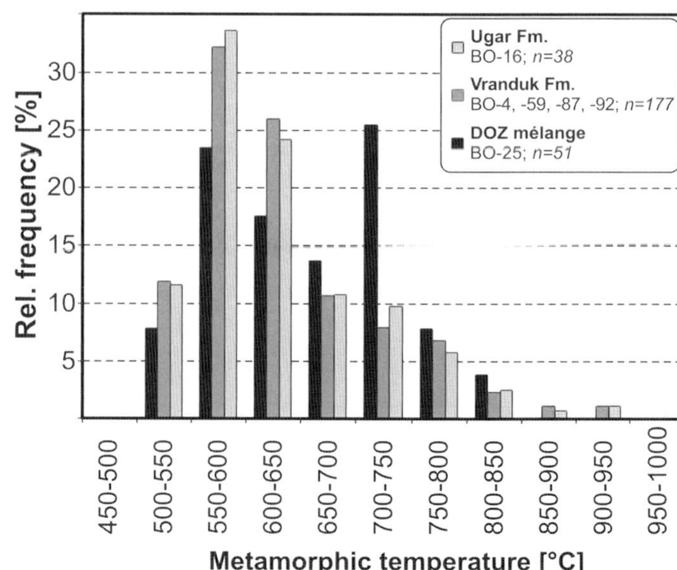

Fig. 9. Thermometry of detrital rutile from the DOZ mélange and the Vranduk and Ugar formations. Histograms show the distribution of metamorphic temperatures recorded by single rutile crystals of metapelitic origin (Cr < Nb) as calculated from their Zr content according to Watson et al. (2006).

87%) mantle-derived spinels and a subordinate magmatic-derived spinel population, based on the criterion by Lenaz et al. (2000), taking 0.20 wt% TiO_2 as a "threshold" value (Fig. 10). The Cr number [Cr# = Cr/(Cr+Al)] of mantle-derived spinels range from 0.20 to 0.75, with more than 50% of the analyses lying between 0.45 and 0.65. This is explained by the dominance of harzburgitic lithologies in a mixed lherzolitic-harzburgitic ophiolite zone, or simply by abundant mantle rocks of transitional harzburgitic character. Occasional across-sample variations in the range of Cr# in the Vranduk and Ugar formations do not affect the overall picture, but reveal local variations in ophiolite petrology.

Geochronology

Zircon U/Pb dating

Detrital zircon U/Pb age data show a marked contrast among the three formations studied, apart from the ubiquitous Permo-Triassic ages (Fig. 11). A Permo-Triassic population, with a minor admixture of Variscan, Caledonian and Pan-African ages, dominates zircon age spectra from the ophiolite mélange. Age spectra of Vranduk zircons display Caledonian, Variscan, Permo-Triassic, and in 3 out of 4 samples, a characteristic Middle/Late Jurassic population with a mode at around 150 Ma. In the Ugar Formation, Permo-Triassic, Variscan and several pre-Variscan populations prevail, with no signs of Jurassic or Cretaceous contribution. The U/Pb ages clearly indicate changing source rocks from the Jurassic mélange formation to Late Jurassic and finally Cretaceous flysch sedimentation.

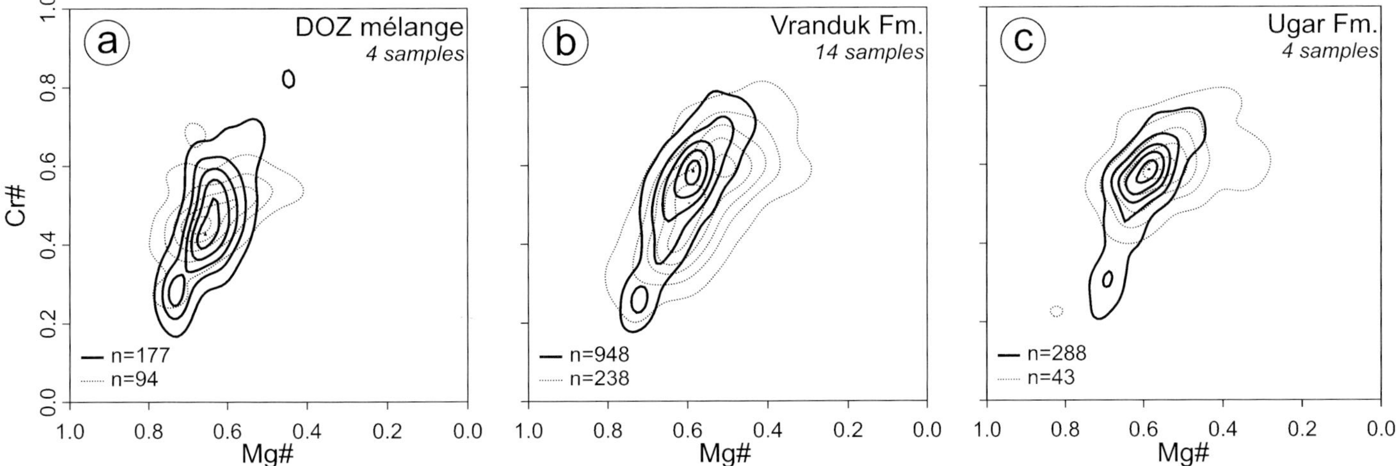

Fig. 10. Geochemistry of detrital Cr-spinel from (a) the DOZ mélange and (b) the Vranduk and (c) Ugar formations, presented by data density contouring of all data points. Contours were derived by two-dimensional kernel density estimation using the *kde2d* function of the software package *MASS* in *R* environment. Contouring covers 90, 70, 50, 30 and 10% of the entire dataset. Bold contours refer to Cr-spinel crystals with TiO_2 concentrations <0.2 wt% (peridotitic origin), Dotted contours account for $TiO_2 > 0.2$ wt% (magmatic origin). A: Cr# = Cr/(Cr + Al), Mg# = Mg/(Mg + Fe^{2+}).

Zircon fission track analysis

Single grain zircon fission track geochronology was performed on selected, zircon-rich arenitic samples, one each from the Ugar and the Vranduk formations. The results are given in Fig. 12. In the Vranduk sample around half of the dated crystals are euhedral, whereas euhedral crystals make up only some 20% in the Ugar sandstone. The majority of the single crystal ages range between 80 and 200 Ma in case of the Vranduk sample. The single grain age distributions (Table T10, electronic supplement) were tested by PopShare computer software (Dunkl & Székely 2002) in order to identify the FT age populations. The youngest and dominant FT age population within the Vranduk Formation lies at 121 ± 21 Ma; an older, more diffuse age group is also present. The age distribution in the Ugar sample is wider compared to the Vranduk sandstone. Here a small and young age component is present at around 80 Ma, while some 90% of the single grain ages form a very diffuse distribution covering an age range between the Early Palaeozoic and the Jurassic. The frequency maximum in the age distribution of the Vranduk sample coincides with a frequency low in the Ugar sample.

Discussion

A large number of new single-grain mineral chemical data (over 3000 electron microprobe analyses), together with clay mineralogical, whole-rock geochemical and geochronological data gathered in this study, allows for a comprehensive characterization of the source area of the two Bosnian Flysch units, as well as that of the sandstone blocks in the DOZ mélange. In the following, we evaluate the biostratigraphic and sedimentary provenance data to assess existing views (e.g. Charvet 1980; Pamić et al. 1998) on the Cretaceous erosion history of the Central Dinaride segment of the Alpine orogen.

Provenance of sandstones incorporated in the DOZ mélange

In addition to low-grade metamorphic (typically Fe- and Mn-rich garnet) and small amounts of ophiolitic (mainly Cr-spinel) detritus, also specific heavy mineral species are identified in the sandstone blocks of the DOZ mélange which proved largely absent in the Vranduk and Ugar formations (Table 1). They include kyanite, two chemical populations of garnet indicating amphibolite facies metabasic and amphibolite to granulite facies metapelitic source lithologies (Fig. 8) and finally, a rutile population revealing higher-temperature (650–700 to 850 °C, i.e. amphibolite to granulite facies) metamorphic conditions of their source rocks (Fig. 9). These mineral signatures can most probably be attributed to the sub-ophiolitic metamorphic soles in the DOZ. On the other hand, zircon U/Pb age spectra (Fig. 11) clearly demonstrate that the detritus was also sourced from the continental margin of Adria exposing Palaeozoic basement rocks and their Permo-Triassic cover. The high proportion of sedimentary lithic fragments among the framework components, and possibly also the presence of "unstable" heavy mineral species (Table 1), hint at relatively short transport distances.

The composition of the sandstone blocks is interpreted in terms of a combined sediment supply into a trench environment both from the basal parts of the overriding ophiolitic slab and from the Adriatic margin, followed by local sediment recycling in the trench.

Provenance of the Vranduk Formation

Apart from abundant quartz and subordinate feldspar, the framework components of the Vranduk sandstones and breccias indicate derivation from a composite source dominated by ophiolitic lithologies, but also comprising carbonates and meta-

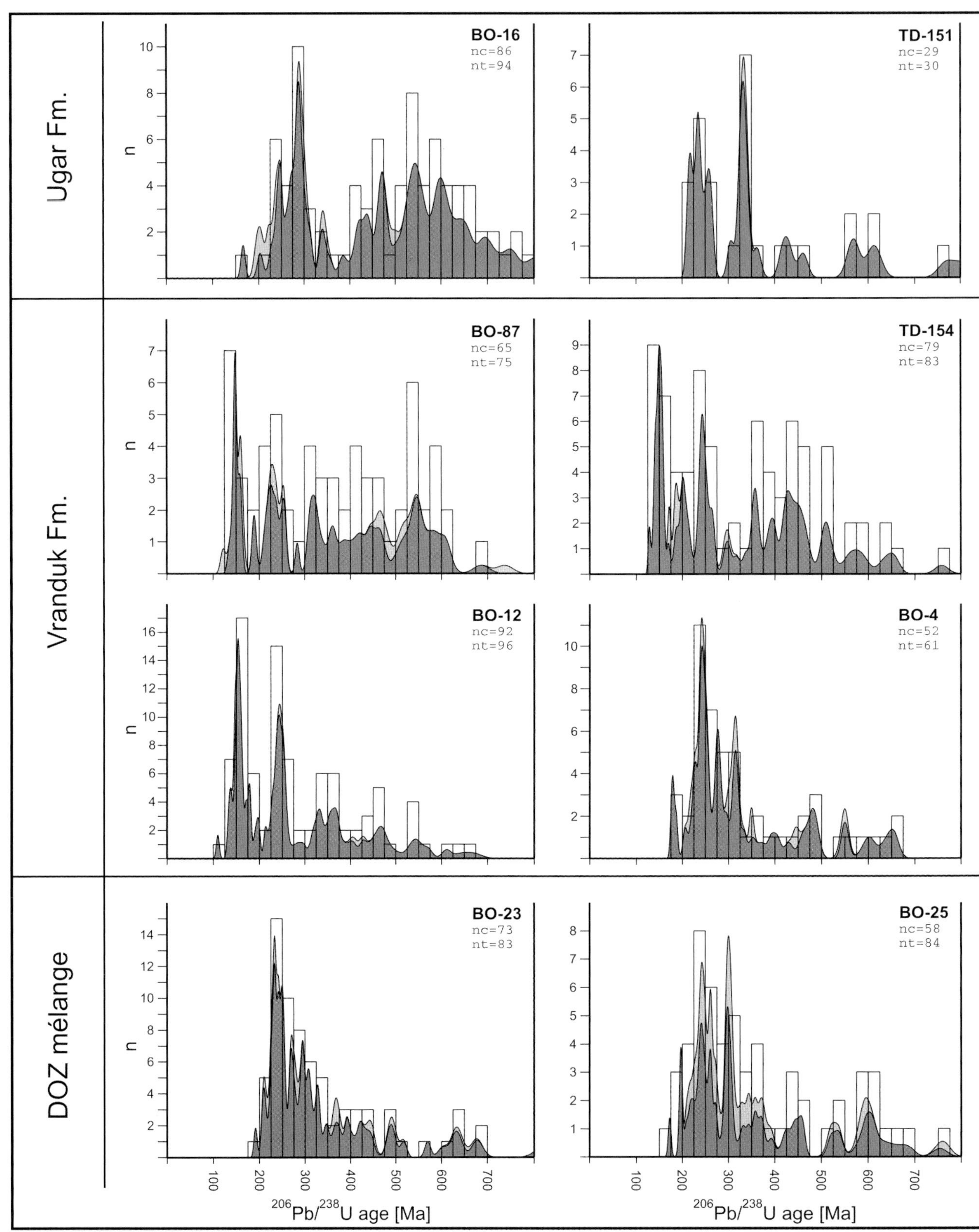

Fig. 11. Zircon U/Pb age distribution in the DOZ mélange, and the Vranduk and Ugar formations. Dark grey area indicates probability density distribution of single-grain age data concordant at the 86–114% level. Histogram also shows the filtered data. Light grey area represents all data. Grains older than 800 Ma are very subordinate and were excluded from the plot. nt: number of all accepted age analyses; nc: number of concordant age data. Calculations according to Sircombe (2004).

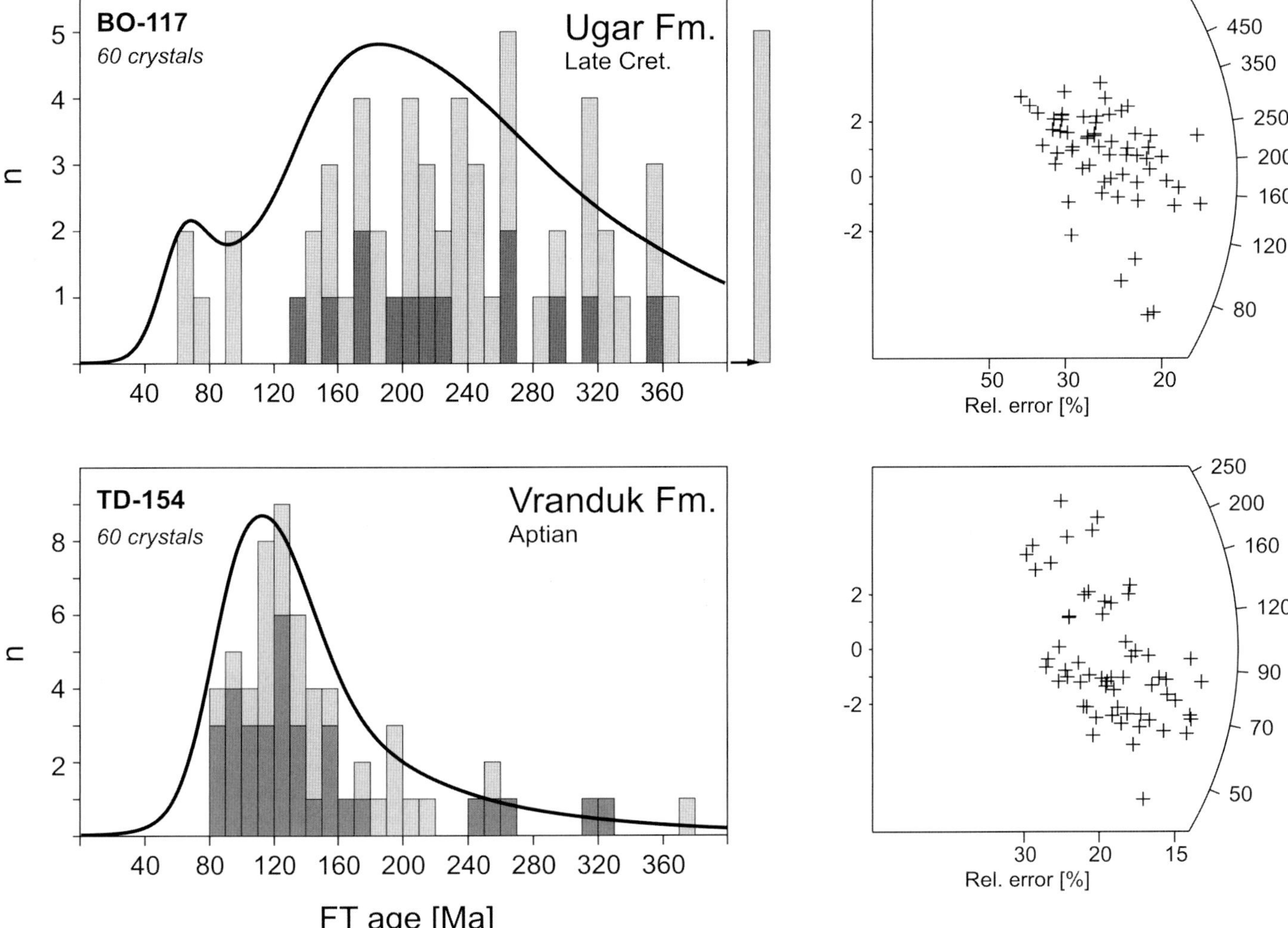

Fig. 12. Zircon single grain fission track age distributions in the Vranduk and Ugar formations. Dark grey bars express the frequency of euhedral zircons, while light grey bars show all grains. The black lines represent age spectra (probability density plots) and they were computed for all crystals according to Hurford et al. (1984). Radial plots are according to Galbraith (1990).

morphic rocks. Lithic fragments mainly include serpentinised mafic/ultramafic rocks, altered basalt fragments, chert, red radiolarite, greywacke and shale (Figs. 4 g & 4j–l). Limestone fragments are also common and yield, among other components, Urgonian facies benthic fauna of a Late Barremian or younger age. One sample from the hemipelagic strata close to the Bosna Valley section yields calcareous nannofossils indicative of an Aptian maximum age (Fig. 3; Table T2, electronic supplement). As supported by petrographic observations, the relatively low carbonate content (mostly <20 wt% CaO) of the Vranduk Formation is partly related to carbonate lithoclasts, and possibly also to intrabasinal carbonate particles.

Throughout the Vranduk Formation, clay mineral assemblages of the pelitic lithologies are dominated by illite/smectite mixed-layer structures. Kaolinite and serpentine were not detected in these samples. This composition conforms to weathering in a mafic-dominated source area, where smectitic soils tend to develop and where lateritic weathering is absent (Thiry 2000). The source rocks of the detrital component of illite were anchi- to epizonal metapelites (e.g. slates, phyllites), eroded together with the ophiolites. Proximity of the exposed ophiolite slices is also shown by trace element systematics (Fig. 6). Pelite Cr/V ratios and carbonate-corrected Cr and Ni concentrations are elevated with respect to the UCC composition and suggest direct input from ophiolitic sources. The Cr/Ni ratios of arenites are within the range of the Dinaride mafic to ultramafic rocks and, thus, preclude significant mineral concentration due to reworking. Finally, a well-represented population of Middle to Late Jurassic ages appears in the zircon U/Pb spectra (Fig. 11), which can likely also be connected to the exhumed ophiolitic sequence and will be discussed later.

In addition to the ophiolites, subordinate amounts of metamorphic source components were also admixed to the Vranduk Formation detritus. This is indicated by: (1) the detrital component of illite, (2) the presence of quartz framework grains, (3) the intermediate position of most samples on the Cr/V–Y/Ni mixing curve which links ultrabasic and UCC composition (Fig. 6a), and (4) heavy mineral occurrences and chemistry.

Zircon, metamorphic tourmaline, the majority of garnet showing Fe- and Mn-rich compositions and accessory amounts of monazite and titanite document the erosion of a diverse suite of low-grade metamorphic lithologies. Zircon U/Pb geochronology shows a predominance of Permo-Triassic, Variscan and pre-Variscan crystallization ages (Fig. 11). These data indicate that the major source of the continental detritus of the Vranduk Formation were Palaeozoic to Permo-Triassic low- to very low-grade metapelitic sequences, probably located on the Adriatic plate. A minor garnet population from amphibolite-facies metabasic source rocks (Fig. 8) may indicate continued erosion of the sub-ophiolitic metamorphic soles, but contribution from the rare amphibolite facies basement units of the Adriatic plate (Pamić & Jurković 2002; Pamić et al. 2004) can not be entirely ruled out.

The comparatively uniform population of zircon fission track ages at around 120 Ma indicates Early Cretaceous, nearly synsedimentary cooling of the source area below mid-crustal temperatures (Fig. 12) which agrees well with the major phase of cooling of the Adriatic basement (Milovanović 1984; Árkai et al. 1995; Belak et al. 1995; Judik et al. 2006).

In summary, our results suggest that the Vranduk Formation records Early Cretaceous exhumation of the Adriatic plate occurring relatively shortly (in less than 20 Ma) after ophiolite obduction. The catchment area included both continental basement and ophiolitic units, capped by short-lived Urgon facies reefs that were immediately redeposited onto the clastic fan.

Provenance of the Ugar Formation

The Ugar Formation is distinguished from the Vranduk Formation by its overall dominance of carbonate clasts, inferred to have been largely derived from the Adriatic Carbonate Platform (AdCP) by many previous workers (e.g. Aubouin 1973). With respect to its rather subordinate siliciclastic source components, several lines of evidence point to a sediment source, which neither entirely matches the eroding DOZ (at least with its structure and composition being as it is known today), nor can it be completely credited to the recycling of the Vranduk Formation. These siliciclastic source components thus also require direct erosion of continental basement units. The signatures of carbonate and siliciclastic detritus are further discussed below.

The high carbonate concentration of the Ugar Formation arenites is distinct from most arenites found in the Vranduk Formation (Fig. 5). The pelites also reveal a similar difference, although Ugar Formation pelites may have been influenced by a higher flux of pelagic carbonate sedimentation beyond their detrital carbonate component. Carbonate-corrected, UCC-normalized concentrations of elements typical for mafic/ultramafic lithologies, such as Cr and Ni, show no marked anomaly for the pelites of the Ugar Formation, and scatter around UCC composition (Fig. 6). These values are 2 to 3 times lower than those of the Vranduk Formation. Combined with low Cr/V ratios and elevated Y/Ni ratios in the Ugar Formation, the trace element data suggest that dilution by carbonate alone is not responsible for the relative scarcity of detrital ophiolitic components. In fact, there is a predominance of felsic components in the siliciclastic portion of the Ugar Formation pelites in comparison with the Vranduk Formation.

Although carbonate debris (orbitolinids, rudist fragments, lithoclasts) are similar in both formations, the following observations suggest that the Ugar Formation was sourced from a carbonate platform with a different setting from that previously supplying the Vranduk Formation with Urgonian facies clasts: (1) orbitolinids agglutinate carbonate particles and Rhaxella spicules instead of quartz, and (2) there is a salient age gap of at least 15 Ma between the bioclasts and the Upper Cretaceous pelagic matrix (Fig. 3). These features, along with the predominance of carbonate detritus in the Ugar Formation, suggest the erosion of a relatively thick carbonate succession, previously deposited on a platform where no siliciclastics were shed to. These data confirm that the AdCP, a thick and isolated carbonate platform located to the SW of the flysch basin and comprising a thick Cretaceous carbonate sequence (Vlahović et al. 2005), represents the source area of the carbonate detritus, in line with earlier models (e.g. Aubouin 1973).

The clay minerals of red marl intercalations in the thin-layered carbonate sequences of the Ugar Formation are dominated by illite/smectite and, contrary to the Vranduk Formation, they contain kaolinite as well. In general, kaolinite in fine-grained hemipelagic sediments indicates highly matured, lateritic soils developed in the source area, although lateritic weathering in tectonically active areas characterized by high erosion rates is not typical (Bárdossy & Aleva 1990; Thiry 2000). However, the Mesozoic sequence of the AdCP, located in a more external paleogeographical position with respect to the site of the deposition of the Ugar Formation, is punctuated by several bauxite and palaeosol horizons (Vlahović et al. 2005), the clay mineral fraction of which is dominantly kaolinite (e.g. Šćavničar 1978). It is thus probable that also the kaolinite in the red shale was derived from the AdCP, as a result of the erosion of its weathering products.

A change in the character of the source area in respect of the felsic detritus is well constrained by the chemistry of detrital garnet and by zircon age data. The Ugar Formation is characterized mostly by a single, almandine-rich garnet population, exclusively derived from low-grade metamorphic sources (Fig. 8). Permo-Triassic zircon crystals and variable amounts of Variscan to pre-Variscan grains are the prominent U/Pb age components in the Ugar Formation (Fig. 11). Although the Vranduk Formation also contains a marked Permo-Triassic group and a rather heterogeneous distribution of pre-Permian

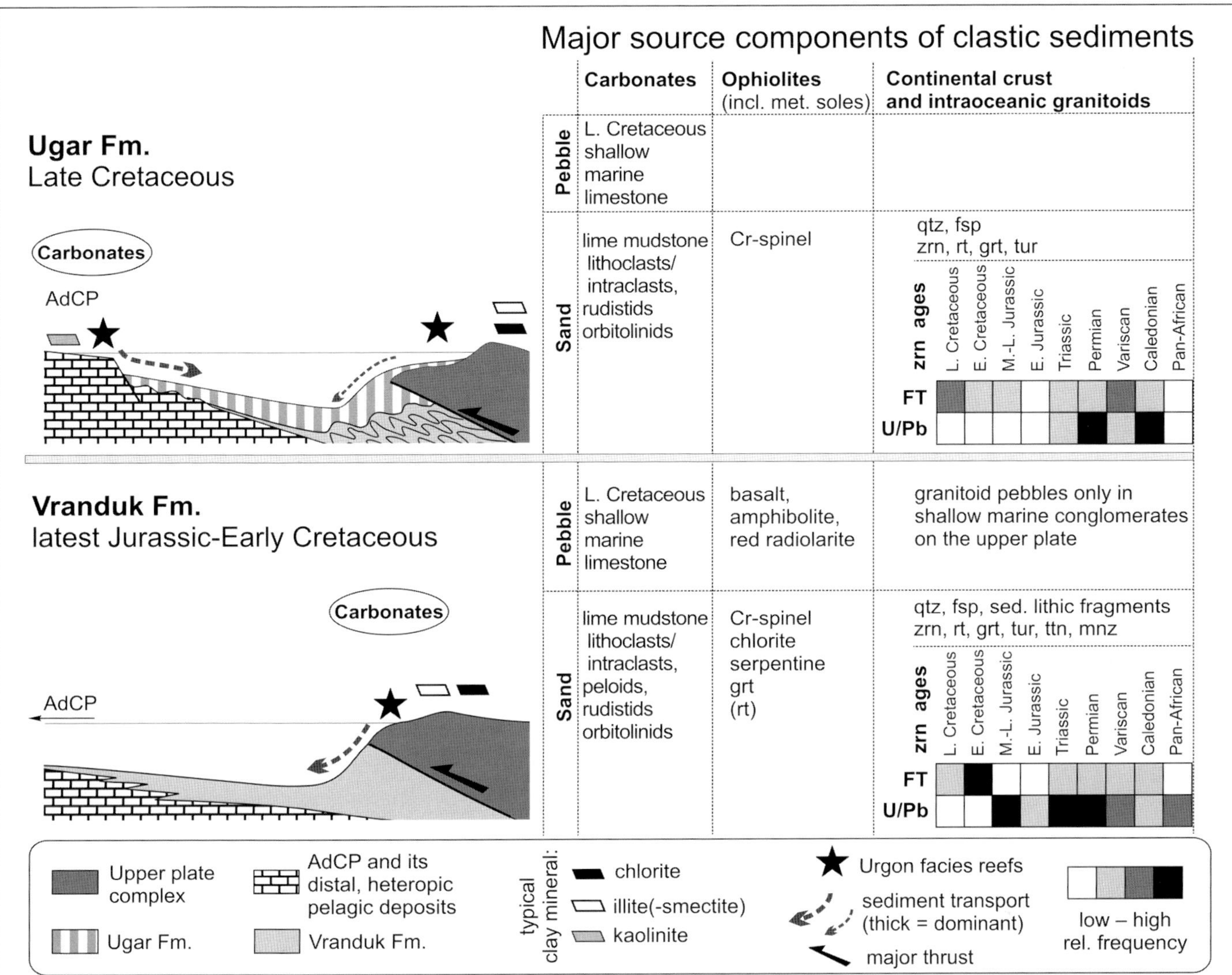

Fig. 13. Sketch summarizing the new provenance data for the Vranduk and Ugar formations. The ophiolitic, continental and carbonate detritus of the Vranduk formation was eroded exclusively from the upper plate, while the sediments of Ugar formation were derived from both sides of the basin (the carbonate gravity flows and the lateritic clay from AdCP, the siliciclastic material from the upper plate).

age components as well, it comprises a significant Middle to Late Jurassic population (Fig. 11) which is absent in the Ugar Formation.

The zircon FT age data (Fig. 12) are well in line with the U/Pb age data: they also call for contrasting source areas for the two formations. Zircon FT age spectrum of the Ugar Formation contains a 80 Ma group (Fig. 12), which could indicate rapid Late Cretaceous exhumation of the Adriatic basement (Fig. 14). This age group could however, from a geochronological point of view, fit well also with other events: (1) the early stage of the Late Cretaceous to Palaeogene acid to intermediate igneous activity in the Sava Zone (Pamić 1993, 1998; Starijaš et al. 2005), (2) connected with HT/LP metamorphism (Pamić 1993, 1998; Balen et al. 2003; Starijaš et al. 2006; Krenn et al. 2008) and in

the (3) 'banatite belt' of the Southern Carpathians and Apuseni Mts. (Kräutner et al. 1984; Wiesinger et al. 2005), as well as (4) with the characteristic Late Cretaceous thermal event recorded in the SE part of the Tisza-Dacia Unit basement rocks (Pamić 1998; Tari et al. 1999; Árkai et al. 2000; Balen et al. 2003; Lelkes-Felvári et al. 2003; Schuller 2004; Starijaš et al. 2005, 2006; Krenn et al. 2008). However, as the detrital zircon U/Pb age spectra record no contribution from Upper Cretaceous igneous rocks, the 80 Ma FT age group of the Ugar Formation more probably reflects a metapelitic rather than magmatic source. Because such a thermal event is not typical for the Adriatic basement, we can not exlude that this minor detrital component is connected to the exhumation of the Sava Zone and/or SE Tisza-Dacia basement units in the Late Cretaceous.

Derivation of the low-grade metamorphic detritus in the Late Cretaceous Ugar Formation from the Bosnian Schist Mountains (BSM) is possible on petrologic grounds, but is contradicted by the structural position of the BSM, which forms the basement of the Pre-Karst (and Bosnian) units (Aubouin et al. 1970; Schmid et al. 2008). Also, the Palaeogene K/Ar and Ar/Ar ages that are frequent in the BSM (Pamić et al. 2004) preclude its erosion in the Late Cretaceous.

In summary, our data suggest that the clastic material of the Ugar Formation was derived from at least three principal sources; (1) dismembered elements of the AdCP, and subordinately (2) ophiolitic units and (3) Variscan low-grade metamorphic basement units (including their Permo-Mesozoic cover), parts of which were affected by a thermal overprint of Late Cretaceous age (Fig. 14). The data also demonstrate that recycled Vranduk Formation sediments did not represent a significant source for the clastic components in the Ugar Formation.

Do the Vranduk and Ugar formations share a common provenance?

The contrasting palaeocurrent directions and lithofacies, as well as the discrepancies reflected by the heavy mineral and geochronological signatures of the Vranduk and Ugar formations suggest that contrasting source rock associations exerted a strong control on the composition of the basin fill. Yet, the provenance signatures with respect to the siliciclastic source components share some comparable aspects, and this is briefly discussed below.

The heavy mineral spectra of both formations include Cr-spinel, a mineral that is characteristic for obducted ophiolites (e.g. Zimmerle 1984). The chemical composition of the spinels indicates a mixed source area exposing lherzolite- and harzburgite-dominated ophiolites, although a single source of "transitional harzburgites", or Type-II peridotites (Dick & Bullen 1984), is equally possible. As shown in Fig. 10, a great part of the spinel compositions exhibit Cr# values between 0.40 and 0.65, and their distribution in the field of Cr# and Mg# is comparable. The proportion of spinels of magmatic origin is similarly low in both formations: about 20% in the Vranduk Formation and 13% in the Ugar Formation (Figs. 10b & 10c). Thus, there is little difference in the ophiolitic source lithologies in time, although the ratio of ophiolitic to continental source rock volumes was smaller during the Late Cretaceous on the basis of whole-rock trace element data (Figs. 6a–b).

The petrology of the felsic crystalline source rock assemblages could also have been, in part, comparable, as can be deduced from the chemical compositions of tourmaline. In both formations, chemical composition and internal texture of tourmaline crystals identify low-grade metamorphic source rocks, similar to those constituting the metapelitic basement of the East-Bosnian–Durmitor (e.g. Rampnoux 1970) and Drina-Ivanjica (e.g. Podubsky 1970; Djoković 1985) units. In addition, the occurrence of a prominent rimward zoning with increasing Fe and decreasing Mg is also a common feature in tourmaline from both formations.

Distribution of calculated metamorphic temperatures using detrital rutile covers a broad range with most data ranging between 500 and 850 °C. This large range of high temperatures may correspond to a uniform contribution from the sub-ophiolitic metamorphic soles, but garnet from such rocks is not detected in the Ugar Formation and is rather subordinate in the Vranduk Formation. Alternatively, and more probably, in both formations most of these rutile crystals were derived from older sediments and low-grade metapelites of the Adriatic plate, which did not experience the temperature conditions necessary for Zr re-equilibration in the rutile lattice, resulting in "inherited" temperatures (Zack et al. 2004a; Triebold et al. 2007).

The above data indicate that the major source of the sand-sized siliciclastic components in both the Vranduk and the carbonate-dominated Ugar formations included comparable rock associations with ophiolites and low- to very low-grade metamorphic continental basement units. The heterogeneity, and contrasting thermal histories of these lithological units on the upper plate, possibly combined with drainage evolution (e.g. the NE-ward shift of the coastline upon the DOZ; Charvet 1980), are reflected by changes in the chemistry of detrital garnet and spinel, and in zircon chronology.

Acid intraoceanic magmatism in the Jurassic

Zircon U/Pb age spectra of the Vranduk Formation (Fig. 11) suggest a pulse of acid to intermediate magmatism in the Middle-Late Jurassic in the Neotethys Ocean. Melt generation was largely contemporaneous with, or closely followed intraoceanic subduction, as reflected by the ages of metamorphic sole formation (about 180 to 160 Ma – Okrusch et al 1978; Parlak & Delaloye 1999; Dimo-Lahitte et al. 2001; Olker et al. 2001; Smith 2006 and references therein) and the age ranges of zircon crystallization (about 165 to 140 Ma). Obvious mechanisms for generation of acid to intermediate magmatic rocks in an intraoceanic setting can be explained by (1) plagiogranite formation at the spreading ridge; (2) island-arc magmatism; and (3) anatexis of metasediments below the ophiolite thrust sheet.

Hitherto there has been no evidence for any voluminous, acid to intermediate Jurassic magmatism that is related to island-arc development in the Dinaride segment of the Neotethys (e.g. Pamić et al. 2002; Dimitrijević et al. 2003) in spite of ample evidence for the existence of intra-oceanic subduction in the Neotethys (see Smith 2006, for a review). Recently however, mafic dykes were reported which intersect the DOZ ophiolites and display island arc geochemical signatures (Lugović et al. 2006).

Plagiogranitic differentiates may be voluminous (e.g. Pamić & Tojerkauf 1970; Bébien et al. 1997) and thus, in principle, capable of producing detritus in such amount that would be sufficiently represented in the turbiditic Vranduk Formation. In this case, an active oceanic ridge would be required at the time of subduction, as melt generation appears to slightly post-date intra-oceanic subduction. Such a scenario has recently been documented in the Vourinos ophiolite (Hellenides) by precise zircon U/Pb geochronology (Liati et al. 2004).

An alternative explanation for the observed age patterns is offered by the anatexis of HT metamorphic rocks at the base of the overriding ophiolite sheets. Jurassic S-type granitoids are known to have intruded many ophiolitic units and their metamorphic soles in the Dinarides–Hellenides (e.g. Borsi et al. 1966; Anders et al. 2005; Resimić-Šarić et al. 2005), that will be discussed in a separate paper (Mikes et al., in prep.).

Sedimentation of Ugar Fm. (from ca. 100 Ma)

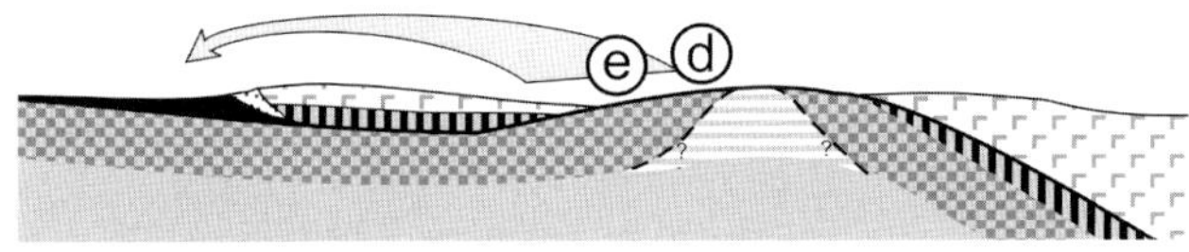

Sedimentation of Vranduk Fm. (from ca. 145 Ma)

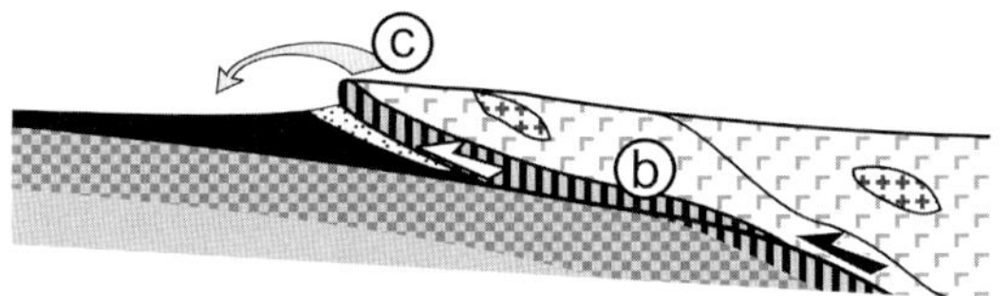

Onset of obduction onto Adria (ca. 150 Ma)

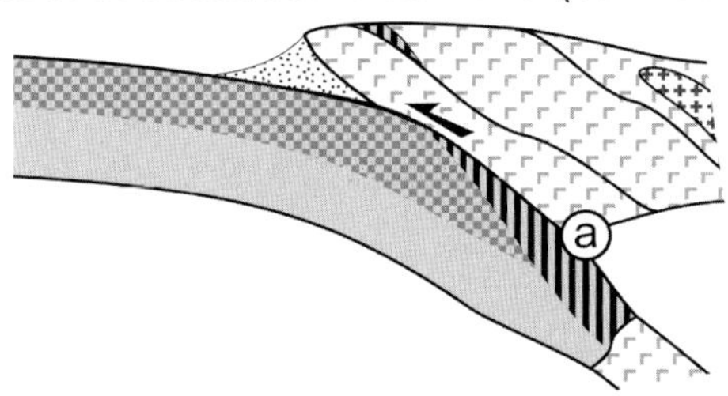

Intra-oceanic subduction (ca. 180-160 Ma)

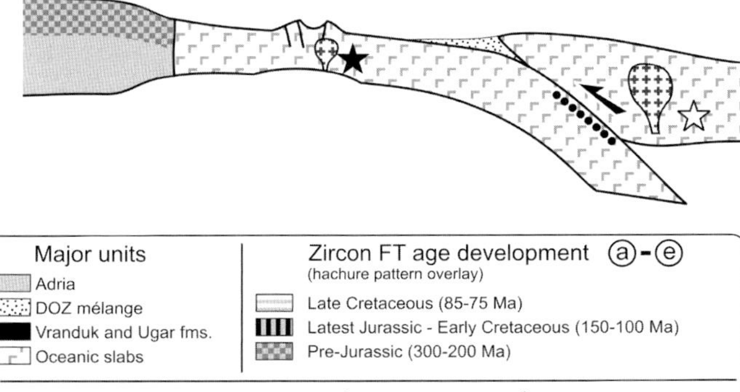

Summary: source area evolution of the Bosnian Flysch

In this chapter we outline several steps that played a major role in the evolution of the Bosnian Flysch basin and its surrounding source areas. Fig. 13 summarizes the most important provenance data for the situation in Early Cretaceous (Vranduk Formation) and Late Cretaceous (Ugar Formation) times. Fig. 14. provides a tentative tectono-thermal synthesis of the zircon-supplying source areas based on thermochronological data published previously and obtained in this study.

Middle to Late Jurassic: intraoceanic subduction

Intraoceanic subduction within the Neotethys resulted in vast amounts of ophiolite bodies exposed at the overriding oceanic plate. Sandstone blocks within the Jurassic DOZ mélange record erosional events linked to the plate convergence. Sandstone composition indicates a dual source with ophiolites and their high-grade metamorphic soles at the upper plate and with detritus derived from the more distally-located continental margin of Adria. The sediments underwent local reworking, consistent with an erosional-depositional scenario in a highly active trench setting. If the oceanic ridge was located continentward of the subduction zone as depicted in Fig 14, then the trench sedimentation could be controlled by detritus from the Adriatic basement only after the oceanic ridge was subducted, that had previously posed a barrier for trenchward sediment transport.

Jurassic–Cretaceous transition: obduction and metamorphism at the continental margin of Adria

Inverse heat transfer from the obducted hot ophiolites caused a regionally widespread, epizonal thermal overprint in the upper part of the Adriatic plate (Fig. 14). Deep crustal levels of both the oceanic plate and the continental margin suffered largely coeval HP/LT metamorphic overprint (Okrusch et al. 1978; Milovanović et al. 1995; Most 2003) related to both intraoceanic subduction and the subsequent obduction onto Adria. The main phase of cooling took place already in the

Fig. 14. Cartoon summarizing the development of the zircon FT ages in tectonic units providing *siliciclastic* detritus to the Vranduk and Ugar formations. (a) Zone of zircon FT reset in the Adriatic basement due to the inverse heat transfer from the overriding hot ophiolites and tectonic load. (b) Obduction is followed by thrusting of the overprinted continental slices that were detached from the topmost level of the lower plate. (c) Early Cretaceous erosion of Adria-derived basement slices is documented by synsedimentary zircon FT ages in the Vranduk Formation. (d) In the Late Cretaceous, the lower plate experienced fast exhumation: the thin upper layer of the Adriatic margin with completely reset zircon FT ages was removed, and a wide range of Jurassic and pre-Jurassic zircon FT ages were supplied. Although these ages are predominant in the Ugar formation, (e) a minor age component with Late Cretaceous zircon FT ages is also present. This can be interpreted by exhumation of deeper-seated, hitherto undated crustal levels of the Adriatic plate, whereas temporally it also agrees with the early stages of collision of Adria with Tisza-Dacia (not shown) implying that the Late Cretaceous catchment already included the docking upper plate. See text for discussion.

Early Cretaceous (Milovanović 1984; Árkai et al. 1995; Belak et al. 1995; Judik et al. 2006). However, basal detachments related to the ophiolite emplacement and nappe advancement possibly gave rise to exhumation of small continental crustal slices and to their incorporation into the thrust wedge already from the latest Jurassic onwards as shown by earliest Cretaceous coarse-grained continental crustal detritus in sediments sealing the DOZ mélange (Charvet 1978; Neubauer et al. 2003).

Early Cretaceous sedimentation: Vranduk Formation

Sediments formed in front of the leading edge of the ophiolite–continental basement thrust wedge complex were subsequently accreted to the base of the thrust wedge, parts of which were exposed to erosion again. Clastic wedge development commenced already at the Jurassic–Cretaceous transition, with the submarine fan deposits prograding SW-ward onto the pelagic sequence of the 'Zone Bosniaque'. Tips of the growing imbricate slices that achievied shallow bathymetry were rapidly colonised by Urgonian facies reefs. These build-ups and their ophiolitic substrate were immediately redeposited onto the clastic fan (Fig. 13).

Erosion, drainage evolution and nappe emplacement could all result in slight changes in source rock composition between incipient obduction and Early Cretaceous flysch deposition. The ophiolitic detritus records an increasing volume of harzburgites exposed to erosion. The obducted oceanic thrust pile contained Jurassic acid to intermediate magmatic bodies as well, interpreted as plagiogranites formed at the ridge, or as S-type granitoids produced by anatexis of subducted oceanic sediment (Fig. 14). Granulite facies metapelitic rocks of the sub-ophiolitic metamorphic soles were no longer eroded, likely due to their extremely small volume relative to the ophiolitic sheet (cf. Karamata et al. 1970, their Fig. 1).

The relatively rapid exhumation and erosion of Adria-derived continental crustal units in the late Early Cretaceous is indicated by a largely synsedimentary age component around 120 Ma in the detrital zircon FT age spectrum of the Vranduk strata (Fig. 14). Parts of the clastic wedge underwent burial and folding in the mid-Cretaceous as suggested by a locally developed erosional unconformity on top of the Vranduk Formation (Figs. 3 & 13; Dimitrijević 1982 p. 14; Csontos et al. 2003; Schmid et al. 2008 p. 25), and by K/Ar age dating of illite from Vranduk Formation shales (Petri 2007).

Late Cretaceous Ugar sedimentation: increasing carbonate contribution and changes in the hinterland

Propagation of the folded imbricate wedge towards the SW led to a platformward migration of the actual depozone. Vast carbonate mass flows released from the AdCP margin were intercalated with Scaglia Rossa-type Upper Cretaceous basin sediments. The origin of the suspended matter can be located on both basin margins, most likely representing a mixture from both areas. Sand-sized siliciclastic components were derived from the DOZ and the underlying, continental crustal thrust sheets of Adria, parts of which experienced fast exhumation. Notably, this *c.* 80 Ma cooling event has been largely unknown in the Adriatic basement so far.

In the light of the largely E-directed palaeocurrent indicators, the site and mechanism of intermingling of the carbonate and sand-sized siliciclastic components prior to the final deposition of the Ugar Formation remains ambiguous. Signs of continued convergence during the Cenozoic, such as fault propagation, basin migration and cannibalistic sediment reworking further toward SW, are seen in the imbricate wedges of Cenozoic flysch of the AdCP (Chorowicz 1977; Mikes et al. 2008 and references therein).

★ ★ ★

The data presented in this paper reinforce previously published tectonic models inasmuch as they relate the deposition of the Vranduk Formation to the adjacent and exhumed ophiolite units (e.g. Aubouin 1973; Charvet 1980; Tari 2002; Schmid et al. 2008). However, although the Vranduk Formation is underlain by pelagic Jurassic strata of the distal Adriatic plate, it represents a SW-ward propagating clastic wedge at the front of the leading edge of the DOZ and a continental thrust sheet complex. Our results thus do not support the model put forward by Pamić (1993) and Pamić et al. (1998) describing the entire Bosnian Flysch in terms of a passive margin sequence of the distal Adriatic plate.

Acknowledgements

The authors are indebted to W. Frisch (Tübingen), L. Csontos, J. Haas and P. Ozsvárt (Budapest), B. Lugović (Zagreb), G. Grathoff (Göttingen/Portland), V. Karius, A. Kronz, R. Tolosana-Delgado, U. Grunewald, I. Ottenbacher and R. Hu (Göttingen) for stimulating discussions and for invaluable field and laboratory support. The reactor staff of the Oregon State University are thanked for the irradiations. The Frljak Family in Breza and the Sehić Family in Vranduk offered very warm hospitality during the field work. All their help is gratefully acknowledged. Careful and very constructive reviews by J. Charvet, an anonymous reviewer and SJG guest editor S.M. Schmid helped to improve the manuscript. The study was supported by the Deutsche Forschungsgemeinschaft (EY 23/4).

REFERENCES

Anders, B., Reischmann, T., Poller, U. & Kostopoulos, D. 2005: Age and origin of granitic rocks of the eastern Vardar Zone, Greece: new constraints on the evolution of the Internal Hellenides. Journal of the Geological Society 162, 857–870.

Árkai, P., Balogh, K. & Dunkl, I. 1995: Timing of low-temperature metamorphism and cooling of the Paleozoic and Mesozoic formations of the Bükkium, Innermost Western Carpathians, Hungary. Geologische Rundschau 84, 334–344.

Árkai, P., Bérczi-Makk, A. & Balogh, K. 2000: Alpine low-T prograde metamorphism in the post-Variscan basement of the Great Plain, Tisza Unit (Pannonian Basin, Hungary). Acta Geologica Hungarica 43, 43–63.

Aubouin, J. 1973: Des tectoniques superposées et de leur signification par rapport aux modèles géophysiques: l'exemple des Dinarides; paléotectonique, tectonique, tarditectonique, néotectonique. Bulletin de la Société géologique de France (7) 15, 426–460.

Babić, Lj., Hochuli, P. A. & Zupanič, J. 2002: The Jurassic ophiolitic melange in the NE Dinarides: Dating, internal structure and geotectonic implications. Eclogae Geologicae Helvetiae 95, 263–275.

Babić, Lj. & Zupanič, J. 1976: Sediments and paleogeography of the Globotruncana calcarata Zone (Upper Cretaceous) in Banija and Kordun, central Croatia. Geološki vjesnik, 29: 49–73.

Bárdossy, Gy. & Aleva, G. J. J. 1990: Lateritic Bauxites. Developments in Economic Geology No. 27, Elsevier, Amsterdam, 624.

Bazylev, B., Popević, A., Karamata, S., Kononkova, N. & Simakin, S. 2006: Spinel peridotites from Zlatibor Massif (Dinaric Ophiolite Belt): Petrological evidences for a supra-subduction origin. In: Mesozoic ophiolite belts of northern part of the Balkan Peninsula. International Symposium, Belgrade-Banja Luka, May 31 – June 6, 2006, Belgrade, 1–4.

Bébien, J., Dautaj, N., Shallo, M., Turku, I. & Barbarin, B. 1997: Diversité des plagiogranites ophiolitiques: l'example albanais. Comptes Rendus de l'Academie des Sciences, sér. IIa 324, 875–882.

Belak, M., Pamić, J., Kolar-Jurkovšek, T., Pécskay, Z. & Karan, D. 1995: The Alpine regional metamorphic complex of Medvednica (NW Croatia) (In Croatian). Proceedings of the First Croatian Geological Congress, Opatija 18.-21.10.1995 67–70.

Belak, M. & Tibljaš, D. 1998: Discovery of blueschists in the Medvednica Mountain (Northern Croatia) and their significance for the interpretation of geotectonic evolution of the area. Geološki vjesnik 51, 27–32.

Bernoulli, D. & Laubscher, H. 1972: The palinspastic problem of the Hellenides. Eclogae geologicae Helvetiae 65, 107–118.

Blanchet, R. 1966: Sur l'âge tithonique-éocrétacé d'un flysch des Dinarides internes en Bosnie: le flysch de Vranduk (Yougoslavie). Compte rendu sommaire des séances de la Société géologique de France 10, 401–403.

Blanchet, R. 1968: Sur l'extension du flysch tithonique-éocrétacé en Bosnie centrale (Yougoslavie). Compte rendu sommaire des séances de la Société géologique de France 3, 97–98.

Blanchet, R. 1970: Données nouvelles sur le flysch bosniaque: la région de Banja Luka, Bosnie septentrionale, Yougoslavie. Bulletin de la Société géologique de France (7) 12, 659–663.

Blanchet, R., Cadet, J.-P., Charvet, J. & Rampnoux, J.-P. 1969: Sur l'existence d'un important domaine de flysch tithonique-crétacé inférieur en Yougoslavie: l'unité du flysch bosniaque. Bulletin de la Société géologique de France (7) 11, 871–880.

Blanchet, R., Durand Delga, M., Moullade, M. & Sigal, J. 1970: Contribution à l'étude du Crétacé des Dinarides internes: la région de Maglaj, Bosnie (Yougoslavie). Bulletin de la Société géologique de France (7) 12, 1003–1009.

Bown, P. R., Rutledge, D. C., Cruy, J. A. & Gallagher, L. T. 1998: Lower Cretaceous. In: Brown, P. R. (Ed.): Calcareous Nannofossil Biostratigraphy. Chapman & Hall, London, 86–131.

Borsi, S., Ferrara, G., Mercier, J. & Tongiorgi, E. 1966: Age stratigraphique et radiométrique jurassique supérieur d'un granite des zones internes des Hellénides (granite de Fanos, Macédoine, Grèce). Revue de Géographie physique et de Géologie dynamique 8, 279–287.

Bown, P. R. & Young, J. R. 1997: Mesozoic calcareous nannoplankton classification. Journal of Nannoplankton Research 19, 21–36.

Burnett, J. A. 1998: Upper Cretaceous. In: Bown, P. R. (Ed.): Calcareous Nannofossil Biostratigraphy. Chapman & Hall, London, 133–199.

Bušer, S. 1987: Development of the Dinaric and the Julian carbonate platforms and of the intermediate Slovenian basin (NW Yugoslavia). Memorie della Società Geologica Italiana 40, 313–320.

Cadet, J.-P. 1968: Sur l'age de flyschs de la haute vallée de la Neretva (région de Ulog, Bosnie Yougoslavie). Compte rendu sommaire des séances de la Société géologique de France 4, 118–120.

Cadet, J.-P. & Sigal, J. 1969: Sur la stratigraphie et l'extension du flysch éocrétacé en Bosnie Hercégovine méridionale. Compte rendu sommaire des séances de la Société géologique de France 2, 52–53.

Carosi, R., Cortesogno, L., Gaggero, L. & Marroni, M. 1996: Geological and petrological features of the metamorphic sole from the Mirdita Nappe, northern Albania. Ofioliti 21, 21–40.

Charvet, J. 1967: Sur un jalon de flysch tithonique-éocrétacé au nord de Sarajevo (Bosnie-Herzégovine, Yougoslavie). Compte rendu sommaire des séances de la Société géologique de France 8, 371–373.

Charvet, J. 1970: Aperçu géologique des Dinarides aux environs du méridien de Sarajevo. Bulletin de la Société géologique de France (7) 12, 986–1002.

Charvet, J. 1973: Sur les mouvements orogéniques du Jurassique-Crétacé dans les Dinarides de Bosnie orientale. Comptes rendus de l'Académie des Sciences Paris, Série D 276, 257–259.

Charvet, J. 1978: Essai sur un orogène alpin: Géologie des Dinarides au niveau de la transversale de Sarajevo (Yougoslavie). Publications de la Société géologique du Nord 2, 1–554.

Charvet, J. 1980: Développement de l'orogène dinarique d'après l'étude du secteur transversal de Sarajevo (Yougoslavie). Revue de Géologie Dynamique et de Géographie Physique 22, 29–50.

Charvet, J. & Termier, G. 1971: Les Nérinéacés de la limite Jurassique-Crétacé de Bjeliš (Nord de Sarajévo, Yougoslavie). Annales – Société Géologique du Nord 91, 187–191.

Chorowicz, J. 1977: Étude géologique des Dinarides le long da la structure transversale Split-Karlovac (Yougoslavie). Publications de la Société géologique du Nord 1, 1–331.

Cousin, M. 1972: Ésquisse géologique des confins italo-yougoslaves; leur place dans les Dinarides et les Alpes méridionales. Bulletin de la Société géologique de France (7) 12, 1034–1047.

Csontos, L., Gerzina, N., Hrvatović, H., Schmid, S. & Tomljenović, B. 2003: Structure of the Dinarides: a working model. Annales Universitatis Budapestinensis, Sectio Geologica 35, 143–144.

Dick, H. J. B. & Bullen, T. 1984: Chromian spinel as a petrogenetic indicator in abyssal and alpine-type peridotites and spatially associated lavas. Contributions to Mineralogy and Petrology 86, 54–76.

Dimitrijević, M. D. 1982: Dinarides: An outline of the tectonics. Earth Evolution Sciences 1, 4–23.

Dimitrijević, M. D. & Dimitrijević, M. N. 1973: Olistostrome mélange in the Yugoslavian Dinarides and late Mesozoic plate tectonics. Journal of Geology 81, 328–340.

Dimitrijević, M. N. & Dimitrijević, M. D. 1968: The multilateral paleotransport on the example of the Durmitor Flysch, Yugoslavia. XXXIII International Geological Congress 3, 249–256.

Dimitrijević, M. N., Dimitrijević, M. D., Karamata, S., Sudar, M., Gerzina, N., Kovács, S., Dosztály, L., Gulácsi, Z., Less, Gy. & Pelikán, P. 2003: Olistostrome/mélanges – an overview of the problems and preliminary comparison of such formations in Yugoslavia and NE Hungary. Slovak Geological Magazine 9, 2–21.

Dimo-Lahitte, A., Monié, P. & Vergély, P. 2001: Metamorphic soles from the Albanian ophiolites: Petrology, ^{40}Ar/^{39}Ar geochronology, and geodynamic evolution. Tectonics 20, 78–96.

Djerić, N., Vishnevskaya, V. S. & Schmid, S. M. 2007: New data on radiolarians from the Dinarides (Bosnia and Serbia). In: Froitzheim, N., Bousquet, R., Fügenschuh, B., Schmid, S. & Tomljenović, B. (Eds.): Abstract Volume, 8th Workshop on Alpine Geological Studies, Davos/Switzerland, 10.–12. October 2007, Bonn, 17–18.

Djoković, I. 1985: The use of structural analysis in determining the fabric of Palaeozoic formations in the Drina-Ivanjica region [in Serbian with English summary]. Geološki anali Balkanskoga poluostrva 49, 11–160.

Dumitru, T. 1993: A new computer-automated microscope stage system for fission-track analysis. Nuclear Tracks and Radiation Measurements 21, 575–580.

Dunkl, I. 2002: Trackkey: a Windows program for calculation and graphical presentation of fission track data. Computers and Geosciences 28, 3–12. Available online at: www.sediment.uni-goettingen.de/staff/dunkl/software/

Dunkl, I. & Székely, B. 2002: Component analysis with visualization of fitting; PopShare, a Windows program for data analysis (abstract). Geochimica et Cosmochimica Acta 66, 201. Available online at: www.sediment.uni-goettingen.de/staff/dunkl/software/

Dunkl, I., Mikes, T., Simon, K. & von Eynatten, H. 2007: Data handling, outlier rejection and calculation of isotope concentrations from laser ICP-MS

Vlahović, I., Tišljar, J., Velić, I. & Matičec, D. 2005: Evolution of the Adriatic carbonate platform; palaeogeography, main events and depositional dynamics. Palaeogeography, Palaeoclimatology, Palaeoecology 220, 333–360.

Watson, E. B., Wark, D. A. & Thomas, J. B. 2006: Crystallization thermometers for zircon and rutile. Contributions to Mineralogy and Petrology 151, 413–433.

Wiesinger, M., Neubauer, F., Berza, T., Jandler, R. & Genser, J. 2005: $^{40}Ar/^{39}Ar$ amphibole and biotite dating of Romanian banatites. Abstracts Book, 7th Workshop on Alpine Geological Studies Opatija, 105–106.

Zack, T., von Eynatten, H. & Kronz, A. 2004a: Rutile geochemistry and its potential use in quantitative provenance studies. Sedimentary Geology 171, 37–58.

Zack, T., Moraes, R. & Kronz, A. 2004b: Temperature dependence of Zr in rutile: empirical calibration of a rutile thermometer. Contributions to Mineralogy and Petrology 148, 471–488.

Zimmerle, W. 1984: The geotectonic significance of detrital brown spinel in sediments. Mitteilungen des dem Geologisch-Paläontologischen Institut der Universität Hamburg 56, 337–360.

Manuscript received February 21, 2008
Revision accepted July 25, 2008
Published Online first November 8, 2008
Editorial Handling: Stefan Schmid & Stefan Bucher

Electronic supplementary material: The online version of this article (DOI: 10.1007/s00015-1289-z) contains supplementary material, which is available to authorized authors.

1661-8726/08/01S055-17
DOI 10.1007/s00015-008-1282-0
Birkhäuser Verlag, Basel, 2008

Late Jurassic tectonics and sedimentation: breccias in the Unken syncline, central Northern Calcareous Alps

HUGO ORTNER[1,*], MICHAELA USTASZEWSKI[1,2] & MARTIN RITTNER[1,3]

Key words: Jurassic tectonics, Jurassic sedimentary rocks, Unken syncline, scarp breccia, Northern Calcareous Alps, Austria

ABSTRACT

This study analyses and discusses well preserved examples of Late Jurassic structures in the Northern Calcareous Alps, located at the Loferer Alm, about 35 km southwest of Salzburg. A detailed sedimentary and structural study of the area was carried out for a better understanding of the local Late Jurassic evolution. The Grubhörndl and Schwarzenbergklamm breccias are chaotic, coarse-grained and locally sourced breccias with mountain-sized and hotel-sized clasts, respectively. Both breccias belong to one single body of breccias, the Grubhörndl breccia representing its more proximal and the Schwarzenbergklamm breccia its more distal part, respectively. Breccia deposition occurred during the time of deposition of the Ruhpolding Radiolarite since the Schwarzenbergklamm breccia is underlain and overlain by these radiolarites. Formation of the breccias was related to a major, presumably north-south

trending normal fault scarp. It was accompanied and post-dated by west-directed gravitational sliding of the Upper Triassic limestone ("Oberrhätkalk"), which was extended by about 6% on top of a glide plane in underlying marls. The breccia and slide-related structures are sealed and blanketed by Upper Jurassic and Lower Cretaceous sediments. The normal fault scarp, along which the breccia formed, was probably part of a pull-apart basin associated with strike slip movements. On a regional scale, however, we consider this Late Jurassic strike-slip activity in the western part of the Northern Calcareous Alps to be synchronous with gravitational emplacement of "exotic" slides and breccias (Hallstatt mélange), triggered by Late Jurassic shortening in the eastern part of the Northern Calcareous Alps. Hence, two competing processes affected one and the same continental margin.

Introduction

The Late Jurassic evolution of the Northern Calcareous Alps (NCA) is a subject of controversy among Alpine geologists. Thereby the position and timing of the closure of the (north) westernmost Tethys embayment (Meliata Ocean) relative to the paleoposition of the NCA is the most contentious issue. Throughout the Triassic, before the opening of the Alpine Tethys, the NCA formed the southeastern continental margin of the European plate against the Meliata Ocean, which was part of the Neotethys (e.g. Stampfli et al. 1998; Schmid et al. 2004). During the Middle and Late Triassic, thick carbonate series accumulated on the inner passive margin, whereas the outer margin was characterized by deep swell deposits (e.g. Mandl 2000) and/or distal periplatform carbonates (Gawlick & Böhm 2000). During Jurassic rifting, the NCA were separated from the European continent by the evolving Piemont-Ligurian Ocean of the Alpine Tethys, and subsequently formed the northwestern margin of the Apulian plate (e.g. Frisch 1979; Faupl & Wagreich

2000). Deposits of the well-studied continental margin and continent-ocean transition crop out in the Austroalpine nappes of eastern Switzerland and westernmost Austria (Eberli 1988; Froitzheim & Eberli 1990; Froitzheim & Manatschal 1996) and in the western part of the Southern Alps (Bertotti et al. 1993). In these areas, half-graben rift basins and low-angle normal faults developed in the Late Triassic to Early Jurassic and the resulting submarine topography persisted to the end of the Middle Jurassic. In response to rifting, the former inner passive margin in the NCA disintegrated and subsided, causing the drowning of carbonate platforms, with the depositional environment changing to a deep water setting. Southeast of the NCA, obduction in the Meliata ocean started in the Middle Jurassic (e.g. see Dimo-Lahitte et al. 2001 for geochronology of the metamorphic sole related to obduction in Albania), followed by obduction onto the eastern margin of the Apulian plate and collision in the Early Cretaceous (Schmid et al. 2008). In the eastern NCA and western Carpathians the obducted unit is not preserved, but documented by ophiolitic detritus in Up-

[1] University of Innsbruck, Institute of Geology and Palaeontology, Innrain 52, A-6020 Innsbruck, Austria.
[2] now at: Department of Geosciences, National Taiwan University, No.1, Sec.4, Roosevelt Road, Taipei 10617, Taiwan.
[3] now at: Department of Geology, Royal Holloway University of London, Egham, Surrey, TW20 0EX, UK.
*Corresponding author: Hugo Ortner. E-mail: hugo.ortner@uibk.ac.at

per Jurassic mélange formations (Kozur & Mostler 1992; Mandl & Ondrejickova 1993). Moreover, pebble analysis (Gruber et al. 1992; Schuster et al. 2007) and chemical analysis of detrital chromium spinel grains (Pober & Faupl 1988) from synorogenic sediments suggest a proximity of the obducted ophiolites to the NCA in the Late Early to Late Cretaceous, when inversion of the passive margin of the NCA and the Eoalpine stacking of thrust sheets began (e.g. Tollmann 1976; May & Eisbacher 1999; Ortner 2003).

Breccias related to the degradation of submarine topography are a common feature within Jurassic deposits of the Eastern Alps (e.g. Bernoulli & Jenkyns 1970; Achtnich 1982; Eberli 1988; Wächter 1987; Froitzheim & Eberli 1990). The tectonic interpretation of breccia deposits is therefore often concerned with the question of how submarine topography did form. In the central part of the Northern Calcareous Alps, Jurassic deep-water carbonates are associated with large slide blocks and blocky breccias, composed of older passive margin sediments (Gawlick et al. 1999a; Mandl 2000). Previous geodynamic interpretations of the breccias have been dependent on the age of the sediments involved. For the Early Jurassic, Channell et al. (1992), Böhm et al. (1995) and Ebli (1997) concluded that sliding and breccia sedimentation were related to block tilting associated with rifting. However, Middle and Late Jurassic (mega-) breccias have also been interpreted as related to compressive tectonics. For example, Tollmann (1987) and Mandl (2000) favour a model of gravitational mobilisation and gliding due to tectonic movements in the hinterland of the

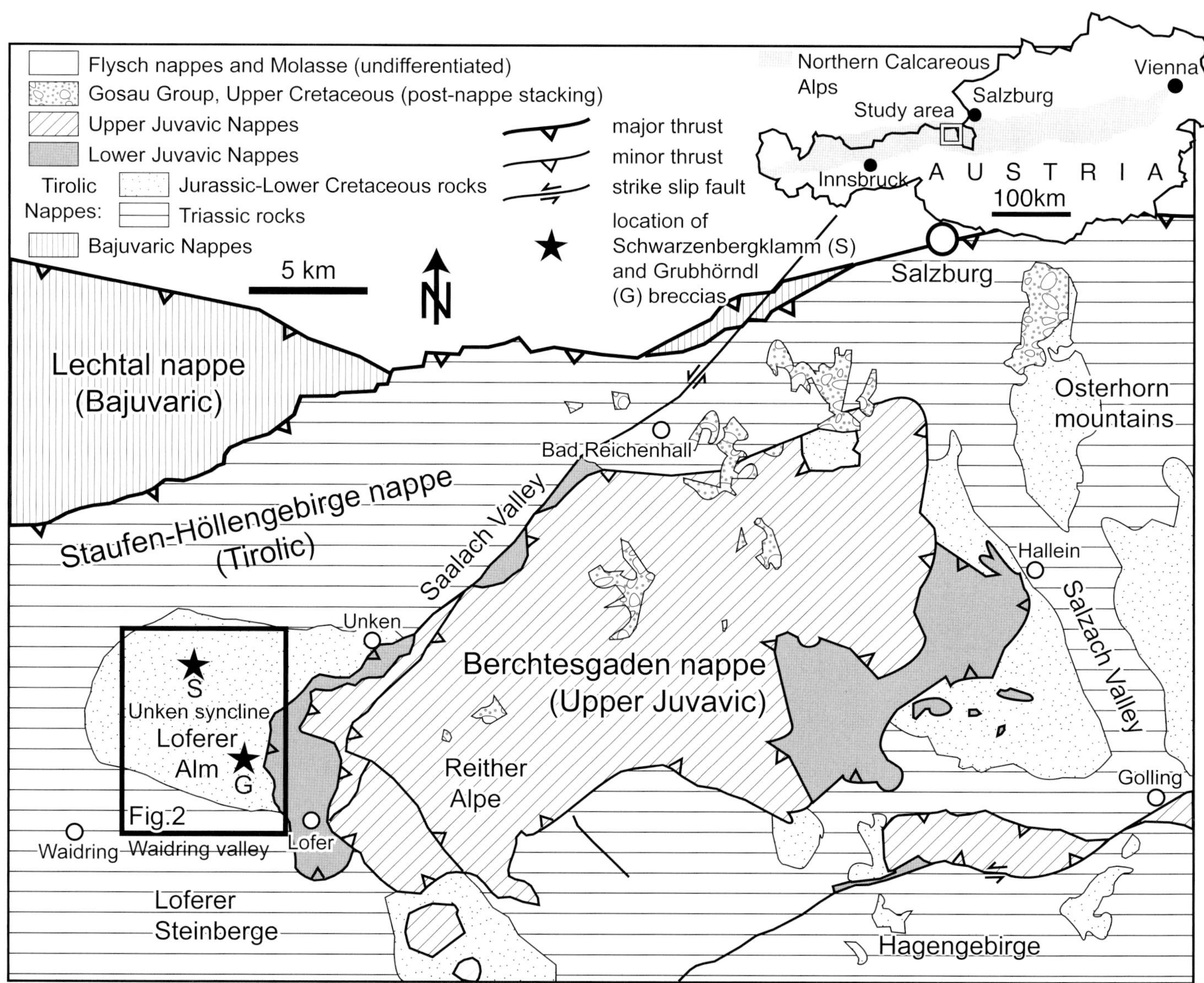

Fig. 1. Tectonic overview of the investigated area based on Schweigl & Neubauer (1997); tectonic subdivision of the thrust sheets of the NCA based on Tollmann (1976).

mega breccias. Frank & Schlager (2006) suggested an interpretation in terms of a depositional domain dissected by major strike-slip faults. Gawlick et al. (1999a) and Frisch & Gawlick (2003) interpret the depositional realm in terms of a foreland basin related to Middle to Upper Jurassic orogeny, which also involved Late Jurassic subduction and high-pressure/low-temperature metamorphism at the southern margin of the Central NCA; however, these authors did not present or discuss associated deformation structures visible in the field.

Previous studies

The Grubhörndl and Schwarzenbergklamm breccias of the Unken syncline (Figs. 1, 2) have been studied for almost a century. Hahn (1910) identified these two occurrences of breccias as Rhaetian and Mid-Liassic sedimentary deposits, respectively. Vortisch (1931) interpreted the Schwarzenbergklamm breccia to be a tectonic breccia attached to a bedding-parallel shear plane. This interpretation was rejected by Fischer (1965), who described the Schwarzenbergklamm breccia as a Jurassic scarp-breccia generated along the western continuation of the "Torrener-Lammer" fault zone. Garrison & Fischer (1969) and Diersche (1980) followed the argument of Fischer (1965) and interpreted the Schwarzenbergklamm breccia as a submarine mass flow deposit generated by a local east-west striking fault scarp, oriented parallel to the Waidring valley. Wächter (1987) gave a detailed sedimentological description of the Schwarzenbergklamm breccia. Channell et al. (1992) presented a paleogeographic reconstruction of the Unken syncline with minor roughly north-south striking Liassic normal faults related to a major E–W striking south-directed normal fault, which separated the future Staufen-Höllengebirge nappe from the Lechtal nappe and was reactivated as a thrust fault during Eoalpine orogeny. Gawlick et al. (1999a, 2002) interpreted the Schwarzenbergklamm breccia as a piece of evidence for the Kimmeridgian orogeny and compared this breccia with deposits in the Tauglboden basin in the Osterhorn Mountains located further to the east (ca. 10 km south/southeast of Salzburg).

Sedimentary succession of the Unken syncline

The Mesozoic of the Unken syncline is characterized by the following succession (Fig. 3). During a tectonically quiet period in the Triassic, shallow water carbonate platform sediments (e.g., Wetterstein Formation, Hauptdolomit Formation, Dachstein limestone) accumulated on the NCA part of the rapidly subsiding Tethyan shelf. In the Rhaetian, reefal limestone ("Oberrhätkalk") interfingered with contemporaneous Kössen marls in an adjacent basin (e.g. Stanton & Flügel 1995). In Early Jurassic times breccias related to tectonic activity were deposited along horst-and-graben structures. These are exposed immediately south of the investigated area (Krainer et al. 1994). However, in the Unken syncline, pre-Jurassic morphology of the Late Rhaetian reef controlled the distribution of facies after a major Earliest Jurassic relative sea-level rise due to subsidence. Lower to Middle Jurassic condensed nodular limestones (i.e. Upper Hettangian Adnet Formation to Bajocian Klaus Formation; Garrison & Fisher 1969) covered the top of the Late Rhaetian reef, while the Kössen basin was filled by Lower Jurassic allodapic limestones and marl-dominated sediments (Kendlbach Formation, Scheibelberg Formation, Allgäu Formation), before Middle Jurassic red nodular limestones of the Klaus Formation covered the former basin (Fischer 1969; Krainer & Mostler 1997). Because the Adnet and Klaus Formations can only be distinguished when biostratigraphic data are available, and because the total thickness of both units is only about 15 meters, we could not distinguish the two formations and mapped "red limestones". A pronounced change in sedimentation occurred at the end of the Middle Jurassic ("Ruhpoldinger Wende"; Schlager & Schöllnberger 1974), as is expressed by widespread deposition of Ruhpolding Radiolarite. This was also the time of deposition of the Grubhörndl and Schwarzenbergklamm breccias, deposited near a rejuvenated submarine topographic gradient, which controlled the distribution of facies in subsequent sedimentary units. On top of the Grubhörndl breccia and southwest of the Grubhörndl, hardgrounds with poorly preserved imprints of ammonites indicate a break in sedimentation, probably even some erosion or dissolution. Contemporaneously, essentially Kimmeridgian-age allodapic radiolarian limestones and siliceous marls of the Tauglboden Formation (sensu Gawlick et al.,1999b) were deposited in the basins. The Tauglboden Formation differs from the Ruhpolding Radiolarite by its content of bio- and coarse lithoclastic material (Schlager & Schlager 1973). Because the radiolarian limestones and siliceous marls overlie both the Ruhpolding Radiolarite and the breccias, and since the latter are contained within the Ruhpolding Radiolarite (see below), we refrain from considering the breccias as part of the Tauglboden Formation in analogy to the stratigraphic succession of the Osterhorn mountains (l.c.; Vecsei et al. 1989). Beyond the area of breccia deposition, pre-Jurassic topography continued to control facies distribution. Upper Jurassic deep water limestones, partly intercalated with bio-/lithoclastic turbidites of the Upper Tithonian to Berriasian Oberalm Formation (Garrison 1967), onlapped onto the breccia and exhumed red limestones and Late Rhaetian limestone, thus levelling out the Early Jurassic relief. Allodapic turbiditic Barmstein beds, which represent a bioclastic shallow water input from a near platform, are intercalated with the Oberalm Formation. The concomitant shallow-water Lärchkogel limestone was accumulated on Late Jurassic topographic highs on top of a mélange of Hallstatt limestones of the Lower Juvavic tectonic unit (e.g. Sanders et al. 2007). Early Cretaceous siliciclastic deposition of the Rossfeld Formation and Lackbach beds preceded overthrusting of the synorogenic sediments (e.g. Darga & Weidich 1986).

Local tectonic setting

The Unken syncline is an open fold structure in the Staufen-Höllengebirge-nappe of the Tirolic nappes of the NCA, located

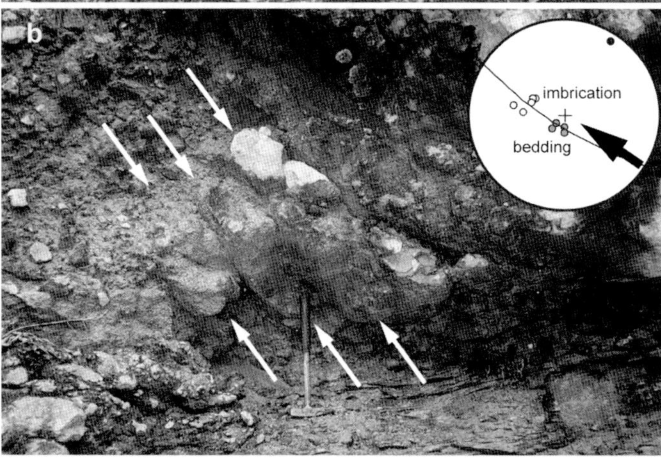

Fig. 4. a) Typical field appearance of the Schwarzenbergklamm and Grubörndl breccias (see Fig. 2 for location). Clast supported breccia composed of unsorted sub-angular carbonate clasts and a marl matrix. b) Imbricate clasts in the Schwarzenbergklamm breccia just above the basal contact to red marly limestones of the Adnet and Klaus Formations. Black arrow in inset indicates sediment transport direction derived from orientation of platy imbricate clasts (white circles: poles to clast surfaces) and bedding (grey circles: poles to bedding). Hammer for scale is 30 cm long.

in the Lackbach beds (Darga & Weidich 1986). The main aim of this contribution is to document structures associated with an Upper Jurassic megabreccia in the Unken syncline. We discuss the sedimentology and three-dimensional distribution of the Grubhörndl and Schwarzenbergklamm breccias and their relation to under- and overlying deposits and adjacent structures. We then compare our data with published results on the Late Jurassic evolution of the NCA.

The Grubhörndl and Schwarzenbergklamm breccias

Description of the breccias

The Grubhörndl breccia and the Schwarzenbergklamm breccia are chaotic, clast- to matrix-supported and lack any kind of sorting (Fig. 4a). The matrix consists of red calcareous micrite. The clasts are angular to sub-angular and consist mainly of "Late Rhaetian" limestone, subordinately of Dachstein limestone, marls of the Kössen Formation, red limestones of the Adnet and Klaus Formations and the Ruhpolding Radiolarite. The clast sizes range from a few millimetres to several meters in diameter. Single blocks measure up to 40 meters in the Schwarzenbergklamm breccia and several hundreds of meters in the Grubhörndl breccia. In the Schwarzenbergklamm breccia, different flow units can be differentiated whereby the base of younger flows cuts into older breccia deposits. Occasionally, imbricated clasts can be found (Fig. 4b). The contact of the breccia with the underlying sediment represents a disconformity or unconformity. In the first case, the breccia was deposited without erosion and gravitationally sank into unconsolidated radiolarite (Fig. 5b). In the second case, an erosive contact of the breccia with the underlying strata could be inferred, whereby the breccia seems to cut stepwise into the underlying sediments at first sight (Fig. 7). However, the presence of a centimetre-thick discontinuous bed of Ruhpolding Radiolarite at the base of the breccia precludes erosion by the breccia and suggests pre-radiolarite slumping during deposition of the youngest parts of the Klaus Formation. Locally, deformation of the underlying red limestones, caused by the overriding breccia, is observed (Unkenbach valley, Fig. 8).

Figure 6a shows the mapped distribution of the Schwarzenbergklamm and Grubhörndl breccias. The Schwarzenbergklamm breccia crops out in the Rottenbach valley, in the Schwarzenbachklamm and in the western Unkenbach valley (Hornsteiner 1991). The Grubhörndl breccia is restricted to an area east of the Loferer Alm. South of Strub Pass, small outcrops of a comparable breccia are found at the Anderlalm. These could represent a southern continuation of the Grubhörndl breccia. A striking feature of the Grubhörndl breccia and the Anderlalm outcrops is their 3 to 5 km north-south extent and their rather narrow east-west extent of less than 500 meters. The Schwarzenbergklamm breccia extends 2 kilometres in a north-south direction, and 2.5 kilometres in an east-west direction. The approximate thickness of the breccias is shown in Figures 6a and 6b. The extraordinary thickness at Grubhörndl and Lärchfeldkopf is a result of the presence of one very large block. The thickness diminishes rapidly towards west and north, but increases slightly in the Schwarzenbergklamm because of a 40 m high block of Late Rhaetian limestone embedded in the breccia.

Sediment transport directions of the Schwarzenbergklamm breccia derived from the long axes of channels and fold axes of slumps were equivocal (Garrison 1964; Wächter 1987). However, these authors inferred north-directed paleoflow for paleogeographical reasons. We measured the imbrication of platy clasts within the breccia (Figs. 4, 6b) and shear structures in red nodular limestones at the base of the breccia near the large block within the Schwarzenbergklamm breccia (Fig. 8). These data indicate sediment transport toward the (W)NW during deposition of the breccia.

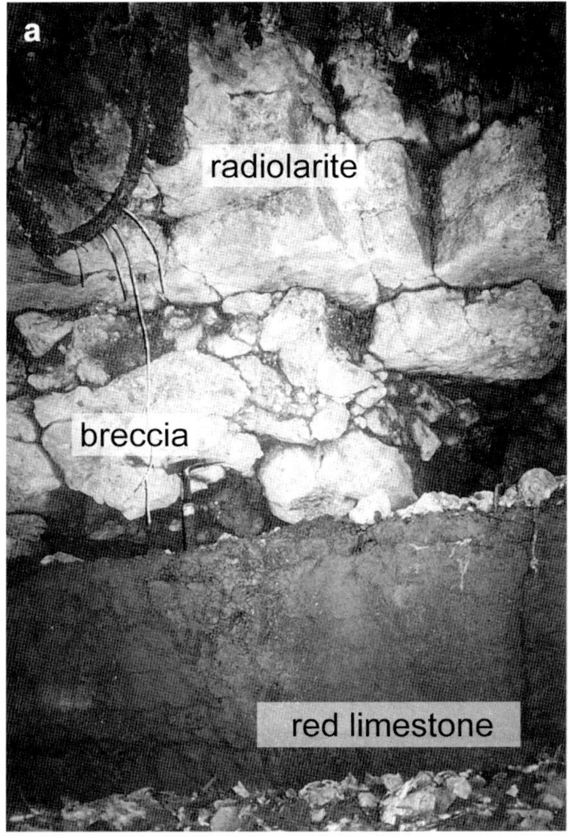

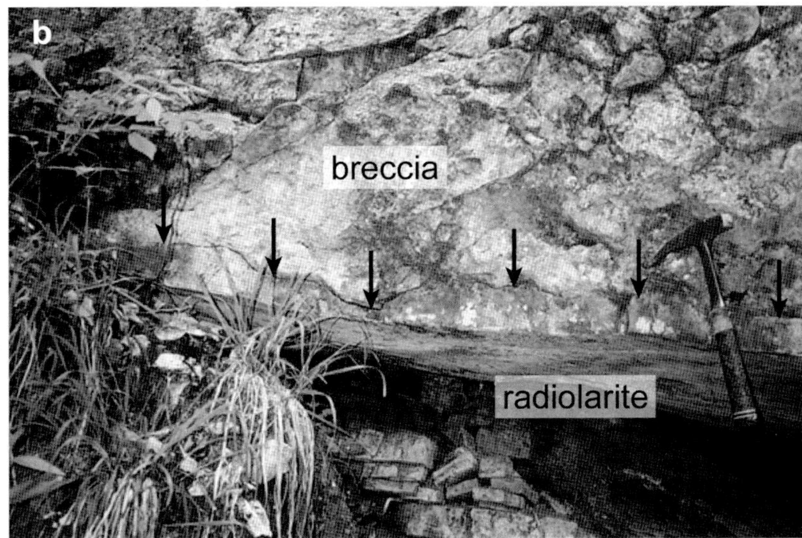

Fig. 5. Relationship of the Schwarzenbergklamm breccia to the Ruhpolding Radiolarite (see Fig. 2 for locations). a) Thin layer of breccia underlain by red limestones of the Adnet or Klaus Formation and overlain by Ruhpolding Radiolarite, 10 m from the outcrop depicted in Fig. 6, where the breccia is at least several meters thick, indicating a pronounced positive topography on top of the breccia body. b) Basal contact of the Schwarzenbergklamm breccia with Ruhpolding Radiolarite. The breccia is gravitationally sunken into the chert after deposition. Together, these outcrops show that the breccia was emplaced during deposition of radiolarian cherts. Hammer for scale is 20 cm high.

Sedimentological interpretation of the breccias

According to Pickering et al. (1986), scarp breccias are characterized by angular components with extreme size variation (micro- to megabreccias), a high percentage of structurally older components, and a matrix devoid of dynamic sedimentary structures. As these characteristics are found in the Grubhörndl breccia, we propose that the Grubhörndl breccia formed as a scarp breccia. The Schwarzenbergklamm breccia is in close geographic vicinity (Fig. 6) and similar in terms of composition and texture, but it shows some evidence for sediment transport, i.e. erosion at the channel basis and imbrications of clasts. We therefore interpret the Schwarzenbergklamm breccia as a portion of the Grubhörndl breccia that was mobilized and transported by debris flows. The large block in the core of the Schwarzenbergklamm breccia can be interpreted as an outrunner block, which became detached from the main breccia mass and which slid further into the basin. This could have occurred by hydroplaning in the frontal part of the debris flow (Prior et al. 1984; Ilstad et al. 2004; De Blasio et al. 2006). The irregular patchy distribution of the Schwarzenbergklamm breccia and its rapid lateral thickness changes (Fig. 6a) are very similar to what is observed between the main body and the outrunner blocks of modern submarine debris flows (Ilstad et al. 2004). The shear structures found at the base of the breccia near the block probably document grounding of the block after water-lubricated transport. The large diameter of the clasts and the rapid decrease of thickness indicate proximity to the scarp, which is assumed to be located east of the line Dietrichshorn-Grubhörndl-Lärchfeldkopf-Anderlkogel (Figs. 2, 6c). It is not possible to reconstruct the exact geometry of the fault, because it was overthrust in the Early Cretaceous, then reactivated in the Miocene with an opposite direction of movement, with at least 500 meters of downfaulting of the eastern block. East of the inferred fault, the former source area of the breccias is covered by the Juvavic nappes.

The mega-block of the Grubhörndl breccia at the Lärchfeldkopf

A block, several hundred meters long and about 350 m high and surrounded by breccia on three sides, is located at the Lärchfeldkopf south of the Grubhörndl (Fig. 9). The block rests on breccia, is covered by it on the western side and overlain by it. The western part of the block consists of Late Rhaetian limestone and the eastern part of Dachstein limestone The sedimentary layering of the block is oriented at 90° to the underlying sedimentary strata (Fig. 9). Below the eastern part of the block, the Late Rhaetian limestone is missing and is replaced by breccia. In the two-dimensional view of Figure 9 the eastern end of the Late Rhaetian limestone is a subvertical step. Within the

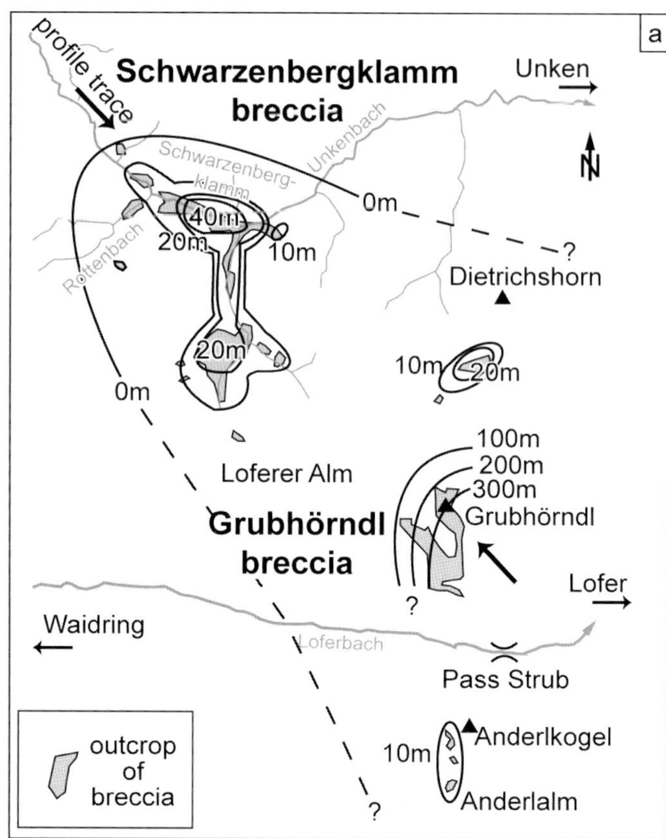

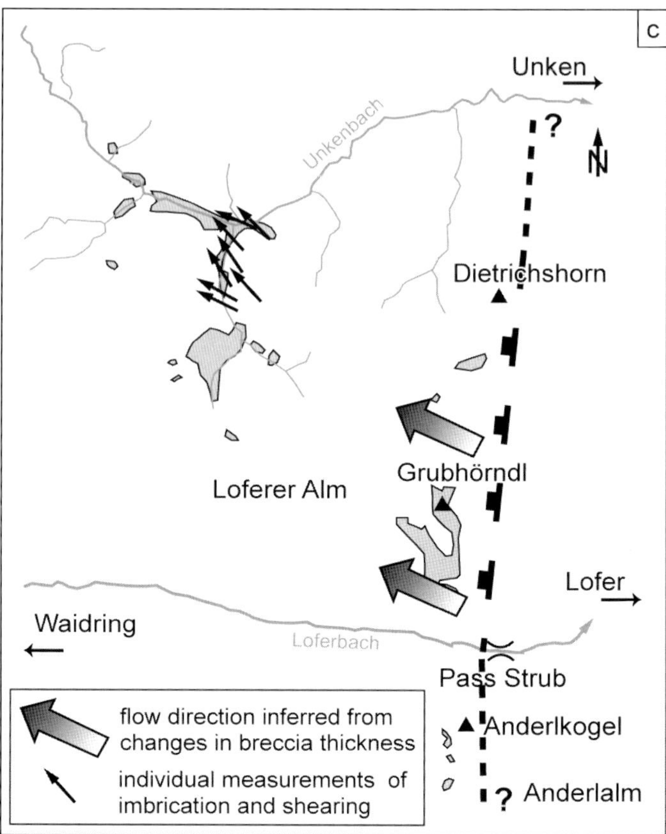

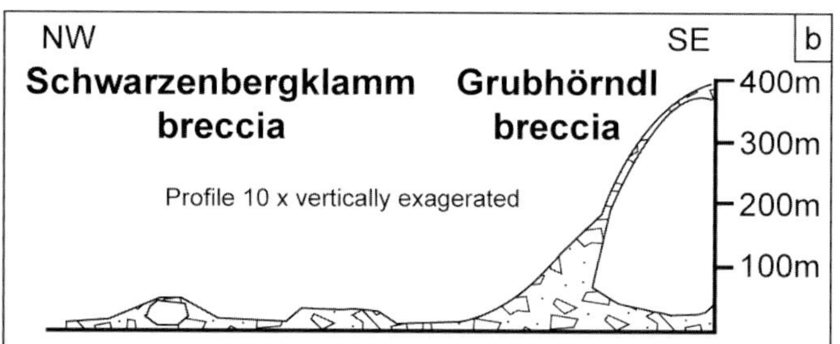

Fig. 6. a) Distribution of thickness of the breccia deposits in the Unken syncline in map view, and, b) in cross section. c) Small arrows: Sediment transport direction deduced from clast imbrication and shear structures at the base of the breccia. Large arrows: Sediment transport inferred from thickness distribution of the breccia deposits.

underlying Dachstein limestone, throughgoing bedding shows that it is not cut by a vertical fault. Therefore, the Late Rhaetian limestone must have moved westward on a bedding-plane-parallel glide surface within the Kössen Formation beneath the Late Rhaetian limestone (Fig. 9).

It is not easy to imagine the scenario in which this block arrived at its present position. One possible scenario would be toppling in the footwall of a major normal fault. In this case the following sequence of events would be necessary: (1) Creation of a major fault scarp, at least as high as the thickness of the block, i.e. several hundred meters, combined with westward tilting. (2) Gliding of the Late Rhaetian limestone in the hanging wall of the detachment, thus creating accommodation space progressively filled by breccia. (3) Gliding of a large block in

the footwall on a bedding plane within the Dachstein limestone across the normal fault into the breccias, such that about half or a third of the block still overlies the footwall. (4) Progressive offset across the normal fault, which leads to progressive tilting of the block into the adjacent basin on the footwall. The latter model is supported by the presence of an array of west-directed normal faults in the Late Rhaetian limestone and red limestones of the Adnet and Klaus Formations of the Sonnwänd (Fig. 2, left side of Fig. 9, Fig. 10), and by half grabens in the Tauglboden Formation in the Unkenbach valley (see below). Examples of large gliding rock masses and allochthonous blocks, as well as different explanations for the development of submarine mega-breccias, have been reported in the literature from different parts of the world, although none of them

Fig. 7. Basal contact of the Schwarzenbergklamm breccia to the red limestones of the Adnet or Klaus Formation (see Fig. 2 for location). A thin discontinuous chert bed of the Ruhpolding Radiolarite underlies the breccia (arrows). The step in the contact must therefore be older than deposition of radiolarian chert and probably represents a slump scar within the red limestones.

includes a block tilted by 90 degrees (compare e.g.: Carrasco 1985; Conaghan et al. 1976; Freeman-Lynde & Ryan 1985; Greb & Weisenfluh 1996; Hesthammer & Fossen 1999; Mullins et al. 1991; Schlager et al. 1984; Surlyk & Ineson 1992; van Weering et al. 1998; Woodcock 1979). The breccia bodies are onlapped by younger sediments. While the thinner part of the Schwarzenbergklamm breccia is overlain and onlapped by the Ruhpolding Radiolarite, the surfaces of the thick parts of the Schwarzenbergklamm breccia and the Grubhörndl breccia were above base level of sedimentation and were only later onlapped by the Oberalm and Ammergau Formations. Southwest of the Grubhörndl (star in Fig. 9), the Grubhörndl breccia body is overlain by a condensation surface, which in turn is overlain by turbidites with coarse-grained shallow water bioclastic debris belonging into the Oberalm Formation (Barmstein beds). Thus, field evidence shows that the submarine topography created by the breccia bodies persisted for several million years and before the base level of sedimentation rose high enough for an onlap of younger formations. The occurrence of large pieces of shallow water organisms indicates the proximity of a carbonate platform, possibly on top of the footwall fault scarp.

Age of the breccias

The age of the breccias was determined by lithostratigraphic correlation because no indicative fossils that would provide a biostratigraphic age were found in the matrix. In the area of the Loferer Alm, the Grubhörndl breccia was deposited on Dachstein and Late Rhaetian limestones (Fig. 9). The breccia consists of reworked Dachstein limestone, Late Rhaetian limestone, marls of the Kössen Formation, red limestones of the Adnet and Klaus Formations and Ruhpolding Radiolarite. Oberalm Formation onlaps the top of the breccia, thus indicating the

end of breccia deposition. The Grubhörndl breccia therefore postdates the onset of deposition of the Ruhpolding Radiolarite and predates deposition of the Oberalm Formation. The Schwarzenbergklamm breccia occurs partly above and partly below the Ruhpolding Radiolarite (Figs. 2, 6, 8). Thus the age of the breccia body can be restricted to the time interval corresponding to the deposition of the Ruhpolding Radiolarite. As both the Grubhörndl and the Schwarzenbergklamm breccia are thought to belong to the same breccia body, this is true for the deposition of both breccias.

The onset of deposition of the Ruhpolding Radiolarite was previously thought to be uniformly Oxfordian (Schlager & Schöllnberger 1974). However, biostratigraphic data from more eastern and southern parts of the central NCA indicate an earlier onset of radiolarite deposition (e.g. Suzuki et al. 2001; review in Gawlick & Frisch 2003). In the westernmost part of the Unken syncline, the youngest rocks below the Ruhpolding Radiolarite are red nodular limestones of the Klaus Formation, which were dated to the Aalenian in two sections at Unkenbach and Scheibelberg (Fischer 1969), and to the Bajocian in the Kammerkehr area (Hahn 1910). The stratigraphic contact to the Ruhpolding Radiolarite is not conformable, however, and is characterized by slump-related scars (Fig. 7) or breccias (Fischer 1969). Direct dating of radiolarian faunas from the Ruhpolding Radiolarite gave an Oxfordian to Mid-Kimmeridgian age (Diersche 1980). The end of deposition of the Ruhpolding Radiolarite is unconstrained because no biostratigraphic data from the Tauglboden Formation of the Unken syncline are available. In the type area in the Osterhorn Mountains further east the Tauglboden Formation ranges from the Late Oxfordian to the Early Tithonian (Gawlick et al., 1999b), overlying Callovian to Oxfordian Ruhpolding Radiolarite (Gawlick et al. 2003). Deposition of the Oberalm Formation in the Unken syncline starts in the Early Tithonian (Garrison 1967), similar to the Osterhorn Mountains (Steiger 1981). The Ruhpolding Radiolarite underlying the Schwarzenbergklamm breccia has a thickness between a few centimetres and one meter. Taking into account the total thickness of about 20 m, the breccia must have been emplaced shortly after the beginning of radiolarite deposition, hence in the Oxfordian. If, however, the onset of deposition of the Ruhpolding Radiolarite is earlier, e.g. Callovian, as in the Osterhorn Mountains (Gawlick et al. 2003), the age of the breccia would also shift to the late Middle Jurassic.

Extensional structures observed in the study area

At the Sonnwänd south of the Loferer Alm (Fig. 1) between Waidringer Steinplatte and Urlkopf (Fig. 10), west-dipping normal faults rooting in the marls of the Kössen Formation cut through Late Rhaetian limestone and Lower to Middle Jurassic red limestones. Upper Jurassic Oberalm Formation seals the faults (Fig. 2, 9). Bedding in the underlying Dachstein limestone is not offset by these faults (Fig. 9, 10). In order to demonstrate the significance of these normal faults, we esti-

Fig. 8. Shear structures at the base of the Schwarzenbergklamm breccia (see Fig. 2 for location). These structures are found below the base of a large block (ø 40 m) within the breccia and are probably related to grounding of the block after transport on top of a water cushion by hydroplaning. Inset shows bedding orientations (white circles: poles to bedding) and transport direction deduced (black arrow).

mated the stretching factor within the Late Rhaetian limestone of the Sonnwänd area (Fig. 10). Due to the low overburden of Lower to Upper Jurassic sediments (total thickness of a few meters) on top of the Rhaetian platform, negligible or no confining pressure acted at the onset of Upper Jurassic extension. The observed normal faults probably were initiated as tensile fractures, orthogonal to the extension direction, i.e. sub-vertical to vertical. Further extension led to domino-style rotation of the fault blocks. As a consequence, bedding in the limestone is perpendicular to the normal faults. As there is no continuous marker bed in the massive Late Rhaetian limestone, arbitrary reference beds have been chosen where fault dips and fault block widths are homogeneous (Fig. 10). Stretching factors calculated from the observed block and fault geometries range from $\beta = 1.06$ to $\beta = 1.12$.

In the Unkenbach valley, fine grained breccias and radiolaria-bearing marly limestone of the Kimmeridgian Tauglboden Formation form a monocline several metres wide (Fig. 11), with a near-horizontal western, and an approximately 20° E-dipping eastern limb. Angular unconformities in the eastern limb document erosion after tilting, followed by deposition parallel to the erosion surface resulting in progressively shallower dips up section. A pronounced angular unconformity marks the boundary between the marly limestones and the marls of the Tithonian Oberalm Formation, which are not tilted and therefore seal the structure.

Although it is tempting to interpret the unconformities as a result of block tilting observed in the underlying Late Rhaetian limestone, the geometry of the structure precludes a direct connection. Progressive domino-style faulting in the subsurface would create basins in which planar bedding would become progressively shallower between angular unconformities, and individual sub basins should be separated by steep faults with diminishing offset up section. The observed monoclinal folding is therefore interpreted as a rollover anticline formed in the hangingwall block of a listric normal fault. The relatively sharp bend in the core of the rollover calls for a shallow detachment, probably within the Tauglboden Formation. Multiple angular unconformities in the Tauglboden Formation indicate progressive tilting of the hanging wall block. The axis of the rollover anticline is oriented north-south and indicates westward normal movement (Fig. 11), which is in concert with the earlier extension phase. The Tithonian Oberalm Formation seals these structures, thus indicating the cessation of extension.

Discussion and conclusion

The detailed documentation of Jurassic sedimentary rocks and related structures in the Unken syncline allows for a reconstruction of the conditions of megabreccia deposition during the Late Jurassic (Fig. 12). Two different controls on facies distribution can be distinguished. During the Early and Middle Jurassic, facies distribution is mainly controlled by the pre-Jurassic topography of the Late Triassic Steinplatte reef with its pronounced slope to the north. In the early Late Jurassic, deposition of megabreccias was linked to tectonic activity, which created a new submarine topography with a west-directed slope along a major N–S-striking fault. The combination of inherited and newly formed slope controlled the observed transport direction of the Schwarzenbergklamm breccia toward the NW. Based on the similarity of clast composition, the two breccia bodies, which previously were described separately, are interpreted as one large breccia body. The Grubhörndl breccia represents the more proximal unit, probably attached to a fault scarp, whereas the Schwarzenbergklamm breccia was transported by debris flows further into the basin. The key observa-

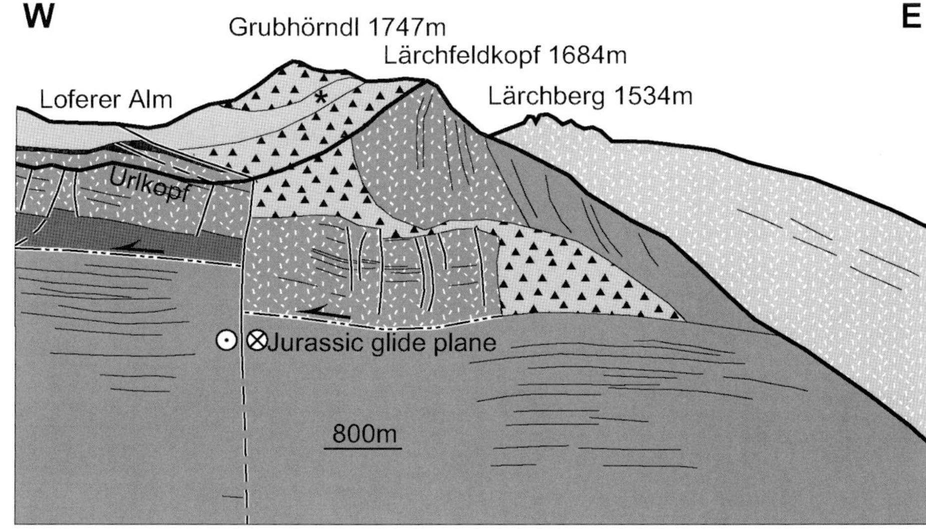

Fig. 9. Aerial view of the southern face of the Lärchfeldkopf and Grubhörndl, showing most of the features described in this paper. The Dachstein limestone is overlain by the Kössen Formation, which interfingers with the Late Rhaetian limestone to the north. Due to a young sinistral fault, a more internal portion of the Late Rhaetian reef, which directly grew on the Dachstein limestone and originally was located further south, is now found east of the fault. The Kössen Formation west of the fault and a bedding plane east of the fault were used as a detachment during gravity-driven westward gliding of the Late Rhaetian limestone. The space created by gliding was filled by Grubhörndl breccia and an embedded mega block. Normal faults crosscutting the Late Rhaetian limestone and rooting in the detachment crosscut red limestones of the Adnet and Klaus Formations, but are sealed by the Ammergau Formation. The Grubhörndl breccia is onlapped by Oberalm and Ammergau Formations. The star in the Oberalm Formation denotes the location of large shallow water fossils in the Oberalm Formation.

tions are (1) the N–S elongation of the several hundred meters thick Grubhörndl breccia, (2) its rapid thinning to the NW, and, (3) the onlap of the younger sedimentary units, demonstrating that the submarine topography was slowly blanketed.

Smaller structures, such as the domino-style faults crosscutting the Late Rhaetian limestone and the roll-over structure in the Tauglboden Formation are attributed to gravity-driven westward gliding of the Late Rhaetian limestone and portions of the Tauglboden Formation in response to westward tilting. Tilting could be a consequence of an E-dipping normal fault to the west and outside the investigated area. As discussed previously, tilting is a necessary precondition for the deposition of the mega block within the Grubhörndl breccia. Therefore,

a N–S-oriented scarp of a major west-dipping normal fault is the most probable setting for the deposition of the Grubhörndl breccia. Fischer (1965) previously proposed N-directed sediment transport of the Schwarzenbergklamm breccia from an E–W-oriented fault scarp along the Waidring valley. However, no major vertical offset is observed in a N–S section across the valley (Lukesch 2003; profile West of Pestal & Hejl 2005).

In the eastern and central NCA Middle to Late Jurassic tectonic processes have been the subject of controversy. The following alternative scenarios have been proposed (1) The NCA are influenced by rifting of the Apulian continental margin adjacent to the Piemont-Ligurian ocean; and this scenario involves extension (Vecsei et al. 1989; Lackschewitz et al. 1991;

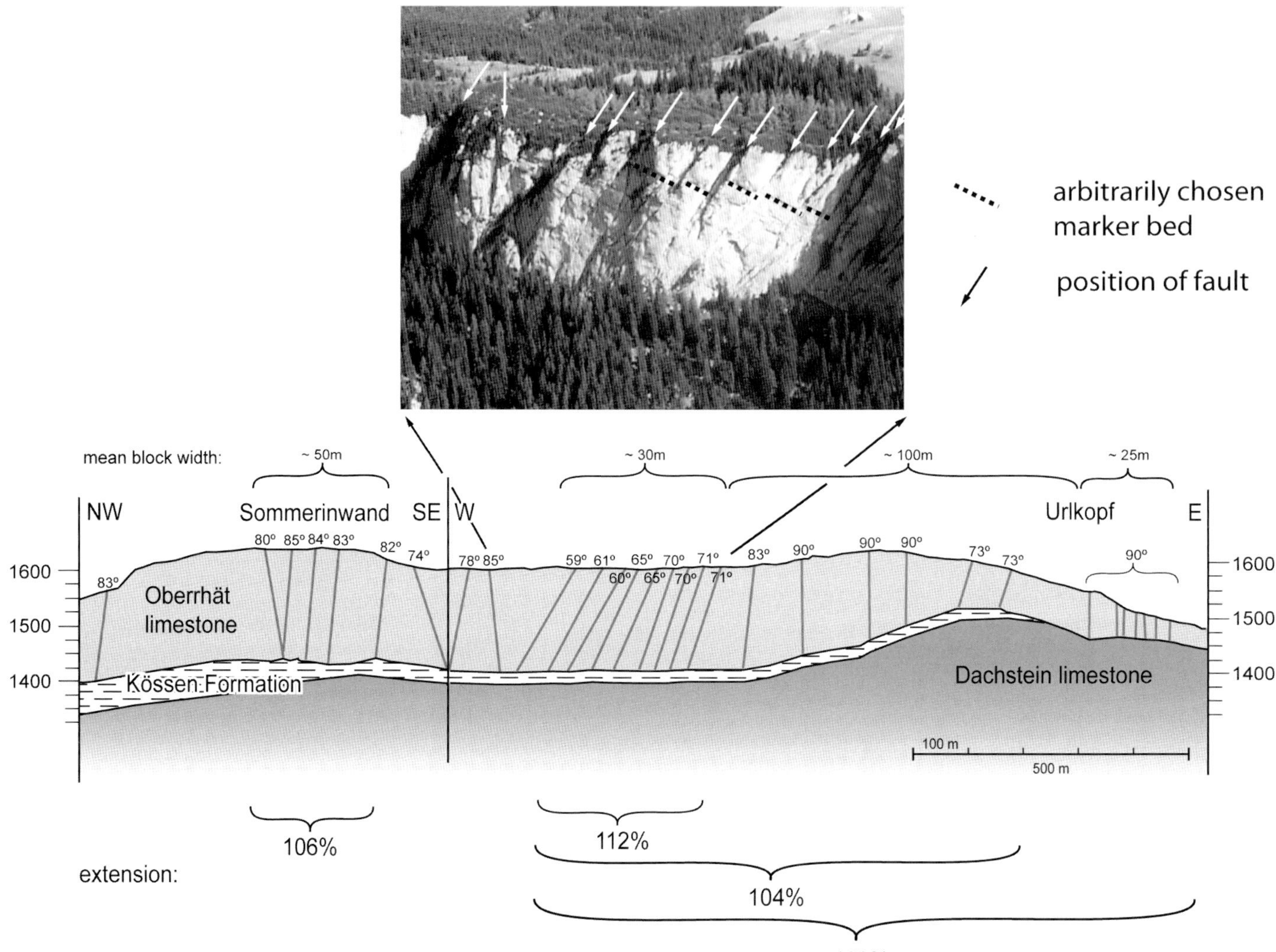

Fig. 10. Normal faults crosscutting the Late Rhaetian limestone (top) and determination of the stretching factor (see Fig. 2 for location). In spite of the clear fault geometry, extension is small.

Hebbeln et al. 1996; May & Eisbacher 1999; Auer & Eisbacher 2003). (2) Nappe stacking and related compression in response to closure of the Meliata Ocean is the main controlling factor (Braun 1998; Gawlick et al. 1999a; Frisch & Gawlick 2003). (3) Jurassic strike-slip faulting in the NCA explains the distribution of Upper Triassic facies, as was proposed by Fischer (1965). Similar concepts were used by e.g. Wächter (1987), Channell et al. (1990), and by Frank & Schlager (2006). The latter also evaluated all the three hypotheses.

We will not repeat this discussion here, except for stating that the observations reported here and for our area of investigation, are in accordance with hypotheses (1) and (3), but not with the thrusting scenario (2). The orientations of early Late Jurassic faults, across which facies changes were reported, vary. Lackschewitz et al. (1991) and Auer & Eisbacher (2003) reported E–W-striking faults, whereas Eberli (1988) and Froitzheim & Eberli (1990) reconstructed N–S-striking faults. Some authors interpreted a complex more or less orthogonal pattern of N–S and E–W-striking faults (Channell et al. 1990; 1992; May & Eisbacher 1999). The structures described in this study are N–S-striking. However, the Grubhörndl breccia and the associated normal fault abruptly end to the north, where the offset must be taken up by some other structure. Such a structure might be a precursor of the so-called Saalachtal fault ("Saalachtal Westbruch"; Hahn 1910), which was later reactivated during the Miocene (Rittner 2006). Large vertical offsets along normal faults, which do not decrease toward the end of the fault, are found within strike-slip systems associated with pull-apart basins (e.g. Aydin & Nur 1982).

On a larger scale, we propose that the above-mentioned three scenarios should perhaps not be treated as mutually exclusive. Each of them describes some other aspect of the complex

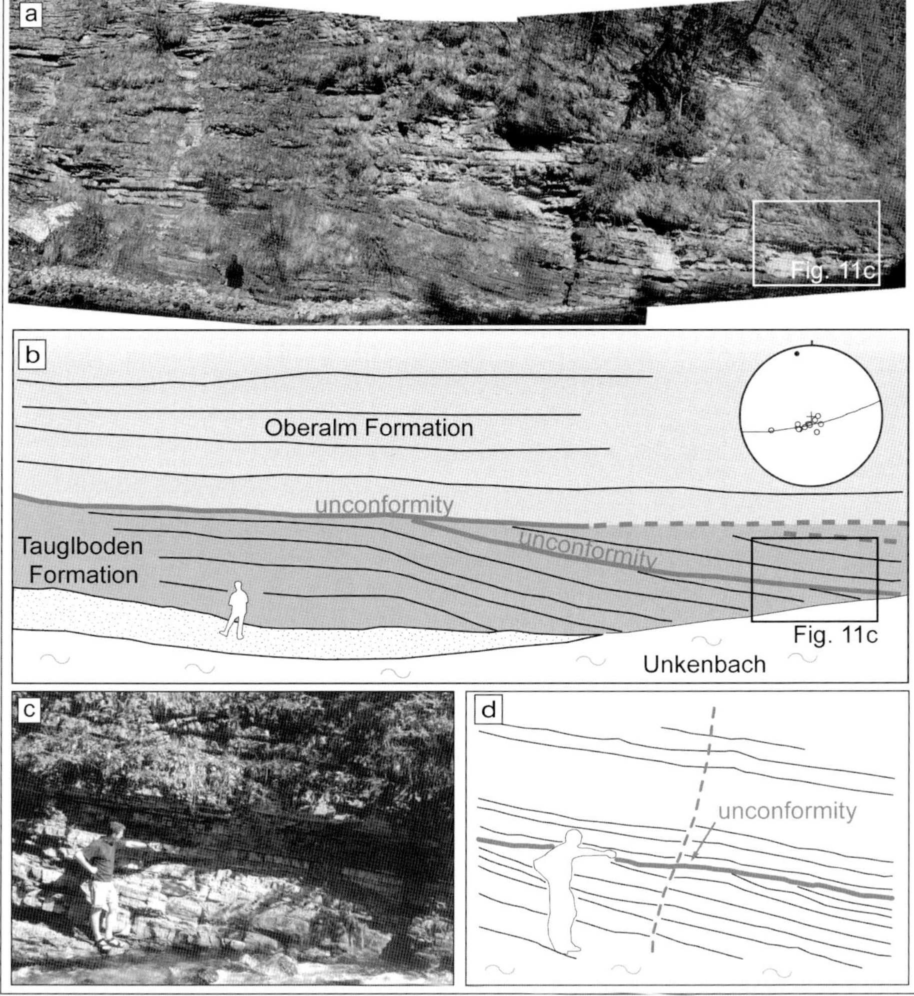

Fig. 11. a) Field photograph and b) sketch of angular unconformities in the Tauglboden Formation, sealed by the Oberalm Formation c) Field photograph and d) sketch of angular unconformity within the Tauglboden Formation All unconformities are erosional, and therefore document deformation, erosion and then sedimentation. See Fig. 2 for location.

evolution of the NCA within the Alps-Carpathian-Dinarides system. During the Jurassic the NCA were located between the opening Piemont-Liguria Ocean to the northwest and the closing Meliata Ocean to the southeast. To the northwest the Jurassic sedimentary succession of the Austroalpine units is controlled by the opening of the Piemont-Liguria ocean: in distal, ocean-near parts of the continental margin (Lower Austroalpine nappes), coarse grained syn-rift sediments of the Lower and early Middle Jurassic Allgäu Formation were deposited in half-graben basins (Eberli 1988). There, these basins and exhumed mantle rocks, formed at the ocean-continent transition, are overlain by post-rift radiolarian cherts and Aptychus/Calpionella pelagic limestones (Dietrich 1970; Weissert & Bernoulli 1985; Froitzheim & Manatschal 1996), which are also abundant in and at the margins of the entire Alpine Tethys (Bernoulli & Jenkyns 1974; Bill et al. 2001).

However, the widespread occurrence of mega-slides and breccias, not only in the syn-rift sediments, but also in Middle to Upper Jurassic post-rift sediments in the central and eastern NCA, was documented in many previous studies (i.e. Tollmann 1987; Mandl 2000; Gawlick et al. 1999a, 2002; Gawlick & Frisch 2003) and points to the activity of another process affecting this same continental margin: Gravitative emplacement of the Hallstatt melange requires tectonic transport of Hallstatt facies sediments, belonging to the outer continental margin facing the Meliata ocean, onto the inner continental margin (Gawlick et al. 1999a, Mandl 2000), represented by the southernmost part of the Tirolic nappe. However, in our opinion, this Jurassic nappe stack is not preserved in the Eastern Alps. As shown by Mandl (2000) and Frisch & Gawlick (2003), late Early Cretaceous thrusting, following the obduction of the Meliata realm (Schmid et al. 2008), superimposed southern parts of the Tirolic nappes out-of-sequence with respect to emplacement of the Hallstatt melange onto northern parts. During this event, the Lower Juvavic nappe, which contains the Hallstatt melange, and the Upper Juvavic nappe, which originally represents a more southern part of the Tirolic unit, both did form. Cretaceous stacking also transported an ophiolithic unit, tec-

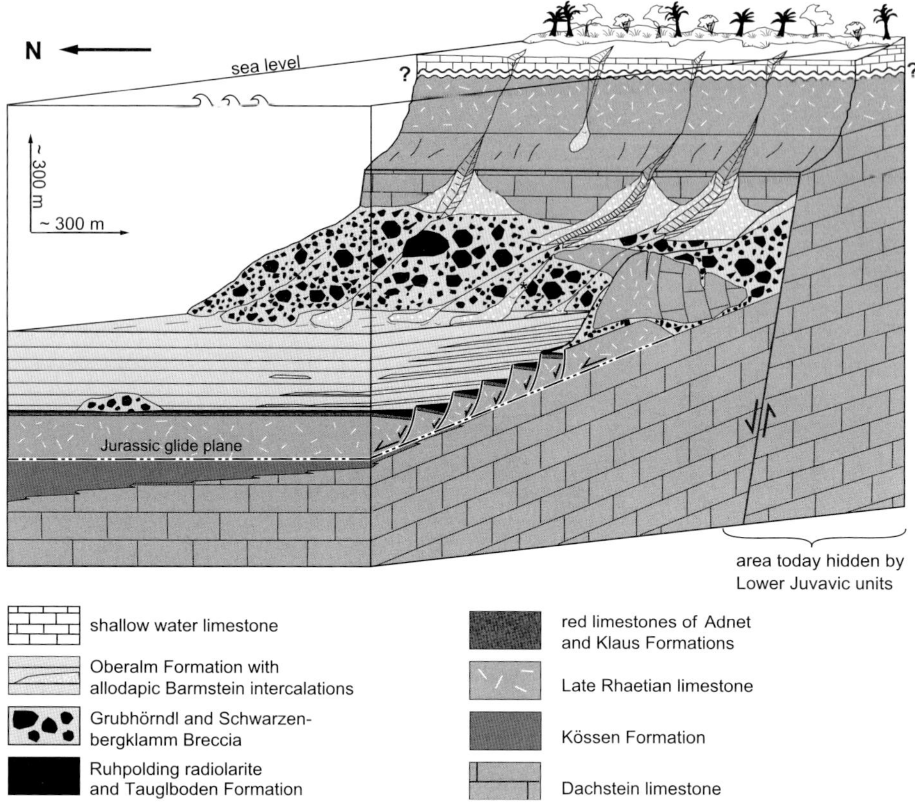

Fig. 12. Tentative reconstruction of the Grubhörndl and Schwarzenbergklamm breccias at the time of deposition of the Oberalm Formation in the Tithonian, giving an impression of the height of the fault scarp. The Grubhörndl breccia (Oxfordian) formed by collapse (and toppling, see text) of the footwall of a normal fault, is overlain by supra-fan lobes fed by a shallow water carbonate platform on top of the footwall. Both the carbonate platform and the normal fault are hidden below the Lower Juvavic nappe (see Fig. 2). The occurrence of Tithonian shallow water carbonates in the footwall is indicated by coarse bioclastic debris in Barmstein beds on top of the Grubhörndl breccia (star). Sub-aerial exposition in the footwall as shown is purely speculative.

tonically positioned on top of the Hallstatt unit, into the immediate vicinity of the present-day NCA, as is documented by ophiolithic detritus in Cretaceous synorogenic deposits of the NCA (e.g. Poper & Faupl 1988). Recent models of the NCA nappe stack that were inspired by the nappe structure described for the Austroalpine nappe system, including the NCA, in general (Schmid et al., 2004; Janak et al., 2004), are, however, not valid for the western NCA. For example, the Inntal thrust sheet represents the western continuation of the Tirolic unit (Tollmann 1976; Ortner et al. 2006). On top it carries several klippen of a tectonically higher thrust sheet, the Krabachjoch nappe, which is equivalent in facies to the underlying Inntal thrust sheet, but equivalent in tectonic position to the Lower Juvavic nappe (Tollmann 1976). This illustrates the non-cylindrical nature of the nappe edifice of the NCA on a large scale. In particular, remnants of the outer continental margin facing the Meliata Ocean are absent in the western NCA; but breccias are commonly found in Middle to Upper Jurassic post-rift sediments (e.g. Eisenspitze breccias, Achtnich 1982; Rofan breccias, Wächter 1987; breccias of the Thiersee syncline, Töchterle 2005). The redeposited material of these breccias consists entirely of debris from local sources. Away from the Hallstatt melange and related breccias, which can readily be explained by sliding from a thrust wedge, the genesis of locally sourced breccias of the western NCA needs an alternative explanation. Plate tectonic reconstructions of the Alpine realm show that the NCA were located in a zone of transform faulting (Weissert & Bernoulli 1985; Trümpy 1988; Channell et al. 1990). Hence, activity of strike-slip faults and formation of pull-apart extensional basins provides a good explanation for the repeated, but not contemporaneous shedding of coarse grained and locally sourced breccias (see above).

The studied area is in close proximity to the Hallstatt melange of the Lower Juvavic nappe and also to the locally sourced breccias within the Tirolic nappe, separated by an early Late Cretaceous thrust (Fig. 2). Cretaceous thrusting superimposed a tectonic unit influenced by the distant effects of Jurassic shortening, causing the gravitative emplacement of the Hallstatt melange (thrusting hypothesis 2) onto another tectonic unit mainly controlled by normal faults related to strike-slip faulting (strike slip hypothesis 3). Thus, Cretaceous thrusting reduced the distance between parts of the NCA that were controlled by different tectonic processes. Even given the evidence for a thrust related fill in southerly adjacent basins (see above), we prefer an interpretation that involves normal faults in a strike-slip scenario for the formation of the breccias of the Unken syncline.

Acknowledgements

The "Fonds zur Förderung der wissenschaftlichen Forschung" (FWF) is gratefully acknowledged for financial support (P-13566-TEC). The Österreichische

Bundesforste and Bayerische Saalforste are thanked for issuing driving permits on their private roads. The reviews of G. Eisbacher and M. Wagreich, and the comments of H.-J. Gawlick and S. Schmid improved the quality of the paper substantially.

REFERENCES

Achtnich, T. 1982: Die Jurabrekzien der Eisenspitze. Geologisch-Paläontologische Mitteilungen der Universität Innsbruck 12, 41–70.

Auer, M. & Eisbacher, G.H. 2003: Deep structure and kinematics of the Northern Calcareous Alps (TRANSALP profile). International Journal of Earth Sciences 92, 210–227.

Aydin, A. & Nur, A. 1982: Evolution of pull-apart basins and their scale independence. Tectonics 1, 91–105.

Bernoulli, D. & Jenkyns, H.C. 1970: A Jurassic basin: the Glasenbach Gorge, Salzburg, Austria. Verhandlungen der Geologischen Bundesanstalt 1970, 504–531.

Bernoulli, D. & Jenkyns, H.C. 1974: Alpine, Mediterranean and central Atlantic Mesozoic facies in relation to the early evolution of the Tethys. In: Dott., R.H. & Shaver, R.H. (Eds.): Modern and ancient geosynclinal sedimentation, SEPM Special Publication 19, 129–160.

Bertotti, G., Picotti, V., Bernoulli, D. & Castellarin, A. 1993: From rifting to drifting: tectonic evolution of the South-Alpine upper crust from the Triassic to the Early Cretaceous. Sedimentary Geology 86, 53–76.

Bill, M., O'Dogherty, L., Guex, J., Baumgartner, P.O. & Masson, H. 2001: Radiolarite ages in Alpine-Mediterranean ophiolites: constraints on the oceanic spreading and the Tethys-Atlantic connection. Geological Society of America Bulletin 113, 129–143.

Böhm, F., Dommergues, J.-L. & Meister, C. 1995: Breccias of the Adnet Formation: indicators of a Mid-Liassic tectonic event in the Northern Calcareous Alps (Salzburg/Austria). Geologische Rundschau 84, 272–286.

Braun, R.1998: Die Geologie des Hohen Gölls. Nationalpark Berchtesgaden, Forschungsbericht 40, Berchtesgaden, 192 pp.

Carrasco, V.B. 1985: Albian sedimentation of submarine autochthonous and allochthonous carbonates, East edge of the Valles-San Luis Potosi platform, Mexico. In: Cook, H.E. & Enos, P. (Eds.): Deep-Water Carbonate Environments, SEPM Special Publication 25, 263–272.

Channell, J.E. T, Brandner, R., Spieler, A. & Stoner, J.S. 1992: Paleomagnetism and paleogeography of the Northern Calcareous Alps (Austria). Tectonics 11, 792–810.

Channell, J.E.T., Brandner, R., Spieler, A. & Smathers, N.P. 1990: Mesozoic paleogeography of the Northern Calcareous Alps – evidence from paleomagnetism and facies analysis. Geology 18, 828–831.

Conaghan, P.J., Mountjoy, E.W., Edgecombe, D.R., Talent, J.A. & Owen, D.E. 1976: Nubrigyn algal reefs (devonian), eastern Australia: Allochtohonous blocks and megabreccias. Geological Society of America Bulletin 87, 515–530.

Darga, R. & Weidich, K.F. 1986: Die Lackbachschichten, eine klastische Unterkreide-Serie in der Unkener Mulde (Nördliche Kalkalpen, Tirolikum). Mitteilungen der bayerischen Staatssammlung für Paläontologie und Historische Geologie 26, 93–112.

De Blasio, F.V., Engvik, L.E. & Elverhøi, A. 2006: Sliding of outrunner blocks from submarine landslides. Geophysical Research Letters 33, L06614.

Diersche, V. 1980: Die Radiolarite des Oberjura im Mittelabschnitt der Nördlichen Kalkalpen. Geotektonische Forschungen 58, 217 pp.

Dietrich, V. 1970: Die Stratigraphie der Platta-Decke. Fazielle Zusammenhänge zwischen Oberpenninikum und Unterostalpin. Eclogae Geologicae Helvetiae 63, 631–671.

Dimo-Lahitte, A., Monié, P. & Vergély, P. 2001: Metamorphic soles from the Albanian ophiolites: Petrology, 40Ar/39Ar geochronology, and geodynamic evolution. Tectonics 20, 78–96.

Dya, M. 1992: Mikropaläontologische und fazielle Untersuchungen im Oberjura zwischen Salzburg und Lofer. Dissertation Technische Universität Berlin, 138 pp.

Eberli, G.P. 1988: The Evolution of the Southern Continental Margin of the Jurassic Tethys Ocean as recorded in the Allgäu Formation of the Austroalpine Nappes of Graubünden (Switzerland). Eclogae Geologicae Helvetiae 81, 175–214.

Ebli, O. 1997: Sedimentation und Biofazies an passiven Kontinentalrändern: Lias und Dogger des Mittelabschnittes der nördlichen Kalkalpen und des Frühen Atlantik (DSDP Site 547B, offshore Marokko). Münchner geowissenschaftliche Abhandlungen, Reihe A 32, 1–255.

Faupl, P. & Wagreich, M. 2000: Late Jurassic to Eocene paleogeography and geodynamic evolution of the Eastern Alps. Mitteilungen der Österreichischen Geologischen Gesellschaft 92, 79–94.

Ferneck, F.A. 1962: Stratigraphie und Fazies im Gebiet der mittleren Saalach und des Reiteralm Gebirges: ein Beitrag zur Deckenfrage in den Berchtesgadener Alpen. Diss. Ludwig-Maximilians-Universität München, 107 pp.

Fischer, A.G. 1965: Eine Lateralverschiebung in den Salzburger Kalkalpen. Verhandlungen der Geologischen Bundesanstalt 1965, 20–33.

Fischer, R. 1969: Roter Ammonitenkalk und Radiolarit aus dem unteren Dogger der Kammerkehr (Nordtirol). Mitteilungen der bayerischen Staatssammlung für Paläontologie und Historische Geologie 9, 93–116.

Frank, W. & Schlager, W. 2006: Jurassic strike slip versus subduction in the Eastern Alps. International Journal of Earth Sciences 95, 431–450.

Freeman-Lynde, R.P. & Ryan, W.B.F. 1985: Erosional modification of Bahama Escarpment. Geological Society of America Bulletin 96, 481–494.

Frisch, W. 1979: Tectonic Progradation and Plate Tectonic Evolution of the Alps. Tectonophysics 60, 121–139.

Frisch, W. & Gawlick, H.-J. 2003: The nappe structure of the central Northern Calcareous Alps and its disintegration during Miocene tectonic extrusion—a contribution to understanding the orogenic evolution of the Eastern Alps. International Journal of Earth Sciences 92, 712–727.

Froitzheim, N. & Eberli, G.P. 1990: Extensional Detachment Faulting in the Evolution of a Tethys Passive Continental Margin (Eastern Alps, Switzerland). Geological Society of America Bulletin 102, 1297–1308.

Froitzheim, N. & Manatschal, G. 1996: Kinematics of Jurassic rifting, mantle exhumation, and passive-margin formation in the Austroalpine and Penninic nappes (eastern Switzerland). Geological Society of America Bulletin 108, 1120–1133.

Garrison, R.E.1964: Jurassic and Early Cretaceous sedimentation in the Unken valley area, Austria. Unpublished PhD dissertation Princeton University, Princeton, N.J., 188 pp.

Garrison, R.E. 1967: Pelagic limestones of the Oberalm beds (Upper Jurassic – Lower Cretaceous), Austrian Alps. Bulletin of Canadian Petroleum Geology 15, 21–49.

Garrison, R.E. & Fischer, A.G. 1969: Deep water limestones and radiolarites of the Alpine Jurassic. In: Friedmann, G.M. (Ed.): Depositional environments in carbonate rocks, SEPM Special Publication 14, 20–56.

Gawlick, H.-J. & Böhm, F. 2000: Sequence and isotope stratigraphy of Late Triassic distal periplatform limestones from the Northern Calcareous Alps (Kälberstein Quarry, Berchtesgaden Hallstatt Zone). International Journal of Earth Sciences 89, 108–129.

Gawlick, H.-J., Frisch, W., Vecsei, A., Steiger, T. & Böhm, F. 1999a: The change from rifting to thrusting in the Northern Calcareous Alps as recorded in Jurassic sediments. Geologische Rundschau 87, 644–657.

Gawlick, H.-J., Suzuki, H., Vortisch, W. & Wegerer, E. 1999b: Zur stratigraphischen Stellung der Tauglbodenschichten an der Typlokalität in der Osterhorngruppe (Nördliche Kalkalpen, Ober-Oxfordium – Unter-Thitonium). Mitteilungen der Gesellschaft der Geologie und Bergbaustudenten Österreichs 42, 1–20.

Gawlick, H.-J. & Frisch, W. 2003: The Middle to Late Jurassic carbonate clastic radiolaritic flysch sediments in the Northern Calcareous Alps: sedimentology, basin evolution, and tectonics – an overview. Neues Jahrbuch für Geologie und Paläontologie, Abhandlungen 230, 163–213.

Gawlick, H.-J., Frisch, W., Missoni, S. & Suzuki, H. 2002: Middle to Late Jurassic radiolarite basins in the central part of the Northern Calcareous Alps as a key for the reconstruction of their early tectonic history – an overview. Memorie Societa' Geologica Italiana 57, 123–132.

Greb, S.F. & Weisenfluh, G.A. 1996: Paleoslumps in coal bearing strata of the Breathitt Group (Pennsylvanian), eastern Kentucky coal field, U.S.A. International Journal of Coal Geology 31, 115–134.

Gruber, P., Faupl, P. & Koller, F. 1992: Zur Kenntnis basischer Vulkanitgerölle aus Gosaukonglomeraten der östlichen Kalkalpen. Mitteilungen der Österreichischen Geologischen Gesellschaft 84, 77–100.

Hahn, F.F. 1910: Die Geologie der Kammerkehr-Sonntagshorngruppe. Jahrbuch der Geologischen Reichsanstalt 60, 637–712.

Hebbeln, D., Henrich, R., Lackschewitz, S. & Ruhland, G. 1996: Tektonische Struktur und fazielle Gliederung der Lechtaldecke am NW-Rand des Tirolischen Bogens in den Chiemgauer Alpen. Mitteilungen der Gesellschaft der Geologie und Bergbaustudenten Osterreichs 39/40, 221–235.

Hesthammer, J. & Fossen, H. 1999: Evolution and geometries of gravitational collapse structures with examples from the Statfjord Fieldnorthern North Sea. Marine and Petroleum Geology 16, 148–170.

Hornsteiner, G. 1991: Die jurassische Entwicklung auf der Waidringer Steinplatte unter besonderer Berücksichtigung der Scheiblberg Schichten an der Typlokalität. Diplomarbeit Universität Innsbruck, 219 pp.

Ilstad, T., De Blasio, F.V., Elverhøi, A., Harbitz, C., Engvik, L., Longva, O. & Marr, J.G. 2004: On the frontal dynamics and morphology of submarine debris flows. Marine Geology 213, 481–497.

Janák, M., Froitzheim, N., Lupták, B., Vrabec, M. & Krogh Ravna, E.J. 2004: First evidence for ultrahigh-pressure metamorphism of eclogites in Pohorje, Slovenia: Tracing deep continental subduction in the Eastern Alps. Tectonics 23, TC5014.

Kozur, H. & Mostler, H. 1992: Erster paläontologischer Nachweis von Meliaticum und Süd-Rudabányaicum in den nördlichen Kalkalpen (Österreich) und ihre Beziehungen zu den Abfolgen in den Westkarpaten. Geologisch-Paläontologische Mitteilungen der Universität Innsbruck 18, 87–129.

Krainer, K. & Mostler, H. 1997: Die Lias-Beckenentwicklung der Unkener Synklinale (nördliche Kalkalpen, Salzburg) unter besonderer Berücksichtigung der Scheibelberg Formation. Geologisch-Paläontologische Mitteilungen der Universität Innsbruck 22, 1–41.

Krainer, K., Mostler, H. & Haditsch, J.G. 1994: Jurassische Beckenbildung in den Nördlichen Kalkalpen bei Lofer (Salzburg) unter besonderer Berücksichtigung der Manganerzgenese. Abhandlungen der Geologischen Bundesanstalt 50, 257–293.

Lackschewitz, K.S., Grützmacher, U. & Henrich, R. 1991: Paleooceanography and rotational block faulting in Jurassic carbonate series of the Chiemgau Alps (Bavaria). Facies 24, 1–24.

Lukesch, M.E. 2003: Die Geologie des Nordwest-Randes der Berchtesgadener Masse bei Lofer (Nördliche Kalkalpen) – Genese oberjurassischer Brekzien und das Westende der Saalachtal Störung. Diplomarbeit Universität Innsbruck, 127 pp.

Mandl, G. 2000: The Alpine sector of the Tethyan shelf – examples for Triassic to Jurassic sedimentation and deformation from the Northern Calcareous Alps. Mitteilungen der Österreichischen Geologischen Gesellschaft 92, 61–77.

Mandl, G.W. & Ondrejickova, A. 1993: Radiolarien und Conodonten aus dem Meliatikum im Ostabschnitt der Nördlichen Kalkalpen (Österreich). Jahrbuch der Geologischen Bundesanstalt 136, 841–871.

May, T. & Eisbacher, G. 1999: Tectonics of the synorogenic "Kreideschiefer basin", northwestern Calcareous Alps, Austria. International Journal of Earth Sciences 92, 307–320.

Mullins, H.T., Dolan, J., Breen, N., Andersen, B., Gaylord, M., Petruccione, J.L., Wellner, R.W., Melillo, A.J., Jurgens, A.D. 1991: Retreat of carbonate platforms: Response to tectonic processes. Geology 19, 1089–1092.

Ortner, H. 2003: Cretaceous thrusting in the western part of the Northern Calcareous Alps (Austria) – evidences from synorogenic sedimentation and structural data. Mitteilungen der Österreichischen Geologischen Gesellschaft 94, 63–77.

Ortner, H., Reiter, F. & Brandner, R. 2006: Kinematics of the Inntal shear zone–sub-Tauern ramp fault system and the interpretation of the TRANSALP seismic section, Eastern Alps, Austria. Tectonophysics 414, 241–258.

Pestal, G. & Hejl, E. 2005: Geologie der österreichischen Bundesländer: Salzburg. Geologische Bundesanstalt, Wien, Geologische Karte 1:200.000

Pickering, K. T, Stow, D.A.V., Watson, M.P. & Hiscott, R.N. 1986: Deep water facies, processes and models: a review and classification scheme for modern and ancient sediments. Earth Science Reviews 23, 75–174.

Piller, W.E., Egger, H., Erhart, C.W., Gross, M., Harzhauser, M., Hubmann, B., Van Husen, D., Krenmayr, H.-G., Krystyn, L., Lein, R., Lukeneder, A.,

Mandl, G.W., Rögl, F., Roetzel, G., Rupp, C., Schnabel, W., Schönlaub, H.-P. Summesberger, H., Wagreich, M. & Wessely, G. (2004): Die stratigraphische Tabelle von Österreich 2004 (sedimentäre Schichtfolgen). Kommission für die paläontologische und stratigraphische Erforschung Österreichs der Österreichischen Akademie der Wissenschaften und Österreichische Stratigraphische Kommission.

Pober, E. & Faupl, P. 1988: The chemistry of detrital chromian spinels and its implications for the geodynamic evolution of the Eastern Alps. Geologische Rundschau 77, 641–670.

Prior, D.B., Bornhold, B.D. & Johns, M.N. 1984: Depositional characteristics of a submarine debris flow. Journal of Geology 92, 707–727.

Rittner, K.M. 2006: Geologie der östlichen Unkener Mulde am Kontakt zur Berchtesgadener Masse – Strukturgeologie und elektronische Verarbeitung geologischer Daten. Diplomarbeit Universität Innsbruck, 104 pp.

Sanders, D., Lukesch, M., Rasser, M. & Skelton, P. 2007: Shell beds of Diceratid rudists ahead of a low-energy gravelly beach (Thitonian, Northern Calcareous Alps, Austria): paleoecology and taphonomy. Austrian Journal of Earth Sciences 100, 186–199.

Schlager, W. & Schlager, M. 1973: Clastic sediments associated with radiolarites (Tauglbodenschichten, Upper Jurassic, Eastern Alps). Sedimentology 20, 65–89.

Schlager, W. & Schöllnberger, W. 1974: Das Prinzip stratigraphischer Wenden in der Schichtfolge der Nördlichen Kalkalpen. Mitteilungen der Österreichischen Geologischen Gesellschaft 66/67, 165–193.

Schlager, W., Austin, J.A., Jr., Corso, W., McNulty, C.L., Flügel, E., Renz, O. & Steinmetz, J.C. 1984: Early Cretaceous platform re-entrant and escarpment erosion in the Bahamas. Geology 12, 147–150.

Schmid, S.M., Fügenschuh, B., Kissling, E. & Schuster, R. 2004: Tectonic map and overall architecture of the Alpine orogen. Eclogae Geologicae Helvetiae 97, 93–117.

Schmid, S.M., Bernoulli, D., Fügenschuh, B., Matenco, L., Schefer, S., Schuster, R., Tischler, M. & Ustaszewski, K. 2008: The Alpine-Carpathian-Dinaride-orogenic system: correlation and evolution of tectonic units. Swiss Journal of Geosciences 101, 139–183.

Schweigl, J. & Neubauer, F. 1997: Structural evolution of the central Northern Calcareous Alps: Significance for the Jurassic to Tertiary geodynamics of the Alps. Eclogae Geologicae Helvetiae 90, 1–19.

Schuster, R., Koller, F. & Frank, W. 2007: Pebbles of upper-amphibolite facies amphibolites of the Gosau Group from the Eastern Alps: relics of a metamorphic sole? In: Froitzheim, N. (Ed.): 8th Workshop on Alpine Geological Studies, Davos, Abstract Volume, p. 74.

Stampfli, G.M., Mosar, J., Marquer, R., Marchant, R., Baudin, T. & Borel, G. 1998: Subduction and obduction processes in the Swiss Alps. Tectonophysics 296, 159–204.

Stanton, R.J. & Flügel, E. 1995: An accretionary distally steepened ramp at an intrashelf basin margin: an alternative explanation for the Upper Triassic Steinplatte "reef" (Northern Calcareous Alps, Austria). Sedimentary Geology 95, 269–286.

Surlyk, F. & Ineson, J.K. 1992: Carbonate Gravity Flow Deposition along a Platform Margin Scarp (Silurian, North Greenland). Journal of Sedimentary Petrology 62, 400–410.

Suzuki, H., Wegerer, E. & Gawlick, H.-J. 2001: Zur Radiolarienstratigraphie im Unter-Callovium in den Nördlichen Kalkalpen – das Klauskogelbachprofil westlich von Hallstatt (Österreich). Zentralblatt für Geologie und Paläontologie, Teil I 2000, 167–184.

Töchterle, A. 2005: Tektonische Entwicklungsgeschichte des Südteiles der Nördlichen Kalkalpen entlang der TRANSALP-Tiefenseismik anhand bilanzierter Profile. Diplomarbeit Universität Innsbruck, 91 pp.

Tollmann, A. 1976: Der Bau der Nördlichen Kalkalpen. Monographie der Nördlichen Kalkalpen, Teil III. Deuticke, Wien, 449 pp.

Tollmann, A. 1985: Geologie von Österreich, Band 2. Deuticke, Wien, 718 pp.

Tollmann, A. 1987: Late Jurassic/Neocomian gravitational tectonics in the Northern Calcareous Alps in Austria. In: Faupl, P. & Flügel, H.W. (Eds.): Geodynamics of the Eastern Alps. Deuticke, Wien, 112–125.

Trümpy, R. 1988: A possible Jurassic-Cretacous transform system in the Alps and the Carpathians. Geological Society of America Special Paper 218, 93–109.

Van Weering, T.C.E., Nielsen, T., Kenyon, N.H., Akentieva, K. & Kuijpers, A.H. 1998: Sediments and sedimentation at the NE Faeroe continental margin; contourites and large-scale sliding. Marine Geology 152, 159–176.

Vecsei, A., Frisch, W., Pirzer, M & Wetzel, A. 1989: Origin and tectonic significance of radiolarian chert in the Austroalpine rifted continental margin. In: Hein, J.R. & Obradovic, J. (Eds.): Siliceous deposits of the Tethys and Pacific regions. Springer, New York, 65–80.

Vortisch, W. 1931: Tektonik und Breccienbildung in der Kammerkehr-Sonntagshorngruppe. Jahrbuch der Geologischen Bundesanstalt 80, 81–96.

Wächter, J. 1987: Jurassische Massflow- und Internbrekzien und ihr sedimentär-tektonisches Umfeld im mittleren Abschnitt der nördlichen Kalkalpen. Bochumer geologische und geotechnische Arbeiten 27, 239 pp.

Weissert, H.J. & Bernoulli, D. 1985: A transform margin in Mesozoic Tethys: evidence from the Swiss Alps. Geologische Rundschau 74, 665–679.

Woodcock, N.H. 1979: Sizes of submarine slides and their significance. Journal of Structural Geology 1, 137–142.

Manuscript received February 26, 2008
Revision accepted July 21, 2008
Published Online first October 22, 2008
Editorial Handling: Stefan Schmid, Stefan Bucher

1661-8726/08/01S073-16
DOI 10.1007/s00015-008-1293-x
Birkhäuser Verlag, Basel, 2008

Swiss J. Geosci. 101 (2008) Supplement 1, S73–S88

The Pfitsch-Mörchner Basin, an example of the post-Variscan sedimentary evolution in the Tauern Window (Eastern Alps)

PETRA VESELÁ & BERND LAMMERER

Key words: post-Variscan sediments, Eastern Alps, Tauern Window, Pfitsch Formation, Windtal Formation, Aigerbach Formation

ABSTRACT

A review of post-Variscan metasedimentary and metavolcanic successions in the western Tauern Window is presented. U/Pb – datations of zircons in metavolcanic rocks reveal ages between 309 and 280 Ma. Deposition of grey conglomerates and black pelites started before 309 Ma in the northernmost basin of the Tauern, the Riffler-Schönach basin. In the more central Pfitsch-Mörchner basin, the onset of conglomerate sedimentation can be dated into the time span between 293 and 280 Ma. The Pfitsch and Windtal Formations are newly defined. The basins were filled with up to 1 km of mainly continen-

tal clastics until Early Triassic. Short marine ingressions in Middle- and Late Triassic times flooded only basinal parts of the area where we suppose a more or less continuous sedimentation until the Late Jurassic. Only the Hochstegen Marble documents a nearly complete submergence in the area of the Tauern Window. In spite of the metamorphic overprint, the tentative interpretations of the sedimentary facies give a reasonable picture and allow correlations to nonmetamorphic areas in South Germany or the External Massifs of Eastern Switzerland.

Introduction

The Inner Tauern Window is considered as a duplex of kilometre-thick slices of external parts of the European crust, which were stacked in the footwall of the Penninic and Austroalpine nappe systems and, finally, uplifted along a deep-reaching reverse fault, the Sub Tauern Ramp (Lammerer et al. 2008). Pre-Variscan basement rocks and Variscan granitoids form the main rock masses in the Tauern Window. They are covered by Late Jurassic marbles (Hochstegen Marble, Silbereck Marble) but, locally, Late Palaeozoic to Early Jurassic rocks are preserved. Primary petrological and sedimentological features are mostly obliterated by Alpine tectonics or metamorphism and the outcrops are scattered over a wide area. Thus, deciphering the pre-orogenic history is far from straightforward.

Numerous earlier studies have already revealed important constraints concerning the stratigraphy (Frasl 1958; Frisch; 1974, 1980a; Thiele 1976, 1980) structure (Thiele 1974; Frisch 1975, 1980b; Lammerer & Weger 1998), metamorphism (Selverstone et al. 1984, 1985), timing of uplift (Fügenschuh et al. 1997; von Blanckenburg 1989; Steenken et al. 2002) and geodynamic evolution of the Tauern window (Ratschbacher et al. 1989, 1991; Frisch et al. 2000; Lammerer et al. 2008). Earlier

descriptions of post-Variscan sediments of the western Tauern Window were given by Frisch (1968), Thiele (1970), Lammerer (1986), Schön & Lammerer (1988), Sengl (1991) and Veselá et al. (2008).

This study provides an overview of the post-Variscan metasedimentary and meta-volcanic rocks within the western Tauern Window. For this purpose some less deformed locations of the relatively well-exposed Pfitsch area in the SW of the Tauern Window were studied in detail. There, major sedimentary structures are still preserved and datations of the meta-volcanic layers were carried out, so that lithostratigraphical correlation of the rock successions is feasible.

Geological setting

The Variscan orogeny left a large mountain chain which crossed the megacontinent Pangea at the end of Westphalian times (von Raumer 1998; Ziegler 1990). The orogenic activity was followed by a collapse of the thickened crust and intraplate reorganisation throughout Western and Central Europe. Normal faulting and strike-slip faulting were prevalent in Late Palaeozoic times (Arthaud & Matte 1977). The extension of parts of the high Andes and Tibet or the Basin-and-Range province may serve

Department of Earth and Environmental Sciences; Ludwig-Maximilians-Universität, Luisenstr. 37, D-80333 München, Germany.
E-mail: petra.vesela@iaag.geo.uni-muenchen.de

as an analogue to the post-collisional stage of the Variscan belt (Ménard & Molnar 1988). In contrast, Ziegler & Dezés (2006) suggest an oblique dextral collision and a wrench-induced collapse of the Variscan orogen. According to McCann et al. (2006) alternating transtensional and transpressional tectonic regimes led to the formation of many basins across the region.

The area of the future Alps was located in Late Carboniferous times close to the warm and humid equatorial zone. Post-orogenic fluvial systems were influenced by the tectonic fracture pattern and basin subsidence. Upper Carboniferous sediments comprise deposits of braided, anastomosing and meandering rivers and swamps. A change towards a drier climate during the Permian led to formation of alluvial fans and playa lakes. The denudation of the Variscan mountain belt continued until the early Mesozoic, when a peneplain formed. As known from other areas in Central Europe (e.g. Wetzel et al. 2003; Ziegler 2005), continental sedimentation prevailed during the Triassic although interrupted by some short marine ingressions. Thereafter, probably during the Jurassic, the area was progressively flooded and the sediments became increasingly calcareous.

The Tauern Window was part of the Moldanubian domain, and its sedimentary history is very similar (e.g. Pfiffner 1998). In the Alpine foreland several intermontane, fault-bounded basins were detected by seismic imaging and drilling beneath the Molasse cover (Fig. 1), e.g. the Permo-Carboniferous Northern Switzerland Basin (Matter 1987), the Lake Constance Trough or the Ries-Salzach Basin close to the Landshut-Neuötting

fault scarp (Lemcke 1988) which continues to a basin within the Zentrale Schwellenzone in Austria (Kröll et al. 2006). Within the Alps, well exposed examples are the Zône Houillère in the French Western Alps (Desmons & Mercier 1993), the Salvan-Dorénaz Trough in the Aiguille Rouge Massif, where coal seams were mined at several locations (Capuzzo et al. 2003; Capuzzo & Wetzel 2004) and the basins within the Aar-Gotthard Massif (Franks 1966; Oberhänsli et al. 1988; Schaltegger & Corfu 1995). Parts of the southernmost post-Variscan basins are the Orobic and Collio Basin in the Southern Alps (Sciunnach 2003) which continue into the Val Gardena Sandstone Plain east of the Giudicarie Line. To the north of the Periadriatic Line, the post-Variscan sediments are incorporated as slices within the Austroalpine nappes (Kreiner 1993). The Sub-Penninic European continental crust sensu Schmid et al. (2004) is in the Eastern Alps exposed only in the Tauern Window. It represents insofar an important link between the basement outcrops of Central Europe and the Variscan crystalline complexes within the Tisza Block in the Pannonian Basin (Haas & Péró 2004). All basins are filled with continental clastics and some volcaniclastic material and are emplaced within the Variscan basement rocks.

The basement

The basement of the Tauern Window is composed of old gneisses or schists ("altes Dach", Frasl, 1958) and late Variscan

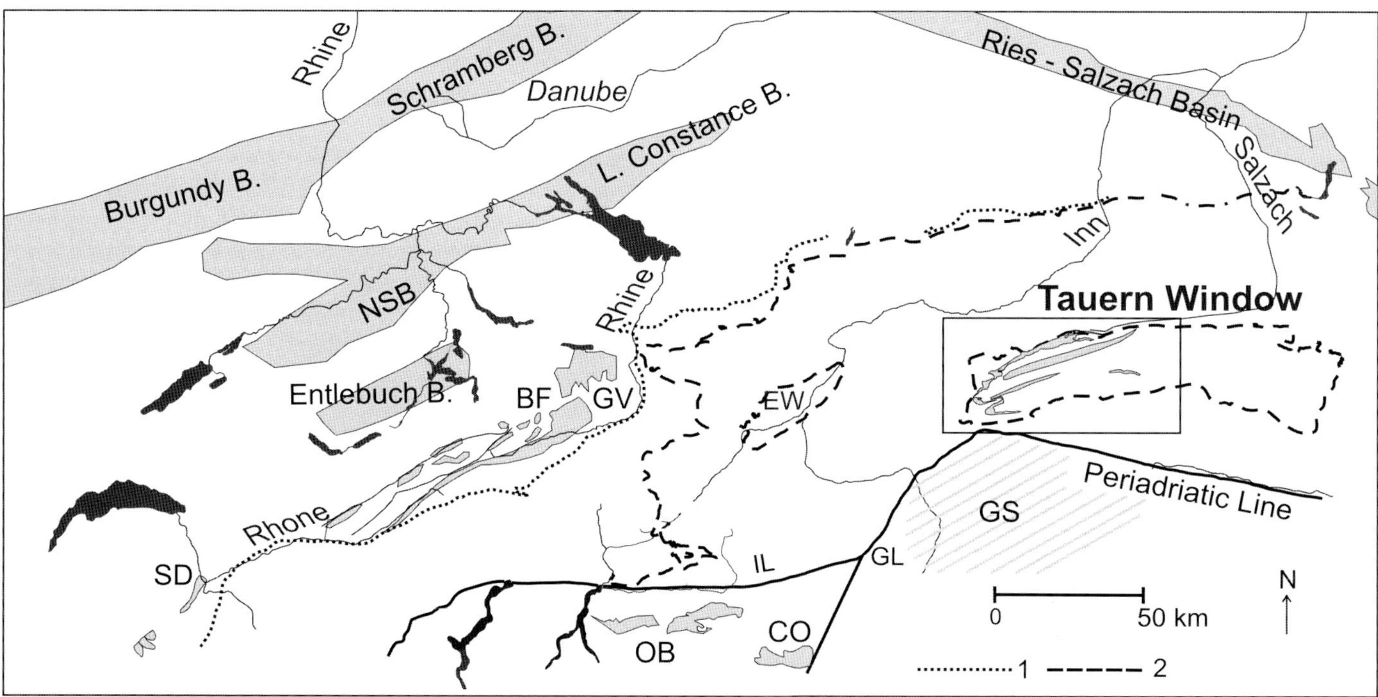

Fig. 1. Post-Variscan basins in Central Europe and Alpine realm (modified after: Lemcke 1988; Ménard & Molnar 1988; Kröll et al. 2006; McCann et al. 2006; Cassinis et al. 2007; Veselá et al. 2008). 1 – Penninic-Helvetic thrust plane, 2 – Austroalpine-Penninic thrust plane, SD – Salvan-Dorénaz Basin, NSB – Northern Swiss Permo-Carboniferous Basin, BF – Bifertengrätli Basin, GV – Glarner Verrucano Basin, OB – Orobic Basin, CO – Collio Basin, GS – Val Gardena Sandstone Plain, IL – Insubric-Tonale Line, GL – Giudicarie Line, EW – Engadine Window. The inset frame shows the position of Figure 2.

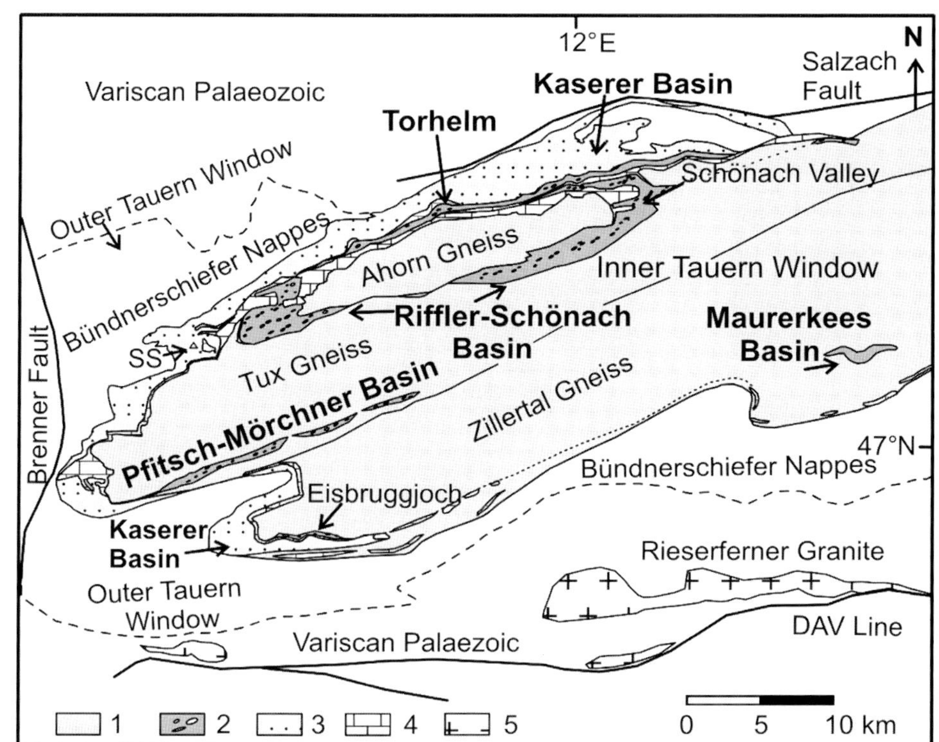

Fig. 2. Geological sketch map of the Tauern Window and the position of the post-Variscan basins (modified after Veselá et al. 2008). 1 – Palaeozoic rocks and Variscan granites, 2 – post-Variscan clastic sediments (Upper Carboniferous-Lower Jurassic), 3 – ? Triassic clastic sediments and carbonates at the base of the Bündnerschiefer, 4 – Hochstegen-Fm. (Jurassic), 5 – Alpine granites (Oligocene), DAV Line – Defereggen-Antholz-Vals Fault, SS – Schöberspitzen.

intrusives (Karl 1959; Finger et al. 1993, 1997). The old Greiner-, Stubach-, Habach-, Storz- and Zwölferzug Gneiss Series are interpreted as pre- or early Variscan terranes derived from island-arcs, back-arc-basins or marginal basins which originated along the Gondwana margin (e.g. Frank et al. 1987; Reitz & Höll 1988; Neubauer et al. 1989; Vavra & Frisch 1989; Frisch et al. 1993; Frisch & Neubauer 1989; Frisch & Raab 1987; v. Quadt 1992; Kupferschmied et al. 1994; v. Raumer 1998). Serpentinites and meta-ophicalcites in mélange-like rocks of the Greiner and Stubach Series may represent remnants of an obducted ocean floor and, hence, mark a pre- or early Variscan suture zone along which the different pieces were amalgamated.

The sedimentary domains within the basement of the Tauern Window experienced metamorphism and even anatexis during complex processes of Variscan nappe-stacking and due to the Variscan intrusive activity (Eichhorn et al. 2000). However, some of the detrital rocks seem to have been metamorphosed only during the Alpine orogeny (Habach Formation, Frasl 1958). Similar to the External Massifs in the Swiss Alps (Schaltegger & Corfu 1995; von Raumer 1998) it appears that the Pfitsch-Mörchner Basin (Fig. 2) could also be a successor basin of an older volcano-sedimentary basin, which was reactivated or continued to exist until the Permian. Kebede et al. (2005) investigated sedimentation time in the central Tauern Window (Zwölferzug, Biotitporphyroblastenschiefer, Habach Series) using detrital zircon U/Pb ages. Graphite-bearing schists and paragneisses, which probably have their continuation in the Greiner Series, contain detrital zircons with ages in the range from Upper Devonian to Lower Carboniferous indicating the maximum age of sedimentation.

Permo-Carboniferous granitoids ("Zentralgneise") intruded into the old gneisses as laccoliths or kilometre-thick sills (Fig. 3; Finger et al. 1997; Lammerer & Weger 1998; Lammerer et al. 2008), but contacts are sometimes overprinted by strike-slip faults (Behrmann & Frisch 1990) or thrusts. The granitic lamellae seem to control the taper of Alpine thrust sheets as contacts of granites to the foliated host rocks serve as plain of weakness and, hence, as detachment horizons (e.g. Tux Gneiss sheet, Eisbruggjoch lamella, Figs. 4, 7).

The plutonic protoliths of the "Zentralgneise" display, from the chemical point of view, features of volcanic arc or continental cordillera magmatism. Series of calc-alkaline magmas with dominantly granitic and tonalitic but also syenitic and monzonitic composition were produced, accompanied and followed by extrusive rhyolitic-dacitic volcanism (Finger & Steyrer 1988; Finger et al. 1993; Eichhorn et al. 2000).

In the western Tauern Window the Late Variscan magmatic activity started at 309 ± 5 Ma with calc-alkaline mafic intrusions, including minor ultramafic cumulates, in the Zillertal Gneiss complex (Cesare et al. 2001). More to the north, meta-rhyodacites of the Riffler-Schönach basin give an age of 309.8 ± 1.5 Ma (Table 1, F. Söllner pers. comm.). The magmatic activity culminated at 295 Ma with emplacement of granodioritic-tonalitic plutons (Cesare et al. 2001) and rhyolite (293 ± 1.9 Ma, Veselá et al. 2008) and ended around 280 Ma with acidic extrusivs (Table 1, F. Söllner, pers. comm.). The evolution is similar in the

Fig. 3. Banded and folded amphibolites of the Greiner Series ("old roof" rocks) are cut by Late-Variscan Tux leuco-granite. Locality Kunerbach water tunnel to the Schlegeis reservoir, Tux Gneiss. Original size is about 2,5 × 4 m.

External Massifs of the western Alps (e.g. Ménot & Paquette 1993; Schaltegger & Corfu 1992, 1995; Capuzzo & Bussy 2000) which supports the paleogeographic interpretation of the Tauern Window as an eastern continuation of the External domains (Thiele 1970; Frisch 1975; Lammerer 1988; Finger et al. 1993; von Raumer 1998).

Sedimentary basins of the western Tauern Window

Within the Inner Tauern Window several elongate, trough-like basins have been identified, representing small remnants of the Late Palaeozoic-Mesozoic sedimentary cover, which survived the post-Variscan uplift, Alpine compression and erosion (Fig. 2). Many palaeogeographic units in the Eastern Alps have their long axes in an ENE–WSW direction, parallel to the strike of the modern orogenic belt and it appears that also in the Tauern Window pre-existing Late Palaeozoic faults and associated rifts strongly affected the tectono-sedimentary development (Frasl & Frank 1966; Arthaud & Matte 1977; Frisch 1977; Kreiner 1993; Frisch et al. 2000). The orientation of Variscan granitoid intrusions follows this trend. The Riffler-Schönach Basin, the Pfitsch-Mörchner Basin, the Maurerkees Basin and the small remains of basins on the southern rim of the Tauern Window, are all confined by tectonic horsts of basement rocks (Ahorn-, Tux-, Zillertal-, and Eisbruggjoch Gneisses). The age and tectonic position of the Kaserer Basin is still not clear.

The Pfitsch-Mörchner Basin

One of the best examples of the Post-Variscan sedimentary successions in the Eastern Alps is found in a narrow syncline which extends from the Pfitsch Valley (Val di Vizze, Italy) in a north-easterly direction to the Mörchenscharte, 2872 m (Austria), (Fig. 5). The Permian-Mesozoic sediments are exposed for about 20 km in an up to 600 m wide zone, covering the Palaeozoic basement rocks of the Greiner Schists and squeezed between Tux and Zillertal Gneiss Horsts (Fig. 7). The Greiner Series comprises hornblende-garbenschists, amphibolites, serpentinites, graphite-biotite schists, quartzites, thin marble layers and migmatites. Dark meta-conglomerates and -breccias have been found in the Haupental (east of Pfitscher Joch) and in contact to the large serpentinite body of the Ochsner (north of Berliner Hütte) and numerous small serpentinite or ophicalcite lenses give the impression of an old mélange.

The Greiner Schists accommodate a large-scale Alpine sinistral shear zone, which affected the whole schist belt and parts of the neighbouring gneisses (Karl & Schmidegg 1979; Behrmann & Frisch 1990). As the syncline plunges 10–15° to the SW, the Greiner Series and the sediments of the Pfitsch-Mörchner basin wedge out close to the Mörchenscharte. All rocks suffered metamorphic recrystallisation and ductile deformation but nevertheless sedimentary structures are still recognizable at many locations in the Permo-Mesozoic rocks. Various lithofacies associations have been distinguished on the basis of lithologic changes, vertical succession, dominant grain size or grading.

An unconformity on top of the Greiner Series is marked by lenses of staurolite-chloritoid-magnetite schists. Because of a high aluminium and iron content, these lenses were interpreted by Barrientos & Selverstone (1987) as erosional remnants of a metamorphosed palaeosol. It documents a period of tectonic quiescence prior to the Permian extensional phase. A Lower Permian meta-rhyolite with an age of 293 ± 1.9 Ma cross-cuts the serpentinite and the other Greiner Schists to the west of the Mörchenscharte (Veselá et al. 2008). It gives the minimal age of the Greiner Series. The rhyolite is, on the other hand, unconformably overlain by the meta-conglomerate of the Pfitsch-Formation which marks its maximum age (Table 2).

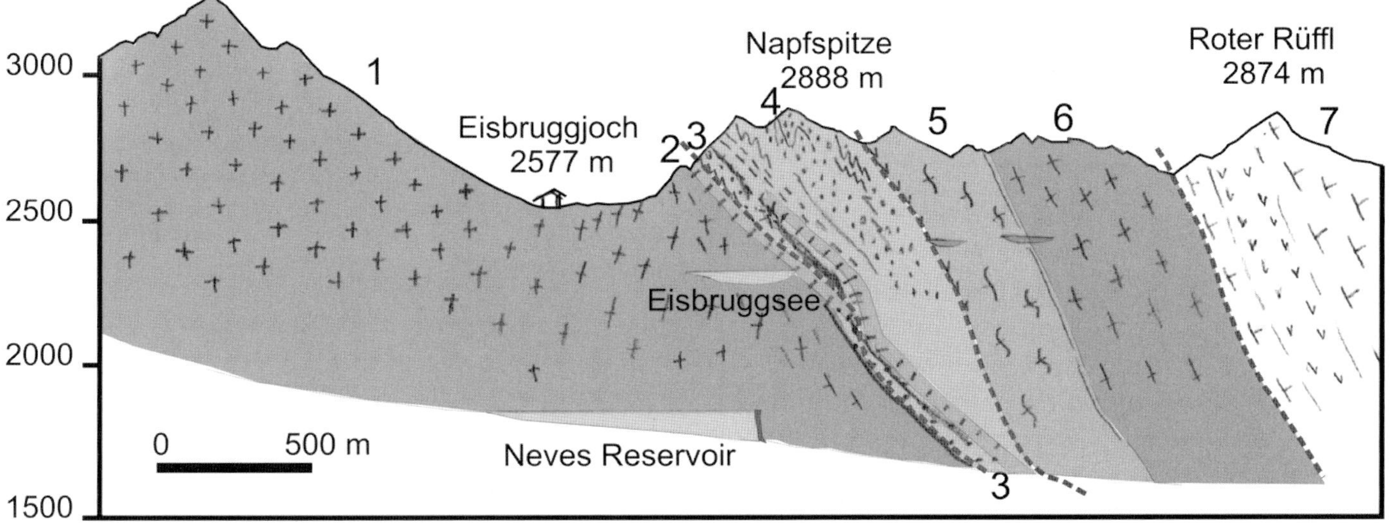

Fig. 4. Eisbruggjoch section. 1 – Zillertal Gneiss, 2 – Hochstegen Marble, 3 – ankerite-chlorite schists, quartzites, 4 – garnet- and graphite-bearing schists, quartzites, mica-schists, 5 – amphibolites, 6 – Eisbruggjoch Lamella (granite gneiss), 7 – Bündnerschiefer and amphibolites of the Glockner Nappe (Outer Tauern Window)

Pfitsch Formation

The Pfitsch Formation (hereafter Pfitsch-Fm.) comprises meta-conglomerates, meta-rhyolite and meta-pelites. The well exposed rock succession in the area of Pfitscher Joch (Passo di Vizze) on the Austrian/Italian border is used as a type section (Fig. 5, Table 2).

Meta-conglomerate (Early Permian)

Sedimentary structures are well preserved in the northern limb of the nearly isoclinal syncline close to the Pfitscher Joch Haus, where strain was relatively low. In protected zones a measurable strain can be more or less absent. Bedding planes are sub-vertical. The rock protolith was a texturally and compositionally immature and poorly sorted coarse-grained polymictic breccia and conglomerate. It was formed by crudely bedded matrix-supported clasts; the matrix consisted of a sandy or silty fraction (Fig. 6 a). Angular to subangular clasts up to 30 cm in size are predominantly aplitic and granitic in origin, but vein-quartz, amphibolites, graphite schists, marbles, greenish calc-silicate-rock pebbles and, very rare, serpentinite clasts occur as well. The base of the unit contains predominantly metamorphic basement rock clasts whereas toward the top granitoid clasts become predominant, which reflects the progressive unroofing of the Variscan "Zentralgneise" (Schön & Lammerer 1988).

Table 1. Radiometric and palaeontologic time markers in the western Tauern Window

Age (method)		Rocks and locations	Reference
309,8 ± 1.5 Ma	(U/Pb Zrn)	meta-rhyodacite, Grierkar, Riffler-Schönach Basin, Tux Alps	F. Söllner pers. comm.
309 ± 5 Ma	(U/Pb Zrn)	ultramafic cumulates, Zillertal Gneiss, Italy	Cesare et al. 2001
295 ± 3 Ma	(U/Pb Zrn)	metagranodiorite, Zillertal Gneiss, Italy	Cesare et al. 2001
293 ± 1.9 Ma	(U/Pb Zrn)	meta-rhyolite, Mörchenscharte, Pfitsch-Mörchner Basin, Zillertal Alps	Veselá et al. 2008
284 + 2/–3 Ma	(U/Pb Zrn)	rhyolitic to andesitic metavolcanic rocks, Porphyrmaterialschiefer – Torhelm Nappe, Tux Alps	Söllner et al. 1991
280.5 ± 2.6 Ma	(U/Pb Zrn)	meta-rhyolite, Pfitscher Joch, Pfitsch-Mörchner Basin Pfitsch Valley, Val di Vizze, Italy	F. Söllner pers. comm.
Fossils			
Late Carboniferous to Early Permian		plant fossils in graphite-bearing schists in the Maurer Kees Basin, southern Venediger Alps	Franz et al. 1991 Pestal et al. 1999
Middle Triassic		crinoids in dolomitic marbles, Kalkwandstange Pfitsch Valley, Val di Vizze, Italy	Frisch 1975
Late Jurassic		ammonite (*Perisphinctes sp.*), Hochsteg, Mayrhofen	Klebelsberg 1940
		belemnite, sponge spicule, radiolaria in the Hochstegen Marble, Tux Valley	Schönlaub et al. 1975 Kiessling 1992

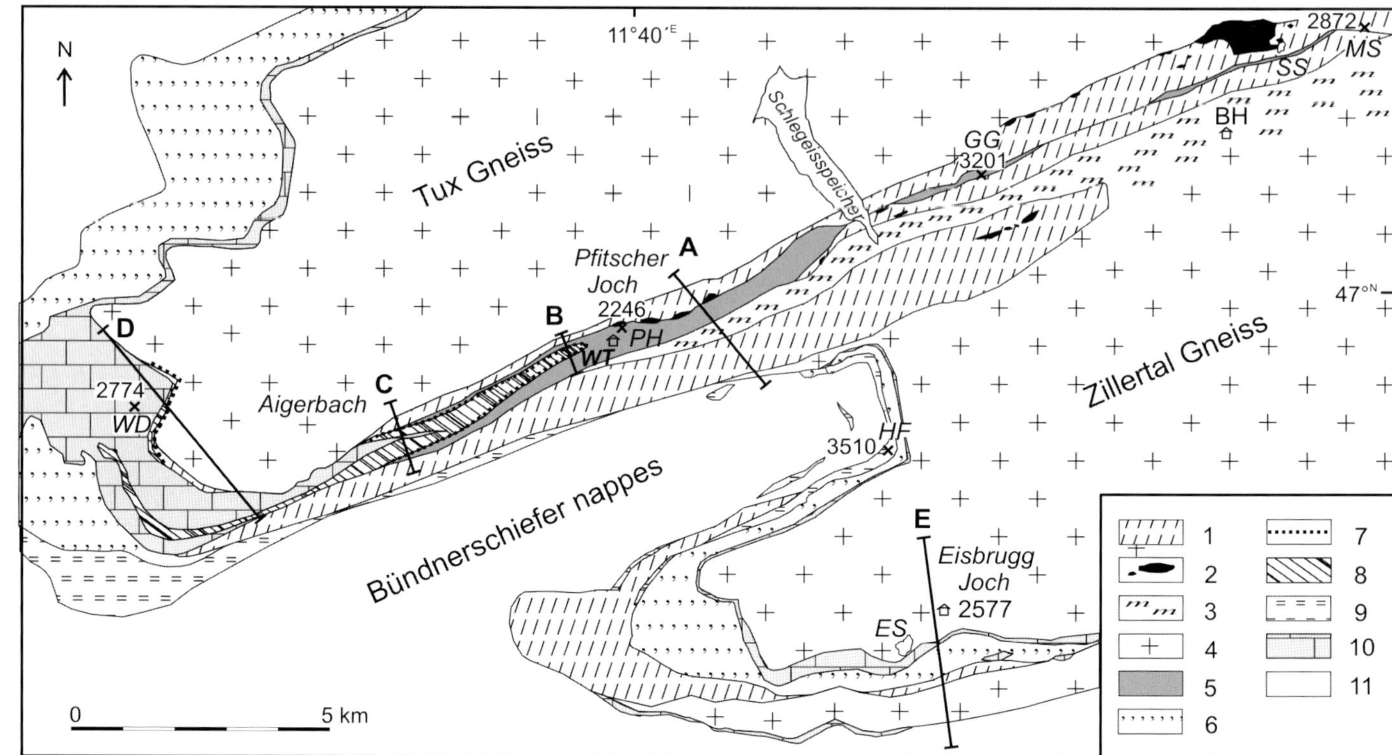

Fig. 5. Geological outline map of the SW Tauern Window. 1 – pre- and Late-Variscan basement rocks (Greiner Schists: amphibolites, hornblende-garbenschists, graphite-bearing schists, quartzites), 2 – serpentinites, 3 – migmatites and sheared gneisses, 4 – Variscan granitoids ("Zentralgneise"), 5 – Pfitsch-Fm., 6 – Kaserer Basin metasediments, 7 – Windtal-Fm., 8 – Aigerbach-Fm., 9 – Middle Triassic carbonates at the base and within the Bündnerschiefer nappes, 10 – Hochstegen-Fm., 11 – Bündnerschiefer nappes, MS – Mörchenscharte, SS – Schwarzsee, BH – Berliner Hütte, GG – Großer Greiner, PH – Pfitscher Joch Haus, WT – Windtal, HF – Hochfeiler, WD – Wolfendorn, ES – Eisbruggsee. Lines show positions of sections shown in figures (A – Fig. 7, B – Fig. 8, C – Fig. 10, D – Fig. 11, E – Fig. 4.).

Other localities show higher strain and clasts are stretched to long prolate bodies north of Berliner Hütte or are strongly flattened, e.g. to the south of the Pfitscher Joch Haus (Fig. 6 b). The average strain ellipsoid measured by the Rf/Φ method in the conglomerate of the northern limb is around x : y : z = 2. 45 : 0.93 : 0.44. In contrast, the southern limb suffered a strong flattening strain of x : y : z = 2.51 : 1.82 : 0.22. The longest axes are gently plunging (10–30°) to the WSW (250–270°), the shortest axes are horizontally NNW–SSE directed.

The actual thickness of the conglomerate member in the northern limb is 60 m, in the southern 94 m (Figs. 8, 9). Considering the strain, the primary thickness should have been around 136 metres in the northern and around 420 metres in the southern limb. Some kilometres more to the east and closer to the fold hinge, nearly plane strain ellipsoids were measured (x : y : z = 2.89 : 1.02 : 0.34) and reconstructed thickness exceeds 400 metres in the southern limb. The age of deposition is limited to the time span between 293 ± 1.9 and 280 ± 2.6 Ma by the unconformity to the meta-rhyolite of the Mörchenscharte and the overlying meta-rhyolite of the Pfitscher Joch (see below).

Interpretation: The meta-conglomerates are interpreted as semi-arid alluvial fans. Some coarse-grained beds give, in spite of the metamorphic overprint, the impression of clast-supported deposits which, in general, represent sieve deposits (Schäfer 2005). Small troughs were incised into the middle part of the alluvial fan, where sediments were partially reworked after heavy rains. They were filled with fining-upward successions.

The rock colour is presently light greyish-greenish from an ubiquitous presence of finely distributed hematite and magnetite and greenish iron-rich phengite. This is taken as a hint that the original rock colour was reddish. The remarkable increase in thickness from north to south and east may indicate a palaeo-relief quickly deepening towards the southeast, or by now undetected faults obliquely cutting the bedding planes in the northern limb.

Meta-rhyolite (280.5 ± 2.6 Ma)

On top of the meta-conglomerates, a 10–50 m thick light grey meta-rhyolite was deposited, Early Permian in age (280.5 ± 2.6 Ma, F. Söllner pers. comm.). It is deformed to gneiss with frequent tourmaline in the foliation plane. The matrix is strongly recrystallized, but euhedral quartz and feldspar grains up to 3 mm are still preserved. The zircon typology (Pupin 1980) presents a bimodal distribution (P2-P5, S10,13,16,17),

Table 2. Lithofacies and stratigraphy of the Pfitsch-Mörchner Basin, not included in the table: 1- Greiner Schists, palaeosol horizon, 2 – Variscan Granitoids, Tux Gneis.

age	lithostratigraphical unit/ rock type	facies/protolith	interpreted depositional environment	section line coordinates
Hochstegen Formation				
Late Jurassic	greyish, bluish calcite marbles, sandy marbles (16)	carbonates	deeper marine environment	11°32′30′′E 46°59′04′′N
? Middle Jurassic	brownish sandy calcite marbles (15), mica- graphite- bearing horizons	sandy impure carbonates	neritic environment	
? Lower Jurassic	kyanite-graphite- phyllite and quarzite-schists (14) graphite-quartzite, pure quartzite (13)	sands, Fe- and Al- rich pelites mud	floodplain deposits, swamps, fluvial sands	11°32′47′′E 46°59′12′′N
Aigerbach Formation				
	chloritoid-, mica-quartzite (12)	sands, impure sands and pelites	supratidal area, coastal flat	11°32′48′′E 46°59′16′′N
	dolomite and cargneuls (11) yellowish sandy calcite marbles (10) dolomitic marbles (9)	carbonates, anhydrite, gypsum	shallow marine to sabhka environment	
Late Triassic	chloritoid-, kyanite-quartzite, mica-, chlorite-schists (12)	fine sands high Fe content, pelites	fluvial sands, muddy floodplain, channels	
	whitish dolomitic marbles, cargneuls (11), thin calcite marble beds (10)	carbonates, anhydrite, gypsum	sabhka environment, lagoon, high evaporation	11°36′18′′E 46°58′34′′N
	calcareous/dolomitic quartzite, kyanite-, chloritoid-, mica-schists, and -quartzite (12)	impure sands and pelites	supratidal area, coastal flat, channels	
	whitish dolomitic marbles and cargneuls (11)	anhydrite, gypsum	sabhka environment, lagoon, high evaporation	
? Middle Triassic	greyish, yellowish sandy calcite marbles (10), yellowish, violet dolomitic marbles (9)	carbonates partly dolomitized	shallow marine environment	11°38′51′′E 46°59′31′′N
Windtal Formation				
? Early Triassic	locally lazulite-kyanite quartzite (8), sericite-, hematite-quartzite (7)	Fe-rich sand deposits, locally P-, Al-rich sediments	braided rivers deposits? / coastal sands ?, palaeosols in vegetated area apart from channel ?	11°39′16′′E 46°59′35′′N
Pfitsch Formation				
? Late Permian	epidote-ankerite schists and quartzite (6)	pelites, mud with sand laminae	distal part of (semi) arid alluvial fan muddy floodplain, playa lake	11°39′14′′E 46°59′29′′N
280,5 ± 2,6 Ma	meta-rhyolite (5)	volcanic deposits	subaerial lava flows	
Early Permian	meta-conglomerate (4) ?discordantly overlying meta-rhyolite (3) W of the Mörchenscharte	crude gravel, matrix/clast supported, sand	proximal, middle part of (semi) arid alluvial fan, sieve deposits	11°50′21′′E 47°02′31′′N

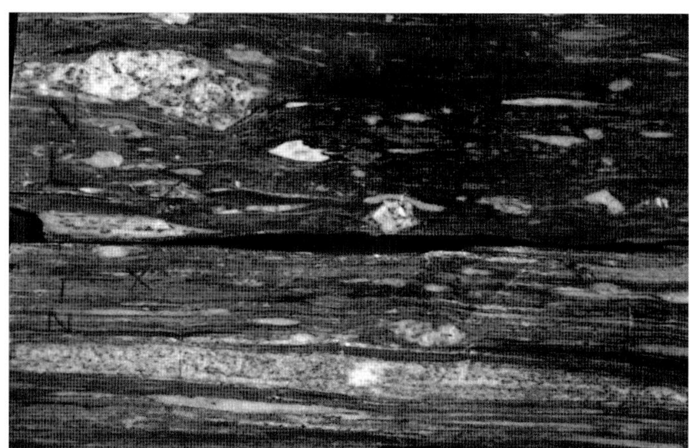

Fig. 6. a) Coarse meta-conglomerate from the Pfitscher Joch, poorly sorted and crudely bedded, consisting mainly of granitic pebbles. Subordinate clasts from graphite-bearing schists, migmatic gneisses and marbles occur. Hammer for scale (middle-right). b) Polished sections of a meta-conglomerate cut in x–z (bottom) and y–z (top) directions of the strain ellipsoid. Flattened pebbles of granites, aplites and quartzites in the highly deformed southern limb of the Pfitsch syncline. Long axis of the specimen is 28 cm.

which suggests derivation from alkaline granitic melt of mainly mantle origin contaminated by tonalitic material of prevalently crustal origin. The layer extends laterally for more than 10 km. Systematic strain analyses have not been carried out, but stretched tourmaline needles point to a similar strain as in the adjacent conglomerates.

Interpretation: As the meta-rhyolite covers terrigenous conglomerates over a large area with a relatively uniform thickness, a subaerial deposition from a pyroclastic flow seems reasonable. The age of the rhyolite is coherent with the Permian volcanic phase in the central Tauern Window (Eichhorn et al. 2000) and many other volcanic domains in the Alps, as the Bozen quartz-porphyry in the Dolomites.

Epidote-ankerite schists

On top of the volcanic bed several metres of finer-grained meta-conglomerates follow. They are arranged in fining-upward cycles and grade into meta-pelites. The sequence consists of epidote-ankerite schists, impure quartzites and mica-schists. In the more pelitic members, occasionally graded quartzite horizons with thicknesses of 2 15 cm are intercalated. Up section the quartz content and grain size increase. The boundary with the overlying Windtal-Fm. is gradual. The high amount of Fe-minerals like iron-epidote (pistacite) and iron-dolomite (ankerite) results probably from the presence of hematite in the original rocks. This points to a warm climate with dry seasons and oxidizing conditions above the groundwater table in a well drained area (e.g. Sheldon 2005). This rather monotonous unit reaches a thickness of 50 m in the northern limb. In the southern limb the thickness increases up to 250 m, which implies a sediment transport towards the south.

Interpretation: A playa lake depositional environment is inferred. Distal parts of alluvial fans delivered fine-grained sediments into the playa lake. The floodplain was transected by a network of channels and sandy beds were deposited by crevasse splays. The stratigraphic position suggests an age in the range of Late Permian to Early Triassic. Several other Variscan basins in Europe display similar lithological characteristics and environmental setting during this time period (e.g. Glarner Verrucano Basin, Trümpy 1966, 1980; German Basin, Hauschke & Wilde 1999).

Windtal Formation

The Windtal Formation is only present to the west of the Pfitscher Joch, where its fold hinge closes. Because of its resistence to erosion it forms prominent cuestas (Fig. 8).

The thickness is about 30–40 metres. Rock types comprise whitish quartzites and muscovite quartzites; a subtle grey shade comes from finely distributed hematite and magnetite. Locally, the quartzites contain albite, lazulite, kyanite, manganese-epidote (thulite), tourmaline and staurolite. Isolated flattened quartz nodules up to 7 cm in diameter occur within the quartz- and muscovite-rich groundmass. Whether they represent originally isolated pebbles or disrupted quartz veins could not be resolved. The top of the Windtal-Fm. is overstepped by the carbonates of the Aigerbach-Fm. Considering the lithological characteristics and the stratigraphical position beneath the Aigerbach Formation, affinities to the Buntsandstein and a Scythian age of the Windtal Formation were already suspected by Lammerer (1986).

Interpretation: The absence of silty horizons and relatively high textural and compositional maturity would correspond to a wide braided river system with sand and gravel bars in channels. Due to the abundant hematite, a reddish colour of the original sandstones is likely and again points to semiarid climate. En-

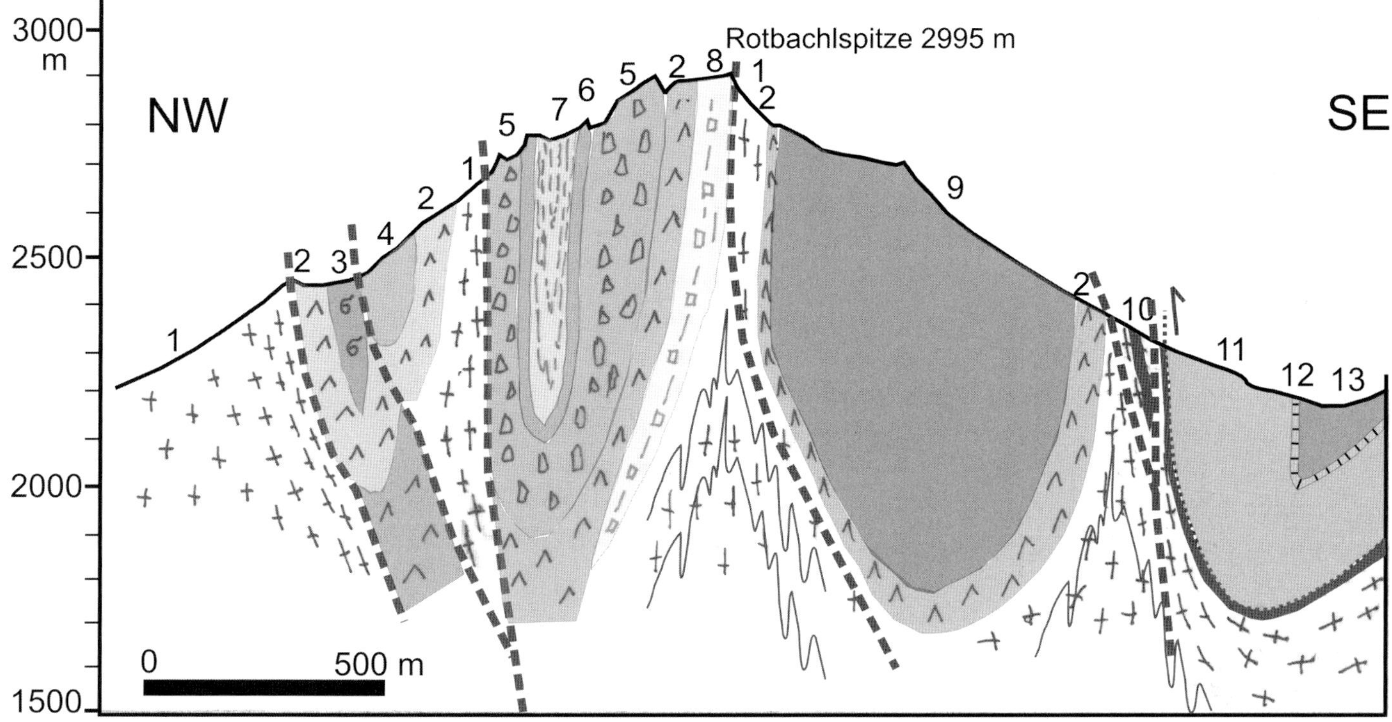

Fig. 7. Cross section through the Pfitsch Mörchner Basin to the east of the Pfitscher Joch. 1 – Tux "Zentralgneise", granite-gneiss lamellae along tectonic contacts Zillertal Gneiss in the SE, 2 – amphibolites or hornblende-garbenschists, 3 – serpentinite, ophicalcite, 4 – northern meta-conglomerates, 5 – coarse breccias and meta-conglomerates, rich in amphiboles in the matrix, 6 – meta-rhyolite, Lower Permian, 7 – Permian meta-conglomerates, quartzites and epidote-ankerite schists, 8 – quartz-pyrite schists, palaeosols, 9 – Palaeozoic graphite-bearing schists, post-Early Devonian, pre-Late Carboniferous, 10 -Hochstegen Marble, Late Jurassic, 11 – Kaserer Series (? Early Cretaceous or ? Late Permian to Early Triassic), 12 – Middle Triassic carbonates, 13 – Bündnerschiefer.

richments of Al and Fe within the sediment could have been caused by weathering effects during the deposition, as the more mobile components were leached away. An alternative explanation is provoked by the local concentration of phosphate minerals, like lazulite. Phosphate concentrations could signify metamorphosed fossil material like bones as a remnant of a bonebed horizon. Such layers are often connected with transgressions. In this case, the quartzites should represent coastal sands and the concentrations of e.g. tourmaline and magnetite could signify enrichment of heavy minerals along the beach. However, this interpretation is highly speculative.

Aigerbach Formation

The Aigerbach-Fm. covers the Windtal-Fm. conformably and continues from the Pfitscher Joch to the west. The boundary between the formations is sharp (Figs. 8, 9). Standard section is the locality Aigerbach, N of St. Jakob (Fig. 10). Its age was long presumed to be Middle and Late Triassic (Baggio et al. 1969). Isotopic studies (δ_{34}S and Sr seawater curves) confirm Late Triassic ages (Brandner et al. 2008). The formation name was given by Brandner et al. (2007).

The lowest part comprises greyish-violet marbles, yellowish calcitic and dolomitic marbles (15 m in thickness). Further up, the main rock portion is composed of thin-bedded white

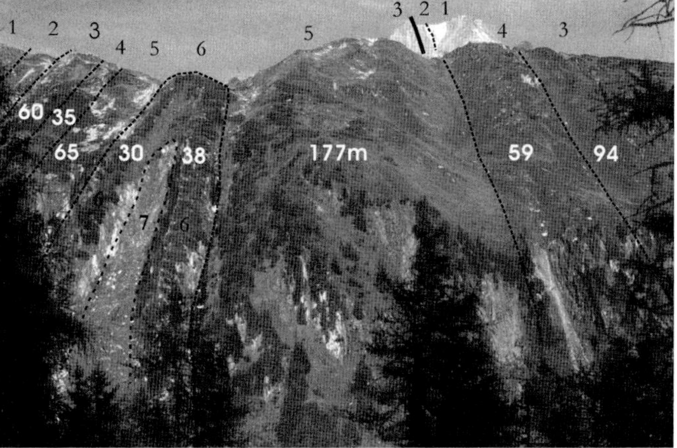

Fig. 8. The Pfitsch syncline in a view from the west towards the Pfitscher Joch and the Rotbachlspitze (2895 m) in the background. Due to an axial plunge steeper than the topography, the nearly isoclinal syncline appears like an anticline. Explanation: 1 – Tux Gneiss and granite-gneiss lamellae at Rotbachlspitze, 2 – amphibolites and serpentinites of the Greiner Series, 3–5: Pfitsch-Fm. (3 – meta-conglomerates, 4 – meta-rhyolite, 5 – epidote-ankerite schists), 6 – quartzites of the Windtal-Fm., 7 – limestones of the Aigerbach-Fm. The thickness of the southern limb (right) is despite higher flattening strain much larger than the northern limb. Bold line between 2 and 3 at the Rotbachlspitze marks the metamorphic soil horizon. White numbers show the true thickness of the beds in metres.

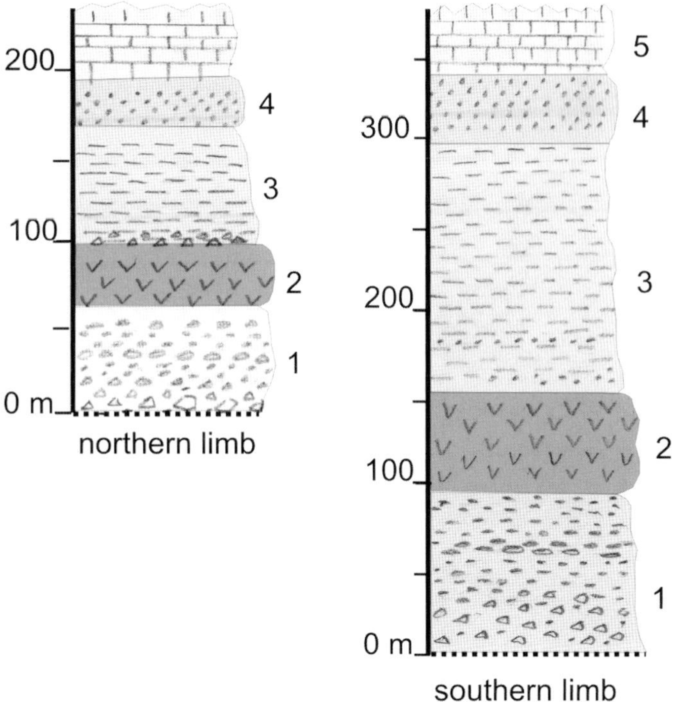

Fig. 9. Actual bed thickness in the northern and southern limb of the Pfitsch syncline, west of the Pfitscher Joch. To the west, thickness increases and has reached up to 1000 m. 1 – meta-conglomerate, 2 – meta-rhyolite (Lower Permian), 3 – epidote-ankerite schists, 4 – Windtal-Fm., 5 – Aigerbach Fm., dotted line – unconformity plane.

fine-grained dolomitic marbles, interlayered with yellowish cargneuls or dolomite which disintegrates surficially to cohesionless sands. Anhydrite and gypsum were discovered in exploration drillings to the Brenner Base Tunnel project (Brandner et al. 2007, 2008). At the surface, cargneuls (cellular dolomite) containing phyllite- and quartzite- fragments are attributed to those leached evaporitic layers. In the lower and in the upper parts of the section calcareous quartzites and various thin layers of kyanite-schists, chloritoid-schists, mica- and chlorite-schists, but also massive micaceous calcitic marble beds are interbedded. The formation reaches a thickness of about 110 m and it is covered by graphite- and kyanite-bearing quartzites, which are attributed to the Hochstegen-Fm.

Because of a thick cover of Pleistocene moraines, the outcrops to the west are not continuous, but can be extrapolated until the Wolfendorn area (Fig. 5). There, the Windtal-quartzites and the marbles of the Aigerbach-Fm. are reduced in thickness and in an onlapping contact to the once elevated area of the Tux Gneiss in the north.

Interpretation: The basal beds document a transition from continental siliciclastic to the lagoonal and shallow marine environment. The lowest part may be ascribed to the Middle Triassic but there is no proof. The higher portion of the Aigerbach-Fm. is interpreted as a sabkha and coastal sedimentary environment. It is characterized by great heterogeneity and alternation of siliciclastic and carbonate lithofacies. Evaporite formation and episodic influx of terrigenous clastics document repeated sea-level fluctuations, which resembles the Keuper facies of the Germanic Basin (Hauschke & Wilde 1999) or in the Helvetic Zone of the Swiss Alps (Frey 1968).

Hochstegen Formation

The Hochstegen-Fm. was first defined by Frisch (1980). It includes graphite-bearing quartzites and dark grey kyaniteschists at the base (10–40 m), some metres of brownish sandy lime marbles in the middle and the Hochstegen Marble on top (10–100 m true thickness, reported 200 m are due to the isoclinal folding).

The lowest unit of the Hochstegen-Fm. is composed of white or grey graphitic quartzites, graphitic kyanite-schists and, locally, a thin horizon of a black calcite marble. Aggregates of kyanite needles ("Rhätizit") are black from numerous tiny inclusions of graphite. An Early Jurassic age for this so-called "Hochstegenquarzit" has been proposed by Frisch (1968, 1980)

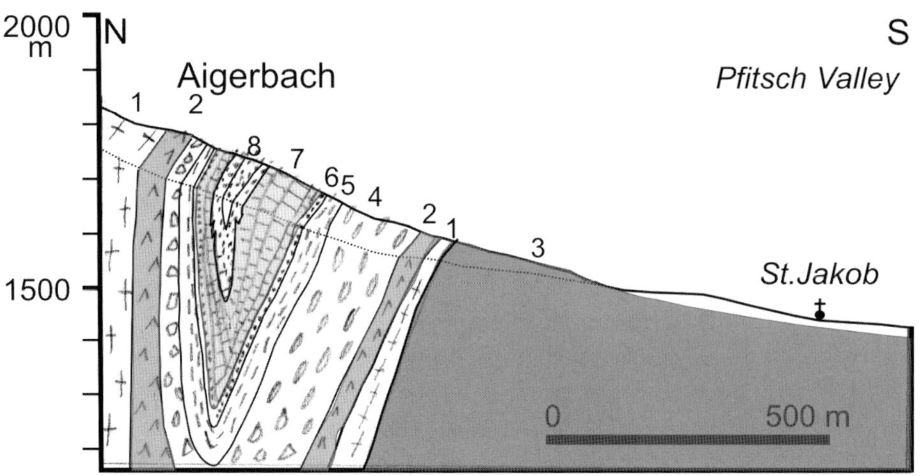

Fig. 10. Aigerbach section. 1 – Tux Gneiss, 2 – amphibolites of the Greiner Series, 3 – graphite-bearing schists of the Greiner Series, 4 – meta-conglomerates, 5 – epidote-ankerite schists, 6 – Windtal-Fm., 7 – Aigerbach Fm., 8 – quartzites of the Hochstegen-Fm.

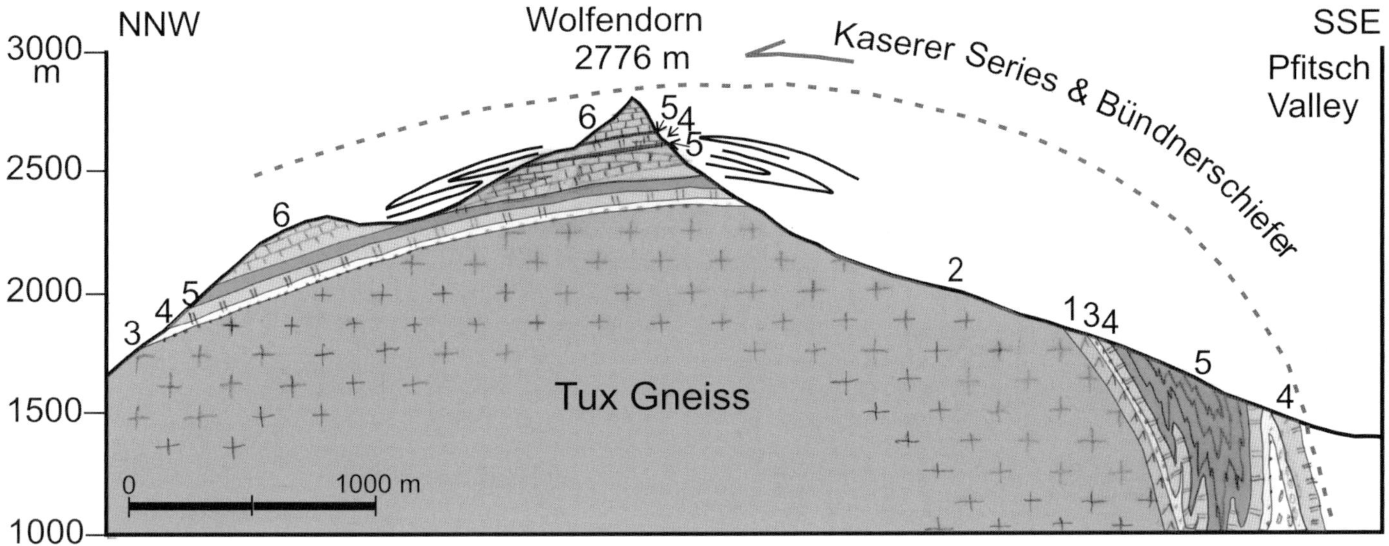

Fig. 11. Wolfendorn section. 1 – amphibolites of the Greiner Series, 2 – Variscan Granites (Tux Gneiss), 3 – Windtal-Fm., 4 – Aigerbach-Fm., 5 – quartzites of the Hochstegen-Fm., 6 – Hochstegen Marble.

in analogy to Liassic blackshales in Germany and Switzerland. It is overlain by only a few metres of brownish sandy calcite marbles, which may be attributed to the Middle Jurassic. On top follows the bluish-grey, fetid Hochstegen Marble. The lower part locally contains boudins of dolomitic beds; in higher horizons cherty nodules are common. An ammonite (*Perisphinctes sp.*), belemnites, sponge spiculae, radiolaria and various open-marine microfossils are described from the Hochstegen Marble (Klebelsberg 1940; Schönlaub et al. 1975; Kiessling 1992). In the equivalent Silbereck Marble of the Eastern Tauern Window also corals could be found (Höfer & Tichy 2005). A detailed description of the Hochstegen-Fm. was given by Kiessling (1992). He found radiolarian and sponge spiculae of Oxfordian and Tithonian ages and stressed the striking similarities to the South German Malm. An affinity to the Helvetic Quinten Limestone of Eastern Switzerland was suspected by Thiele (1970). Frisch (1975b) made a comparative study with the Ultrahelvetic Grestener Zone in Austria.

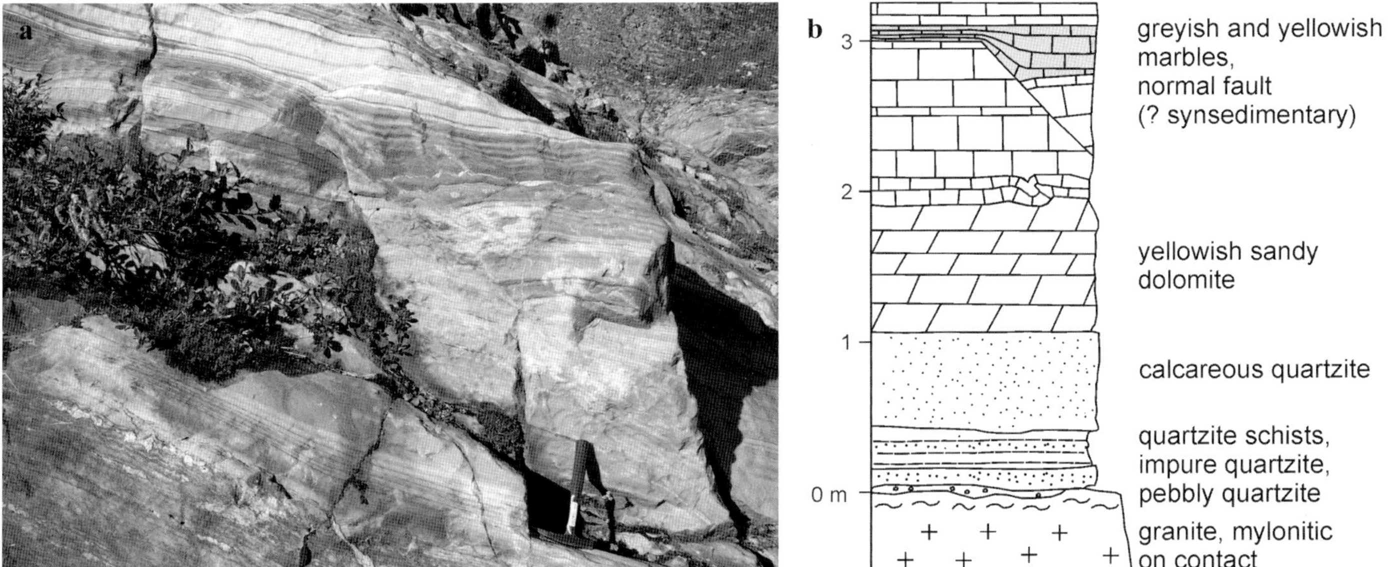

Fig. 12. a) The basal part of the Hochstegen-Fm. with ? synsedimentary normal fault, outcrop Grosser Kunerbach (11° 39' 38" E, 47° 05' 06" N). Fig. 12b) Detailed profile of basal units of Hochstegen-Fm., the same outcrop as Fig. 12a).

The Upper Jurassic carbonates overly in the Tauern Window post-Variscan sediments in the basins and granites or other basement rocks in the horst areas. A notable feature is the mylonitic shear deformation of the "Zentralgneise" at some contacts while the Hochstegen Marble is much less deformed. This implies that the sediments transgressed over a tectonically exhumed basement, which reminds of a Basin-and-Range-like situation. Continuing extensional processes during deposition of the Hochstegen Limestone are visible from synsedimentary normal faults (Fig. 12).

Interpretation: A coastal plain depositional environment with organic-rich sands and mudstones is inferred for the quartzites and the kyanite-schists. The local massive light quartzites may represent fluvial sand bodies or sandy deltaic horizons. A short episode of submergence is documented by the single calcite marble horizon. The more finely laminated quartzite and graphite-bearing schists resemble a fan-delta environment with succession of mudstones and sandstones. The overlying brownish sandy calcite marbles mark the widespread marine transgression during the Middle Jurassic which can be traced far to the north under the Molasse basin (Lemcke 1988). The deposition of the Hochstegen Marble took place under shallow water conditions in the lower parts (dolomitic horizons), but the higher horizons are attributed to deeper water conditions. Kiessling (1992) proposed an outer shelf environment under semi-reducing conditions because of a frequent H_2S content and microfossils which were pyritized during early diagenetic processes.

The well exposed Wolfendorn section is a matter of debate since decades. Tollmann (1963) proposed a Palaeozoic age of the graphite-bearing quartzites (? Lower Jurassic) and, consequently, assumed a thrust plane here. Because the quartzites apparently rest also over the Hochstegen Marble, due to a recumbent isoclinal fold (Fig. 11), he drew another thrust. Tollmann (1963) and Frisch (1975 a) describe both carbonates as Hochstegen Marble. The present authors, on the contrary, attribute the lowest carbonates to the Aigerbach-Fm. because of the remnants of Windtal-Fm. beneath and the sedimentary contact to the Hochstegen-quartzite of presumed Early Jurassic age above which was already recognized by Frisch (1975a).

Other occurrences of post-Variscan sediments

The well exposed Riffler-Schönach Basin forms an elongate belt between the Ahorn- and Tux Gneisses. Detailed descriptions are already given by Thiele (1974), Sengl (1991) and Veselá et al. (2008). Therefore we present here only a new age datum from a meta-rhyodacite of Westphalian age (309.8 ± 1.5 Ma, Table 1) from the Hoher Riffler area and show its position in the stratigraphic column (Table 3).

Around Mayrhofen, the "Porphyrmaterialschiefer Series" frames as a thin ribbon the northern Tauern Window and represents the northern continuation of the Tux Gneiss thrust sheet (Veselá et al. 2008). It contains layers of meta-rhyolites with porphyritic textures. In addition, graphite-bearing phyllites, quartzites, arcoses and amphibolites occur. The Porphyrmaterialschiefer is unconformably covered by calcite marbles which are attributed to the Hochstegen-Formation. Deformation of the rocks is too severe for detailed sedimentological investigation (e.g. Dietiker 1938; Beil-Grzegorczyk 1988). A layer of rhyolitic to andesitic metavolcanic rocks has been dated at 284 +2/-3 Ma (Söllner et al. 1991, Table 1). The Lower Permian meta-rhyolites cover directly the Tux Gneiss at several locations in the western Tauern Window (e.g. the Venntal meta-rhyolite gives an age of about 293 Ma, F. Söllner pers. comm). It documents Early Permian subaerial volcanic activity after the exhumation of the Tux Gneiss and the age maximum of the Porphyrmaterialschiefer Series. The Porphyrmaterialschiefer Series forms a thrust sheet which is sometimes described as "Porphyrmaterialschieferschuppe" (Frisch 1968, 1974; Thiele 1974). Because of the unspeakable name, Thiele (1976) proposed to rename it Torhelm nappe, after the Torhelm (2452 m), a mountain east of Mayrhofen (Fig 2).

The Maurerkees Basin is situated on the southern margin of the Tauern Window and it is a part of the so-called "Mi-

Table 3. Lithostratigraphy of the Riffler-Schönach Basin, Hoher Riffler area

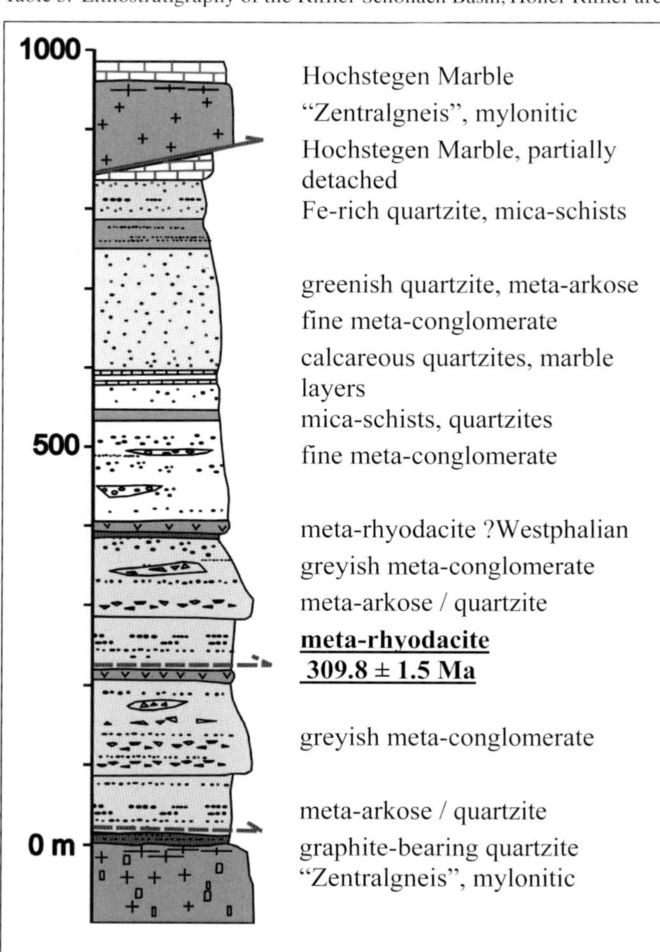

caschist unit", which is folded into the migmatic basement rocks ("Old Gneiss Series") (Schmidegg 1961; Raith et al. 1980). It comprises graphite-garnet schists, quartzites, meta-arkoses and meta-conglomerates. The sedimentary succession displays characteristics of an anastomosing river system and a shallow lacustrine depositional environment, where peat deposits developed. Although the metamorphism reached amphibolite facies, the graphitic schists yielded plant fossils proving Stephanian to Early Permian age (Franz et al. 1991; Pestal et al. 1999).

In the Eisbruggjoch area (Ponte di Ghiaccio) meta-sediments have been found between the two lamellae of the Upper Jurassic Hochstegen Marble. The marbles were folded and internally thrust, so that the subjacent chlorite-schists, banded mica-schists and quartzite-schists are emplaced in between. All these rocks are characterized by a varying amount of finely disseminated ankerite. Thin marble horizons are intercalated and Baggio et al. (1982) report also meta-conglomerates with amphibolite- clasts and lenses within this layer. The age of the meta-sediments is unknown. However, considering the presence of basement-derived rocks (amphibolite) and carbonate horizons, a Permian-Triassic age can be presumed.

The stratigraphic position of the Kaserer Series is still debated due to the complete lack of datable fossils. A Cretaceous age was deduced from apparent conformable contacts to the Hochstegen marble by Frisch (1974), Thiele (1974) and Rockenschaub et al. (2003). In contrast, a Permo-Triassic age was suspected by Dietiker (1938), Tollmann (1963), Fenti & Friz (1974), Lammerer (1998) and Veselá et al. (2008). The older age is supported by the occurrence of sheared anhydrite (of ? Triassic age) within the Kaserer Series and along the contact to the Upper Jurassic Hochstegen Marble which was found in drill cores from the Brenner base tunnel project (Brandner et al. 2007, 2008). In addition, there is an apparent oscillating sedimentary transition to the Middle Triassic carbonates of the Schöberspitzen. The Kaserer Series is composed of a variegated succession of fluvial sediments and shallow marine deposits with evaporites, coarse- to fine-grained quartzites, arkoses, meta-conglomerates, black phyllites, mica-schists and dolomites. In our interpretation, the Kaserer Basin represents the southernmost rift-related trough, which evolved later to the Penninic Ocean. Several serpentinite lenses which are incorporated into the Kaserer metasediments indicate that the once shallow basin lost its substratum by large-scale low-angle extensional faults and came into contact with mantle rocks. During the Alpine convergence phase, the sediments of the Kaserer Basin were stacked and thrust together with a basement slice of a Cambrian meta-gabbro over the Tauern area together with the whole stack of the Penninic Bündnerschiefer nappes of the Outer Tauern Window (Veselá et al. 2008).

Conclusions

In spite of the metamorphic overprint, the tentative interpretation of the sedimentary history of the Carboniferous to Jurassic strata gives a reasonably consistent picture which fits into the geodynamic history of the Alps. The metasediments of the Tauern Window exhibit striking similarities with coeval non-metamorphic deposits within the Germanic Basin. The study area is therefore considered to have been part of the Vindelician Land until the Middle Jurassic.

The earliest post-Variscan sediments of the western Tauern Window could be dated into the Westphalian. These are greyish conglomerates and minor blackschists, which cover the undated but almost certainly Upper Carboniferous Ahorn Gneiss and are topped by metavolcanics of Westphalian age (309.8 ± 1.5 Ma, Moscowian stage after Gradstein et al. 2004). These deposits are interpreted as sediments of alluvial fans and braided rivers. The blackschists, which locally contain plant fossils, reflect formation of peat deposits within a low energy environment of anastomosing rivers. Together with the rhyodacitic and rhyolitic lava flows, they mark periods of accelerated tectonic subsidence in the basin at the end of Carboniferous and beginning of Permian. The relatively short time span between the intrusion of the granitoids and the onset of the basin formation requires fast uplift rates and active tectonic exhumation. In the Pfitsch-Mörchner Basin, the volcanic activity lasted until the Artinskian stage of Early Permian, accompanied by the formation of fanglomerates. The sedimentary fill nicely documents the stepwise denudation of the Variscan orogen and concomitant subsidence of the basins.

The Mesozoic history of the basins may be interpreted as follows: The increasing amount of playa-lake deposits in Late Permian and Early Triassic times documents the lowering of the continental relief. Carbonates and cargneuls are attributed to the ? Middle- and Upper Triassic. They document that the elevation of the basin floors was close to sea level where even a small relative sea-level rise led to flooding of vast areas. A coastal to deltaic depositional environment prevailed but sediment delivery from continental sources persisted until the Jurassic. In response to continued crustal extension and relative sea-level rise, marine conditions were established from the ? Middle Jurassic which is contemporaneous to the break-up of the Penninic-Ligurian Ocean (e.g. Bill et al. 2001) leading to rapid subsidence and the submergence of the adjacent continental margins. The Late Jurassic Hochstegen-Fm. was deposited when the entire area (or at least most parts) of the Tauern Window was drowned and, locally, even deeper marine conditions established. Unambigeously dated Cretaceous sediments are not known from the Inner Tauern Window, but are debated for the Kaserer Serie. On the other hand, there is no need for a continuous sedimentation because an erosional surface on top of Upper Jurassic strata is widespread also in the Molasse foreland.

Acknowledgments

We are grateful to Andreas Wetzel for his sedimentological advice and inspiring discussions in the field. We acknowledge the kind permission from F. Söllner (Munich) to use some unpublished radiometric data on the meta-rhyolites. The manuscript could be considerably improved by the careful reviews and helpful comments of W. Frisch and D. Sciunnach. We further are grateful to the Deutscher Akademischer Austauschdienst (DAAD), who financially supported Petra Veselá.

REFERENCES

Arthaud, F. & Matte, P.H. 1977: Late Palaeozoic strike-slip faulting in southern Europe and northern Africa. Geological Society of America Bulletin 88, 1305–1320.

Baggio, P., De Vecchi, G.P. & Mezzacasa, G. 1982: Carta geologica della media ed alta Valle di Vizze e regione vicine (Alto Adige). Bollettino Società Geologica Italiana 101, 89–116.

Baggio, P., Bosellini, A., Braga, G., Castiglioni, G.B., Corsi, M., Dal Cin, R., De Vecchi, G.P., Friz, C., Gatto, G.O., Gatto, P., Gregnanin, A., Mezzacasa, G., Sassi, F.P., Zirpoli, G., Zulian, T. 1969: Note illustrative della carta geologica d'Italia alla scala 1:100000, Foglio 1 Passo del Brennero, Foglio 4a Bressanone. Servizio Geologico d'Italia, Napoli (Ercolano), 120 pp.

Barrientos, X. & Selverstone, J. 1987: Metamorphosed soils as stratigraphic indicators in deformed terranes: An example from Eastern Alps. Geology 15, no. 1, 841–844.

Behrmann, J. & Frisch, W. 1990: Sinistral ductile shearing associated with metamorphic decompression in the Tauern Widow, Eastern Alps. Jahrbuch der Geologischen Bundesanstalt Wien 133, no. 2, 135–146.

Beil-Grzegorczyk, F. 1988: Petrographie, Genese und stratigraphische Stellung des "Porphyrmaterialschiefers" am Nordrand des Tauernfensters zwischen Hintertux und Gerlospass. Jahrbuch der Geologischen Bundesanstalt Wien 131, 219–230.

Bill, M., O'Dogherty, L., Guex, J., Baumgartner, P.O. & Masson, H., 2001: Radiolarite ages in Alpine-Mediterranean ophiolites: Constraints on the oceanic spreading and the Tethys–Atlantic connection, GSA Bulletin 113, 129–43.

v. Blanckenburg, F., Villa, I., Baur, H., Morteani, G. & Steiger, R.H. 1989:. Time calibration of a PT-path in the Western Tauern Window, Eastern Alps: The problem of closure temperatures. Contributions to Mineralogy and Petrology 101, 1–11.

Brandner, R., Reiter, F. & Töchterle, A. 2007: Geologische Prognose des Brenner Basistunnels – ein Überblick. In: Schneider et al. (Eds.): BBT 2007, Internationales Symposium Brenner Basistunnel und Zulaufstrecken. Innsbruck University Press, 13–23.

Brandner, R., Reiter, F. & Töchterle, A. 2008: Überblick zu den Ergebnissen der Geologischen Vorerkundung für den Brenner-Basistunnel. Geo.Alp 5, 165–174.

Cassinis, G., Durand, M. & Ronchi, A. 2007: Remarks on the Permian-Triassic transition in Central and Eastern Lombardy (Southern Alps, Italy). Journal of Iberian Geology 33, no. 2, 143–162.

Capuzzo, N. & Bussy, F. 2000: High-precision dating and origin of synsedimentary volcanism in the Late Carboniferous Salvan Dorénaz basin (Aiguilles-Rouges Massif, Western Alps). Schweizerische Mineralogische und Petrographische Mitteilungen 80, no. 2, 147–167.

Capuzzo, N., Handler, R., Neubauer, F. & Wetzel, A. 2003: Post-collisional rapid exhumation and erosion during continental sedimentation: the example of the late Variscan Salvan-Dorénaz basin (Western Alps). International Journal of Earth Sciences 92, 364–379.

Capuzzo, N. & Wetzel, A. 2004: Facies and basin architecture of the Late Carboniferous Salvan-Dorénaz continental basin (Western Alps, Switzerland/France). Sedimentology 51, 675–697.

Cesare, B., Rubatto, D., Hermann, J., & Barzi, L. 2001: Evidence for Late Carboniferous subduction type magmatism in mafic – ultramafic cumulates of the Tauern window (Eastern Alps). Contributions to Mineralogy and Petrology 142, 449–464.

Desmons, J. & Mercier, D. 1993: Passing through the Briançon zone. In: von Raumer, J. & Neubauer, F. (Eds.): Pre-Mesozoic geology in the Alps. Springer, Heidelberg, 279–295.

Dietiker, H. 1938: Der Nordrand der Hohen Tauern zwischen Mayrhofen und Krimml (Gerlostal, Tirol). PhD Thesis, ETH Zürich, 131 pp.

Eichhorn, R., Loth, G., Höll, R., Finger, F., Schermaier, A. & Kennedy, A. 2000: Multistage Variscan magmatism in the central Tauern Window (Austria) unveilen by U/Pb SHRIMP zircon data. Contributions to Mineralogy and Petrology 139, 418–435.

Fenti, V. & Friz, C. 1974: Il progetto della galleria ferroviaria Vipiteno-Innsbruck (versante italiano); I, Ricerche geostrutturali sulla regione del Brennero. The Vipiteno-Innsbruck railroad tunnel project; I, Structural geology of the Brenner region. Memorie del Museo Tridentino di Scienze Naturali 20, no.1, 5–59.

Finger, F. & Steyrer, H.P. 1988: Granite-types in the Hohe Tauern (Eastern Alps, Austria) – some aspects on their correlation to Vaiscan plate tectonic processes. Geodiynamica Acta 2, 75–87.

Finger F., Frasl G., Haunschmid B., Lettner H., Schermaier A., von Quadt A., Schindlmayr A. & Steyrer H.P. 1993: The Zentralgneise of the Tauern Window (Eastern Alps) – insight into an Intra-Alpine Variscan batholith. In: von Raumer, J. & Neubauer, F. (Eds.): Pre-Mesozoic Geology in the Alps. Springer, Heidelberg, 375–391.

Finger, F., Roberts M.P., Haunschmid, B., Schermaier, A. & Steyrer H.P. 1997: Variscan granitoids of central Europe, their typology, potential sources and tectonothermal relations. Mineralogy and Petrology 61, 67–96.

Frank, W., Höck, V. & Miller, Ch. 1987: Metamorphic and tectonic history of the Central Tauern Window. In: Flügel, H.H. & Faupl, P. (Eds.): Geodynamics of the Eastern Alps. Deuticke, Wien, 34–54.

Franks, G.D. 1966: The development of the limnic Upper Carboniferous of the eastern Aar Massif. Eclogae Geologicae Helvetiae 59, 943–950.

Franz, G., Mosbrugger, V. & Menge, R. 1991: Carbo-Permian pteridophyll leaf fragments from an amphibolite facies basement, Tauern Window, Austria. Terra Nova 3, no. 2, 137–141.

Frasl, G. 1958: Zur Seriengliederung der Schieferhülle in den Mittleren Hohen Tauern. Jahrbuch der Geologischen Bundesanstalt Wien 101, 323–472.

Frasl, G. & Frank, W. 1966: Einführung in die Geologie und Petrographie des Penninikums im Tauernfenster (mit besonderer Berücksichtigung des Mittelabschnittes im Oberpinzgau, Land Salzburg). Der Aufschluß, Heidelberg, Sonderheft 15, 30–58.

Frey, M. 1968: Quartenschiefer, Equisetenschiefer und germanischer Keuper – ein lithostratigraphischer Vergleich. Eclogae Geologicae Helvetiae 61, 141–156.

Frisch, W. 1968. Geologie des Gebietes zwischen Tuxbach und Tuxer Hauptkamm bei Lanersbach (Zillertal, Tirol). Mitteilungen der Geologie- und Bergbaustudenten 18, 287–336.

Frisch, W. 1974: Die stratigraphisch-tektonische Gliederung der Schieferhülle und die Entwicklung des penninischen Raumes im westlichen Tauernfenster (Gebiet Brenner – Gerlospaß). Mitteilungen der Österreichischen Geologischen Gesellschaft 66/67, 9–20.

Frisch, W. 1975: Ein Typ-Profil durch die Schieferhülle des Tauernfensters: Das Profil am Wolfendorn (westlicher Tuxer Hauptkamm, Tirol). Verhandlungen der Geologischen Bundesanstalt Wien 2–3, 201–221.

Frisch, W. 1975: Hochstegen-Fazies und Grestener Fazies – ein Vergleich des Jura. Neues Jahrbuch für Geologie und Paläontologie Monatshefte, Stuttgart, 82–90.

Frisch, W. 1977: Der alpidische Internbau der Venedigerdecke im westlichen Tauernfenster (Ostalpen). Neues Jahrbuch für Geologie und Paläontologie Monatshefte, Stuttgart, 11, 675–696.

Frisch, W. 1980: Post-Hercynian formations of the western Tauern window: sedimentological features, depositional environment and age. Mitteilungen der Österreichischen Geologischen Gesellschaft 71/72, 49–63.

Frisch, W. 1980: Tectonics of the western Tauern Window. Mitteilungen der Österreichischen Geologischen Gesellschaft 71/72, 65–71.

Frisch, W. & Neubauer, F. 1989: Pre-Alpine terranes and tectonic zoning in the Eastern Alps. In: Dallmeyer, R.D. (Ed.): Terranes in the Circum-Atlantic Paleozoic orogens. Special Paper, Geological Society of America 230, 91–100.

Frisch, W. & Raab, D. 1987: Early Palaeozoic back-arc and island-arc settings in greenstone sequences of the central Tauern Window (Eastern Alps). Jahrbuch der Geologischen Bundesanstalt Wien 129, 545–566.

Frisch, W., Vavra, G. & Winkler, M. 1993: Evolution of the Penninic Basement of the Eastern Alps. In: von Raumer, J. & Neubauer, F. (Eds.): Pre-Mesozoic Geology in the Alps. Springer, Heidelberg, 349–360.

Frisch, W., Dunkl, I. & Kuhlemann, J. 2000: Post-collisional orogen-parallel large-scale extension in the Eastern Alps. Tectonophysics 327, 239–265.

Fügenschuh, B., Seward, D. & Mancktelow, N. 1997: Exhumation in a convergent orogen: the western Tauern window. Terra Nova 9, 213–217.

Gradstein, F.M., Ogg, J.G., and Smith, A.G., Agterberg, F.P., Bleeker, W., Cooper, R.A., Davydov, V., Gibbard, P., Hinnov, L.A., House, M.R., Lourens, L., Luterbacher, H.P., McArthur, J., Melchin, M.J., Robb, L.J., Shergold, J., Villeneuve, M., Wardlaw, B.R., Ali, J., Brinkhuis, H., Hilgen, F.J., Hooker, J., Howarth, R.J., Knoll, A.H., Laskar, J., Monechi, S., Plumb, K.A., Powell, J., Raffi, I., Röhl, U., Sadler, P., Sanfilippo, A., Schmitz, B., Shackleton, N.J., Shields, G.A., Strauss, H., Van Dam, J., van Kolfschoten, T., Veizer, J., and Wilson, D. 2004: A Geologic Time Scale 2004. Cambridge University Press, 589 pp.

Haas, J. & Péró, C. 2004: Mesozoic evolution of the Tisza Mega-unit. International Journal of Earth Sciences (Geologische Rundschau) 93, 297–313.

Hauschke, N. & Wilde, V. (Eds.) 1999: Trias, eine ganz andere Welt: Mitteleuropa im frühen Erdmittelalter. Verlag Dr. Friedrich Pfeil, München, 636 pp.

Höfer, C.G. & Tichy, G. 2005: Fossilfunde aus dem Silbereckmarmor des Silberecks, Hafnergruppe (Hohe Tauern, Salzburg). Journal of Alpine Geology 47, 145–158.

Karl, F. 1959: Vergleichende petrographische Studien an den Tonalit-Graniten der Hohen Tauern und den Tonalit-Graniten einiger periadriatischer Intrusivmassive. Jahrbuch der Geologischen Bundesanstalt Wien 102, 1–192.

Karl, F. & Schmidegg, O. 1979: Geologische Spezialkarte Blatt Krimml (151) 1:50'000. Geologische Bundesanstalt Wien.

Kebede, T., Kloetzli, U., Kosler, J. & Skiold, T. 2005: Understanding the pre-Variscan and Variscan basement components of the central Tauern Window, Eastern Alps, Austria; constraints from single zircon U-Pb geochronology. International Journal of Earth Sciences 94, no. 3, 336–353.

Kiessling, W. 1992: Palaeontological and facial features of the Upper Jurassic Hochstegen Marble (Tauern Window, Eastern Alps). Terra Nova 4, 184–197.

v. Klebelsberg, R. 1940: Ein Ammonit aus dem Hochstegenkalk des Zillertales (Tirol). Zeitschrift der Deutschen Geologischen Gesellschaft Berlin 92, 582–586.

Kreiner, K. 1993: Late-and post-Variscan sediments of the eastern and southern Alps. In: von Raumer, J. & Neubauer, F. (Eds.): Pre-Mesozoic Geology in the Alps. Springer, Heidelberg, 537–564.

Kröll, A., Meurers, B., Oberlercher, G., Seiberl, W., Slapansky, P., Wagner, L., Wessely, G. & Zych, D. 2006: Erläuterungen zu den Karten Molassezone Salzburg – Oberösterreich. Geologische Bundesanstalt Wien, 24 pp.

Kupferschmied, M., Höll, R., Miller, H. 1994: Lithologische und strukturgeologische Untersuchungen in der Krimmler Gneiswalze (Tauernfenster/Ostalpen) und in ihrem Umfeld. Jahrbuch der Geologischen Bundesanstalt Wien 137, 155–170.

Lammerer, B. 1986: Das Autochthon im westlichen Tauernfenster. Jahrbuch der Geologischen Bundesanstalt Wien 129, 51–67.

Lammerer, B. 1988: Thrust-regime and transpression-regime tectonics in the Tauern Window. Geologische Rundschau 71, 143–156.

Lammerer, B. 1998: Bericht 1997 über geologische Aufnahmen im Tauernfenster auf Blatt 149 Lanersbach. Jahrbuch der Geologischen Bundesanstalt Wien 147, 3–4, 297.

Lammerer, B. & Weger, M. 1998: Footwall uplift in an orogenic wedge; the Tauern Window in the Eastern Alps of Europe. Tectonophysics 285, no. 3–4, 213–230.

Lammerer, B., Gebrande, H., Lüschen, E. & Veselá, P. 2008: A crustal-scale cross section through the Tauern Window (eastern Alps) from geophysical and geological data. In: Siegesmund, S. et al. (Eds.): Tectonic Aspects of the Alpine-Dinaride-Carpathian System. Geological Society, London, Special Publications 298, 219–229.

Lemcke, K. 1988: Geologie von Bayern 1, I.: Das bayerische Alpenvorland vor der Eiszeit. Schweizerbart'sche Verlagsbuchhandlung, 175 pp.

Matter, A. 1987: Faziesanalyse und Ablagerungsmilieus des Permokarbons im Nordschweizer Trog. Eclogae Geologicae Helvetiae 80, 345–367.

McCann, T., Pascal, C., Timmerman, M.J., Krzywiec, P., López-Gómez, J., Wetzel, A., Krawczyk, C.M., Rieke, H. & Lamarche, J. 2006: Post-Variscan (End Carboniferous – Early Permian) basin evolution in Western and Central Europe. In: Gee, D.G. & and Stephenson, R.A. (Eds.): European Lithosphere Dynamics. Geological Society London, Memoirs no. 32, 355–388.

Ménard G. & Molnar P. 1988: Collapse of Hercynian Tibetan Plateau into a late Palaeozoic European Basin and Range province. Nature 334, 235–237.

Ménot R.P. & Paquette J.L. 1993: Geodynamic Significance of Basic and Bimodal Magmatism in the External Domain. In: von Raumer, J. & Neubauer, F. (Eds.): Pre-Mesozoic Geology in the Alps. Springer, Heidelberg, 241–254.

Neubauer, F., Frisch, W., Schmerold, R. & Schlöser, H. 1989: Metamorphosed and dismembered ophiolite suites in the basement of the eastern Alps. Tectonophysics 164, 49–62.

Oberhänsli, R., Schenker, F. & Mercolli, I. 1988: Indications of Variscan nappe tectonics in the Aar Massif. Schweizerische Mineralogische und Petrographische Mitteilungen 68, 509–520.

Pestal, G., Brueggemann-Ledolter, M., Draxler, I., Eibinger, D., Eichberger, H., Reiter, C., Scevik, F., Fritz, A. & Koller, F. 1999: Ein Vorkommen von Oberkarbon in den mittleren Hohen Tauern. Jahrbuch der Geologischen Bundesanstalt Wien 141, 491–502.

Pfiffner, A. 1998: Palinspastic reconstruction of the Pre-Triassic Basement Units in the Alps: the Central Alps. In: von Raumer, J. & Neubauer, F. (Eds.): Pre-Mesozoic Geology in the Alps. Springer, Heidelberg, 29–39.

Pupin, J.P. 1980: Zircon and Granite petrology. Contributions to Mineralogy and Petrology 73, 207–220.

von Quadt, A.H.F.C. 1992: U-Pb zircon and Sm-Nd geochronology of mafic and ultramafic rocks from the central part of the Tauern Window (Eastern Alps). Contributions to Mineralogy and Petrology 110, 57–67.

Raith, M., Mehrens, C. & Thöle, W. 1980: Gliederung, tektonischer Bau und metamorphe Entwicklung der penninischen Serien im südlichen Venediger-Gebiet, Osttirol. Jahrbuch der Geologischen Bundesanstalt Wien 123, 1–37.

Ratschbacher, L., Frisch, W., Neubauer, F., Schmid, S.M. & Neugebauer, J. 1989: Extension in compressional orogenic belts: The Eastern Alps. Geology 17, 404–407.

Ratschbacher, L., Frisch, W., Linzer, H.G. & Merle, O. 1991: Lateral extrusion in the Eastern Alps, part 2: Structural analysis. Tectonics 10, 257–271.

von Raumer, J.F. 1998: The Palaeozoic evolution in the Alps: from Gondwana to Pangea. Geologische Rundschau 87, 407–435.

Reitz, E. & Höll, R. 1988: Jungproterozoische Mikrofossilien aus der Habachformation in den mittleren Hohen Tauern und im nordostbayirischen Grundgebirge. Jahrbuch der Geologischen Bundesanstalt Wien 133, 611–618.

Rockenschaub, M., Kolenprat, B. & Nowotny, A. 2003. Das westliche Tauernfenster. In: Rockenschaub, M. (Ed.): Brenner, Arbeitstagung 2003. Geologische Bundesanstalt Wien, 7–38.

Sciunnach, D. 2003: Fault-controlled stratigraphic architecture and magmatism in the western Orobic Basin (Lower Permian, Lombardy Southern Alps). Bollettino Società Geologica Italiana, Volume Speciale 2, 49–58.

Schäfer, A. 1989: The Permo-Carboniferous Saar-Naahe basin, south-west Germany and northeast France: basin formation and deformation in a strike-slip regime. Geologische Rundschau 78, 499–524.

Schäfer, A. 2005: Klastische Sedimente, Fazies und Sequenzstratigraphie. Elsevier, Heidelberg, 414 pp.

Schaltegger, U. & Corfu, F. 1992: The age and source of Late Hercynian magmatism in the Central Alps: evidence from precise U-Pb ages and initial Hf isotopes. Contributions to Mineralogy and Petrology 111, 329–344

Schaltegger, U. & Corfu, F. 1995: Late Variscan "Basin and Range" magmatism and tectonics in the Central Alps: evidence from U-Pb geochronology. Geodinamica Acta 8, 82–98.

Schmid, S.M., Fügenschuh, B., Kissling, E. & Schuster, R. 2004: Tectonic map and overall architecture of the Alpine orogen. Eclogae geologicae Helvetiae 97, 93–117.

Schmidegg, O. 1961: Geologische Übersicht der Venedigergruppe. Verhandlungen der Geologischen Bundesanstalt Wien 1, 35–56.

Schön, Ch. & Lammerer, B. 1988: Die Post-Variszischen Metakonglomerate des westlichen Tauernfensters, Österreich. Mitteilungen der Österreichischen Geologischen Gesellschaft 81, 219–232.

Schönlaub, H.P., Frisch, W. & Flajs, G. 1975: Neue Fossilfunde aus dem Hochstegenmarmor (Tauernfenster, Österreich). Neues Jahrbuch für Geologie und Paläontologie, Monatshefte, Stuttgart 2, 111–128.

Selverstone, J., Spear, F.S., Franz, G. & Morteani, G. 1984: High pressure metamorphism in the SW Tauern Window, Austria: PT Paths from hornblende-kyanite-staurolite schists. Journal of Petrology 25, 501–531.

Selverstone, J. 1985: Petrologic constraints on imbrication, metamorphism and uplift in the SW Tauern window, Eastern Alps. Tectonics 4, no. 7, 687–704.

Sengl, F. 1991: Geologie und Tektonik der Schönachmulde (Zillertaler Alpen, Tirol). Unpublished PhD Thesis, Ludwig-Maximilians-Universität München, 183 pp.

Sheldon, N.D. 2005: Do red beds indicate paleoclimatic conditions? A Permian case study. Palaeogeography, Palaeoclimatology, Palaeoecology 228, 305–319.

Söllner, F., Höll, R. & Miller, H. 1991: U-Pb-Systematik der Zirkone in Meta-Vulkaniten ("Porphyroiden") aus der Nördlichen Grauwackenzone und dem Tauernfenster (Ostalpen, Österreich). Zeitschrift der deutschen geologischen Gesellschaft, 142, 285–299.

Steenken, A., Siegesmund, S., Heinrichs, T. & Fügenschuh, B. 2002: Cooling and exhumation of the Rieserferner pluton (Eastern Alps, Italy/Austria). International Journal of Earth Sciences (Geologische Rundschau) 91, 799–817.

Thiele, O. 1970: Zur Stratigraphie und Tektonik der Schieferhülle der westlichen Hohen Tauern (Zwischenbericht und Diskussion über Arbeiten auf Blatt Lanersbach, Tirol). Verhandlungen der Geologischen Bundesanstalt Wien, 230–244.

Thiele, O. 1974: Tektonische Gliederung der Tauernschieferhülle zwischen Krimml und Mayrhofen. Jahrbuch der Geologischen Bundesanstalt Wien 117, 55–74.

Thiele, O. 1976: Der Nordrand des Tauernfensters zwischen Mayrhofen und Inner Schmirn (Tirol). Geologische Rundschau, Stuttgart, 65, 410–421.

Thiele, O. 1980: Das Tauernfenster. In: Oberhauser, R. (Ed.): Der Geologische Aufbau Österreichs. Springer Verlag Wien – New York, 300–314.

Tollmann, A. 1963: Ostalpensynthese. Deuticke, Wien, 265 pp.

Trümpy, R. 1966: Considérations générales sur le "Verrucano des Alpes suisses". In: Tongiorgi, M. & Rau, A. (Eds.): Atti del symposium sul Verrucano, Pisa 1965. Società Toscana di Scienze Naturali in Pisa, 212–232.

Trümpy, R. 1980: Geology of Switzerland. Wepf, Basel, 104 pp.

Vavra, G. & Frisch, W. 1989: Pre-Variscan back-arc and island-arc magmatism in the Tauern Window (Eastern Alps). Tectonophysics 169, 271–280.

Veselá, P., Lammerer, B., Wetzel, A., Söllner, F. & Gerdes, A. 2008: Post-Variscan to Early Alpine sedimentary basins in the Tauern Window (eastern Alps). In: Siegesmund, S. et al. (Eds.): Tectonic Aspects of the Alpine-Dinaride-Carpathian System. Geological Society, London, Special Publications 298, 83–100.

Wetzel, A., Allenbach, R. & Allia, V. 2003: Reactivated basement structures affecting the sedimentary facies in a tectonically "quiescent" epicontinental basin: an example from NW Switzerland. Sedimentary Geology 157, 153–172.

Ziegler, P.A. 1990: Geological Atlas of Western and Central Europe, 2nd edn. Shell Internationale Petroleum Maatschappij, The Hague, 239 pp.

Ziegler, P.A. 2005: Permian to Recent Evolution. In: Selley, R.C., Cocks, L.R. & Plimer, I.R. (Eds.) The Encyclopedia of Geology, Elsevier, World Wide Web Address: http://www.encyclopediaofgeology.com/samples/europe2.pdf

Ziegler, P.A. & Dezés, P. 2006: Crustal evolution of Western and Central Europe. In: Gee, D.G. & Stephenson, R.A. (Eds.): European lithosphere dynamics. Memoirs of the Geological Society of London 32, pp. 43–56.

Manuscript received January 1, 2008
Revision accepted June 5, 2008
Published Online first November 13, 2008
Editorial Handling: Stefan Schmid & Stefan Bucher

1661-8726/08/01S089-22
DOI 10.1007/s00015-008-1280-2
Birkhäuser Verlag, Basel, 2008

Swiss J. Geosci. 101 (2008) Supplement 1, S89–S110

Multistage shortening in the Dauphiné zone (French Alps): the record of Alpine collision and implications for pre-Alpine restoration

THIERRY DUMONT[1], JEAN-DANIEL CHAMPAGNAC[2], CHRISTIAN CROUZET[3] & PHILIPPE ROCHAT[4]

Key words: External Western Alps, multistage deformation history, Hercynian fabric reactivation, Tethyan rifting, inversion tectonics, lateral escape, 3D modelling

ABSTRACT

Three-dimensional modelling tools are used with structural and palaeo-magnetic analysis to constrain the tectonic history of part of the Dauphiné zone (external Western Alps). Four compressive events are identified, three of them being older than the latest Oligocene. Deformation D1 consists of W–SW directed folds in the Mesozoic cover of the study area. This event, better recorded in the central and southern Pelvoux massif, could be of Eocene age or older. Deformation D2 induced N-NW-oriented basement thrusting and affected the whole southern Dauphiné basement massifs south of the study area. The main compressional event in the study area (D3) was WNW oriented and occurred before 24 Ma under a thick tectonic load probably of Penninic nappes. The D2-D3 shift corresponds to a rapid transition from northward propagation of the Alpine collision directly driven by Africa-Europe convergence, to the onset of westward escape into the Western Alpine arc. This Oligocene change in the collisional regime is recorded in the whole Alpine realm, and led to the activation of the Insubric line. The last event (D4) is late Miocene in age and coeval with the final uplift of the Grandes Rousses and Belledonne external massifs. It produced strike-slip faulting and local rotations that significantly deformed earlier Alpine folds and thrusts, Tethyan fault blocks and Hercynian structures. 3D modelling of an initially horizontal surface, the interface between basement and Mesozoic cover, highlights large-scale basement involved asymmetric folding that is also detected using structural analysis. Both, Jurassic block faulting and basement fold-and-thrust shortening were strongly dependent on the orientation of Tethyan extension and Alpine shortening relative to the late Hercynian fabric. The latter's reactivation in response to oblique Jurassic extension produced an en-échelon syn-rift fault pattern, best developed in the western, strongly foliated basement units. Its Alpine reactivation occurred with maximum efficiency during the early stages of lateral escape, with tectonic transport in the overlying units being sub-perpendicular to it.

RESUME

L'histoire des deformations dans un secteur du Dauphiné (zone externe des Alpes occidentales) est précisée en utilisant à la fois l'analyse structurale, le paléomagnétisme et la modélisation 3D. On y reconnaît quatre évènements compressifs, dont trois antérieurs à l'Oligocène terminal. La deformation D1 se marque par des plis à vergence W à SW dans la couverture mésozoïque. Ce premier évènement, mieux connu dans la partie sud des massifs cristallins dauphinois, est antérieur à l'Eocène supérieur. La déformation D2 a produit des chevauchements de socle orientés vers le N et le NW dans les massifs dauphinois situés au Sud du secteur étudié. Mais ce dernier a été principalement marqué par le raccourcissement D3 orienté ESE–WNW, qui s'est produit avant 24 Ma sous une importante couverture tectonique de nappes penniques. Ce changement entre D2 et D3 marque une évolution importante de la collision alpine, d'abord dominée par la subduction continentale en conséquence directe de la convergence N–S entre Afrique et Europe, puis par l'échappement latéral vers l'ouest qui a généré l'arc des Alpes occidentales. Cette réorganisation oligocène a aussi provoqué l'activation du décrochement insubrien. La dernière déformation D4, d'âge Miocène supérieur, a accentué les bombements de socle des massifs des Grandes Rousses et de Belledonne, entre lesquels des mouvements décrochants et des rotations ont distordu à la fois les structures alpines antérieures, les failles et blocs téthysiens et la foliation hercynienne. L'interface socle hercynien/couverture mésozoïque, qui était initialement plan et horizontal, a été construit en 3D. Le réseau des blocs téthysiens y apparaît déformé par des failles et des plis de socle. Les observations de terrain montrent que l'orientation du grain hercynien a largement influencé la nature et la localisation des structures distensives et compressives. Les dépocentres jurassiques se seraient disposés en échelon à cause de l'obliquité de l'extension téthysienne par rapport à la foliation du socle. Durant la collision, cette dernière a été le plus intensément réactivée lorsque la contraction lui était subperpendiculaire, c'est à dire à partir de l'Oligocène quand le régime en échappement latéral vers l'Ouest a été établi.

1 Introduction

The origin of the arcuate shape of the Western Alpine fold belt is still debated. Some models regard it as inherited either from the shape of the Jurassic margins (Lemoine et al. 1989) or from the shape of the Adriatic indenter (Coward & Dietrich 1989). Other authors propose that the arc was created during Alpine collision due to indentation and lateral escape (e.g. Tapponnier

[1] CNRS, Université Joseph Fourier, Laboratoire de Géodynamique des Chaînes Alpines (UMR 5025), 1381 rue de la Piscine, F-38400 St Martin d'Hères. Email: Thierry.Dumont@ujf-grenoble.fr
[2] Institute of Mineralogy, University of Hannover, Callinstrasse 3, D-30167 Hannover.
[3] allée des Bayardines, 38530 Pontcharra.
[4] Total, BP730, F-92007 Nanterres cedex.

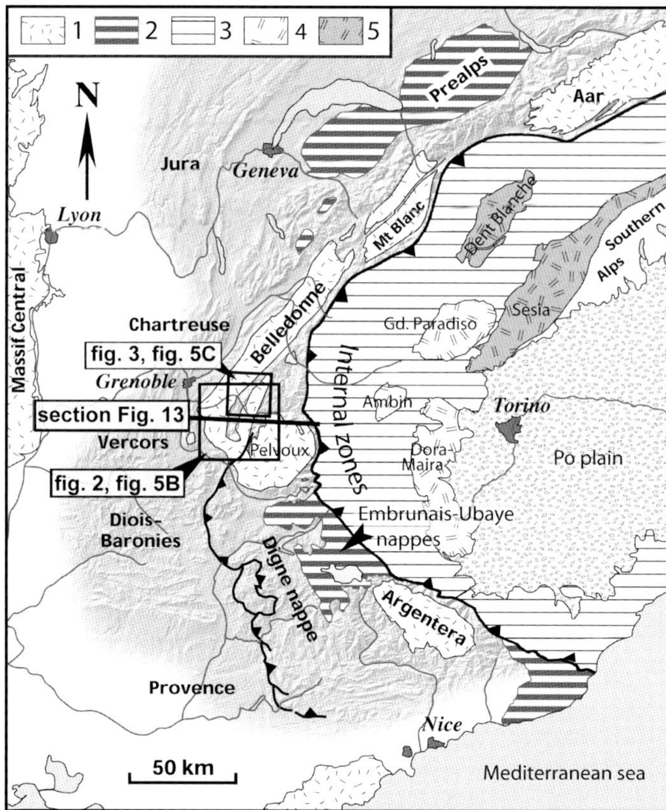

Fig. 1. Location of the study area. Shaded relief: deformed Meso-Cenozoic cover of the External zone; 1: External basement massifs (Hercynian basement); 2: Exotic flysch and klippe; 3: nappes of the Internal zones; 4: Internal basement massifs, European origin; 5: Crystalline basement, Austro-Alpine origin.

1977; Vialon et al. 1989). However, various associated driving mechanisms are involved: counter-clockwise rotation of the indenter and/or of the Penninic foreland (Ricou & Siddans 1986; Choukroune et al. 1986; Laubscher 1988; Thomas et al. 1999), change in relative motion of the indenter (Schmidt & Kissling 2000; Lickorish et al. 2002; Ford et al. 2006), or indenter-induced body forces with constant plate motion (radial outward model of Platt et al. 1989a, Ratschbacher 1989). The arcuate trend is outlined by the trace of the Internal Alpine units and their western boundary, the so-called Frontal Pennine thrust (Butler 1992), Briançonnais Frontal Thrust (Sue & Tricart 2002), Penninic Basal Contact (Ceriani & Schmid 2004), and Basal Penninic Fault (Ford et al. 2006). By contrast, the trend of the external crystalline massifs shows an abrupt change from NE–SW to NW–SE, where the study area is located (Fig. 1). Belonging to the Dauphiné zone, it shows high Hercynian basement massifs with Mesozoic series pinched in roughly N–S trending synclines (Fig. 2). In the literature these structures have been described as inverted Jurassic basins (Barféty et al. 1979; Lemoine et al. 1981). Alpine shortening was strongly influenced by the Mesozoic rift pattern (Tricart & Lemoine 1986), which needs a

detailed understanding of compressional deformation history to be restored. Because the orientation of rift structures (tilted blocks) and the transport direction along the major thrusts changed through time (Graciansky & Lemoine 1988; Claudel & Dumont 1999; Ceriani et al. 2001), the final geometry must be studied in three dimensions, and maintaining consistency in the deformation history between large-scale and small-scale structures is of utmost importance. In this study field data and palaeomagnetic records are considered in a regional framework improved by the use of synthetic imagery (synthetic aerial views and 3D geological maps). Additionally, 3D modelling of the Hercynian basement/Mesozoic cover interface was carried out because this initially flat reference surface provides a marker for finite deformation during the Alpine cycle (Tethyan extension plus Alpine shortening). This multidisciplinary approach leads to a reappraisal of the relative effects of Hercynian heritage, Mesozoic rifting and Alpine shortening on the present geometry. It also facilitates proposing time constraints for some stages of Alpine exhumation in this area.

2 Structural and stratigraphic framework of the northern Dauphiné

Many previous studies focused on structures and deformation processes (Vernet 1965, 1974; Vialon 1974, 1986; Gratier et al. 1973, 1978; Gratier & Vialon 1975, 1980; Lamarche 1987; Gillcrist 1988; Grand 1988; Crouzet et al. 1996). The northern Dauphiné region shows N–S to NE–SW trending massifs composed of Cambro-Ordovician to early Carboniferous metamorphic basement and intruded by late Variscan granites (Guerrot & Debon 2000), locally with late Carboniferous cover of molasse and conglomerates. The Palaeozoic basement together with these molasse sediments are truncated by a regional unconformity overlain by thin but widespread Triassic shallow marine carbonates, and by deeper marine Liassic to Upper Jurassic limestones and marls (Barféty 1988; Dumont 1998). The Mesozoic series are pinched between the basement massifs (Fig. 2; Fig. 3) and deformed under zeolite to lower greenschist metamorphic facies during Alpine shortening (see references in Crouzet et al. 2001a). The overall structure was first regarded as Alpine folds involving the basement (Vernet 1965), then later as remnants of Tethyan tilted blocks (Barféty et al. 1979; Lemoine et al. 1981; Barféty & Gidon 1983, 1984) more or less affected by tectonic inversion processes in a brittle setting (Grand et al. 1985; Lemoine et al. 1986; Gillcrist et al. 1987). However, the occurrence of large-scale basement folds, already noted by previous authors (e.g. Vernet 1965), has been highlighted more recently (Ford 1996; Dumont et al. 1997).

The tectono-sedimentary analysis of the Mesozoic record documents Alpine and pre-Alpine deformation in the following ways:

– The sharp Triassic transgression seals Hercynian structures and makes it possible to distinguish between Hercynian and Alpine cleavages;

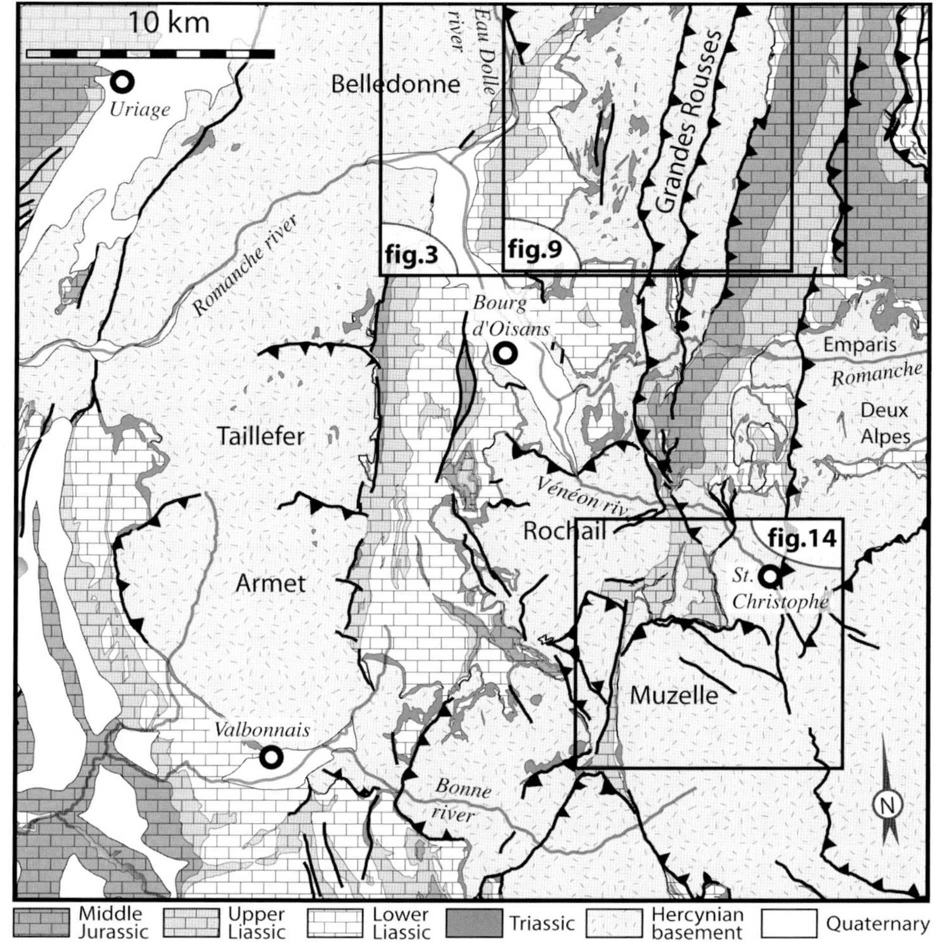

Fig. 2. Structural map of the central Dauphiné region, corresponding to the 3D model of Fig. 5b.

– The constant thickness and facies of the widespread lower Triassic dolomites, which were deposited in peritidal environments, demonstrate that the whole domain was very flat and horizontal at that time. Thus, the present geometry of the transgression surface is the result of Mesozoic extension combined with Alpine shortening;

– The distribution, stratal patterns and palaeotectonic features within Liassic syn-rift sediments help to reconstruct the shape of Jurassic extensional blocks (Barféty & Gidon 1983; Lemoine et al. 1986; Barféty 1988, and refs. herein). The vertical movements and extensional fault kinematics can be deduced from analysis of palaeobathymetry and sequence stratigraphy (Roux et al. 1988; Chevalier et al. 2003). Since the Triassic series contain few evaporites, the lower part of the syn-rift sequence generally remains attached to the basement, such that the contribution of Mesozoic rift deformation on the geometry of the basement blocks of this area can be evaluated.

Despite Alpine shortening, the Mesozoic rift structures are still clearly visible on E–W profiles across the Bourg d'Oisans half-graben (Fig. 4; Fig. 5a). However, the 3D structure is poorly understood. Due to its present rhomb shape, it has been tentatively interpreted as a pull-apart basin (Grand et al. 1985; Dumont & Grand 1987), but the amount of Alpine deformation was probably underestimated.

3 Compressional structures and deformation history

Multistage deformation is documented by Gratier & Vialon (1980), with strong reorientation of fold axis near the Tethyan-inherited Taillefer-Belledonne buttress due to increasing strain intensity (Gratier et al. 1978). Here we propose a sequence of tectonic events which is similar to the one identified by Sue et al. (1998) in a nearby area, but with different ages (see discussion in § 5). Our outcrop observations focus on the basement-cover interface, whose 3D regional-scale shape is described in § 4. We investigated the structural relationships of this interface with two types of structures: (a) Alpine compressive structures in the overlying Mesozoic sedimentary cover, and (b) the late Hercynian foliation in the underlying basement, which is locally involved in Alpine folds but which can be restored together with the originally horizontal basement-cover interface.

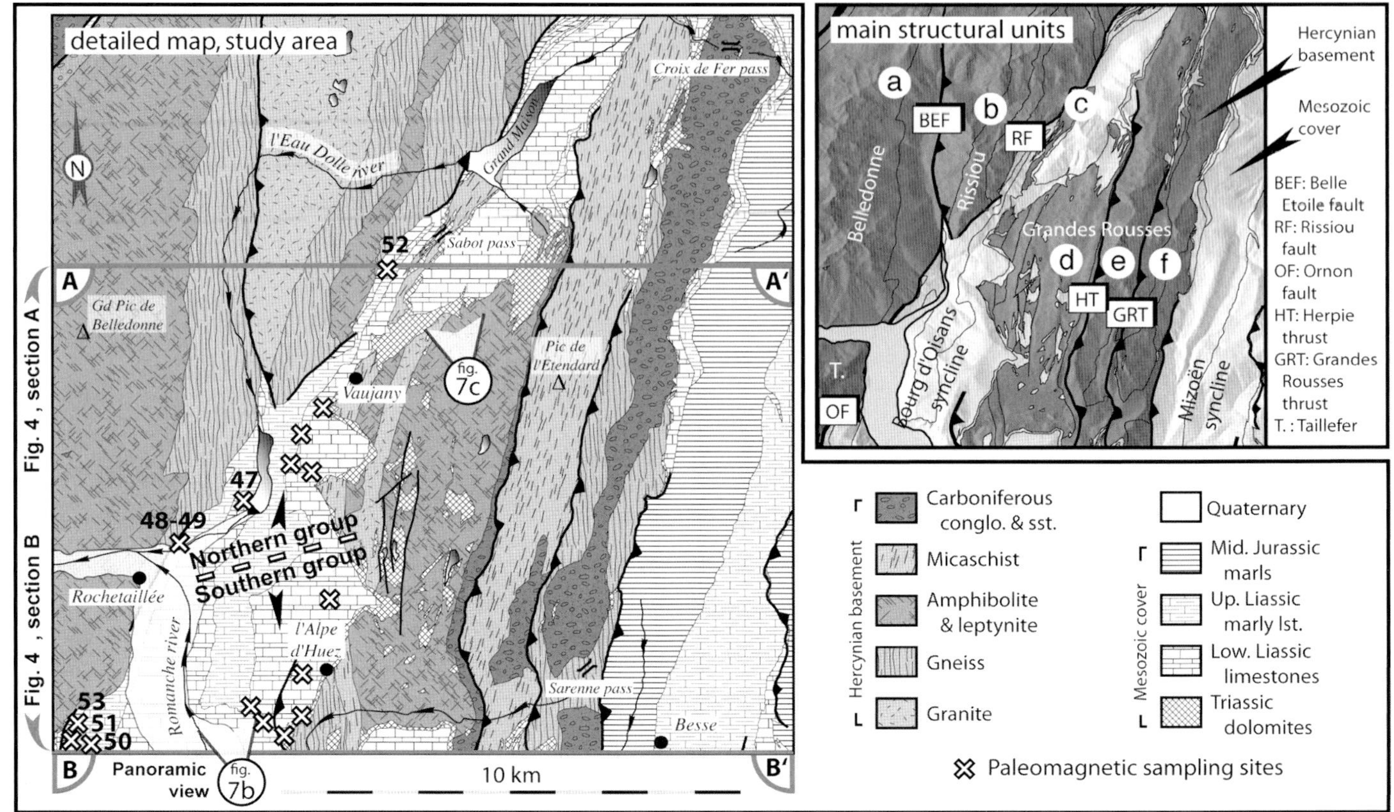

Fig. 3. Left: Geological map of the main study area, corresponding to the 3D model of Fig. 5c. Locations of Fig. 4 (sections) and Fig. 7 (panoramic views) also shown. Paleomagnetic sampling sites located: numbers refer to previously unpublished data (Table 1 and text), while other sites are described in Crouzet et al. (1996, 1999 & 2001a). Right: cartoon of the same area showing the main structural units: (a): Belledonne; (b): Rissiou; (c): Sabot; (d): Huez-western Grandes Rousses; (e): central Grandes Rousses; (f): eastern Grandes Rousses (same units as in Fig. 4, 5c & 12a).

3.1 Deformation sequence recorded in the Mesozoic cover

Alpine deformation was polyphase and the cleavages developed during folding under low to medium grade metamorphic conditions (Gratier & Vialon 1980; Aprahamian 1988; Nziengui 1993; Crouzet et al. 2001a). Fold trends and cleavages measured within the Triassic and lower Liassic formations around the Grandes Rousses bulge (Fig. 6 and examples in Fig. 7) document four folding and thrusting events (D1 to D4, in order of increasing age):

– D1 produced WSW- to SW-recumbent folds only visible in the southern part of the study area (Auris fold; site 13, Fig. 6). It is locally overprinted by north-directed D2 crenulation, and by D3 westward shortening (Fig. 7a). A similar deformation history is known further south (Sue et al. 1998 and §6.4).

– D2 involves only the southern part of the study area and is much more developed further south (central and southern Pelvoux). It consists of north- to NW-directed, high-angle basement thrusts, which affect the Liassic cover only locally. These features indicate N–S to NW–SE shortening, which is parallel to the trend of the Alpine chain in this area. This particular feature will be documented and discussed below (§6.4).

– D3 is responsible for many of the observed structures, with both small-scale folding of Triassic beds attached to the basement and large-scale ductile deformation of the lower Liassic wedge, disharmonically folded above basement thrusts (Fig. 7b). D3 features are enhanced by the reactivation of basement foliation, which was suitably oriented for high-angle reverse faulting (e.g. the Herpie and La Garde thrusts: HT & LGT, Fig. 4b & 5a). D3 is top-W to top-NW oriented, but the orientation of fold axes vary significantly from south to north (Fig. 6). This curved path is due to D4 rotation, as shown by structural and palaeomagnetic data (§3.2 and §5).

– D4 consists of open folds with steep axial planes and kink geometries, with local backward (eastward) fold and thrust features (Fig. 7b & 7c). The associated cleavage S4 clearly overprints D3 structures along the road from l'Alpe d'Huez to Villard Reculas (Fig. 7b), in agreement with Vialon (1968) and Gratier et al. (1973). D4 folding is consistent with conjugate strike-slip faulting on both sides of the Rissiou basement block (Fig. 6, sites 4 to 6) and on the western slope of the Grandes Rousses massif, indicating an EW to N100° oriented maximum stress axis. D4 is probably coeval with the final uplift of the Grandes Rousses massif, which occurred recently according fission-track data (Sabil 1995).

3.2 Alpine compressional structures in the shallow part of the basement

Alpine folding affected both the basement-cover interface and the Hercynian foliation. In the hanging wall of the La Garde thrust (LGT, Fig. 7b), the Hercynian foliation is involved in a ramp anticline, as shown by the scatter of poles of Hercynian foliation along a small circle of the stereogram (Fig. 8). This small circle indicates a primary angle of no more than 20° between the strike of Hercynian foliation before folding and the D3 Alpine fold axis measured in the sedimentary cover immediately above the interface. This low-angle obliquity is repeatedly observed in five locations from south to north (Fig. 9a, Fig. 10), with Hercynian foliation and D3 fold axis rotated together but maintaining about the same angle between both. This rotation of the D3 fold axis together with the Hercynian foliation, about 40° clockwise from south to north, is due to the later Alpine event D4. This renewed E–W shortening also involved (Fig. 9b) conjugate strike-slip motion of the Rissiou and Alpe d'Huez faults (RF and AHF) and updoming of the Grandes Rousses bulge (Fig. 9c). Older Alpine structures such as D3 fold trends and basement thrusts (HT, GRT; Fig. 3, Fig. 9b) are rotated and involved in the bulge. Before rotation, the D3 structures and the Hercynian fabric were more rectilinear, and were slightly oblique to each other.

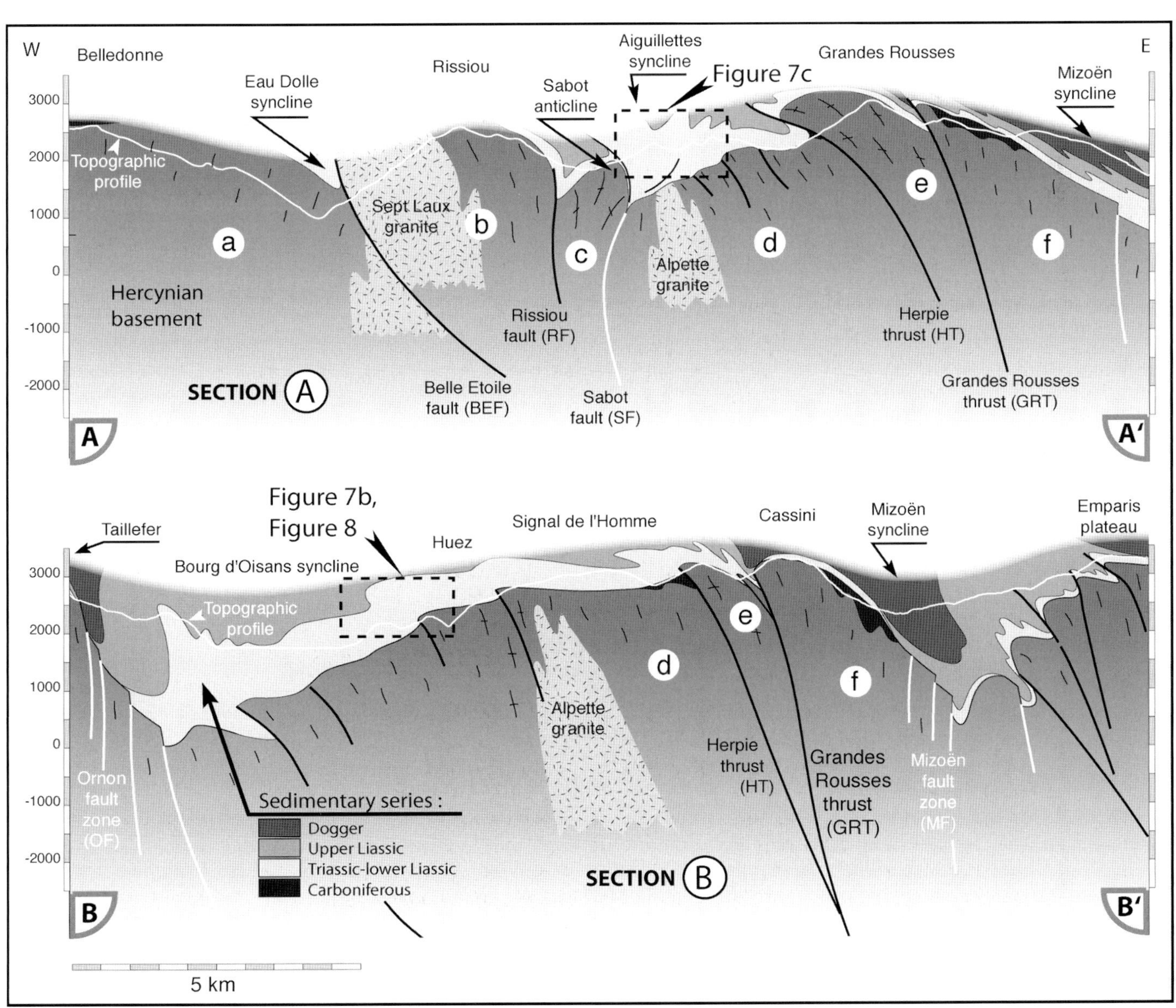

Fig. 4. E–W interpretative cross sections in the Bourg d'Oisans area (location Fig. 3).

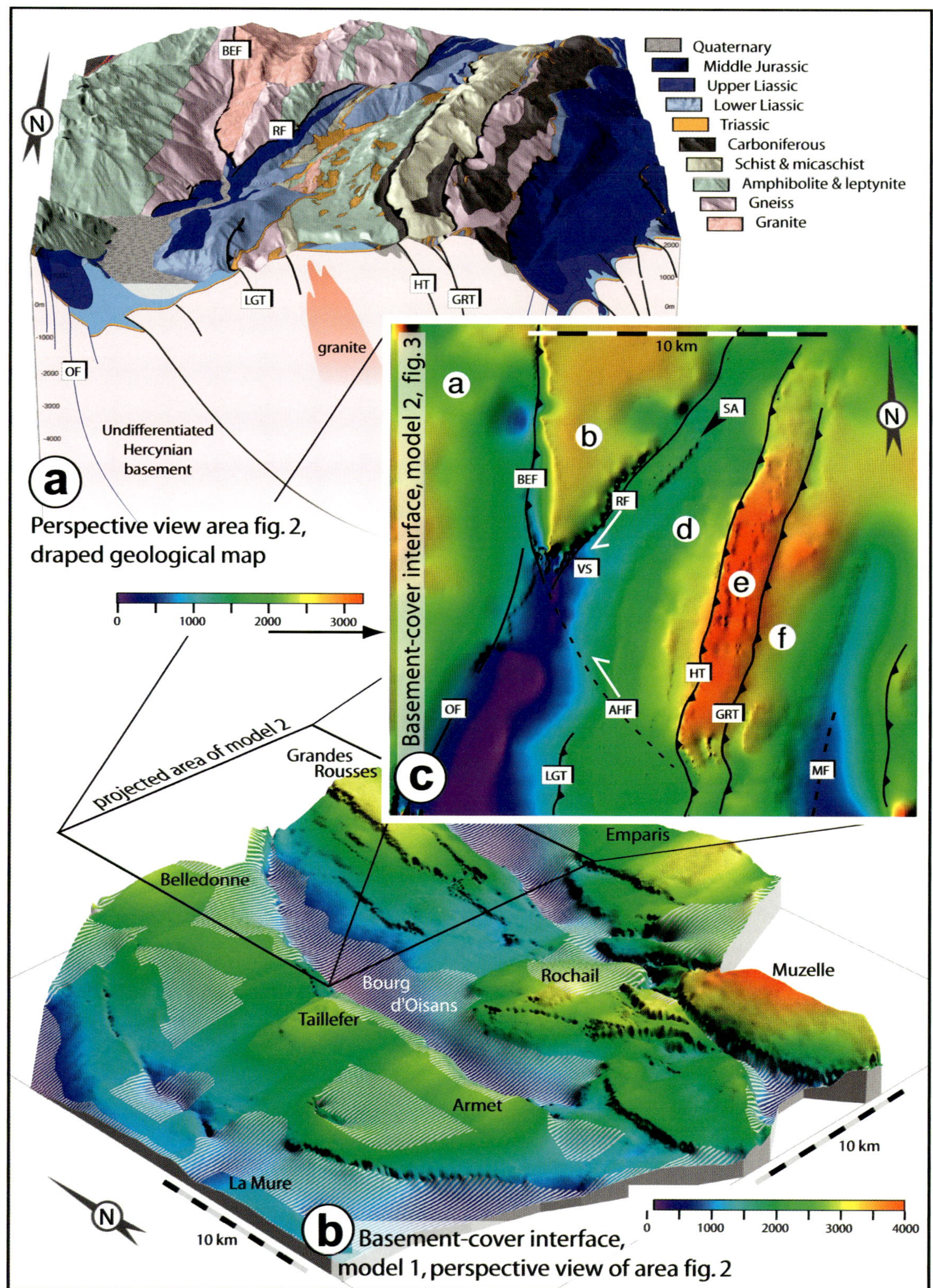

Fig. 5. a) Perspective geological map view from the South (same area as Fig. 3) combined with the interpretative cross-section of Fig. 4b. b) 3D model of the Basement-cover interface (BCI) in the area corresponding to Fig. 2, perspective view from SW. Shaded relief combined with rainbow colour scale for altitude. c) Detailed 3D model of the BCI in the main area of interest (corresponding to Figs. 3 and 5a). Altitude scale is different from fig. 5b.
Main Alpine features: AHF: Alpe d'Huez fault; BEF: Belle Etoile fault; GRT: Grandes Rousses thrust; HT: Herpie thrust; LGT: La Garde thrust; RF: Rissiou fault; SA: Sabot anticline; VS: Vaujany syncline; *Main pre-Alpine (Tethyan) features:* OF: Ornon fault zone; MF: Mizoën fault zone.

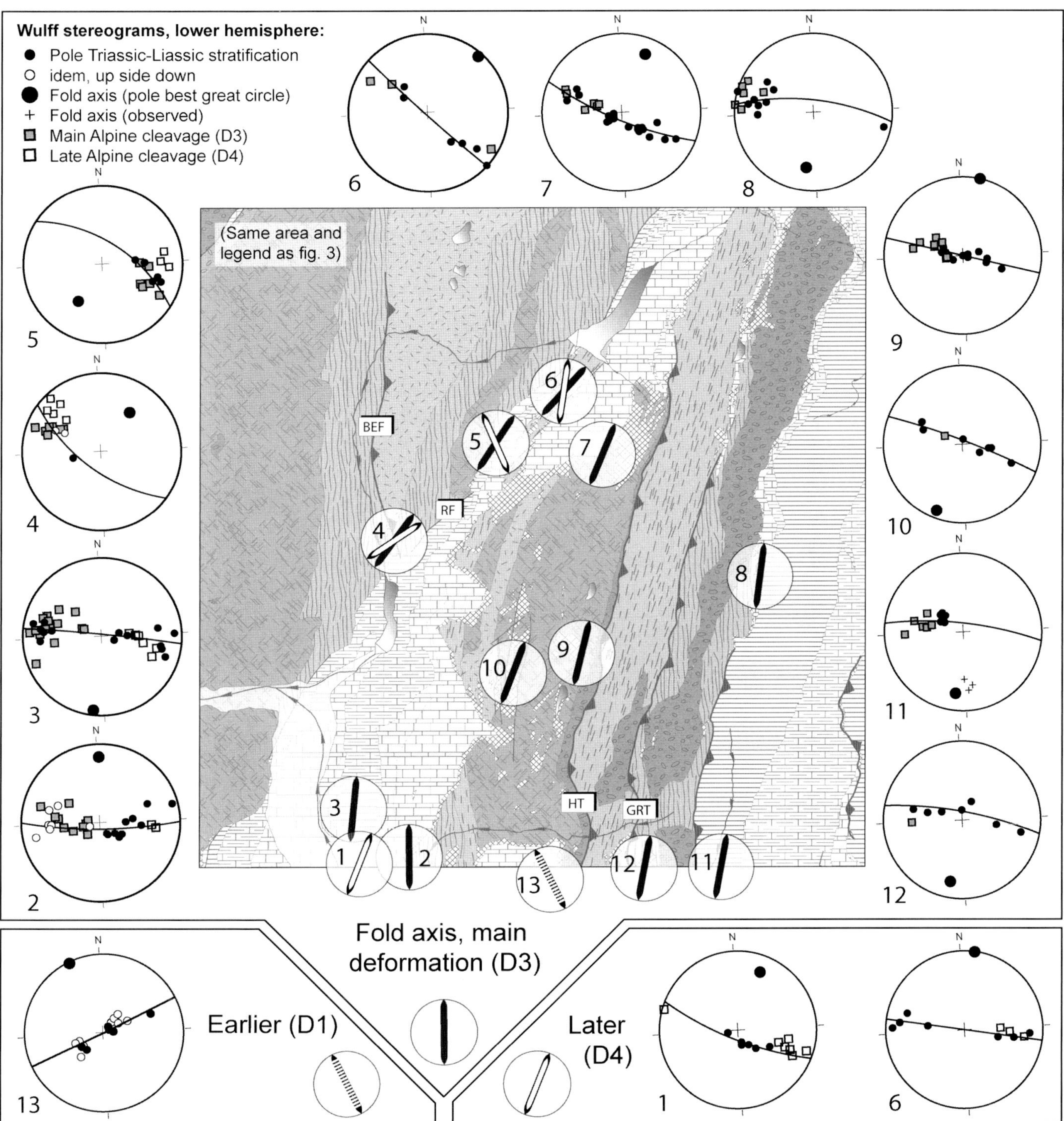

Fig. 6. Kinematic data from the Mesozoic sedimentary cover, mostly from Triassic and lower Liassic strata (sites numbers are different from Fig. 9). Outcrop pictures are given in Fig. 7. Most tectonic marks refer to deformation D3, locally overprinted by D4. A gradual clockwise rotation of F3 fold trends is observed from south to north, as in the basement structures (Fig. 9). This D4 deformation in the Grandes Rousses massif may be related with conjugate strike-slip motions of the southern BEF and RF suggested by steeply dipping F4 fold axes on both sides of the Rissiou block (sites 4 & 5).
Sites location: (1) Côte Alamelle cliff, (2) road to l'Alpe d'Huez, (3) road from Huez to Villard Reculas, (4) 1 km north from Le Verney, (5) below Les Grandes Côtes, SW from the Sabot pass, (6) 0,5 km west from the Sabot pass, (7) near the Couard pass, (8) southern slope of Crête des Sauvages, (9) between Blanc lake and Milieu lake, (10) above Besson lake, (11) southern shoulder of the Chambon dam, below high water level (projected), (12) north of Sarenne pass, (13) between Les Courts, Combe Gillarde and Cluy pass (projected).

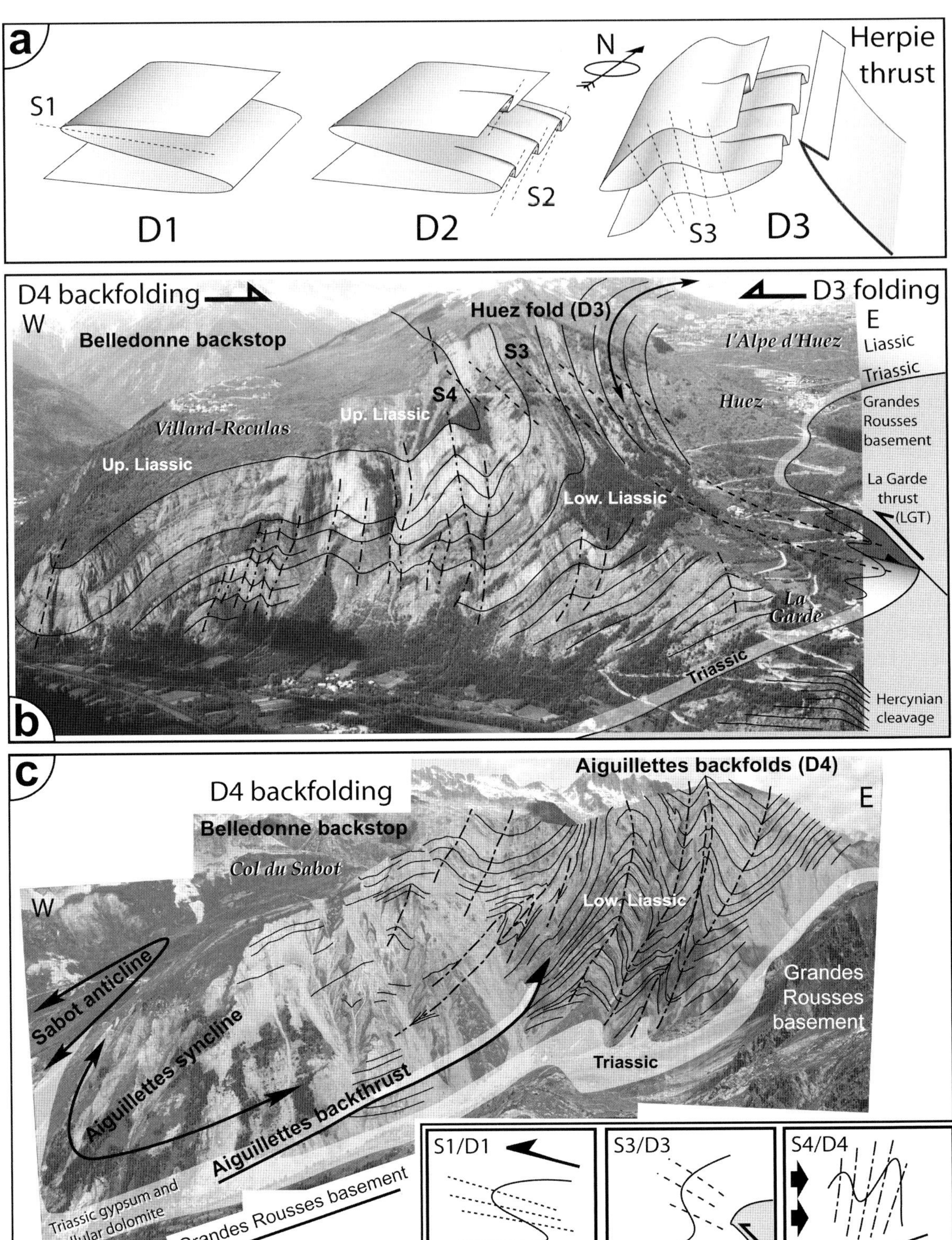

Fig. 7. Examples of multistage folding in the Liassic sedimentary cover of the study area, from north to south: a) Sketch of interferences observed in the southern part of the study area (Auris; site 13 of Fig. 6): the reverse limb of a D1 recumbent fold was deformed by D2 N-directed crenulation before D3 folding in the footwall of the Herpie W-directed thrust. b) The West-directed F3 Huez fold, surrounding the La Garde basement thrust and ramp anticline, is overprinted by D4 backward folding with kink geometry. The S4/S3 overprint can be observed along the Huez to Villard Reculas road. This area corresponds to sites 1, 2 and 3 of Fig. 6. c) Backward D4 fold and thrust structures in the Aiguillettes thick lower Liassic series, East from the Sabot pass. The lower Liassic series surrounding the Sabot anticline and onlapping the Belledonne backstop is very condensed (Fig. 4a), which implies the occurrence of a hidden Liassic palaeofault between the Sabot anticline and the Aiguillettes syncline.

4 Large-scale present geometry of the Hercynian basement-Mesozoic cover interface

4.1 modelling the basement-cover interface in 3D

As stated above, the present shape of the basement-cover interface represents the combined effects of Tethyan and Alpine deformations, because this surface was flat and horizontal 230 My ago before Jurassic extension (normal faulting, tilting) and Alpine shortening (folding, thrusting). This surface frequently has no Mesozoic sedimentary cover and is intersected by valleys, making it easy to model. Two 3D models of this basement-cover interface were made using DEMs from the French Institut Géographique National (BD-Alti database) and 1/50 000 scale geological maps from Bureau de Recherches Géologiques et Minières (BRGM), modified and georeferenced in NTF Lambert III (Vizille sheet, Barféty et al. 1972; La Grave sheet, Barbier et al. 1973; La Mure sheet, Barféty et al. 1988; St Christophe sheet, Barféty, Pêcher & coll. 1984). The areas covered by these two models correspond to Fig. 2 and Fig. 3, respectively. The first larger model covers the Dauphiné and northern Oisans (Fig. 5b; 37×37 km). The second model focuses on the Grandes Rousses-Rissiou area (Fig. 5c; 18×18 km). We used the following method:

- E–W topographic profiles were extracted every 350 m (model 1, Fig. 5b) or 250 m (model 2, Fig. 5c) from the DEM. Along each profile, the basement-cover interface was located from outcrops or its depth was estimated considering the thickness of the sedimentary cover. Gaps were preserved when such estimation was unreasonably constrained (e.g. central Bourg d'Oisans valley) or when a significant part of the basement was suspected to have been removed by erosion (e.g. Belledonne massif). In the latter case, the envelope surface of the highest preserved relief was taken as a minimum altitude value for the interface.
- The profiles were included in a 3D modelling box. The traces of interface from each profile were connected, giving patches which cover about 40% of the total area.
- Finally, an automatic interpolation was carried out (minimum curvature method, Surfer software) in order to fill the areas without data. These interpolated areas, lacking in geological constraints, are striped on Fig. 5b.

The models are presented in colour/altitude scale to show elevation (probably underestimated in the Belledonne massif), and lightened to enhance the relief. They can be rendered as a block-diagram (Fig. 5b) or map view (Fig. 5c).

4.2 Characteristic features of the 3D models: Tethyan vs. Alpine influence

The large model (Fig. 5b) shows relicts of the Jurassic half-grabens (Bourg d'Oisans and Mizoën basins) which are N–S oriented and which are best preserved from Alpine shorten-

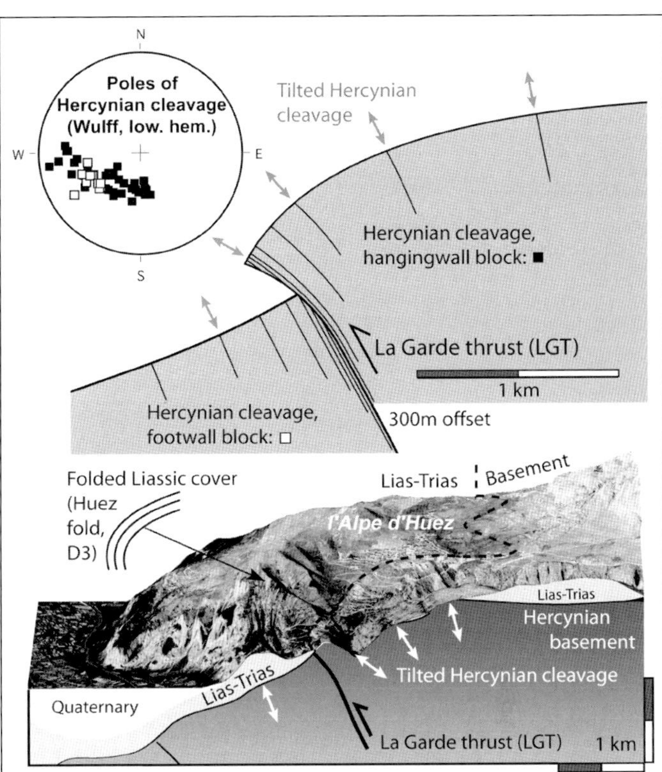

Fig. 8. Top: schematic cross-section of the La Garde thrust, with Alpine folding of Hercynian cleavage measured in the hanging wall block (stereogram). Bottom: block-diagram of the Huez fold-La Garde thrust area (satellite view over DEM).

ing in the northern and western parts of the modelled area. The strongest compressional deformation of the interface together with its highest altitude is observed in the southeastern part of the area, in the vicinity of the Pelvoux massif. Cross sections extracted from this model in different directions illustrate the variable relative influence of Tethyan and Alpine deformation in its present shape (Fig. 11): the tilted blocks geometry is visible on E–W and ENE–WSW profiles n° 1 and 3, whereas the E–W shortening component is more prominent to the south (profiles n° 2, 4 and 5). The latter produced basement uplift and forward short cuts, which are contractional faults slicing through the basement in the footwall of the normal faults. Along N–S profiles, trending parallel to the Jurassic fault blocks, the changes in elevation of the basement/cover interface are significant and show the occurrence of a N–S Alpine shortening component. The associated N-directed thrusts (Fig. 11) are D2 structures (Sue et al. 1999, and personal data). The compressional deformation is decreasing from the SE (A, profile 6) to the NW (C, profile 8). The A, B, C and D uplifted massifs lie in the hanging wall of both north-directed and west-directed thrusts, which suggests that this SE–NW gradient is due to partitioning between N–S and E–W shortening events. This is in good agreement with the field data presented above, and corresponds to

D2 and D3 events, respectively. The Grandes Rousses uplift (E) is mainly due to E–W shortening, a conjugate result of D3 and D4 events (§3.2). This model shows that the southern closure of the Mizoën Jurassic half-graben is due to Alpine shortening, namely westward and northward thrusting. In the Bourg d'Oisans half graben, stratigraphic data suggest that the offset of the Ornon Jurassic fault was decreasing southwards (Gidon 1980; Bas 1985; Barféty 1988). Our field investigations and modelling concentrated in the northern termination of this half-graben. The Grandes Rousses massif shows both a horizontal curvature on its western side and a gradual uplift towards the central part of the massif. This uplift occurs within a conjugate strike-slip system, that is between (i) the Alpe d'Huez fault zone (AHF, Fig. 5c) including several N120° sinistral faults visible along the road to l'Alpe d'Huez, and the Belle Etoile reverse fault cutting the Belledonne basement (BEF), and (ii) the Rissiou dextral fault (RF). The Bourg d'Oisans syncline splits northwards into two branches on each side of the Rissiou block (unit b, Fig. 5c). The Rissiou fault (RF) is not connected to the Jurassic Ornon fault (OF) that runs northwards in the footwall of the BEF. Here is actually the northern termination of the Bourg d'Oisans graben which therefore cannot be interpreted as a pull-apart basin as proposed by Grand et al. (1985).

5 Thermochronological constraints

5.1 Age of peak metamorphism

Alpine metamorphic peak conditions are estimated at about 320 °C to 370 °C and 2 to 3.5 kbar, corresponding to a maximum overburden of 7 to 12 km (see discussion in Crouzet et al. 2001a). Surprisingly, the age of peak metamorphism in the studied area is not clearly constrained. It can be estimated from some debatable data in the literature: the K/Ar method gives a reset apparent age of 24 Ma from biotites of the Belledonne massif (Demeulemeester et al 1986); the <2μm clay fraction of Liassic rocks dated using the same method gives 26.5 Ma (Nziengui 1993). The thermo-palaeomagnetic record starts at 24.1 Ma in the Bourg d'Oisans area (Crouzet et al. 2001a), thus, the metamorphic peak was reached earlier, and it may correspond to the thermal overprint at 26.5 Ma or may be even older.

5.2 Rotation tested using re-interpretation of palaeomagnetic data

The palaeomagnetic interpretation of Crouzet et al. (1996) was focused on tilting around a horizontal axis. The magnetic mineralogy, the procedure and treatments used are extensively

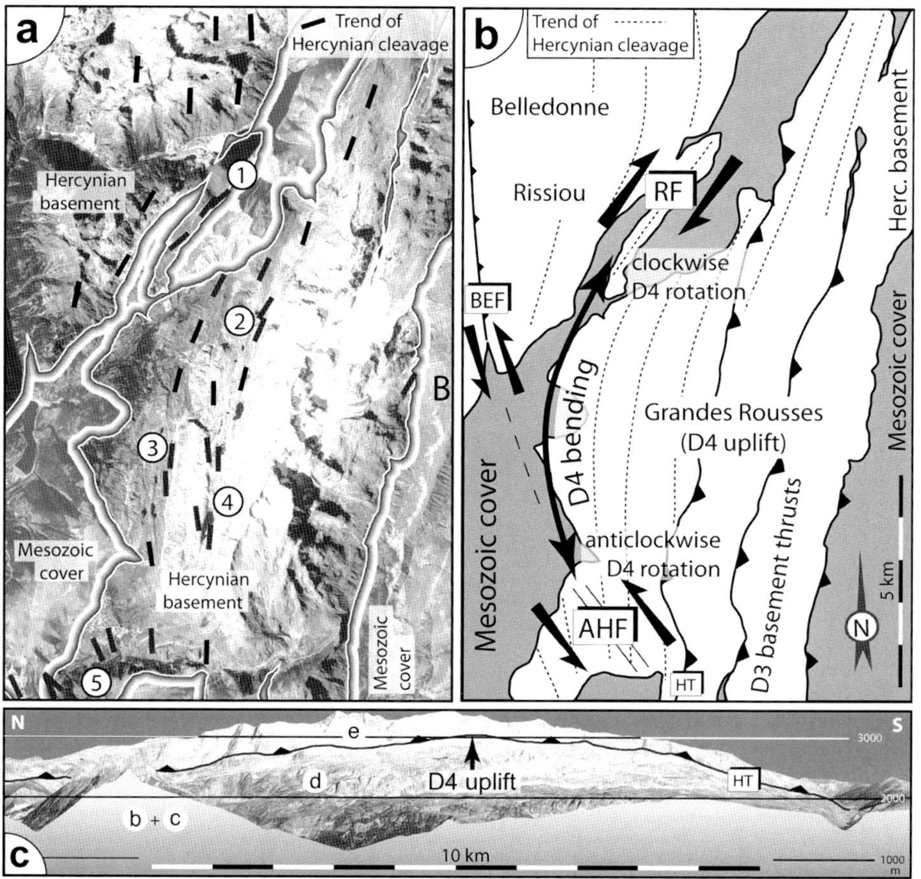

Fig. 9. Alpine fold trends and Hercynian grain between the Grandes Rousses and Belledonne massifs. a) Satellite view with location of five sites showing a gradual change in orientation of the late Hercynian cleavage from north to south (stereoplots in Fig. 10). Sites location: (1) Sabot pass, (2) from Couard pass to Jasse lake, (3) above Besson lake, (4) Blanc lake, (5) road to l'Alpe d'Huez. b) Simplified sketch map showing the trend of late Hercynian cleavage (white areas: basement outcrops), bended within two late Alpine conjugate strike-slip fault boundaries (RF: Rissiou fault; AHF: Alpe d'Huez fault; BEF: Belle Etoile fault). c) Synthetic, horizontal perspective view of the Grandes Rousses bulge from West, with no vertical exaggeration. Updoming is probably associated with bending during D4, as a compensation of westward displacement of the Grandes Rousses massif within the conjugate RF and AHF/BEF fault zones, coeval with the curvature of its western side (Fig. 9b). The D3 Herpie thrust (HT) is involved in bulging.

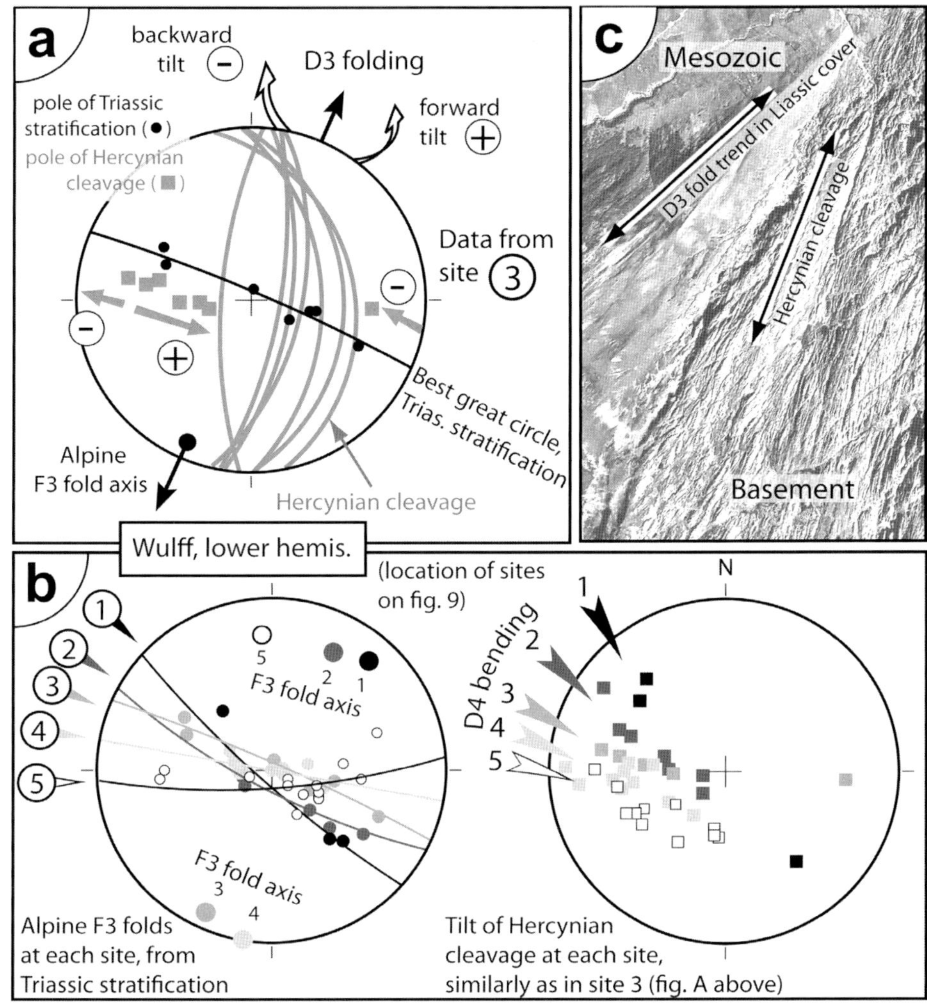

Fig. 10. a) Microstructures at site 3 (Fig. 9a): the plotted data show an Alpine (D3) folding of Hercynian cleavage, whose strike was initially different from the fold axis (initial obliquity of about 15°), so that the poles of Hercynian cleavage are scattered along small circles. b) Repeated observation from the five sites: both Hercynian cleavage and Alpine (D3) fold axes display a gradual clockwise rotation from south to north, approaching the Sabot wrench zone. Thus this is a late Alpine feature, assigned to D4. c) Detail aerial view of the northern Grandes Rousses area, showing about ~20° discordance between the strike of Hercynian cleavage and Alpine fold trends in the Liassic cover.

described in other papers (Crouzet et al. 1996, 1999 and 2001b). Here these data are re-interpreted and some new data are presented (Table 1) in order to test the hypothesis of recent rotation obtained from interpretation of our 3D model and structural data. The investigated sites are split into a northern and a southern group (location Fig. 3) corresponding to areas 1 and 3 of Crouzet et al. (1996), respectively. These sites were selected using the following criteria: (1) sites definitely tilted (Crouzet et al. 1996) and sites with k < 10 were not taken into consideration; (2) only sites from the lower Liassic formation were used (except area 4 of Crouzet et al. 1996), because they are closer to the basement/cover interface. The mean data for these two groups are the following:

– Southern group (Fig. 3; Tables 1 & 2, area 3): the mean direction (D = 350.0°, I = 52.5°, k = 76, α95 = 6.0) shows a significant apparent counter-clockwise rotation of 17.7° ± 8.5° (Table 2) with respect to the expected direction at ~25 Ma (D = 7.7 ± 3.9; I = 58.3 ± 2.7; a95 = 2.0) calculated after the apparent polar wander path for stable Europe (Besse and Courtillot 1991).

– Northern group (Fig. 3; Tables 1 & 2, area 1): the mean palaeomagnetic direction is D = 359.5°, I = 53.2°, k = 559, α95 = 2.6. It also shows a significant rotation of 8.2° ± 4.6° from the expected direction (Table 2).

Table 1. New paleomagnetic site characteristics. X, Y Lambert zone III coordinates; Z: altitude in m; N: number of samples used; Dec (Inc): Geographic declination (inclination) of the characteristic remanent magnetisation; k: precision parameter; α95: confidence cone for a 95% level; areas are defined according to Crouzet et al., (1996). A new area corresponding to Villard Notre Dame is introduced (sites BA, BB, BC).

Site	X	Y	Z	N	Dec	Inc	K	α95	area
47	891.34	3321.12	780	6	1.0	49.8	492.6	3.0	1
48+49	889.77	3320.00	740	13	3.9	56.4	13.8	11.6	1
50	887.31	3314.92	1400	4	169.7	−58.2	179.1	6.9	4
51	887.38	3315.08	1550	5	173.9	−24.9	4.8	38.8	4
52	894.94	3326.43	1740	10	358.9	52.5	52.8	6.7	1
53	888.00	3314.92	1445	7	190.7	−62.3	120.0	5.5	4
BA	891.80	3307.95	1465	5	187.3	−52.5	18.9	18.1	6
BB	891.81	3307.90	1500	5	193.5	−50.7	57.1	10.2	6
BC	891.83	3307.85	1480	7	191.1	−57.5	87.3	6.5	6

The differential rotation between the southern and northern groups is $9.5° \pm 8.6°$. This rotation is significant at 95% confidence. Therefore, the northern sites underwent a clockwise rotation with respect to the southern sites, which is consistent with the structural and kinematic observations discussed above. Additional sampling was carried out on the SW part of the study area:

– The mean directions for areas 4 (Oulles road) and 5 (Villard-Reymond), located in the western part of the Bourg d'Oisans syncline, do not show any significant rotation with respect to stable Europe since ~25 Ma. Thus D4 rotations occurred only in front of the Grandes Rousses bulge, as expected from the model discussed in § 4.2.
– Area 6 samples were taken in the lower Liassic beds in the footwall of the D2 Villard Notre Dame thrust (Desthieux & Vernet 1968), in order to test the influence of such structure on the magnetization direction. The samples from both limbs of a D2 fold show no significant rotation with respect to the expected direction of stable Europe (Table 2). This shows that northward thrusting associated with D2 occurred before the acquisition of magnetization.

5.3 Age of deformation stages

It has been demonstrated that the magnetization carried by pyrrhotite in the Bourg d'Oisans area is of thermal origin and was acquired during cooling in a temperature/time range between >320 °C at 24.1 Ma to around 220 °C at 20 Ma (Crouzet 1997; Crouzet et al. 1999, 2001a and b). As stated above, magnetization postdates D2 deformation. The fold test is also negative concerning the Huez F3 fold (Crouzet et al. 1996), therefore, magnetization also postdates D3. Therefore, deformations D1 to D3 must be older than 24.1 Ma. The Zijderveld diagrams of most of the thermally demagnetized samples are linear between the pyrrhotite Curie temperature of 320 °C and 150 °C. Consequently, no deformation seems to have occurred during this cooling interval. The 150 °C isotherm may have been reached at about 9 to 13 Ma according to Sabil (1995) and Crouzet et al. (1996). Rotation and tilting which have affected the palaeomagnetic record and which correspond to D4 deformation, correlated with the Grandes Rousses bulge uplift and bending, must then be younger than 13 Ma. The final exhumation is upper Miocene in age (9.3 ± 1 Ma and 5.8 ± 0.7 Ma AFT ages: Sabil 1995).

6 Discussion

6.1 Mesozoic fault pattern: a result of oblique extensional reactivation of Hercynian fabric

The structural and kinematic data demonstrate that the shape of the Bourg d'Oisans Jurassic basin must have been significantly distorted by Alpine events. The western sidewall, corresponding to the N–S Ornon boundary fault (OF) south of the Romanche valley, can be followed further north, in the footwall of the BEF (Eau Dolle syncline, Fig. 4a). Its N–S trend parallels the late Hercynian foliation, but conjugate strike-slip and/or reverse faulting on both sides of the Rissiou basement massif (area b, Fig. 3 & 4), has strongly distorted the pre-Alpine geometry of the northern Bourg d'Oisans basin. Evidence of syn-rift extensional deformation is found between the Rissiou and the northern Grandes Rousses massifs (Col du Sabot area c, Fig. 3 & 4): an east-dipping Liassic normal fault was located between unit c (Col du Sabot highly condensed lower Liassic series; Chevalier et al. 2003) and unit d (Aiguillettes thick series, Fig. 7c). This Sabot Liassic fault (SF, Fig. 4a) coincides with highly deformable basement (Sabot anticline schists, Fig. 4a & 5c), located between two late-Hercynian granites (southern termination of Sept Laux granite and Alpette granite), which favoured the distortion of Tethyan structures in this area. A tentative pre-Alpine restoration is given in Fig. 12a. Basement deformation in the Sabot unit (c) and the curvature of the western Grandes Rousses unit (d) were taken into consideration. E–W shortening of about 50% is in agreement with the deformation measured in the Jurassic series (Gratier & Vialon 1980). In this model, the Sabot paleofault (SF) died out southwards beneath the Liassic cover of the Bourg d'Oisans basin, and extension was then transferred to the Ornon paleofault (OF) further west (Fig. 12b). The syn-rift (early Liassic) depocentres were then aligned from the NE (Aiguillettes depocentre) to the SW (Bourg d'Oisans depocentre, and further on the Beaumont basin: Bas, 1988). Because there is evidence that the Sabot, Ornon and La Mure palaeofaults developed by extensional reactiva-

Table 2. Mean palaeomagnetic directions for the different areas. Dec (Inc): Geographic declination (inclination) of the area, calculated using sites with k > 10; k: precision parameter; $\alpha 95$: confidence cone for a 95% level; areas are defined according to Crouzet et al. (1996); N: number of sites used; R: rotation in respect to the stable Europe pole at 25 Ma calculated from Besse and Courtillot (1991); dR: error on the rotation, calculated after Demarest (1983). While sites from areas 3 to 6 are mainly reverse, the mean directions are given as a normal polarity in order to allow comparison between different areas. For areas 1 to 5, the mean is calculated using data of Crouzet et al. (1996) and data presented in Table 1. Area 6 is newly defined in this paper.

Area	Name	Dec	Inc	k	A95	N	R	dR
1	Allemond-Oz-Vaujany	359.5	53.2	559	2.6	7	–8.2	4.6
3	Huez	350.0	52.5	76	6.0	9	–17.7	8.5
4	Oulles	3.6	56.5	88	4.1	15	–4.1	6.7
5	Villard Reymond	8.6	55.3	38	3.6	7	0.9	5.9
6	Villard Notre Dame	10.7	53.6	410.4	6.1	3	3.0	8.8

tion of N–S Hercynian foliation, we propose that this NE–SW distribution of depocenters is due to en-échelon pattern of the N–S master faults, and we interpret this feature as a response of N–S Hercynian foliation to obliquely oriented Jurassic extension (NW–SE; Lemoine et al. 1989). Such a fault relay geometry is observed in other natural examples (Cornfield & Sharp 2000) and in analogue experiments (McClay & White 1995).

6.2 Regional restoration: major half-grabens in foliated basement

A larger scale restored cross-section perpendicular to the Jurassic tilted blocks and to the Hercynian foliation is proposed in Fig. 13. The eastern half of this reconstruction is significantly different from previously published models (Lemoine et al. 1986; Gillcrist et al. 1987; Trift & de Graciansky 1988; Chevalier 2002). Only two major boundary faults are involved, the La Mure and the Ornon faults. The syn-rift sequence is clearly pinching eastwards on the Grandes Rousses block, but further east the two limbs of the Mizoën syncline are not regarded as conjugate normal palaeofaults because they both display condensed, but conformable syn-rift sediments. Two small half-grabens are now buried beneath the Mizoën syncline and are exposed further south (see § 6.4, Fig. 14). We found no evidence to support the interpretation of the Meije and Combeynot massifs as syn-rift tilted blocks (Lemoine et al. 1986; Gillcrist et al. 1987), because all the areas east of the Grandes Rousses block and around the Pelvoux massif bear condensed syn-rift lower Liassic series (Barféty 1988; Corna et al. 1997; Chevalier 2002). Thus the eastern Dauphiné zone represented a large plateau without major fault blocks. This suggests that the offsets of the western Dauphiné faults (La Mure and Ornon) were transferred at depth to a flat detachment underlying the whole eastern Dauphiné area (Fig. 13), as proposed in different models (Gillcrist et al. 1987; Chevalier 2002). Considering the width of the western tilted blocks (~15 km), this detachment may have occurred in the middle crust, either close to the brittle-ductile transition, or as a reactivation of a Hercynian low-angle thrust.

According to this reconstruction, Jurassic extension was accommodated in different ways in the western and eastern parts of the Dauphiné zone, corresponding to different types of basement. In the western area (central Oisans), the metamorphic units with synkinematic intrusions of early Carboniferous granites are extensively foliated. This regionally consistent N–S striking, east-dipping foliation was conveniently oriented for extensional reactivation (negative inversion), and the major fault blocks developed here. In contrast, the eastern basement represented by the Pelvoux massif shows extensive late Carboniferous, late-orogenic granitic plutons (300 Ma; Guerrot & Debon 2000) intruded at shallow depth. There, the early syn-rift series are always condensed, indicating that no major fault blocks developed. Further southeast, the Hercynian fabric is oriented NW–SE (south-eastern Pelvoux massif, Argentera massif), similarly to the syn-rift Jurassic basins in the southern French Alps (Dardeau 1983). This also supports the interpretation that Hercynian foliation was a primary controlling factor in the development of tilted blocks.

6.3 Multistage shortening: A proposed tectonic and geodynamic scenario

The following sequence of events is proposed to explain the data (Fig. 15a):

Earliest compressional structures: a consequence of Pyrenean-Provence orogeny?

The Mesozoic cover of central Oisans was locally affected by top southwest transport and recumbent folding (D1), locally re-oriented along the Ornon Jurassic fault buttress. This early compressional event is consistent with the SW-directed basement-thrusting event well documented in the southern Pelvoux area (Gidon 1979; Ford 1996; Lazarre et al. 1996). There, this event pre-dates the deposition of late Eocene limestones. Thus this early phase D1 must also predate the emplacement of Alpine nappes and be older than the peak of low grade metamorphism in the Dauphiné.

Early collision stage: Northward to NNW-ward thrusting (Early Oligocene):

This event is hardly visible in the study area, but is a key feature for understanding the kinematics of the Western Alpine external zone. The Alpine realm suffered important N–S contraction in response to Adria-Europe collision in late Eocene-early Oligocene times (Lacassin 1989; Schmid & Kissling 2000; Dèzes et al. 2004). Early Alpine northward to NW-ward thrusting and folding is documented in many places throughout the Alps: in the Helvetic zone (Dietrich & Durney 1986; Ramsay 1989; Wildi & Huggenberger 1993), in the Internal Zones (Caby 1973; Maury & Ricou 1983; Merle & Brun 1984; Merle et al. 1989; Platt et al. 1989b; Schmid et al 1996; Ceriani & Schmid 2004) and in the Austro-Alpine realm (Froitzheim et al. 1994). N-directed thrusting is also known in the Pelvoux realm (Barbier 1956; Desthieux & Vernet 1968; Bartoli et al. 1974; Gidon 1979; Gillcrist 1988; Barféty & Gidon 1990; Pêcher et al. 1992; Ford 1996; Sue et al. 1998; Gidon 1999) but it has been generally confused with Pyrenean-Provence events. We propose that this D2 episode, which predates the onset of westward escape in the Western Alps but involves the late Eocene sediments, occurred in early Oligocene times, consistently with our palaeomagnetic record (§ 5.3). We also propose that this event is linked with the emplacement of the exotic early Alpine Embrunais-Ubaye nappes to the SE of the study area (2, Fig. 1). There, the Eocene flysch was buried to more than 7 km depth (south of the Pelvoux massif; Waibel 1990) and suffered local isoclinal folding in the footwall of these nappes (Prapic area, Orcières geological sheet; Debelmas et al. 1980). According to Kerckhove et al. (1978), the associated tectonic transport direction was towards the NW. Both these D2 structures and the nappes are overprinted by D3 SW-directed tectonic transport. The Penninic Romand and Chablais Prealps nappes were also emplaced in early Oligocene times towards the N to NW onto the external European units (Mosar et al. 1996; Bagnoud et al. 1998).

W-directed basement stacking and early stage of exhumation (Oligocene, >24 Ma):

A sharp kinematic shift must have occurred between D2 and D3, since all the post-D2 deformations correspond to westward propagation of shortening. The eastern part of the Grandes Rousses area is affected by D3 high-angle basement thrusts that reactivate the N–S trending late Hercynian lithological boundaries and cleavage (e.g. the Herpie thrust, HT, Fig. 4). D3 shortening is properly oriented to produce both short cuts of the Tethyan horsts and pinching of the syn-rift basins along the Belledonne-Taillefer, Jurassic-inherited buttress (Dumont et al. 1996). Greater amounts of inversion are observed in the Helvetics (Morcles nappe, Badertscher & Burkhard 1998). In the study area, the offset of basement thrusts is generally less than 1 km. They form ramp anticline features in the shallow part of the basement and their displacement is transferred to large-scale folds in the lower Liassic cover (Huez fold, Fig. 7b). This deformation still occurred at elevated temperatures, probably >320 °C, because it predates the pyrrhotite magnetization (Crouzet et al. 1996 and 2001a). Thus D3 represents an early stage of the exhumation history, soon after the metamorphic peak which may have occurred in Oisans aroud 26 Ma or earlier (§5). The latter is very likely related with the westward emplacement of the Internal Nappes stack onto the External zone along the Briançonnais Frontal Thrust (*sensu* Sue & Tricart 2002; D3 activation of Roselend thrust according Ceriani et al. 2001 or D2 stage of Fügenschuh & Schmid 2003) which occurred in early and middle Oligocene times, between 33 Ma and 27.5 Ma (Tricart et al. 2000; Ceriani et al. 2001; Tricart 2004). The D3 westward tectonic transport of the Internal Nappes was strongly influenced by the Pelvoux basement bulge, previously thickened during D1 and D2 episodes (Gamond 1980; Tricart 1980). The propagation of this tectonic overload is also recorded in the Mt Blanc and the Argentera external massifs, which were buried to over 14 km at 22 Ma (Leloup et al. 2005; Corsini et al. 2004).

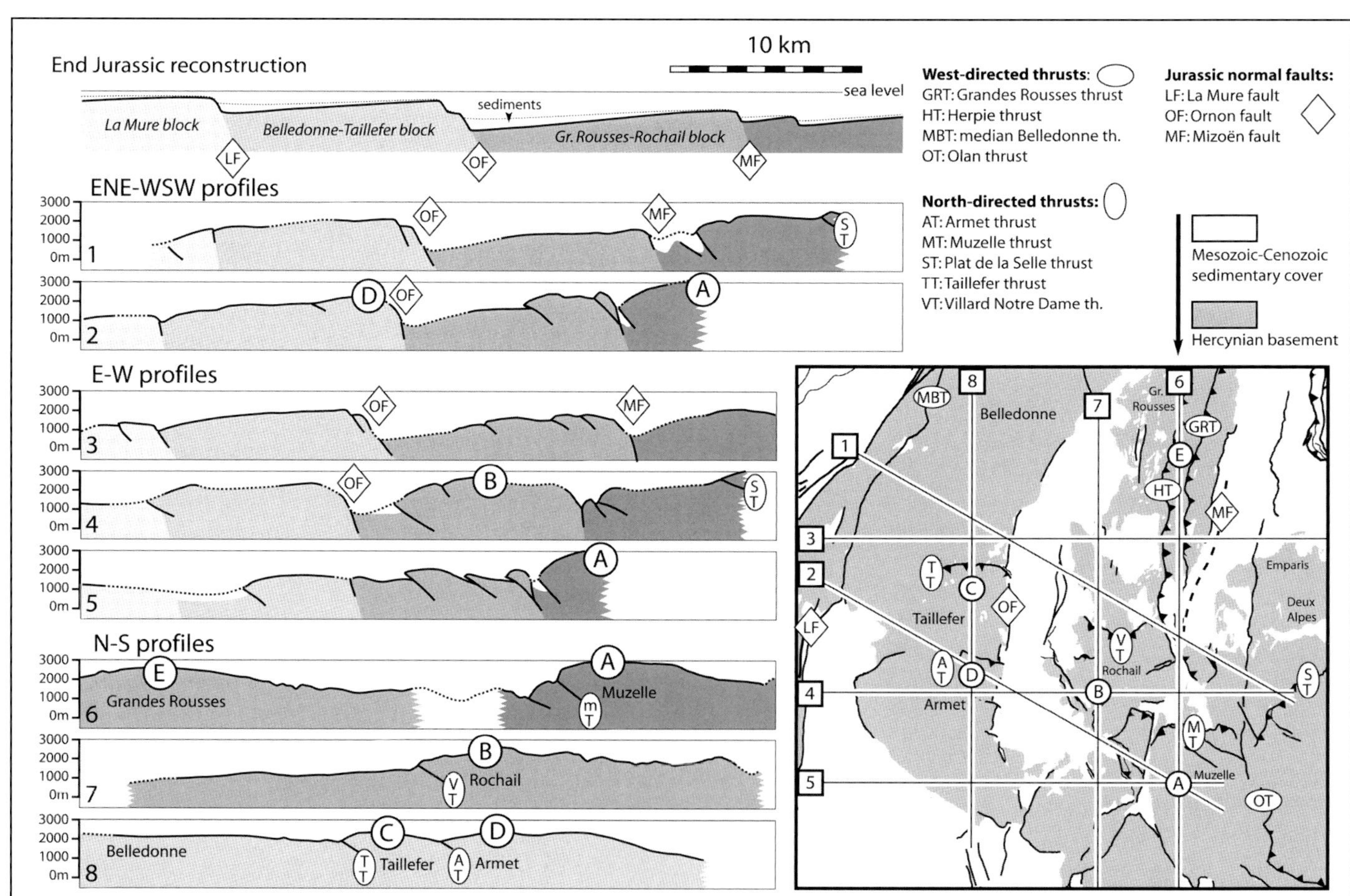

Fig. 11. Vertical cross-sections of the basement-cover interface extracted from the 3D model of Fig. 5b, with location shown in the sketch map corresponding to Fig. 2. On top is shown a reconstructed ENE–WSW profile at the end of Jurassic rifting.

Buttressing and forward propagation of uplift
(Middle-Late Miocene, <13 Ma):

According Crouzet et al. (1996 and 2001a), no deformation occurred during the main cooling interval from 320 °C to 200 °C–
150 °C. Structural and palaeomagnetic records indicate that
post-D3 clockwise rotation and strike-slip faulting occurred between the Grandes Rousses and Belledonne basement blocks.
D4 also produced long wavelength uplift of the Grandes Rousses massif, later than 10 Ma. The large scale anticlinal shape
of this massif (Vernet 1965) formed at that time, as shown by
opposite tilting of the palaeomagnetic vectors in the Liassic
sedimentary cover (West side: Crouzet et al. 1996; East side:
Lamarche et al. 1988). D4 in the Bourg d'Oisans and Grandes
Rousses areas was enhanced by the uplift of the Belledonne-
Taillefer buttress, which increased shortening of sedimentary
series with local backward folding. This recent exhumation of
the Belledonne massif is documented by:

–   The lack of basement rocks from the Belledonne massif in
    the Middle Miocene conglomerates of the proximal delta
    near Grenoble (Bocquet 1966).
–   The upper Middle Miocene to Pliocene range of apatite fis-
    sion-track data from the eastern Belledonne near the study
    area (14.3 ± 2.0 to 4.2 ± 0.5 Ma; Sabil 1995).

The uplift and subsequent increasing buttress effect to the east
of the Belledonne-Taillefer massif is probably a consequence
of the late Miocene transport of these external massifs above
a crustal ramp, which accommodated the westward propagation of the Alpine orogen (Mugnier et al. 1990; Deville &
Chauvières 2000).

6.4 Interpretation of D2 shortening: an early, N-directed stage of Alpine collision

The Dauphiné crystalline basement in the study area is shortened and uplifted by several E–W, southward dipping high-
angle thrusts (Fig. 2, Fig. 11). Most of the highest peaks in the
region are in the hanging wall of such structures. Our kinematic
data from the Mesozoic cover indicate that this remarkable
event occurred between D1 and D3, and the Eocene flysch is
involved. Further south in the Lanchatra valley (Fig. 14, location Fig. 2), the Liassic strata in the footwall of the Muzelle E–
W thrust (MT, Fig. 11 & 14) are affected by both E–W and N–S
shortening. Northward recumbent folding below this thrust
affects an earlier cleavage and is overprinted by a younger
one, namely D1 and D3, respectively. Thus MT is a D2 feature.
Other D2 thrusts (ST, MT, VT, TT; Fig. 11) bear evidence for
dextral strike-slip reactivation and/or deformation by E–W D3
shortening, and several of them crosscut older compressional
structures. For example (Fig. 14), the hanging wall block above
MT transports a huge D1 granitic-core anticline and previously
shortened Tethyan tilted blocks are cut in its footwall.

Such perpendicular shifts in orientation of shortening were
already noticed by Gillcrist (1988) and Sue et al. (1998). Together with Ford (1996) and Gidon (1999), they regarded all
the N–S shortening features as related to pre-Late Eocene Pyrenean-Provence orogeny. Bravard & Gidon (1979) proposed
that the north-directed thrust system of the northeastern Pelvoux was active during Late Eocene, but the synsedimentary
tectonic evidence is highly questionable according Butler (1992).
In our opinion, two episodes with a N–S shortening component
must be distinguished: a pre-Late Eocene event, so-called "Pyrenean-Provence", and an Early Oligocene, early Alpine event.

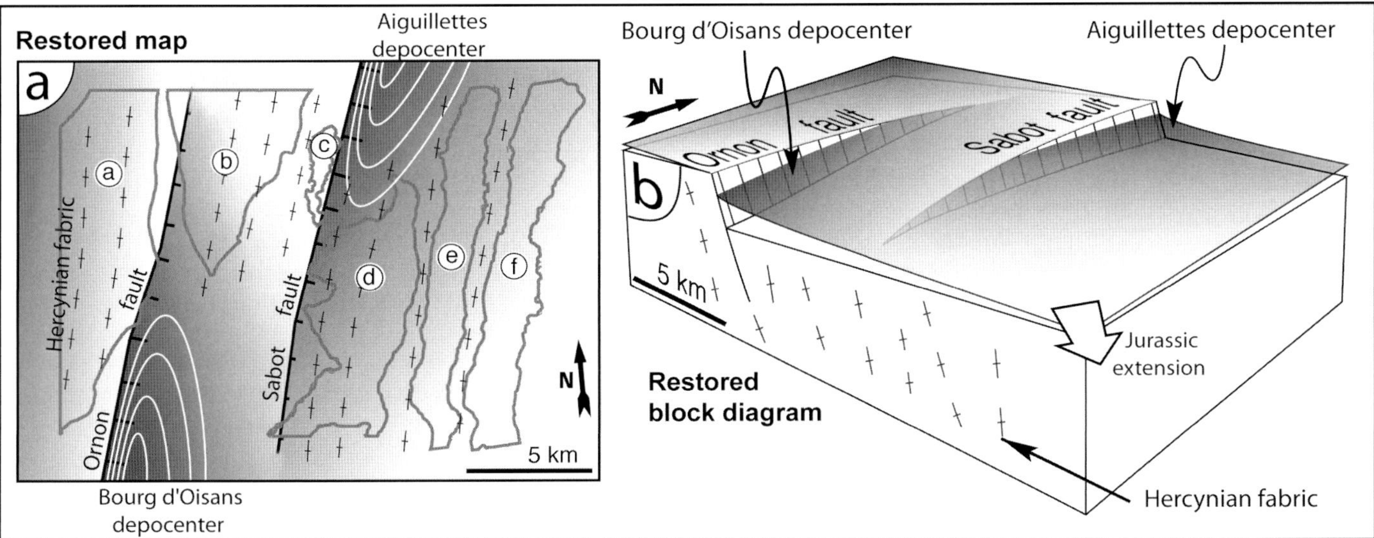

Fig. 12. The main structural units (a) to (f) corresponding to Fig. 3 & 5c are restored in a pre-Alpine position based on semi-quantitative estimation of Alpine deformation (left). The superposition of the main Liassic faults and depocentres over this restored map leads to reconstruction b, right: the faults were subparallel to the late Hercynian fabric and close to N–S oriented. Extension was transferred laterally from the Sabot fault to the Ornon fault. This could be due to the initial obliquity between the involved Hercynian fabric and NW–SE Jurassic extension.

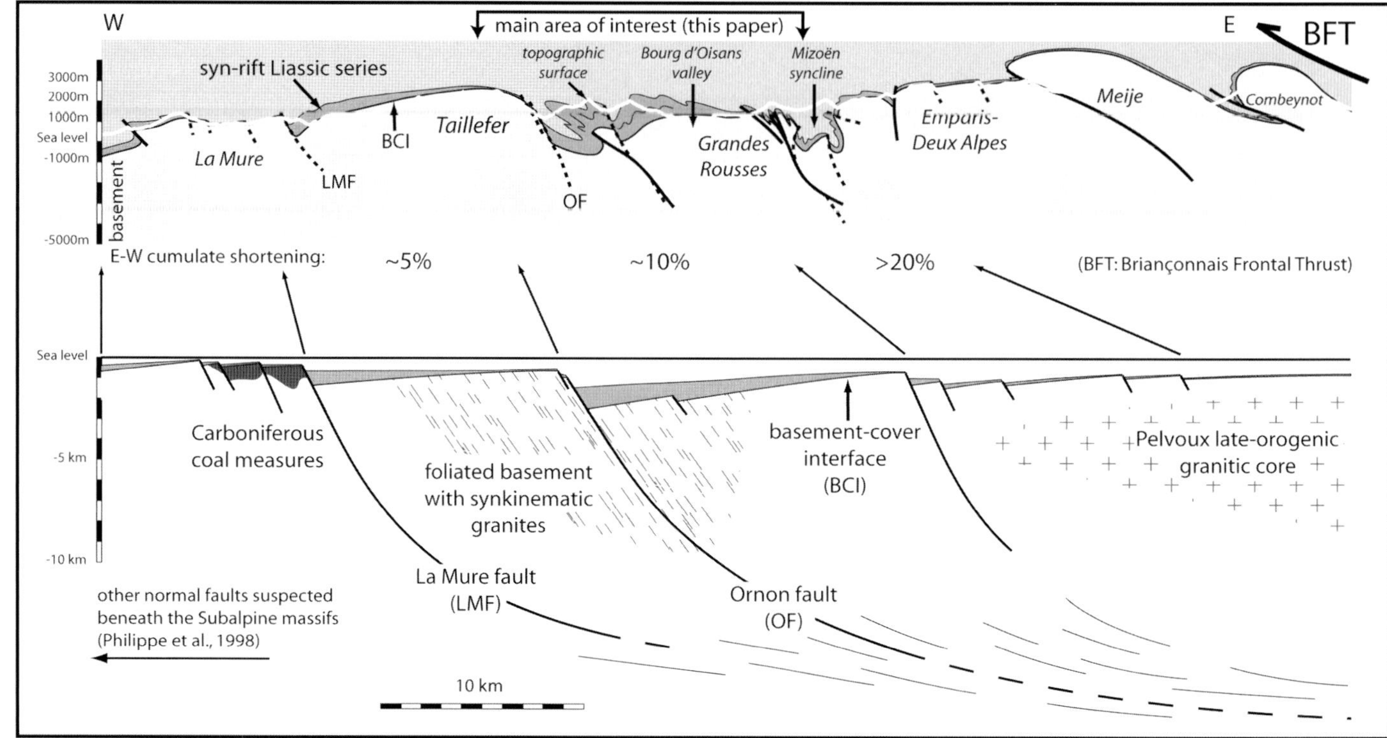

Fig. 13. E–W cross-section of the whole Dauphiné zone, with proposed reconstruction during the early Jurassic, based on the distribution and facies of early syn-rift sediments. Large-scale blocks tilting occurred only in the western part, as a suspected consequence of specific basement lithologies and postulated connection of the major normal faults with a mid-crustal detachment level.

The former (D1) is dominantly S-directed and is sealed by the Priabonian beds in southern Pelvoux (Gidon 1979; Ford 1996; Gupta 1997). The latter (D2) overprints D1 features in many places, it is north to NW-directed and is nowhere sealed by the Late Eocene sediments. It clearly deformed the Nummulitic formations both in the eastern Pelvoux (Bravard 1982, and personal data) and in the southern Pelvoux area (Gidon & Pairis 1981).

6.5 Basement folding

Despite metamorphic conditions lower than greenschist facies, the Dauphiné basement accommodated a significant amount of E–W shortening by large-scale folding, as shown by the contorted shape of the initially flat basement-cover interface in our 3D models. This feature has been considered of minor importance in some thin-skinned type models that emphasize brittle deformation (Butler 1984; Platt 1984), but kilometric-scale folds occur in the hanging wall of high-angle basement ramps, and even granitic blocks show large wavelength folds: the Grandes Rousses massif is a 10 km-wide basement anticline, and the Rochail massif is a granitic dome in the hanging wall of both D2 and D3 thrusts. The Pelvoux massif itself is a 5 km-high basement bulge created by D1 and D2 N–S shortening, which suffered D3 underthrusting below the westward-propagating internal nappes (Gamond 1980; Tricart 1980). Since the conditions for ductility of granitic minerals were not reached in this region (§5.1), the «semi-ductile» deformation of basement must involve the following processes

(i) pressure-solution for a prolonged period; and/or (ii) simple shear reactivation of the Hercynian foliation when it was optimally oriented (e.g. eastward dip favoured reverse reactivation during E–W shortening). The features observed here represent an incipient stage of basement folding much more developed in recumbent structures of some other external crystalline massifs (Aar-Gotthard massif, Pfiffner et al. 1990).

6.6 Integration of our observations with the Alpine collisional evolution

Two different nappes stacks are found to the S-SE and to the east of the Dauphiné basement massifs (Fig. 15b): (1) the Embrunais-Ubaye nappes which propagated north-westwards during the Early Oligocene (Kerckhove et al. 1978) and which experienced a 90° anticlockwise rotation in transport direction (Merle & Brun 1984), and (2) the Internal nappes transported westwards since the Early Oligocene (Ceriani et al. 2001; Tricart 2004). The latter clearly crosscuts the former south of the city of Briançon, and a sharp kinematic change is needed to explain this regional interference. We propose that this change is recorded in the footwall of both Alpine nappe systems (the Dauphiné basement massifs and sedimentary cover) by the D2/D3 shift. This shift occurred in the lowermost Oligocene according to some preliminary dating of synkinematic minerals in the eastern Pelvoux basement. D3 buried the study area to ~10 km during Early Oligocene times. Palaeomagnetic data

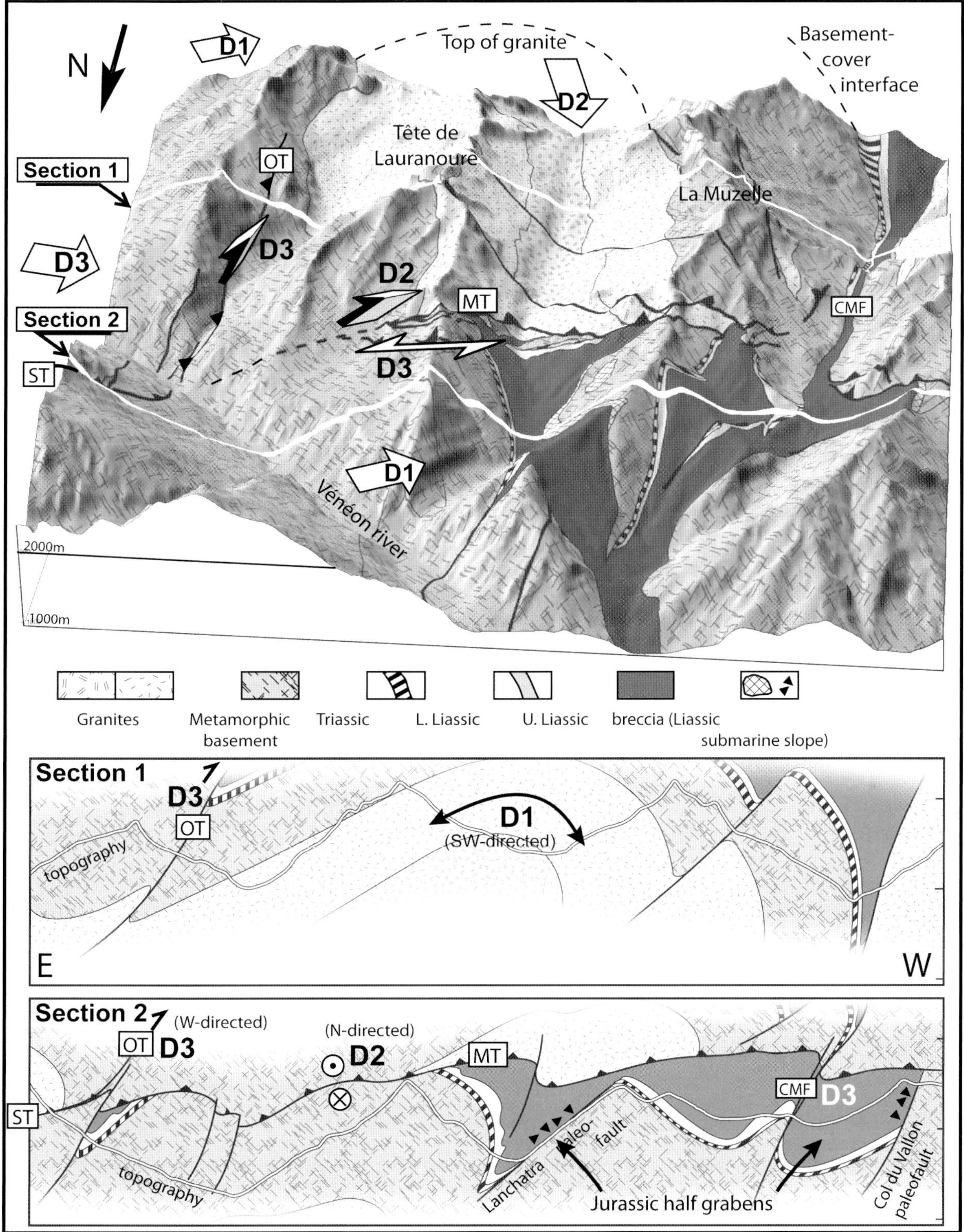

Fig. 14. Map scale evidence for deformation D2 in the Vénéon valley, to the south of the study area. Top: 3D geological map of the Tête de Lauranoure-Muzelle area, with no vertical exaggeration; view from the North with 45° incidence. Southward dipping thrusts (ST: Selle thrust; MT: Muzelle thrust) look roughly aligned and rectilinear. They clearly cut a huge D1 basement anticline uplifted in the hanging wall block (bottom, section 1). They also cut two small-scale Jurassic half-grabens still visible in the footwall (bottom, section 2), and which had been previously pinched as already stated by Gillcrist (1988). So deformation D2 postdates an earlier, D1 basement involved shortening event. ST and MT are in turn overprinted by a younger E–W shortening event (D3), which caused dextral strike-slip reactivation and crosscutting by more recent Olan thrust (OT) and Col de la Muzelle fault (CMF).

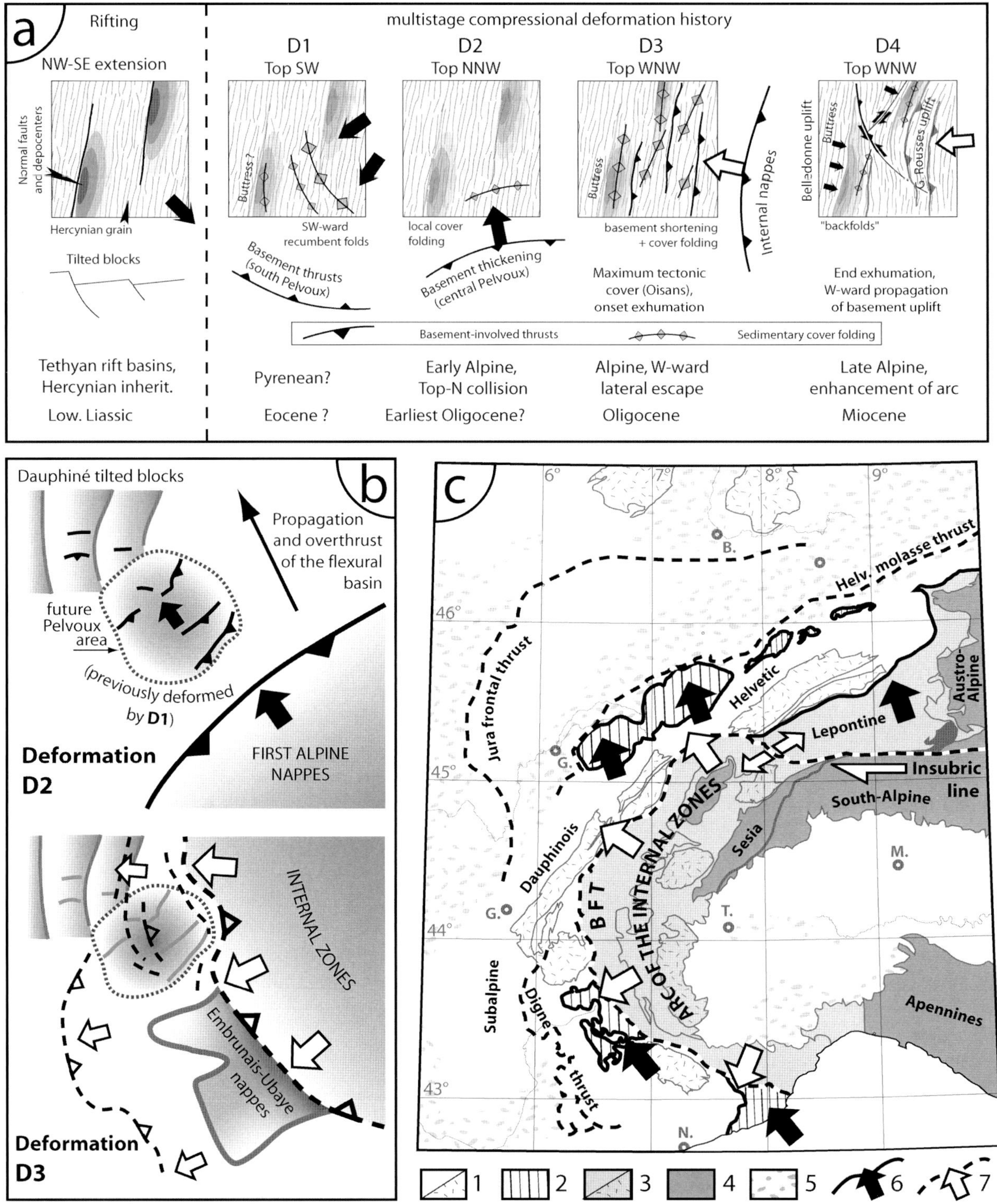

Fig. 15. a) Summary sketch of local deformation history. b) Regional interpretation including the Pelvoux area: an early Alpine nappe build-up was emplaced from the SE soon after the closure of the Late Eocene flexural basin; D2 shortening features (black thrusts) occurred in the footwall of this stack, represented by the Embrunais-Ubaye klippen. The latter was later on crosscut by the Internal Nappes stack propagating westwards in the study area (D3, dashed thrusts). c) Proposed integration in the Western Alps kinematics: D3 corresponds to the onset of lateral extrusion which gave birth to the arc of the Internal zones, linked with dextral motion along the Insubric line. 1: External zone, Mesozoic cover (white) and external basement massifs (dotted); 2: Early Alpine exotic units, from N to S: Prealpine nappes, Embrunais-Ubaye nappes & Ligurian flyschs (first nappes stack, D2); 3: Internal zones (European margin and Ligurian Tethys derived units; internal basement massifs dotted); 4: Apulian-Adriatic derived units; 5: Tertiary foreland basins; 6: Early Alpine thrusts, related with the N–NW-directed collision phase (black arrows, D2); 7: Thrusts and faults related with the Oligocene westward extrusion phase (white arrows, D3); Briançonnais Frontal Thrust (BFT) dashed.

(this paper and Crouzet et al., 2001a) recorded the initiation of unroofing in the footwall of the Internal zones during early Late Oligocene times, consistent with extensional reactivation of the Briançonnais Frontal Thrust (Tricart et al. 2000). Exhumation occurred with little deformation in the central Dauphiné until the lower Miocene, coeval with westward propagation of subsidence and deformation in the foreland molasse basins (Ménard 1988; Guellec et al. 1990; Deville et al. 1994). During late Miocene times, the exhumation of the Belledonne massif resulted in the study area in backward buttressing, renewed shortening at shallow structural levels (D4) and final exhumation and bending of the Grandes Rousses massif. Considering the whole Alpine realm (Fig. 15c), the major D2/D3 kinematic change can be correlated with the shift from N-directed (orogen-perpendicular) crustal stacking to westward (orogen-parallel) extrusion, which developed the western Alpine arc (Schmid & Kissling 2000; Ford et al. 2006). The latter is marked since the Early Oligocene in the central Alps by transcurrent motion along the Insubric line (Schmid et al. 1989; Müller et al. 2001). In the Western Alps, this late collisional extrusion regime has extensively overprinted the initial build-up, but interferences with earlier stages are still visible in the external zone.

7 Conclusions

Basement behaviour and Hercynian heritage:

The different Alpine compressive stages reactivated the late-Hercynian foliation depending on their relative orientation. The strongest reactivation was caused by maximum stress axis subperpendicular to the Hercynian grain: In the northern part of the Oisans region, the early Oligocene WNW-ward transport of Penninic nappes (D3) produced reverse faulting involving reactivation of the N–S trending, steeply east-dipping Hercynian cleavage, and subsequent basement thickening. D3 maximum stress axis was probably not strictly perpendicular to this cleavage, so that the coeval structures are trending ~20° W with respect to the Hercynian fabric. Further southeast, D3 inversion is more strike-slip because the orientation of Hercynian fabric changes to NW–SE (southern Pelvoux and Argentera massifs). Large scale Hercynian structures such as late-orogenic pinched synclines and magmatic contacts (boundaries of granitic plutons) have also been reactivated during the Alpine orogeny. The late Hercynian granitoid massifs are mainly located in the hanging wall of north-directed D2 thrusts in the central Oisans region. The influence of granitoids distribution on the development of Alpine structures is also seen in the Aar massif (Pfiffner et al. 1990).

Tethyan structures and inversion:

Extensional reactivation of the Hercynian grain detected in the foreland (Roure & Colletta 1996) was somehow underestimated in the Western Alps. Our reconstruction suggests that the syn-rift depocenters have been distributed en-échelon as a response to the obliquity of Jurassic extension (NW–SE) in the N–S foliated basement. The Alpine compressional reactivation of Tethyan fabric consists of incipient short cuts in the shallow basement, mostly important during the D3 deformation stage. There is little evidence for compressional reactivation of first-order normal faults which were probably too steep (since they developed over the steeply dipping late Hercynian cleavage).

Multistage shortening: lateral escape since middle Oligocene:

The present structure results from four individual events separated by drastic changes in local shortening direction, which took place over a long time span. Three of them occurred before Neogene times: D1, D2 and D3, respectively directed SW-wards, N-wards and WNW-wards. The interpretation of the earliest event D1 is still speculative, but in the eastern and southern Pelvoux areas similar deformations occurred before Priabonian sedimentation, that is, before Alpine flexural subsidence. Thus a link with Pyrenean deformations in Provence is suspected. By contrast, we propose that D2 corresponds to an early stage of north-directed Alpine collision following the overthrusting of the Paleogene flexural basin during earliest Oligocene. The D3 stage occurred after a sharp kinematic change which probably marks the onset of westward extrusion in the W-Alpine arc.

Acknowledgements

Financial support from Total company under supervision of Jean-Claude Chermette is gratefully acknowledged. Interpretations were improved by discussions in the framework of GéoFrance3D Alpes. Structural data were processed using "Stem" software from Arnaud Pêcher. Gilles Ménard is acknowledged for discussions and help during palaeomagnetic sampling, and Pierre Rochette for giving access to one of us (CC) to the palaeo¬magnetic laboratory of the University of Aix Marseille III. One of us (JDC) would like to warmly acknowledge the Swiss National Science Foundation for several grants. The manuscript was improved by discussions with Stéphane Guillot and english reviewed by Matthias Bernet. Michel Guiraud and Mark Handy are gratefully acknowledged for their fruitful and thorough review.

REFERENCES

Aprahamian, J. 1988: Cartographie du métamorphisme faible à très faible dans les Alpes françaises externes par l'utilisation de la cristallinité de l'Illite. Geodynamica Acta 2, 25–32.
Badertscher, N. & Burkhard, M. 1998: Inversion alpine du graben Permo-Carbonifère de Salvan-Dorénaz et sa relation avec le chevauchement de la nappe de Morcles sus-jacente. Eclogae Geologicae Helvetiae 91, 359–373.
Bagnoud, A., Wernli, R. & Sartori, M. 1998: Découverte de foraminifères planctoniques paléogènes dans la zone de Sion-Courmayeur à Sion (Valais, Suisse). Eclogae Geologicae Helvetiae 91, 421–429.
Barbier, R. 1956: L'importance de la tectonique anténummulitique dans la zone ultradauphinoise au nord du Pelvoux: la chaîne Arvinche. Bulletin de la Société géologique de France 6, 355–370.
Barbier, R., Barféty, J.C., Bocquet, A., Bordet, P., Le Fort, P., Meloux, J., Mouterde, R., Pêcher, A. & Petiteville, M. 1973: Carte géologique de la France (1/50000), feuille La Grave (798), Bureau de Recherches géologiques et minières, Orléans.
Barféty, J.C. 1988: Le Jurassique dauphinois entre Durance et Rhône. Etude stratigraphique et géodynamique. Documents du Bureau de Recherches géologiques et minières 131, 655 pp.

Barféty, J.C., Bordet, P., Carme, F., Debelmas, J., Meloux, M., Montjuvent, G., Mouterde, R. & Sarrot-Reynauld, J. 1972: Carte géologique de la France (1/50000), feuille Vizille (797), Bureau de Recherches géologiques et minières, Orléans.

Barféty, J.C. & Gidon, M. 1983: La stratigraphie et la structure de la couverture dauphinoise au Sud de Bourg d'Oisans. Leurs relations avec les déformations synsédimentaires jurassiques. Géologie Alpine 59, 5–32.

Barféty, J.C. & Gidon, M. 1984: Un exemple de sédimentation sur un abrupt de faille fossile: le Lias du versant Est du massif du Taillefer (Zone dauphinoise, Alpes occidentales). Revue de Géographie Physique et Géologie Dynamique 25, 267–276.

Barféty, J.-C. & Gidon, M. 1990: La tectonique alpine du massif cristallin du Taillefer (Alpes occidentales françaises): découverte de chevauchements vers le nord. Géologie Alpine 66, 1–9.

Barféty, J.-C., Gidon, M., Lemoine, M. & Mouterde, R. 1979: Tectonique syn-sédimentaire liasique dans les massifs cristallins de la zone externe des Alpes occidentales françaises: la faille du col d'Ornon. Comptes Rendus de l'Académie des Sciences Paris 289, 1207–1210.

Barféty, J.C., Montjuvent, G., Pêcher, A. & Carme, F. 1988: Carte géologique de la France (1/50000), feuille La Mure (821), Bureau de Recherches géologiques et minières, Orléans.

Barféty, J.C., Pêcher, A. & coll. 1984: Carte géologique de la France (1/50000), feuille St. Christophe (822), Bureau de Recherches géologiques et minières, Orléans.

Bartoli, F., Pêcher, A. & Vialon, P. 1974: Le chevauchement Meije-Muzelle et la répartition des domaines structuraux alpins du massif de l'Oisans, partie nord du Haut-Dauphiné cristallin. Géologie Alpine 50, 17–26.

Bas, T. 1985: Caractéristiques du rifting liasique dans un secteur d'une marge passive de la Téthys: le haut-fond de La Mure et le bassin du Beaumont (Alpes occidentales). Unpublished PhD Thesis, Grenoble, 193pp.

Bas, T. 1988: Rifting liasique dans la marge passive téthysienne: le haut-fond de La Mure et le bassin du Beaumont (Alpes occidentales). Bulletin de la Société géologique de France 4, 717–724.

Besse, J. & Courtillot, V. 1991: Revised and synthetic apparent polar wander paths of the african, eurasian, north american and indian plates, and true polar wander since 200 Ma. Journal of Geophysical Research 96, 4029–4050.

Bocquet, J. 1966: Le delta miocène de Voreppe. Etude des faciès conglomératiques du Miocène des environs de Grenoble. Travaux du Laboratoire de Géologie de Grenoble 42, 54–75.

Bravard, C. 1982: Données nouvelles sur la stratigraphie et la tectonique de la zone des Aiguilles d'Arves au nord du col du Lautaret. Géologie Alpine 58, 5–14.

Bravard, C. & Gidon, M. 1979: La structure du revers oriental du massif du Pelvoux: observations et interprétations nouvelles. Géologie Alpine 55, 23–33.

Butler, R.W.H. 1984: Balanced cross-sections and their implications for the deep structure of the northwest Alps: reply. Journal of Structural Geology 5, 607–612.

Butler, R.W.H. 1992: Thrust zone kinematics in a basement-cover imbricate stack; eastern Pelvoux Massif, French Alps. Journal of Structural Geology 14, 29–40.

Caby, R. 1973: Les plis transversaux dans les Alpes occidentales: implications pour la genèse de la chaîne alpine. Bulletin de la Société géologique de France 15, 624–634.

Ceriani, S., Fügenschuh, B. & Schmidt, S. 2001: Multi-stage thrusting at the « Penninic Front » in the Western Alps between Mont Blanc and Pelvoux massifs. Geologische Rundschau 90, 685–702.

Ceriani, S. & Schmid, S. 2004: From N–S collision to WNW-directed post-collisional thrusting and folding: Structural study of the Frontal Penninic Units in Savoie (Western Alps, France). Eclogae Geologicae Helvetiae 97, 347–369.

Chevalier, F. 2002: Caractérisation du fonctionnement d'une faille normale et de son impact sur l'enregistrement stratigraphique à partir de l'étude d'un hémigraben jurassique (Bourg d'Oisans, France). Unpublished PhD Thesis, Dijon.

Chevalier, F., Guiraud, M., Garcia, J.P., Dommergues, J.L., Quesne, D., Allemand, P. & Dumont, T. 2003: Calculating the long-term displacement rates of a normal fault from the high-resolution stratigraphic record (early Tethyan rifting, French Alps). Terra Nova 15, 410–416.

Corna, M., Dommergues, J.L., Meister, C. & Page, K. 1997: Les faunes d'ammonites du Jurassique inférieur (Hettangien, Sinémurien et Pliensbachien) au nord du massif des Ecrins (Oisans, Alpes occidentales françaises). Revue de Paléobiologie, Genève 16, 321–409.

Choukroune, P., Ballèvre, M., Cobbold, P., Gauthier, Y., Merle, O. & Vuichard, J.-P. 1986: Deformation and motion in the Western Alpine arc. Tectonics 5, 215–226.

Claudel, M.E. & Dumont, T. 1999: A record of multistage continental break-up on the Briançonnais marginal plateau (Western Alps): Early and Middle-Late Jurassic rifting. Eclogae Geologicae Helvetiae 92, 45–61.

Cornfield, S & Sharp, I.R. 2000: Structural style and stratigraphic architecture of fault propagation folding in extensional settings; a seismic example from the Smorbukk area, Halten Terrace, Mid-Norway. Basin Research 12, 329–341.

Corsini, M., Ruffet, G. & Caby, R. 2004: Alpine and late-Hercynian geochronological constraints in the Argentera Massif (Western Alps). Eclogae Geologicae Helvetiae 97, 3–15.

Coward, M.P. & Dietrich, D. 1989: Alpine tectonics, an overview. In: Coward, M.P., Dietrich, D. & Park, R.G. (eds.), Alpine Tectonics, Geological Society of London, Special Publication n°45, 1–29.

Crouzet, C. 1997: Thermopaléomagnétisme: principe et applications (tectoniques, thermiques et géochronologique) à la zone dauphinoise interne (Alpes occidentales, France). Géologie Alpine, Mémoire Hors Série n°27, 197 pp.

Crouzet, C., Ménard, G. & Rochette, P. 1996: Post-Middle Miocene rotations recorded in the Bourg d'Oisans area, Western Alps, France by paleomagnetism. Tectonophysics 263, 137–148.

Crouzet, C., Ménard G. & Rochette, P. 1999: High precision three dimensional paleothermometry derived from paleomagnetic data in an Alpine metamorphic unit, Geology 27, 503–506.

Crouzet, C., Ménard, G. & Rochette, P. 2001a: Cooling history of the Zone dauphinoise (Western Alps, France) deduced from the thermopaleomagnetic record: geodynamic implications. Tectonophysics 340, 79–93.

Crouzet, C., Rochette, P. & Ménard, G. 2001b: Experimental evaluation of thermal recording of polarity reversals during metasediments uplift. Geophysical Journal International 145, 771–785.

Dardeau, G. 1983: Le Jurassique des Alpes Maritimes (France). Stratigraphie, paléogéographie, évolution du contexte structural à la jonction des dispositifs dauphinois, briançonnais et provençal. Unpublished Thèse d'Etat, Nice, 391 pp.

Debelmas, J., Kerckhove, C., Monjuvent, G., Mouterde, R. & Pêcher, A. 1980: Carte géologique de la France (1/50000), feuille Orcières (846), Bureau de Recherches géologiques et minières, Orléans.

Demarest, H. 1983: Error analysis for the determination of tectonic rotation from paleomagnetic data. Journal of Geophysical Research 88, 4321–4328.

Demeulemeester, P., Roques, M., Giraud, P., Vivier, G. & Bonhomme, M.G. 1986: Influence du métamorphisme alpin sur les âges isotopiques des biotites des massifs cristallins externes (Alpes françaises). Géologie Alpine 62, 31–44.

Desthieux, F. & Vernet, J. 1968: Les failles inverses du flanc nord du Rochail. Géologie Alpine 44, 113–115.

Deville, E., Blanc, E., Tardy, M., Beck, C., Cousin, M. & Ménard, G. 1994: Thrust propagation and syntectonic sedimentation in the Savoy Tertiary molasse basin (Alpine foreland). In: Mascle, A. (Ed.), Hydrocarbon and Petroleum Geology of France. Springer, Berlin, 269–280.

Deville, E. & Chauvières, A. 2000: Thrust tectonics at the front of the western Alps: constraints provided by the processing of seismic reflection data along the Chambéry Transect. Comptes Rendus de l'Académie des Sciences Paris 331, 583–585.

Dèzes, P., Schmid, S. & Ziegler, P.A. 2004: Evolution of the European Cenozoic Rift System: interaction of the Alpine and Pyrenean orogens with their foreland lithosphere. Tectonophysics 389, 1–33.

Dietrich, D. & Durney, D.W. 1986: Change of direction of overthrust shear in the Helvetic nappes of western Switzerland. Journal of Structural Geology 8, 389–398.

Dumont, T. 1998: Sea-Level changes and early rifting of a European Tethyan margin in the western Alps and Southeastern France. In. Graciansky, P.C. de, Hardenbol, J., Jacquin, T. & Vail, P.R. (Eds): Mesozoïc and Cenozoic sequence stratigraphy of European Basins. Society of Economic Petrologists and Mineralogists, Special Publication 60, 623–642.

Dumont, T. & Grand, T. 1987: Caractères communs entre l'évolution précoce d'une portion de marge passive fossile (marge européenne de la Téthys ligure, Alpes occidentales) et celle du rift de Suez. Comptes Rendus de l'Académie des Sciences Paris 305, 1369–1373.

Dumont, T., Vidal, G., Ustal, M. & Tricart, P. 1997: L'interface socle-couverture dans la région de Bourg d'Oisans: modélisation numérique 3D. Documents du Bureau de Recherches géologiques et minières 274, 27–30.

Dumont, T., Wieczorek, J. & Bouillin, J.P. 1996: Inverted Mesozoic rift structures in the Polish Western Carpathians (High-Tatric units). Comparison with similar features in the Western Alps. Eclogae Geologicae Helvetiae 89, 181–202.

Ford, M. 1996: Kinematics and geometry of early Alpine, basement involved folds, SW Pelvoux Massif, SE France. Eclogae Geologicae Helvetiae 89, 269–295.

Ford, M., Duchêne, S., Gasquet, D. & Vanderhaeghe, O. 2006: Two-phase orogenic convergence in the external and internal SW Alps. Journal of the Geological Society of London 163, 1–12.

Froitzheim, N., Schmid, S.M. & Conti, P. 1994: Repeated change from crustal shortening to orogenparallel extension in the Austroalpine units of Graubunden. Eclogae Geologicae Helvetiae 87, 559–612.

Fügenschuh, B. & Schmid, S. 2003: Late stages of deformation and exhumation of an orogen constrained by fission-track data: a case study in the Western Alps. Geological Society of America Bulletin 115, 1425–1440.

Gamond, J.F. 1980: Direction de déplacement et linéation: cas de la couverture sédimentaire dauphinoise orientale. Bulletin de la Société géologique de France 22, 429–436.

Gidon, M. 1979: Le rôle des étapes successives de déformation dans la tectonique alpine du massif du Pelvoux. Comptes Rendus de l'Académie des Sciences Paris 288, 803–806.

Gidon, M. 1980: Carte géologique de la France (1/50 000), notice de la feuille St Bonnet (845). Bureau de Recherches géologiques et minières, Orléans, 43pp.

Gidon, M. 1999: L'origine des abrupts septentrionaux du Taillefer, massifs cristallins externes, Isère, France. Géologie Alpine 75, 103–109.

Gidon, M. & Pairis, J.L. 1981: Nouvelles données sur la structures des écailles de Soleil Bœuf (bordure sud du massif du Pelvoux). Bull. Bureau de Recherches géologiques et minières 2, 35–41.

Gillcrist, R. 1988: Mesozoic basin development and structural inversion in the French Western Alps. Unpublished PhD thesis, Imperial College, London, 430 pp.

Gillcrist, R., Coward, M. & Mugnier, J.-L. 1987: Structural inversion and its controls: examples from the Alpine foreland and the French Alps. Geodynamica Acta 1, 5–34.

Graciansky, P.C. de, & Lemoine, M. 1988: Early Cretaceous extensional tectonics in the southwestern French Alps: a consequence of North-Atlantic rifting during Tethyan spreading. Bulletin de la Société géologique de France 5, 733–738.

Grand, T. 1988: Mesozoic extensionnal inherited structures on the European margin of the Ligurian Tethys. The example of the Bourg d'Oisans half-graben, western Alps. Bulletin de la Société géologique de France 4, 613–621.

Grand, T., Dumont, T. & Pinto-Bull, F. 1985: Distensions liées au rifting téthysien et paléochamps de contrainte associés dans le bassin liasique de Bourg d'Oisans (Alpes occidentales). Bulletin de la Société géologique de France 4, 699–704.

Gratier, J.-P., Lejeune, B. & Vergne, J.-L. 1973: Etude des déformations de la couverture et des bordures sédimentaires des massifs cristallins externes de Belledonne, des Grandes Rousses et du Pelvoux. Unpublished PhD thesis, Grenoble, 289 pp.

Gratier, J.-P., Pêcher, A. & Vialon, P. 1978: Relations entre déformation interne et déplacement-glissement dans les roches. In: Mém. Bureau de Recherches géologiques et minières 91, 207–214.

Gratier, J.-P. & Vialon, P. 1975: Clivage schisteux et déformations: analyse d'un secteur clé du bassin mésozoïque de Bourg d'Oisans, Alpes dauphinoises. Géologie Alpine, Grenoble 51, 41–50.

Gratier, J.-P. & Vialon, P. 1980: Deformation pattern in a heterogeneous material: folded and cleaved sedimentary cover immediatly overlying a cristalline basement, Oisans, French Alps. Tectonophysics 65, 151–180.

Guellec, S., Mugnier, J.-L., Tardy, M. & Roure, F. 1990: Neogene evolution of the Western Alpine foreland in the light of Ecors data and balanced cross section. Mémoire Société géologique de France, nouvelle série 156, 165–184.

Guerrot, C. & Debon, F. 2000: U-Pb zircon dating of two contrasting Late Variscan plutonic suites from the Pelvoux massif (French Western Alps). Schweizerische Mineralogische und Petrographische Mitteilungen 80, 249–256.

Gupta, S. 1997: Tectonic control on paleovalley incision at the distal margin of the Early Tertiary Alpine foreland basin, south-east France. Journal of Sedimentary Research 67, 1030–1043.

Kerckhove, C., Cochonat, P. & Debelmas, J. 1978: Tectonique du soubassement parautochtone des nappes de l'Embrunais-Ubaye sur leur bordure occidentale, du Drac au Verdon. Géologie Alpine, Grenoble 54, 67–82.

Lacassin, R. 1989: Plate-scale kinematics and compatibility of crustal shear zones in the Alps. In: Coward, M.P., Dietrich, D. & Park, R.G. (Eds.), Alpine Tectonics, Geological Society of London, Special Publication 45, 339–352.

Lamarche, G. 1987: Analyse microstructurale et fabrique magnétique. L'exemple des calcschiste et des flyschs de la zone dauphinoise. Alpes françaises. Unpublished PhD thesis, Université Joseph Fourier Grenoble, France, 168 pp.

Lamarche, G., Ménard, G. & Rochette, P. 1988: Données paléomagnétiques sur le basculement tardif de la zone dauphinoise interne (Alpes occidentales). Comptes Rendus de l'Académie des Sciences Paris 306, 711–716.

Laubscher, H.P. 1988: The arcs of the Western Alps and the Northern Appennines: an updated view. Tectonophysics 146, 67–78.

Lazarre, J., Tricart, P., Courrioux, G. & Ledru, P. 1996: Héritage téthysien et polyphasage alpin: réinterprétation tectonique du «synclinal» de l'Aiguille de Morges (massif du Pelvoux, Alpes occidentales, France). Comptes Rendus de l'Académie des Sciences Paris 323, 1051–1058.

Leloup, P.H., Arnaud, N., Sobel, E.R. & Lacassin, R. 2005: Alpine thermal and structural evolution of the highest external cristalline massif: the Mont Blanc. Tectonics 24, TC4002, 26 p.

Lemoine, M., Gidon, M. & Barféty, J.-C. 1981: Les Massifs Cristallins Externes des Alpes occidentales: d'anciens blocs basculés nés au Lias, lors du rifting téthysien. Comptes Rendus de l'Académie des Sciences Paris 292, 917–920.

Lemoine, M., Bas, T., Arnaud-Vanneau, A., Arnaud, H., Dumont, T., Gidon, M., Bourbon, M., de Graciansky, P.C., Rudckiewicz, J.L., Megard-Galli, J. & Tricart, P. 1986: The continental margin of the Mesozoic Tethys in the Western Alps. Marine and Petroleum Geology 3, 179–199.

Lemoine, M., Dardeau, G., Delpech, P.Y., Dumont, T., de Graciansky, P.C., Graham, R., Jolivet, L., Roberts, D. & Tricart, P. 1989: Extension syn-rift et failles transformantes jurassiques dans les Alpes occidentales. Comptes Rendus de l'Académie des Sciences Paris 309, 1711–1716.

Lickorish, W.H., Ford, M., Bürgisser, J. & Cobbold, P.R. 2002: Arcuate thrust systems in sandbox experiments: A comparison to the external arcs of the Western Alps. Geological Society of America Bulletin 114, 1089–1107.

Maury, P. & Ricou, L.-E. 1983: Le décrochement subbriançonnais: une nouvelle interprétation de la limite externe-interne des Alpes franco-italiennes. Revue de Géographie Physique et Géologie Dynamique 24, 3–22.

McClay, K.R. & White, M.J. 1995: Analogue modelling of orthogonal and oblique rifting. Marine and Petroleum Geology 12, 137–151.

Ménard, G. 1988: Structure et cinématique d'une chaîne de collision: les Alpes occidentales et centrales. Unpublished Doctorate thesis, Grenoble, 268 pp.

Merle, O. & Brun, J.-P. 1984: The curved translation path of the Parpaillon nappe (French Alps). Journal of Structural Geology 6, 711–719.

Merle, O., Cobbold, P.R. & Schmid, S. 1989: Tertiary kinematics in the Lepontine dome. In: Coward, M.P., Dietrich, D. & Park, R.G. (Eds.): Alpine Tectonics, Geological Society of London, Special Publication n°45, 113–134.

Mosar, J., Stampfli, G.M. & Girod, F. 1996: Western Préalpes Médianes Romandes: Timing and structure. A review. Eclogae Geologicae Helvetiae 89, 389–425.

Mugnier, J.L., Guellec, S., Ménard, G., Roure, F., Tardy, M. & Vialon, P. 1990: A crustal scale balanced cross-section through the external Alps deduced from the Ecors profile. Mémoire Société Géologique de France, nouvelle série 156, 203–216.

Müller, W., Prosser, G., Mancktelow, N., Villa, I., Kelly, S.P., Viola, G. & Oberli, F. 2001: Geochronological constraints on the evolution of the Periadriatic Fault system (Alps). International Journal of Earth Sciences 90, 623–653.

Nziengui, J.-J. 1993: Excès d'argon radiogénique dans les quartz des fissures tectoniques: implications pour la datation des séries métamorphiques. L'exemple de la coupe de la Romanche, Alpes occidentales françaises. Unpublished PhD Thesis, Grenoble, 209 pp.

Pecher, A., Barféty, J.C. & Gidon, M. 1992: Structures est-ouest anténummulitiques à la bordure orientale du masif des Ecrins-Pelvoux (Alpes françaises). Géologie alpine, Série spéciale Résumés de colloques n°1, 72–73.

Pfiffner, A.O., Klaper, E.M., Mayerat, A.M. & Heitzmann, P. 1990: Structure of the basement-cover contact in the Swiss Alps. In: Roure, F., Heitzmann, P. & Polino, R., Eds., Deep Structure of the Alps. Memoire Société Géologique de France 156, 247–262.

Philippe, Y., Deville, E. & Mascle, A. 1998: Thin-skinned inversion tectonics at oblique basin margin: examples of the western Vercors and Chratreuse subalpine massifs (SE France). In: Mascle A., Puigdefabregas C., Luterbacher H.P. & Fernandez M. (Eds.): Cenozoic Foreland basins of Western Europe. Geological Society of London, Special Publication 134, 239–262.

Platt, J.P. 1984: Balanced cross-sections and their implications for the deep structure of the northwest Alps: discussion. Journal of Structural Geology 5, 603–606.

Platt, J.P., Behrmann, J.H., Cunningham, P.C., Dewey, J.F., Helman, M., Parrish, M., Shepley, M.G., Wallis, S. & Weston, P.J. 1989a: Kinematics of the Alpine arc and the motion history of Adria. Nature 337, 158–161.

Platt, J.P., Lister, G.S., Cunningham, P., Weston, P., Peel, F., Baudin, T. & Dondey, H. 1989b: Thrusting and backthrusting in the Briançonnais domain of the western Alps. In: Coward, M.P., Dietrich, D. & Park, R.G. (Eds.): Alpine Tectonics. Geological Society of London, Special Publication 45, 135–152.

Ramsay, J.G. 1989: Fold and fault geometry in the western Helvetic nappes of Switzerland and France and its implications for the evolution of the arc of the Western Alps. In: Coward, M.P., Dietrich, D. & Park, R.G. (Eds.), Alpine Tectonics, Geological Society of London, Special Publication 45, 33–46.

Ratschbacher, L. 1989: Discussion on kinematics of the Alpine arc. Journal of the Geological Society of London 147, 572–575.

Ricou, L.-E. & Siddans, A.-W.-B. 1986: Collision tectonics in the Westyern Alps. In: Coward, M.-P. & Ries, A.-C., Eds., Collision tectonics, Geological Society of London, Special Publication 19, 229–244.

Roure, F. & Colletta, B. 1996: Cenozoic inversion structures in the foreland of the Pyrenees and Alps. In: Ziegler, P.A. & Horvath, F. (Eds.): Structure and prospects of Alpine basins and forelands, Peri-Tethys Memoire 2, Memoires du Museum d'Histoire Naturelle Paris 170, 173–209.

Roux, M., Bourseau, J.P., Bas, T., Dumont, T., De Graciansky, P.C. & Rudckiewicz, J.L. 1988: Bathymetric evolution of the Tethyan margin in the Western Alps, data from stalked crinoids: a reappreisal of eustatism problems during the Jurassic. Bulletin de la Société géologique de France 4, 633–641.

Sabil, N. 1995: La datation par traces de fission: aspects méthodologiques et applications thermochronologiques en contexte alpin et de marge continentale. Unpublished PhD Thesis, Grenoble, 238 pp.

Schmid, S.M., Aebli, H.R., Heller, F. & Zingg, A. 1989: The role of the Periadriatic Line in the tectonic evolution of the Alps. In: Coward, M.P., Dietrich, D. & Park, R.G. (Eds.): Alpine Tectonics, Geological Society of London, Special Publication 45, 153–171.

Schmid, S.M. & Kissling, E. 2000: The arc of the Western Alps in the light of geophysical data on deep crustal structure. Tectonics 19, 62–85.

Schmid, S.M., Pfiffner, A., Froitzheim, N., Schönborn, G. & Kissling, E. 1996: Geophysical-geological transect and tectonic evolution of the Swiss-Italian Alps. Tectonics 15, 1036–1064.

Sue, C. & Tricart, P. 2002: Widespread post-nappe normal faulting in the Internal Western Alps: a new constraint on arc dynamics. Journal of the Geological Society of London 159, 61–70.

Sue, C., Tricart, P., Dumont, T. & Pêcher, A. 1998: Raccourcissement polyphasé dans le massif du Pelvoux, Alpes occidentales: exemple du chevauchement de Villard Notre Dame. Comptes Rendus de l'Académie des Sciences Paris 324, 847–854.

Tapponnier, P. 1977: Evolution tectonique du système alpin en Méditerrannée: poinçonnement et écrasement rigide-plastique. Bulletin de la Société géologique de France 19, 437–460.

Thomas, J.C., Claudel, M., Collombet, M., Tricart, P., Chauvin, A. & Dumont, T. 1999: First paleomagnetic data from the sedimentary cover of the French penninic Alps: evidence for Tertiary counterclockwise rotations in the Western Alps. Earth and Panetary Science Letters 171, 561–574.

Tricart, P. 1980: Tectoniques superposées dans les Alpes occidentales, au sud du Pelvoux. Evolution structurale d'une chaîne de collision. Thèse de Doctorat d'Etat, Strasbourg, 407 pp.

Tricart, P. 2004: From extension to transpression during final exhumation of the Pelvoux and Argentera massifs, Western Alps. Eclogae Geologicae Helvetiae 97, 429–439.

Tricart, P. & Lemoine, M. 1986: From faulted blocks to megamullions and megaboudins. Tethyan heritage in the structure of the Western Alps. Tectonics 5, 95–110

Tricart, P., Schwartz, S., Sue, C., Poupeau, G. & Lardeaux, J.M 2000: La dénudation tectonique de la zone ultradauphinoise et l'inversion du front briançonnais au sud-est du Pelvoux (Alpes occidentales): une dynamique miocène à actuelle. Bulletin de la Société géologique de France 172, 49–58.

Trift, M. & de Graciansky, P.C. 1988: Aspects du rifting téthysien: petits grabens et mégabrèches du Domérien-Toarcien sur le plateau d'Emparis (Isère et Hautes Alpes). Bulletin de la Société géologique de France 4, 643–650.

Vernet, J. 1965: La zone Pelvoux-Argentera. Bull. Serv. Carte géol. France, 275, 131–424.

Vernet, J. 1974: Sur la tectonique alpine des massifs cristallins dauphinois dans leur région culminante (Pelvoux, Grandes Rousses et leurs abords) et l'histoire de leur édification. Géologie Alpine 50, 195–236.

Vialon, P. 1968: Clivages schisteux et déformations: répartition et genèse dans le bassin mésozoïque de Bourg d'Oisans. Géologie Alpine 44, 353–366.

Vialon, P. 1974: Les déformation synschisteuses superposées en Dauphiné. Leur place dans la collision des éléments du socle préalpin. Bulletin Suisse de Minéralogie et Petrographie 54, 663–690.

Vialon, P. 1986: Les déformations alpines de la couverture sédimentaire de blocs du socle cristallin basculés de Belledonne, Grandes Rousses et Pelvoux dans la région de Bourg d'Oisans. Réunion Extraordinaire de la Société Géologique de France: De la marge océanique à la chaîne de collision dans les Alpes du Dauphiné. Bulletin de la Société géologique de France 8, 197–231.

Vialon, P., Rochette, P. & Ménard, G. 1989: Indentation and rotation in the Western Alpine arc. In: Coward, M.-P., Dietrich, D. & Park, R.-G., (Eds.): Alpine Tectonics, Geological Society of London, Special Publication 45, 329–338.

Waibel, A.F. 1990: Sedimentology, petrographic variability and very-low-grade metamorphism of the Champsaur sandstone (Paleogene, Hautes Alpes, France). PhD Thesis, Geneva, 140 pp.

Wildi, W. & Huggenberger, P. 1993: Reconstitution de la plate-forme européenne anté-orogénique de la Bresse aux Chaînes subalpines; éléments de cinématique alpine (France et Suisse orientale). Eclogae Geologicae Helvetiae 86, 47–64.

Manuscript received July 30, 2007
Revision accepted July 22, 2008
Published Online first November 1, 2008
Editorial Handling: Stefan Schmid, Stefan Bucher

1661-8726/08/01S111-16
DOI 10.1007/s00015-008-1290-0
Birkhäuser Verlag, Basel, 2008

Swiss J. Geosci. 101 (2008) Supplement 1, S111–S126

The metamorphic evolution of migmatites from the Ötztal Complex (Tyrol, Austria) and constraints on the timing of the pre-Variscan high-T event in the Eastern Alps

WERNER F. THÖNY [*,1], PETER TROPPER [1], FRIEDERIKE SCHENNACH [1], ERWIN KRENN [2], FRIEDRICH FINGER [2], REINHARD KAINDL [1], FRANZ BERNHARD [3] & GEORG HOINKES [3]

Key words: pre-Variscan, high-T, migmatite, monazite, geochronolgy, Eastern Alps

ABSTRACT

Within the Ötztal Complex (ÖC), migmatites are the only geological evidence of the pre-Variscan metamorphic evolution, which led to the occurrence of partial anatexis in different areas of the complex. We investigated migmatites from three localities in the ÖC, the Winnebach migmatite in the central part and the Verpeil- and Nauderer Gaisloch migmatite in the western part. We determined metamorphic stages using textural relations and electron microprobe analyses. Furthermore, chemical microprobe ages of monazites were obtained in order to associate the inferred stages of mineral growth to metamorphic events. All three migmatites show evidence for a polymetamorphic evolution (pre-Variscan, Variscan) and only the Winnebach migmatite shows evidence for a P-accentuated Eo-Alpine metamorphic overprint in the central ÖC. The P-T data range from 670–750 °C and <2.8 kbar for the pre-Variscan event, 550–650 °C and 4–7 kbar for the Variscan event and 430–490 °C and ca. 8.5 kbar for the P-accentuated Eo-Alpine metamorphic overprint. U-Th-Pb electron microprobe dating of monazites from the leucosomes from all three migmatites provides an average age of 441 ± 18 Ma, thus indicating a pervasive Ordovician-Silurian metamorphic event in the ÖC.

Introduction

Polymetamorphic crystalline complexes provide a window into the metamorphic evolution of large-scale orogenic belts, such as the Alps. Deciphering the metamorphic history by means of petrographic, mineral chemical and geochronological data will help to unravel the geodynamic evolution of these belts and increase our knowledge and understanding of large-scale orogenic processes. Due to the mineralogical and textural complexities in polymetamorphic rocks, it is imperative to apply advanced analytical techniques with high spatial resolution to obtain reliable information on the P-T-t path.

The Austroalpine Ötztal Complex (ÖC) in the Eastern Alps provides an excellent opportunity to study a metamorphic complex which underwent several episodes of metamorphic overprint. Although extensive research has been performed on the two predominant orogenic episodes in the Eastern Alps, namely the Variscan (Hercynian) and Eo-Alpine orogenic events, very little attention has been paid to the pre-Variscan metamorphic history so far. Compared to the Variscan and Eo-Alpine events, which can be traced throughout the whole ÖC, the pre-Variscan history is manifested only in localized migmatite occurrences within the ÖC. Thus these migmatites provide the opportunity to study polymetamorphic rocks that underwent at least three metamorphic episodes (Neubauer et al. 1999; Hoinkes et al. 1999). Within the ÖC, several migmatite occurrences were described from the Ötz valley, the Kauner valley, the Reschenpass area and the Stubai valley (Schindlmayer 1999). We will focus our investigations on three migmatite bodies. Two of these migmatite bodies were only recently discovered, the Verpeil migmatite and the migmatite from the Nauderer Gaisloch (Schweigl 1993; Bernhard 1994) and of these two, only the latter was studied to some extent petrographically and geochronologically (Klötzli-Chowanetz 2001). The third is the Winnebach migmatite from the central ÖC, but despite of its relatively large size (25 km²) and importance as a locality to study the pre-Variscan metamorphic history, only limited information on its petrology and mineral chemistry is available (e.g. Hoinkes et al. 1972; Hoinkes 1973; Klötzli-Chowanetz 2001). Although the electron microprobe has been the major tool to perform micro-analytical techniques at small scales over the last thirty years, only very few mineral chemical data are avail-

[1] University of Innsbruck, Faculty of Geo- and Atmospheric Sciences, Institute of Mineralogy and Petrography, Innrain 52, A-6020 Innsbruck, Austria.
[2] University of Salzburg, Department of Material Sciences, Division of Mineralogy, Hellbrunnerstr. 34, A-5020 Salzburg, Austria.
[3] University of Graz, Institute of Geo Sciences, Division of Mineralogy and Petrography, Universitätsplatz 2, A-8010 Graz, Austria.
*Corresponding author: Werner Thöny. E-mail: Werner.Thoeny@uibk.ac.at

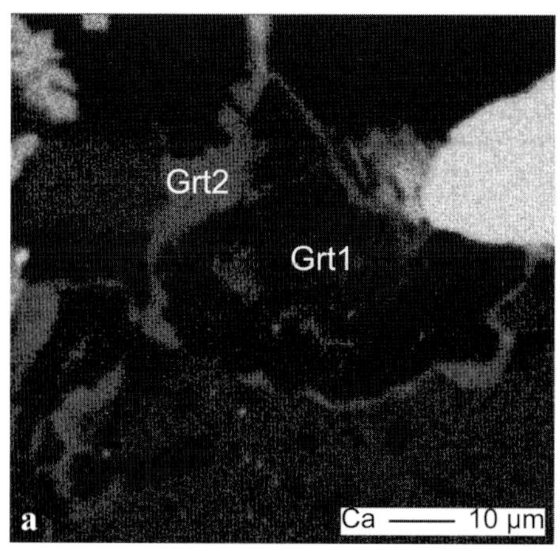

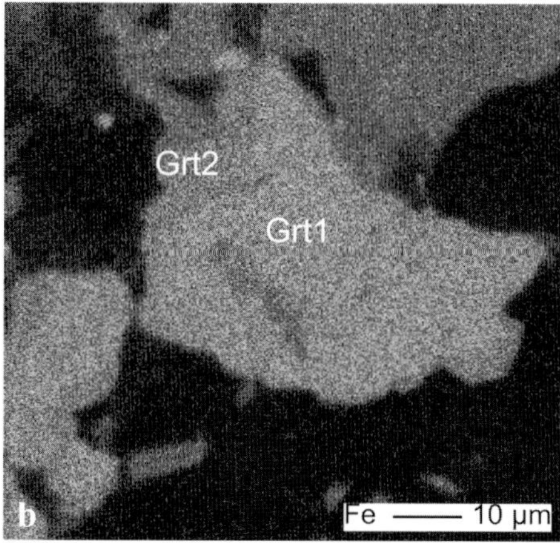

Fig. 3. (a): X-ray distribution image for the element Ca showing two generations of garnet (Grt_1 + Grt_2) from a leucosome sample of the Winnebach migmatite (sample WB 70). Ca is distinctly enriched at the rims. (b): X-ray distribution image for the element Fe also showing the two generations of garnet (Grt_1 + Grt_2) and a decrease in Fe at the rims of the grain.

± garnet ± kyanite ± silimanite ± andalusite ± cordierite. Previous petrographic investigations from the three localities by Hoinkes et al. (1972), Söllner et al. (1982), Chowanetz (1991), Schweigl (1993), Bernhard (1994), and Klötzli-Chowanetz (2001) provided evidence for several stages of mineral growth of plagioclase, biotite and quartz during the metamorphic evolution of these migmatite bodies.

In the Winnebach migmatite, garnet forms hypidiomorphic to idiomorphic grains ranging from 50 to 100 µm in size. Backscatter electron (BSE) images reveal two stages of plagioclase and garnet growth (Figs. 3a-b). Biotite texturally also shows evidence for two generations. Older biotites are large, idiomorphic crystals which are surrounded by fine-grained younger biotites (Fig. 4). Similarly, two generations of kyanite occur (Fig. 5a), which were distinguished by micro-Raman spectroscopy since the younger fine-grained kyanite shows a strongly fluorescing Raman spectrum (Figs. 5b-c). Textures which indicate melting rarely occur (Figs. 6a-c). These textures show a haplogranitic assemblage and indicate melting according to the model reactions (Boettcher & Wyllie 1968; Spear et al. 1999):

Plagioclase + K-feldspar + Biotite + Quartz = Melt (1)

Muscovite + K-feldspar + Quartz + H_2O = Melt (2)

In addition, large pseudomorphs containing muscovite + chlorite are interpreted as pinite pseudomorphs after cordierite. Chloritoid occurs as fine-grained needles along former fractures.

In contrast to the other two migmatites, the Verpeil migmatite still contains relicts of cordierite as shown in Figures 7a–b, which is actually the first report of cordierite from migmatites from the ÖC. These cordierites are partly replaced by kyanite, biotite, and plagioclase (Figs. 7a–c) indicating a reaction such as:

Cordierite + K-feldspar + H_2O = Biotite + Kyanite (3)

Garnet grows around biotite (Fig. 8) and kyanite forms either needle-shaped grains or relict large crystals, surrounded by later formed andalusite (Figs. 9a–b), as identified with Raman spectroscopy (Fig. 9c). In addition, cordierite is also replaced by

extremely fine-grained muscovite and chlorite (pinitization), which formed during a later stage.

The mineral assemblage of the Nauderer Gaisloch migmatite is very similar to the one of the Verpeil migmatite and is comprised of garnet + quartz + biotite + plagioclase + muscovite + $kyanite_{1,2}$. $Kyanite_1$ occurs as large grains, and $kyanite_2$ occurs as small needles intergrown with biotite (Fig. 10), which could also have replaced former cordierite according to reaction (3).

Analytical methods

Electron microprobe analysis (EMPA) of all minerals except monazite was performed using the JEOL 8100 SUPERPROBE at the Institute of Mineralogy and Petrography at the University of Innsbruck. Operating conditions were 15 kV acceleration voltage and 20 nA beam current. A defocused beam with a

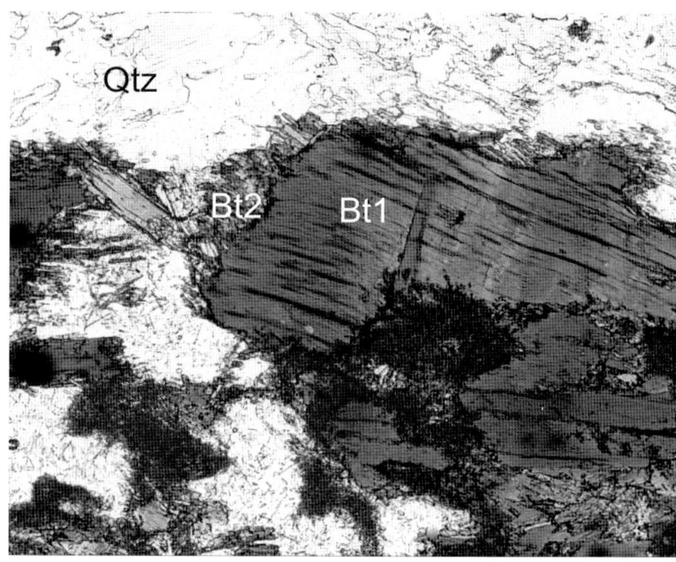

Fig. 4. Photograph of a thin section of sample WB9A2 showing the two, texturally different, generations of biotite. (//P, length of image = 1.13 mm, Bt1 = biotite1, Bt2 = biotite2, Qtz = quartz)

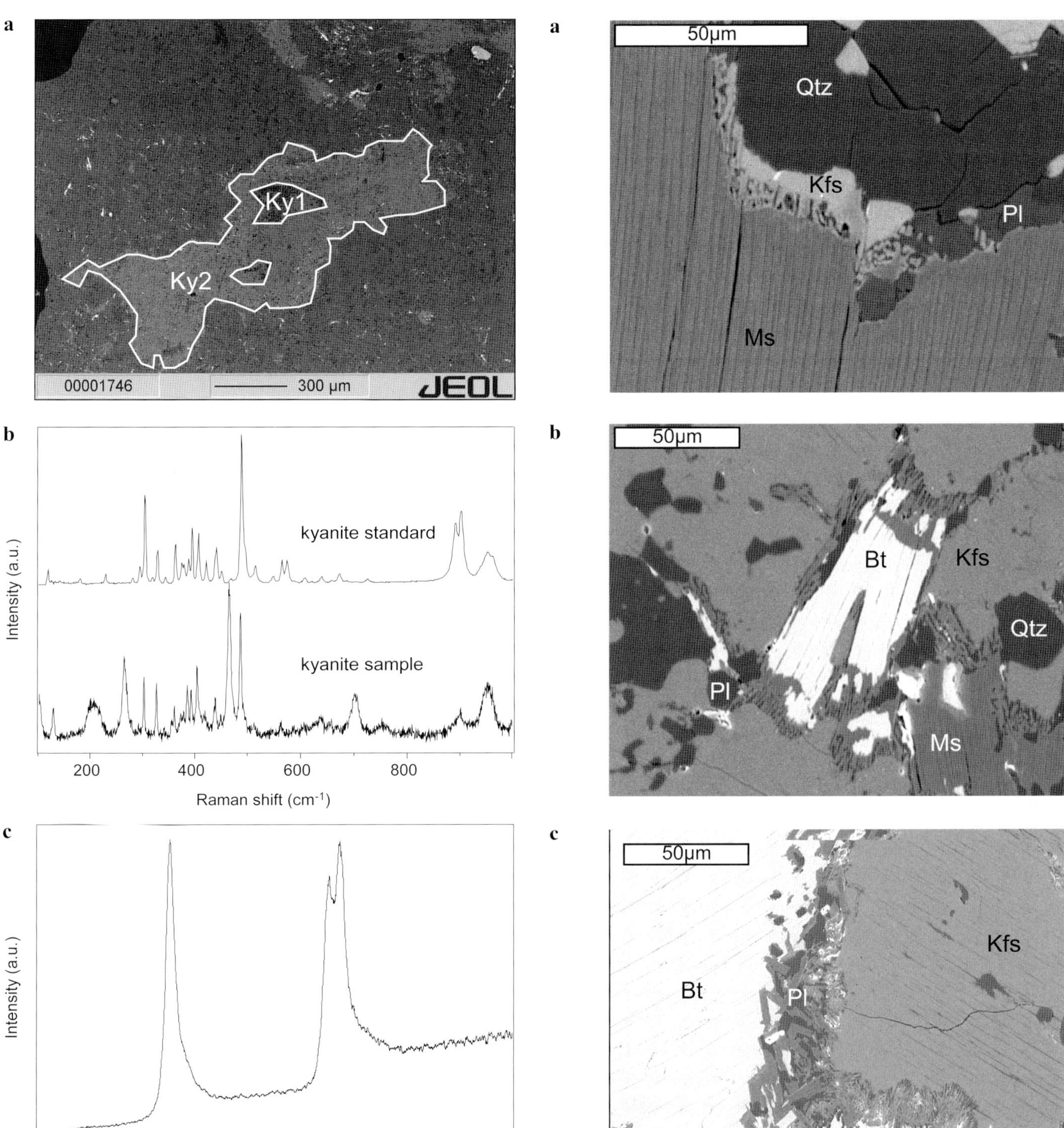

Fig. 5. (a): Backscatter electron (BSE) image showing the two generations of kyanite (Ky$_1$ + Ky$_2$) from the Winnebach migmatite (sample WB 70). (b): Raman spectrum of the texturally older kyanite$_1$. The dark grey spectrum is from sample WB9A2, the light grey is a standard kyanite spectrum. The spectrum also includes peaks from adjacent minerals (quartz) and adhesive but clearly identifies the aluminium silicate as kyanite. (c): Raman spectrum of the fluorescing kyanite$_2$ (WB9A2).

Fig. 6. BSE images of melting textures from the Winnebach migmatite (a): melting textures involving the assemblage plagioclase (Pl) + K-feldspar (Kfs) + muscovite (Ms) + quartz (Qtz) (sample WB9A2). (b): melting textures involving the assemblage biotite (Bt) + muscovite (Ms) + K-feldspar (Kfs) + plagioclase (Pl) + quartz (Qtz) (sample WB9A2). (c): melting textures involving melting of the assemblage biotite (Bt) + K-feldspar (Kfs) + quartz (Qtz) + plagioclase (Pl) (sample WB9A2).

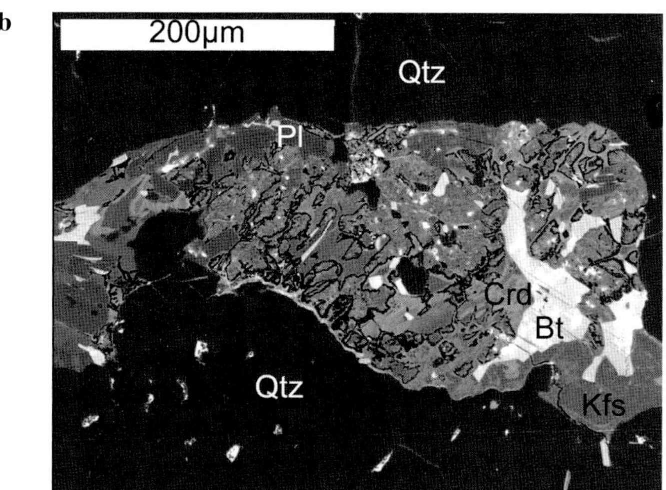

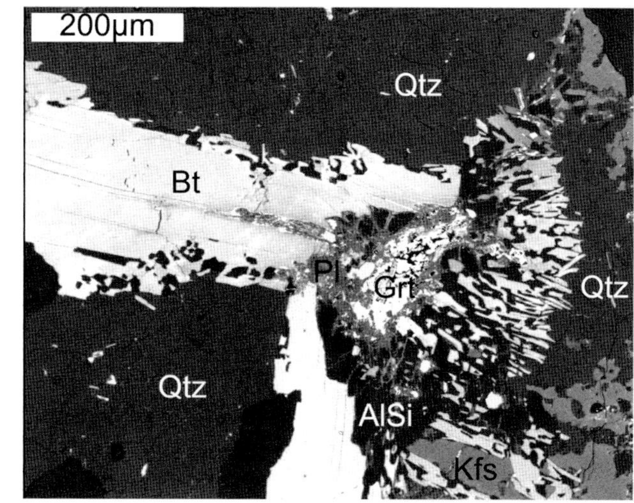

diameter of 5 μm was used for the analysis of feldspars and micas to prevent loss of alkalis. Mineral formulae were calculated using the software NORM II (Ulmer 1993, written comm.). Microprobe analyses of monazite were obtained in wavelength dispersive (WD) mode on a JEOL JX 8600 electron microprobe at the University of Salzburg. Operating conditions were 15 kV acceleration voltage, 150–200 nA beam current. Mα lines were chosen for Th, U and Pb; Lβ lines for Pr, Nd, Sm, Eu and Gd; Lα lines for Sr, Y, La, Ce, Dy, Er and Yb and Kα lines for Si, Al, Ca, P, Fe, Ti, Mn. With exception of Pb, which was counted between 160–320 s all other elements were rated 10–50 s. Related 3 sigma detection limits were 0.025 wt.% (for Th), 0.023 wt.% (U), 0.011–0.015 wt.% (Pb) and <0.1 wt.% for all other elements. For acquisition of Pb an exponential background model was used. Details on the background modelling, analytical setting (element lines, crystals, and standards), detection limits and interference correction are given in Krenn et al. (2008).

Standards with contrasting U and Th concentrations were routinely used to control dead time adjustment and ZAF factors. As a further control of the analytical precision and the reliability of chemical dating, a monazite standard with well known age (341 ± 2 Ma, Friedl 1997) was routinely measured prior to and after 5 to 8 standard monazite analyses. In addition, standard monazites with Eo-Alpine and Proterozoic ages (~90 Ma, ~1385 Ma, ~1865 Ma) were also analysed together with the sample monazites.

Age and error calculations were carried out using the technique of Montel et al. (1996). Weighted average ages were calculated by internal error propagation (95% confidence) using Isoplot 2.49e (Ludwig 2001). Errors of the isochron ages were calculated using the least-squares fitting method of York (1966).

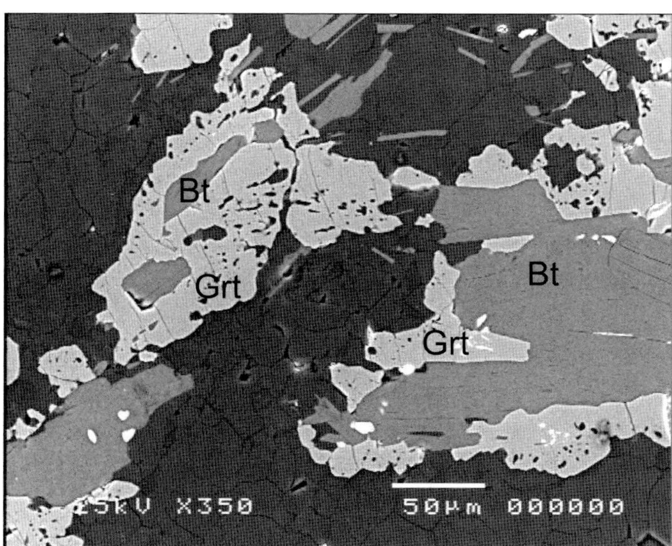

Fig. 7. BSE images of biotite + kyanite pseudomorphs after cordierite from the Verpeil migmatite. Relict cordierite is still present in the samples FB56E and VP16 (a, b). (c): garnet + K-feldspar also occur within the pseudomorphs according to reaction (5). Abbreviations: Crd = cordierite, Bt = biotite, Ky = kyanite, Ms = muscovite, Pl = plagioclase, Qtz = quartz, Grt = garnet, Kfs = K-feldspar, AlSi = aluminium silicate.

Fig. 8. BSE image from the Verpeil migmatite showing the formation of garnet around biotite (sample FB56E). Abbreviations: Grt = garnet, Bt = biotite. Note that garnet contains an inclusion-free inner rim and an inclusion-rich outer rim.

Confocal micro-Raman spectra were obtained with a HORIBA JOBIN YVON LabRam-HR 800 micro-Raman spectrometer. Samples were excited at room temperature with the 633 nm emission line of a 17 mW He-Ne-laser through an OLYMPUS 100X objective. The laser spot on the surface had a diameter of approximately 1 µm and a power of about 5 mW. Light was dispersed by a holographic grating with 1800 grooves/mm. Spectral resolution of about 1.8 cm^{-1} was experimentally determined by measuring the Rayleigh line. The dispersed light was collected by a 1024×256 open electrode CCD detector. Confocal pinhole was set to 1000 µm. Several spectra of single crystals of aluminium silicates in thin sections were recorded without polarizers for the exciting laser and the scattered Raman light. The spectra were baseline-corrected by subtracting linear or squared polynomial functions and fitted to Voigt functions. Peak shifts were calibrated by regular adjusting the zero-order position of the grating and controlled by measuring the Rayleigh line of a (100) polished single crystal silicon-wafer. Accuracy of Raman peak shifts was better than 0.5 cm^{-1}. The detection range was 100–4000 cm^{-1}.

Mineral chemistry

Garnet: Within the Winnebach migmatite polyphase garnet growth is evident in BSE images. The older garnets (garnet$_1$) (X_{Mg} = 0.07–0.09, X_{Fe} = 0.70–0.76, X_{Mn} = 0.12–0.13, X_{Ca} = 0.04–0.06) are almandine-rich with relatively constant Mg- and Mn-contents. The second generation (garnet$_2$) (X_{Mg} = 0.02–0.07, X_{Fe} = 0.52–0.56, X_{Mn} = 0.06–0.10, X_{Ca} = 0.25–0.26) forms grossular-rich rims around garnet$_1$ (Table 1, Fig. 3a). The garnets of the Verpeil migmatite are almandine-rich with high spessartine- and relatively low amounts of grossular- and pyrop components (X_{Mg} = 0.09, X_{Fe} = 0.6–0.62, X_{Mn} = 0.22–0.25, X_{Ca} = 0.01–0.02; Table 1, Fig. 8). Though two texturally different generations of garnet, an older, inclusion-free and a younger inclusion-rich garnet growing around biotite were detected, no chemical difference was observed. Garnet from the Nauderer Gaisloch migmatite shows a uniform composition of X_{Mg} = 0.08–0.11, X_{Fe} = 0.64–0.67, X_{Mn} = 0.20–0.23, X_{Ca} = 0.03–0.05 (Table 1).

Aluminium silicates: Two kyanite generations were identified in the Winnebach migmatite and the Nauderer Gaisloch migmatite using micro-Raman spectroscopy. The large, prismatic kyanites (kyanite$_1$) are chemically pure aluminium silicates. Fine-grained needles of kyanite 2 (kyanite$_2$) show evidence for Fe contents since they have a different micro-Raman spectrum showing fluorescence (Fig. 5c). These data are in agreement with the data of Klötzli-Chowanetz (2001) who identified two kyanite generations based on X-ray diffraction data. The fine-grained felt of kyanite$_2$ often occurs as a replacement product of kyanite$_1$ (Fig. 5a). Micro-Raman investigations on samples of the Verpeil migmatite also resulted in the identification of two generations of aluminium silicates, namely andalusite and kyanite, which can also be texturally discerned (Fig. 9). While kyanite forms larger prismatic crystals, andalusite appears as

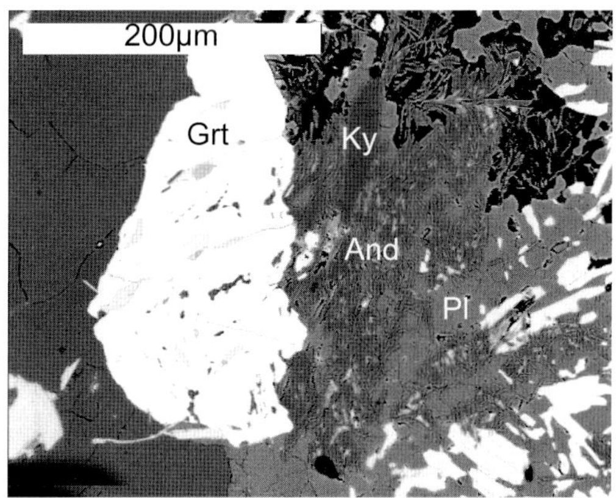

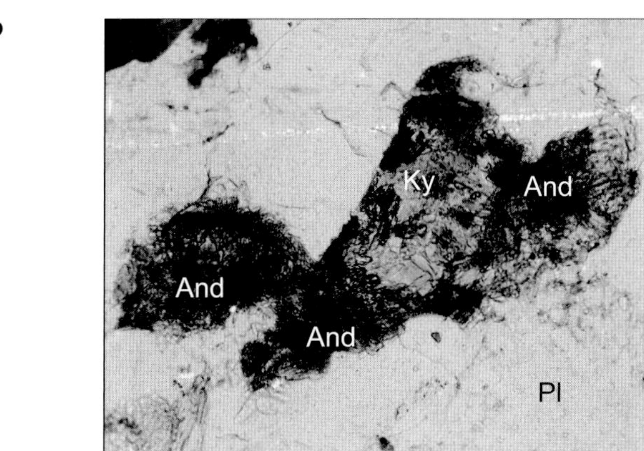

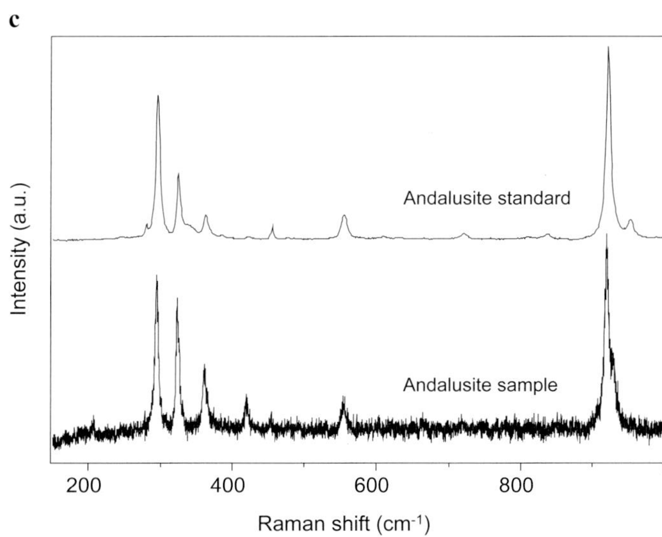

Fig. 9. a) BSE image of the Verpeil migmatite showing a large grain of kyanite surrounded by later formed andalusite adjacent to garnet (sample FB56E). (b): Microphotograph of a thin section from the Verpeil migmatite showing kyanite and later formed andalusite. (sample VP16, //P, length of bottom edge = 1.13 mm). c) Raman spectrum of andalusite from the Verpeil migmatite (sample FB56E).

fine-grained aggregates. The large kyanite grains of the Nauderer Gaisloch migmatite contain 0.018 apfu Fe on average and therefore show a fluorescing Raman spectrum. The fine-grained kyanite felt within the muscovite-biotite-quartz aggre-

Table 1. Representative electron microprobe analyses of garnet: Basis of formula calculation: 12 O and 8 cations; Fe3+ was calculated based on charge balance considerations. 1: garnet core (WB70); 2: garnet core (WB70); 3: garnet rim (WB49A); 4: garnet rim (WB49A); 5: garnet core (FB56A); 6: garnet rim (FB56A); 7: garnet core (NA50); 8: garnet rim (NA50);

Analysis	1	2	3	4	5	6	7	8
SiO_2	36.62	36.64	37.02	37.22	36.63	36.52	36.70	36.40
Al_2O_3	21.16	20.57	21.12	20.89	20.76	20.70	21.59	20.74
Fe_2O_3	1.82	0.55	1.81	3.05	1.29	1.42	1.20	1.55
FeO	29.96	31.58	24.58	22.70	27.35	27.03	28.99	28.37
MnO	5.31	5.51	2.97	2.67	9.49	10.38	10.00	8.37
MgO	1.88	1.79	1.51	1.76	2.29	2.21	1.90	2.62
CaO	3.27	2.70	10.92	11.88	2.15	1.71	0.91	1.57
Σ	100.02	99.34	99.93	100.17	99.96	99.96	101.36	99.89
Si	2.954	2.993	2.953	2.947	2.969	2.966	2.946	2.953
Al	2.012	1.980	1.985	1.949	1.983	1.981	2.043	1.983
Fe^{3+}	0.110	0.034	0.109	0.182	0.078	0.087	0.073	0.095
Fe^{2+}	2.021	2.157	1.639	1.503	1.854	1.836	1.946	1.924
Mn	0.363	0.381	0.201	0.179	0.652	0.714	0.680	0.575
Mg	0.226	0.218	0.180	0.208	0.277	0.268	0.227	0.317
Ca	0.283	0.236	0.933	1.008	0.187	0.149	0.078	0.137
Σ Cat.	7.969	7.999	8.000	7.976	8.000	8.001	7.993	7.984
Mg#	0.075	0.072	0.059	0.068	0.091	0.088	0.076	0.104
Grossular	0.09	0.08	0.30	0.33	0.06	0.05	0.03	0.05
Almandine	0.67	0.71	0.53	0.49	0.61	0.60	0.64	0.63
Pyrope	0.08	0.07	0.06	0.07	0.09	0.09	0.08	0.10
Spessartine	0.12	0.13	0.07	0.06	0.21	0.23	0.23	0.19
Andradite	0.04	0.01	0.04	0.05	0.03	0.03	0.02	0.03

Table 2. Representative electron microprobe analyses of feldspars: Basis of formula calculations: 8 O and 5 cations; 1: plagioclase core (WB70); 2: plagioclase core (WB70); 3: plagioclase rim (WB43); 4: plagioclase (FB56A); 5: plagioclase (NA49a); 6: K-feldspar (VP16); 7: K-feldspar (WB70); 8: K-feldspar (NA49); n.d.: not detected

Analysis	1	2	3	4	5	6	7	8
SiO_2	64.17	63.58	64.30	62.85	68.35	64.06	62.63	63.29
Al_2O_3	20.83	22.49	20.22	23.50	20.08	18.72	19.54	18.89
CaO	2.26	3.26	1.62	4.36	0.38	0.16	n.d.	0.22
Na_2O	10.37	8.64	9.98	9.15	10.99	0.81	0.19	1.23
K_2O	0.13	0.31	0.23	0.34	n.d.	16.14	15.93	14.88
Σ	98.65	98.64	96.35	100.54	99.80	99.89	98.29	98.51
Si	2.863	2.863	2.936	2.776	2.998	2.932	2.922	2.921
Al	1.095	1.194	1.088	1.219	1.038	1.009	1.074	1.028
Ca	0.108	0.157	0.079	0.206	0.018	0.008	n.d.	0.011
Na	0.897	0.754	0.884	0.781	0.935	0.072	0.017	0.110
K	0.007	0.018	0.013	0.019	n.d.	0.942	0.948	0.876
Σ Cat.	4.970	4.986	5.000	5.001	4.989	4.963	4.961	4.946
Albite	0.88	0.81	0.91	0.78	0.98	0.07	0.02	0.11
Anorthite	0.11	0.17	0.08	0.20	0.02	0.01	n.d.	0.01
K-feldspar	0.01	0.02	0.01	0.02	n.d.	0.92	0.98	0.88

gates shows a normal Raman spectrum but was too small for chemical analyses using the electron microprobe.

Plagioclase: Plagioclase is generally albite or oligoclase (Table 2). Plagioclase analyses of the Winnebach migmatite samples chemically show two generations similar to garnet. The Ca-rich cores (plagioclase$_1$; $X_{Ca} = 0.11–0.16$, $X_{Na} = 0.81–0.89$, $X_K \leq 0.02$) of the plagioclase are attributed to an older metamorphic event together with the Fe/Mn-rich garnet cores. The Na-rich rims of the plagioclase (plagioclase$_2$; $X_{Ca} = 0.08$, $X_{Na} = 0.91–0.92$, $X_K \leq 0.01$) and the Ca-rich rims of the garnet represent a younger metamorphic event. The plagioclase compositions from the Verpeil migmatite are uniform with $X_{Ca} = 0.20–0.22$, $X_{Na} = 0.78$, $X_K \leq 0.02$. Plagioclase from the Nauderer Gaisloch shows $X_{Ca} = 0.02–0.05$, $X_{Na} = 0.95–0.98$, $X_K \leq 0.01$.

K-feldspar: K-feldspar analyses from the Verpeil migmatite show X_K = 0.88–0.92, X_{Na} = 0.07–0.11, $X_{Ca} \leq 0.01$ and analyses from the Winnebach migmatite show compositions of X_K = 0.98–1.00, $X_{Na} \leq 0.02$. The K-feldspar compositions of the Nauderer Gaisloch migmatites are very similar to the samples from the Verpeil migmatite and are X_K = 0.88, X_{Na} = 0.11, X_{Ca} = 0.01 (Table 2).

Biotite: Although biotites from the Winnebach migmatite appear as two texturally different generations, chemically they are identical (Fe = ~1.30 apfu, Mg = 1.15 apfu, Ti = ~0.13 apfu). Biotites from the Verpeil migmatite show Fe = 1.28 apfu, Mg = 1.1 apfu and Ti = 0.15 apfu. Biotites of the Nauderer Gaisloch migmatite are also not zoned and show Ti contents between 0.1–0.18 apfu. Fe is around 1.25 apfu and Mg is between 0.95 and 1.05 apfu (Table 3).

Muscovite: White mica analyses from the three migmatite bodies show very similar results indicating a slight Tschermak's substitution which is most distinct in the Winnebach migmatite. White mica from the Winnebach migmatite shows Si = 3.19 apfu and Al = 2.43 apfu. The mica analyses from the other two migmatites show Si = 3.02–3.06 apfu and Al = 2.77–2.78 apfu. The celadonite component (Si – 3) is around 0.19 for the Winnebach migmatite and between 0.10 and 0.04 for the Verpeil migmatite. The Nauderer Gaisloch migmatite shows the lowest celadonite component of around 0.02 (Table 4).

Cordierite: Within the samples of the Verpeil migmatite cordierite is the only relict of a pre-Variscan metamorphic event.

Table 3. Representative electron microprobe analyses of biotite: Basis of formula calculations: 11 O; 1: biotite (WB70); 2: biotite (WB70); 3: biotite (FB56A); 4: biotite (FB56A); 5: biotite (NA49a); 6: biotite (NA49a); n.d.: not detected.

Analysis	1	2	3	4	5	6
SiO_2	34.48	35.13	36.03	35.80	35.34	35.33
TiO_2	1.74	1.61	0.28	0.13	2.40	2.27
Al_2O_3	19.46	19.08	21.48	21.22	20.24	20.17
FeO	21.31	21.30	16.91	18.06	18.30	18.90
MnO	0.29	n.d.	n.d.	0.32	0.42	0.43
MgO	9.21	9.27	11.18	11.35	9.14	9.02
Na_2O	n.d.	0.21	n.d.	n.d.	n.d.	0.29
K_2O	9.15	9.51	9.96	9.45	9.47	9.16
H_2O	3.92	3.94	4.02	4.03	4.50	3.88
Σ	99.68	100.17	99.86	100.47	99.75	99.80
Si	2.639	2.675	2.685	2.665	2.694	2.679
Ti	0.100	0.092	0.016	0.007	0.138	0.129
Al	1.713	1.713	1.887	1.862	1.819	1.803
Fe^{2+}	1.364	1.357	1.054	1.125	1.167	1.199
Mn	0.019	n.d.	n.d.	0.020	0.027	0.028
Mg	1.051	1.052	1.242	1.260	1.003	1.020
Na	n.d.	0.031	n.d.	n.d.	n.d.	0.043
K	0.893	0.924	0.947	0.898	0.921	0.917
H	2.000	2.000	2.000	2.000	2.000	2.000
Σ Cat.	7.830	7.847	7.830	7.845	7.783	7.809
Mg#	0.435	0.437	0.541	0.528	0.462	0.459

Table 4. Representative electron microprobe analyses of muscovite: Basis of formula calculations: 11 O; 1: muscovite (WB54); 2: muscovite (WB54); 3: muscovite (FB56A); 4: muscovite (FB56A); 5: muscovite (NA49a); 6: muscovite (NA49a); n.d.: not detected

Analysis	1	2	3	4	5	6
SiO_2	47.82	47.25	46.64	45.78	45.67	45.56
TiO_2	0.26	0.50	0.30	0.11	0.78	0.78
Al_2O_3	30.52	30.52	34.81	35.74	36.00	36.54
FeO	3.11	2.42	1.51	1.47	1.14	1.20
MgO	2.21	1.92	0.88	0.88	0.56	0.78
Na_2O	0.22	0.19	0.34	0.41	0.40	0.44
K_2O	11.33	11.35	10.87	11.39	11.01	10.78
H_2O	4.49	4.43	4.51	4.52	3.94	3.82
Σ	99.96	98.58	99.86	100.30	99.50	99.90
Si	3.193	3.195	3.098	3.039	3.022	2.995
Ti	0.013	0.025	0.015	0.006	0.046	0.039
Al	2.402	2.432	2.725	2.796	2.807	2.831
Fe^{2+}	0.156	0.123	0.076	0.073	0.057	0.059
Mg	0.220	0.194	0.087	0.087	0.055	0.076
Na	0.029	0.025	0.044	0.053	0.051	0.056
K	0.965	0.979	0.921	0.964	0.929	0.904
H	2.000	2.000	2.000	2.000	2.000	2.000
Σ Cat.	6.978	6.973	6.966	7.018	6.967	6.960

Table 5. Representative electron microprobe analyses of cordierite: Basis of formula calculations: 18 O + OH; Fe^{3+} was calculated based on charge balance considerations; 1–4: cordierite (FB56E).

Analysis	1	2	3	4
SiO_2	47.90	48.13	48.40	48.47
Al_2O_3	32.39	32.71	32.96	31.69
FeO	7.54	7.32	6.90	7.30
MnO	0.28	0.20	0.44	0.19
MgO	7.45	7.21	7.54	7.94
Na_2O	1.08	0.91	1.16	0.98
K_2O	0.05	0.08	0.02	0.05
Σ	96.69	96.56	97.42	96.62
Si	5.012	5.031	5.007	5.066
Al	3.995	4.030	4.019	3.903
Fe^{2+}	0.659	0.641	0.597	0.638
Mn	0.025	0.017	0.038	0.017
Mg	1.162	1.123	1.163	1.236
Na	0.219	0.186	0.232	0.198
K	0.006	0.009	0.002	0.007
Σ Cat.	11.078	11.037	11.058	11.065
Mg#	0.638	0.637	0.661	0.659

Strongly pinitized grains of cordierite could be measured and yielded X_{Mg} = 0.64–0.66 (Table 5).

Discussion

Textural relations between mineral assemblages and the sequence of metamorphic overprints

Since these migmatites underwent three metamorphic events, a pre-Variscan high-T, a Variscan medium P-T amphibolite-facies, and a weak, Eo-Alpine, greenschist-facies overprint, and due to the fine-grained nature of the rocks and the development of subsequent mineral assemblages, detailed petrography involving the scanning electron microscope was performed to identify equilibrium mineral assemblages and attribute them to the metamorphic events. It is well known that in the case of the ÖC, the dominant mineral assemblages can be attributed to the Variscan metamorphic event and that the Eo-Alpine metamorphic overprint increases from NW to SE (e.g. Hoinkes et al. 1997; Neubauer et al. 1999). According to the metamorphic map of the ÖC, the Winnebach migmatite shows a strong Eo-Alpine metamorphic overprint since it is in the vicinity of the chloritoid-in isograd and the other two migmatites show a greenschist-facies metamorphic overprint. In the case of the Winnebach migmatite, the minerals plagioclase$_{1,2}$ + K-feldspar + garnet$_{1,2}$ + pinites + biotite + kyanite$_{1,2}$ + zircon + apatite + muscovite + chloritoid + quartz are present. We interpret garnet cores (garnet$_1$) as being part of the Variscan mineral assemblage while the Ca-rich rims (garnet$_2$) are attributed to the Eo-Alpine metamorphic event. Pre-Variscan cordierite, which occurs only as pinite relicts, could have formed by the model reaction:

Biotite + Sillimanite = Cordierite + K-feldspar + Melt   (4)

During the Variscan event garnet$_1$ + kyanite$_1$ + plagioclase$_1$ + biotite formed, which led to the transformation of cordierite into kyanite and biotite by reactions such as (3) and into garnet$_1$ by reactions such as:

Cordierite + Biotite = Garnet + K-feldspar + H$_2$O   (5)

No association of kyanite and crystallized melt domains could be observed and we propose that kyanite formed later, most probably during the Variscan metamorphic event, which also led to the formation of amphibolite-facies staurolite (Chowanetz 1991). This interpretation is in contrast to Klötzli-Chowanetz (2001) who attributed kyanite formation to the migmatization. During the Eo-Alpine metamorphic event garnet$_2$ (Ca-rich) + plagioclase$_2$ (Na-rich) + kyanite$_2$ + biotite + chloritoid formed in small microdomains along former fractures.

The mineral assemblage in the Verpeil migmatite was also determined by SEM and micro-RAMAN spectroscopy and is: K-feldspar + plagioclase + garnet + biotite + muscovite + cordierite + andalusite + kyanite + clinozoisite + zircon + apatite + rutile + quartz. Cordierite is interpreted to be the only remaining pre-Variscan relict. The dominant Variscan mineral assemblage is garnet + biotite + kyanite + K-feldspar + quartz. Again, cordierite most likely reacted to form garnet, biotite and kyanite by model reactions (3) and (5) as shown in Figures 7a–c. Andalusite formed during the retrograde Variscan evolution as described further south by Tropper & Hoinkes (1996). The Eo-Alpine mineral assemblage is chlorite + muscovite + albite, consistent with the Eo-Alpine greenschist-facies conditions in this area (Hoinkes et al. 1997).

The mineral assemblage of the Nauderer Gaisloch migmatite is garnet + quartz + biotite + plagioclase + muscovite +

kyanite$_{1,2}$. In contrast to Verpeil, no relict cordierite is present anymore and only pseudomorphs after cordierite containing biotite and kyanite occur. Kyanite$_1$, which occurs as large grains in the matrix, was probably formed during the Variscan event, which also led to the transformation of cordierite into pseudomorphs containing biotite + kyanite$_2$ according to reaction (3) where kyanite$_2$ occurs as small needles (Fig. 10).

Thermobarometry

Multi-equilibrium calculations: The simultaneous calculation of all possible reactions within a defined chemical system has been done by using the program THERMOCALC v. 2.7 (Holland 1999, written comm.) and the data set of Holland & Powell (1998). The natural composition of coexisting minerals is taken into account using the activity models for garnet, plagioclase, muscovite and biotite from the set of proposed activity models from the program MacAX (Holland 1999, written comm.).

Inverse equilibrium approach (WEBINVEQ): In this approach, by giving the activities of the end-members of the coexisting minerals, a least squares estimate of the pressure and temperature of the equilibration of a mineral assemblage is calculated, based on the activities of the participating phases. Instead of using a set of independent equilibria, P and T estimates are found by finding the best-fit hyperplane to the partial molar free energies of all phase components. The basic principles of the method are described by Gordon (1992).

Pseudosections: Equilibrium phase diagrams for unmelted protolith paragneiss samples adjacent to the leucosomes, with the given bulk-rock composition were calculated in the chemical system SiO$_2$–TiO$_2$–Al$_2$O$_3$–FeO–MnO–MgO–CaO–Na$_2$O–

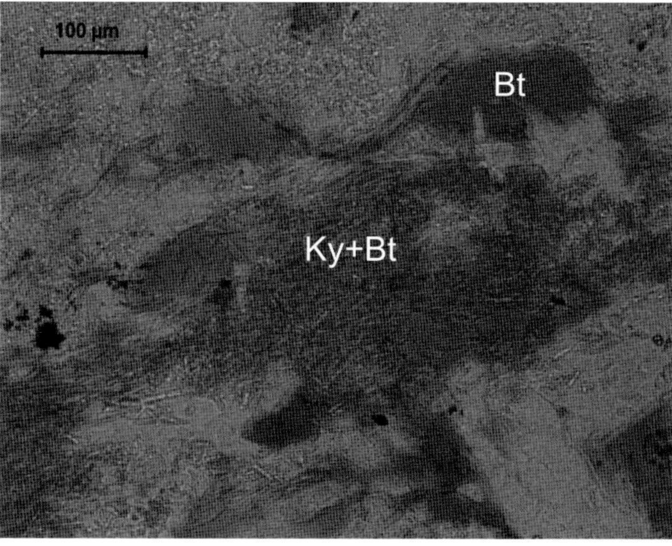

Fig. 10. Photograph of a thin section of sample NA49a from the Nauderer Gaisloch migmatite showing the intergrowth of biotite and kyanite (//P).

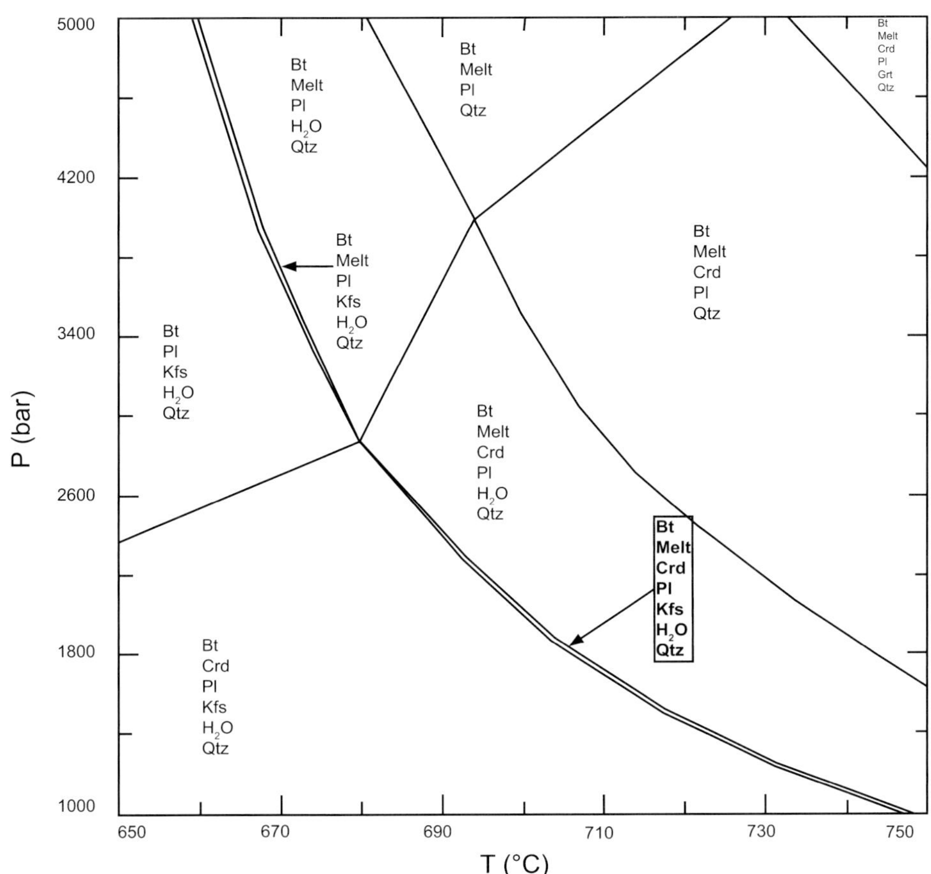

Fig. 11. Pseudosection calculated for sample VP2b from the Verpeil migmatite. The boxed assemblage biotite + cordierite + plagioclase + K-feldspar + melt + quartz is the observed mineral assemblage in the migmatite and is stable over a T-range of 670–750 °C and <2.8 kbar. Abbreviations: Crd: cordierite, Bt: biotite, Kfs: K-feldspar, Pl: plagioclase, Qtz: quartz.

K_2O–H_2O with the software PerpleX (Connolly 2005, written comm.) and the thermodynamic database of Holland & Powell (1998).

In addition to the programs discussed above, conventional thermobarometry using Fe-Mg exchange reactions and net-transfer equilibria was done with the program P-T-t-path by Spear & Kohn (1999 written comm.). For temperature calculations the garnet – biotite thermometer with the calibrations of Kleemann & Reinhardt (1994), Patiño Douce (1993), Perchuk & Lavrent'eva (1983) and Ferry & Spear (1978) with Berman (1990) were used. Pressure calculations were done with the the garnet – kyanite – plagioclase – quartz (GASP) barometer with the calibration of Koziol & Newton (1989) and the garnet – plagioclase – biotite – muscovite barometer using the calibration of Chun-Ming Wu (2004).

Pre-Variscan thermobarometry

Verpeil migmatite: The P-T conditions for this event were calculated with a pseudosection for a biotite-plagioclase host rock sample with the bulk-rock composition of $SiO_2 = 77.51$, $TiO_2 = 0.61$, $Al_2O_3 = 11.17$, $Fe_2O_3 = 2.43$, $MnO = 0.05$, $MgO = 1.42$, $CaO = 1.99$, $Na_2O = 2.99$, $K_2O = 1.62$, $P_2O_5 = 0.19$, Total = 99.96 and LOI = 1.11. The calculations yielded about 670–750 °C and pressures <2.8 kbar for the maximal baric stability of the assemblage biotite + cordierite + plagioclase + K-feldspar + quartz + melt as shown in Figure 11. These low P estimates also indicate that garnet was not stable during the high-T event.

Winnebach migmatite: Textural evidence from the Winnebach migmatite shows that during the pre-Variscan event the water saturated granite solidus was overstepped indicating temperatures of at least 650 °C at pressures <5 kbar (Boettcher & Wyllie 1968). Hoinkes (1973) experimentally obtained temperatures of 685 °C and pressures ≥4 kbar for the pre-Variscan event. Similar P-T conditions as for the Verpeil migmatite were also calculated using the composition of a biotite-plagioclase host rock sample from the Winnebach migmatite given by Hoinkes et al. (1972). The obtained pressure estimates of <2.8 kbar are also in agreement with the petrogenetic grid of Spear et al. (1999) since the presence of K-feldspar in the leucosomes indicates pressures <4 kbar, which is also consistent with the absence of garnet during migmatization. In contrast to these data, Klötzli-Chowanetz (2001) assumes partial anatexis in the kyanite stability field, since she interprets kyanite$_1$ to be a product of anatectical reactions above the invariant point of the system KASH, which is above 4 kbar and hence attributes kyanite$_1$ to be of pre-Variscan origin. She assumes that garnet$_1$ is also of pre-Variscan origin and in combination with the petrogenetic grid of Spear et al. (1999) this indicates pressures of at least 7–8 kbar for the partial anatexis.

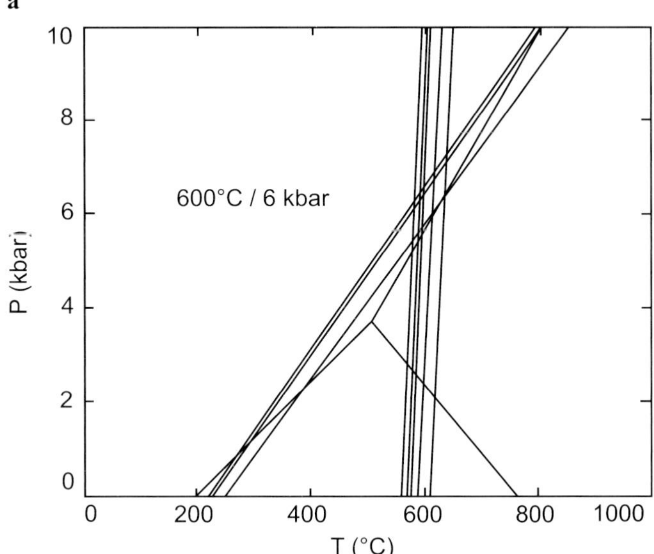

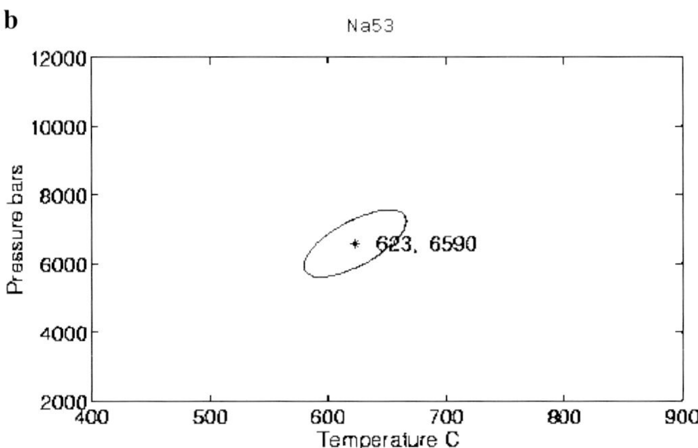

Fig. 12. a) Variscan P-T conditions of the Verpeil migmatite calculated with the program P-T-t-path (sample VP16). b) Variscan P-T conditions of the Nauderer Gaisloch migmatite calculated with the program WEBINVEQ (sample NA 53).

Variscan thermobarometry

Verpeil migmatite: Garnet-biotite temperatures from the Verpeil migmatite range from 470 to 650 °C (at 5 kbar). The most uniform data set were obtained using the calibration of Kleemann & Reinhardt (1994), which yielded 530–600 °C. The highest results were obtained with the calibration Patiño Douce (1993) and yielded approximately 660 °C. The lowest temperatures were calculated with the calibration of Ferry and Spear (1978) with the garnet activity model of Berman (1990). The calibration of Perchuk & Lavrent'eva (1983) also yielded uniform temperatures of 580–615 °C for garnet-biotite pairs from the Verpeil migmatite. Calculation with THERMOCALC v 2.7 yielded the following reactions for the equilibrium assemblage garnet + plagioclase + K-feldspar + quartz:

Phlogopite + Anorthite
  = Pyrope + Grossular + Eastonite + Quartz   (6)

Phlogopite + Annite + Anorthite
  = Grossular + Almandine + Eastonite + Quartz   (7)

Pyrope + Annite + Anorthite =
  Grossular ⊦ Almandine + Eastonite + Quartz   (8)

Almandine + Phlogopite = Pyrope + Annite   (9)

Due to the small grossular component in the garnets, the P-estimates have to be treaded with caution (Todd 1998). Nonetheless, these calculations yield P-T conditions of 6–7 kbar and 550–650 °C. These data are in very good agreement with the data of Tropper & Hoinkes (1996), who calculated metamorphic conditions of 600–640 °C and 5–6 kbar for the Variscan event in the western ÖC. Calculations with the program P-T-t-path using the garnet – biotite thermometer and the garnet – kyanite – plagioclase – quartz (GASP) barometer yielded the same P-T conditions of 600 °C and 6 kbar (Fig. 12a). Calculations using WEBINVEQ yielded 5.8 ± 0.7 kbar and 612 ± 32 °C.

Winnebach migmatite: Garnet-biotite thermometry, using the calibration of Perchuk & Lavrent'eva (1983) yielded temperatures of 580–620 °C (at 5 kbar), which is a much smaller temperature range than from Chowanetz (1991) and Klötzli-Chowanetz (2001) who obtained temperatures of 480–700 °C for the Variscan event. Calculation with the program THERMOCALC resulted in the following reactions using the mineral assemblage garnet$_1$ + biotite + plagioclase$_1$ + K-feldspar + quartz:

Phlogopite + Anorthite + Quartz
  = Pyrope + Grossular + Sanidine + H$_2$O   (10)

Pyrope + Grossular + Muscovite
  = Phlogopite + Anorthite   (11)

Grossular + Muscovite + Quartz
  = Anorthite + Sanidine + H$_2$O   (12)

Phlogopite + Muscovite + Quartz
  = Pyrope + Sanidine + H$_2$O   (13)

The intersection of these reactions yielded P-T conditions of 5.7 ± 2.6 kbar and 676 ± 56 °C.

Nauderer Gaisloch migmatite: Calculations with the program P-T-t-path using the garnet biotite thermometer, the GASP barometer and the garnet – plagioclase – biotite – muscovite barometer according to reaction (11) using the calibration of Chun-Ming Wu (2004), resulted in P-T conditions of 550–570 °C and 4–7 kbar. Calculations with WEBINVEQ resulted in a temperature range of 500–650 °C and pressures of 5–6.5 kbar (Fig. 12b). THERMOCALC v 2.7 yielded similar conditions of 610–660 °C and 5.1–5.3 kbar.

Eo-Alpine thermobarometry

Calculation of the Eo-Alpine P-T conditions was only attempted in the Winnebach migmatite. For these calculations, the assemblage garnet$_2$ (Ca-rich rims) + plagioclase$_2$ (albite-rich rims) + biotite (small grains adjacent to Ca-rich garnet$_2$, Fig. 4) + kyanite + quartz + chloritoid was used. The calculations using the program P-T-t-path yielded P-T conditions of 485 °C and 8.5 kbar as shown in Figure 13. These results are in agreement with thermobarometric results from this area (Tropper & Recheis 2003). Calculations with THERMOCALC v 2.7 resulted in pressures of 8.6 ± 0.2 kbar and slightly lower temperatures of 433 ± 12 °C.

Geochronology

U-Th-Pb electron microprobe dating of monazites from three samples from the leucosome of the Winnebach migmatite yielded ages ranging from 408 ± 46 Ma to 472 ± 36 Ma for the partial anatexis as shown in Table 6. These ages show a higher spread and are also slightly younger than previously reported ages from the Winnebach migmatite which were around 490 Ma (Klötzli-Chowanetz 2001). The geochronological data from monazites of four leucosome samples from the Verpeil migmatite range from 409 ± 50 Ma to 457 ± 42 Ma for the anatectic event, which is similar within the error with the data obtained from the Winnebach migmatite. Geochronological investigations on monazites from two leucosome samples of the Nauderer Gaisloch migmatite resulted in ages ranging from 431 ± 37 Ma to 472 ± 31 Ma (Fig. 14), again in good agreement with the data from the other two migmatites. In addition, a second generation of monazite which formed during the subsequent Variscan metamorphic event was detected. These grains yielded ages ranging from 305 ± 43 Ma to 336 ± 42 Ma (Table 6; Fig. 14).

Discussion

Although our P-T data of the pre-Variscan migmatization yield somewhat lower pressures of <2.8 kbar than previous investigations, the data are in agreement with petrogenetic grids and textural observations. Hoinkes (1973) estimated the P-T conditions of the migmatization, based on experiments, to be 660–685 °C and ≥4 kbar. Söllner et al. (1982) obtained similar P-T results by using petrogenetic grids from the literature. Klötzli-Chowanetz et al. (2001) deduced significantly higher pressures and temperatures of 8 kbar and ca. 750 °C for the anatexis based on petrographic evidence and comparison to the petrogenetic grid by Spear et al. (1999). The latter pressure estimates strongly depend on the interpretation of kyanite as being present during anatexis, which could not be verified in our investigation.

Previous thermobarometric investigations of the Variscan P-T conditions by Veltman (1986), Tropper & Hoinkes (1996), and Tropper & Recheis (2003) from the ÖC yield P-T conditions of 570–750 °C and 5.8–8 kbar. Veltman (1986) obtained from the northern kyanite zone 570–650 °C and 4.3–7.8 kbar. For the sillimanite zone, he reported P-T conditions of 600–750 °C and 4.2–6.8 kbar. Tropper & Hoinkes (1996) obtained 570–640 °C and 5.8–7.5 kbar for the central andalusite zone and Tropper & Recheis (2003) obtained P-T conditions of 469–630 °C and 4.2–7.3 kbar for the northern kyanite zone and 578 °C and 7.2 kbar for the southern kyanite zone. The data from the Verpeil- and the Nauderer Gaisloch migmatites are well in accordance with these results, only the temperatures of the Winnebach migmatite are slightly lower, thus possibly indicating a later Eo-Alpine rejuvination.

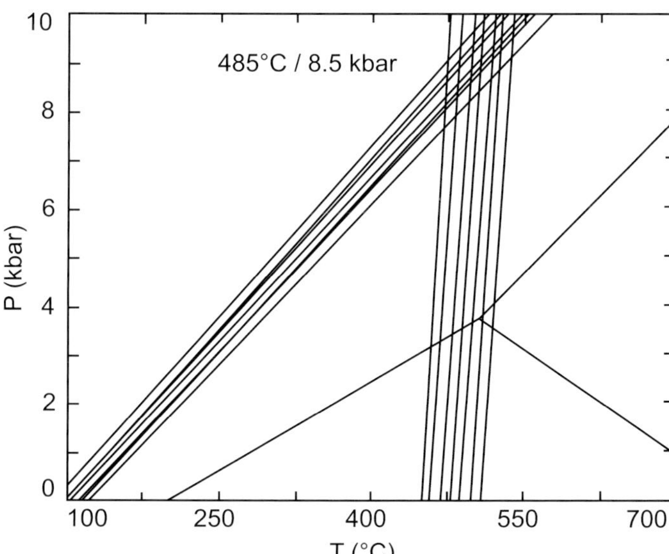

Fig. 13. Eo-Alpine P-T conditions calculated with the program P-T-t-path for the Winnebach migmatite (sample WB 70).

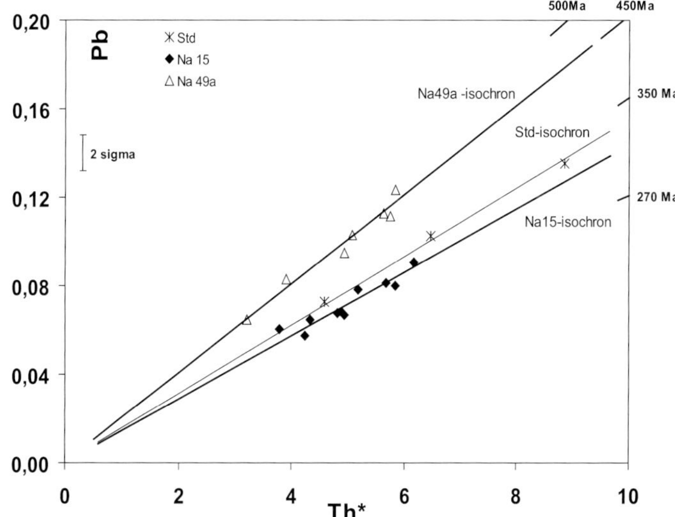

Fig. 14. Th*–Pb diagram showing the isochron ages of the two different monazite generations from the Nauderer Gaisloch migmatite (samples Na49a, Na15). Na49a (open triangles) yields pre-Variscan ages and Na15 (black diamonds) Variscan ages as discussed in the text.

Table 6. U-Th-Pb electron microprobe analyses data of monazites: Th. U and Pb contents. Th* values (Suzuki et al. 1991) and chemical ages (calculated after Montel et al. 1996) of monazites from samples WB70B1, WB29/2, VP28, FB56A, NA15, NA49. Errors are 2σ for the single points and at the 95% confidence level for the weighted average age.

	Y	Th	U	Pb	Th*	Age	Error (2σ)	
WB 70 B1 Mon1	1,306	3,702	0,371	0,095	4,917	433	33	
WB 70 B1 Mon2	1,304	3,525	0,780	0,116	6,081	426	27	
WB 70 B1 Mon3	1,268	3,822	0,241	0,098	4,614	472	36	MSWD = 1.8
WB 70 B1 Mon4	1,702	3,821	1,245	0,164	7,911	463	21	
WB 70 B1 Mon5	1,515	3,759	0,401	0,100	5,073	441	32	
WB 29/2 Mon1	1,639	3,702	0,943	0,138	6,798	455	27	
WB 29/2 Mon2	1,627	3,485	0,985	0,132	6,716	442	27	
WB 29/2 Mon3	1,387	2,246	0,602	0,084	4,219	444	43	
WB 29/2 Mon4	1,038	2,548	0,501	0,086	4,192	457	43	MSWD = 0.74
WB 29/2 Mon5	1,282	2,174	0,544	0,072	3,953	408	46	
WB 29/2 Mon6	1,141	3,203	0,371	0,087	4,418	438	41	
VP 28 Mon1	1,335	2,121	0,353	0,060	3,275	409	50	
VP 28 Mon2	1,616	1,765	0,664	0,080	3,945	457	42	
VP 28 Mon3	1,204	4,280	0,428	0,115	5,685	452	29	MSWD = 1.6
VP 28 Mon4	0,824	3,937	0,314	0,101	4,969	456	33	
VP 28 Mon5	1,189	2,809	0,421	0,076	4,185	408	39	
FB 56A Mon1	1,763	3,726	0,729	0,118	6,114	433	30	
FB 56A Mon2	2,056	7,259	0,781	0,200	9,824	454	18	
FB 56A Mon3	1,881	3,530	0,759	0,115	6,015	427	30	
FB 56A Mon4	1,127	2,332	0,280	0,061	3,249	419	56	MSWD = 1.6
FB 56A Mon5	1,749	3,214	1,379	0,143	7,728	416	24	
FB 56A Mon6	1,927	3,591	1,532	0,164	8,610	428	21	
NA 15 Mon1	1,968	2,887	0,615	0,068	4,885	313	37	
NA 15 Mon2	2,204	3,056	0,652	0,079	5,178	340	35	
NA 15 Mon3	2,056	2,606	0,678	0,068	4,807	316	38	
Na 15 Mon4	2,231	3,788	0,632	0,080	5,838	308	31	
NA 15 Mon5	1,531	4,329	0,566	0,091	6,169	330	29	
NA 15 Mon6	1,882	2,343	0,611	0,065	4,329	336	42	MSWD = 0.75
NA 15 Mon7	1,978	2,848	0,641	0,067	4,928	305	37	
NA 15 Mon8	1,697	2,311	0,454	0,061	3,791	358	48	
NA 15 Mon9	1,886	3,392	0,702	0,081	5,674	321	32	
NA 15 Mon10	1,967	2,088	0,662	0,057	4,237	305	43	
NA 49 Mon1	1,036	2,053	0,355	0,065	3,217	449	56	
NA 49 Mon2	1,137	2,705	0,364	0,083	3,901	475	46	
NA 49 Mon3	1,484	3,242	0,517	0,095	4,935	431	37	
NA 49 Mon4	1,424	3,516	0,471	0,103	5,061	455	36	MSWD = 0.88
NA 49 Mon5	1,736	4,310	0,464	0,123	5,836	472	31	
NA 49 Mon6	1,536	3,993	0,536	0,112	5,749	435	31	
NA 49 Mon7	1,331	4,054	0,480	0,113	5,630	448	32	

The Eo-Alpine P-T conditions of this investigation from the Winnebach migmatite agree with P-T estimates of the chloritoid-in isograd of 421–495 °C and 6.7–9.4 kbar for samples from the Ortler-Campo Crystalline Complex (Mair et al. 2006) as well as with the Eo-Alpine P-increase from NW to SE (e.g. Veltman 1986; Tropper & Recheis, 2003).

Although the obtained U-Th-Pb ages of monazite cluster around 430–450 Ma in all three migmatites and thus confirm a pervasive pre-Variscan event throughout the ÖC, which has not been confirmed yet, still large discrepancies concerning previously published age data remain. Söllner & Hansen (1987) and Söllner (2001) obtained Pan-African ages of 607–670 Ma on zircons, which they interpret as the age of the anatexis. Klötzli-Chowanetz et al. (1997) obtained an age of the migmatization of 490 ± 9 Ma and minimum cooling ages, based on Rb-Sr isochron ages of muscovite, of 461 ± 4 Ma (Chowanetz 1991; Klötzli-Chowanetz 2001). On the other hand, even the age of the pre-Variscan event in the Winnebach migmatite is still not unambiguous since Klötzli-Chowanetz et al. (1997) also provided some evidence for even older thermal events at 560 Ma and 635 Ma. Similarly, highly variable pre-Variscan ages have also been obtained from the Nauderer Gaisloch migmatite, which yielded several stages of zircon growth, namely 531 ± 11 Ma, which is thought to be the age of the anatexis, 585 ± 8 Ma (Pan-African), and 430 ± 6 Ma (Silurian high-T event). The age of the tonalite is constrained to be 487 ± 11 Ma and that of a pegmatite crosscutting the migmatites at the Nauderer Gaisloch is 472 ± 26 Ma (Klötzli-Chowanetz et al. 2001; Schweigl 1993). These data are also somewhat older than the ages we obtained from this migmatite complex (Table 6). The only ages that were so far available from the Verpeil area are Rb-Sr ages from two different types of orthogneisses, which are in the vicinity of the migmatite body (Bernhard 1994). The first type, a hedenbergite-hornblende orthogneiss, shows a minimum intrusion age of 417 ± 9 Ma, which is similar to the age of the second type of intrusion, a hornblende-bearing tonalitic gneiss, which shows an age of 408 ± 20 Ma. Both ages are considerable younger than the ages of the monazites obtained in this study. The majority of intrusives in the central ÖC shows ages in the range between 420 and 485 Ma (Thöni 1999). These ages seem to correlate well with the monazite ages of the ÖC migmatites, thus placing a rather reliable time constraint on the formation of the migmatites during the early Ordovician.

Acknowledgments

The authors wish to thank the FWF for partial financial support in the course of project P17878-N10. Edgar Mersdorf and Bernhard Sartory are thanked for their help with the electron microprobe. The manuscript was considerably improved by the careful reviews of Thorsten Nagel and Igor Petrik. Niko Froitzheim is thanked for the editorial handling.

REFERENCES

Berman, R.G. 1990: Mixing properties of Ca-Mg-Fe-Mn garnets, American Mineralogist 75, 328–344.

Bernhard, F. 1994: Zur magmatischen und metamorphen Entwicklung im westlichen Ötztal-Stubai Kristallin (Bereich Feichten-Verpeil, mittleres Kaunertal). Unpublished Masters Thesis, University Graz, 314 pp.

Bernhard, F., Klötzli, U., Thöni, M. & Hoinkes, G. 1996: Age, origin and geodynamic significance of a polymetamorphic felsic intrusion in the Ötztal Crystalline Basement, Tirol, Austria. Mineralogy and Petrology 58, 171–196.

Boettcher, A.L. & Wyllie, P.J. 1968: Melting of granite with excess water to 30 kbar pressure. Journal of Geology 76, 235.

Chowanetz, E. 1991: Strukturelle und geochronologische Argumente für eine altpaläozoische Anatexis im Winnebachmigmatit (Ötztal/Tirol, Österreich). Mitteilungen der Gesellschaft der Geologie und Bergbaustudenten Österreichs 37, 15–36.

Cocherie, A., Mezeme, E.B., Legendre, O., Fanning, C.M., Faure, M. & Rossi, P. 2005: Electron-microprobe dating as a tool for determining the closure of Th-U-Pb system in migmatitic monazites. American Mineralogist 90, 607–618, 2005.

Drong, H-J. 1959: Das Migmatitgebiet des „Winnebachgranits" (Ötztal-Tirol) als Beispiel einer petrotektonischen Analyse. Tschermaks Mineralogische Petrographische Mitteilungen 7, 1–69.

Ferry, J.M. & Spear, F.S. 1978: Experimental calibration of the partitioning of Fe and Mg between biotite and garnet. Contributions to Mineralogy and Petrology 66, 113–117.

Finger, F. et al. 1996: Altersdatierungen von Monaziten mit der Elektronenstrahlmikrosonde – Eine wichtige neue Methode in den Geowissenschaften. TSK 6, 118–122.

Finger, F., Broska, I., Roberts, M.P. & Schermaier, A. 1998: Replacement of primary monazite by apatite-allanite-epidote coronas in amphibolite facies granite gneiss from the eastern Alps. American Mineralogist 83, 248–258.

Finger, F. & Helmy, H. 1998: Composition and total-Pb model ages of monazite from high-grade paragneisses in the Abu Swayel area, southern Eastern Desert, Egypt. Mineralogy and Petrology 62, 269–289.

Goncalves, P., Nicollet, C. & Montel, J.M. 2004: Petrology and in situ U-Th-Pb Monazite Geochronology of Ultrahigh-Temperature Metamorphism from the Andriamena Mafic Unit, North-Central Madagascar. Significance of a Petrographical P-T Paths in a Polymetamorphic Context. Journal of Petrology 10, 1923–1957.

Gordon, T.M. 1992: Generalized thermobarometry: solution of the inverse chemical equilibrium problem using data for individual species. Geochimica et Cosmochimica Acta 56, 1793–1800.

Hammer, W. 1925: Cordieritführende metamorphe Granite aus den Ötztaler Alpen. Tschermaks Mineralogische Petrographische Mitteilungen 38, 67–87.

Hoinkes, G., Purtscheller, F. & Schantl, J. 1972: Zur Petrographie und Genese des Winnebachgranites (Ötztaler Alpen, Tirol). Tschermaks Mineralogische Petrographische Mitteilungen 18, 292–311.

Hoinkes, G. 1973: Die Anatexis des Winnebachgranites (Ötztaler Alpen, Österreich) am Beispiel eines Aufschlusses. Tschermaks Mineralogische Petrographische Mitteilungen 20, 225–239.

Hoinkes, G., Thöni, M., Bernhard, F., Kaindl, R., Lichem, C., Schweigl, J., Tropper, P. & Cosca, M. 1997: Metagranitoids and associated metasediments as indicators for the pre-Alpine magmatic and metamorphic evolution of the Western Austroalpine Ötztal Basement (Kaunertal, Tirol). Schweizerische Mineralogische und Petrographische Mitteilungen 77, 299–314.

Hoinkes, G., Koller, F., Rantitsch, G., Dachs, E., Höck, V., Neubauer, F. & Schuster, R. 1999: Alpine metamorphism of the Eastern Alps. Schweizerische Mineralogische Petrographische Mitteilungen 79, 155–181.

Holland, T.J. B. & Powell, R. 1998: An internally-consistent thermodynamic data set for phases of petrological interest. Journal of Metamorphic Geology 8, 89–124.

Jarosevich, E.J., Nelen, J.A. & Norberg, J.A. 1980: Reference samples for electron microprobe analysis. Geostandards Newsletter 4, 87–133.

Jarosevich, E.J. & Boatner, L.A. 1991: Rare earth element reference samples for electron microprobe analysis. Geostandards Newsletter 15, 397–399.

Jercinovich, M.J. & Williams, M.L. 2005: Analytical perils (and progress) in electron microprobe trace element analysis applied to geochronology: Background acquisition, interferences and beam irradiation effects. American Mineralogist 90, 526–546.

Kleemann, U. & Reinhardt, J. 1994: Garnet – biotite thermometry revisited; the effect of Al (super VI) and Ti in biotite. European Journal of Mineralogy 6, 925.

Klötzli-Chowanetz, E., Klötzli, U. & Koller, F. 1997: Lower Ordovician migmatization in the Ötztal crystalline basement (Eastern Alps, Austria): linking U-Pb and Pb-Pb dating with zircon morphology. Schweizerische Mineralogische Petrographische Mitteilungen 77, 315–324.

Klötzli-Chowanetz, E. 2001: Migmatite des Ötztalkristallins – Petrologie und Geochronologie. Unpublished PhD Thesis University of Vienna, 155 pp.

Koziol, A.M. & Newton, R.C. 1989: Grossular activity – composition relationships in ternary garnets determined by reversed displaced equilibrium experiments. Contributions to Mineralogy and Petrology 103, 423.

Mair, V., Tropper, P., Schuster, R. 2006: The P-T-t evolution of the Ortler-Campo Crystalline (South-Tyrol/Italy). PANGEO 2006, Innsbruck University Press, 186–187.

Mehnert, K.R. 1968: Migmatites and the origin of granitic rocks. Elsevir, Amsterdam.

Miller, C. & Thöni, M. 1995: Origin of eclogites from the Austroalpine Ötztal Basement (Tirol, Austia): geochemistry and Sm-Nd vs. Rb-Sr isotope systematics. Chemical Geology 137, 283–310.

Neubauer, F., Hoinkes, G., Sassi, F.P., Handler, R., Höck, V., Koller, F. & Frank, W. 1999: Pre-Alpine metamorphism in the Eastern Alps. Schweizerische Mineralogische Petrographische Mitteilungen 79, 41–62.

Ohnesorge, T. 1905: Die vorderen Kühtaier Berge. Verhandlungen der Geologischen Reichsanstalt Wien. Jahrgang 1905, 175–182.

Patiño Douce, A.E. 1993: Titanium substitution in biotite: an empirical model with applications to thermometry, O_2 and H_2O barometries, and consequences for biotite stability. Chemical Geology 108, 133–162.

Perchuk, L.L. & Lavrent'eva, I.V. 1983: Experimental investigations of exchange equilibria in the system cordierite – garnet – biotite. In: Saxena (Ed). Kinetics and Equilibrium in Mineral Reactions Berlin (Springer Verlag), 199–239.

Schindlmayer, A. 1999: Granitoids and plutonic evolution of the Ötztal-Stubai Massif: a key for understanding the Early Paleozoic history of the Austroalpine crystalline basement in the western Eastern Alps. Unpublished Ph.D. Thesis Universität Salzburg, 287 pp.

Schweigl, J. 1993: Kristallingeologische Untersuchungen in den Nauderer Bergen (Westliche Ötztaler Alpen, Tirol). Unpublished Masters Thesis Universität Wien, 87 pp.

Söllner, F., Schmidt, K., Bumann, A. & Hansen, B.T. 1982: Zur Altersstellung des Winncbach-Migmatits im Ötztal (Ostalpen). Verhandlungen der Geologischen Bundesanstalt 1982, 95–106.

Söllner, F. & Hansen, B.T. 1987: „Pan-afrikanisches" und „kaledonisches" Ereignis im Ötztal-Kristallin der Ostalpen: Rb-Sr- und U-Pb-Altersbestimmungen an Migmatiten und Metamorphiten. Jahrbuch der Geologischen Bundesanstalt 130/4, 529–569.

Söllner, F. 2001: The Winnebach migmatite (Ötz-Stubai crystalline unit) – evidence for a Pan-Africanic metamorphism in an overthrust nappe sequence in the Eastern Alps. Geologisch Paläontologische Mitteilungen der Universität Innsbruck 25, 199–200.

Spear, F.S., Kohn, M.J. & Cheney, J.T. 1999: P-T paths from anatectic pelites. Contributions to Mineralogy and Petrology 134, 17–32.

Thöni, M. 1999: A review of geochronological data from the Eastern Alps. Schweizerische Mineralogische Petrographische Mitteilungen 79, 209–230.

Todd, C.S. 1998: Limits on the precision of geobarometry at low grossular and anorthite content. American MIneralogist 83, 1161–1168.

Tropper, P. & Hoinkes, G. 1996: Thermobarometry in Al_2SiO_5-bearing metapelites in the western Austroalpine Ötztal-basement. Mineralogy and Petrology 58, 145–170.

Tropper, P. & Recheis, A. 2003: Garnet zoning as a window into the metamorphic evolution of a crystalline complex: the northern and central Austroalpine Ötztal-Complex as a polymorphic example. Mitteilungen der Österreichischen Geologischen Gesellschaft 94, 27–53.

Veltman, C. 1986: Zur Polymetamorphose metapelitischer Gesteine des Ötztal Stubaier Altkristallins. Unpublished Ph.D. Thesis Universität Innsbruck, 164 pp.

Williams, M.L., Jercinovic, M.J., Goncalves, P. & Mahan, K. 2006: Format and philosophy for collecting, compiling, and reporting microprobe monazite ages. Chemical Geology 225, 1–15.

Wu, C.-M., Zhang, J. & Ren, L.-D. 2004: Empirical garnet – muscovite – plagioclase – quartz – geobarometry in medium- to high-grade metapelites. Lithos 78, 319.

Manuscript received January 18, 2008
Revision accepted June 3, 2008
Published Online first November 13, 2008
Editorial Handling: Nikolaus Froitzheim & Stefan Bucher

1661-8726/08/01S127-29
DOI 10.1007/s00015-008-1289-6
Birkhäuser Verlag, Basel, 2008

Swiss J. Geosci. 101 (2008) Supplement 1, S127–S155

From subduction to collision:
thermal overprint of HP/LT meta-sediments in the north-eastern Lepontine Dome (Swiss Alps) and consequences regarding the tectono-metamorphic evolution of the Alpine orogenic wedge

MICHAEL WIEDERKEHR[1,2,*], ROMAIN BOUSQUET[2], STEFAN M. SCHMID[1] & ALFONS BERGER[3]

Key words: Lepontine dome, meta-sediments, Fe-Mg carpholite, Barrovian metamorphism, high-pressure metamorphism, Alpine tectonics

ABSTRACT

The Cenozoic-age metamorphic structure of the Alps consists of a through-going pressure-dominated belt (blueschists and eclogites) that strikes parallel to the orogen and was later truncated by two thermal domes characterised by Barrow-type metamorphism (Lepontine dome and Tauern window). This study documents for the first time that relics of Fe-Mg carpholite occur also within meta-sedimentary units that are part of the north-eastern Lepontine structural and metamorphic dome, where so far exclusively Barrovian assemblages were found. They occur in meta-sediments of both Valais Ocean-derived Lower Penninic Bündnerschiefer and structurally lower Europe-derived Sub-Penninic cover nappes and slices. These high-pressure units were subsequently overprinted by a thermal event, as is documented by the growth of new minerals typical for Barrovian metamorphism.

We present evidence for a two-stage metamorphic evolution in the northern part of the Lepontine dome: (1) Early subduction-related syn-D1 (Safien phase) HP/LT metamorphism under blueschist facies conditions (350–400 °C and 1.2–1.4 GPa) was immediately followed by "cold" isothermal (or cooling) decompression during D2 nappe-stacking (Ferrera phase). (2) Collision-related Barrovian overprint (500–570 °C and 0.5–0.8 GPa) postdates the D3 nappe-refolding event (Domleschg phase) and represents a late heating pulse, separated by D2 and D3 from the D1 high-pressure event. It occurred before and/or during the initial stages of D4 (Chièra phase) representing a second nappe-refolding event.

In discussing possible heat sources for the late Barrow-type heating pulse it is argued that heat release from radioactive decay of accreted material may play an important role in contributing much to heat production. Based on the field evidence, we conclude that heat transfer was essentially conductive during these latest stages of the thermal evolution.

1. Introduction

The zoning of Alpine metamorphism is rather complex, evolving over a very long period of time before, during and after the collision of Europe with Adria, i.e. from Late Cretaceous to Late Cenozoic times. Mapping of metamorphic facies in the Alps started with early pioneering studies based on the spatial distribution of index minerals and mineral assemblages (Wenk 1962; Niggli & Niggli 1965; Trommsdorff 1966; Frey 1969; Fox 1975; Frey et al. 1980). Metamorphic maps at the scale of the Alpine orogen, showing the spatial arrangement of the different metamorphic facies types, were repeatedly synthesised and improved (Ernst 1971; Niggli & Zwart 1973; Frey et al. 1999; Oberhänsli et al. 2004). The Cenozoic-age metamorphic pattern is characterised by a pressure-dominated belt (blueschists and eclogites) that strikes orogen-parallel but is interrupted by two thermal domes, the Lepontine dome in the Central Alps and the Tauern window in the Eastern Alps (Oberhänsli et al. 2004).

Our area of investigation is located at the NE border of the Lepontine thermal dome. There, along strike of the tectonic units, a remarkable metamorphic field gradient that ranges from pressure-dominated blueschist facies in the NE to temperature-dominated Barrovian metamorphism in the SW is observed within an amazingly short distance (<10 km, Fig. 1 & 2). This allows for a clear correlation between the two metamorphic events and structures that resulted from a polyphase deformation history. Hence, the area is well suited for studying spatial and temporal relationships between these two types of metamorphism, including their relative timing in respect to discrete deformation phases linked to particular geodynamical stages.

The availability of meta-sediments all along strike facilitates the reconstruction of the metamorphic and structural evolution

[1] Geologisch-Paläontologisches Institut, Universität Basel, Bernoullistrasse 32, CH-4056 Basel, Switzerland. E-mail: m.wiederkehr@unibas.ch
[2] Institut für Geowissenschaften, Universität Potsdam, Karl-Liebknecht-Strasse 24/25, D-14476 Potsdam/Golm, Germany.
[3] Institut für Geologie, Universität Bern, Baltzerstrasse 3, CH-3012 Bern, Switzerland.
*Corresponding author: Michael Wiederkehr. E-mail: m.wiederkehr@unibas.ch

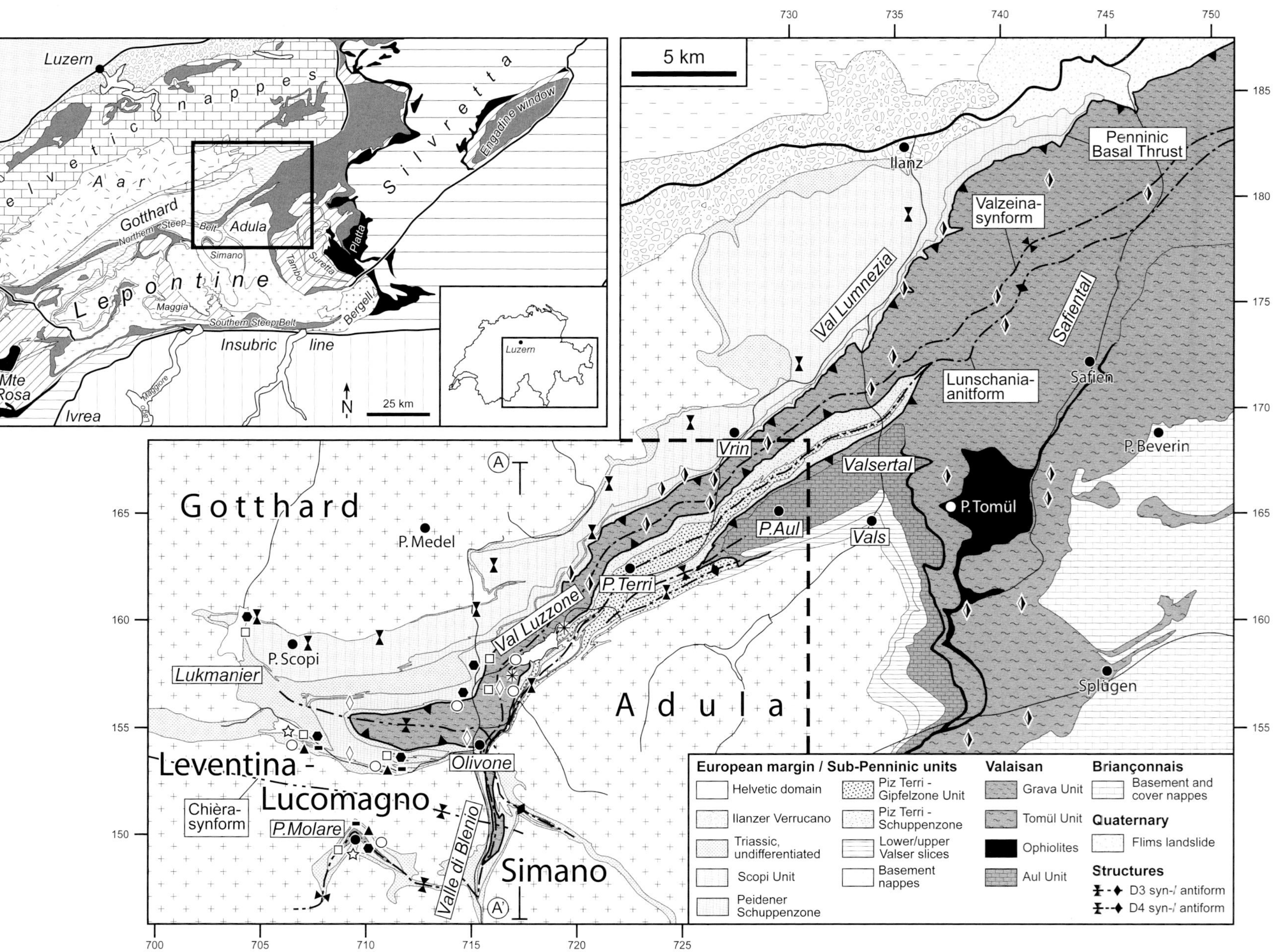

Fig. 1. Tectonic sketch map of the study area showing the main geographic localities mentioned in the text as well as traces of axial planes of major D3 and D4 folds, and the main occurrences of index minerals and mineral assemblages found in the meta-sedimentary units (light-grey: Sub-Penninic/European units, dark-grey: Lower Penninic/Valaisan units; symbols are explained in Figure 2). The tectonic map of the Central Alps in the upper left is after Schmid et al. (2004), the frame shows the location of the study area. Letters A–A' mark the trace of the composite cross section shown in Figure 3. The dashed line delineates the cut-out shown in greater detail in Figure 2.

in the working area. These meta-sediments contain widespread occurrences of Fe-Mg carpholite within an orogen-parallel HP/LT-metamorphic belt in eastern Switzerland (Grisons), characterised by blueschist facies conditions (Goffé & Oberhänsli 1992; Oberhänsli 1994; Oberhänsli et al. 1995; Bousquet et al. 2002). Approaching the Lepontine dome, the same meta-sediments become increasingly affected by a temperature-dominated, Barrovian metamorphic event, as is documented by amphibolite facies mineral assemblages characterised by garnet, biotite, staurolite and kyanite (Chadwick 1968; Frey 1969; Fox 1975; Engi et al. 1995; Frey & Ferreiro Mählmann 1999). No evidence is available, so far, that this part of the Lepontine dome, characterised by this Barrow-type MP/MT metamorphism, could have been previously also affected by HP/LT metamorphism.

This tectono-metamorphic study primarily aims to document this transition from HP/LT blueschist facies metamorphism in the east to amphibolite-grade Barrow-type metamorphism within the Lepontine dome further west. This will allow deducing whether these two contrasting types of metamorphism evolved at the same time but differently in the different parts of the study area, or alternatively, whether they evolved during consecutive stages of the evolution of the Alpine orogen. In the second case the pressure-dominated metamorphism represents an early stage related to subduction, followed by a temperature-dominated event, as proposed by Bousquet et al. (2008). The latter, i.e. a two-stage metamorphic evolution, was also proposed for more southerly located parts of the Lepontine dome. However, the question whether the Barrow-type overprint is associated with a second and discrete heating pulse that followed high-pressure metamorphism (e.g. Engi et al. 2001), or alternatively, simply represents a late stage during isothermal decompression (e.g. Nagel et al. 2002; Keller et al. 2005) is of fundamental importance from a geodynamic point of view. Furthermore, our study will address the important key question concerning the heat source of Barrow-type Lepontine metamorphism, debated since the pioneering metamorphic studies in the Alps. While Niggli (1970) concluded that regional metamorphism in the Lepontine region is caused by tectonic burial during nappe stacking, Wenk (1970) proposed late-stage thermal doming induced by an additional (magmatic) heat source underneath a pre-existing overburden of nappes. The results presented in this study will provide insights that are of fundamental importance for understanding the thermo-mechanical evolution of collisional orogens in general.

2. Geological setting and major tectonic units

The investigated area is located at the north-eastern edge of the Lepontine structural dome and extends from the Lukmanierpass and Pizzo Molare areas in the west to the Safiental area in the east (Fig. 1). The Sub-Penninic nappes, interpreted as derived from the distal European margin (Milnes 1974; Schmid et al. 2004), originally occupied a lower tectonic position within the working area. Among these nappes are basement nappes that predominantly consist of pre-Mesozoic igneous and meta-sedimentary rocks, and cover nappes forming an orogen-parallel belt of Mesozoic meta-sediments. The cover nappes are not interrupted by oblique tectonic contacts (thrusts or faults) and they overlie the pre-Mesozoic basement units or nappes (Gotthard "Massif", Leventina-Lucomagno and Simano nappes). These Sub-Penninic nappes are structurally overlain by Lower Penninic cover nappes that originated from the Valais Ocean, largely consisting of Mesozoic meta-sediments referred to as Bündnerschiefer. The front of the Adula nappe complex only reaches the southern rim of the working area. The occurrences of oceanic remnants that are imbricated with typical continental crustal rocks in the overlying Misox Zone (e.g. Partzsch 1998), and according to some authors also within the Adula-Cima Lunga nappe complex itself (e.g. Trommsdorff 1990, and references therein), indicate that the Adula nappe complex contains slivers from the continent-ocean transition between the European margin and the Valais Ocean (lithospheric mélange; Trommsdorff 1990). The Penninic Basal Thrust represents an early-stage first-order thrust along which the Valaisan Bündnerschiefer were originally thrust onto the Europe-derived Sub-Penninic units. However, this thrust was subsequently isoclinally refolded and hence penetratively overprinted by later structures. The tectonic units, subdivided following the schemes proposed by Schmid et al. (2004) and Berger et al. (2005), are mapped in Figures 1 and 2, as well as in cross section view (Fig. 3). In the following they are further described.

2.1. Sub-Penninic basement nappes

The Gotthard-"massif" is the lowermost thrust sheet of the Sub-Penninic nappe pile (Fig. 3) and represents a backfolded nappe front (Milnes 1974) rather than a par-autochthonous massif. This unit consists of pre-Mesozoic crystalline basement (Steiger 1962; Mercolli et al. 1994) overlain in stratigraphic contact only by an extremely thin veneer of Early to Middle Triassic quartzites, occasionally also containing dolomitic marbles and/or meta-evaporites (Frey 1967).

The Lucomagno-Leventina nappe structurally overlays the Gotthard-"massif", from which it is separated by the Piora Zone, which represents an intervening Sub-Penninic cover nappe (Fig. 3). The southern realms of the Lucomagno-Leventina nappe predominantly consist of Variscan orthogneiss (Leventina gneiss; Casasopra 1939; Köppel et al. 1981; Rütti et al. 2008), forming the deepest outcropping parts of the Ticino sub-dome within the Lepontine dome (Merle et al. 1989). The northern part of the Leventina-Lucomagno nappe reaches the working area and consists of Pre-Mesozoic poly-metamorphic meta-sedimentary complexes (Lucomagno crystalline; Bossard 1925, 1929; Chadwick 1968).

The next higher Sub-Penninic basement nappe, i.e. the Simano nappe, contains Caledonian to Variscan orthogneisses and pre-Mesozoic poly-metamorphic pelitic to psammitic meta-sediments (Jenny et al. 1923; Niggli et al. 1936; Keller 1968; Köppel et al. 1981). It is separated from the underlying

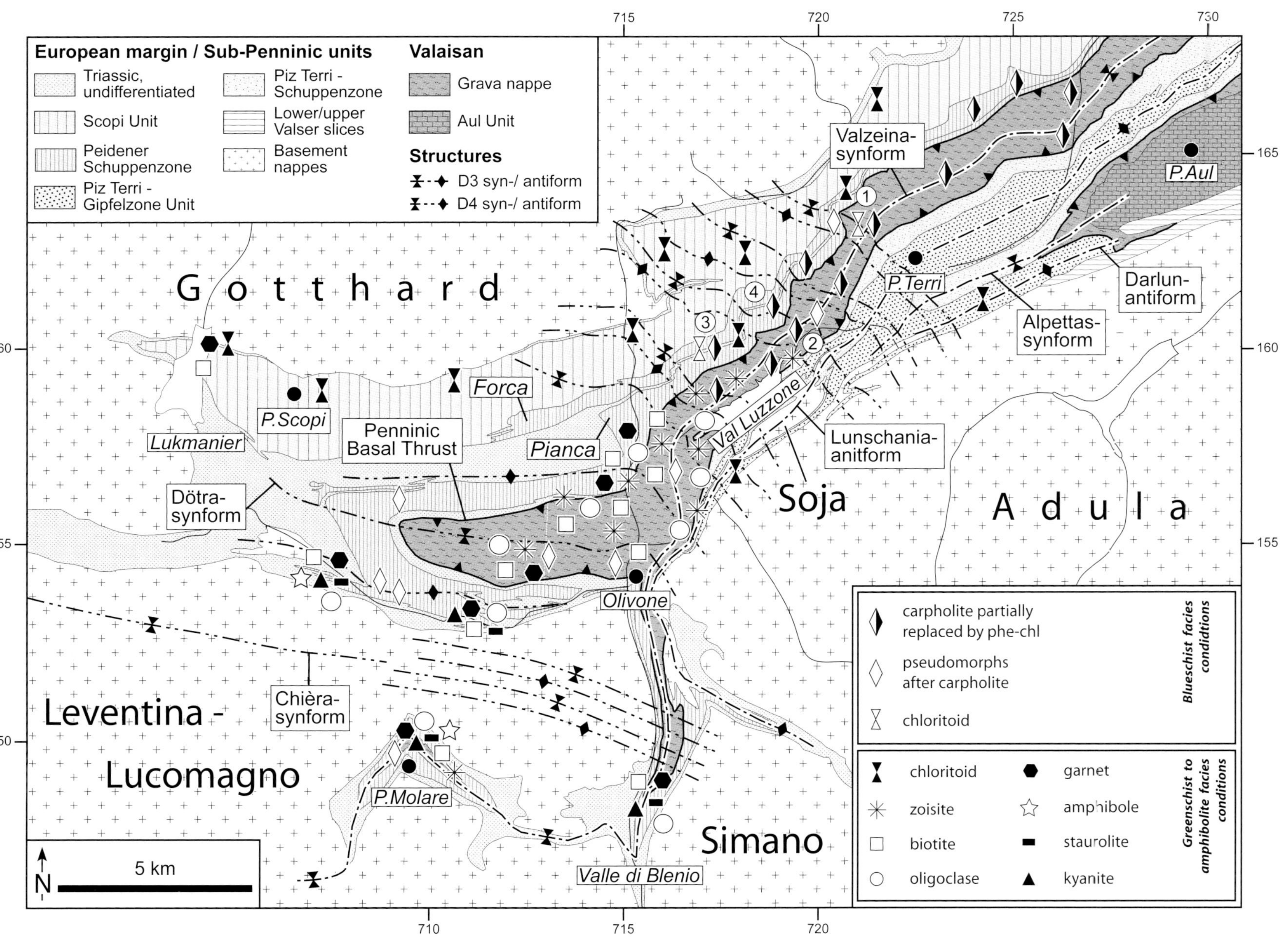

Fig. 2. Blow-up of the south-western part of the study area (bordered by the dashed line in Fig. 1) showing the traces of axial planes of major D3 and D4 folds, as well as the main occurrences of index minerals and mineral assemblages found in the meta-sedimentary units. Numbers 1–4 refer to the locations of the investigated samples presented in Table 1; coordinates are given in Swiss map coordinates (1: LUZ 0432, 720'938/162'275, 2600 m; 2: LUZ 0416, 719'235/160'052, 1670 m; 3: LUZ 047, 717'435/160'282, 1850 m; 4: LAR 061, 718'5317160'727, 1960 m).

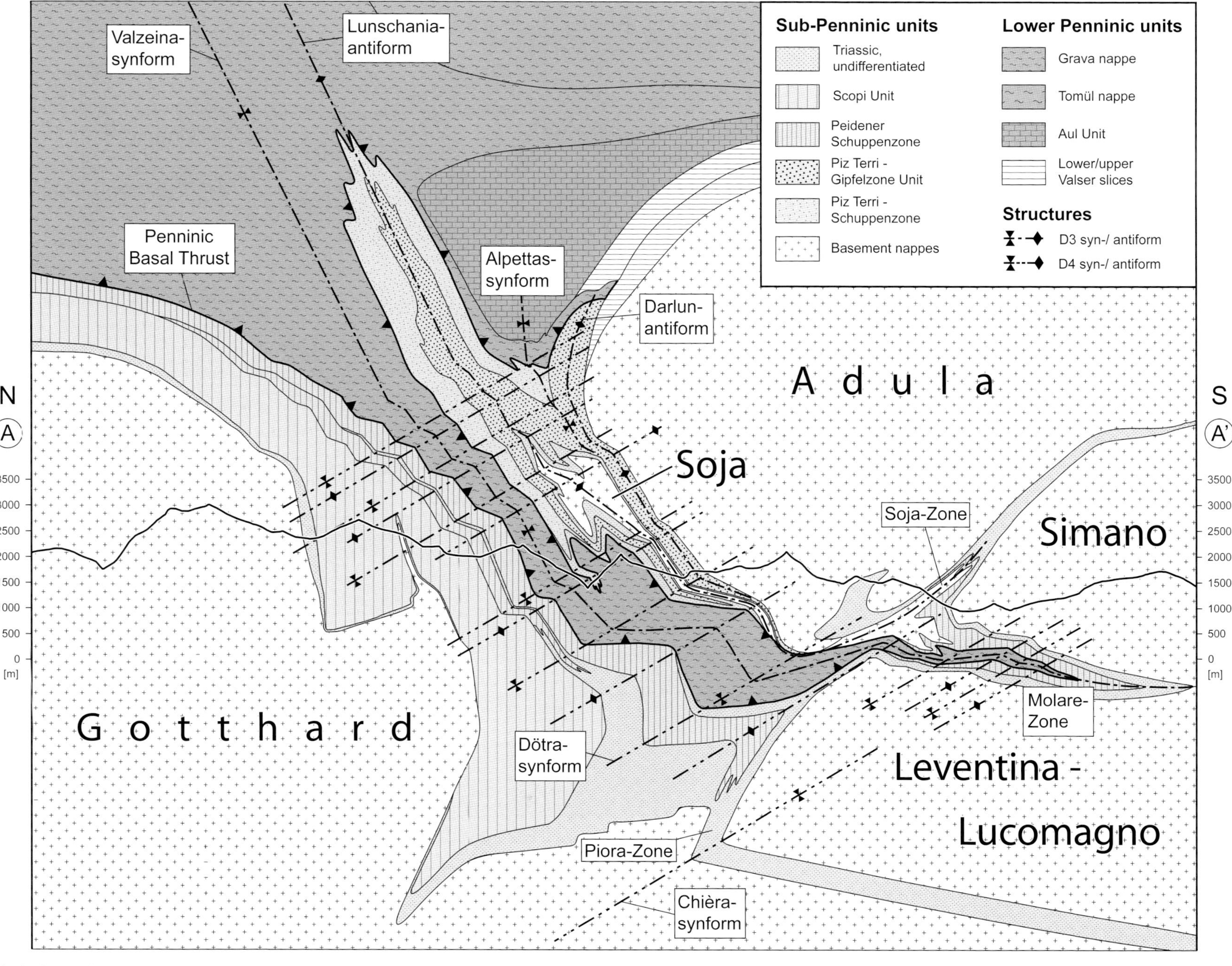

Fig. 3. Composite cross section, established on the basis of a series of cross sections by axial projection, showing the overall architecture of the working area (see Figure 1 for the trace of the cross-section into which laterally adjacent profiles were projected).

Lucomagno-Leventina basement nappe by the Molare Zone, again consisting of cover nappes (Fig. 3).

A thin sliver of pre-Mesozoic meta-sediments (locally Verrucano-type meta-conglomerates of presumed Permian age), that is part of the so-called Soja nappe (Jenny et al. 1923; Egli 1966) crops out in Val Luzzone at the front of the Adula nappe complex. This Soja nappe can be followed southwards where it is seen to separate the Simano nappe from the Adula nappe complex (Soja Zone of Fig. 3). Most authors correlate the Soja nappe with the Lebendun nappe west of the Lepontine dome (Burckhardt 1942; Egli 1966).

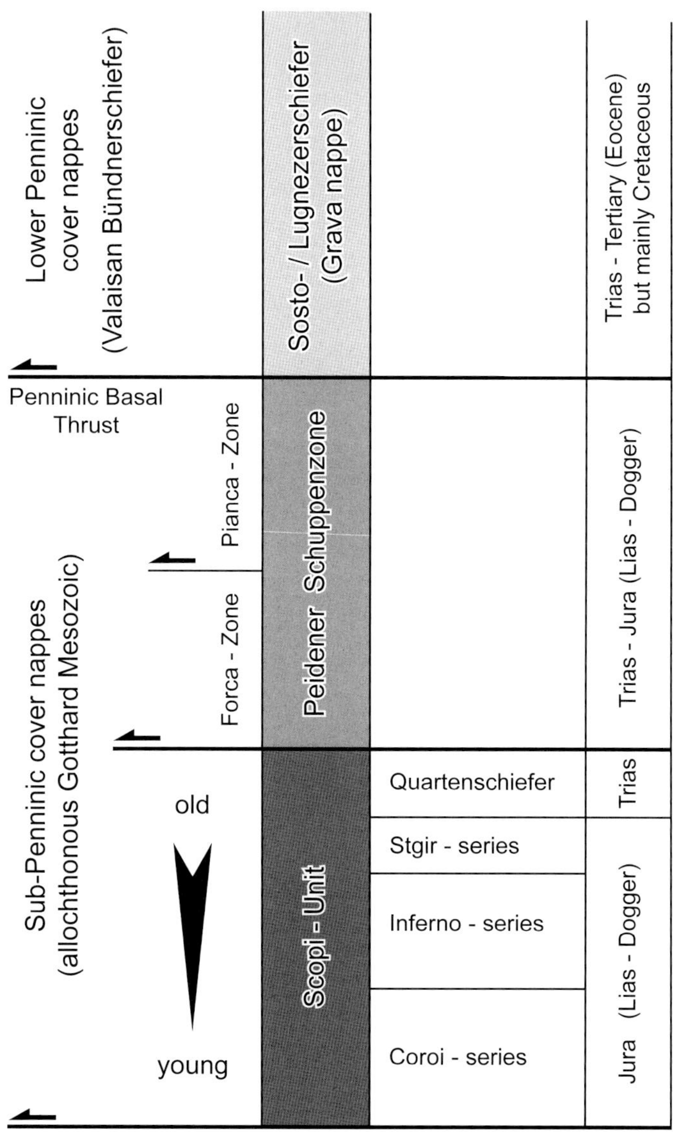

Gotthard - "Massif" (Sub-Penninic basement) covered with autochthonous Lower Triassic

Fig. 4. Tectono-stratigraphic scheme of the Sub-Penninic – Lower Penninic transition after Etter (1987).

The Adula nappe complex represents the highest structural unit within the Sub-Penninic nappe pile. It is generally overlain by the Lower Penninic Bündnerschiefer, but in the working area the contact between the Adula and overlying units is subvertical due to a late stage tectonic event. This nappe complex is not a coherent basement sliver but consists of several thin basement slices, separated from each other by Mesozoic slivers ("internal Mesozoic"; Löw 1987) and thin sediment-bearing mélange units, including meta-basalts and ultramafics, i.e. remnants of oceanic (presumably Valaisan) crust (Jenny et al. 1923; Trommsdorff 1990; Berger et al. 2005). Since it predominantly consists of continental basement rocks, we attribute it to the Sub-Penninic nappe complex (Schmid et al. 2004).

The Adula nappe complex is well known for its high-pressure metamorphism, showing a progressive increase from blueschist facies conditions (1.2 GPa and 500 °C) in the north to eclogite facies conditions (800 °C and >3 GPa) in the south (Heinrich 1982; Heinrich 1986; Löw 1987; Meyre et al. 1997; Nimis & Trommsdorff 2001; Nagel et al. 2002). Slightly different P-T conditions are given by Dale & Holland (2003) who estimated 1.7 GPa and 640 °C in the north and 2.5 GPa and 750 °C in the south. High-pressure metamorphism is interpreted to be due to Eocene subduction of the distal European margin beneath the Adriatic continent (Becker 1993; Froitzheim et al. 1996; Schmid et al. 1996). A pressure-dominated upper blueschist facies event is also reported from the Simano nappe (1.2–1.4 GPa / 500 °C; Rütti et al. 2005; Bousquet et al. 2008). Adula nappe complex and Simano nappe, together with the rest of the Sub-Penninic nappe stack from which so far no pressure-dominated metamorphism is reported, were overprinted by Barrow-type metamorphism reaching lower amphibolite facies conditions within the investigated area, i.e. 500–550 °C and 0.5–0.8 GPa (Engi et al. 1995; Todd & Engi 1997; Frey & Ferreiro Mählmann 1999).

2.2. Sub-Penninic cover nappes and slices

The sedimentary sequences found in these nappes and tectonic slices have strong affinities to non-metamorphic sequences of the southern Helvetic paleogeographic domain (Trümpy 1960). Hence, they are interpreted to represent the sedimentary cover of the most distal European margin (Froitzheim et al. 1996).

The Scopi Unit represents a lowermost cover nappe and is characterised by a coherent sedimentary stack in an overturned position. It tectonically overlays a thin veneer of Lower and Middle Triassic stratigraphic cover (Melser Sandstone Formation and Röti Dolomite Formation) of the Gotthard-"massif" basement nappe (Fig. 3; Baumer et al. 1961; Jung 1963; Baumer 1964; Frey 1967; Etter 1987). The Scopi Unit, together with the structurally higher Forca- and Pianca Zones (Fig. 4), forms what is often referred to as Gotthard-Mesozoic. These tectonic units are built up of sedimentary units detached along the evaporites of the Middle Triassic Röti Dolomite Formation from their former crystalline substratum that has to be looked for south of the Gotthard "massif" (Etter 1987). The Scopi Unit

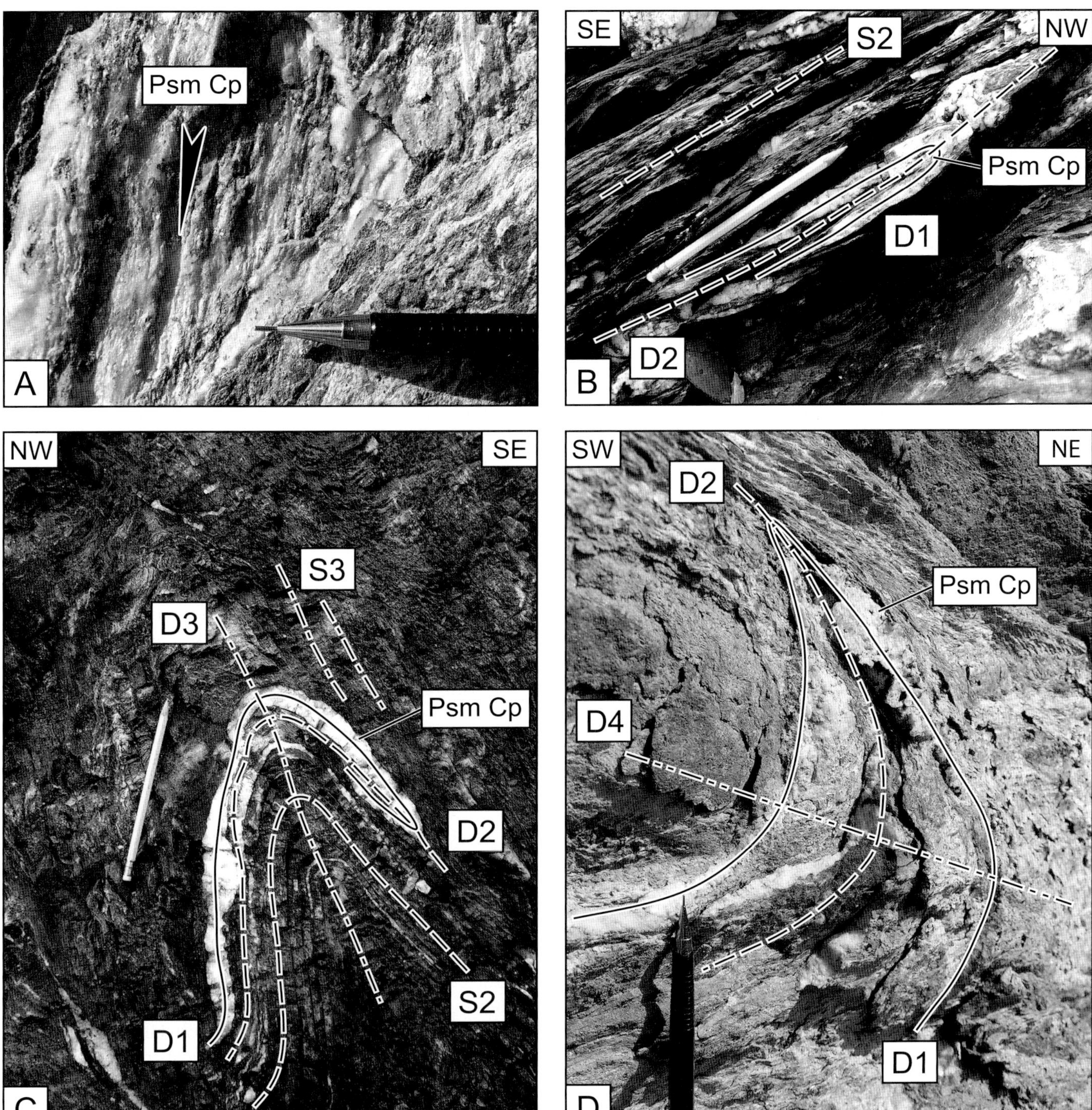

Fig. 5. Photographs of quartz-calcite veins representing pseudomorphs after Fe-Mg carpholite (Psm Cp). (A) Black arrow marks the orientation of characteristic fibrous growth of quartz, typical for pseudomorphs after Fe-Mg carpholite (Val Luzzone, 716'865/158'363, 1780 m). (B) Pseudomorphs after carpholite (related to D1 deformation, Safien phase) are folded by D2 (Ferrera phase) and aligned parallel to the main foliation S2 in the fold limbs (Safiental, 739'942/160'530, 2350 m). (C) Pseudomorph after carpholite, refolded by D2 and D3 (Domleschg phase; Valsertal, 738'292/174'969, 1580 m). Generally a new axial planar schistosity S3 ("spaced cleavage") evolved. In D3 fold hinges, S2 and S3 can easily be distinguished. (D) Pseudomorphs after carpholite, refolded by D2 (Ferrera phase) and D4 (Chièra phase; Val Luzzone, 716'865/158'363, 1780 m).

comprises a series of Late Triassic meta-pelites and meta-marls (Quartenschicfer), stratigraphically overlain by Jurassic sediments. The Lower to Middle Jurassic cover in Ultrahelvetic facies consists of carbonates, calc-schists and meta-pelites that can be subdivided into three mappable formations, referred to as Stgir, Inferno and Coroi Series (Baumer et al. 1961).

The Peidener Schuppenzone (imbricate zone), which tectonically overlies the Scopi Unit, consists of incoherent and chaotic sedimentary slices. This imbricate zone can be further subdivided into two major parts (Forca and Pianca Zones, Fig. 4; Frey 1967). We interpret the Peidener Schuppenzone as a sedimentary accretionary complex, which contains lithologies that are identical with those of the Scopi Unit, and they directly underlie the Lower Penninic cover nappes along the Penninic Basal Thrust (Figs. 3 & 4). The southern and western parts of the Gotthard-Mesozoic, which underwent metamorphism under lower to middle amphibolite facies conditions (0.5–0.8 GPa and 500–550 °C; Chadwick 1968; Frey 1969; Engi et al. 1995; Frey & Ferreiro Mählmann 1999), represent a classical area of regional Barrow-type metamorphism (Frey 1969; Niggli 1970; Wenk 1970; Fox 1975; Frey 1978).

Another pile of Sub-Penninic cover nappes and slices, referred to as Piz Terri-Lunschania Unit, forms the core of a large isoclinal antiform (the so-called Lunschania antiform; Figs. 1 & 3), located in front of the Adula nappe complex. The Piz Terri-Lunschania Unit is considered as Sub-Penninic because it is structurally in the footwall of the folded Penninic Basal Thrust. The Piz Terri-Lunschania Unit originally represents the sedimentary cover of the basement of the Soja nappe and hence roots below the Adula nappe complex, i.e. also below the Penninic Basal Thrust (Figs. 1 & 3; Probst 1980). The structurally lower cover nappe, the so-called Piz Terri Gipfelzone Unit, consists of thick and rather homogenous black and sandy calc-schists, which often resemble the Bündnerschiefer of the Lower Penninic units. The Piz Terri Gipfelzone Unit was overlain by heterogeneous sedimentary slices consisting of dolomitic marbles, meta-pelites, black shales, quartzitic micaschists and layered shaly-sandy calcareous sediments (Piz Terri Schuppenzone). These are now found along both limbs of the Lunschania antiform. They have been interpreted as tectonic imbricates (Kupferschmid 1977; Probst 1980; Uhr unpubl.) and they are in direct tectonic contact with the Lower Penninic Grava nappe. The stratigraphy of the sediments of the Piz Terri-Lunschania Unit is ill defined due to intense deformation and scarcity of fossils. A Triassic to Middle Jurassic age is inferred, based on lithological criteria and fossil record (Kupferschmid 1977; Probst 1980). So far little was known regarding the metamorphic overprint of this zone.

2.3. Lower Penninic cover nappes (Valais Bündnerschiefer)

In the eastern part of the study area (Fig. 1) the lower Penninic cover nappes represent an up to 15 km thick volume of limestones, shales, marls and calc-schists originally deposited in the Valaisan Ocean. This thickness diminishes towards the west and around Olivone only very reduced series are conserved (Figs. 1, 2 & 3). Further to the east (Grisons area) the Lower Penninic Bündnerschiefer can be subdivided into, from top to bottom, the Tomül and Grava nappes, consisting of a Cretaceous- to Eocene-age sedimentary sequence (Nänny 1948; Steinmann 1994a, b), and three imbricate zones: Aul-Unit,

Upper and Lower Valser Schuppenzone (Steinmann 1994a, b). In contrast to the Sub-Penninic cover nappes all these units are rooted above the Adula nappe complex (Probst 1980; Steinmann 1994a, b).

Only the Grava nappe reaches the Val Luzzone and northern Valle di Blenio in the central and western parts of the working area, respectively, where the base of this cover nappe (here referred to as Sosto and Lugnez schists; Probst 1980) forms the Penninic Basal Thrust (Fig. 4). In the east, i.e. in the area around Vals, Jurassic-age (dating based on stratigraphic criteria; Steinmann 1994a) mafic and ultramafic rocks are associated with meta-sediments, both forming the Aul Unit (Figs. 1 & 3). In some places the meta-basalts preserve pillow structures (Steinmann 1994a, b) and are locally associated with serpentinites (Piz Aul; Nabholz 1945). This indicates that parts of the Valais Bündnerschiefer were deposited on oceanic crust (Steinmann 1994a, b).

The metamorphism of the Valais Bündnerschiefer units is characterised by the occurrence of Fe-Mg carpholite (Goffé & Oberhänsli 1992; Bousquet et al. 2002), i.e. a typical index mineral for HP/LT conditions in meta-sediments (Goffé & Chopin 1986; Bousquet et al. 2008). Interestingly, Fe-Mg carpholite is described from both sides of the Lepontine dome, documenting blueschist facies conditions for the Petit St. Bernard area in the west (1.7 GPa, 350–400 °C; Goffé & Bousquet 1997) as well as in the Grisons including the Engadine window in the east (1.2–1.3 GPa, 350–400 °C; Bousquet et al. 1998). Both these high-pressure occurrences follow a northern suture zone between Briançonnais micro-continent and distal European margin that is formed by tectonic units attributed to the former Valais Ocean (Bousquet et al. 2002).

3. Structural evolution

In the following, we will first describe the four major deformation phases observed in the studied area. These are documented by clearly observable overprinting patterns. We will then correlate structural and metamorphic evolution in a second step and finally discuss and compare the results obtained in the working area with the large-scale geological context concerning the geodynamic evolution of the Alps.

3.1. First deformation phase (D1)

The first phase of deformation led to the formation of widespread, often carpholite-bearing calcite, quartz and quartz-calcite veins mainly found in calcareous schists in both the Sub-Penninic and Lower Penninic meta-sediments (Voll 1976). These veins represent oblique fibrous veins that opened in a transtensive manner by re-precipitation from hydrous solutions, which led to the growth of the fibres (Weh & Froitzheim 2001). The veins typically resemble fibrous carpholite pseudomorphs described in the literature (Fig. 5a; Goffé & Chopin 1986; Agard et al. 2001; Rimmelé et al. 2003; Trotet et al. 2006). Such fibres of Fe-Mg carpholite, indicative of HP/LT metamorphic conditions, are typically found only within fibrous segregations and rarely in the surrounding rock matrix. Since no major folding structures formed during D1 and since the surrounding rock matrix is occasionally found virtually undeformed (Fig. 6), we infer semi-ductile behaviour during D1, largely characterised by solution and re-precipitation processes.

Since the HP/LT mineral assemblage carpholite – chlorite – phengite – quartz ± chloritoid is found in meta-sediments of both Lower Penninic (Grava and Tomül nappes) and Sub-Penninic units (Peidener Schuppenzone), the tectonic contact between the most distal European margin and the Valais Ocean, the Penninic Basal Thrust, must have already formed during D1. Although the more external Sub-Penninic units may also have been affected by D1, these units lack carpholite-bearing veins.

3.2. Second deformation phase (D2)

This deformation post-dates D1, since the carpholite-bearing quartz-calcite veins are isoclinally folded and overprinted by the S2 penetrative main and axial planar schistosity (Fig. 5b). Due to polyphase overprinting during later phases of folding, the orientations of D2 structures, such as fold axes and fold axial planes, show a large spread in orientation. The D2 schistosity completely transposes bedding, and possibly, relics of an earlier D1 schistosity that may have existed in pelitic lithologies.

This penetrative, main D2 foliation is largely formed by phengite and chlorite, which indicates greenschist facies conditions during D2 deformation. Hence D2 was associated with early exhumation of the blueschist-facies rocks. In analogy with the findings of Bucher & Bousquet (2007) and Bucher et al. (2003, 2004) in the Western Alps, we interpret this main phase of nappe stacking to be associated with the exhumation of high-pressure rocks. Thereby relatively more internal and higher-pressure units were thrust onto relatively more external and lower-pressure units.

3.3. Third deformation phase (D3)

D3 deformation produces tight mega-folds such as the Lunschania antiform (Figs. 1 & 3), as well as folds observable at the mesoscopic and microscopic scale. A second strong axial planar cleavage S3 is associated with D3 folding. However, the distinction between S2 and S3 can only be made in D3 fold hinges where S3 represents a spaced cleavage (Fig. 5c), while S3 completely transposes S2 in F3 fold limbs. In most places the main foliation represents a composite S2/S3 schistosity.

Due to D4 overprint in the western part of the working area, the D3 fold axes and fold axial planes often show variable orientations. Further east (east of Vrin), D3 fold axes strike SW–NE, plunging gently either NE or SW; fold axial planes

Fig. 6. Photograph showing pseudomorphs after carpholite, together with almost undeformed crinoids in the Forca slice of the Peidener Schuppenzone (Sub-Penninic sediments; Val Luzzone, 717'364/160'358, 1860 m). The black arrow in the lower picture is oriented parallel to the quartz fibres.

steeply dip SE to SSE. There a crenulation is often associated with D3 deformation, the crenulation lineation being oriented parallel to D3 fold axes.

3.4. Fourth deformation phase (D4)

D4 deformation is only observed in the SW part of the investigated area (Figs. 1 & 2), strain intensity rapidly decreasing towards the NE. D4 deformation sets in east of Piz Terri while intensive folding affects the area around Pizzo Molare and between the southern Lukmanier area and Olivone (Figs. 2 & 3). D4 folds are open, often without an axial planar schistosity. They refold the S2/S3 composite schistosity (Fig. 5d). Typically, D4 folds have an undulating and wavy appearance, producing a staircase-like set of syn- and antiforms on the macroscopic scale, striking E–W to ESE–WNW (Figs. 2 & 3).

However, a new axial plane schistosity S4 locally evolves in fold hinges of tighter D4 folds and overprints the S2/S3 composite foliation. In such cases this new S4 foliation represents a pressure solution cleavage producing microlithons within which overprinting of the S2/S3 composite foliation is well preserved on a microscopic scale.

The E–W striking D4 fold axes dip moderately (20–40°) towards the E in most of the working area. Only around Pizzo Molare and south of Olivone the fold axes dip towards SE-ESE. It is important to note that the D4 fold axial planes generally plunge with 30–50° to the NE. Hence the D4 folds represent back-folds, which are typical for the so-called Northern Steep Belt at the northern rim of the Lepontine dome (Milnes 1974).

4. New data regarding the metamorphic evolution of the area

A remarkable metamorphic field gradient is deduced for the investigated area, ranging from low-temperature (≤400 °C) blueschist facies metamorphism in the east all the way to classical Barrow-type amphibolite facies overprint (up to 570 °C) further west (Figs. 1 & 2). Moreover, for the first time we are able to distinguish between two separate metamorphic stages in the Sub-Penninic and Lower Penninic meta-sedimentary units in the north-eastern Lepontine: an earlier HP/LT event is followed by a MP/MT Barrovian event.

4.1. Spatial distribution of index minerals and mineral parageneses

The HP/LT event is documented by the mineral assemblage Fe-Mg carpholite – chlorite – phengite – quartz ± chloritoid, relics of which are widespread in the Lower Penninic Grava and Tomül nappes east of the Lepontine dome (Bousquet et al. 2002). Remnants of this assemblage, notably relics and/or pseudomorphs after Fe-Mg carpholite, are even found in the westernmost exposures of the Grava nappe as far as Pizzo Molare, hence in an area well inside the thermally overprinted realm of the Lepontine dome (Fig. 2). Relics and pseudomorphs af-

ter Fe-Mg carpholite were also found further north within the imbricated Sub-Penninic Peidener Schuppenzone. There, such pseudomorphs occur as fibrous quartz-calcite veins, which crosscut meta-sedimentary units containing almost undeformed crinoids (Fig. 6). This documents for the first time that HP/LT metamorphic conditions also affected Ultrahelvetic meta-sediments, derived from the former distal European margin. This clearly points to the existence of a second and more northerly-located Alpine subduction zone with respect to the Upper Penninic Suture Zone, as pointed out by several authors (e.g. Stampfli 1993; Stampfli et al. 1998; Bousquet et al. 2002; Nagel et al. 2002; Froitzheim et al. 2003; Pleuger et al. 2003).

The classical Barrow-type amphibolite facies overprint is associated with a pronounced metamorphic field gradient. Towards the south-west a gradual succession of newly growing minerals indicates an increase in temperature from greenschist to lower/middle amphibolite facies conditions. This is deduced from the progressive appearance of chloritoid, zoisite, plagioclase, titanite, biotite, garnet, staurolite, kyanite and finally amphibole (Fig. 2). Furthermore, in the south-western part of the working area around Pizzo Molare and between Olivone and the southern Lukmanier area, where pseudomorphs of Fe-Mg carpholite are also present, the mineral assemblage staurolite – kyanite – garnet – plagioclase – biotite – phengite ± amphibole clearly indicates a thermal overprint under lower to middle amphibolite facies conditions (Chadwick 1968; Frey 1969; Thakur 1971; Engi et al. 1995). Towards the north-east the mineral parageneses chloritoid – phengite – chlorite – quartz and zoisite/clinozoisite – chlorite – phengite – quartz – calcite/ dolomite indicate greenschist facies conditions related to the same Barrow-type event (Jung 1963; Frey 1967; Frey & Ferreiro Mählmann 1999). In contrast to the earlier HP/LT event, restricted to the Grava nappe and the Peidener Schuppenzone, the Barrow-type overprint affected all the units of the working area (Fig. 2). In the following the new data concerning the two metamorphic events will be described in more detail.

4.2. Data on the earlier HP/LT metamorphic event

Occurrences or relics of the HP/LT mineral assemblage Fe-Mg carpholite – chlorite – phengite – quartz ± chloritoid are only found within quartz- and calcite-bearing veins or segregations. Fibrous mesoscopic appearance and characteristic light green silvery colour of such veins or segregations resemble the typical Fe-Mg carpholite pseudomorphs described in the literature (Fig. 5a; Goffé & Chopin 1986; Goffé et al. 1989; Fournier et al. 1991). It is important to emphasise that in the study area Fe-Mg carpholite was never found in the matrix of the rocks; it exclusively occurs within veins and/or segregations. This, together with the fact that undeformed crinoids are found in the neighbouring rocks next to these veins (Fig. 6), indicates that carpholite growth occurred in the context of veining and dehydration, i.e. probably during subduction.

However, Fe-Mg carpholite is only preserved in the form of microscopic-scale relics (Fig. 7a) within quartz-calcite segrega-

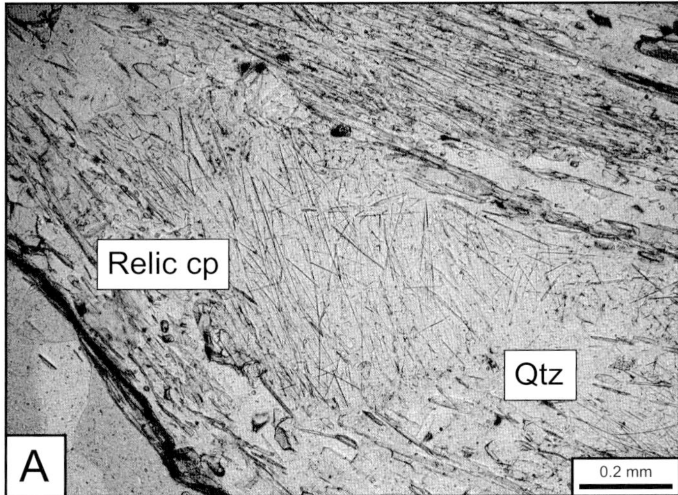

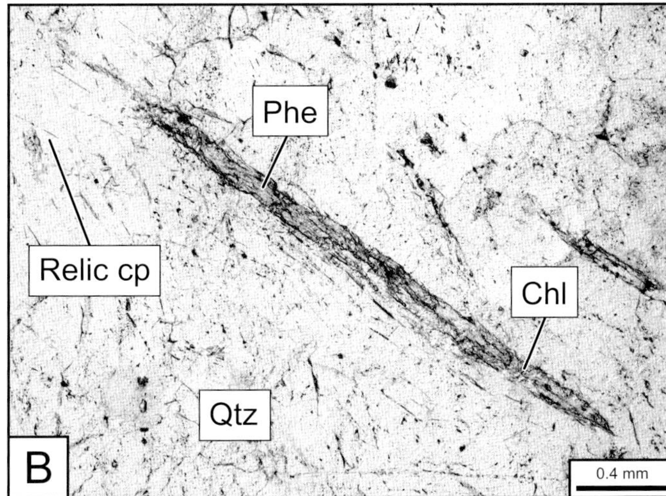

Fig. 7. Photomicrographs of thin sections showing mineral assemblages related to the HP/LT event, preserved as inclusions in quartz-calcite fibrous veins generally interpreted as pseudomorphs after Fe-Mg carpholite. (A) Hair-like fibres represent relics of Fe-Mg carpholite (cp) as inclusions in quartz (SW of Vrin, 723'990/166'197, 1890 m). (B) Phengite, quartz and chlorite define a needle-shaped pseudomorph after carpholite. Still preserved relics of carpholite can be found as hair-like fibres (Val Luzzone, 720'938/162'275, 2600 m).

tions or veins that represent macroscopically visible pseudomorphs after large former Fe-Mg carpholite crystals (Fig. 5a). No Fe-Mg carpholite crystals preserved on a macroscopic scale, such as described from the Engadine window (Bousquet et al. 1998), were found. The macroscopic pseudomorphs are mainly built up by fibrous quartz (Fig. 5a). Its shimmering silver-green lustre is due to a thin layer of chlorite and phengite. In thin section, fibrous quartz is full of inclusions of chlorite, phengite and paragonite. This assemblage often forms needle-shaped pseudomorphs after Fe-Mg-carpholite (Fig. 7b). Rarely chloritoid has also been found as inclusions in such fibrous quartz-calcite veins.

In order to estimate peak-pressure conditions, the chemical compositions of HP/LT minerals were determined by wavelength-dispersive X-ray analysis (WDS) using a CAMECA SX-100 electron microprobe at the GeoForschungsZentrum (GFZ) Potsdam. The analytical conditions included an acceleration voltage of 15 kV, a beam current of 20 nA and beam diameters of 1–10 μm; PAP corrections were applied. Natural and synthetic minerals were used as standards. Peak counting times were 10–20 s for major and 20–40 s for minor elements; backgrounds were counted for 5–20 s.

Relics of Fe-Mg carpholite are hair-thin micro-fibres (10 to 100 μm long, 0.5 to 10 μm wide) embedded in quartz (Fig. 7a) documented by Raman spectroscopy and microprobe analysis. In order to avoid the effect of contamination by the surrounding quartz, the chemical composition of Fe-Mg carpholite $[(Mg,Fe,Mn)Al_2Si_2O_6(OH,F)_4]$ was calculated on the basis of a fixed atomic number of cations (Goffé & Oberhänsli 1992). The value of X_{Mg} [Mg/(Mg+Fe+Mn)] is rather constant, ranging from 0.44 to 0.55 (mean value 0.49) in the Penninic Bündnerschiefer of the Grava nappe, and from 0.39 to 0.57 (mean value 0.49) in the Sub-Penninic Peidener Schuppenzone (Table 1).

Generally the fluorine content is very low and varies from 0.0 to 0.99 wt%.

White mica is also found as inclusions in quartz grains. Both phengite and paragonite are present and associated with chlorite and quartz. They form also part of the pseudomorphs after Fe-Mg carpholite (Fig. 7b). There, phengite and paragonite occur as fine-grained flakes without any shape preferred orientation. The Si^{4+} content, reflecting the phengitic substitution in white mica replacing Fe-Mg carpholite, ranges from 3.24 in the Valais Bündnerschiefer to 3.12 p.f.u in the Sub-Penninic Peidener Schuppenzone (Table 1).

Chlorite occurs as randomly arranged grains, together with white mica and quartz, but also within pseudomorphs after Fe-Mg carpholite (Fig. 7b). In the Valais Bündnerschiefer the Tschermak substitution in chlorites is around 2.57–2.70 (Si p.f.u.), X_{Mg} ranging from 0.43 to 0.60. In the Sub-Penninic sediments between 2.52 and 2.61 Si p.f.u. and X_{Mg} from 0.44 to 0.48 were measured (Table 1).

Only in a few cases chloritoid was found inside quartz-calcite segregations containing relics of Fe-Mg carpholite. Such chloritoid forms small prisms with X_{Mg}-values varying between 0.12 in the Valais Bündnerschiefer and 0.20 in the Sub-Penninic meta-sediments (Table 1). This chloritoid is interpreted to have formed during prograde metamorphism by reaction from Fe-Mg carpholite, as shown by Vidal et al. (1992); i.e. during the high-pressure stage rather than during a greenschist facies event (see discussion in Oberhänsli et al. 2003).

4.3. Data on the subsequent temperature-dominated event

While remnants of the HP/LT metamorphic event arc restricted to quartz-calcite segregations within the Lower Penninic Bündnerschiefer and the Sub-Penninic Peidener Schuppenzone, the

Table 1. Representative microprobe mineral analyses of HP/LT assemblages found as inclusions in quartz-calcite veins, interpreted as pseudomorphs after carpholite of Valaisan (Grava nappe) and Sub-Penninic (Peidener Schuppenzone) meta-sedimentary units given in weight-percents. The locations of the samples are shown in Figure 2. The deviations from 100% are mainly due to the OH-content not detected by microprobe analyses. The sums of Fe-Mg carpholite analyses have been corrected for fluorine content; the values given under "Total corr." take into account that fluorine occupies an oxygen site. The structural formulae for carpholite were calculated by using 5 cations for Si and 3 cations for Al, Fe, Mn and Mg, following Goffé & Oberhänsli (1992). For chlorite we used 14 oxygens, for white mica 11 and for chloritoid 12, following Chopin et al. (1992). The following abbreviations have been used: Cp = Fe-Mg carpholite, Ctd = chloritoid, Chl = chlorite, Phe = phengite, Pg = paragonite.

Unit	Valaisan Bündnerschiefer (Grava nappe)								Sub-Penninic sediments (Peidener Schuppenzone)						
									Forca slice				Pianca slice		
Sample	LUZ 0432					LUZ 0416			LUZ 047				LAR 061		
Minerals	**Cp**	**Ctd**	**Chl**	**Phe**	**Pg**	**Cp**	**Chl**	**Phe**	**Cp**	**Ctd**	**Chl**	**Phe**	**Cp**	**Chl**	**Phe**
Analysis (wt-%)	*Cp828*	*Ctd215*	*Chl219*	*Phe221*	*Pg220*	*Cp849*	*Chl207*	*Phe224*	*Cp9B*	*Ctd16*	*Chl5*	*Phe1*	*Cp12*	*Chl4*	*Phe12*
SiO_2	37.40	24.46	25.36	49.46	48.43	37.63	25.30	48.60	37.85	25.32	23.17	47.18	38.00	23.63	47.20
TiO_2	0.19	0.01	0.04	0.20	0.04	0.25	0.06	0.17	0.00	0.00	0.06	0.13	0.27	0.03	0.00
Al_2O_3	31.11	39.36	22.91	33.52	40.30	31.31	22.40	34.08	32.41	38.80	23.57	35.94	30.02	22.60	36.34
Cr_2O_3	0.03	0.00	0.00	0.00	0.00	0.00	0.00	0.00	0.00	0.00	0.00	0.00	0.03	0.00	0.00
FeO	11.44	25.42	27.37	1.34	0.20	12.64	26.51	1.24	10.92	22.55	27.27	0.81	13.90	27.05	0.49
MnO	0.25	0.06	0.02	0.00	0.01	0.30	0.05	0.00	0.06	0.27	0.01	0.00	0.03	0.08	0.01
MgO	6.32	1.85	12.73	1.39	0.12	5.68	13.39	1.20	6.27	2.83	11.98	0.85	6.49	12.62	0.62
CaO	0.00	0.00	0.00	0.00	0.07	0.02	0.01	0.00	0.01	0.00	0.00	0.00	0.00	0.03	0.02
Na_2O	0.00	0.00	0.04	0.67	7.26	0.00	0.00	0.76	0.01	0.00	0.01	0.84	0.04	0.03	0.92
K_2O	0.04	0.00	0.02	8.82	1.11	0.03	0.01	9.12	0.02	0.01	0.01	8.75	0.00	0.02	8.59
BaO	0.00	0.00	0.00	0.00	0.00	0.00	0.00	0.00	0.00	0.00	0.00	0.00	0.00	0.00	0.00
F	0.65	0.00	0.00	0.00	0.00	0.77	0.00	0.00	0.49	0.00	0.00	0.00	0.00	0.00	0.00
Total	87.43	91.15	88.48	95.39	97.54	88.64	87.72	95.16	88.03	89.78	86.07	94.50	88.76	86.08	94.19
Total corr.	**87.17**	**91.15**	**88.48**	**95.39**	**97.54**	**88.32**	**87.72**	**95.16**	**87.83**	**89.78**	**86.07**	**94.50**	**88.76**	**86.08**	**94.19**
Si	2.00	2.05	2.67	3.24	3.02	2.00	2.67	3.20	2.00	2.13	2.52	3.12	2.00	2.57	3.12
Ti	0.01	0.00	0.00	0.01	0.00	0.01	0.00	0.01	0.00	0.00	0.00	0.01	0.01	0.00	0.00
Al	1.96	3.90	2.84	2.59	2.96	1.96	2.79	2.64	2.02	3.85	3.02	2.80	1.86	2.90	2.83
Cr	0.00	0.00	0.00	0.00	0.00	0.00	0.00	0.00	0.00	0.00	0.00	0.00	0.00	0.00	0.00
Fe^{3+}	0.04	0.05	–	–	–	0.04	–	–	0.00	0.02	–	–	0.14	–	–
Fe^{2+}	0.47	1.74	2.41	0.07	0.01	0.52	2.34	0.07	0.52	1.56	2.48	0.04	0.47	2.46	0.03
Mn	0.01	0.00	0.00	0.00	0.00	0.01	0.00	0.00	0.00	0.02	0.00	0.00	0.00	0.01	0.00
Mg	0.50	0.23	1.99	0.14	0.01	0.45	2.11	0.12	0.49	0.35	1.94	0.08	0.51	2.04	0.06
Ca	0.00	0.00	0.00	0.00	0.00	0.00	0.00	0.00	0.00	0.00	0.00	0.00	0.00	0.00	0.00
Na	0.00	0.00	0.01	0.08	0.88	0.00	0.00	0.10	0.00	0.00	0.00	0.11	0.00	0.01	0.12
K	0.00	0.00	0.00	0.74	0.09	0.00	0.00	0.77	0.00	0.00	0.00	0.74	0.00	0.00	0.72
Ba	0.00	0.00	0.00	0.00	0.00	0.00	0.00	0.00	0.00	0.00	0.00	0.00	0.00	0.00	0.00
F	0.11	0.00	0.00	0.00	0.00	0.13	0.00	0.00	0.08	0.00	0.00	0.00	0.00	0.00	0.00
XMg	0.51	0.12	0.45			0.46	0.47		0.49	0.18	0.44		0.52	0.45	

temperature-dominated Barrovian event affected all tectonic units (Fig. 2). From north-east to south-west, i.e. from greenschist to lower/middle amphibolite facies overprint, the following mineral assemblages are described.

Greenschist facies: Chloritoid growing during the second metamorphic event and at the expense of pyrophyllite and chlorite can be texturally distinguished from that produced by the breakdown of Fe-Mg carpholite. Whereas the scarce occurrences of HP/LT chloritoid are restricted to microscopic-scale inclusions in quartz-calcite veins containing preserved relics of Fe-Mg carpholite, LP/LT chloritoid occurs in the rock-matrix as idiomorphic rosettes, bundles and prisms, together with quartz, white mica and chlorite. It is common in shaly formations of the Stgir and Coroi Series of the Gotthard-Mesozoic (Scopi Unit and Peidener Schuppenzone), as well as in quartzitic formations of the Piz Terri-Lunschania Unit (Jung 1963; Frey 1967; Probst 1980; Fig. 2). The mineral assemblage chloritoid-white mica-chlorite-quartz is typical for the Sub-Penninic sedimentary units in the study area. The overlying Lower Penninic

Bündnerschiefer are characterised by the mineral assemblage phengite – chlorite – quartz – calcite/dolomite. However, chloritoid has not been identified with the exception of rare occurrences in quartz-calcite veins interpreted as relics of the HP/LT stage, as discussed above. The lack of chloritoid is most probably due to the bulk rock composition in the Lower Penninic Bündnerschiefer, which is generally Ca-rich and Al-poor.

Upper greenschist facies: In the Val Luzzone area, the mineralogical composition of the Lower Penninic Bündnerschiefer changes dramatically (Fig. 2). Newly formed zoisite/clinozoisite, plagioclase and titanite indicate an increase in temperature. This is corroborated by the mineral assemblage plagioclase – zoisite/clinozoisite – titanite – phengite – chlorite – quartz – calcite/dolomite found in marly or calcareous schists of the Grava nappe. When entering the upper greenschist facies stability field, zoisite/clinozoisite is the first newly grown mineral observed. Based on mineral shape and composition the following two zoisite/clinozoisite types are found within the same rocks (Frank 1983; Kuhn et al. 2005): (1) Fine-grained, needle-shaped, prismatic zoisite/clinozoisite, enriched in Fe^{3+} relative to Al^{VI} with an X_{Ep} [$Fe^{3+}/(Fe^{3+} + Al-2)$] of 0.15–0.20. Some grains show distinct compositional zoning, with cores being richer in Fe^{3+} compared to the rims (X_{Ep} core = 0.63; rim = 0.20), and, (2) an almost pure Al-end member with X_{Ep} of 0.03–0.05 forming relatively large crystals arranged as rosettes and sheaves, up to 5 cm in diameter (Fig. 8a). Plagioclase forming black plates up to several mm in size is present as oligoclase, with an An-content varying between 0.19 and 0.36. Titanite, another characteristic mineral of this zone, forms bi-pyramidal crystals up to 3 mm in size.

Upper greenschist to amphibolite facies transition: Biotite first appears west of the retaining wall of Lago di Luzzone within the Lower Penninic Bündnerschiefer (Fig. 2). This site roughly coincides with the north-eastern border of the Lepontine thermal dome as mapped by Spicher (1980). The assemblage biotite – plagioclase – zoisite/clinozoisite – titanite – phengite – quartz – calcite/dolomite ± chlorite is characteristic for the upper greenschist to amphibolite facies transition. The Jurassic sediments of the Sub-Penninic units (Stgir Series of the Peidener Schuppenzone) are less calcareous and more pelitic than the overlying Lower Penninic Bündnerschiefer (Fox 1975). This difference in chemical composition led to the formation of garnet in the Stgir Series, which is absent in the Lower Penninic Bündnerschiefer in the study area. Garnet first appears north of Olivone (northern Valle di Blenio; Figs. 1 & 2) within the assemblage garnet – biotite – plagioclase – phengite – quartz ± chlorite ± zoisite/clinozoisite. The co-existence of garnet and biotite is typical for the upper greenschist to amphibolite facies transition (Bucher & Frey 2002). The almandine-rich garnets ($Alm_{0.71}Prp_{0.06}Grs_{0.21}Sps_{0.02}$) are chemically more or less unzoned and form crystals up to 1 cm in size.

Lower to middle amphibolite facies: Kyanite, staurolite and amphibole appear in addition further SW (Valle di Blenio,

Pizzo Molare, and south of the Lukmanier pass; Figs. 1 & 2). Pelitic meta-sediments of the Sub-Penninic units are characterised by the mineral assemblage staurolite – kyanite – garnet – biotite – plagioclase – phengite – quartz (Baumer 1964; Chadwick 1968; Frey 1969; Thakur 1971; Fox 1975; Probst 1980), typically indicating lower to middle amphibolite facies conditions. In general, garnet is almandine-rich and shows a normal zoning pattern with increasing Mg- and decreasing Mn-content from core to rim. Garnets from the Pizzo Molare area yield $Alm_{0.62}Prp_{0.06}Grs_{0.23}Sps_{0.09}$ for the core and $Alm_{0.68}Prp_{0.12}Grs_{0.20}Sps_{0.00}$ for the rim. The An-content of plagioclase ranges between 0.16 and 0.30. The more calcareous chemistry of the Valaisan Bündnerschiefer does not allow for the growth of these new minerals; the assemblage biotite – plagioclase – zoisite/clinozoisite – titanite – phengite – quartz – calcite/dolomite ± chlorite still persists in this metamorphic zone.

4.4. Correlations between the structural and metamorphic evolution

The early HP/LT event was associated with the formation of quartz-calcite veins containing pseudomorphs after Fe-Mg carpholite indicating blueschist facies conditions. We infer that these veins formed during D1 since they were folded by isoclinal F2 folds (Figs. 5b, c & d). These isoclinal folds are associated with a D2 penetrative axial planar foliation that formed under greenschist-facies conditions. From this we deduce that greenschist-facies conditions were already established during D2.

Porphyroblasts related to Barrow-type thermal overprint, such as chloritoid, zoisite/clinozoisite, titanite, plagioclase, biotite, garnet, kyanite, staurolite and amphibole, all clearly overgrow the S2/S3 composite main foliation and have no shape preferred orientation (Figs. 8a & b). Big flakes of biotite typically grow across the S2/S3 main foliation ("Quer-Biotit"). This implies a temporal hiatus between syn-D1 HP/LT metamorphism and post-D2/D3 Barrow-type overprint. Moreover, the random orientation of porphyroblasts related to Barrow-type thermal overprint, as well as the conservation of an unfolded internal S2/S3 compositional foliation in the cores of garnets (Fig. 8b), both indicate that at least the initial stages of the Barrovian overprint occurred under static conditions in most parts of the working area. The famous syn-D4 snowball garnets (Chadwick 1968; Fox 1975; Robyr et al. 2007) represent an exception and are restricted to a specific level of the Stgir Series.

In most places the Barrovian mineral assemblages have been overprinted by D4 crenulation deformation; needles and prisms of chloritoid, zoisite/clinozoisite and kyanite are kinked, bent or broken (Figs. 8c & d). Most garnet porphyroblasts show some rotation of the internal S2/S3 foliation towards the rims (Fig. 8b), indicating that the last stages of the garnet growth occurred during D4. However the D2/D3 composite foliation is strongly crenulated in the rock-matrix outside of the garnet, indicating that most of the D4 crenulation post-dates the growth of garnet.

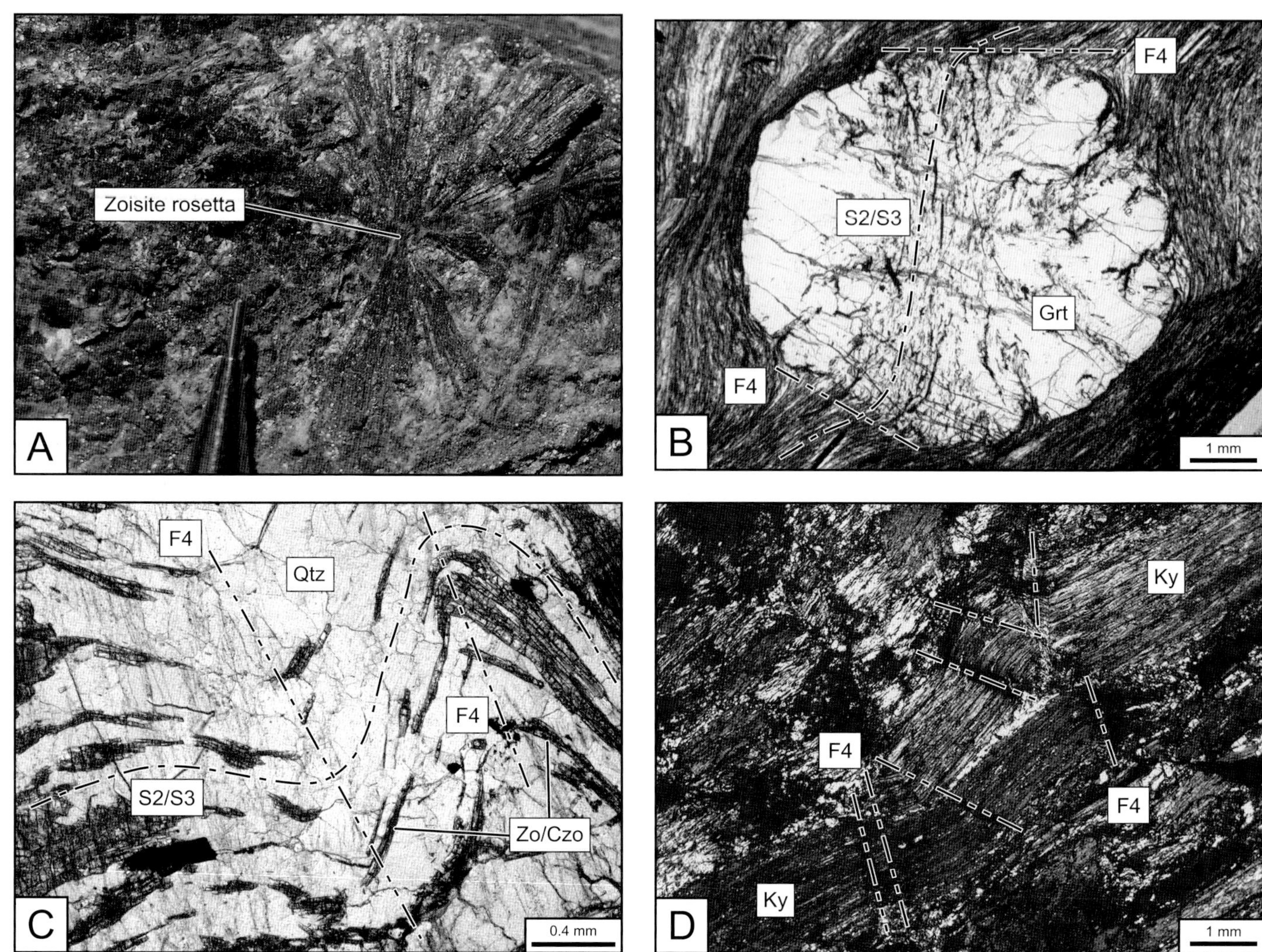

Fig. 8. Photographs showing microstructural relationships between porphyroblasts related to Barrowian overprint and deformation phases. (A) Rosetta of zoisite growing over the S2/S3 composite main foliation (Val Luzzone, 716'469/158'124, 1450 m). (B) Straight internal S2/S3 composite foliation inside a garnet porphyroblast (Grt), slightly curving at the rim; deflection of the S2/S3 foliation around the garnet and its relative rotation are the effects of subsequent D4 deformation (Val Luzzone, 715'029/156'841, 1190 m). (C) Zoisite/clinozoisite needles (Zo/Czo) oriented parallel to the S2/S3 composite main foliation are broken and bent by D4 deformation (Val Luzzone, 716'093/157'771, 1390 m). (D) Needles of kyanite (Ky) kinked by D4 folds (S of Pizzo Molare, 709'541/149'308, 2310 m).

In summary (see Figs. 9 & 10), the Barrovian-type thermal overprint definitely post-dates D3 and started during a period without any significant deformation. The HP/LT event, however, was syn-D1 and terminated before D2. This implies that greenschist facies conditions were already established during decompression from the HP/LT stage and before the Barrow-type heating event.

4.5. Relations between HP/LT and MP/MT metamorphism: Significance of a metamorphic field gradient

Pressure-dominated metamorphic event

Peak-pressure and -temperature conditions can be estimated from the composition of coexisting phengite, Fe-Mg carpholite, and chloritoid according to Bousquet et al. (2002). P-T calculations were carried out with the GEO-CALC software (Brown et al. 1988), by using the updated JAN92.RGB thermodynamic database (Berman 1988), Mg-carpholite data from Vidal et al. (1992), Mg-chloritoid data of B. Patrick (listed in Vidal & Theye 1996), and alumino-celadonite data from Massonne (1995). The mineral activities used are listed in Bousquet et al. (2002). In the working area, the measured mineral compositions of Fe-Mg carpholite, phengite, and chloritoid are similar to those described by Goffé & Oberhänsli (1992), Oberhänsli et al. (1995) and Bousquet et al. (2002). The pressure conditions for carpholite-bearing rocks are defined by the location of the equilibrium (Fig. 11):

2 phengite + chlorite + 5 Quartz + 2 H_2O = 3 carpholite + 2 phengite (R1)

while the stability field of Fe-Mg carpholite towards higher temperature is limited by the equilibrium:

carpholite = chloritoid + quartz + 2 H_2O (R2)

From the mineral chemistry of the observed mineral assemblage and the position of the above-described equilibria, peak metamorphic conditions of 1.2–1.4 GPa and 350–400 °C are estimated for both Lower Penninic Bündnerschiefer and Sub-Penninic meta-sediments of the Peidener Schuppenzone (Fig. 11), similar to the P-T conditions estimated in the Safiental further to the east (Bousquet et al. 2002).

Most of the meta-sediments show retrogressed greenschist-facies assemblages documented by the widespread mineral assemblage phengite – paragonite – chlorite – quartz – calcite/dolomite. However, the preservation of Fe-Mg carpholite relics, as well as a retrograde path mainly characterised by the decay of Fe-Mg carpholite to chlorite and phengite (following R1; Fig. 11) forming pseudomorphs after carpholite, implies a

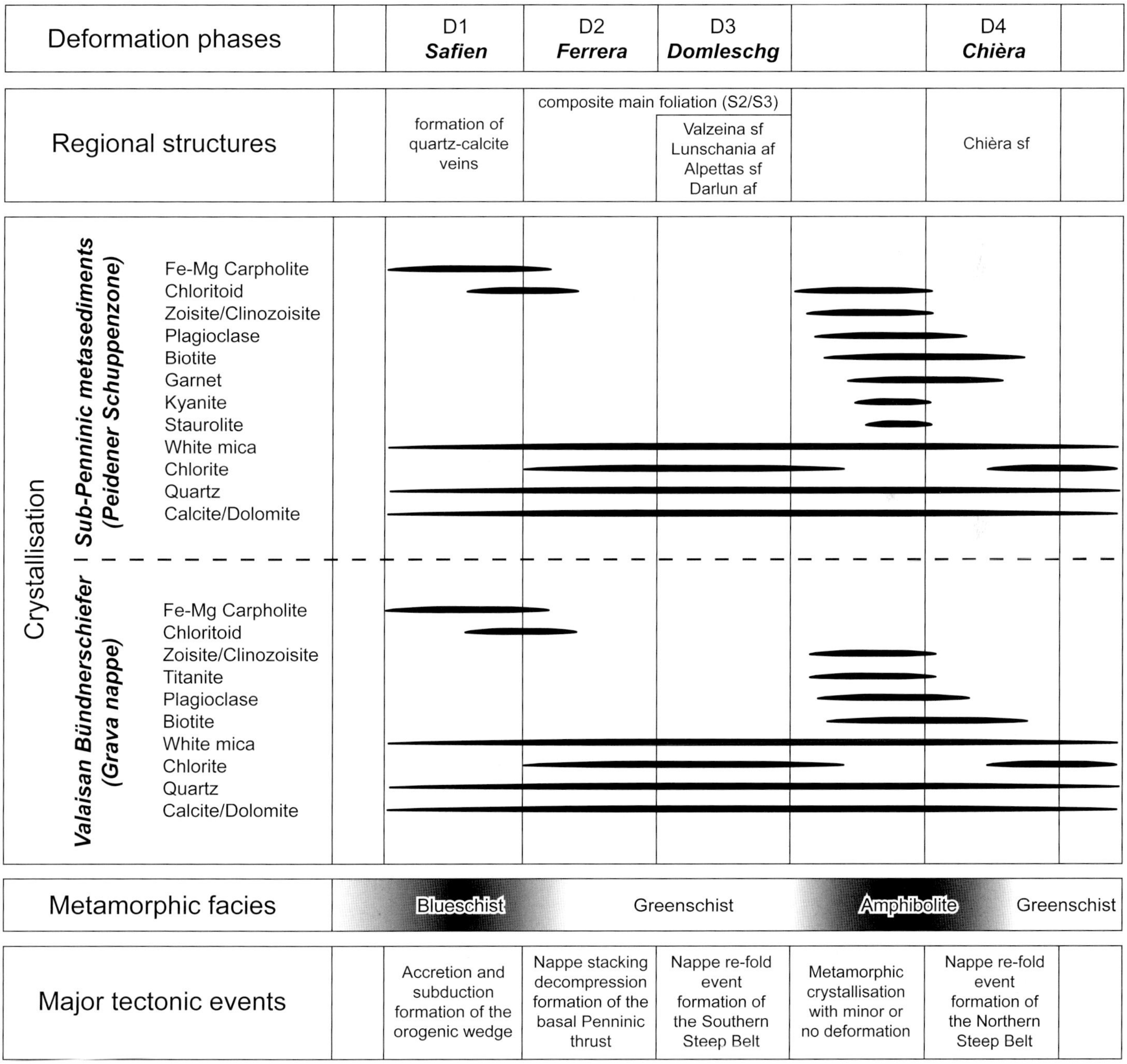

Deformation phases		D1 **Safien**	D2 **Ferrera**	D3 **Domleschg**		D4 **Chièra**	
Regional structures		formation of quartz-calcite veins	composite main foliation (S2/S3)			Chièra sf	
				Valzeina sf Lunschania af Alpettas sf Darlun af			

Fig. 9. Summary of relationships between crystallisation and deformation in the Lower Penninic and Sub-Penninic meta-sediments; sf: synform, af: antiform.

Metamorphic facies: Blueschist — Greenschist — Amphibolite — Greenschist

Major tectonic events:
- D1: Accretion and subduction formation of the orogenic wedge
- D2: Nappe stacking decompression formation of the basal Penninic thrust
- D3: Nappe re-fold event formation of the Southern Steep Belt
- D3/D4: Metamorphic crystallisation with minor or no deformation
- D4: Nappe re-fold event formation of the Northern Steep Belt

Crystallisation — Sub-Penninic metasediments (Peidener Schuppenzone): Fe-Mg Carpholite, Chloritoid, Zoisite/Clinozoisite, Plagioclase, Biotite, Garnet, Kyanite, Staurolite, White mica, Chlorite, Quartz, Calcite/Dolomite.

Crystallisation — Valaisan Bündnerschiefer (Grava nappe): Fe-Mg Carpholite, Chloritoid, Zoisite/Clinozoisite, Titanite, Plagioclase, Biotite, White mica, Chlorite, Quartz, Calcite/Dolomite.

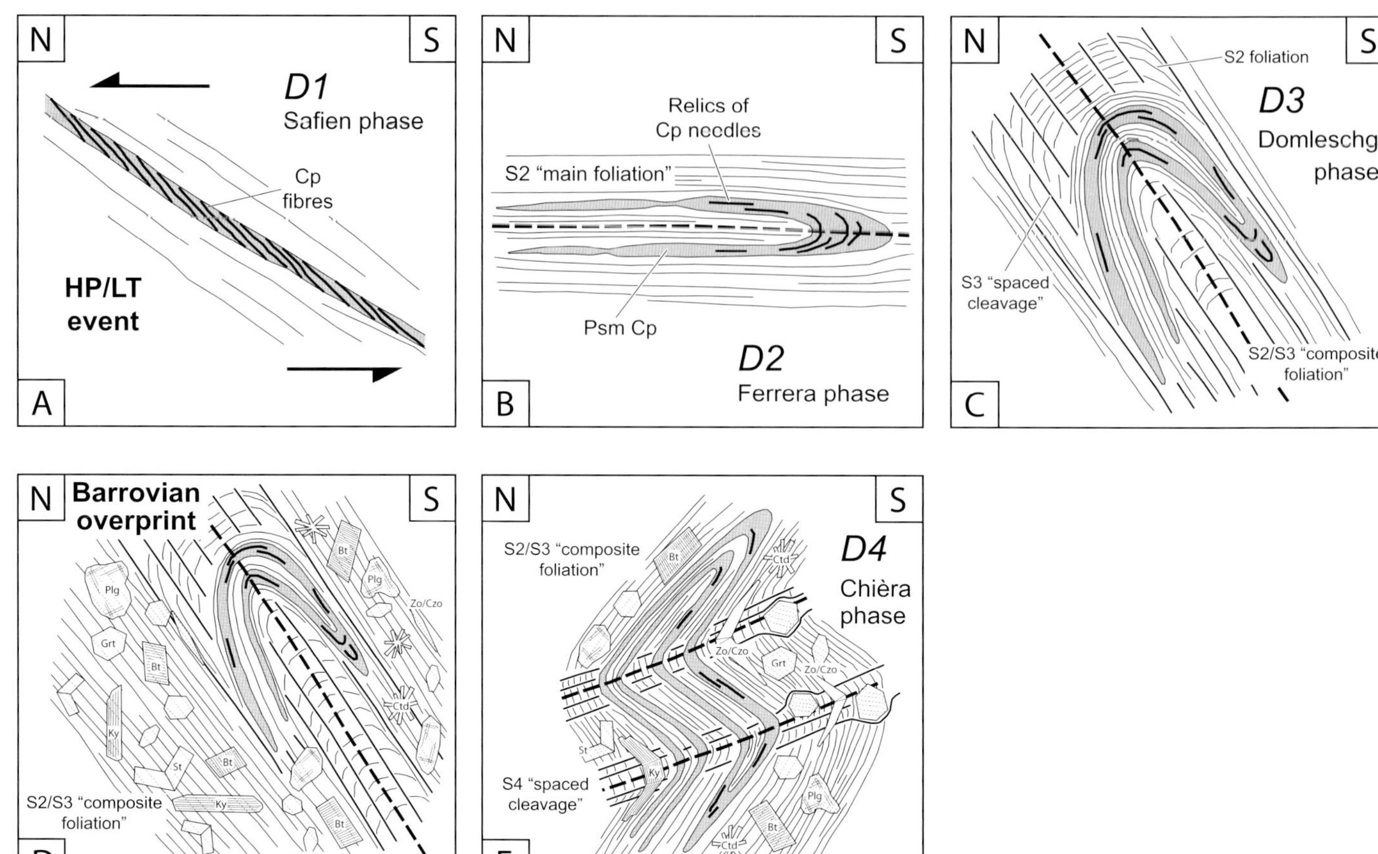

Fig. 10. Schematic sketches illustrating the tectono-metamorphic evolution in the study area. (A) Formation of fibrous quartz-calcite veins and Fe-Mg carpholite related to the HP/LT event, i.e. subduction (D1, Safien phase). (B) Pseudomorphs after carpholite, refolded by D2 (Ferrera phase), associated with the formation of the main penetrative foliation S2. (C) main foliation and earlier formed pseudomorphs after carpholite, refolded by D3 (Domleschg phase) and overprinted by a new spaced cleavage S3. (D) Amphibolite-facies Barrovian overprint leading to the growth of new porphyroblasts over the pre-existing S2/S3 composite foliation under static conditions. (E) D4 (Chièra phase), refolding the quartz-calcite veins a third time and deforming the amphibolite-facies mineral assemblages.

cold (or fast) decompression path after the HP/LT metamorphic stage (Gillet & Goffé 1988). In the Engadine window and in Safiental, no re-heating during this decompression can be evidenced from the observed mineral assemblages (Bousquet et al. 1998).

Temperature-dominated, Barrow-type overprint: Results based on graphite-thermometry

The P-T conditions are only well constrained for the pelitic rocks of the Sub-Penninic sediments in the south-western part of the working area, around Pizzo Molare and the area between Olivone and southern Lukmanier. There, earlier investigations yielded 500–550 °C and 0.5–0.8 GPa (Frey 1969; Fox 1975; Engi et al. 1995; Todd & Engi 1997; Frey & Ferreiro Mählmann 1999). In order to provide more information on the temperature gradient from NE to SW associated with the Barrow-type amphibolite facies overprint, we performed graphite-thermometry following the procedure proposed by Beyssac et al. (2002). The method is based on the degree of crystallisation of organic

material, which is mainly temperature dependent (Buseck & Bo-Jun 1985). Relationships between grade of crystallisation and metamorphic conditions are empirically calibrated (Beyssac et al. 2002). Since graphitisation of organic matter is strictly irreversible (Pesquera & Velasco 1988) this geothermometer always records the peak temperature reached by a rock specimen along its P-T loop, with a relative accuracy in the order of 10–15 °C (Beyssac et al. 2004). Here we only present the main results of this analysis of the "field thermal gradient" (Bollinger et al. 2004); details on method and results will be published elsewhere.

The peak temperatures derived from the Raman spectra obtained from over 140 samples collected between the Lucomagno/Pizzo Molare area in the west and Safiental in the east continuously increase from 350 °C in Safiental to 570 °C at the Pizzo Molare over an amazingly short distance along strike (Fig. 12). Most of this increase in temperature occurs in the Val Luzzone, i.e. between Piz Terri and Olivone. Further east a fairly homogeneous temperature between 350 and 400 °C, with only a moderate gradient, has been deduced.

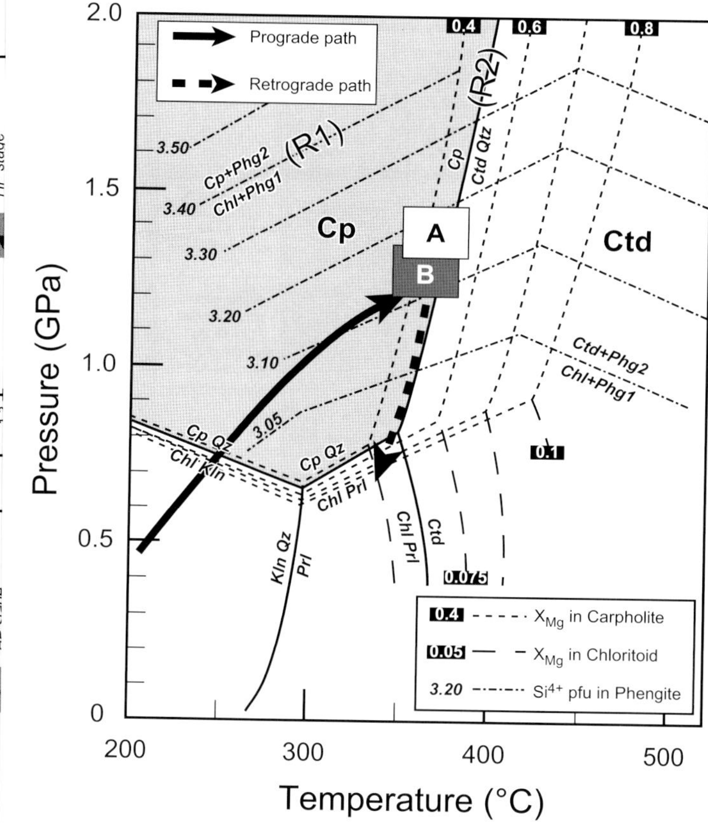

Fig. 11. Estimated P-T conditions for the North Penninic Bündnerschiefer (A) and the Sub-Penninic (European) meta-sediments (Peidener Schuppenzone, B). P-T conditions have been estimated for Fe-Mg carpholite-phengite-chlorite-quartz±chloritoid assemblages (only preserved in quartz-calcite veins) using the chemistry of the different minerals (Table 1) according to reactions R1 & R2 (see explanations in the text). Peak-pressure conditions were established very close to the position of the reaction R2, as is documented by the scarce occurrences of chloritoid that formed by the breakdown of Fe-Mg carpholite along the prograde path, i.e. during the HP/LT stage. The preservation of both Fe-Mg carpholite and associated rare occurrences of HP/LT chloritoid imply near-isothermal decompression from the HP/LT stage (Bousquet et al. 1998). Petrogenetic grid for HP/LT metapelites after Bousquet et al. (2002, 2008).

Comparison of the resulting temperature distribution pattern with the geological structures yields the following observation: In the south-west, the "isotherms" clearly cut all D2 nappe contacts and D3 post-nappe mega-folds, while in the north-east the "425 °C isotherm" is folded around the large scale D3 Lunschania antiform (Fig. 12). This shows that the peak temperatures were reached at different times and under different metamorphic conditions in the east and west. The post-D3 temperature increase in the SW is related to the later Barrow-type overprint. The folded isotherms in the NE, however, indicate that the temperatures derived for this area are "older", i.e. they pre-date the onset of the Barrovian overprint and are hence related to the high-pressure event and/or greenschist facies overprint that followed isothermal decompression. This in turn implies that a Barrow-type overprint possibly did not exist at all in the north-east, or was only associated with temperatures

lower than 425 °C, i.e. temperature ranges previously reached during the blueschist and/or greenschist facies event.

The temperatures determined by Raman microscopy of carbonaceous rocks are in excellent agreement with previously published temperature estimates based on traditional methods in the SW part of the working area (500–550 °C; e.g. Frey & Ferreiro Mählmann 1999). They also indicate temperatures of 500–550 °C, reaching 570 °C in the Pizzo Molare area. In the north-east the inferred temperatures (<425–375 °C) are near those inferred for the blueschist facies peak-pressure with traditional methods, ranging between 350–400 °C according to our study and that of Bousquet et al. (2002), as well as near those obtained for the greenschist facies overprint in the Grisons area (400 °C; Rahn et al. 2002).

The superposition of a Barrow-type over an earlier HP/LT evolution clearly indicates that the notion of a metamorphic field gradient can lead to misinterpretations. Strictly, a metamorphic field gradient primarily reflects the present-day distribution of pressure and/or temperature and cannot a priori be interpreted in terms of a particular geodynamic evolution.

5. Discussion and interpretation of the results

In the following we discuss the results obtained within the working area in a regional context and address the timing of the geodynamical evolution of the Alps. Then, in a qualitative way, we discuss possible heat sources that could be held responsible for Barrovian metamorphism in the north-eastern part of Lepontine thermal dome.

5.1. Regional correlations of the tectono-metamorphic evolution established in the working area and timing constraints

The D3 and D4 deformation phases led to the major features that are visible in map (Figs. 1 & 2) and cross-section (Fig. 3) view. D4 deformation resulted in the cascade-like geometry formed by a set of parasitic syn- and antiforms in the western part of the working area (Fig. 3), related to the formation of the Chièra synform which is well-developed only west of the working area (Milnes 1976; Etter 1987). There, the composite D2/D3 main foliation generally steeply dips northward and represents the overturned nappe-stack characterising the Northern Steep Belt. In our study area (Val Luzzone, Piz Terri, Val Lumnezia) the Chièra synform is only weakly developed, and instead, a series of cascade-like D4 syn- and antiforms overprint the D3 Lunschania antiform; these gradually fade out further to the east. The overall geometry of the cross section in Figure 3 is characterised by progressive steepening of the main foliation from a sub-horizontal orientation in the south to a generally moderate southward dip in the north that is produced by this D4 folding; subvertical and overturned composite D2/D3 foliations and nappe contacts are restricted to the structurally lowest levels.

(Milnes & Schmutz 1978; Schmid et al. 1990, 1996; Schreurs 1993), based on the fact that both these events are related to nappe stacking and that both are responsible for the formation of the first penetrative foliation. Weh and Froitzheim (2001) traced the Ferrera phase into the area of the Lower Penninic Bündnerschiefer (their D1a, b phase). It has to be emphasised, however, that the Ferrera phase was defined in a structurally higher level, i.e. the Middle Penninic Suretta nappe (Milnes & Schmutz 1978). Hence, the correlation with our D2 in a geometrical and kinematic sense does not imply that deformation in this structurally lower level was contemporaneous with the Ferrera phase, active during the 56–35 Ma age interval in the area of the Middle Penninic nappes (Schmid et al. 1996) as also documented by radiometric dating in the Suretta nappe (46 ± 5 Ma; Challandes et al. 2003). We emphasise that not all structural correlations presented in Table 2 imply that deformation producing these structures was contemporaneous at the scale of the Alps.

In the Lower and Middle Penninic units east of our area of investigation, the Ferrera phase has been severely overprinted by the Niemet-Beverin phase, which represents a first nappe refolding stage, resulting in large scale back-folding and inverting the nappe pile in the upper limb of the recumbent Niemet-Beverin mega-fold (Milnes & Schmutz 1978; Schmid et al. 1990; Schreurs 1993; Mayerat Demarne 1994; Weh & Froitzheim 2001; Pleuger et al. 2003). Interestingly, no effects of this Niemet-Beverin phase (35–30 Ma; Schmid et al. 1996) were found in our area of investigation. Therefore, it is theoretically possible that the D2-event (Ferrera phase) could have lasted until some 30 Ma ago in our study area, i.e. at a much deeper structural level and in units occupying a more external paleogeographical position.

D1-deformation related to formation of quartz-calcite segregations, pre-dating D2 of the working area, corresponds to the sub-stage D1a of Weh and Froitzheim (2001; Table 2). These authors proposed that the formation of tight to isoclinal folds (their D1b; our D2) post-dates the formation of Fe-Mg carpholite-bearing veins associated with the formation of a penetrative foliation during their D1a (our D1). An estimate on the age of D1 in the working area may be obtained by considering the fact that D1 is linked with thrusting along the Penninic Basal Thrust, whose age is constrained by the age of the youngest sediments involved. Sedimentation in the Lower Penninic Bündnerschiefer realm lasted until Lowest Eocene times (i.e. some 50 Ma ago) according to Weh & Froitzheim (2001), but until Bartonian times (i.e. some 40 Ma) in the Sardona Unit (Lihou & Allen 1996). The paleogeographical position of the Sardona Unit is considered Ultra-Helvetic by some authors (e.g. Lihou & Allen 1996) but Penninic by others (mainly based on sedimentological and age criteria; e.g. Trümpy 1980; Hsü & Briegel 1991). Regardless of its precise paleogeographical position, it is extremely unlikely that the Sardona Flysch is of more external origin in respect to the Ultrahelvetic sediments of the Peidener Schuppenzone, given its high content of siliciclastic detritus partly shed from the "North Prättigau High" (Lihou & Allen 1996). Hence, D1-deformation, sediment-accretion and blueschist facies overprint are unlikely to have started before the Bartonian, i.e. before some 40 Ma ago.

Interestingly, the main nappe stacking Ferrera phase also post-dates an early thrusting event formed under HP/LT conditions, known as the Avers phase (Milnes & Schmutz 1978; Schmid et al. 1997b; Wiederkehr 2004) in the Briançonnais-Piemont-Liguria Ocean contact area in the Avers. However, since the Avers phase is related to the closure of the Piemont-Liguria Ocean during the Late Paleocene (Schmid et al. 1996), it must substantially pre-date the D1 event in our working area, which is related to the closure of the more northerly Valais Ocean.

In summary, this study could, for the first time in the investigated area, decipher the existence of an early blueschist-facies tectono-metamorphic event related to subduction and sediment-accretion. We refer to this event as the Safien phase (Table 2, Figs. 9 & 10). In contrast to Weh & Froitzheim (2001; their D1a and D1b), we emphasise a clear separation between the D1 and D2 events. This separation is supported by the fact that D2 formed under greenschist facies conditions. The age of the HP/LT Safien phase is constrained to post-date Bartonian times, i.e. 40 Ma. D2 deformation in our area is probably younger than the Ferrera phase in the Schams area; it probably lasted until the onset of D3-deformation, i.e. some 30 Ma ago. The onset of Barrow-type overprint, which post-dates D3 (i.e. 25 Ma according to the correlation of D3 with the age of the Domleschg phase; Schmid et al. 1996) is likely to be very much younger in respect of the high-pressure event. Given an almost static interval of mineral growth and the new radiometric dating of Barrow-type metamorphism in the area (Allaz et al. 2007; Janots et al. 2007), this heating pulse post-dates 20 Ma.

5.2. P-T-d-t path and reconstruction of the regional tectono-metamorphic evolution

The complex metamorphic evolution characterised by an early HP/LT stage (350–400 °C, 1.2–1.4 GPa), later overprinted by a Barrow-type amphibolite facies event (500–570 °C, 0.5–0.8 GPa) can be reconciled with either of two different P-T path trends: (1) A single P-T loop whereby the amphibolite facies overprint results from heating during decompression after HP/LT metamorphism, or alternatively, (2) a two-stage P-T evolution, whereby the amphibolite facies Barrovian overprint represents a separate heating pulse that follows earlier isothermal or cooling decompression from HP/LT conditions (Fig. 13). For the reconstruction of the regional tectono-metamorphic evolution and the interpretation of the geodynamic scenario, it is crucial to obtain constraints on the shape of the P-T path and its timing.

The following facts argue for isothermal or slightly cooling decompression of the Lower Penninic Valaisan Bündnerschiefer and the Sub-Penninic Peidener Schuppenzone after the HP stage: (1) very good preservation of Fe-Mg carpholite east of the study area (Engadine window), (2) its replacement exclusively by a lower pressure mineral assemblage (pheng-ite-chlorite-quartz) within the studied area, and (3) the preservation of both pseudomorphs and relics of Fe-Mg carpholite within the north-eastern Lepontine dome. Hence, the breakdown of carpholite in the Valaisan of the study area is entirely pressure-controlled and not associated with a temperature increase.

The analysis of the relationships between deformation and metamorphism and the timing constraints provide additional arguments in favour of a two-stage P-T-evolution characterised by a separate heating pulse that followed isothermal or cooling decompression from earlier HP/LT metamorphic conditions:

1) Both metamorphic events are separated from each other by two deformation phases implying a considerable time gap between them. The HP/LT event predates D2 and was estimated to have started ca. 40 Ma ago, certainly before 30 Ma (onset of D3-deformation). Barrow-type amphibolite facies overprint post-dates D2 nappe stacking and a first nappe re-folding event D3 (30–25 Ma), and hence, was younger than 20 Ma (Figs. 9 & 10).

2) Substantial decompression was associated with D2 nappe stacking and therefore predates the heating pulse that took place after D3.

3) The increase in temperature took place under more or less static conditions between D3 and D4.

4) Graphite thermometry documents a two-stage temperature distribution pattern. The D3 Lunschania antiform folded an older HP/LT-related pattern and a younger, onion-shaped pattern cutting the D3 Lunschania antiform was associated with Barrovian amphibolite facies overprint (Fig. 12).

We propose the following 5-stage scenario regarding the tectono-metamorphic evolution in the area (Fig. 14):

(1) *Subduction and sediment-accretion stage (Safien phase):* The Lower Penninic Bündnerschiefer of the Grisons area (mainly Grava and Tomül nappes) that today build up a 20 km thick accretionary wedge of meta-sediments (Hitz & Pfiffner 1997) formed during Cenozoic subduction of the Valais Ocean and the distal European margin beneath the Briançonnais micro-continent. Deeper parts of this sedimentary accretionary wedge experienced pressure-dominated metamorphism under blueschist facies conditions (350–400 °C, 1.2–1.4 GPa), including parts of the sedimentary cover in Ultrahelvetic facies (Peidener Schuppenzone), which were detached from their crystalline basement and incorporated into the HP/LT part of the accretionary wedge. The associated deformation (D1 Safien phase) was semi-ductile and led to the formation of shear fibre veins consisting of quartz, calcite and Fe-Mg carpholite. Exact timing of this HP/LT subduction and sediment-accretion stage is not yet possible. It is, however, constrained to have occurred during the Late Eocene (after 40 but before 30 Ma ago; see Berger & Bousquet 2008), hence substantially after similar but Late Paleocene to Middle Eocene high-pressure stages pro-

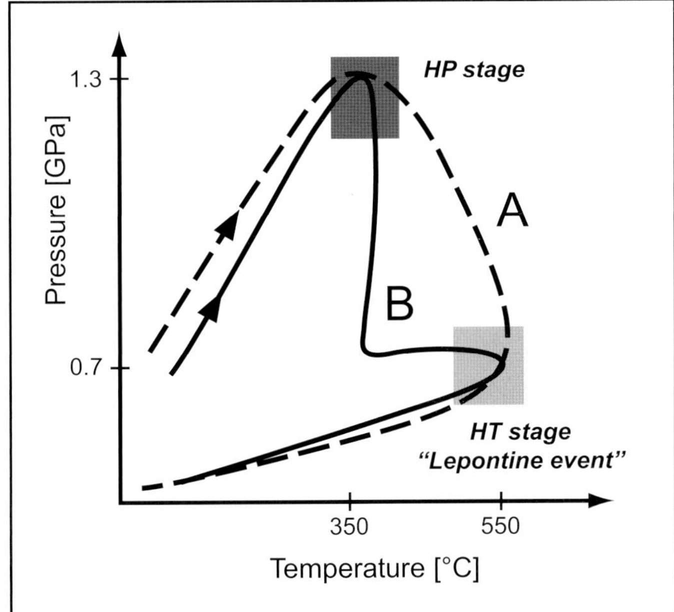

Fig. 13. Alternative P-T paths connecting the HP/LT event and the amphibolite-facies Barrovian overprint. Path A is a single P-T path, decompressional heating of the HP/LT stage leading to Barrovian overprint. Path B is characterised by a two-stage P-T path, Barrovian overprint represented by a separate heating pulse that follows isothermal or cooling decompression.

posed for the Middle and Upper Penninic units (see overviews given by Froitzheim et al. 1996; Schmid et al. 2004; Berger & Bousquet 2008; Bousquet et al. 2008).

(2) *Nappe stacking and decompression stage (Ferrera phase):* Nappe stacking was associated with substantial decompression of the blueschist-facies rocks. The presence of relics of Fe-Mg carpholite indicates decompression under isothermal or cooling conditions. This thermal regime, as well as ongoing accretion of the rest of the Sub-Penninic sediments (Scopi Unit, Fig. 3), suggests that also European continental basement rocks, such as preserved in the Adula, Simano and Leventina-Lucomagno units, became involved in ongoing subduction and accretion during the Ferrera phase. We propose that the Ferrera phase in our working area may have outlasted earlier stages of the nappe-stacking Ferrera phase that affected the Middle Penninic units (e.g. Baudin et al. 1993), and probably was active until 30 Ma ago.

(3) *Nappe re-folding (Domleschg phase):* This D3 nappe folding event substantially modified the Penninic nappe stack in the working area. It post-dates an earlier nappe re-folding phase established only within structurally higher North and Middle Penninic nappes east of the working area (Niemet-Beverin phase; e.g. Schreurs 1993; Baudin et al. 1993; Marquer et al. 1996; Weh & Froitzheim 2001). D3 deformation produced tight to isoclinal mega-folds with amplitudes up to some 10 km (Figs. 3 & 14): the most prominent Lunschania antiform, but

also the Valzeina and Alpettas synforms, as well as the Darlun antiform (Voll 1976; Kupferschmid 1977; Probst 1980; Steinmann 1994a, b; Weh & Froitzheim 2001; Uhr unpubl.; Figs. 1, 2 & 3). On the scale of the entire Alpine orogen the Domleschg phase, characterised by far less intense folding at higher structural levels, is interpreted as contemporaneous with back thrusting along the Insubric mylonite belt (Schmid et al. 1987), which occurred during the 30–25 Ma time interval (see discussion given in Schmid et al. 1997b). Note also that this phase is associated with ongoing accretion of continental basement (i.e. Lucomagno-Leventina nappe, Fig. 14). Regarding the eastern part of the investigated area, the tectono-metamorphic evolution essentially came to a halt after D3 deformation (Fig. 12). The following metamorphic and tectonic events only affected the western part of the working area.

(4) *Barrow-type thermal overprint:* This thermal pulse occurred during a tectonically quiescent phase within the working area (but not necessarily elsewhere, i.e. in the more external parts of the Alps), initiating shortly after some 20 Ma ago. Increasing temperatures led to the formation of porphyroblasts related to classical Barrow-type amphibolite facies overprint. This thermal overprint was sustained until the beginning stages of the last tectonic (D4) event (Figs. 12 & 14).

(5) *Back-folding in the Northern Steep Belt (Chièra phase):* This second nappe re-folding event leads to back-folding within the Gotthard "massif" and adjacent areas. It is associated with the formation of the Northern Steep Belt of the Penninic realm that is well developed only west of our area of investigation (Milnes 1974). D4 deformation is intense in the south-west but gradually becomes weaker towards the north-east and, finally, fades out somewhere east of the Piz Terri-Vrin area (Figs. 2 & 3). A relatively tight synform, the Chièra synform (Milnes 1974; Milnes 1976; Etter 1987), brings the Lower Penninic and Sub-Penninic nappe stack into an overturned, steeply north dipping position (Northern Steep Belt) at the deepest structural levels. Within most of the working area, a set of parasitic syn- and antiforms develops, structurally located between the Chièra synform and the more northerly located corresponding Greina or Gotthard antiform (Thakur 1973) which brings the overturned nappe pile back into a normal position (Fig. 3). Hence, back-folding is much less pronounced in our working area compared to further west (Figs. 2 & 3). This folding outlasted Barrovian overprint (18–20 Ma; Allaz et al. 2007; Janots et al. 2007) and hence is very young (probably post-18 Ma) and contemporaneous with the N-directed thrusting in the Aar massif in the more external parts of the Alps (Grindelwald phase; Burkhard 1988; Schmid et al. 1996; Pfiffner et al. 1997) and movements along the Simplon line associated with back folding west of the Lepontine dome (Steck 1984, 1990; Marquer & Gapais 1985; Mancktelow 1992; Mancktelow & Pavlis 1994; Steck & Hunziker 1994; Keller et al. 2006).

We conclude that Barrovian overprint in the working area, representing a separate heating pulse (Fig. 14) is surprisingly young (18–20 Ma; Allaz et al. 2007; Janots et al. 2007) when compared to the timing of a similar separate heating pulse proposed for the Southern Steep Belt at around 30–27 Ma (Engi et al. 2001; Berger et al. 2005; Brouwer et al. 2005; Brouwer & Engi 2005). Barrow-type overprint at the western edge of the Lepontine dome, which occurred along a single continuous P-T path, occurred before some 20 Ma ago according to Keller et al. (2005).

Barrow-type metamorphism in the Lepontine area is often referred to as the Lepontine metamorphic event. The term "event" is totally misleading since Lepontine Barrow-type metamorphism, rather than representing one single event, is diachronous; thermal overprint becomes progressively younger towards the north (Köppel et al. 1981; Engi et al. 1995). Consequently, different relations between deformation and crystallisation are commonly observed (Berger et al. 2005). We emphasise that the separate heating pulse described in this study is characteristic for the north-eastern part of the Lepontine dome only and that no direct inferences in terms of timing, geodynamic setting or nature of the heat source should be drawn regarding the rest of the Lepontine area, particularly its southern part.

5.3. Discussion of potential heat sources of Barrow-type overprint and thermal evolution

We now discuss possible heat sources that could potentially be responsible for Barrovian metamorphism in the north-eastern part of the Lepontine thermal dome. We do this in a qualitative way, being aware of the complexities of the subject. We first briefly introduce the presently known potential heat sources and then qualitatively discuss which of these heat sources could best explain the observations.

The existence of a separate heating pulse raises the old and still widely debated question after the heat sources for Barrovian overprint in collisional orogens (e.g. Jamieson et al. 1998). The following potential heat sources have been proposed for Barrow-type thermal overprint: (1) Shear or viscous heating (Burg & Gerya 2005). (2) Advective heat transfer by rising magma, i.e. plutons and dykes (Engi et al. 1995; Frey & Ferreiro Mählmann 1999), possibly induced by up-welling of hot asthenosphere due to slab break-off (von Blanckenburg & Davies 1995). (3) Advective heat transfer to the upper crust by exhumation of hot eclogitic slices within a subduction channel (Becker 1993; Engi et al. 2001), or alternatively, by extension-related exhumation of hot high-pressure rocks (Platt 1986; Ballèvre et al. 1990). (4) Accretion of continental crustal rocks characterised by high radioactive heat production (Chamberlain & Sonder 1990; Bousquet et al. 1997; Huerta et al. 1998; Roselle et al. 2002; Goffé et al. 2003).

Discussions on shear heating need to (1) evaluate the expected spatial and temporal distribution of shearing-induced heat and (2) quantify the amount of heat produced, which depends on strain-rates and deformation mechanisms (e.g. Peacock 1996). In the case of the Alpine orogen such shear heating

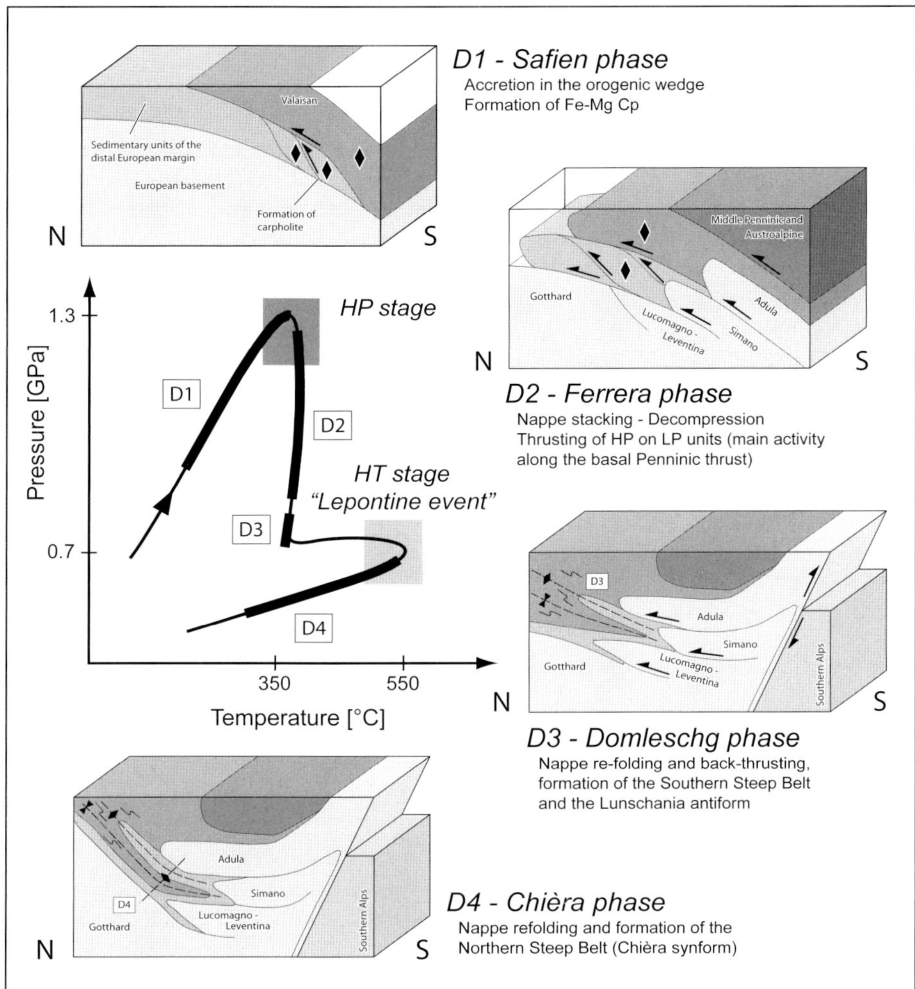

Fig. 14. P-T-d path for the Lower Penninic and Sub-Penninic (Peidener Schuppenzone) metasedimentary units and sketches of the geodynamic scenario during the various stages of the tectono-metamorphic evolution.

would be expected to lead to a thermal zonation, which parallels the strike of the orogen, i.e. parallel to the strike of the potential high-strain zones (in general nappe contacts and shear zones associated with the formation of the nappe-stack) that produce this heat. This, however, is not the case within the study area since isotherms are perpendicular to important structural elements, such as the Penninic Basal Thrust and other nappe contacts (Fig. 12). Moreover, plate convergence rates in the Alps are considered too low (in the order of 1 cm per year; Schmid et al. 1996) to produce enough heat at a nappe or orogen-wide scale, while this heat source may play a role in case of localised shear zones. Hence, only at high convergence rates can shear heating be an important heat source at a large scale. This, however, is not the case regarding the thermal overprint that occurred at a very late stage, i.e. when the Alpine edifice was essentially established. Moreover microstructural observations of porphyroblasts related to Barrovian metamorphism indicate that thermal overprint took place under more or less static conditions and was not at all associated with deformation. Based on field evidence we conclude that shear heating probably had a rather limited influence on the thermal evolution in our working area.

The effect of advective heat transport by magma and/or local melt is also negligible during Barrovian overprint within our working area. While the intrusion of the Bergell pluton and segregating migmatitic melts (Berger et al. 1996, 2007; Burri et al. 2005) may additionally contribute to the heat budget in case of the Southern Steep belt, these effects can probably be ignored in case of our working area at the northern rim of the Lepontine dome.

Exhumation of hot eclogitic slices, combined with radioactive heat production by accreted continental crust, was proposed as a model for Barrow-type overprint within the Southern Steep Belt (Tectonic Accretion Channel model, e.g. Engi et al. 2001, Roselle et al. 2002). Rising high-temperature eclogites as a potential advective heat source are indeed available outside the working area, i.e. in form of the eclogite facies Adula nappe complex. Moreover, in the Western Alps, characterised by the occurrence of large volumes of high-pressure and ultra-high-pressure units, such as the Gran Paradiso and Dora-Maira

Internal Massifs, no Barrovian thermal overprint is observed (Oberhänsli et al. 2004). It seems that the rising of eclogites is not efficient enough to explain Barrovian overprint in our working area, since such eclogites are restricted to more southerly areas within the Lepontine dome, and also in view of the large temporal hiatus between high-pressure event and Barrovian overprint.

The strongest argument against advective heat transport by exhumation of eclogitic material in the northern Lepontine dome and the Tauern window is that the distribution of the HP/LT metamorphic units in the Alps is completely different from that of the areas characterised by a Barrovian overprint (Bousquet et al. 2008). The latter are restricted to dome-shaped areas, such as the Lepontine dome and the Tauern window (Bousquet et al. 2008). Both the Lepontine dome and the Tauern window are characterised by massive accretion of granitoid basement units derived from the distal European margin (Sub-Penninic nappe stack; Milnes 1974; Schmid et al. 2004). Hence, Barrovian overprint is spatially coupled with exposures of large nappe-stacks of continental material, characterised by high radioactive heat production. However, one might argue that, since both these domes represent structural highs, similar Barrovian-metamorphism would be expected at depth outside these domes. A simple consideration of the volume available between the earth's surface and the Moho, which is at approximately constant depth along strike (Waldhauser et al. 1998), excludes along-strike doming of the entire crust. Hence, this doming is related to the localised accretion of large volumes of upper European basement, as is documented by the stacking of the Sub-Penninic basement nappes; in other words, doming is the direct isostatic response of such localised accretion of European upper crust (see Bousquet et al. 2008 for a more detailed discussion).

The effect of radioactive heat production is only relevant if such material is accreted at certain depths (locations with primary lower radioactive heat productions, see Jamieson et al. 1998). The most likely way to add such heat sources is the combination of subduction and subsequent thickening (e.g. Jamieson et al. 1998; Engi et al. 2001; Roselle et al. 2002; Goffé et al. 2003). Our field observations fulfil these prerequisites since subduction was followed by late-stage heating. Hence, we propose that the thermal regime during Barrovian overprint is determined by the thermal structure during the final stages of subduction and the additional heat release from radioactive decay of accreted material. However, these considerations do not quantitatively explain the observed distribution of temperatures and additional model calculations to those of Bousquet et al. (1997), Roselle et al. (2002) and Goffé et al. (2003) are necessary.

Nevertheless the field data can be discussed within the frame of the thermal evolution of the Alpine orogenic wedge (see Fig. 15). The low-temperature regime is associated with an early subduction and sediment accretion stage and led to the formation of mineral assemblages that are typical for subduction processes and related down-folding of the iso-

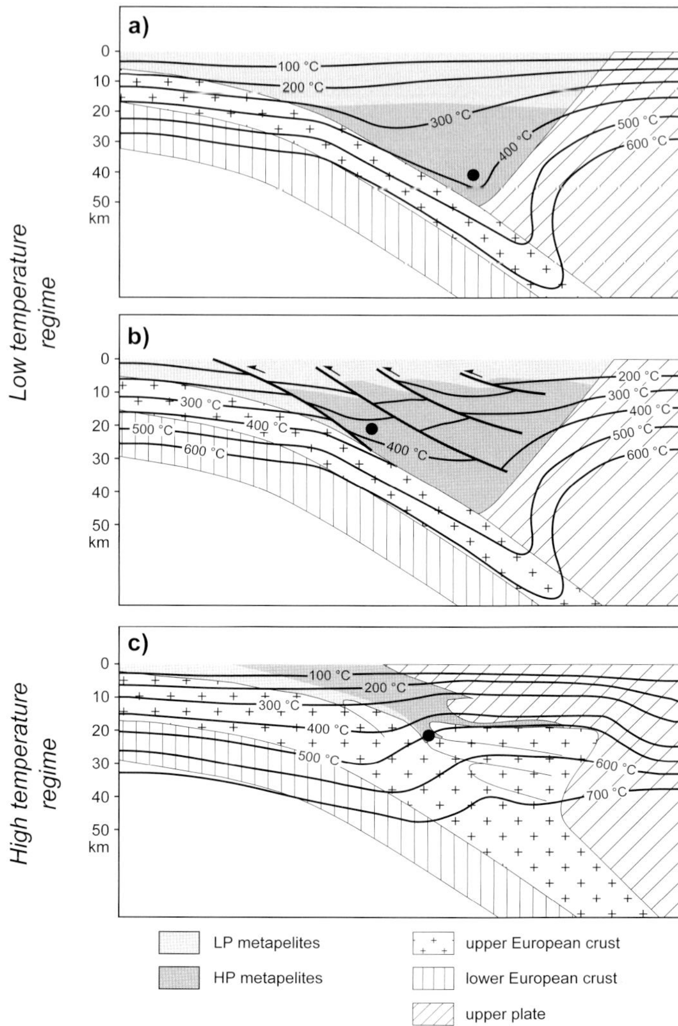

Fig. 15. Very schematic sketches, illustrating the evolution of isotherms in an orogenic wedge (inspired by Goffé et al. 2003) such as the Alps during three geodynamic stages of the orogen. Two strongly differing thermal regimes are distinguished: a low temperature (a & b) and a high temperature regime (c). The black dot represents the location of the meta-sediments of the study area. (a) Typical pattern of isotherms during the subduction stage. The wedge is dominated by accretion of large amounts of sediments. The deeper parts of the wedge are characterised by pressure-dominated metamorphism under blueschist facies conditions. (b) Isothermal or cooling decompression during nappe-stacking, bringing HP onto LP units; the isotherms remain downbent. (c) Due to massive accretion of continental crustal material after collision, the rock composition within the wedge changes dramatically: Large amounts of upper-crustal European granitoid rocks were accumulated within the wedge (Sub-Penninic nappe stack). This accumulation of heat-producing crustal material is responsible for increasing temperatures by the up-bending of isotherms inducing the late-stage amphibolite-facies Barrovian overprint observed in the working area.

therms (Fig. 15a). During ongoing subduction, deeper parts of the orogenic wedge were thrust onto lower pressure units, a process that is accompanied with nearly isothermal or cooling decompression (Fig. 15b). The high-temperature regime (Fig. 15c) occurs after the accretion of additional continental middle crust that led to the formation of the Sub-Penninic nappe-stack. This temperature regime leads to the up-doming of isotherms, which cut through nearly all the structural units (Fig. 12). Therefore, we infer that the observable crosscutting isotherms are most likely related to a late stage of purely conductive heat transfer. Note that these rising isotherms, in the absence of mass transport, elegantly explain the Barrow-type amphibolite facies overprint of HP/LT units that already experienced substantial decompression before, as is the case in our study area.

6. Conclusions

We provided evidence for a two-stage metamorphic evolution of meta-sedimentary units derived from the Valaisan Ocean (Grava nappe) and the distal European margin (Peidener Schuppenzone) in the north-eastern part of the Lepontine dome. A first HP/LT metamorphic event under blueschist facies conditions (350–400 °C and 1.2–1.4 GPa) was associated with subduction and sediment-accretion. It was immediately followed by "cold" isothermal or cooling decompression during nappe stacking. Continent-collision-related classical Barrow-type amphibolite facies overprint (500–570 °C and 0.5–0.8 GPa) represents a separate heating pulse that post-dates the D3 nappe-refolding event. It was induced by post-collisional accretion of continental crust, and it largely occurred under static conditions, partly during the initial stages of the D4 back-folding event that led to the formation of the Northern Steep Belt of the Penninic nappe pile. The two metamorphic events are separated by a time gap within our working area, estimated to be in the order of 20 Ma.

Amongst the various possible heat sources of Barrovian metamorphism we regard radiogenic heat production by accretion of continental crust during the collisional and post-collisional stages of Alpine orogeny, associated with rising isotherms, to be mainly responsible for this separate late-stage heating event at the north-eastern rim of the Lepontine dome. We propose that the Lepontine and Tauern thermal and structural domes both largely resulted from the local accretion of massive volumes of Sub-Penninic basement nappes derived from the distal European margin. This well explains the substantial Barrow-type thermal gradient observed at the north-eastern rim of the Lepontine dome, cutting across former nappe contacts almost perpendicular to strike. We emphasise, however, that in the southern parts of the Lepontine dome (Southern Steep Belt) other heat sources such as heat advection by rising eclogitic bodies and melts are probably also very important (e.g. Frey & Ferreiro Mählmann 1999; Nagel et al. 2002; Keller et al. 2005; Berger et al. 2007).

The new data from the north-eastern rim of the Lepontine dome provide strong evidence for the former existence of a contiguous HP/LT belt, representing a second northern suture zone associated with the closure of the Valais Ocean. Moreover, relative timing constraints indicate that both HP/LT metamorphism and Barrow-type overprint were diachronous at the scale of the Alpine orogen; hence all indicators of metamorphic zonation such as index mineral zone boundaries must be strongly diachronous.

Acknowledgements

Excellent preparation of numerous samples by W. Tschudin, as well as hauling masses of rocks down the mountains by G. Derungs, including great support in the field, are gratefully acknowledged. O. Appelt and Dr. D. Rhede from the GeoForschungsZentrum Potsdam are thanked for help with microprobe analyses. A supporting field visit of A. Riemann from the University of Potsdam is appreciated. Dr. M.A. Ziemann is thanked for the introduction and support in the Raman laboratory of Potsdam University. We also thank the two reviewers J. Pleuger and M. Janak for their constructive comments and suggestions for improving this paper, and the editor N. Froitzheim for his careful handling of the manuscript. Substantial funding by the Swiss National Science Foundation (project NF-200020-113585 and precursor project NF-200020-103585) is gratefully acknowledged.

REFERENCES

Agard, P., Jolivet, L. & Goffé, B. 2001: Tectonometamorphic evolution of the Schistes Lustrés complex: implications for the exhumation of HP and UHP rocks in the Western Alps. Bulletin de la Société Géologique de France 172, 617–636.

Allaz, J., Janots, E., Engi, M., Berger, A. & Villa, I.M. 2007: Understanding Tertiary metamorphic ages in the northern Central Alps. Geophysical Research Abstracts 9, EGU2007-A-07684.

Ballèvre, M., Lagabrielle, Y. & Merle, O. 1990: Tertiary ductile normal faulting as a consequence of lithospheric stacking in the Western Alps. Mémoires de la Société Géologique Suisse 1, 227–236.

Baudin, T., Marquer, D. & Persoz F. 1993: Basement-cover relationships in the Tambo nappe (Central Alps, Switzerland) – geometry, structure and kinematics. Journal of Structural Geology 15, 543–553.

Baumer, A. 1964: Geologie der gotthardmassivisch-penninischen Grenzregion im oberen Bleniotal. Geologie der Blenio-Kraftwerke. Beiträge zur Geologie der Schweiz, Geotechnische Serie 39, 105 pp.

Baumer, A., Frey, J.D., Jung, W. & Uhr, A. 1961: Die Sedimentbedeckung des Gotthard-Massivs zwischen oberen Bleniotal und Lugnez (Vorläufige Mitteilung). Eclogae geologicae Helvetiae 54, 478–491.

Becker, H. 1993: Garnet peridotite and eclogite Sm-Nd mineral ages from the Lepontine dome (Swiss Alps) – New evidence for Eocene high-pressure metamorphism in the Central Alps. Geology 21, 599–602.

Berger, A. & Bousquet, R. 2008: Subduction related metamorphism in the Alps: Review of isotopic ages based on petrology and their geodynamic consequences. In: Siegesmund, S. et al. (Eds.): Tectonic Aspects of the Alps-Dinarides-Carpathians system. Geological Society Special Publications 298, 117–144.

Berger, A., Rosenberg, C. & Schmid, S.M. 1996: Ascent, emplacement and exhumation of the Bergell pluton within the Southern Steep Belt of the Central Alps. Schweizerische Mineralogische und Petrographische Mitteilungen 76, 357–382.

Berger, A., Mercolli, I. & Engi, M. 2005: The central Lepontine Alps: Notes accompanying the tectonic and petrographic map sheet Sopra Ceneri (1:100'000). Schweizerische Mineralogische und Petrographische Mitteilungen 85, 109–146.

Berger, A., Burri T., Alt-Epping, P. & Engi, M. 2007: Tectonically controlled fluid flow and water-assisted melting in the middle crust: An example from the Central Alps. Lithos, doi:10.1016/j.lithos.2007.07.027.

Berman, R.G. 1988: Internally-consistent thermodynamic data for minerals in the system $Na_2O\text{-}K_2O\text{-}CaO\text{-}MgO\text{-}FeO\text{-}Fe_2O_3\text{-}Al_2O_3\text{-}SiO_2\text{-}TiO_2\text{-}H_2O\text{-}CO_2$. Journal of Petrology 29, 445–522.

Beyssac, O., Goffé, B., Chopin, C. & Rouzaud, J.N. 2002: Raman spectra of carbonaceous material in metasediments: a new geothermometer. Journal of Metamorphic Geology 20, 859–871.

Beyssac, O., Bollinger, L., Avouac, J.P. & Goffé, B. 2004: Thermal metamorphism in the lesser Himalaya of Nepal determined from Raman spectroscopy of carbonaceous material. Earth and Planetary Science Letters 225, 233–241.

Bollinger, L., Avouac, J.P., Beyssac, O., Catlos, E.J., Harrison, T.M., Grove, M., Goffé, B. & Sapkota, S. 2004: Thermal structure and exhumation history of the Lesser Himalaya in central Nepal. Tectonics 23, TC5015, doi:10.1029/2003TC001564.

Bossard, L. 1925: Der Bau der Tessiner Kulmination. Eclogae geologicae Helvetiae 19, 504–521.

Bossard, L. 1929: Zur Petrographie unterpenninischer Decken im Gebiet der Tessiner Kulmination. Schweizerische Mineralogische und Petrographische Mitteilungen 9, 47–106.

Bousquet, R., Goffé, B., Henry, P., Le Pichon, X. & Chopin, C. 1997: Kinematic, thermal and petrological model of the Central Alps: Lepontine metamorphism in the upper crust and eclogitisation of the lower crust. Tectonophysics 273, 105–127.

Bousquet, R., Oberhänsli, R., Goffé, B., Jolivet, L. & Vidal, O. 1998: High-pressure-low-temperature metamorphism and deformation in the Bündnerschiefer of the Engadine window: implications for the regional evolution of the eastern Central Alps. Journal of Metamorphic Geology 16, 657–674.

Bousquet, R., Goffé, B., Vidal, O., Oberhänsli, R. & Patriat, M. 2002: The tectono-metamorphic history of the Valaisan domain from the Western to the Central Alps: New constraints on the evolution of the Alps. Geological Society of America Bulletin 114, 207–225.

Bousquet, R., Oberhänsli, R., Goffé, B., Wiederkehr, M., Koller, F., Schmid, S.M., Schuster, R., Engi, M., Berger, A. & Martinotti, G. 2008: Metamorphism of metasediments in the scale of an orogen: A key to the Tertiary geodynamic evolution of the Alps. In: Siegesmund, S. et al. (Eds.): Tectonic Aspects of the Alps-Dinarides-Carpathians system. Geological Society Special Publications 298, 393–411.

Brouwer, F.M. & Engi, M. 2005: Staurolite and other aluminous phases in Alpine eclogite from the Central Swiss Alps: Analysis of domain evolution. Canadian Mineralogist 43, 105–128.

Brouwer, F.M., Burri, T., Engi, M. & Berger, A. 2005: Eclogite relics in the Central Alps: PT-evolution, Lu-Hf ages and implications for formation of tectonic mélange zones. Schweizerische Mineralogische und Petrographische Mitteilungen 85, 147–174.

Brown, T.H., Berman, R.G. & Perkins, E.H. 1988: GEO-CALC – Software package for calculation and display of pressure-temperature-composition phase diagrams using IBM or compatible personal computer. Computer and Geosciences 14, 279–289.

Bucher, K. & Frey, M. 2002: Petrogenesis of metamorphic rocks. 7th edition, Springer, Berlin, 341 pp.

Bucher, S. & Bousquet, R. 2007: Metamorphic evolution of the Briançonnais units along the ECORS-CROP profile (Western Alps): New data on metasedimentary rocks. Swiss Journal of Geosciences 100, 227–242.

Bucher, S., Schmid, S.M., Bousquet, R. & Fügenschuh, B. 2003: Late-stage deformation in a collisional orogen (Western Alps): nappe refolding, back-thrusting or normal faulting? Terra Nova 15, 109–117.

Bucher S., Ulardic, C., Bousquet R., Ceriani, S., Fügenschuh B. & Schmid, S.M. 2004: Tectonic evolution of the Briançonnais units along a transect (ECORS-CROP) through the Italian-French Western Alps. Eclogae geologicae Helvetiae 97, 321–345.

Burckhardt, C.E. 1942: Geologie und Petrographie des Basodino-Gebietes (nordwestliches Tessin). Schweizerische Mineralogische und Petrographische Mitteilungen 22, 99–188.

Burg, J.P. & Gerya, T.V. 2005: The role of viscous heating in Barrovian metamorphism of collisional orogens: thermomechanical models and application to the Lepontine Dome in the Central Alps. Journal of Metamorphic Geology 23, 75–95.

Burkhard, M. 1988: L'Helvétique de la bordure occidentale du massif de l'Aar (évolution tectonique et métamorphique). Eclogae geologicae Helvetiae 81, 63–114.

Burri, T., Berger, A. & Engi, M. 2005: Tertiary migmatites in the Central Alps: Regional distribution, field relations, conditions of formation, and tectonic implications. Schweizerische Mineralogische und Petrographische Mitteilungen 85, 215–232.

Buseck, P.R. & Bo-Jun, H. 1985: Conversion of carbonaceous material to graphite during metamorphism. Geochimica et Cosmochimica Acta 49, 2003–2016.

Casasopra, S. 1939: Studio petrografico dello gneiss granitico Leventina. Schweizerische Mineralogische und Petrographische Mitteilungen 19, 449–709.

Chadwick, B. 1968: Deformation and Metamorphism in the Lukmanier Region, Central Switzerland. Geological Society of America Bulletin 79, 1123–1150.

Challandes, N., Marquer, D & Villa, I.M. 2003: Dating the evolution of C-S microstructures: a combined $^{40}Ar/^{39}Ar$ step heating and UV laser-probe analysis of the Alpine Roffna shear zone. Chemical Geology 197, 3–19.

Chamberlain, C.P. & Sonder, L.J. 1990: Heat-producing elements and the thermal and baric patterns of metamorphic belts. Science 250, 763–769.

Chopin, C., Seidel, E., Theye, T., Ferraris, G., Ivaldi, G. & Catti, M. 1992: Magnesiochloritoid, and the Fe-Mg series in the chloritoid group. European Journal of Mineralogy 4, 67–76.

Dale, J. & Holland, T.J.B. 2003: Geothermobarometry, P-T paths and metamorphic field gradients of high-pressure rocks from the Adula Nappe, Central Alps. Journal of Metamorphic Geology 21, 813–830.

Egli, W. 1966: Geologisch-petrographische Untersuchungen an der NW-Aduladecke und in der Sojaschuppe (Bleniotal, Kanton Tessin). Unpublished PhD Thesis, ETH Zürich, 160 pp.

Engi, M., Todd, C.S. & Schmatz, D.R. 1995: Tertiary metamorphic conditions in the eastern Lepontine Alps. Schweizerische Mineralogische und Petrographische Mitteilungen 75, 347–369.

Engi, M., Berger, A. & Roselle, G.T. 2001: Role of the tectonic accretion channel in collisional orogeny. Geology 29, 1143–1146.

Ernst, W.G. 1971: Metamorphic zonations on presumably subducted lithospheric plates from Japan, California and the Alps. Contributions to Mineralogy and Petrology 34, 43–59.

Etter, U. 1987: Stratigraphische und strukturgeologische Untersuchungen im gotthardmassivischen Mesozoikum zwischen dem Lukmanierpass und der Gegend von Ilanz. Unpublished PhD Thesis, Universität Bern, 162 pp.

Fournier, M., Jolivet, L., Goffé, B. & Dubois, R. 1991: Alpine Corsica metamorphic core complex. Tectonics 10, 1173–1186.

Fox, J.S. 1975: Three-dimensional isograds from the Lukmanier-Pass, Switzerland, and their tectonic significance. Geological Magazine 112, 547–564.

Frank, E. 1983: Alpine metamorphism of calcareous rocks along a cross-section in the Central Alps: occurrence and breakdown of muscovite, margarite and paragonite. Schweizerische Mineralogische und Petrographische Mitteilungen 63, 37–93.

Frey, J.D. 1967: Geologie des Greinagebietes (Val Camadra – Valle Cavalasca – Val di Larciolo – Passo della Greina). Beiträge zur Geologischen Karte der Schweiz – NF 137, 112 pp.

Frey, M. 1969: Die Metamorphose des Keupers vom Tafeljura bis zum Lukanier-Gebiet. Beiträge zur Geologischen Karte der Schweiz NF 137, 160 pp.

Frey, M. 1978: Progressive low-grade metamorphism of a black shale formation, Central Swiss Alps, with special reference to Pyrophyllite and Margarite bearing assemblages. Journal of Petrology 19, 95–135.

Frey, M. & Ferreiro Mählmann, R. 1999: Alpine metamorphism of the Central Alps. Schweizerische Mineralogische und Petrographische Mitteilungen 79, 135–154.

Frey, M., Bucher, K., Frank, E. & Mullis, J. 1980: Alpine metamorphism along the geotraverse Basel-Chiasso – a review. Eclogae geologicae Helvetiae 73, 527–546.

Frey, M., Desmons, J. & Neubauer, F. 1999: Metamorphic map of the Alps. 1:500'000. Schweizerische Mineralogische und Petrographische Mitteilungen 79.

Froitzheim, N., Schmid, S.M. & Conti, P. 1994: Repeated change from crustal shortening to orogen-parallel extension in the Austroalpine units of Graubünden. Eclogae geologicae Helvetiae 87, 559–621.

Froitzheim, N., Schmid, S.M. & Frey, M. 1996: Mesozoic paleogeography and the timing of eclogite-facies metamorphism in the Alps: A working hypothesis. Eclogae geologicae Helvetiae 89, 81–110.

Froitzheim, N., Pleuger, J., Roller, S. & Nagel, T. 2003: Exhumation of high- and ultrahigh-pressure metamorphic rocks by slab extraction. Geology 31, 925–928.

Gillet, P. & Goffé, B. 1988: On the signifiance of aragonite occurrence in the Western Alps. Contributions to Mineralogy and Petrology 99, 70–81.

Goffé, B. & Bousquet, R. 1997: Ferrocarpholite, chloritoïde et lawsonite dans les métapelites des unités du Versoyen et du Petit St Bernard (zone valaisanne, Alpes occidentales). Schweizerische Mineralogische und Petrographische Mitteilungen 77, 137–147.

Goffé, B. & Chopin, C. 1986: High-pressure metamorphism in the Western Alps: zoneography of metapelites, chronology and consequences. Schweizerische Mineralogische und Petrographische Mitteilungen 66, 41–52.

Goffé, B. & Oberhänsli, R. 1992: Ferro- and magnesiocarpholite in the "Bündnerschiefer" of the eastern Central Alps (Grisons and Engadine Window). European Journal of Mineralogy 4, 835–838.

Goffé, B., Michard, A., Garcia-Dueñas, V., Gonzalez-Lodeiro, F., Monié, P., Campos, J., Galindo-Zaldivar, J., Jabaloy, A., Martinez-Martinez, J.M. & Simancas, J.F. 1989: First evidence of high-pressure, low-temperature metamorphism in the Alpujarride nappes, Betic Cordilleras (SE Spain). European Journal of Mineralogy 1, 139–142.

Goffé, B., Bousquet, R., Henry, P. & Le Pichon, X. 2003: Effect of the chemical composition of the crust on the metamorphic evolution of orogenic wedges. Journal of Metamorphic Geology 21, 123–141.

Heinrich, C.A. 1982: Kyanite-eclogite to amphibolite facies evolution of hydrous mafic and pelitic rocks, Adula-nappe, Central Alps. Contributions to Mineralogy and Petrology 81, 30–38.

Heinrich, C.A. 1986: Eclogite facies regional metamorphism of hydrous mafic rock in the Central Alpine Adula Nappe. Journal of Petrology 27, 123–154.

Hitz, L. & Pfiffner, O.A. 1997: Geologic interpretation of the seismic profiles of the Eastern Traverse (lines E1–E3, E7–E9): eastern Swiss Alps. In: Pfiffner et al. (Eds.): Deep Structure of the Swiss Alps – Results of NRP 20, Birkhäuser, Basel, 73–100.

Hsü, K.J. & Briegel, U. 1991: Geologie der Schweiz – Ein Lehrbuch für den Einstieg, und eine Auseinandersetzung mit den Experten. Birkhäuser, Basel, 219 pp.

Huerta, A.D., Royden, L.H. & Hodges, K.V. 1998: The thermal structure of collisional orogens as a response to accretion, erosion, and radiogenic heating. Journal of Geophysical Research Solid Earth 103, 15287–15302.

Jamieson, R.A., Beaumont, C., Fullsack, P. & Lee, B. 1998: Barrovian regional metamorphism: where's the heat? In: Treloar, P.J. & O'Brian, P.J. (Eds.): What drives metamorphism and metamorphic reactions? Geological Society Special Publications 138, 23–51.

Janots, E., Engi, M., Berger, A., Rubatto, D. & Gregory, C. 2007: Texture, chemistry and age of monazite and allanite in the northern Central Alps. Geophysical Research Abstracts 9, EGU2007-A-08582.

Jenny, H., Frischknecht, G. & Kopp, J. 1923: Geologie der Adula. Beiträge zur Geologischen Karte der Schweiz NF 51, 123 pp.

Jung, W. 1963: Die mesozoischen Sedimente am Südostrand des Gotthard-Massivs (zwischen Plaun la Greina und Versam). Eclogae geologicae Helvetiae 56, 653–754.

Keller, F. 1968: Mineralparagenesen und Geologie der Campo Tencia-Pizzo Forno-Gebirgsgruppe. Beiträge zur Geologischen Karte der Schweiz NF 135, 71 pp.

Keller, L.M., Hess, M., Fügenschuh, B. & Schmid, S.M. 2005: Structural and metamorphic evolution of the Camughera – Moncucco, Antrona and Monte Rosa units southwest of the Simplon line, Western Alps. Eclogae geologicae Helvetiae 98, 19–49.

Keller, L.M., Fügenschuh, B., Hess, M., Schneider, B. & Schmid, S.M. 2006: Simplon fault zone in the Western and Central Alps: Mechanism of Neogene faulting and folding revisited. Geology 34, 317–320.

Köppel, V., Günthert, A. & Grünenfelder, M. 1981: Patterns of U-Pb zircon and monazite ages in polymetamorphic units of the Swiss Central Alps. Schweizerische Mineralogische und Petrographische Mitteilungen 61, 97–119.

Kuhn, B.K., Reusser, E., Powell, R. & Günther, D. 2005: Metamorphic evolution of calc-schists in the Central Alps, Switzerland. Schweizerische Mineralogische und Petrographische Mitteilungen 85, 175–190.

Kupferschmid, C. 1977: Geologie auf der Lugnezer Seite der Piz Aul-Gruppe. Eclogae geologicae Helvetiae 70, 1–58.

Lihou, J.C. & Allen, P.A. 1996: Importance of inherited rift margin structures in the early North Alpine Foreland Basin, Switzerland. Basin Research 8, 425–442.

Löw, S. 1987: Die tektono-metamorphe Entwicklung der nördlichen Adula-Decke. Beiträge zur Geologischen Karte der Schweiz NF 161, 84 pp.

Mancktelow, N.S. 1992: Neogene lateral extension during convergence in the Central Alps: evidence from interrelated faulting and backfolding around the Simplonpass (Switzerland). Tectonophysics 215, 295–317.

Mancktelow, N.S. & Pavlis, T.L. 1994: Fold-fault relationships in low-angle detachment systems. Tectonics 13, 668–685.

Marquer, D. & Gapais, D. 1985: Les massifs cristallins externes sur une transversale Guttannen-Val Bedretto (Alpes centrales): structure et histoire cinématique. Comptes Rendues de l'Académie des Sciences de Paris 301, 543–546.

Marquer, D., Challandes, N. & Baudin, T. 1996: Shear zone patterns and strain distribution at the scale of a Penninic nappe: The Suretta nappe (eastern Swiss Alps). Journal of Structural Geology 18, 753–764.

Massonne, H.-J. 1995: Experimental and petrogenetic study of UHPM. In: Coleman, R.G. & Wang, X. (Eds): Ultra-high pressure metamorphism. Cambridge University Press, 159–181.

Mayerat Demarne, A.-M. 1994: Analyse structurale de la zone frontale de la nappe de Tambo (Pennique, Grisons, Suisse). Beiträge zur Geologischen Karte der Schweiz NF 165, 68 pp.

Mercolli, I., Biino, G.G. & Abrecht, J. 1994: The lithostratigraphy of the pre-Mesozoic basement of the Gotthard massif – a review. Schweizerische Mineralogische und Petrographische Mitteilungen 74, 29–40.

Merle, O., Cobbold, P.R. & Schmid, S.M. 1989: Tertiary kinematics in the Lepontine Alps. In: Coward, M.P. et al. (Eds.): Alpine Tectonics. Geological Society Special Publications 45, 113–134.

Meyre, C., De Capitani, C. & Partzsch, J.H. 1997: A ternary solid solution model for omphacite and its application to geothermobarometry of eclogites from the Middle Adula nappe (Central Alps, Switzerland). Journal of Metamorphic Geology 15, 687–700.

Milnes, A.G. 1974: Structure of the Pennine zone (Central Alps) – new working hypothesis. Geological Society of America Bulletin 85, 1727–1732.

Milnes, A.G. 1976: Strukturelle Probleme im Bereich der Schweizer Geotraverse – das Lukmanier-Massiv. Schweizerische Mineralogische und Petrographische Mitteilungen 56, 615–618.

Milnes, A.G. & Schmutz, H.-U. 1978: Structure and history of the Suretta nappe (Pennine zone, Central Alps) – a field study. Eclogae geologicae Helvetiae 71, 19–33.

Nabholz, W.K. 1945: Geologie der Bündnerschiefergebirge zwischen Rheinwald, Valser- und Safiental. Eclogae geologicae Helvetiae 38, 1–119.

Nagel, T., de Capitani, C., Frey, M., Froitzheim, N., Stünitz, H. & Schmid, S.M. 2002: Structural and metamorphic evolution during rapid exhumation in the Lepontine dome (southern Simano and Adula nappes, Central Alps, Switzerland). Eclogae geologicae Helvetiae 95, 301–321.

Nänny, P. 1948: Zur Geologie der Prättigauschiefer zwischen Rhätikon und Plessur. Mitteilungen aus dem Geologischen Institut der Eidgenössischen Technischen Hochschule und der Universität Zürich 30, 127 pp.

Niggli, E. 1970: Alpine Metamorphose und alpine Gebirgsbildung. Fortschritte der Mineralogie 47, 16–26.

Niggli, E. & Niggli, C. 1965: Karten der Verbreitung einiger Mineralien der alpidischen Metamorphose in den Schweizer Alpen (Stilpnomelan, Alkali-Amphibol, Chloritoid, Staurolith, Disthen, Sillimanit). Eclogae geologicae Helvetiae 58, 335–368.

Niggli, E. & Zwart, H.J. 1973: Metamorphic Map of the Alps, scale 1:1'000'000. Subcommission for the Cartography of the Metamorphic Belts of the World. Sheet 17 of the Metamorphic Map of Europe. Leiden/UNESCO, Paris.

Niggli, P., Preiswerk, H., Grütter, O., Bossard, L. & Kündig, E. 1936: Geologische Beschreibung der Tessineralpen zwischen Maggia und Bleniotal. Beiträge zur Geologischen Karte der Schweiz NF 71, 190 pp.

Nimis, P. & Trommsdorff, V. 2001: Revised thermobarometry of Alpe Arami and other garnet peridotites from the Central Alps. Journal of Petrology 42, 103–115.

Oberhänsli, R. 1994: Subducted and obducted ophiolites of the Central Alps: Paleotectonic implications deduced by their distribution and metamorphic overprint. Lithos 33, 109–118.

Oberhänsli, R., Goffé, B. & Bousquet, R. 1995: Record of a HP-LT metamorphic evolution in the Valais zone: Geodynamic implications. In: Lombardo, B. (Ed.): Studies on metamorphic rocks and minerals of the western Alps. A Volume in Memory of Ugo Pognante. Bolletino Museo Regionale di Scienze Naturali, Torino, 13, 221–239.

Oberhänsli, R., Bousquet, R. & Goffé, B. 2003: Comment to «Chloritoid composition and formation in the eastern Central Alps: a comparison between Penninic and Helvetic occurrences» by M. Rahn, M. Steinmann & M. Frey. Schweizerische Mineralogische und Petrographische Mitteilungen 83, 341–344.

Oberhänsli, R., Bousquet, R., Engi, M., Goffé, B., Gosso, G., Handy, M., Höck, V., Koller, F., Lardeaux, J.-M., Polino, R., Rossi, P., Schuster, R., Schwartz, S. & Spalla, M.I. 2004: Metamorphic structure of the Alps. In: Oberhänsli, R. (Ed.): Explanatory note to the map «Metamorphic structure of the Alps, 1:1'000'000», Commission for the Geological Map of the World, Paris. Mitteilungen der Österreichischen Mineralogischen Gesellschaft, 149.

Partzsch, J.H. 1998: The tectono-metamorphic evolution of the middle Adula nappe, Central Alps, Switzerland. Unpublished PhD Thesis, Universität Basel, 142 pp.

Peacock, S.M. 1996: Thermal and petrologic structure of subduction zones. In: G.E. Bebout et al. (Eds.): Subduction from Top to Bottom. American Geophysical Union, Washington D.C., 119–133.

Pesquera, A. & Velasco, F. 1988: Metamorphism of the Paleozoic Cinco Villas massif (Basque Pyrenees) – Illite crystallinity and graphitization degree. Mineralogical Magazine 52, 615–625.

Pfiffner, O.A. 1977: Tektonische Untersuchungen im Infrahelvetikum der Ostschweiz. Mitteilungen aus dem Geologischen Institut der Eidgenössischen Technischen Hochschule und der Universität Zürich 217, 432 pp.

Pfiffner, O.A., Sahli, S. & Stäuble, M. 1997: Compression and uplift of the external massifs in the Helvetic zone. In: Pfiffner et al. (Eds.): Deep Structure of the Swiss Alps – Results of NRP 20, Birkhäuser, Basel, 139–153.

Platt, J.P. 1986: Dynamics of orogenic wedges and the uplift of high-pressure metamorphic rocks. Geological Society of America Bulletin 97, 1037–1053.

Pleuger, J., Hundenborn, R., Kremer, K., Babinka, S., Kurz, W., Jansen, E. & Froitzheim, N. 2003: Structural evolution of Adula nappe, Misox zone, and Tambo nappe in the San Bernardino area: Constraints for the exhumation of the Adula eclogites. Mitteilungen der Österreichischen Geologischen Gesellschaft 94 (2001), 99–122.

Pleuger, J., Nagel, T.J., Walter, J.M., Jansen, E. & Froitzheim, N. 2008: On the role and importance of orogen-parallel and –perpendicular extension, transcurrent shearing, and backthrusting in the Monte Rosa nappe and the Southern Steep Belt of the Alps (Penninic zone, Switzerland and Italy). In: Siegesmund, S. et al. (Eds.): Tectonic Aspects of the Alps-Dinarides-Carpathians system. Geological Society Special Publications 298, 251–280.

Probst, P. 1980: Die Bündnerschiefer des nördlichen Penninikums zwischen Valser Tal und Passo di San Giacomo. Beiträge zur Geologischen Karte der Schweiz NF 153, 64 pp.

Rahn, M.K., Steinmann, M. & Frey, M. 2003: Chloritoid composition and formation in the eastern Central Alps: a comparison between Penninic and Helvetic occurrences. Schweizerische Mineralogische und Petrographische Mitteilungen 82, 409–426.

Rimmelé, G., Jolivet, L., Oberhänsli, R. & Goffé, B. 2003: Deformation history of the high-pressure Lycian Nappes and implications for tectonic evolution of SW Turkey. Tectonics 22, doi:10.1029/2001TC901041.

Robyr, M., Vonlanthen, P., Baumgartner, L.P. & Grobety, B. 2007: Growth mechanism of snowball garnets from the Lukmanier Pass area (Central Alps, Switzerland): a combined mCT/EPMA/EBSD study. Terra Nova 19, 240–244.

Roselle, G.T., Thüring, M. & Engi, M. 2002: MELONPIT: A finite element code for simulating tectonic mass movement and heat flow within subduction zones. American Journal of Science 302, 381–409.

Rütti, R., Maxelon, M. & Mancktelow, N.S. 2005: Structure and kinematics of the northern Simano Nappe, Central Alps, Switzerland. Eclogae geologicae Helvetiae 98, 63–81.

Rütti, R., Marquer, D. & Thompson, A.B. 2008: Tertiary tectono-metamorphic evolution of the European margin during Alpine collision: Example of the Leventina Nappe (Central Alps, Switzerland). Swiss Journal of Geosciences doi: 10.1007/s00015-008-1278-9.

Schmid, S.M., Zingg, A. & Handy, M. 1987: The kinematics of movements along the Insubric line and the emplacement of the Ivrea zone. Tectonophysics 135, 47–66.

Schmid, S.M., Rück, P. & Schreurs, G. 1990: The significance of the Schams nappes for the reconstruction of the paleotectonic and orogenic evolution of the Penninic zone along the NFP-20 East traverse (Grisons, eastern Switzerland). In: Roure, et al. (Eds.): Deep structure of the Alps, Mémoire de la Société de France, Paris, 156, 263–287.

Schmid, S.M., Pfiffner, O.A., Froitzheim, N., Schönborn, G. & Kissling, E. 1996: Geophysical-geological transect and tectonic evolution of the Swiss-Italian Alps. Tectonics 15, 1036–1064.

Schmid, S.M., Pfiffner, O.A., Schönborn, G., Froitzheim, N. & Kissling, E. 1997a: Integrated cross section and tectonic evolution of the Alps along the Eastern Traverse. In: Pfiffner et al. (Eds.): Deep Structure of the Swiss Alps – Results of NRP 20, Birkhäuser, Basel, 289–304.

Schmid, S.M., Pfiffner, O.A. & Schreurs, G. 1997b: Rifting and collision in the Penninic zone of eastern Switzerland. In: Pfiffner et al. (Eds.): Deep Structure of the Swiss Alps – Results of NRP 20, Birkhäuser, Basel, 160–185.

Schmid, S.M., Fügenschuh, B., Kissling, E. & Schuster, R. 2004: Tectonic map and overall architecture of the Alpine orogen. Eclogae geologicae Helvetiae 97, 93–117.

Schreurs, G. 1993: Structural analysis of the Schams nappes and adjacent tectonic units: implications for the orogenic evolution of the Pennine zone in eastern Switzerland. Bulletin de da Société Géologique de France 164, 415–435.

Spicher, A. 1980: Tektonische Karte der Schweiz, 1:500'000. Schweizerische Geologische Kommission, Bern.

Stampfli, G.M. 1993: Le Briançonnais: terrain exotique dans les Alpes? Eclogae geologicae Helvetiae 86, 1–45.

Stampfli, G.M., Mosar, J., Marquer, D., Marchant, R., Baudin, T. & Borel, G. 1998: Subduction and obduction processes in the Swiss Alps. In: Vauchez, A. & Meissner, R. (Eds): Continents and their mantle root. Tectonophysics 296, 159–204.

Steck, A. 1984: Structures de deformations tertiaires dans les Alpes centrales (transversale Aar-Simplon-Ossola). Eclogae geologicae Helvetiae 77, 55–100.

Steck, A. 1990: Une carte des zones de cisaillement ductile des Alpes Centrales. Eclogae geologicae Helvetiae 83, 603–627.

Steck, A. & Hunziker, J.C. 1994: The Tertiary structural and thermal evolution of the Central Alps – Compressional and extensional structures in an orogenic belt. Tectonophysics 238, 229–254.

Steiger, R.H. 1962: Petrographie und Geologie des südlichen Gotthardmassivs zwischen St. Gotthard- und Lukmanierpass. Schweizerische Mineralogische und Petrographische Mitteilungen 64, 381–577.

Steinmann, M. 1994a: Die nordpenninischen Bündnerschiefer der Zentralalpen Graubündens: Tektonik, Stratigraphie und Beckenentwicklung. Unpublished PhD Thesis, ETH Zürich, 220 pp.

Steinmann, M. 1994b: Ein Beckenmodell für das Nordpenninikum der Ostschweiz. Jahrbuch der Geologischen Bundesanstalt 137, 675–721.

Thakur, V.C. 1971: The structural and metamorphic history of the Mesozoic and pre-Mesozoic basement rocks of the Molare region, Ticino, Switzerland. Unpublished PhD Thesis, Imperial college of London, 229 pp.

Thakur, V.C. 1973: Events in the Alpine deformation and metamorphism in the northern Pennine zone and southern Gotthard massif regions, Switzerland. Geologische Rundschau 62, 549–563.

Todd, C.S. & Engi, M. 1997: Metamorphic field gradients in the Central Alps. Journal of Metamorphic Geology 15, 513–530.

Trommsdorff, V. 1966: Progressive Metamorphose kieseliger Karbonatgesteine in den Zentralalpen zwischen Bernina und Simplon. Schweizerische Mineralogische und Petrographische Mitteilungen 46, 431–460.

Trommsdorff, V. 1990: Metamorphism and tectonics in the Central Alps: The Alpine lithospheric mélange of Cima Lunga and Adula. Memorie della Societa Geologica Italiana 45, 39–49.

Trotet, F., Goffé, B., Vidal, O. & Jolivet, L. 2006: Evidence of retrograde Mg-carpholite in the Phyllite-Quartzite nappe of Peloponnese from thermobarometric modelisation – geodynamic implications. Geodinamica Acta 19, 323–343.

Trümpy, R. 1960: Paleotectonic evolution of the Central and Western Alps. Geological Society of America Bulletin 71, 843–908.

Trümpy, R. 1980: Geology of Switzerland – a guide-book. Part A: An outline of the Geology of Switzerland. Wepf, Basel, 104 pp.

Uhr, A. Unpublished: Geologische Untersuchungen im Gebiet des Piz Terri (Kt. Tessin und Graubünden). Textmanuskript, hinterlegt beim BWG, Bern-Ittigen.

Vidal, O., Goffé, B. & Theye, T. 1992: Experimental study of the stability of sudoite and magnesiocarpholite and calculation of a new petrogenetic grid for the system FeO-MgO-Al$_2$O$_3$-SiO$_2$-H$_2$O. Journal of Metamorphic Geology 10, 603–614.

Vidal, O. & Theye, T. 1996: Comment on "Petrology of Fe-Mg-carpholite-bearing metasediments from northeast Oman". Journal of Metamorphic Geology 14, 381–386.

Voll, G. 1976: Structural Studies of the Valser Rhine Valley and the Lukmanier Region and their Importance for the nappe Structure of the Central Swiss Alps. Schweizerische Mineralogische und Petrographische Mitteilungen 56, 619–626.

von Blanckenburg, F. & Davies, J.H. 1995: Slab breakoff: A model for syncollisional magmatism and tectonics in the Alps. Tectonics 14, 120–131.

Waldhauser, F., Kissling, E., Ansorge, J. & Mueller, S. 1998: Three-dimensional interface modelling with two-dimensional seismic data: the Alpine crust-mantle boundary. Geophysical Journal International 135, 264–278.

Weh, M. & Froitzheim, N. 2001: Penninic cover nappes in the Prättigau half-window (Eastern Switzerland): Structure and tectonic evolution. Eclogae geologicae Helvetiae 94, 237–252.

Wenk, E. 1962: Plagioklas als Indexmineral in den Zentralalpen. Schweizerische Mineralogische und Petrographische Mitteilungen 42, 139–152.

Wenk, E. 1970: Zur Regionalmetamorphose und Ultrametamorphose im Lepontin. Fortschritte der Mineralogie 47, 34–51.

Wiederkehr, M. 2004: Strukturelle und metamorphe Entwicklung der Metasedimente am Kontakt Briançonnais – Piemont-Liguria, Val Madris, Avers/GR. Unpublished Diploma Thesis, Universität Basel. 145 pp.

Manuscript received February 1, 2008
Revision accepted May 26, 2008
Published Online first November 13, 2008
Editorial Handling: Nikolaus Froitzheim & Stefan Bucher

1661-8726/08/01S157-15
DOI 10.1007/s00015-008-1278-9
Birkhäuser Verlag, Basel, 2008

Swiss J. Geosci. 101 (2008) Supplement 1, S157–S171

Tertiary tectono-metamorphic evolution of the European margin during Alpine collison: example of the Leventina Nappe (Central Alps, Switzerland)

ROGER RÜTTI[1,2], DIDIER MARQUER[3] & ALAN BRUCE THOMPSON[1,4]

Key words: Central Alps, Leventina Nappe, deformation history, P-T evolution

ABSTRACT

The Leventina Nappe represents one of the lowermost exposed units in the Alpine nappe stack and corresponds to a slice of the European margin that was entrained into the Alpine continental accretionary prism during the Tertiary tectonic event. This study yields details regarding the tectonic and metamorphic history of the Leventina Nappe, through detailed analysis of structures and shear zone patterns, and the examination of the Si-content of white mica along a north-south profile. The Leventina Nappe underwent three phases of ductile deformation. Foliation S1 is mostly sub-parallel to the regionally dominant structural fabric (the S2 foliation). S2 foliation is penetratively developed in the structurally higher portions of the Leventina Nappe toward the Simano Nappe, while it is only weakly developed in the core of the Leventina Nappe. A 50 to 200 m wide mylonite zone, with a D2 top-to-NW sense of shear marks the boundary to the Simano Nappe. Throughout the Leventina Nappe only small-scale D2 shear bands (mm to cm wide) are observed, showing a top-to-NW sense of shear. Deformation phase D3 locally generated a vertical axial plane foliation (S3) associated with the large-scale D3 Leventina antiform.

Microtextural evidence and phengite geobarometry were used to constrain the temperature and pressure conditions of equilibration of the Leventina Gneisses. Highest Si (pfu) values are preserved in the core of phengitic micas and reflect pressure and temperature conditions of around 8 kbar at 550 °C and 10 kbar at 650 °C in the northern and southern parts of the Leventina Nappe, respectively. Lower Si (pfu) values from the rims of white micas correspond to a metamorphic pressure of ca. 5 kbar during the exhumation of the unit. These metamorphic conditions are related to the underthrusting of the thinned European margin into the continental accretionary prism during late Eocene time. These new data allow us to propose a kinematic model for the Leventina Nappe during the Tertiary Alpine tectonics.

ZUSAMMENFASSUNG

Die Leventina-Decke stellt eine der tiefsten Einheiten dar, welche im alpinen Deckenstapel aufgeschlossen ist. Sic entspricht einem Teil des europäischen Kontinentalrandes, welcher während der alpinen Kontinentalkollision in den Akkretionskeil einbezogen wurde.

Die detaillierte Analyse der Strukturelemente und der Scherzonenverteilung sowie die Untersuchung von Si-Gehalten in metamorphen Hellglimmern entlang eines Nord-Süd-Profiles ergeben neue Erkenntnisse zur tektonischen und metamorphen Geschichte der Leventina-Decke. Drei Deformationsphasen werden in der Leventina-Decke beobachtet. Die Schieferung S1 ist mehrheitlich sub-parallel zur regional vorherrschenden Schieferung (S2). S2 ist in den höheren Teilen der Leventina-Decke zur Simano-Decke hin penetrativ ausgebildet, wogegen sie im Kern der Leventina-Decke nur schwach ausgeprägt ist. Ein 50 bis 200 m mächtiger Mylonit-Horizont mit einem NW-gerichteten D2-Schersinn stellt im Hangenden die Grenze zur Simano-Decke dar. In der ganzen Leventina-Decke werden nur feine D2-Scherbänder (mm bis cm) mit einem top-nach-NW Schersinn beobachtet. Die D3-Deformationsphase erzeugte lokal eine vertikale Achsenebenenschieferung S3, welche im Zusammenhang mit der grossmassstäblichen D3-Leventina-Antiform steht.

Mikrotexturen und Phengit-Barometrie wurden benutzt, um Temperatur- und Druckbedingungen für die Metamorphose der Leventina Gneise abzuschätzen. Hohe Si (pfu)-Gehalte der Kerne von Hellglimmer widerspiegeln metamorphe Druck- und Temperaturbedingungen von ungefähr 8 kbar bei 550 °C im Norden und 10 kbar bei 650 °C im Süden der Leventina-Decke. Tiefere Si (pfu)-Gehalte von Hellglimmer-Rändern entsprechen einem metamorphem Druck von ca. 5 kbar während der Heraushebung der Einheit. Diese metamorphen Bedingungen widerspiegeln das "Underthrusting" der ausgedünnten europäischen Kruste in den kontinentalen Akkretionskeil im späten Eozän. Diese neuen Daten erlauben es uns, ein kinematisches Modell für die tertiäre alpine Orogenese zu formulieren.

1. Introduction

The studied area is located in the Central Lepontine Alps, south of the External Crystalline Massifs and north of the Insubric line (see review by Berger et al. 2005 and references therein). This part of the Penninic domain corresponds to the internal part of the Alpine Belt and consists of different tectonic nappes derived, from bottom to top of the nappe stack, of the European

[1] Institute for Mineralogy and Petrology, ETH Zurich, Clausiusstrasse 25, 8092 Zurich, Switzerland.
[2] Present address: Dr. M. Kobel & Partner AG, Büro für Technische Geologie, Grossfeldstrasse 74, 7320 Sargans, Switzerland.
E-mail: roger.ruetti@btgeo.ch
[3] UMR 6249 Chrono-Environnement-Geosciences, Université de Franche-Comté, 16, Route de Gray, 25030 Besançon-Cedex, France.
[4] Also at Faculty of Science, University of Zürich, Zürich, Switzerland.

margin, the Valaisan Basin, the Briançonnais Units and parts of the Liguro-Piedmontais Ocean, respectively (Stampfli et al. 1998; Berger et al. 2005). The Leventina Nappe (Tektonische Karte der Schweiz 1:500000, 2005) is amongst the structurally deepest exposed units in the Central Alps of Switzerland (Fig. 1 and 2). It is part of the lowermost units in the Penninic Nappe stack referred to as Subpenninic units (Milnes 1974; Schmid et al. 2004). This unit therefore is derived from the European margin that was entrained into the Alpine continental accretionary prism. Although the Leventina gneisses have been the subject of a detailed lithologic and petrographic study (Casasopra 1939) and several hypotheses have been formulated regarding their structural position within the Alpine Nappe Stack (e.g. Bossard in Niggli et al. 1936, Berger et al. 2005), its kinematic indicators and pressure-temperature path during the Alpine metamorphic event are barely known and have not been the subject of research in recent times. Note, however, that the relationships of the Leventina Nappe with the northerly adjacent Lucomagno Nappe were not a subject of this study.

To investigate kinematics at a larger scale, shear band patterns (e.g. Gapais et al. 1987) are generally very useful and readily interpretable in regionally sheared granitic rocks that were initially homogeneous and isotropic (e.g. Aar Massif: Choukroune & Gapais 1983; St. Cast granite in Brittany: Gapais et al. 1987; Gotthard Massif: Marquer 1990; Tambo and Suretta Nappes: Marquer 1991; Marquer et al. 1996). These shear criteria in ductily deformed crystalline rocks exist at all scales within the crust. Micro- to mesoscale deformation features were investigated to detect and characterize individual shear zones within the Leventina Nappe.

The present study thus aimed at understanding the tectonic processes governing the Leventina Nappe during the Tertiary Alpine event. It uses mapping of the different structures, analysis of shear zone patterns, and investigation of the Si (pfu) content of white micas in gneisses along a north-south profile, to yield details regarding the tectonic and metamorphic history of the Leventina Nappe. These new data will be integrated into a kinematic model for this unit during the Tertiary Alpine tectonics.

2. Previous Work and Geologic Setting

The Leventina Nappe represents the structurally deepest of the Subpenninic units in the Central Alps of Switzerland and forms the valley floor and cliffs of Valle Leventina in southern Switzerland (Figs. 1 & 2). In 1925, Bossard first mapped and described the rocks of the so-called "Tessiner Kulmination". His work was subsequently used in the classic studies by Preiswerk et al. (1934) and Niggli et al. (1936). Bossard (1925) already postulated that the Leventina Gneisses form an independent nappe – contrary to many geologists working in the area at that period (see various tectonic schemes in Niggli et al. 1936; Berger et al. 2005). This metagranitic body of unspectacular and homogeneous appearance was later thoroughly mapped and described by Casasopra (1939). After these pioneering studies little re-

search has been conducted in this unit, probably due to the homogeneous character of the unit. The Leventina Gneisses are important from a geotechnical point of view as they support the Gotthard highway and some 20 km of the new railroad tunnel (Gotthard Base Tunnel GBT of the "Neue Eisenbahn-Alpen-Transversale" NEAT) currently in construction.

The Leventina Gneisses consist of a Variscan metagranite with trondhjemitic-leucogranitic chemical bulk composition, with some scarce lenses of different composition such as paragneiss, micaschist and amphibolite (Casasopra 1939). These lenses, which represent remnants of the country rocks of the Leventina metagranites at the time of its magmatic intrusion, are now completely integrated into the Leventina Gneisses (Hiss 1975). In the north of the unit, the Leventina Gneisses contain the so-called "Intercalazione Centrale di Chironico – Faido – Piottino" (Casasopra 1939). This 80 m wide zone consists of various rock types such as quartzite, micaschist, paragneiss, aplite and amphibolite. The Leventina Gneisses have intrusion ages of ca. 270 Ma, as indicated by their zircon age pattern (Allègre et al. 1974; Köppel et al. 1980; Köppel 1993).

The Leventina Nappe is separated from the Simano Nappe by metasediments of presumed but unproven Mesozoic age (Niggli et al. 1936; Bianconi 1971; see Figs. 1 & 2). According to some studies, the Lucomagno unit is interpreted as the metasedimentary country rock into which the magmatic protoliths of the Leventina Gneisses intruded and accordingly Leventina Gneisses and Lucomagno metasediments are viewed as two lithologies of a coherent Alpine Nappe referred to as the Lucomagno-Leventina Nappe (Fig. 1; Milnes 1976; Spicher 1980; Etter 1992). Based on the existence of nappe-dividing Mesozoic metasediments, the Lucomagno (Bianconi 1971) and Leventina Nappes are locally separated by Mesozoic metasediments, and can actually be divided into two distinct nappes north of the area of this study (Berger et al. 2005; Rütti et al. 2005).

Toward the south, the metasediments found between Leventina and the overlying Simano Nappe are rare and the boundary has been defined by lithological criteria. This boundary is formed by a strongly deformed mylonitic horizon already described by Casasopra (1939) and confirmed by several other studies (Irouschek 1983; Merle et al. 1989; Timar-Geng et al. 2004; Rütti et al. 2005). Whether the lower boundary of the Leventina Nappe is represented by an anhydrite-bearing zone known from the subsurface (Biaschina Bore Hole) remains an open question, as below the anhydrite-bearing zone Leventina Gneisses are found again (Hiss 1975). Recently, Timar-Geng et al. (2004) used samples from the Leventina Gneisses to constrain the latest exhumation stage of this unit by Apatite Fission Track ages dated around 10 to 3.7 Ma.

3. Tertiary Deformation and Kinematics

3.1 Deformation History

Although the Leventina Gneisses have not previously been the subject of a detailed structural study, there are data regarding

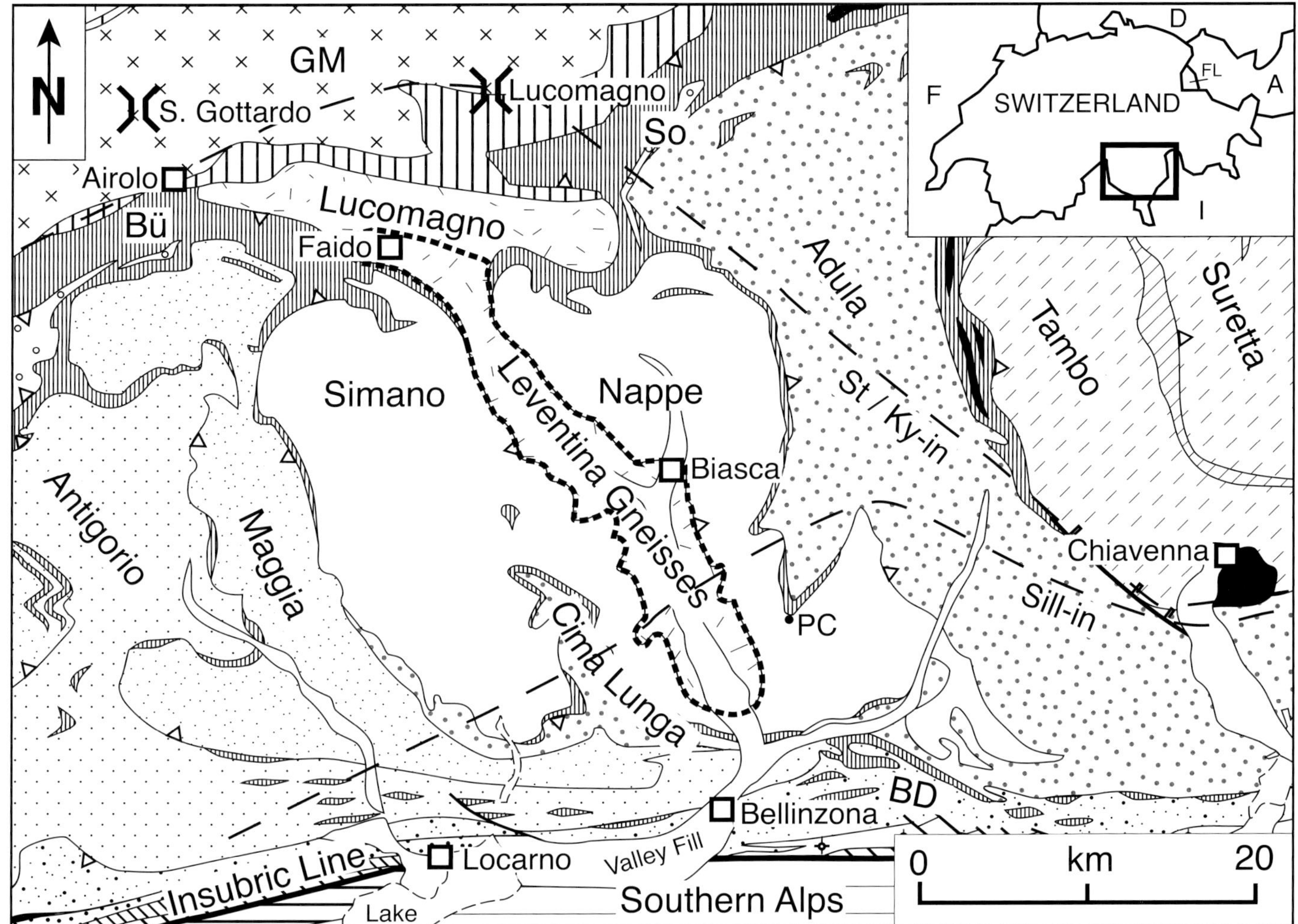

Fig. 1. Map of the Central Alps of Switzerland modified from Spycher (1980). The study area is enclosed by the dashed line. The staurolite/kyanite-in and sillimanite-in isograds are from Niggli and Niggli (1965) and Thompson (1976). Abbreviations: BD = Bellinzona-Dascio Zone, GM = Gotthard Massif, So = Soja Zone, Bü = Bündnerschiefer, PC = Pizzo di Claro.

the orientation of the main foliation outlining the contours of the Toce and Ticino culminations (Wenk 1955). Structural data were also compiled by Merle et al. (1989) and more recently Maxelon & Mancktelow (2005). During our study, structural data were collected in order to define the internal structure of the Leventina Nappe and the structural relationships with the surrounding units.

Deformation Phase D1

In much of the Leventina Gneisses, S1 is sub-parallel to the regionally dominant gently NW-dipping S2 foliation and in many outcrops it is not possible to distinguish between these two foliations since both are associated with north-directed shearing (Rütti et al. 2005 and references therein). However, the relation-

ships between the S1 and the S2 foliations are well preserved in the frontal part of the Leventina Nappe. For example, in the Dazio Grande Gorge (Swiss Coordinates* 700883/149665), the S1 foliation of the Leventina Gneisses is affected by the subsequent deformation phase D2 (Fig. 3a). In this place, the S1 foliation is partly preserved between anastomosing D2 shear zones.

Deformation Phase D2

The main foliation S2 is axial planar to mesoscopic folds affecting S1, and is more or less penetrative throughout the entire unit. Towards the core of the nappe, the main foliation S2 becomes less penetrative as some of the large alkali feldspar augen are not aligned along S2, but show a more random distribution. Also white micas and biotites in these rocks are less

---

*Throughout this paper references to Swiss map Coordinates are made with the letters S.C.

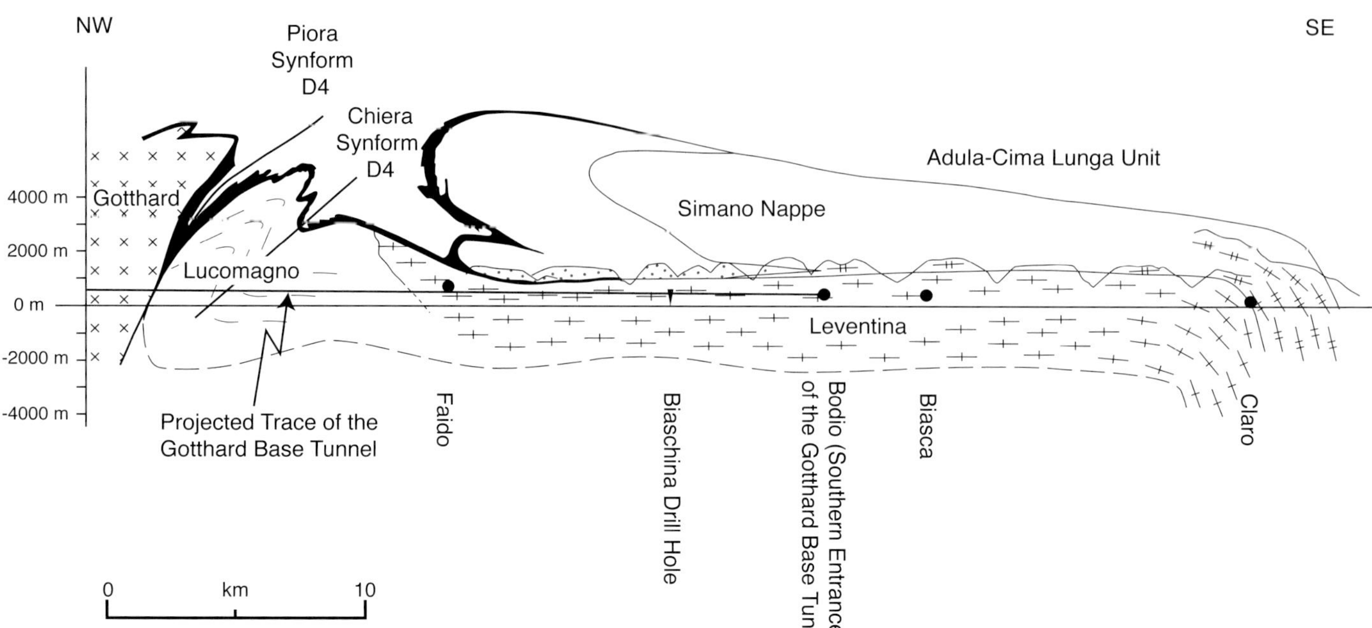

Fig. 2. Structural cross-section across the Leventina valley (Lucomagno-Leventina Nappe). Geological distinctions: in black, Triassic rocks, elongated crosses, mainly orthogneisses; short lines, essentially paragneisses (after Preiswerk et al. 1934, Casasopra 1939, Heitzmann 1991, Etter 1992 and S. Schmid unpublished). The dashed line represents a suggested tectonic contact with a deeper non-exposed unit at the base of the Leventina Nappe.

rigorously oriented along S2. Close to the overlying Simano Nappe, a 50 to 200 m wide mylonitic horizon (Fig. 3b) is observed. This mylonite crops out along both the eastern and western side of Valle Leventina. The protolith of the mylonite, as determined from the mineral assemblage and modal estimations, corresponds to the Leventina Gneisses. In some cases a strongly sheared leucocratic rock containing quartz, alkali feldspar, plagioclase, biotite and muscovite (e.g. a leucogranite) is associated with the mylonite and can macroscopically be mistaken for quartzite. Asymmetric feldspar porphyroclasts as well as shear bands indicate a top-to-NW sense of shear (Fig. 3b), which is assigned to the D2 deformation. Figure 4 shows a map of the trend of the main foliation S2. The field measurements were averaged in order to produce a better legible map. The overall flat-lying D2 foliation shows orientation deflections and inceasing angles of dip near the boundaries of the mapped area, due to the superimposition of D3 and D4 deformations described below.

Deformation Phase D3

During D3, large-scale regional folds with small amplitude-wavelength ratios (1 to 2 km amplitude, wavelength 8 to 10 km) affected the whole nappe stack in this internal part of the Alps. The Leventina Antiform is one of these large-scale N–S oriented folds and observed near or at the eastern boundary between Leventina Nappe and overlying Simano Nappe (Fig. 4). East of its fold axial trace, S2 dips east. In the Dazio Grande gorge-outcrop (S.C. 700883/149665), the overprinting relationships between N–S oriented D3 folds and the S2 foliation are

observed at the large scale (20 to 30 meters). A gentle undulation (amplitude of 6 to 8 meters) also affects all the structures observed in the gorge.

Table 1 correlates structures and deformation phases of the Leventina Nappe with those of the Simano Nappe in the hanging wall to the east and the west according to the relative age of deformation, as well as the metamorphic conditions deduced by several studies in the Simano Nappe. Table 1 shows that the deformation history of the two units is very similar. However, the directional kinematics during a same deformation phases differ when going east and structurally upwards in the Alpine Nappe pile as will be discussed in the next section.

Following D3 deformation, all the structures are steepened in the northern part of the Leventina Nappe by the D4-Chiera synform (see Fig. 2; Milnes 1974; Etter 1992), related to backfolding induced by Oligo-Miocene backthrusting of the Gotthard and Aar massifs (Marquer 1990). Toward the south the main foliation S2 is also strongly steepened due to the formation of the Southern Steep Belt associated with Oligocene dextral and reverse displacement along the Insubric Line (Milnes 1974).

3.2 Kinematic History

The distribution of shear band patterns was studied in numerous quarries of Valle Leventina. The quarries represent the bulk of the accessible 3D-outcrops of the Leventina Gneisses on the valley floor. Shear bands within the Leventina Gneisses were generally found to be small, the bands being a few cm wide (Figs. 3c & 3d). They are associated with and deflect the

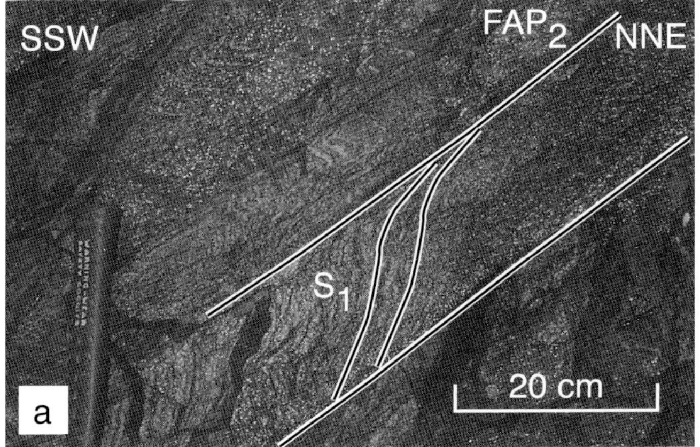

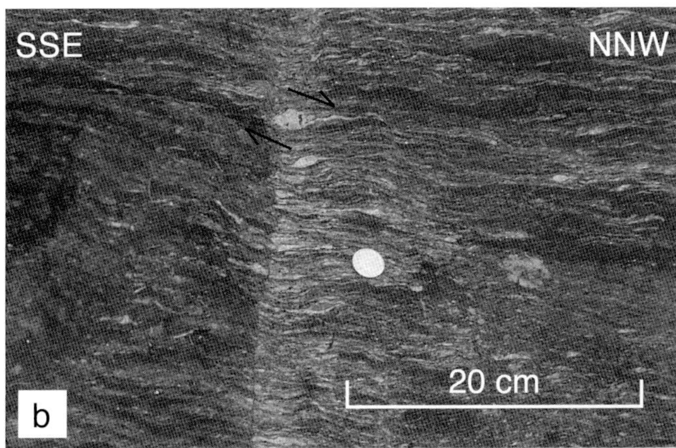

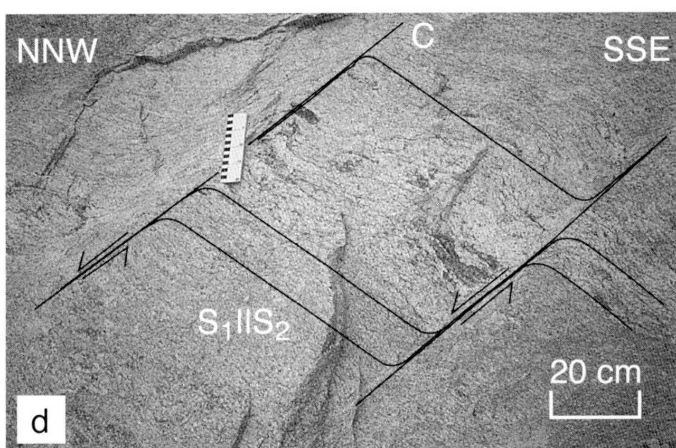

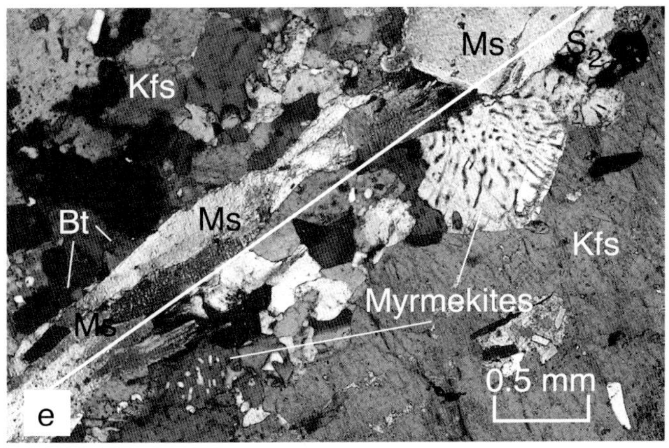

Fig. 3. Field pictures and microphotographs of structural features observed in the Leventina Gneisses. a) D1/D2 interference in the Leventina Gneisses in the Dazio Grande gorge (S.C. 700883/149665). b) Asymmetric feldspar clast showing top-to-NW sense of shear near Faido (S.C. 705520/147010). c) cm-wide shear bands in a quarry near Iragna (S.C. 718290/130120). d) cm-wide shear zone in the Leventina Gneisses showing top-to-NW sense of shear (S.C. 718200/129925). e) Myrmekite-structures present at the rim of plagioclase grain exposed to the principal strain axis Z as described by Simpson (1985) and Simpson & Wintsch (1989) (Sample LEV012, S.C. 710250/138975).

composite S1-S2 foliation. The C-type shear bands (Passchier & Trouw, 1996) therefore formed during D2 deformation (Fig. 3d). The shear bands dip toward the NW and the shear sense is consistently top-to-NW (Fig. 3d). Figure 5 shows plots of the main foliation S2, the associated stretching lineation L2, and the orientation of shear bands measured in the Leventina Gneisses. The main foliation is mainly horizontal throughout the unit. L2 stretching lineations (Ls) consistently strike NW–

SE, which accords well with observations in the overlying units (see Rütti et al. 2005 for data on the Simano Nappe,). Shear bands (C planes) have the same strike as S2 planes but show higher angles of dip toward the NW. The stretching lineation (Lc) associated with the shear bands is also oriented in a NW–SE direction, except for the measurement in Cresciano (Fig. 5). The geometrical relationships between the shear planes and the schistosity reveal a bulk top-to-NW sense of shear (Fig. 5).

Table 1. Correlation of Mesoscopic structural elements in the Subpenninic Domain of the Central Alps of Switzerland (Leventina Nappe and Simano Nappe) and inferred ages after Becker (1993), Gebauer (1996), Hurford et al. (1989), Nievergelt et al. (1996), Villa & von Blanckenburg (1991) von Blanckenburg (1992) and the references cited in the Table.

Correlation of structural elements in outcrop

Age/Nappe	western Simano Nappe	Leventina Nappe	eastern Simano Nappe
	Grond et al. (1995); Pfiffner (1999); Rütti et al. (2005).	This study	Partzsch (1998); Rütti (2001); Nagel et al. (2002).
Eocene to Oligocene (40 to 35 Ma)	D_1 *(Nappe formation):* Subhorizontal S_1, NW–SE-oriented L_1, top-to-NW thrusting, isoclinal folds (relict, rare). T: ~500 °C; P: 9 to 11 kbar.	D_1 *(Nappe formation):* Subhorizontal S_1, except in the north toward the NSB, NW–SE oriented L_1, top-to-NW-thrusting. T: >550 °C; P: 8 to 10 kbar.	D_1 *(Zapport phase):* Foliation S_1, N–NNE-dipping L_1, top-to-N shear sense, isoclinal, similar folds. T: >600 °C; P: ~12 kbar.
Oligocene (pre-Bergell, 35 to 30 Ma)	D_2 *(Nappe folding):* subhorizontal "main" foliation S_2, NW–SE-dipping L_2 with top-to-NW in lower, and top-to-SE shear sense of in upper parts. T: ~650 °C; P: 8 to 10 kbar.	D_2 *(Nappe folding):* subhorizontal "main" foliation S_2, NW-SE-dipping lineation L_2, top-to-NW-sense of shear in the upper parts. T: ~550 to 650 °C; P: 8 to 10 kbar.	D_2 *(Niemet-Beverin phase):* New axial planar foliation S_2, stretching lineation L_2, top-to-SE shear sense, open to tight isoclinal folds, T: 650 to 700 °C; P: 10 to 12 kbar.
Oligocene (syn- to post-Bergell, 30 to 25 Ma)	D_3 *(Cross folding):* Open folds with steep FAP_3, very frequent around P. Campo Tencia, T: 650 to 700 °C; P: 6 to 8 kbar.	D_3 *(Cross folding):* Open folds with steep FAP_3 (rare). T: 550 to 650 °C; P: 5 to 8 kbar.	D_3 *(Cressim phase):* Lineation L_3, open folds, E-W striking in the south, N–S stiking in the north, syn-magmatic with the Bergell intrusion T: 650 to 750 °C; P: 4 to 8 kbar.
Miocene to today		*post D_3:* Ultracataclasites and pseudotachylytes in the upper part, kinking of the "main" foliation	

Apart from the mylonite zone at the boundary with the Simano Nappe, no other major macroscopic shear zones within the Leventina Nappe were observed at the surface within the internal parts of the Leventina Gneisses. This is intriguing as recent published observations from the construction of the new railroad tunnel (Gotthard Base Tunnel, NEAT) show that ductile shear zones at hectometric scale exist in the subsurface (Bonzanigo & Oppizi 2006).

Heterogeneous deformation occurs at various scales in the Leventina Nappe as shown by the different shear zone patterns. The sense of shear during D2 is consistently top-to-NW in the Leventina Nappe. This is in agreement with D2 deformation in the footwall and the core of the Simano Nappe (Rütti et al. 2005). This shear sense, however, contrasts with shear senses observed in the footwall of the Adula-Cima Lunga Nappe, which are consistently top-to-SE (Table 1 and references therein). This implies that the D1 to D2 deformations in the Subpenninic Nappes, we assume to be contemporaneous, correspond to a continuous and progressive process during which a large-scale change in the sense of shearing occurred, depending on the position of the nappe unit within the Alpine Nappe pile (see Meyre et al. 1998, for kinematics-time relationships for the upper units). For example, the base of the Adula Nappe exhibits top-SE shearing while the Leventina-Simano Nappes shows top-NW shearing during the same D1 and D2 nappe stacking and folding process (Table 1).

4. Metamorphism

4.1 Rock Types and Mineral Chemistry

The Leventina Gneisses have a "granitic" bulk rock composition (trondhjemitic-leucogranitic Casasopra 1939; Hiss 1975). The main minerals in the samples investigated during this study are quartz (Qtz), plagioclase (Pl) and alkali feldspar (Kfs) varying considerably in amount, and additionally, biotite (Bt) and muscovite (Ms); no paragonite was found. In addition chlorite (Chl) or garnet (Gt) occur with accessory phases such as apatite (Ap), zircon (Zr), clinozoisite (Clz), epidote (Ep), zoisite (Zs) and calcite (Cc). Hiss (1975) further mentions accessory phases such as scapolite, kyanite, staurolite and hornblende, which are found in lenses or horizons representing remnants of the country rocks of the Leventina Gneisses at the time of intrusion. Such a lens of amphibolite is observed SE of Faido (S.C. 705560/147070). It is a garnet–bearing amphibolite (Hbl + Pl + Gt) with structures that are parallel to the main foliation S2. Also the rocks of the «intercalazione centrale» (e.g. road outcrops at Valbona; S.C. 706000/146600), show more semi- to metapelitic and amphibolitic mineralogical compositions. The mineral textures evolve in a general manner from more or less foliated in the core of the metagranite to strongly deformed at the boundary of the Leventina Nappe with the overlying Si-

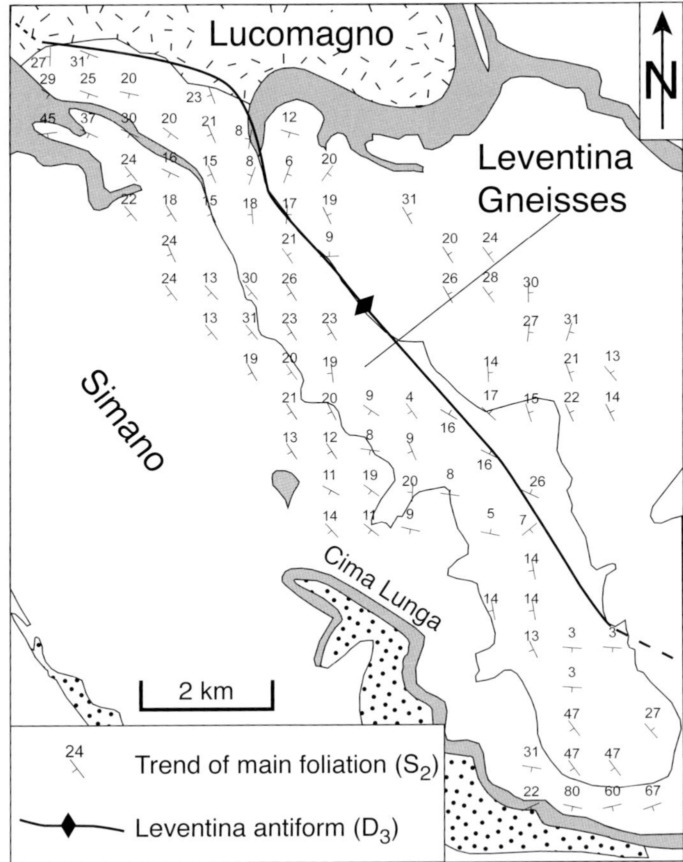

Fig. 4. Map of S2 foliations in the study area. 257 field measurements of S2 were averaged through inverse distance weighted spatial averaging with the program SpheriStat for Windows (Pangaea Scientific, Brookville, Ontario). The spacing between two points measures 1.5 km. The D3 Leventina antiform is indicated for reference.

mano Nappe, where they show mylonitic textures (Rütti et al. 2005).

In the core of the Leventina Nappe, quartz has a large grain size (mm to several mm), decreasing strongly towards the mylonitic horizon of the Leventina Gneisses. In most cases, quartz shows high-energy grain boundaries, subgrains and undulose extinction.

Feldspars often form large porphyroclasts (up to several cm in length) in the central parts of the Leventina Gneisses. Generally they are strongly altered. Exsolution textures (exsolution blebs) were observed in almost all investigated samples. The overall grain size of feldspar depends on the structural position of the sample with respect to shear zones. Near the boundary towards the Simano Nappe, the grain size is generally small (mm to several mm).

Alkali Feldspar shows cross-hatched twinned microcline, and Carlsbad twins are frequent at least in the central part of the Leventina Gneisses. Flame perthites occur also quite frequently. A wide range of alkali feldspar-compositions is observed (X_{Or}-content ranges between 0.84 and 0.97). No systematic pattern of variation (e.g. core to rim) is observed; the varia-

tion may be explained by cation exchange during retrograde metamorphism. In some cases the cross-hatched microcline twins are deformed.

Plagioclase often shows deformed polysynthetic twins. Myrmekite structures associated with plagioclase are frequent in most of the samples. Zoning of plagioclase is very frequent and readily observed in thin section. Plagioclase-composition also shows a range of X_{An} (0.09 continuous to 0.21), representing compositions at the albite-oligoclase miscibility gap and in the oligoclase field. Higher X_{An}-values (0.18 to 0.21) are recorded in the cores of the plagioclase grains and lower values at the rim (average around 0.16).

Micas are all recrystallized in the investigated samples. Some of the micas are deformed and show undulose extinction. Biotite growing obliquely to the main foliation (Querglimmer) is frequently observed. Biotite has a typical value for X_{Mg} around 0.36, but X_{Mg} as low as 0.12 (LEV30) and as high as 0.51 (LEV35) have been measured. Chemical compositions of white micas used to deduce metamorphic pressure estimates are described in the next section (Table 2).

Generally the degree of chloritization of the samples substantially varies, ranging from almost no retrogression to nearly complete alteration of biotite. Some of the chlorite grains appear to be in equilibrium with the surrounding minerals and might have grown during the retrograde path in the chlorite stability field. X_{Mg} of chlorite varies less compared to the X_{Mg} of biotite, but also shows a considerable range from 0.34 to 0.49 (average X_{Mg} is about 0.45).

Sample LEV30 additionally contains garnet grains with a skeletal or atoll habit and closely associated with quartz and feldspar. This garnet is very rich in Mn with X_{sps} 0.25, X_{alm} 0.67, X_{pyp} 0.02 and X_{grs} 0.06.

4.2 P-T Evolution

In order to decipher the P-T evolution of the Leventina Nappe, six samples collected along the structural NW–SE trend of the Leventina Gneisses (Fig. 6) have been studied in detail. The main objectives were to see whether the Leventina Gneisses show any change in deduced pressure of equilibration from north to south and to determine a quantitative P-T path for the unit. Equilibration temperatures in orthogneisses were estimated using observations from mineral textures. Equilibration pressures in orthogneisses were derived through the Si-content of white mica assuming equilibrium with Kfs, Pl, Qtz and Bt, using the calibrations of Velde (1965) and Massonne & Schreyer (1987), and the calculated grid of Simpson et al. (2000).

Metamorphic Temperature Estimates

An independent estimate of the temperature of metamorphic equilibration is needed in order to apply the phengite-geobarometer to metagranitic rocks, because of its significant temperature dependence. An estimate of the temperature was derived by studying the microtextures of metagranitoids.

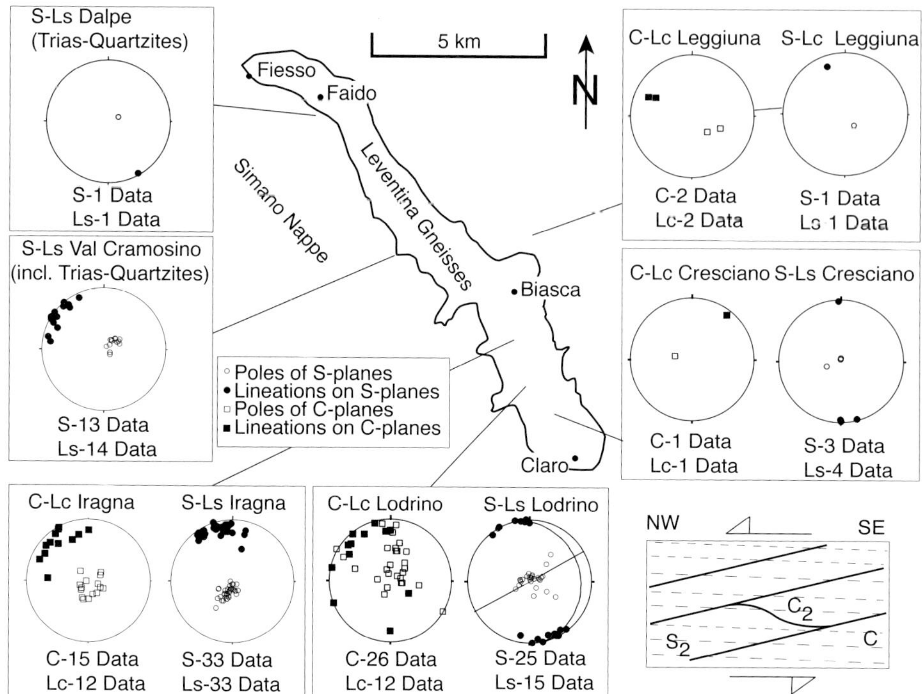

Fig. 5. Map showing stereoplots (lower hemisphere, equal area) of foliation planes (s) and lineations (L) and the associated shear bands (c) planes; Lc lineation on c planes) of selected quarries in Valle Leventina. The sketch to the right shows the field relations of the shear bands in these locations (shear sense: top-to-NW)

Minerals of all the Leventina Gneisses samples are completely recrystallized. In the core of the unit, large porphyroclasts of alkali feldspar recrystallized into alkali feldspar and plagioclase. Carlsbad twins with irregular twin faces are also frequent. The polysynthetic twins of plagioclase are usually deformed and irregularly shaped. A large number of plagioclase grains are irregularly zoned. Subgrain rotation and grain boundary migration are the processes associated with dynamic recrystallization of feldspars in the Leventina Gneisses. These processes are believed to indicate that the deformation probably took place at temperatures around or greater than 550 °C (e.g. Olsen & Kohlstedt 1985; Pryer 1993).

Myrmekites, bulbous symplectitic intergrowth of vermicular quartz in plagioclase, are quite frequent in the Leventina Gneisses. Myrmekites are common in high-grade metamorphic and igneous rocks. Their presence in metagranitic rocks indicates equilibration temperatures of at least 550 °C (Simpson 1985; Simpson & Wintsch 1989; Pryer 1993). Myrmekites are normally associated with plagioclase in Leventina Gneisses and two types can be distinguished: (a) Myrmekites, which are characterized by bulbous outlines and (b) myrmekite-structures present at the rims of plagioclase grains exposed to the principal strain axis Z, as described in Simpson (1985) and Simpson & Wintsch (1989). All myrmekites are thus observed along the main foliation, regardless of the type of myrmekite (Fig. 3c). The temperature of 550 °C is therefore considered as a minimum estimate for the metamorphism of the Leventina Gneisses during D2. Based on the shape and the location of the myrmekites within the section, the processes leading

to the formation of these structures is a combination of replacement of plagioclase to albite and quartz and the accommodation of deformation in the rocks (Simpson & Wintsch 1989).

Engi et al. (1995) calculated metamorphic temperatures for the eastern part of the Lepontine zone of the Central Alps by using the multi-equilibria method of Berman (1991) and analysing metamorphic assemblages in metapelites equilibrated during the main thermal peak (syn- to post-D2). In the northern Leventina Valley the calculated temperatures range between 550 and 575 °C, in the southern part of the valley (south of Biasca) temperatures of 650 °C and higher are obtained.

Metamorphic Pressure Estimates

The main constituents of metagranitoids, alkali feldspar, plagioclase, quartz, biotite and white mica are stable over a wide range of pressure and temperature conditions and show only little variation with changing P-T conditions. Nevertheless, it is possible to closely define the metamorphic conditions in terms of continuous reactions involving changes in composition of coexisting mineral phases along both the Tschermak [$Al_2(Fe,Mg)_{-1}Si_{-1}$] and $FeMg_{-1}$ exchange vectors (J.B. Thompson 1979). Numerous workers have reported that the Si-content in natural phengite increases (via the anti-Tschermak substitution $Si(Fe,Mg)Al_{-2}$) with increasing pressure and temperature in the model system $K_2O\text{-}MgO\text{-}Al_2O_3\text{-}SiO_2\text{-}H_2O$ (KMASH; Velde 1965; Massonne & Schreyer 1987). Several petrological studies (e.g. Massonne & Chopin 1989; Baudin et

Table 2. Representative electron microprobe analyses of micas from the Leventina Gneisses (wm = white mica, Bt = coexisting bioitite). All iron is reported as FeO. Electron microprobe analyses (EMP) were carried out on the Cameca SX50 of the Institute for Mineralogy and Petrology at ETH Zurich. All analyses were measured with an accelerating voltage of 15 kV and a 20 nA beam current. Synthetic and natural standards were used. All the EMP-analyses of micas were recalculated on the base of 11 oxygens. The X_{Or}-content of alkali feldspars in the samples represented in the Table is: LEV010 = 0.84 to 0.91; LEV011 = 0.90; LEV029 = 0.90 to 0.95; LEV030 = 0.85 to 0.93; LEV035 = 0.89 to 0.96; LEV037 = 0.90 to 0.94.

Mineral Analyses – Leventina Gneisses (weight %)

Sample Nr.	LEV010	LEV010	LEV011	LEV011	LEV029	LEV029	LEV030	LEV030	LEV035	LEV035	LEV037	LEV037	LEV037
Analysis Nr.	18	4	46	23	2	7	47	7	45	34	29	30	14
Mineral	Wm	Bt	Wm	Bt	Wm	Bt	Wm	Bt	Wm	Bt	Wm	Wm	Bt
Location	core	rim	rim	rim	core	rim	core	rim	core	rim	core	rim	rim
SiO_2	50,37	35,13	48,43	34,57	50,54	34,85	50,71	33,09	51,49	35,20	51,26	48,39	35,27
TiO_2	0,72	2,62	0,52	4,46	0,90	2,13	0,48	2,92	0,65	4,03	0,72	0,89	2,83
$Al2O_3$	31,33	17,58	34,68	17,09	28,10	15,80	29,06	15,32	28,69	16,58	31,31	35,21	18,25
FeO	2,48	21,77	1,65	21,69	5,02	23,35	5,38	28,83	4,49	20,52	2,08	1,86	18,48
MnO	0,02	0,31	0,00	0,26	0,02	0,36	0,07	0,32	0,05	0,43	0,04	0,03	0,27
MgO	1,92	7,78	0,84	7,48	2,09	7,75	1,44	3,52	2,40	7,64	1,94	0,84	9,83
CaO	0,00	0,05	0,01	0,00	0,01	0,03	0,05	0,05	0,00	0,07	0,02	0,01	0,01
Na_2O	0,32	0,10	0,26	0,13	0,32	0,10	0,23	0,06	0,24	0,15	0,30	0,46	0,12
K2O	9,77	9,36	9,67	9,06	10,17	9,55	9,75	8,84	10,11	9,27	9,58	9,83	9,35
Total	96,93	94,70	96,07	94,74	97,18	93,92	97,16	92,94	98,12	93,89	97,25	97,52	94,40

Cations Calculated on the Basis of 11 Oxygens

	LEV010	LEV010	LEV011	LEV011	LEV029	LEV029	LEV030	LEV030	LEV035	LEV035	LEV037	LEV037	LEV037
Si	3,2842	2,7315	3,1706	2,6879	3,3441	2,7687	3,3460	2,7362	3,3566	2,7499	3,3162	3,1327	2,7039
Ti	0,0353	0,1531	0,0254	0,2605	0,0447	0,1275	0,0239	0,1814	0,0317	0,2367	0,0349	0,0434	0,1631
Al	2,4078	1,6111	2,6758	1,5661	2,1911	1,4799	2,2597	1,4932	2,2044	1,5264	2,3871	2,6863	1,6488
Fe_2	0,1351	1,4155	0,0904	1,4104	0,2778	1,5517	0,2969	1,9936	0,2451	1,3403	0,1124	0,1008	1,1850
Mn	0,0012	0,0206	0,0000	0,0168	0,0011	0,0240	0,0042	0,0222	0,0029	0,0286	0,0020	0,0016	0,0175
Mg	0,1866	0,9022	0,0824	0,8674	0,2062	0,9178	0,1419	0,4341	0,2333	0,8892	0,1870	0,0806	1,1236
Ca	0,0000	0,0040	0,0010	0,0000	0,0007	0,0024	0,0033	0,0045	0,0000	0,0060	0,0014	0,0008	0,0006
Na	0,0399	0,0156	0,0335	0,0203	0,0413	0,0154	0,0288	0,0089	0,0303	0,0223	0,0380	0,0579	0,0178
K	0,8128	0,9281	0,8077	0,8986	0,8586	0,9684	0,8203	0,9325	0,8405	0,9239	0,7907	0,8116	0,9146
Total Cations	6,9029	7,7817	6,8868	7,7280	6,9656	7,8558	6,9250	7,8066	6,9448	7,7233	6,8697	6,9157	7,7749

al. 1993) have also applied the phengite geobarometer since the pioneering work of Velde (1965), e.g. Powell & Evans (1983) and Bucher-Nurminen (1987). Unfortunately, these studies give no compositional data of the white micas and the coexisting phases.

Simpson et al. (2000) have presented diagrams of calculated isopleths for $[Al_2(Fe,Mg)_{-1}Si_{-1}]$ and $FeMg_{-1}$ in phengite relative to KMASH and KFASH end-members for the metamorphic assemblage phengite, chlorite, biotite, alkali feldspar, quartz and H_2O, which can be used for thermobarometry purposes. Figure 7 combines these calculated isopleths by Simpson et al. (2000) as well as the loci of the experiments carried out by Velde (1965), Massonne & Schreyer (1987, 1989) to give a P-T calibration for various white mica assemblages as functions of $[Al_2(Fe,Mg)_{-1}Si_{-1}]$ in terms of Si per formula unit (pfu).

For metamorphic equilibration temperatures around 550 °C in the north (3.36 Si (pfu) for sample LEV35) and around 650 °C in the south (3.32 Si (pfu) for sample LEV37) the cal-culated grid of Simpson et al. (2000; their Figure 1a), yields a range of equilibration pressures between 8 and 10 kbar (Fig. 7). These P-T conditions correspond well with those of the Simano Nappe during D2 (Rütti et al. 2005). As the deformation along the Leventina-Simano nappe boundary occurred during D2, and hence these two rock units were in direct contact, this pressure estimate for the Leventina Gneisses is considered reasonable. The experimental fit by Massonne & Szpurka (1997) gives higher pressures of equilibration (by up to 3 kbar at 550 °C, Fig. 7) compared to the thermodynamic calculations by Simpson et al. (2000) using the Holland & Powell (1998) database.

The present understanding of Fe-Mg on the white mica Si-content is ambiguous – the Holland & Powell (1998) dataset would lower the pressure of equilibration of intermediate X_{Fe} mica by up to 2 kbar, whereas other evidence (Massonne & Szpurka 1997; Simpson et al. 2000) would favor increasing the pressure of equilibration with increasing X_{Fe}. We have not specifically corrected for X_{Fe} to obtain equilibration pressures

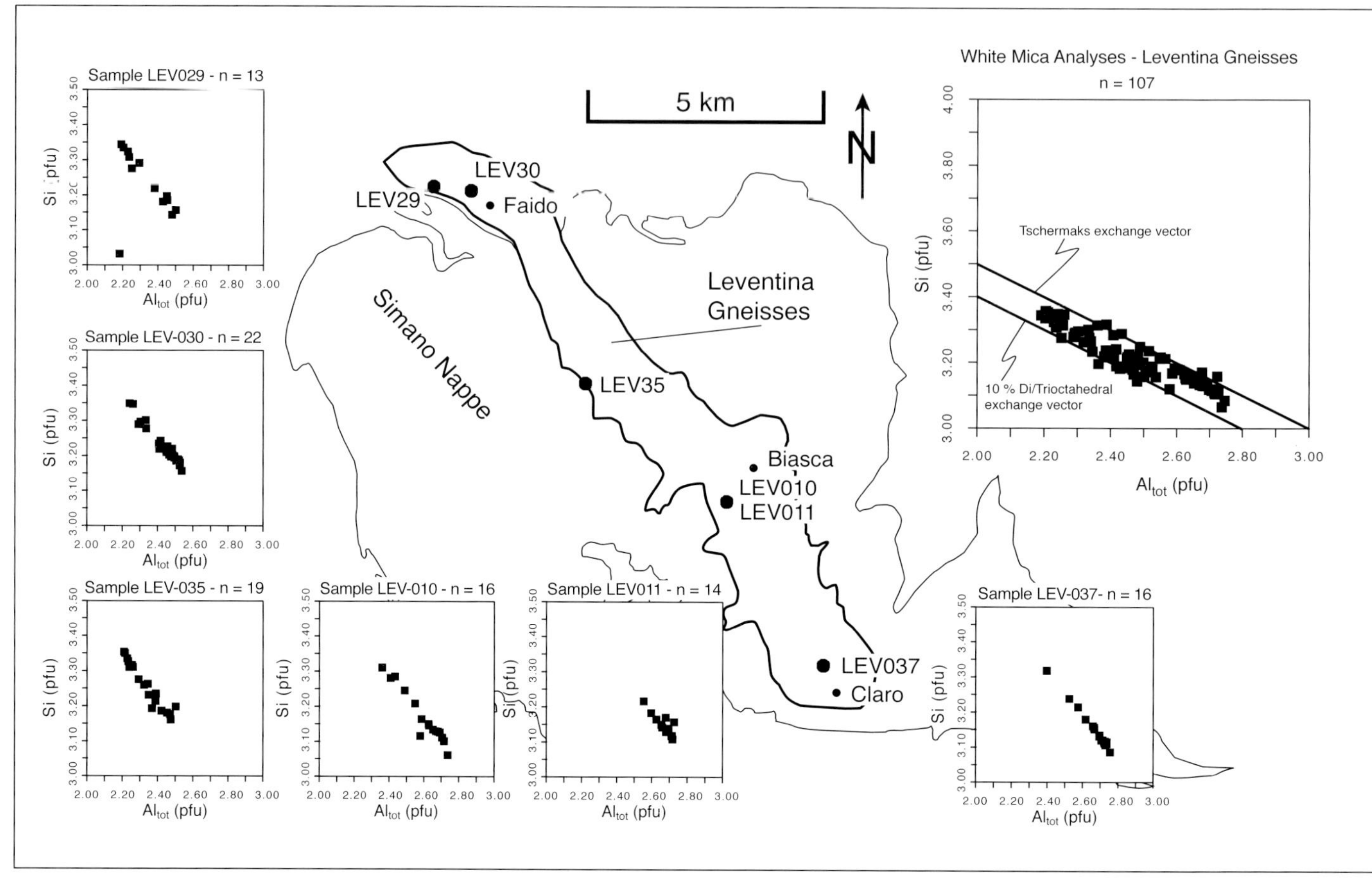

Fig. 6. Map showing the sample locations of this study. The map also shows the Si vs. Al$_{tot}$ ratio of all the measurements for white mica in each of all investigated samples (all with Bt + Kfs + Qtz) as well as a Si vs. Al$_{tot}$ ratio diagram with all the measurements plotted (upper right).

for the Leventina Gneisses (for equilibration temperatures between 550 and 650 °C in Fig. 7).

The white mica compositions were also examined to determine whether differences in relative metamorphic pressure were recorded by white micas oriented in the main foliation along a profile through the Leventina Gneisses, all with the common mineral assemblage. The samples LEV29, LEV30 and LEV35 represent outcrops from the northern and middle part of the Leventina Nappe (Fig. 6) and give a comparable maximum Si (pfu) content of 3.34 to 3.35 (Fig. 6). The samples LEV10, LEV11, LEV37 show an average Si (pfu) value (3.15 to 3.18) that is lower than the one of the first group (3.23 to 3.26, Fig. 6) and could indicate re-equilibration at higher temperature (ca. 650 °C; e.g. Engi et al. 1995; Kuhn et al. 2005) during the exhumation path in the southern part of the Leventina Nappe. Based on the assumed metamorphic temperatures in this study, the central and southern samples originating from the Ticino plain between Biasca and Claro, were plotted at higher temperature in Figure 8. The resulting metamorphic pressure estimate (ca. 10 kbar for samples of the south and ca. 8 kbar vs. for the samples in the north) therefore appears to re-

flect the position of the samples within the Leventina Gneisses. The continental subduction during the collision of the Adriatic and European plates was south-directed; therefore for the samples in the south, the deduced pressures (Fig. 8) could reflect a gentle dip of the Leventina Nappe within the crustal accretionary prism.

Lower Si-values for mica rim compositions (3.14 to 3.17 for the southern samples, 3.21 for the northern samples) allow us to obtain a metamorphic pressure estimate of 5 kbar for all the investigated samples during the exhumation path. This part of the "retrograde" path (Fig. 8) is correlated with the Tertiary exhumation of the Lepontine Nappes and could be related to D2 and D3 deformations in the Simano Nappe.

5. Tectonic Evolution

In the eastern Lepontine area, deduced metamorphic peak pressures are higher in the higher Subpenninic and Penninic units compared to those inferred for the lowermost Subpenninic unit (e.g. the Leventina Nappe) located in the center of the Lepontine Gneiss Region (see review in Berger et al. 2005

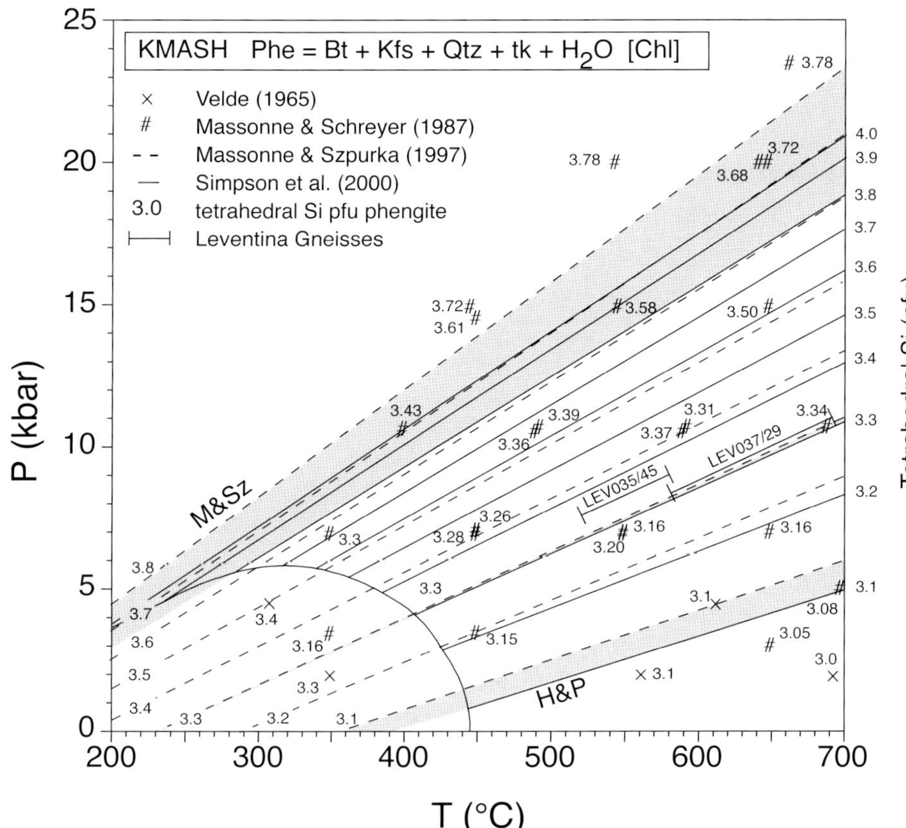

Fig. 7. P-T projection (Simpson et al. 2000) showing the results of calculations (solid lines and numbers on the right side of the diagram) using the thermodynamic database of Holland & Powell, (1998, H & P); isopleths calculated by Massonne & Szpurka, 1997 (M & Sz, dashed lines); and experimental data (hash symbols: Massonne & Schreyer, 1987; crosses: Velde, 1965) for the KMASH chlorite-absent assemblage phengite + biotite + alkali feldspar + quartz. The shaded regions in the P-T projection indicate the difference between Simpson et al. (2000, with H&P database) and M & Sz calculations for 10, 30 and 80% celadonite.

and references therein). This has been explained by a subduction-related regime (Jamieson & Beaumont 1989; Beaumont et al. 1996; Schmid et al. 1996). In such a model the Leventina Nappe stays longer in front of the rest of the Subpenninic and Penninic Nappes during progressive entrainment into the Tertiary subduction zone. Subsequently the Leventina Nappe was thrust by units that came from deeper levels in the subduction zone and that were exhumed during Tertiary collision, thus creating the present-day Alpine stack.

Figure 9 (a–d) illustrates the major stages of the Alpine evolution of the Leventina Nappe. The restored initial geometry of the basement of the thinned European margin before the closure of the Valais Ocean (before some 45 Ma) is shown in Figure 9a. After 45 Ma underthrusting of the European Margin initiated due to the closure of the Valais Ocean. This phase generates D1 structures, nappe formation and stacking. The Adula Nappe (southernmost European Margin) is underthrusted and will later re-ascend in the accretionary prism through extrusion. At this point, the Briançonnais Tambo and Suretta Nappes have already reached their maximum burial depth (pressure peak at ca. 10 to 13 kbar equivalent to 35 to 42 km; Marquer et al. 1994, Nussbaum et al. 1998, Challandes et al. 2003). The Subpenninic Adula, Simano and Leventina Nappes, corresponding to the thinned European margin, are progressively individualized and stacked from 40 to 35 Ma (Stampfli et al. 1998; Meyre et al. 1998; Engi et

al. 2001; Berger et al. 2005). Around 35 Ma the metamorphic pressure peak, corresponding to the deformation phase D2 in both Simano and Leventina Nappes, is attained in the depth range of 30 to 35 km. Subpenninic and Penninic units are juxtaposed and afterwards followed together a common P-T path (Fig. 9c). The upper part of the accretionary prism thins by normal fault movements directed to the E-SE (Stampfli et al. 1998; Challandes et al. 2003). However, in the lower part of the pile there is coeval NW-directed thrusting. As a result the kinematics vary across the nappe pile: decompression at the top of the nappe pile is associated with shearing towards SE (top-to-SE), whereas stacking at the lower part of the nappe pile yields top-to-NW sense of shear (around 35 Ma during the metamorphic pressure peak). From 30 Ma onwards, the Lepontine Gneiss region followed a Barrovian-type P-T path. During the D3 deformation phase the nappe pile is folded into large-scale open folds (wavelength 8 to 10 km) without any further internal thrusting. The underthrusting of the thick European margin (Aar and Gotthard Massifs) from the north and the Adriatic plate from the south only causes the frontal- and the southernmost parts of the Penninic nappes to be backthrusted and backfolded along the Insubric Line and south of the Gotthard Massif during late Oligocene and early Miocene, respectively (Berger et al. 2005; D4, Fig. 2). The formation of the steeply dipping Northern Steep belt occurred at around 20 Ma in response to the underthrusting of the ex-

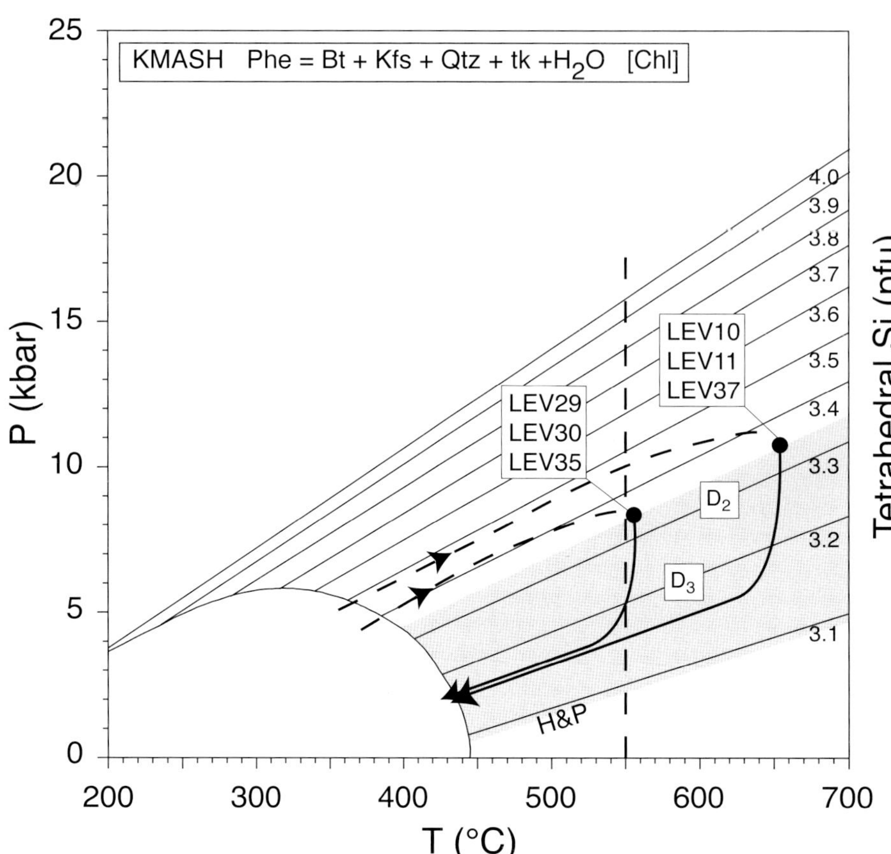

Fig. 8. Suggested P–T paths for the Leventina Gneisses with solid line isopleths from Figure 7. On the prograde path, Si (pfu) values in the core of white micas up to 3.36 are reached at temperatures of circa 550 °C where myrmekite is stable. The samples LEV10, LEV11, LEV37 originate from the central and southern part of the Leventina Gneisses and are therefore plotted at higher temperature of 650 °C according to Engi et al. (1995). On the retrograde path, the chlorite stability field is reached, because biotite in the main foliation is replaced by chlorite (e.g. in sample LEV035).

ternal crystalline massifs (e.g. the Aar and Gotthard Massifs; Challandes et al. 2008).

6. Conclusions

The Leventina Gneisses show evidence for three phases of ductile deformation. Foliation S1 is mostly sub-parallel to the regionally dominant structural fabric, the S2 foliation. The composite S1-S2 foliation is penetratively developed in the hanging wall toward the Simano Nappe, whereas in the core of the Leventina Gneisses, S2 is only weakly developed. A 50 to 200 m wide mylonite zone, showing a top-to-NW sense of shear for D2 marks the boundary to the Simano Nappe. At the surface, no major shear zones were detected within the interior of the Leventina Nappe, but mesoscopic shear zones are reported underground from the Gotthard Base Tunnel (Bonzanigo & Oppizi 2006). In local outcrops only small shear bands (D2, mm to cm wide) are commonly observed. They show a top-to-NW sense of shear throughout the unit. At the outcrop scale, deformation phase D3 only locally generated a new axial plane foliation S3. However, the large-scale effects of this phase are clearly observed with the D3 Leventina antiform.

The nappe stack indicates that the Leventina Nappe represents one of the lowermost units in the Central Alps of Switzerland. Structural data of this study and Rütti et al. (2005) show that D2 deformation was the same both in the Leventina Nappe and the overlying Simano Nappe. At the end of the D2 deformation phase, the superposition of these two units was as observed today, later they were folded together by the subsequent D3 deformation phase.

The metamorphic microstructures and temperature calibrations quoted in the literature for the Leventina Gneisses (e.g. Engi et al. 1995) associated with the dominant foliation S2, constrain metamorphic temperatures for this deformation phase at 550 to 650 °C for the northern and southern parts of the Leventina Nappe, respectively. Si (pfu) values of white micas oriented along the main foliation are always relatively higher in the core compared to with the rim. The higher Si (pfu) values at different temperatures indicate maximum metamorphic pressure conditions at around 8 and 10 kbar for the north and south within the nappe, respectively. The lower Si (pfu) values at the rim of white micas reflect pressures of 5 kbar reached during the exhumation path.

The inferred metamorphic P-T-path shows that the estimates for the Leventina Nappe concur with the conditions estimated for the Simano Nappe during D2 and D3 (Rütti 2003), implying a common metamorphic and deformation history during these two phases of the Alpine orogeny. They are related to the underthrusting of the thinned European margin into the crustal accretionary prism that initiated during late Eocene to early Oligocene times.

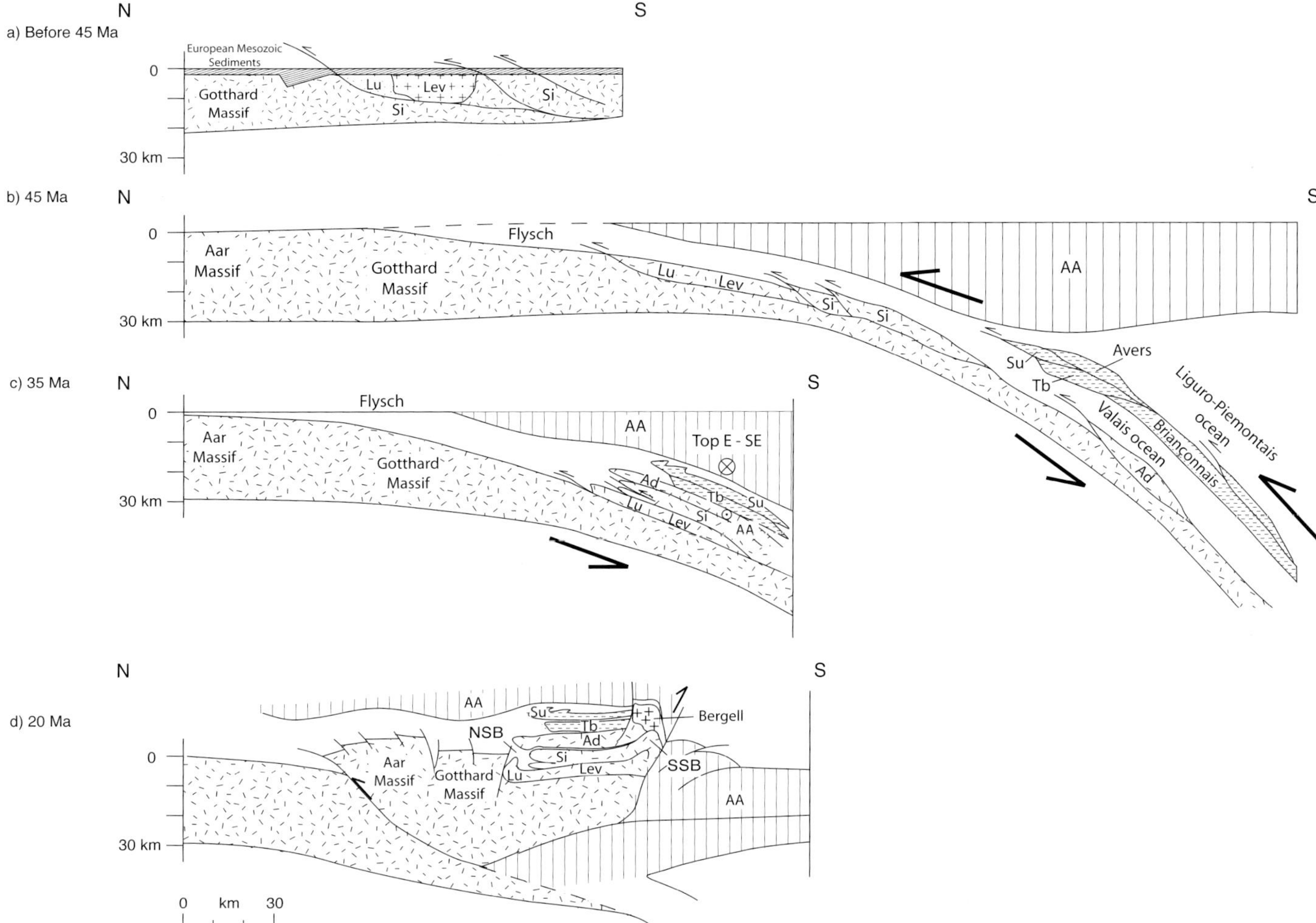

Fig. 9. Sketches tracing the major stages of the Alpine tectonic evolution of the Leventina Gneisses. Abbreviations: Lu = Lucomagno Nappe, Le = Leventina Gneisses, Si = Simano Nappe, Su = Suretta Nappe, Tb = Tambo Nappe, Ad = Adula Nappe, AA: Austroalpine domain, NSB = Northern Steep Belt, SSB = Southern Steep Belt. a) presumed initial configuration of the European margin. b) Geometry during the underthrusting of the European plate. The Adula Nappe remains deep in the subduction zone, while the Suretta and Tambo Nappes are already ascending. c) Situation when the Penninic units are being stacked at 30 to 35 km depth. From this point onwards the Leventina Gneisses and the Simano Nappe have a common history (D2). d) Configuration of the Alpine Nappe pile during the last stage of collision, backfolding in the north and south generating the Northern und Southern Steep Belts, respectively.

Acknowledgements

This contribution is part of the doctorate thesis work of RR and funding by ETH Zürich (Grant No. 0-20738-99) is acknowledged. RR is grateful to Guy Simpson for help with white mica microscopy and mineral chemistry. The authors thank Thorsten Nagel and Djordje Grujic for the helpful reviews and Stefan Schmid for careful editing.

REFERENCES

Allègre, C.J., Albarède, F., Grünenfelder, M. & Köppel, V. 1974: ^{238}U/^{206}Pb-^{235}U/^{207}Pb-^{232}Th/^{208}Pb Zircon geochronology in Alpine and non-Alpine environment. Contributions to Mineralogy and Petrology 43, 163–194.

Baudin, T., Marquer, D. & Persoz, F. 1993: Basement-cover relationships in the Tambo nappe (Central Alps, Switzerland): geometry, structure and kinematics. Journal of Structural Geology 15, 543–553.

Beaumont, C., Ellis, S., Hamilton, J. & Fullsack, P. 1996: Mechanical model for subduction-collision tectonics of Alpine-type compressional orogens. Geology 24, 675–678.

Becker, H. 1993: Garnet peridotite and eclogite Sm-Nd mineral ages from the Lepontine dome (Swiss Alps): New evidence for Eocene high-pressure metamorphism in the Central Alps, Geology 21, 599–602.

Berger, A., Mercolli, I., & Engi, M. 2005: The central Lepontine Alps: Notes accompanying the tectonic and petrographic map sheet Sopra Ceneri (1:100 000). Schweizerische Mineralogische und Petrographische Mitteilungen 85, 109–146.

Berman, R.G. 1991: Thermobarometry suing multi-equilibrium calculations: a new technique, with petrological applications. Canadian Mineralogist 29, 833–855.

Bianconi, F. 1971: Geologia e petrografia della regione del Campolungo. Beiträge zur Geologie der Schweiz NF 142, 238 pp.

Bonzanigo, L. & Oppizzi, P. 2006: Low angle fault zones and TBM excavation in Bodio section of Gotthard Base Tunnel. In Löw S., Ed. Geologie und Geotechnik der Basistunnels am Gotthard und am Lötschberg, 155–165. vdf Hochschulverlage AG, Zürich.

Bossard, L. 1925: Der Bau der Tessinerkulmination. Eclogae Geologicae Helvetiae 19, 504–521.

Bucher-Nurminen, K. 1987: A recalibration of the chlorite-biotite-muscovite geobarometer. Contributions to Mineralogy and Petrology 96, 519–522.

Casasopra, S.F. 1939: Studio petrografico dello Gneiss granitico Leventina (Valle Riviera e Valle Leventina, Canton Ticino). Schweizerische Mineralogische und Petrographische Mitteilungen 19, 449–708.

Challandes, N., Marquer, D. & Villa, I.M. 2003: Dating the evolution of C-S microstructures: a combined $^{40}Ar/^{39}Ar$ step-heating and UV laserprobe analysis of the Alpine Roffna shear zone. Chemical Geology 197, 3–19.

Challandes N., Marquer D. & Villa I. 2008: P-T-t modeling, fluid circulation, and ^{39}Ar-^{40}Ar and Rb-Sr mica ages in the Aar Massif shear zones (Swiss Alps). Swiss Journal of Geosciences. DOI 10.1007/S00015-009-1260-6.

Choukroune, P. & Gapais, D. 1983: Strain pattern in the Aar granite (Central Alps): orthogneiss developed by bulk inhomogeneous flattening. Journal of Structural Geology 9, 411–418.

Engi, M., Todd, C.S. & Schmatz, D. R. 1995: Tertiary metamorphic conditions in the eastern Lepontine Alps. Schweizerische Mineralogische und Petrographische Mitteilungen 75, 347–369.

Engi M., Berger A. & Roselle G.T. 2001: Role of the tectonic accretion channel in collisional orogeny. Geology 29, 1143–1146.

Etter, U. 1992: Die Chierà-Synform. Bulletin der Vereinigung Schweizerischer Petroleum-Geologen und -Ingenieure 59, 93–99.

Gapais, D., Bale, P., Choukroune, P., Cobbold, P.R., Mahjoub, Y. & Marquer, D. 1987: Bulk kinematics from shear zone patterns: some field examples. Journal of Structural Geology 9, 635–646.

Gebauer, D. 1996: A P-T-t path for an (ultra?) High-pressure ultramafic/mafic rock association and its felsic country rocks based on SHRIMP-dating of magmatic and metamorphic Zircon domains. Example: Alpe Arami (Central Alps). In: Earth Processes: Reading the Isotopic Code. Geophysical Monograph 95, 307–329.

Grond, R., Wahl, F. & Pfiffner, M. 1995: Mehrphasige alpine Deformation und Metamorphose in der nördlichen Cima–Lunga–Einheit, Zentralalpen (Schweiz). Schweizerische Mineralogische und Petrographische Mitteilungen 75, 371–386.

Heitzmann, P., Frei, W., Lehner, P. & Valasek, P. 1991: Crustal indentation in the Alps – an overview of reflection seismic profiling in Switzerland. Geodynamics 22, 161–176.

Hiss, B.M. 1975: Petrographische Untersuchungen der SBB-Sondierbohrung Biaschina (TI). Schweizerische Mineralogische und Petrographische Mitteilungen 55, 201–215.

Holland, T.J.B. & Powell, R. 1998: An internally consistent thermodynamic data set for phases of petrological interest. Journal of Metamorphic Geology 16, 309–343.

Hurford, A.J., Flisch, M. & Jäger, E. 1989: Unravelling the thermo-tectonic evolution of the Alps: a contribution from fission track analysis and mica dating. In: Coward, M.P., Dietrich, D. & Park, R.G. (Eds.): Alpine Tectonics. Geological Society Special Publication 45, 369–398.

Irouschek, A. 1983: Mineralogie und Petrographie von Metapeliten der Simano-Decke mit besonderer Berücksichtigung cordieritführender Gesteine zwischen Alpe Sponda und Biasca. Ph.D Thesis, Universität Basel, 205 pp.

Jamieson, R.A. & Beaumont, C. 1989: Deformation and metamorphism in convergent orogens: a model for uplift and exhumation of metamorphic terrains. In Daly, J.S., Cliff, R.A. & Yardley, B.W.D. (Eds.): Evolution of metamorphic Belts. Geological Society Special Publication 43, 117–129.

Köppel, V., Günthert, A. & Grünenfelder, M. 1980: Patterns of U-Pb zircon and monazite ages in polymetamorphic units of the Swiss Central Alps. Schweizerische Mineralogische und Petrographische Mitteilungen 61, 97–119.

Köppel, V. 1993: The Lepontine area, a geochronological summary. In von Raumer J.F. & Neubauer, F., Eds. Pre-Mesozoic Geology in the Alps, 345–348. Springer-Verlag, Berlin.

Kuhn, B.K., Reusser, E., Powell, R. & Günther, D. 2005: metamorphic evolution of calc-schists in the Central Alps, Switzerland. Schweizerische Mineralogische und Petrographische Mitteilungen 85, 175–190.

Marquer, D. 1990: Structures et déformations alpines dans les granites hercyniens du massif du Gothard (Alpes centrales suisses) Eclogae Geologicae Helvetiae 83, 77–97.

Marquer, D. 1991: Structures et cinématique des déformations alpines dans le granite de Truzzo (Nappe de Tambo: Alpes Centrales Suisses). Eclogae Geologicae Helvetiae 84, 107–123.

Marquer, D., Baudin, T., Peucat, J.-C. & Persoz, F. 1994: Rb-Sr mica ages in the Alpine shear zones of the Truzzo granite: Timing of the Tertiary alpine P-T-deformations in the Tambo nappe (Central Alps, Switzerland). Eclogae Geologicae Helvetiae 87, 225–239.

Marquer, D. Challandes, N. & Baudin, T. 1996: Shear zone patterns and strain distribution at the scale of a Penninic nappe: the Suretta nappe (Eastern Swiss Alps). Journal of Structural Geology 18, 753–764.

Massonne, H.J. & Schreyer, W. 1987: Phengite geobarometry based on the limiting assemblage with K-feldspar, phlogopite, and quartz. Contributions to Mineralogy and Petrology 96, 212–224.

Massonne, H.J., Schreyer, W. 1989: Stability field of the high-pressure assemblage talc + phengite and two new phengite barometers. European Journal of Mineralogy 1, 391–410.

Massonne, H.J., & Chopin, C. 1989: P-T history of the Gran Paradiso (Western Alps) metagranites based on phengite geobarometry. In Daly, J.S., Cliff, R.A. & Yardley, B.W.D., Eds. Evolution of Metamorphic Belts, Geological Society Special Publication 43, 545–549.

Massonne, H.J. & Szpurka, Z. 1997: Thermodynamic properties of white micas on the basis of high pressure experiments in the systems K2O-MgO-Al2O3-SiO2-H2O and K2O-FeO-Al2O3-SiO2-H2O. Lithos 41, 229–250.

Maxelon, M. & Mancktelow, N.S. 2005: Three-dimensional geometry and tectonostratigraphy of the Pennine zone, Central Alps, Switzerland and Northern Italy. Earth Science Reviews 71, 171–227.

Merle, O., Cobbold, P.R. & Schmid, S. 1989: Tertiary kinematics in the Lepontine dome. In Coward, M.P., Dietrich, D. & Park, R.G. (Eds): Alpine Tectonics. Geological Society Special Publication 45, 113–134.

Meyre, C., Marquer, D., Schmid, S.M. & Ciancaleoni, L. 1998: Syn-orogenic extension along the Forcola fault: Correlation of Alpine deformations in the Tambo and Adula nappes (Eastern Penninic Alps). Eclogae Geologicae Helvetiae 91, 409–420.

Milnes, A.G. 1974: Structure of the Pennine zone (Central Alps): A new working hypothesis. Geological Society of America Bulletin 85, 1727–1732.

Milnes, A.G. 1976: Strukturelle Probleme im Bereich der Schweizer Geotraverse – das Lukmanier-Massiv. Schweizerische Mineralogische und Petrographische Mitteilungen 56, 615–618.

Nagel, T., de Capitani, C., Frey, M., Froitzheim, N., Stünitz, H. & Schmid, S.M. 2002: Structural and metamorphic evolution during rapid exhumation in the Lepontine dome (southern Simano and Adula nappes, Central Alps, Switzerland). Eclogae Geologicae Helvetiae 95, 301–321.

Niggli, P., Preiswerk, H., Grütter, O., Bossard, L. & Kündig, E. 1936: Geologische Beschreibung der Tessiner Alpen zwischen Maggia- und Bleniotal. Beiträge zur Geologie der Schweiz NF 71, 190 pp.

Niggli, E. & Niggli, C.R. 1965: Karten der Verbreitung einiger Mineralien der alpidischen Metamorphose in den Schweizer Alpen (Stilpnomelan, Alkali-Amphibol, Chloritoid, Staurolith, Disthen, Sillimanit). Eclogae Geologicae Helvetiae 58, 335–368.

Nussbaum, C., Marquer, D. & Biino, G.G. (1998) Two subduction events in a polycyclic basement: Alpine and pre-Alpine high-pressure metamorphism in the Suretta nappe, Swiss eastern Alps. Journal of Metamorphic Geology 16, 591–605.

Olsen, T.S. & Kohlstedt, D.L. 1985: Natural deformation and recrystallization of some intermediate plagioclase feldspars. Tectonophysics 111, 107–131.

Partzsch, J.H. 1998: The tectono-metamorphic evolution of the middle Adula nappe, Central Alps, Switzerland. Ph. D. Thesis, Universität Basel, Switzerland.

Passchier, C.W. & Trouw, R.A.J. 1996: Microtectonics. Springer-Verlag, Berlin, 289 pp.

Pfiffner, M. 1999: Genese der hochdruckmetamorphen ozeanischen Abfolge der Cima Lunga Einheit (Zentralalpen). Ph. D. Thesis, ETH Zürich, 248 pp.

Powell, R. & Evans, J.A. 1983: A new geobarometer for the assemblage biotite-muscovite-chlorite-quartz. Journal of Metamorphic Geology 1, 331–336.

Preiswerk, H., Bossard, L., Grütter, O., Niggli, P., Kündig, E. & Ambühl, E. 1934: Geologische Karte der Tessineralpen zwischen Maggia- und Blenio-

Tal. Geologische Spezialkarte Nr. 116, 1:50 000. Schweizerische Geologische Kommission, Bern.

Pryer, L.L. 1993: Microstructures in feldspars from a major crustal thrust zone: The Grenville Front, Ontario, Canada. Journal of Structural Geology 15, 21–36.

Rütti, R. 2001: Tectono–metamorphic evolution of the Simano–Adula nappe boundary, Central Alps, Switzerland. Schweizerische Mineralogische und Petrographische Mitteilungen 81, 115–129.

Rütti, R. 2003: The tectono-metamorphic evolution of the northwestern Simano Nappe (Central Alps, Switzerland). Ph. D Thesis, ETH Zürich, 112 p.

Rütti, R., Maxelon, M. & Mancktelow, N.S. 2005: Structure and kinematics of the northern Simano Nappe, Central Alps, Switzerland. Eclogae Geologicae Helvetiae 98, 63–81.

Schmid, S.M., Pfiffner, O.A., Froitzheim, N., Schönborn, G. & Kissling, E. 1996: Geophysical-geological transect and tectonic evolution of the Swiss-Italian Alps. Tectonics 15, 1036–1064.

Schmid, S.M., Fügenschuh, B., Kissling, E. & Schuster, R. 2004: Tectonic map and overall architecture of the Alpine orogen. Eclogae Geologicae Helvetiae 97, 93–117.

Simpson, C. 1985: Deformation of granitic rocks across the brittle-ductile transition. Journal of Structural Geology 7, 503–511.

Simpson, C. & Wintsch, R.P. 1989: Evidence for deformation-induced K-feldspar replacement by myrmekite. Journal of Metamorphic Geology 7, 261–275.

Simpson, G.D.H., Thompson, A.B. & Connolly, J.A.D. 2000: Phase relations, singularities and thermobarometry of metamorphic assemblages containing phengite, chlorite, biotite, K-feldspar, quartz and H2O. Contributions to Mineralogy and Petrology 139, 555–569.

Spicher, A. 1980: Tektonische Karte der Schweiz 1:500000. Schweizerische Geologische Kommission.

Stampfli, G.M., Mosar, J., Marquer, D., Marchant, R., Baudin, T. & Borel, G. 1998: Subduction and obduction processes in the Swiss Alps. Tectonophysics 296, 159–204.

Tektonische Karte der Schweiz 1:500000, 2005: Bundesamt für Landestopografie, Wabern.

Thompson, J.B. 1979: The Tschermak substitution and reactions in pelitic schists. In: Zharikov, V.A., Fonarev, V.I. & Karikovskii, S.P. (Eds.): Problems in physicochemical petrology (in Russian), 149–159. Academy of Sciences, Moskow.

Thompson, P.H. 1976: Isograd pattern and pressure-temperature distribution during regional metamorphism. Contributions to Mineralogy and Petrology 57, 277–295.

Timar-Geng, Z., Grujic, D. & Rahn, M. 2004: Deformation at the Leventina-Simano nappe boundary, Central Alps, Switzerland. Eclogae Geologicae Helvetiae 97, 265–278.

Velde, B. 1965: Experimental determination of muscovite polymorph stabilities. American Mineralogist 50, 436–449.

Villa, I. & von Blanckenburg, F. 1991: A hornblende ^{39}Ar-^{40}Ar age traverse of the Bregaglia tonalite (southeast Central Alps). Schweizerische Mineralogische und Petrographische Mitteilungen 71, 73–87.

von Blanckenburg, F. 1992: Combined high precision chronometry and geochemical tracing using accessory minerals: applied to the Central-Alpine Bergell intrusion. Chemical Geology 100, 19–40.

Wenk, E. 1955: Eine Strukturkarte der Tessineralpen. Schweizerische Mineralogische und Petrographische Mitteilungen 35, 311–319.

Manuscript received 25 October, 2006
Revision accepted 21 July, 2008
Published Online first November 1, 2008
Editorial Handling: Stefan Schmid & Stefan Bucher

1661-8726/08/01S173-17
DOI 10.1007/s00015-008-1292-y
Birkhäuser Verlag, Basel, 2008

Swiss J. Geosci. 101 (2008) Supplement 1, S173–S189

Lu-Hf garnet geochronology of eclogites from the Balma Unit (Pennine Alps): implications for Alpine paleotectonic reconstructions

DANIEL HERWARTZ [1,*], CARSTEN MÜNKER [1,2], ERIK E. SCHERER [2], THORSTEN J. NAGEL [1], JAN PLEUGER [1] & NIKOLAUS FROITZHEIM [1]

Key words: Lu-Hf geochronology, Western Alps, Balma Unit, Monte Rosa, tectonic evolution

ABSTRACT

Three samples of eclogite from the Balma Unit, an ophiolite sheet on top of the Monte Rosa Nappe in the Pennine Alps, were investigated in terms of their P-T evolution, geochemistry, and Lu-Hf geochronology. The paleogeographic origin of this unit is controversial (North Penninic vs. South Penninic). It has been interpreted as a piece of Late Cretaceous oceanic crust, on the basis of ca. 93 Ma U-Pb SHRIMP ages of synmagmatic zircon cores in an eclogite. Trace element and isotope data suggest a mid ocean ridge (MOR) rather than an intraplate or OIB setting for the protoliths of the eclogites. Electron microprobe analyses of representative garnets show typical prograde zoning profiles. Estimated peak metamorphic temperatures of 550–600 °C most likely did not exceed the closure temperature of the Lu-Hf system. Hence, Lu-Hf ages most likely reflect garnet growth in the studied samples. To minimize inclusion effects on age determinations, a selective digestion procedure for garnet was applied, in which zircon and rutile inclusions are not dissolved. The ages obtained for three samples, 42.3 ± 0.6 Ma (MSWD: 0.47), 42 ± 1 Ma (MSWD: 3.0) and 45.5 ± 0.3 Ma (MSWD: 0.33), are younger than all Lu-Hf ages reported so far for South Penninic Units. Metamorphic zircon domains of the 42.3 Ma sample (PIS1) were previously dated by U-Pb SHRIMP at 40.4 ± 0.7 Ma, indicating that the growth of metamorphic zircon post-dated the onset of garnet growth.

These new data put important constraints on the paleogeographic reconstruction of the Alps. The MORB character of the rocks, together with their previously published protolith age, imply that oceanic spreading was still taking place in the Late Cretaceous. This supports a North Penninic origin for our samples because plate tectonic models predict Cretaceous spreading in the North Penninic but not in the South Penninic Ocean. If the Balma Unit is indeed North Penninic, the new Lu-Hf data, in combination with published geochronological data, require that two independent subduction zones consumed the South and North Penninic oceans.

1. Introduction

Despite a wealth of geological and geochronological data, there are still conflicting models for the tectonic evolution of the Alps. Current rapid progress in dating metamorphic minerals such as garnet has provided a new tool to date tectonometamorphic events in high pressure terrains. In this study, we apply Lu-Hf garnet geochronology to high-pressure rocks of the Balma Unit in the Swiss-Italian Pennine Alps. This area is formed by a stack of nappes, the Penninic Nappes (Figs. 1, 2), which originate partly from continental crust (Variscan basement with Permian and Mesozoic cover rocks), and partly from Mesozoic oceanic crust. These rocks were imbricated from the latest Cretaccous through the Early Tertiary and affected by several phases of folding and shearing. This modified the initial geometry of the nappes to such an extent that restoring the deformation back to the pre-orogenic arrangement of the units is difficult and controversial.

Most authors agree that in the Cretaceous, two partly oceanic basins existed between Europe to the Northwest and Adria (Apulia) to the Southeast: the South Penninic or Piemont-Ligurian Ocean, and the North Penninic or Valais Ocean. The former opened in the Middle and Late Jurassic, and the latter opened in the Cretaceous (Stampfli et al. 1998). The Briançonnais continental peninsula represented an eastward-tapering promontory of the Iberian continent between these two oceanic basins (Frisch 1979; Stampfli 1993). Due to the sinistrally transtensive opening of the North Penninic Ocean along a trace oblique to the South Penninic Ocean, Jurassic crust of the South Penninic Ocean was captured in the North Penninic basin, in addition to the new, Cretaceous-age oceanic crust (Fig. 2b; Liati et al. 2005).

Continental crustal rocks of the Sesia and Dent Blanche nappes occupy the highest position in the nappe stack. They originated – according to an interpretation preferred by the present authors – from a continental fragment in the South

[1] Steinmann-Institut, Universität Bonn, Germany. E-mail: danielherwartz@gmx.de
[2] Institut für Mineralogie, Universität Münster, Germany.

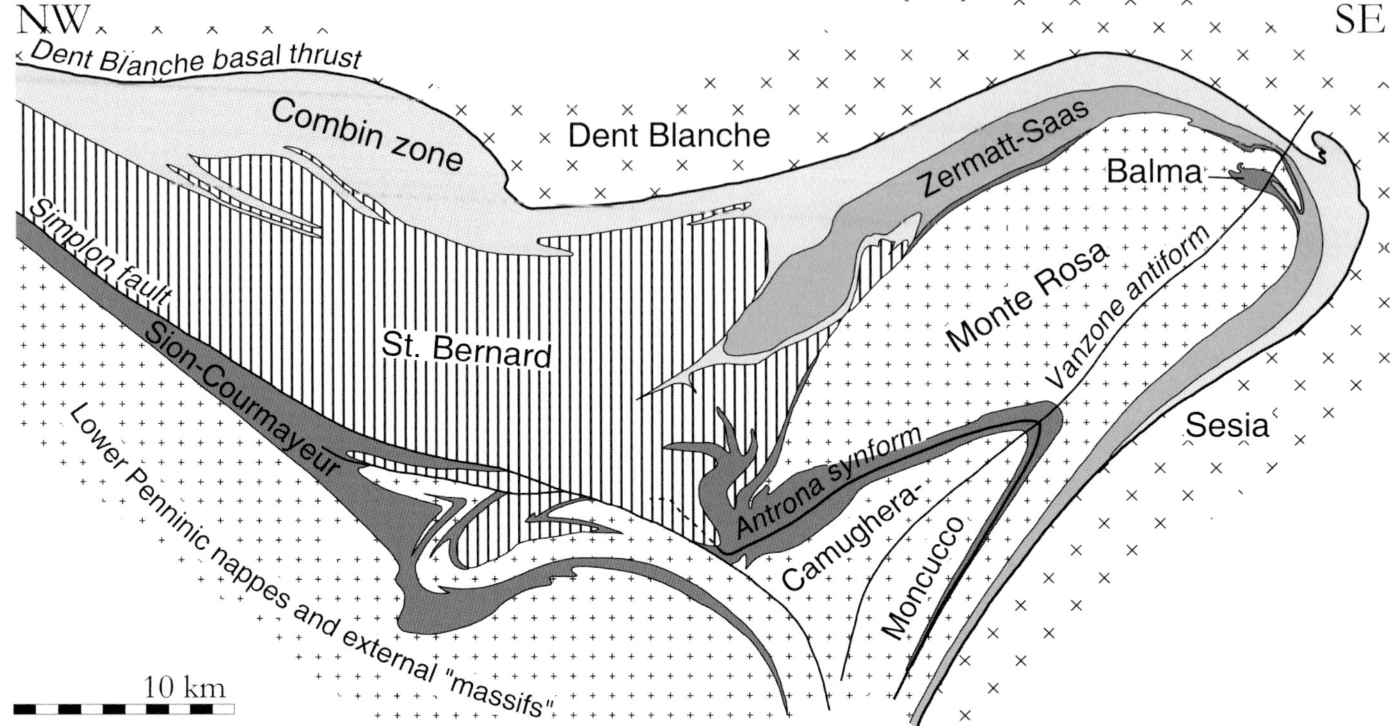

Fig. 1. Cross section through the Swiss–Italian Western Alps, modified after Escher et al. (1993). In this profile and in Fig. 2a, small crosses indicate former European continental crust, dark grey North Penninic ophiolites, vertical ruling Briançonnais continental crust, middle grey South Penninic ophiolites from the Zermatt-Saas Basin, light grey South Penninic ophiolites from the Tsaté Basin, and "x" pattern units from the Cervinia continental fragment. The paleogeographic affiliation shown here follows the interpretation of Froitzheim (2001) and Pleuger et al. (2007).

Penninic Ocean, called the Margna-Sesia fragment or Cervinia (Froitzheim & Manatschal 1996; Froitzheim et al. 1996; Schmid et al. 2004; Pleuger et al. 2007). According to other workers, these nappes were derived from the Adriatic continental margin proper (Stampfli 1993; Avigad et al. 1993). Ophiolite units occur below the Dent Blanche-Sesia nappe system. These are subdivided into two large units, the structurally higher Combin Zone and the structurally deeper Zermatt-Saas Zone. The Combin Zone records lower pressures during the subduction-related metamorphism (blueschist facies; Bousquet et al. 2004) than the Zermatt-Saas Zone (eclogite facies, locally with coesite; Reinecke 1991, 1998; Bousquet et al. 2004).

The eclogite-facies ophiolites of the Zermatt-Saas Zone overlie continental basement rocks of the Monte Rosa Nappe. The latter include paragneisses, Late Variscan to post-Variscan granites partly transformed to orthogneisses, and amphibolite boudins that in some places contain eclogite relics. In the area under consideration, on the southern flank of the Monte Rosa massif, the ophiolites of the Zermatt-Saas Zone do not rest directly on the Monte Rosa gneisses, but on a thin layer of gneiss (Stolemberg Unit), which overlies an ophiolite layer (Balma Unit), which in turn rests on the Monte Rosa Nappe (Pleuger et al. 2005). This geometry is modified by three phases of folding postdating the emplacement of the nappes (Figs. 3, 4). The Balma Unit is made up mostly of serpentinite, eclogite,

and minor amphibolite. The paleogeographic origins of these units are controversial. According to Froitzheim (2001); Liati & Froitzheim (2006), and Pleuger et al. (2005, 2007), the Monte Rosa Nappe represents the European continental margin, the overlying Balma Unit the North Penninic Ocean, the Stolemberg Unit the Briançonnais, and the rest of the Zermatt-Saas Zone the South Penninic Ocean. According to other authors (Keller & Schmid 2001; Kramer et al. 2003), the Monte Rosa Nappe originates from the Briançonnais continental crust and the overlying units (Balma, Stolemberg, Zermatt-Saas) all represent the South Penninic Ocean. These controversial views result from different ways of retrodeforming the complex geometry of the Penninic nappes.

Eclogite from the Balma Unit (sample PIS1) has recently been the subject of U-Pb SHRIMP geochronology on zircon (Liati & Froitzheim 2006). Magmatic zircon domains yielded a Cretaceous protolith age of 93.4 ± 1.7 Ma (Cenomanian-Turonian) and metamorphic domains a Late Eocene age of 40.4 ± 0.7 Ma, interpreted as the age of high-pressure metamorphism. In the present study, we studied the geochemistry, P-T evolution, and Lu-Hf geochronology of the same sample PIS1 and two other eclogites from the Balma Unit to put additional constraints on the origin and metamorphic history of these rocks.

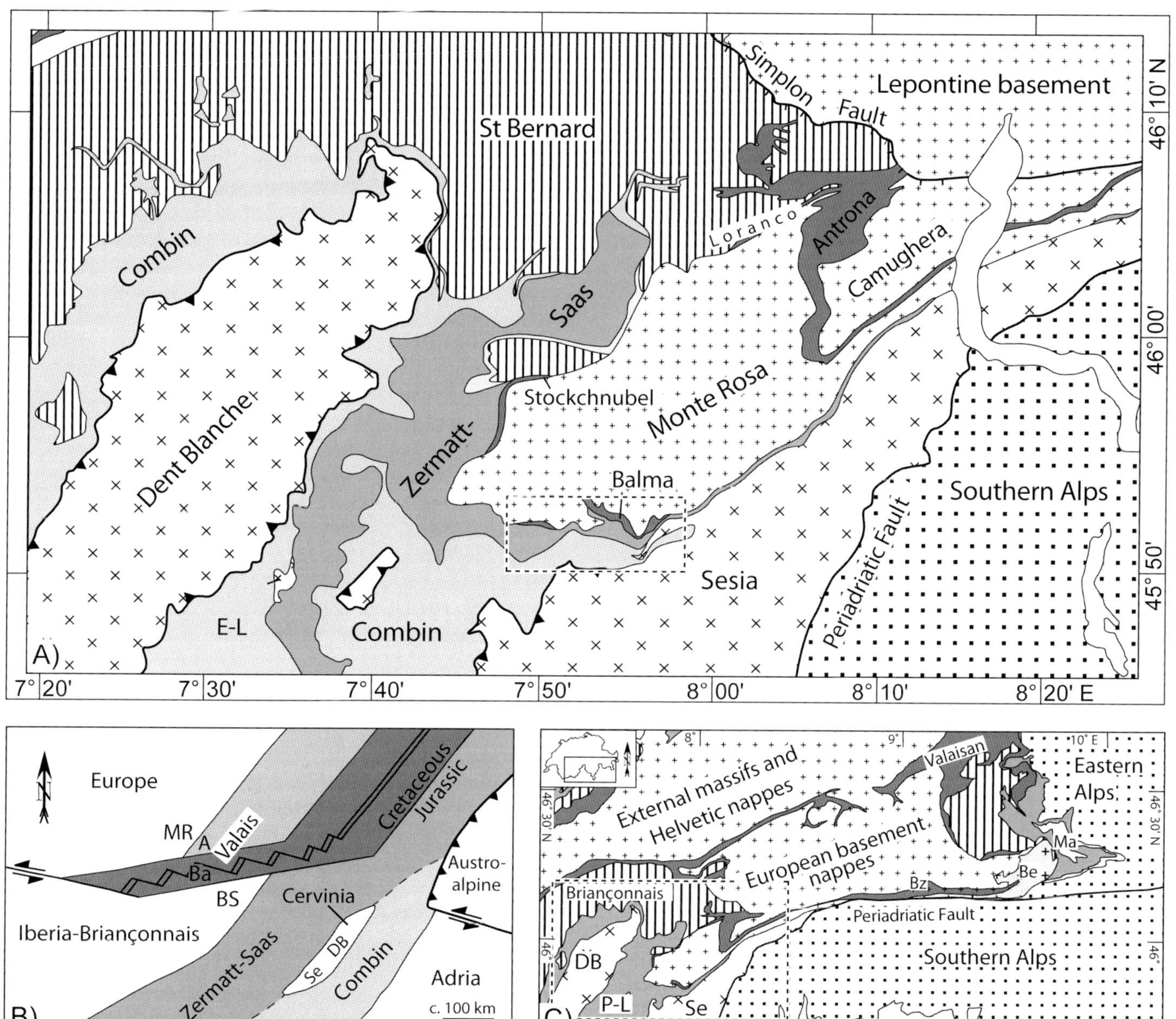

Fig. 2. (A) Tectonic map of the Penninic nappe stack between the Sesia-Dent-Blanche nappe system and the Simplon line, modified after Steck et al. (1999). E-L Etirol-Levaz sliver. (B) Paleogeographical sketch map of continental and oceanic domains for early Late Cretaceous time. MR Monte Rosa, A Antrona, Ba Balma, BS St. Bernard and Stolemberg, DB Dent Blanche, Se Sesia. (C) Overview map of the Central Alps. Bz Bellinzona-Dascio Zone, Be Bergell Pluton, Ma Margna Nappe, P-L Piemont-Ligurian (South Penninic) ophiolites.

2. Sample description, petrography and P-T evolution

Three partially retrogressed eclogites, all sampled close to Alpe la Balma, NW of Alagna Valsesia (Figs. 3 and 4), have been investigated in detail. All three eclogites contain garnet, amphibole, clinozoisite, quartz, paragonite, albite, and rutile ± epidote, ± chlorite, ± zircon and ± opaques. Sample MR56 also displays a small amount of phengite allowing the application of the Grt-Cpx-Phg barometer (Ravna & Terry 2004). Although there is a weak retrograde overprint, the peak-pressure assemblage (Grt + Cpx1 + Pg Qtz + Rt + Am1 ± Phg) is well preserved in all samples investigated (Fig. 5).

Garnets are usually 0.5–3 mm in diameter and euhedral. They often have inclusions in their cores and wide, almost inclusion-free rims (Fig. 5a). Inclusions in garnet comprise clinopyroxene, amphibole, quartz, clinozoisite, rutile, sphene. There are also rare plagioclase inclusions, but it is not clear whether these are primary or products of retrograde reactions. Garnets display typical prograde compositional patterns with elevated grossular and spessartine components in the cores and a bell-

Table 1. Representative microprobe analyses of eclogite phases in wt% and p.f.u. *All Fe is calculated as FeO.

Sample: Mineral:	MR56 Grt (rim)	MR56 Grt (core)	MR56 Cpx 1	MR56 Am 1	MR56 Am 2	MR56 Ab	MR56 Pg	MR1 Grt (rim)	MR1 Grt (core)	MR1 Cpx 1	MR1 Cpx 2	MR1 Ab	PIS1 Grt (rim)	PIS1 Grt (core)
SiO_2	37.9	37.5	55.5	53.7	40.4	67.4	51.1	38.9	39.3	56.3	53.3	67.4	38.3	38.3
TiO_2	0.073	0.152	0.033	0.135	0.032	<DL	0.175	0.072	0.191	0.024	0.093	0.002	0.051	0.103
Al_2O_3	21.8	21.6	10.5	7.02	18.1	19.7	27.7	22.1	22.0	11.8	1.85	20.0	22.2	22.1
FeO^*	29.6	28.1	6.18	7.83	14.9	0.178	1.63	29.3	19.9	3.32	5.85	0.234	28.8	19.4
MnO	0.404	1.37	0.041	0.093	0.183	<DL	0.012	0.155	4.96	<DL	0.163	<DL	0.610	8.57
MgO	3.89	2.21	7.31	16.3	9.10	<DL	3.54	2.82	0.803	7.71	13.2	0.022	3.58	0.697
CaO	6.65	8.79	11.3	8.80	10.0	0.207	0.009	7.37	14.0	12.0	22.3	0.454	6.55	10.8
Na_2O	0.013	0.012	8.00	3.50	3.57	11.7	0.677	<DL	0.026	7.86	1.58	11.4	0.039	0.044
K_2O	0.029	0.016	<DL	0.172	0.743	0.061	9.725	0.040	0.043	0.014	<DL	<DL	<DL	<DL
Cr_2O_3	0.002	0.032	<DL	0.012	<DL	0.014	0.023	<DL	0.115	0.061	0.033	<DL	0.0000	0.030
Sum	100.28	99.82	98.86	97.42	96.35	99.27	94.57	100.77	101.18	99.10	98.32	99.47	100.04	100.12
Si	5.96	5.97	2.01	7.53	6.31	2.97	6.82	6.07	6.09	2.01	2.00	2.96	6.01	6.05
Ti	0.009	0.018	0.001	0.014	0.004	0.00	0.018	0.008	0.022	0.001	0.003	0.000	0.006	0.012
Al	4.04	4.05	0.449	1.16	3.33	1.03	4.36	4.07	4.01	0.497	0.081	1.04	4.15	4.12
Fe	3.90	3.74	0.187	0.919	1.95	0.007	0.182	3.83	2.58	0.099	0.183	0.009	3.78	2.55
Mn	0.054	0.185	0.001	0.011	0.024	0.000	0.001	0.021	0.651	0.000	0.005	0.000	0.081	1.15
Mg	0.912	0.524	0.395	3.41	2.12	0.000	0.703	0.655	0.185	0.409	0.735	0.001	0.837	0.164
Ca	1.12	1.50	0.438	1.32	1.67	0.010	0.001	1.23	2.32	0.456	0.894	0.021	1.10	1.83
Na	0.004	0.004	0.562	0.951	1.08	0.998	0.175	0.000	0.008	0.542	0.115	0.976	0.012	0.014
K	0.006	0.003	0.00	0.031	0.148	0.003	1.65	0.080	0.008	0.001	0.000	0.000	0.000	0.000
Cr	0.000	0.004	0.00	0.001	0.000	0.000	0.000	0.000	0.014	0.002	0.001	0.000	0.000	0.004
Sum	16.01	15.99	4.04	15.36	16.64	5.02	13.90	15.89	15.87	4.01	4.01	5.01	15.94	15.89
O	24.00	24.00	6.00	23.00	23.00	8.00	22.00	24.00	24.00	6.00	6.00	8.00	24.00	24.00

Mg-Ca-Na-O system). There are currently no widely accepted solution models available for clinopyroxene and especially for amphibole. For both minerals, we have used an arbitrary ideal solution model. Further limitations arise from the fact that individual domains of the sample obviously equilibrated at different metamorphic stages, calling into question the use of bulk rock XRF major element data as a viable input composition. Nevertheless, the equilibrium phase diagram allows some conclusions about the P-T evolution. For the given bulk composition, the observed peak-pressure assemblage (Grt + Cpx1 + Qtz + Pg + Am) is predicted to be stable over a fairly large P-T range (500–700 °C and 12–20 kbar). Assuming that the pressure range inferred from barometry is correct, peak conditions should be around 550–600 °C and 17–19 kbar. The stability field of the peak-pressure assemblage is limited towards higher temperatures through the breakdown of paragonite which should decompose to form kyanite and clinopyroxene1 at higher pressures (assemblage 1 in Fig. 7) and plagioclase and amphibole at lower pressures (assemblage 3). The locations of the paragonite breakdown reactions in P-T space are relatively insensitive to the bulk composition and are in any case associated with dehydration. Thus we consider the presence of paragonite to be a robust constraint on P and T conditions. Hence, the equilibrium phase diagram indicates that pressures did not exceed 20 kbar and that decompression was associated with cooling, a conclusion that is also supported by the compositions of retrograde amphiboles and feldspar. The equilibrium phase diagram fur-

ther predicts the appearance of clinozoisite at conditions below 500 °C and 12 kbar. The peak P-T conditions proposed here agree well with independent P-T calculations on sample PIS1 (500–590 °C and 13–14.5 kbar minimum P) by Liati & Froitzheim (2006).

Some studies have inferred a reheating of samples from the southern flank of the Monte Rosa massif based on late generations of hornblende and plagioclase (e.g., Alta Luce, Borghi et al. 1996). In our samples, we observe the hornblende rims but could not identify Ca-bearing plagioclase. Our calculations predict decreasing glaucophane components in amphibole towards lower pressures even during cooling along the proposed P-T path. Given the limited coverage of this study, we do not exclude the possibility of a reheating event at low pressures. However, the persistence of large paragonite crystals, which were apparently part of the peak metamorphic assemblage, should constrain the temperatures to less than 500 °C at pressures below 10 kbar. Along the P-T path, garnet is most abundant at peak-pressure conditions. Because the chemical zonation of garnet is of a typical prograde nature, we propose that garnet growth in our samples occurred as pressures and temperatures increased during subduction to peak pressure conditions.

3. Analytical methods

Samples were crushed in a steel mortar and divided into two splits. One split was powdered in an agate mill and used for major- and trace element analyses. The second split was used for mineral separation after the clay-size fraction was removed. Following separation of a garnet-rich fraction by a Franz LB-1 magnetic separator, visibly inclusion free garnet separates were hand picked mostly from the 128–180 µm size range. To avoid biasing the bulk garnet samples toward cores or rims, magnetic separator settings were chosen such that only a few impure garnet grains remained in the 'garnet-poor' fraction. However, because inclusion-free garnet fragments were preferred during handpicking, a bias towards the inclusion-poor garnet rims relative to the inclusion-rich garnet cores is likely. Sample PIS1 had already been crushed at ETH-Zürich and was used for zircon separation (Liati & Froitzheim 2006). From this sample, only the 63–128 µm size fraction was available for garnet picking. Garnet separates (100–150 mg) were cleaned in an ultrasonic bath with deionised water and dried. Whole-rock analyses were performed on representative sample powders.

All samples were spiked with mixed ^{176}Lu-^{180}Hf and ^{149}Sm-^{150}Nd tracers before digestion. Two different digestion procedures were applied. (1) The rutiles and one set of whole rock powders were digested in steel-jacketed PARR bombs with HF-HNO$_3$-HClO$_4$ for 24 hours at 180 °C. (2) To selectively dissolve the garnet fractions without digesting microscopic grains of Hf-bearing phases such as zircon, a tabletop procedure was applied, whereby samples are digested with HF-HNO$_3$-HClO$_4$ in closed Teflon vials on 120 °C hotplates as described in Lagos et al. (2007). Both digestion methods continue with samples being dried down on a hotplate, evaporating virtually all of the HClO$_4$, and re-dissolving in 6 M HCl. In most cases, the digestion procedure had to be repeated at least once to achieve a visibly clear sample solution. At this point, the sample was assumed to be fully equilibrated with spike. To screen for the presence of inherited zircon, which would affect the whole-rock Hf compositions, an additional set of whole rock samples was run through the tabletop digestion procedure. Sufficient amounts of significantly older inherited zircon would cause the bombed whole rock fraction to lie below – rather than on – a table-top digested whole rock – garnet isochron. All sample solutions were dried down, re-dissolved in 2.5 M HCl and centrifuged to remove any newly-formed precipitates or undigested minerals prior to loading onto cation exchange columns. A single-column separation procedure using Eichron Ln-Spec resin was used to separate Lu and Hf from the rock matrix (Münker et al. 2001). Apart from the Lu and Hf cuts, a matrix cut that included the LREE and MREE was collected. Two further column separation steps were carried out to purify Sm and Nd. The LREE and MREE were first separated from the matrix elements by using cation exchange resin (AG 50 W ×12, 200–400 mesh) and 2.5 M HCl. The resulting REE fraction was collected in 6 M HCl. Samarium and Nd were then purified on a third column using HDEHP- coated Teflon beads after Richard et al. (1976).

Lutetium and Hf measurements were carried out in static mode using the Micromass Isoprobe MC-ICPMS at Universität Münster. Mass bias for Hf was corrected by using ^{179}Hf/^{177}Hf of 0.7325 (Patchett & Tatsumoto 1980) and the exponential law.

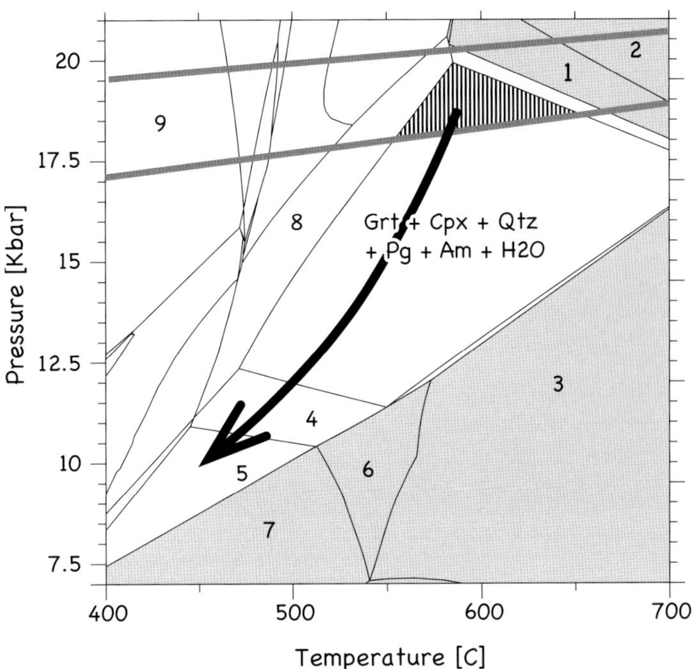

Fig. 7. Equilibrium phase diagram (De Capitani 1994) calculated for the bulk composition of sample MR56. Chemical system is Si-Al-Fe-Mg-Ca-Na-O. Light grey lines indicate range of Grt-Cpx-Phg barometry in the same sample. Dark arrow indicates inferred P-T path. Grey shaded area indicates paragonite-free assemblages posing a robust constraint on the P-T path of the sample. Black and white shaded area indicates constraints on peak P-T conditions. Selected assemblages: (1) Grt + Cpx1 + Am + Qtz + Ky + H$_2$O, (2) Grt + Cpx + Qtz + Ky + H$_2$O, (3) Grt + Fsp + Cpx + Am + Qtz + H$_2$O, (4) Grt + Cpx + Qtz + Pl + Am + Czo + H$_2$O, (5) Cpx + Qtz + Pl + Am + Czo + H$_2$O, (6) Grt + Fsp + Cpx + Czo + Am + Qtz + H$_2$O, (7) Fsp + Cpx + Am + Qtz + H$_2$O, (8) Grt + Cpx + Qtz + Pl + Am + Lws + H$_2$O, (9) Cpx + Chl + Lws + Qtz + H$_2$O. Mineral abbreviations are after Kretz (1983); except: Am = amphibole.

Measured ^{176}Hf/^{177}Hf values are reported relative to ^{176}Hf/^{177}Hf = 0.282160 for the Münster Ames Hf standard, which is isotopically identical to the JMC-475 standard. For the purpose of plotting isochrons, the external reproducibility of analyses was estimated using the empirical relationship 2σ (external 2 s.d.) = ~2 × 2σ$_m$ (internal 2 s.e. run statistic) for replicate measurements of different concentrations of Hf standard solutions (see Bizzarro 2003). For interference corrections on ^{176}Hf and ^{180}Hf, the ^{173}Yb, ^{175}Lu, ^{181}Ta, and ^{182}W signals were monitored. For Lu measurements, mass bias correction and correction of the ^{176}Yb interference was achieved by monitoring the naturally occurring Yb in the Lu cuts and using the trend defined by ln(^{176}Yb/^{171}Yb) vs. ln(^{173}Yb/^{171}Yb) of Yb standard analyses that were interspersed with samples during the run sessions (e.g., Blichert-Toft et al. 2002; Albarède et al. 2004; Vervoort et al. 2004). This procedure typically results in an external reproducibility of ~0.2% (2σ) for the ^{176}Lu/^{177}Hf values of ideally spiked sample solutions. Blanks for Lu and Hf were <10 and <50 pg respectively.

Neodymium isotope ratios of whole rock samples were determined by MC-TIMS (Finnigan Triton) in Münster. The iso-

Table 2. XRF and LA-ICP-MS analyses for Balma eclogite samples PIS1, MR1, and MR56.

	MR1	MR56	PIS1
Major elements in weight percent oxides			
SiO_2	48.5	49.0	46.5
TiO_2	1.42	2.04	1.63
Al_2O_3	16.2	15.9	16.0
Fe_2O_3	9.62	10.8	15.0
MnO	0.18	0.18	0.28
MgO	7.84	6.56	6.62
CaO	10.6	10.6	10.9
Na_2O	3.46	3.08	2.47
K_2O	0.030	0.16	0.030
P_2O_5	0.302	0.186	0.067
SO_3	0.031	0.021	0.093
Trace elements (XRF) in ppm			
Sc	40	37	53
V	250	311	264
Cr	262	205	205
Co	40	39	47
Ni	133	96	83
Cu	39	49	18
Zn	141	83	83
Ga	12	22	19
L.O.I.	0.82	0.67	0.03
Total (in %)	99.09	99.28	99.82
Trace elements (LA-ICP-MS) in ppm			
Cs	0.044	0.11	0.67
Rb	0.33	2.4	0.22
Ba	10	6.6	7.3
Th	0.20	0.24	0.19
U	0.11	0.15	0.17
Nb	2.4	3.6	3.4
Ta	0.16	0.21	0.20
La	7.5	8.1	6.2
Ce	13	18	12
Nd	13	19	12
Zr	112	168	171
Hf	2.8	4.0	4.0
Pr	2.2	3.3	2.0
Sm	4.2	6.0	3.7
Eu	1.5	2.0	1.3
Gd	5.0	7.5	5.3
Tb	0.77	1.1	1.3
Ho	1.2	1.8	2.6
Y	29	45	67
Er	3.4	5.0	7.6
Tm	0.45	0.71	1.1
Yb	3.1	5.1	7.4
Lu	0.41	0.7	1.1

baric interference on ^{144}Nd was corrected by monitoring ^{147}Sm and using the natural $^{147}Sm/^{144}Sm$. Mass fractionation was corrected using a $^{146}Nd/^{144}Nd$ of 0.7219 and the exponential law. The typical reproducibility of $^{143}Nd/^{144}Nd$ values is ± 50 ppm. A value of 0.511847 was obtained for the LaJolla Nd standard during the course of this study.

X-ray fluorescence analyses of major and trace elements in whole rock samples were carried out at the Steinmann-Institut Bonn, and electron microprobe work was conducted at the Institut für Geologie und Mineralogie in Cologne. Further trace element abundances in whole rock samples were analysed by melting the samples with lithium tetraborate. The resulting fused disks were analysed directly using the Thermo Finnigan Element2 LA-SF-ICP-MS at the Max-Planck-Institut für Chemie in Mainz. Analyses of Nist 612 SRM reference glass measured together with the samples indicate, that the 95% confidence level of trace element concentration is usually better than 5–10% of the sample concentration (Jochum et al. 2007).

4. Results

4.1 Major and trace elements

Major and trace element data for the eclogite samples are given in Table 2. All samples have similar basaltic compositions with 46.4–49.0 wt% SiO_2, Al2O3 of 15.9–16.2 wt% and 6.56–7.84 wt% MgO. Titanium contents are high (1.42–2.04 wt%) and Zr contents are moderately high for mafic rocks (113–173 ppm). Chondrite-normalized REE patterns reveal that sample PIS1 is clearly depleted in LREE, whereas samples MR1 and MR56 have rather flat patterns with a slight enrichment in LREE (Fig. 8a). Extended trace element patterns normalized to primitive mantle are similar to those of typical N-MORB (Hofmann 1988). There appears to be evidence for U and LREE mobility, as indicated by a U-enrichment in all samples and a positive Zr-Hf anomaly in sample PIS1 (Fig 8b). Element concentrations are given in Table 2.

All measured garnets are typical almandine-rich, eclogitic garnets. Element profiles across representative garnets show characteristic zoning patterns of Fe, Mg, Mn, and Ca. Bell-shaped manganese profiles are generally interpreted to indicate original prograde growth zoning in garnet (Spear 1991; Kohn 2003). In addition to this observation, molar $Fe^{2+}/(Fe^{2+}+Mg^{2+})$ decreases from core to rim in all samples. Some garnets show slight enrichment of spessartine component at the outer garnet rim which could be interpreted either as the result of garnet dissolving on a retrograde PT path or by breakdown of a Mn-rich phase during a late stage of garnet growth.

4.2 Lu-Hf geochronology

Figure 9 illustrates the $^{176}Hf/^{177}Hf$ and Lu-Hf results for the three analysed samples in Lu-Hf isochron space. Isochron regressions were calculated using ISOPLOT v. 2.49 (Ludwig 2001), using the 2σ uncertainties in $^{176}Lu/^{177}Hf$ (Table 3), and 2σ uncertainties in $^{176}Hf/^{177}Hf$ estimated from the $2\sigma_m$ internal run statistics as previously described. Calculated ages are based on the decay constant of $\lambda^{176}Lu = 1.865 \times 10^{-11}$ yr^{-1} (Scherer et al. 2001; corroborated by Söderlund et al. 2004).

Hafnium contents in the whole rocks range from 0.4 to 1.3 ppm. The garnets have Hf contents of 30 to 93 ppb and

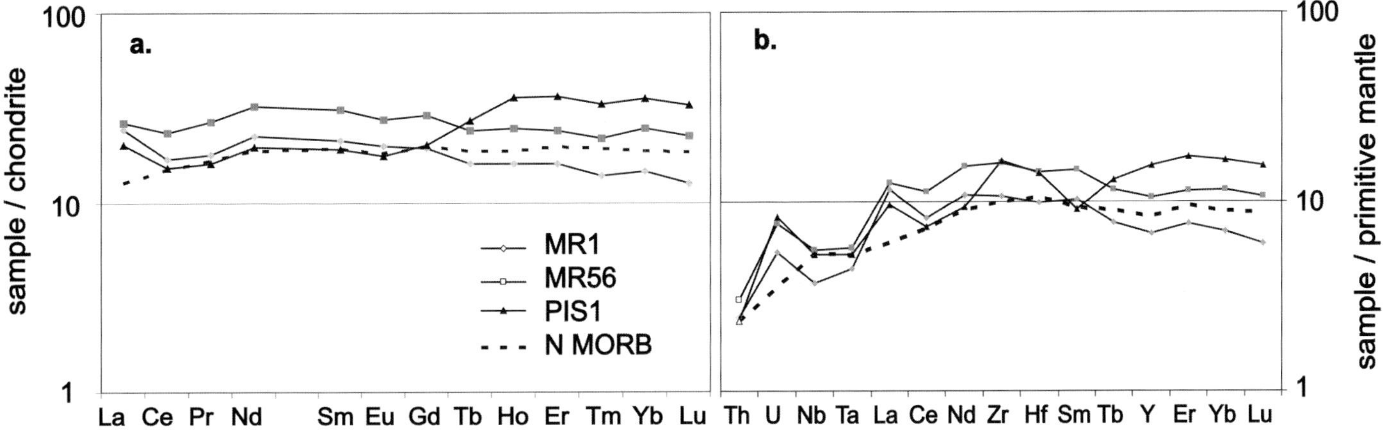

Fig. 8. (a) Chondrite-normalized REE diagram (values from Boynton 1984), showing compositions of the Balma eclogites in comparison to those of N-MORB (values are taken from Hofmann 1988). (b) Incompatible trace element diagram, normalized to primitive mantle (McDounough & Sun 1995).

Table 3. Lu and Hf concentrations and Hf-Nd isotope data for Balma eclogite samples PIS1, MR1 and MR56. 'Whole rock 1' fractions were digested via the selective tabletop digestion procedure; 'whole rock 2' fractions were digested in bombs. Uncertainties on the last decimal places (in parentheses) are estimated 2σ external reproducibility for $^{176}Lu/^{177}Hf$ and $^{147}Sm/^{144}Nd$, and $2\sigma/\sqrt{n}$ internal run statistics for $^{176}Hf/^{177}Hf$ and $^{143}Nd/^{144}Nd$. For the purpose of plotting Lu-Hf isochrons, the 2σ uncertainties on $^{176}Hf/^{177}Hf$ are estimated using the method of Bizzarro et al. (2003). See text for details. The 93 Ma zircon protolith age (Liati & Froitzheim 2006) is used to calculate $\varepsilon Hf(t)$ and $\varepsilon Nd(t)$.

fraction	ppm Lu	ppm Hf	$^{176}Lu/^{177}Hf$	$^{176}Hf/^{177}Hf$	$\varepsilon Hf(0)$	$\varepsilon Hf(t)$
sample: PIS1						
rutile 1	0.0212	2.02	0.001491 (8)	0.283242 (20)		
rutile 2	0.0162	0.326	0.00706 (2)	0.282980 (32)		
whole-rock 1	0.597	0.441	0.1922 (3)	0.283314 (25)	19.2	10.7 (9)
whole-rock 2	0.977	1.19	0.1170 (2)	0.283230 (8)	16.2	11.8 (5)
garnet 1	1.88	0.113	2.361 (4)	0.284977 (25)		
garnet 2	2.04	0.101	2.858 (5)	0.285395 (21)		
garnet 3	1.68	0.0927	2.576 (5)	0.285190 (22)		
sample: MR1						
rutile 1	0.0158	1.11	0.00201 (1)	0.283258 (23)		
rutile 2	0.0108	0.260	0.00587 (2)	0.283448 (57)		
whole-rock 1	0.492	0.400	0.1747 (3)	0.283325 (15)	19.5	12.0 (5)
whole-rock 2	0.410	0.872	0.06671 (13)	0.283190 (8)	14.8	13.0 (5)
garnet 1	1.26	0.0433	4.128 (8)	0.286436 (56)		
garnet 2	1.30	0.0321	5.752 (10)	0.287797 (22)		
garnet 3	1.28	0.0297	6.126 (11)	0.288053 (47)		
sample: MR56						
rutile	0.0162	0.208	0.01104 (3)	0.283015 (37)		
whole-rock 1	0.620	0.433	0.2031 (4)	0.283294 (16)	18.5	9.5 (11)
whole-rock 2	0.659	1.33	0.07025 (15)	0.283179 (8)	14.4	12.5 (6)
garnet 1	1.85	0.0534	4.922 (9)	0.287246 (36)		
garnet 2	1.78	0.0465	5.437 (10)	0.287747 (21)		
garnet 3	1.85	0.0358	7.344 (14)	0.289348 (36)		

fraction	ppm Sm	ppm Nd	$^{143}Nd/^{144}Nd$	$^{147}Sm/^{144}Nd$	$\varepsilon Nd(0)$	$\varepsilon Nd(t)$
sample: PIS1						
whole-rock 1	4.548	14.27	0.513001 (158)	0.1927 (4)	7.1	7.1 (31)
whole-rock 2	3.448	10.489	0.513088 (10)	0.1990 (4)	8.8	8.8 (2)
sample: MR1						
whole-rock 1	3.787	11.47	0.513095 (6)	0.1996 (4)	8.9	8.9 (2)
whole-rock 2	3.863	11.67	0.513093 (9)	0.2001 (4)	8.9	8.9 (2)
sample: MR56						
whole-rock 1	4.914	14.76	0.513094 (12)	0.2014 (4)	8.9	8.9 (2)
whole-rock 2	5.549	16.62	0.513095 (10)	0.2019 (4)	8.9	8.9 (2)

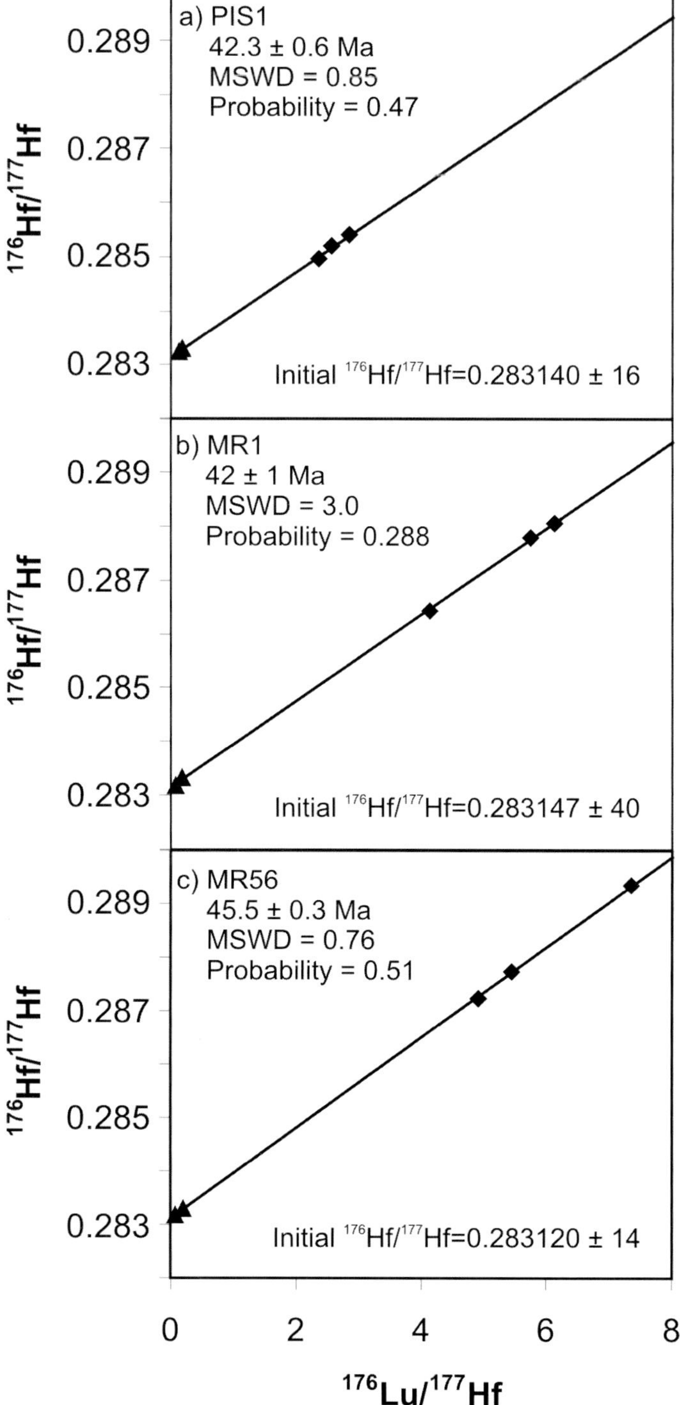

Fig. 9. Lu-Hf garnet (diamonds) whole rock (triangle) isochron plots for samples a) PIS1, b) MR1 and c) MR56. 2σ uncertainties used in regressions are always smaller than symbol sizes. Calculated initial values and ages are based on λ^{176}Lu = 1.865 × 10^{-11} yr^{-1} (Scherer et al. 2001).

^{176}Lu/^{177}Hf ranging from 2.36 to 7.34. For each sample, both whole rock splits and all three garnet separates measured define isochrons, suggesting that sample-spike equilibration was achieved during the selective digestion procedure used for garnets and whole rock split. The two rutile separates analysed for each sample however, do not plot on these isochrons, possibly indicating the lack of full isotopic equilibrium between the rutile fraction and their host rocks at the time of their crystallization or that the rutile crystallized at a different time than garnet. The Lu-Hf garnet-whole-rock ages that have been obtained for eclogites from the Balma unit (Fig. 9) are 42.3 ± 0.6, 42 ± 1 and 45.5 ± 0.3 Ma. These ages are younger than all Lu-Hf ages established so far for South Penninic Units (Fig. 10).

5. Discussion

5.1 Significance of Lu-Hf Ages

If Lu-Hf garnet geochronology is to be applied to determine prograde growth ages, the following criteria have to be met: (1) garnet rims must have been in isotopic equilibrium with the whole rock matrix during garnet growth, (2) the garnets must have remained closed systems for Lu and Hf since their formation, and (3) the garnet separates analysed must not contain any significantly older inherited components, such as zircon (e.g., Scherer et al. 2000). The Lu-Hf closure temperature (Tc) after Dodson (1973) depends on multiple factors such as peak temperature, cooling rate, mineral composition, and grain size and shape. The latter, along with presence of inclusions and their distribution, affect the effective diffusion radii. Only one direct estimate for Tc of Lu-Hf isotopic system in garnets has been proposed: about 720–755 °C for rapid cooling rates (Skora et al. 2006b). Most other estimates were made relative to the Tc of the Sm-Nd isotope system in garnet (e.g., Scherer et al. 2000; Lapen et al. 2003). These authors predict that Tc for the Lu-Hf system is higher than for the Sm-Nd isotope system in the same garnet. Numerous Tc estimates for Sm-Nd in garnet have been published (Jagoutz 1988; Cohen et al. 1988; Mezger et al. 1992; Burton et al. 1995; Hensen & Zou 1995; Becker 1997; Scherer et al. 2000; Van Orman et al. 2002; Thoeni 2002; Tirone et al. 2004), ranging between 490 °C for 0.2 mm size garnets and low cooling rates to 1050 °C for 3 mm size garnets and high cooling rates. The garnet porphyroblasts studied here have experienced peak temperatures of only 550–600 °C and are therefore unlikely to have undergone Lu or Hf exchange with their surroundings after their growth. This, together with the preserved prograde growth zoning in Mn^{2+}, which, according to Van Orman et al. (2002), would diffuse faster than 3$^+$ ions (e.g., Lu) and probably also 4$^+$ ions (e.g., Hf), suggests that the Lu-Hf ages reflect garnet growth rather than cooling ages.

The effects of trace mineral inclusions on Lu-Hf garnet geochronology were investigated by Scherer et al. (2000). Hafnium-rich inclusions, such as zircon and rutile, can lower the Lu/Hf and ^{176}Hf/^{177}Hf of a bulk garnet separate relative to those of pure garnet. Zircon is the most problematic of these minerals for Lu-Hf garnet dating because it may contain a significantly older inherited component that did not isotopically equilibrate with the whole rock at the time of garnet nucleation. In such a case, measured ^{176}Lu/^{177}Hf and ^{176}Hf/^{177}Hf values would lie on

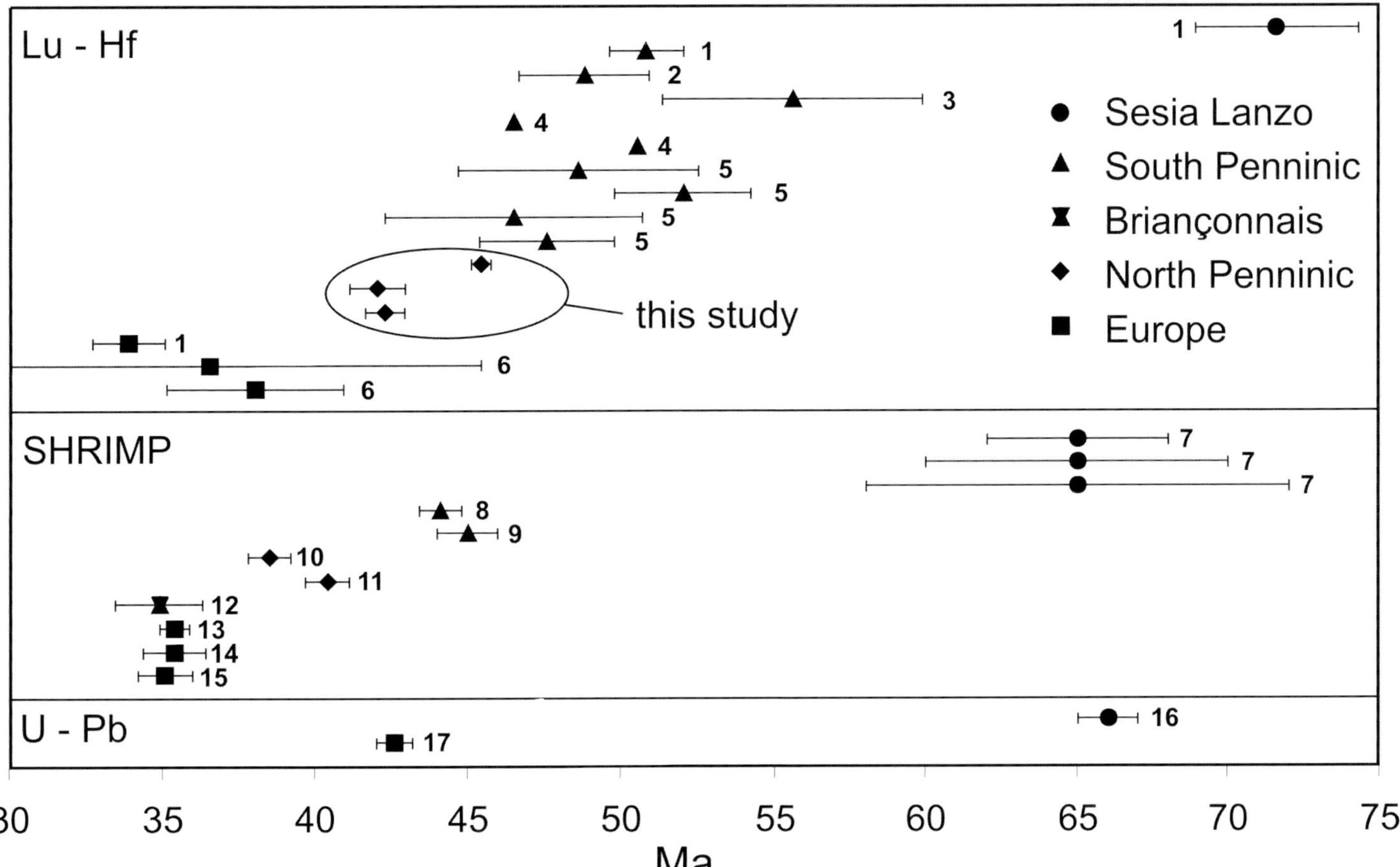

Fig. 10. Ages of HP and UHP metamorphism established in the western and central Alps over the last 10 years by Lu-Hf garnet, U-Pb SHRIMP-zircon, and U-Pb-rutile dating (1) Duchêne et al. (1997) recalculated with the decay constant of Scherer et al. (2001). (2) Lapen et al. (2003), (3) Mahlen et al. (2003), (4) Mahlen et al. (2005), (5) Mahlen et al. (2006), (6) Brouwer et al. (2005), (7) Rubatto et al. (1999), (8) Rubatto et al. (1998), (9) Rubatto & Hermann (2003), (10) Liati et al. (2005), (11) Liati & Froitzheim (2006), (12) Rubatto & Gebauer (1999), (13) Gebauer et al. (1996), (14) Gebauer et al. (1997), (15) Rubatto & Hermann (2001), (16) Inger et al. (1996), (17) Lapen et al. (2007). The Lu-Hf ages of 36.6 ± 8.9 and 38.1 ± 2.9 Ma (Brouwer et al. 2005) come from Alpe Arami and Gorduno, respectively. Further Lu-Hf garnet ages by Brouwer et al. (2005) are not discussed, as their paleogeographic origin is difficult to interpret and/or they contain older components.

mixing lines between pure garnet and zircon and whole rock (minus its zircon) and zircon, thus potentially producing an age bias. Zircon inclusions are commonly submicroscopic, and even the most carefully handpicked garnet separates may contain them. Rutile is part of the high pressure assemblage but may also carry an inherited Hf isotope signature, perhaps in the form of zircon inclusions. This might be the case for the Balma samples as some of the rutile separates plot below their respective isochrons. Other rutile separates plot above the garnet-whole rock isochrons, suggesting that their Hf may be derived from the breakdown of more radiogenic minerals. In the case of initial isotopic equilibrium among all mineral phases, the resulting lower Lu/Hf of zircon- or rutile-bearing garnet separates merely affect the precision of the age calculated and not its accuracy. To minimize inclusion effects, (1) the separates were carefully hand picked and (2) a selective tabletop digestion procedure was applied, as previously described. For all three cases, all garnet separates plot on isochrons with both table-top- and bomb-digested whole rock fractions, suggesting that full sample-spike equilibrium was achieved and that any zircon present contains little if any significantly older inherited Hf component.

Despite the fact that all three samples originate from the same tectonic unit, the calculated Lu-Hf age of MR56 clearly differs from that of the other two samples. The cause of this age difference may include one or both of the following explanations: (1) If the garnet porphyroblasts grew over a long time interval (several million years), it is possible that different core-to-rim Lu distributions would bias Lu-Hf ages of different samples towards different parts of the growth intervals. Such Lu zoning can be produced by Rayleigh fractionation during garnet growth (Lapen et al. 2003) or by diffusion-limited garnet growth (Skora et al. 2006). Modelling by Skora et al. (2006) suggests that at low temperatures, diffusion-limited garnet growth might even lead to a situation where Lu concentrations in garnet rims are in fact higher than in garnet cores.

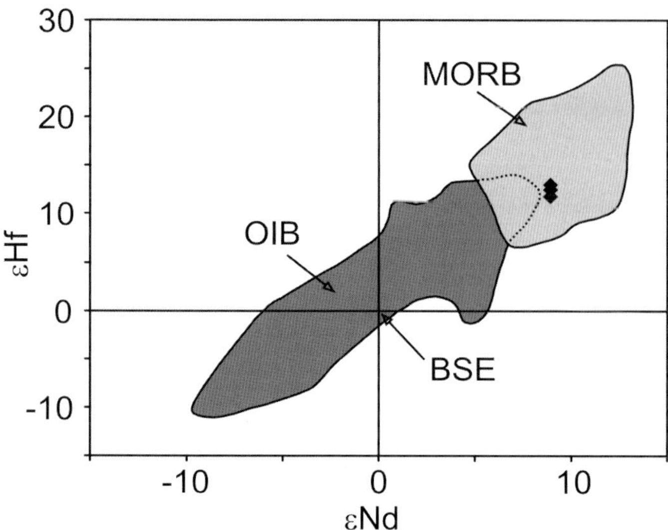

Fig. 11. Initial Hf-Nd isotope compositions (t at 93 Ma) of eclogites from the Balma Unit compared to compositions of MORB and OIB (Salters & White 1998; Chauvel & Blichert-Toft 2001 and references therein). Whole rock samples plotted here were all dissolved in PARR Bombs. 2σ errors are approximately symbol size. BSE values for Hf are taken from Blichert-Toft & Albarède (1997); values for Nd from (Jacobsen & Wasserburg 1980).

In the absence of trace element profiles through the garnets in this study, neither of these two explanations can be excluded. (2) As inclusion-free garnet fragments were preferred during hand picking, there might be a systematic bias towards inclusion-poor rim material. We therefore tentatively interpret the garnet ages to be weighted towards the later part of the garnet growth interval.

Even with the small age difference among the three eclogites, all of the Lu-Hf ages are consistently older than the U-Pb SHRIMP age of a metamorphic zircon rim in sample PIS1 (40.4 ± 0.7 Ma, Liati & Froitzheim 2006). The offset of two to five Myr between the two chronometers is not unusual and can be seen in almost all Alpine units (Fig. 10), except for the Dora-Maira Massif, where coesite-bearing quartzites, rather than eclogites, were used for zircon dating (Gebauer et al. 1997; Duchêne et al. 1997). In addition, the Lu-Hf age established by Duchêne et al. (1997) for Dora Maira was based on only a single two-point isochron. The consistent age differences between U-Pb zircon ages and Lu-Hf ages of garnet growth indicate that zircon growth most likely postdates garnet nucleation, which occurred during subduction-driven prograde metamorphism.

5.2 Characterisation of the protolith

An assessment of the tectonic setting (MORB vs. OIB origin) of the Balma metabasites (especially sample PIS1) can contribute to the paleogeographic assignment of the Balma unit.

Rare earth element patterns presented in Fig. 8a show a consistent depletion of LREE relative to HREE, suggesting a MORB origin. However, a depletion of LREE might also reflect fluid-rock interaction due to subduction-related dehydration

(e.g., John et al. 2004). Light REE and U are likely to be mobilized, whereas HREE, Nb, Ta, Zr, and Hf are often relatively immobile. This appears to be the case for sample PIS1, which is strongly depleted in LREE and shows a markedly positive Zr-Hf anomaly. In contrast, samples MR1 and MR56 do not show Zr-Hf anomalies and exhibit flat REE patterns. Hence, the combined trace element evidence from the three eclogite samples suggests an origin of the eclogites from MORB protoliths. Consistently high Zr/Nb ratios (46–50) also confirm a depleted mantle source.

Initial Hf and Nd isotope ratios obtained for the whole rocks clearly plot inside the field of MORB (Salters & White 1998; Chauvel & Blichert-Toft 2001 and references therein, Fig. 11). Even though trace element patterns show that Nd may have been mobile during subduction, the εNd and εHf values of all three samples broadly overlap, indicating that the Nd isotope composition was not significantly disturbed. Because zircon is not effectively dissolved by the table top digestion procedure, only Hf data from whole rocks digested in PARR-bombs were plotted in Fig. 11.

Collectively, the combined trace element and isotope evidence strongly suggest that the protolith for the studied eclogites was a MORB, thus precluding the possibility that the Cretaceous protolith age of sample PIS1 (Liati & Froitzheim 2006) reflects late ocean island magmatism.

5.3. Previous geochronological results from the Penninic nappes

Ophiolite protolith ages: Jurassic protolith ages of ca. 164 Ma (U-Pb SHRIMP on zircon) were determined for metagabbros of the Zermatt-Saas Zone by Rubatto et al. (1998). All protolith ages from South Penninic ophiolites in the Central and Western Alps range between ca. 142 and ca. 166 Ma (e.g., Kaczmarek et al. 2008; review of data in Liati et al. 2003). Jurassic SHRIMP ages of ca. 155, 158, and 156 Ma, were also determined for metagabbro, amphibolitized eclogite, and amphibolite of the Antrona Unit (Fig. 1, 2) by Liati et al. (2005). The Antrona ages were interpreted to indicate that this unit is a piece of South Penninic Ocean floor that was geometrically captured in the North Penninic Basin by sinistral movement (Fig. 2; Liati et al. 2005). As mentioned above, eclogite sample PIS1 from the Balma Unit yielded a Cretaceous protolith age of 93.4 ± 1.7 Ma (Cenomanian-Turonian; Liati & Froitzheim 2006). The age from the Balma Unit is identical to two ages (93.0 ± 2.0 and 93.9 ± 1.8 Ma, Liati et al. 2003) determined on amphibolites of the Chiavenna Ophiolite in the eastern Central Alps, for which an origin from the North Penninic Ocean is widely accepted (e.g., Schmid et al. 1996).

High-pressure metamorphism: Until ca. 1995; the high-pressure metamorphism in the Penninic Nappes was generally assumed to be Cretaceous in age, mostly based on K-Ar and ^{40}Ar-^{39}Ar geochronology (compromised by excess argon). However, the application of U-Pb, Sm-Nd, and Lu-Hf geochronology has shown that the high-pressure metamorphism is Tertiary in age, except for the Sesia Nappe, which already had experienced eclogite-facies metamorphism in the Latest Cretaceous. The ages decrease from the upper to the lower tectonic units, that is, from southeast to northwest in terms of paleogeography. More recent [40]Ar-[39]Ar work has confirmed the Tertiary ages (Dal Piaz et al. 2001; Agard et al. 2002; Bucher et al. 2003). Lutetium-Hf and U-Pb ages are compiled in Fig. 10 and discussed below with our tentative tectonic model (Fig. 12).

5.4 Tectonic implications

Our study provides additional support for a MORB origin of the protoliths. Together with the previously published ca. 93 Ma protolith age, this strengthens the evidence that oceanic spreading was still active in the Late Cretaceous. According to a plate tectonic model for the Alpine paleogeography (Stampfli & Borel 2004), spreading in the South Penninic Ocean took place in the Jurassic (Fig. 12a) and ceased by the Early Cretaceous (Barremian to Aptian), whereas the North Penninic Ocean opened in the Cretaceous (Fig 12b). Geochronological data from South Penninic ophiolites and biochronological data from their cover units indicate that spreading began during the Bajocian and ended in the Kimmeridgian (Bill et al. 2001). This reflects the northward propagation of the Atlantic opening past Iberia. In the Jurassic, the northern tip of the North Atlantic ridge was at the latitude of Gibraltar, and Atlantic opening was transferred into the South Penninic (Piemont-Ligurian) Ocean through a strike-slip zone south of Iberia. During the Early Cretaceous, Atlantic opening propagated west of Iberia and was transferred to the North Penninic (Valais) Ocean through a transtensional zone between Iberia and Europe. Therefore, our geochemical results, together with the Late Cretaceous protolith age, support the interpretation of the Balma Unit as North Penninic.

Figure 12 shows a tentative reconstruction of the paleotectonic evolution assuming that the Balma Unit is indeed North Penninic. In this case, the South Penninic and North Penninic oceans must have been consumed by two separate subduction zones. If only one subduction zone had existed, the paleogeographic units should have arrived in this subduction zone one after the other, starting with the South Penninic units and ending with the European margin. The presently available Lu-Hf data (Fig. 10), however, suggest that subduction of the North Penninic Ocean started when subduction of the South Penninic Ocean was still going on or had just ended. This leaves no time for the subduction of the lithosphere which carried the Briançonnais units. Our tentative model presented in Fig. 12 therefore includes two partly contemporaneously active subduction zones: one consuming the South Penninic and the other subducting the North Penninic Ocean. The Lu-Hf ages presented here, as well as all other published U-Pb SHRIMP and Lu-Hf data for the western and central Alps, fit well into this model. As pointed out before and illustrated in Fig. 10, U-Pb SHRIMP ages seem to be offset relative to Lu-Hf garnet

ages by roughly 4 Myr. This is interpreted to indicate that metamorphic zircon crystallized after garnet had already nucleated along the prograde path. If our garnets are indeed biased by hand-picking towards inclusion free rims, this would place zircon growth in the latter part of – or even after – the garnet growth interval.

Eclogites from the Sesia Nappe yielded ages of 71.6 Ma using Lu-Hf (Duchêne et al. 1997 recalculated with the decay constant of Scherer et al. 2001) and of ca. 65 Ma using U-Pb on sphenes (Inger et al. 1996) and U-Pb SHRIMP on zircon (Rubatto et al. 1999) (Fig 12c). In units from the South Penninic Ocean (Zermatt-Saas Zone, Monviso Unit) eclogite-facies metamorphism has been dated at 60 ± 12 and 62 ± 9 Ma (Sm-Nd garnet ages by Cliff et al. 1998) (Fig. 12d). However, most Lu-Hf ages cluster around 49 Ma (Duchêne et al. 1997 recalculated with the decay constant of Scherer et al. 2001; Lapen et al. 2003; Mahlen et al. 2003, 2005, 2006) (Fig 12e), and U-Pb SHRIMP ages are again slightly younger (44.1 ± 0.7 and 45 ± 1 Rubatto et al. 1998; Rubatto & Hermann 2003) (Fig. 12f). Lutetium-Hf ages of 45.5 ± 0.3, 42 ± 1 and 42.3 ± 0.6 Ma (this study) for North Penninic Ophiolites (Fig 12f, g), suggest that subduction of the North Penninic Ocean started roughly at around 50 Ma (Fig 12e). Once again, U-Pb SHRIMP ages (37.1 ± 1.9, 38.5 ± 0.7 and 40.4 ± 0.7 Ma, Liati et al. 2003; 2005 and Liati & Froitzheim 2006) (Fig. 12h) are younger than the Lu-Hf ages, not only for the same unit but in this case even for the same sample: Sample PIS1 was analyzed by both methods and yields a U-Pb SHRIMP age of 40.4 ± 0.7 Ma (Liati & Froitzheim 2006) and a Lu-Hf age of 42.3 ± 0.6 Ma (this study). The first HP ages in the European margin, more precisely in the Monte Rosa Nappe, are 42.6 Ma (Lapen et al. 2007) using U-Pb in rutile (Fig. 12g). However, it must be stressed that the paleogeographic position of the Monte Rosa Nappe is controversial. A ca. 35 Ma U-Pb SHRIMP age has been published for the Gornergrat series (Rubatto & Gebauer 1999), which was attributed to the Monte Rosa Nappe by these authors, but according to our tectonic interpretation it rather belongs to the St. Bernard Nappe (Briançonnais). This age is therefore difficult to interpret. Lu-Hf garnet ages of 36.6 ± 8.9 and 38.1 ± 2.9 Ma (Brouwer et al. 2005) from the Adula Nappe (European margin) are in good agreement with 35.4 ± 0.5 Ma U-Pb SHRIMP ages from the same area determined by Gebauer et al. (1996) (Fig 12i). Similar ages were determined for UHP metamorphism in the Dora-Maira Unit (which also belongs to the European margin according to our interpretation): 35.4 ± 1 and 35.1 ± 0.9 Ma using U-Pb SHRIMP (Gebauer et al. 1997; Rubatto & Hermann 2001) and 34.1 ± 1.2 Ma using Lu-Hf (Duchêne et al. 1997 recalculated with the decay constant of Scherer et al. 2001). This is the only occasion where a Lu-Hf age (two point isochron) is not significantly older than associated U-Pb SHRIMP ages, which could be due to the fact that coesite-bearing quartzites rather than eclogites were analyzed (Gebauer et al. 1997; Duchêne et al. 1997). Fig. 12j shows our tentative model for the Western Alps, taking into account the slab extraction model after Froitzheim et al. (2003). Such an event could have triggered ca. 33–30 Ma

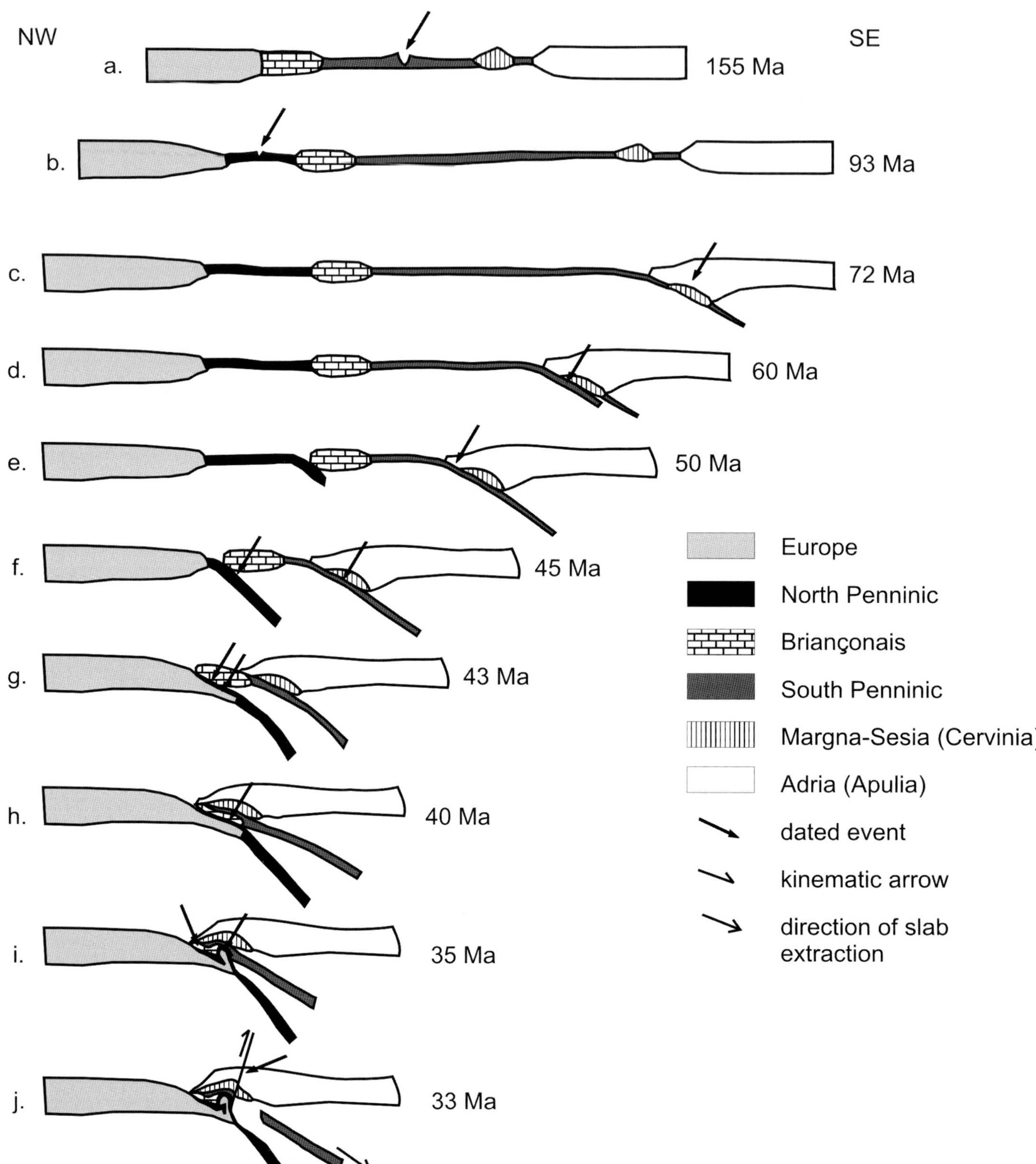

Fig. 12. Tentative model for the paleotectonic evolution of the western Alps, showing the spreading and the subduction of the South and North Penninic Oceans. All published Lu-Hf and U-Pb SHRIMP ages are consistent with this model as discussed in the text, and indicated by arrows. (j) is taking into account the slab extraction model after Froitzheim et al. (2003) (lower arrow). Such an event could have triggered ca. 33–30 Ma old magmatism along the Insubric lineament (kinematic arrow). Sketches have not been scaled for convergence rate or subduction depth.

old magmatism along the Insubric lineament (e.g. Oberli et al. 2004).

6. Conclusions

Our combined major- and trace element and Lu-Hf data for eclogites from the Balma Unit provide new constraints on the tectonic evolution of the western Alps. Previous models can now be modified as follows:

1) Trace elements and initial Hf-Nd isotope data strongly suggest that MORB-type ocean crust provides the protolith for the Balma eclogites. Consequently, the Cretaceous U-Pb SHRIMP age of synmagmatic zircon cores established from sample PIS1 (Liati & Froitzheim 2006) does not relate to late OIB magmatism in the South Penninic Ocean. Rather, the Balma unit was apparently derived from the North Penninic Ocean because that ocean was still actively spreading at the time.
2) Lutetium-Hf garnet ages obtained for three eclogites from the Balma unit range from 42 to 45.5 Ma and are systematically younger than those reported for South Penninic units (46.5–55.6 Myr).
3) In most cases Lu-Hf ages for eclogites in the Alps are consistently two to six Myr older than U-Pb SHRIMP ages for the same units. For example, the Lu-Hf garnet age of PIS1 is ca. 2 Myr older than the U-Pb SHRIMP age determined for metamorphic zircon in the same sample. This may reflect garnet growth during prograde metamorphism and zircon growth at or after the peak pressure conditions.
4) Our tentative paleotectonic model suggests simultaneous subduction of two oceanic basins in the Western Alps.

Acknowledgments

We thank reviewers Jan Kramers and Alfons Berger as well as guest editor Stefan Schmid for their constructive and very helpful criticism. We would also like to thank Klaus Peter Jochum and Brigitte Stoll for supervising the LA-ICP-MS analyses. Work of N.F., T.J.N., and J.P. was supported by DFG project FR700/6. Special thanks goes to Angelika and Thomas Herwartz for their selfless support of my Education.

REFERENCES

Agard, P., Monié, P., Jolivet, L. & Goffé, B. 2002: Exhumation of the Schistes Lustrés complex: in situ laser probe ^{40}Ar/^{39}Ar constraints and implications for the Western Alps. Journal of Metamorphic Geology 20, 599–618.

Albarède, F., Telouk, P., Blichert-Toft, J., Boyet, M., Agranier, A., Nelson, B., 2004: Precise and accurate isotopic measurements using multiple-collector ICPMS. Geochemica et Cosmochemica Acta 68, 2725–2744.

Avigad, D., Chopin, C., Goffe, B. & Michard, A. 1993: Tectonic model for the evolution of the Western Alps. Geology 21, 659–662.

Becker, H. 1997: Sm-Nd garnet ages and cooling history of high-temperature garnet peridotite massifs and high-pressure granulites from lower Austria. Contributions to Mineralogy and Petrology 127, 224–236.

Bill, M., O'Dogherty, L., Guex, J., Baumgartner, P. O. & Masson, H. 2001: Radiolarite ages in Alpine-Mediteranean ophiolites: Constraints on the oceanic spreading and the Tethys-Atlantic connection. Geological Society of America Bulletin 113, 129–143.

Bizzarro, M., Baker, J.A., Haack, H., Ulfbeck, D. & Rosing, M. 2003: Early history of Earth's crust-mantle system inferred from hafnium isotopes in chondrites. Nature 421, 931–933.

Blichert-Toft J. & Albarède F. 1997: The Lu-Hf isotope geochemistry of chondrites and the evolution of the mantle-crust system. Earth and Planetary Science Letters 148, 243–258.

Blichert-Toft, J., Boyet, M., Télouk, P., Albarède, F. 2002: ^{147}Sm-^{143}Nd.and ^{176}Lu-^{176}Hf in eucrites and the differentiation of the HED parent body. Earth and Planetary Science Letters 204, 167–181.

Borghi, A., Compagnoni, R. & Sandrone, R. 1996: Composite P-T paths in the internal Penninic Massifs of the western Alps: Petrological constraints to their thermo-mechanical evolution, Eclogae geologicae Helvetiae 89, 345–367.

Bousquet, R., Engi, M., Gosso, G., Oberhänsli, R., Berger, A., Spalla, M.I., Zucali, M. & Goffé, B. 2004: Explanatory notes to the map: Metamorphic structure of the Alps Transition from the Western to the Central Alps. Mitteilungen der Österreichischen Mineralogischen Gesellschaft 149, 145–156.

Boynton, W.V. (1984): Cosmochemistry of the rare earth elements; meteorite studies, in Henderson, P. (Ed.), Rare Earth Element Geochemistry, Elsevier, Amsterdam, 1984, pp. 63–102.

Brouwer, F.M., Burri, T., Engi, M., Berger, A. 2005: Eclogite relics in the Central Alps: PT – evolution, Lu-Hf ages and implications for formation of tectonic mélange zones. Schweizerische Mineralogische und Petrographische Mitteilungen 85, 147–174.

Bucher, S. 2003: The Briançonnais units along the ECORS-CROP transect (Italian-French Alps): structures, metamorphism and geochronology. Unpublished PhD thesis, University Basel, 175 pp.

Burton, K.W., Kohn, M.J., Cohen A.S. & O'Nions, R.K. 1995: The relative diffusion of Pb, Nd, Sr and O in garnet. Earth and Planetary Science Letters 133, 199–211.

Chauvel, C. & Blichert-Toft, J. 2001: A hafnium isotope and trace element perspective on melting of the depleted mantle. Earth and Planetary Science Letters 190, 137–151.

Cliff, R.A., Barnicoat, A.C., Inger, S. 1998: Early Tertiary eclogite facies metamorphism in the Monviso Ophiolite. Journal of Metamorphic Geology 16, 447–455.

Cohen, A.S., O'Nions, R.K., Siegenthaler, R. & Griffin, W.I. 1988: Chronology of the pressure-temperature history recorded by a granulite terrain. Contributions to Mineralogy and Petrology 98, 303–311.

Dal Piaz, G. V., Cortiana, G., Del Moro, A., Martin, S., Pennacchioni, G. & Tartarotti, P. 2001: Tertiary age and paleostructural inferences of the eclogitic imprint in the Austroalpine outliers and Zermatt-Saas ophiolite, western Alps. Internationl Journal of Earth Sciences 90, 668–684.

De Capitani C. 1994: Gleichgewichts-Phasendiagramme: Theorie und Software. Beihefte zum European Journal of Mineralogy 72. Jahrestagung der Deutschen Mineralogischen Gesellschaft 6, 48.

Dodson, M.,H. 1973: Closure Temperature in Cooling Geochronological and Petrological Systems. Contributions to Mineralogy and Petrology 40, 259–274.

Duchêne, S., Blichert-Toft, J., Luais, B., Téluk, P., Lardeaux, J.M. & Albarède, F. (1997): The Lu-Hf dating of garnets and the ages of the Alpine high-pressure metamorphism. Nature 387, 586–589.

Escher, A., Masson, H. & Steck, A. 1993: Nappe geometry in the western Swiss Alps. Journal of Structual Geology 15, 501–509.

Frisch, W. 1979: Tectonic progranation and plate tectonic evolution of the Alps Tectonophysics 60(3–4), 121–139.

Froitzheim, N. & Manatschal, G. 1996: Kinematics of Jurassic rifting, mantle exhumation, and passive-margin formation in the Austroalpine and Penninic nappes (eastern Switzerland). Geological Society of American Bulletin 108, 1120–1133.

Froitzheim, N., Schmid, S.M. & Frey, M. 1996: Mesozoic paleogeography and the timing of eclogitfacies metamorphism in the Alps: A working hypothesis. Eclogae geologicae Helvetiae 89(1), 81–110.

Froitzheim, N. 2001: Origin of the Monte Rosa nappe in the Pennine Alps: A new working hypothesis. Geological Society of America Bulletin 113, 604–614.

Froitzheim, N., Pleuger, J., Roller, S. & Nagel, T. 2003: Exhumation of high and ultrahigh-pressure metamorphic rocks by slab extraction. Geology, 31, 925–928.

Gebauer, D. 1996: A P–T–t path for an (ultra?-) high-pressure ultramafic/mafic rock-association and its felsic country-rocks based on SHRIMP-dating of magmatic and metamorphic zircon domains. Example: Alpe Arami (Cetral Swiss Alps). In: Basu, A., Hart, S. (Eds.), Earth Processes: Reading the Isotopic Code. Geophys. Monogr. AGU. Washington DC 95, 307–329.

Gebauer, D., Schertl, H.P., Brix, M. & Schreyer, W. 1997: 35 Ma old ultrahigh-pressure metamorphism and evidence for very rapid exhumation in the Dora Maira massif, Western Alps. Lithos 41, 5 –24.

Hensen, B.J. & Zhou, B. 1995: A Pan-African granulite facies metamorphic episode in Prydz Bay, Antarctica: evidence from Sm-Nd garnet dating. Australian Journal of Earth Science 42(3), 249–258.

Hofmann, A.W. 1988: Chemical differentiation of the Earth: the relationship between mantle, continental crust and oceanic crust. Earth and Planetary Science Letters 90, 297–314.

Inger, R., Ramsbotham, W., Cliff, R.A. & Rex, D.C. 1996: Metamorphic evolution of the Sesia-Lanzo Zone, Western Alps: time constraints from multi-system geochronology. Contributions to Mineralogy and Petrology 126, 152–168.

Jacobsen, S.B., Wasserburg, G.J. 1980: Sm-Nd isotopic evolution of chondrites. Earth and Planetary Science Letters 50, 139–155.

Jagoutz, E. 1988: Nd and Sr systematics in an eclogite xenolith from Tanzania: Evidence for frozen mineral equilibria in the continental lithosphere. Geochimica et Cosmochimica Acta 52, 1285–1293.

Jochum K.P., Stoll B., Herwig K. & Willbold M. 2007: Validation of LA-ICP-MS trace element analysis of geological glasses using a new solid-state 193 nm Nd: YAG laser and matrix-matched calibration. Journal of Analytical Atomic Spectrometry 22(2), 112–121.

John, T., Schere, E.E., Haase, K. & Schenk, V. 2004: Trace element fractionation during fluid-induced eclogitization in a subduction slab: trace element and Lu-Hf-Sm-Nd isotope systematics. Earth and Planetary Science Letters 277, 441–456.

Kaczmarek, M.A., Müntener, O. & Rubatto, D. 2008: Trace element chemistry and U-Pb dating of zircons from oceanic gabbros and their relationship with whole rock composition (Lanzo, Italian Alps). Contributions to Mineralogy and Petrology 155, 295–312.

Keller, L.M. & Schmid, S.M. 2001: On the kinematics of shearing near the top of the Monte Rosa nappe and the nature of the Furgg zone in Val Loranco (Antrona valley, N. Italy): Tectonometamorphic and paleogeographical consequences. Schweizerische Mineralogische und Petrographische Mitteilungen 81, 347–367.

Kohn, M.J. 2003: Geochemical zoning in metamorphic minerals. In: Rudnik, R.L. (Ed.), The Crust. Treatise on Geochemistry 3, 229–261.

Kramer J., Abart, R., Müntener, O., Schmid, S.M., Stern, W. 2003: Geochemistry of metabasalts from ophiolitic and adjacent distal continental margin units: Evidence from the Monte Rosa region (Swiss and Italian Alps). Schweizerische Mineralogische und Petrographische Mitteilungen 83, 217–240.

Kretz, R. 1983: Symbols for rock-forming minerals. American Mineralogist 68, 277–279.

Lagos, M., Scherer, E.E., Tomaschek, F., Münker, C., Keiter, M., Berndt, J., Ballhaus, C. 2007: High precision Lu-Hf geochronology of Eocene eclogite-facies rocks from Syros, Cyclades, Greece. Chemical Geology 243, 16–35.

Lapen, T.J., Johnson, C.M., Baumgartner, L.P., Mahlen, N.J., Beard, B.L. & Amato, J.M. 2003: Burial rates during prograde metamorphism of an ultra-high-pressure terrane: an example from Lago di Cignana, Western Alps, Italy. Earth and Planetary Science Letters 215, 57–72.

Lapen, T.J., Johnson, C.M., Baumgartner, L.P., Dal Piaz, G.V., Skora, S. & Beard, B.L. 2007: Coupling of oceanic and continental crust during Eocene eclogite facies metamorphism: evidence from the Monte Rosa nappe, western Alps. Contributions to Mineralogy and Petrology 153, 139–157.

Liati, A., Gebauer, D. & Fanning, M.C. 2003: The youngest basic oceanic magmatism in the Alps (Late Cretaceous; Chiavenna unit, Central Alps): geochronological constraints and geodynamic significance. Contributions to Mineralogy and Petrology 146(2), 144–158.

Liati, A., Froitzheim, N. & Fanning, C.M. 2005: Jurassic Ophiolithes within the Valais domain of the western and central Alps: Geochronological evidence for re-rifting of oceanic crust, Contributions to Mineralogy and Petrology 149(4), 446–461.

Liati, A. & Froitzheim, N. 2006: Assessing the Valais ocean, Western Alps: U-Pb SHRIMP zircon geochronology of eclogite in the Balma unit, on top of the Monte Rosa nappe. European Journal of Mineralogy 18(3), 299–308.

Ludwig, K.R. 2001: Isoplot/Ex version 2.49, Geochronological Toolkit for Microsoft Excel, Berkeley Geochronology Center Special Publications 1a.

Mahlen, N.J., Lapen, T.J., Johnson, C.M., Baumgartner, L.P. & Beard, B.L. 2003: Duration of Prograde Metamorphism as constrained by high-precision Lu-Hf Geochronology of HP/UHP Eclogites from the western Alps. Geological Society of America Abstracts with Programs 35(6), 638.

Mahlen, N.J., Skora, S., Johnson, C.M., Baumgartner, L.P., Lapen, T.J., Beard, B.L. & Pilet, S. 2005: Lu-Hf geochronology of eclogites from Pfulwe, Zermatt-Saas Ophiolithe, western Alps, Switzerland. Geochimica et Cosmochimica Acta 69, A305-A305 Suppl. S.

Mahlen, N.J., Johnson, C.M., Baumgartner, L.P., Lapen, T.J., Skora, S. & Beard, B.L. 2006: Protracted Subduction History and HP/UHP Metamorphism of the Zermatt-Saas-Ophiolithe, Western Alps, as Constrained by Lu-Hf Geochronology. Eos Trans. AGU 87 (52), Fall Meeting Supplement.

McDounough, W.F., & Sun, S. s. 1995: The composition of the Earth. Chemical Geology 120, 223–253.

Mezger, K., Essene, E.J. & Halliday, A.N. 1992: Closure temperatures of the Sm-Nd system in metamorphic garnets. Earth and Planetary Science Letters 113, 397–409.

Münker C., Weyer S., Scherer E. & Mezger K. 2001: Separation of high field strength elements (Nb, Ta, Zr, Hf) and Lu from rock samples for MC-ICPMS measurements. Geochemistry Geophysics Geosystems 2, Nr. 2001GC000183

Oberli, F., Meier, M., Berger, A., Rosenberg, C.L., Gieré, R. 2004: U-Th-Pb and $^{230}Th/^{238}U$ disequilibrium isotope systematics: Precise accessory mineral chronology and melt evolution tracing in the Alpine Bergell intrusion. Geochimica et Cosmochimica Acta, 68 (11), 2543–2560.

Patchett, P.J. & Tatsumoto, M. 1980: A routine high precision method for Lu-Hf isotope geochemistry and chronology. Contributions to Mineralogy and Petrology 75, 263–267.

Pleuger, J., Froitzheim, N. & Jansen, E. 2005: Folded continental and oceanic nappes on the southern side of Monte Rosa (Western Alps, Italy): anatomy of a double collision suture. Tectonics 24, DOI 10.1029/2004TC001737

Pleuger, J., Roller, S., Walter, J.M., Jansen, E. & Froitzheim, N. 2007: Structural evolution of the contact between two Penninic nappes (Zermatt-Saas zone and Combin zone, Western Alps) and implications for the exhumation mechanism and palaeogeography. International Journal of Earth Science 96, 229–252.

Ravna, E.J.K. & Terry, M.P. 2004: Geothermobarometry of UHP and HP eclogites and schists – an evaluation of equilibria among garnet-clinopyroxene-kyanite-phengite-coesite/quartz. Journal of Metamorphic Geology 22(6), 579–592.

Reinecke, T. 1991: Very-high-pressure metamorphism and uplift of coesite-bearing metasediments from the Zermatt-Saas zone, Western Alps. European Journal of Mineralogy 3, 7–17.

Reinecke, T. 1998: Prograde high- to ultrahigh-pressure metamorphism and exhumation of oceanic sediments at Lago di Cignana, Zermatt-Saas Zone, western Alps. Lithos 42, 147–189.

Richard, P., Shimizu, N., Allègre, C.J. 1976: $^{143}Nd/^{144}Nd$ a natural tracer: An application to oceanic basalts. Earth and Planetary Science Letters 31, 269–278.

Rubatto, D., Gebauer, D. & Fanning, M. 1998: Jurassic formation and Eocene subduction of the Zermatt-Saas Fee Ophiolithes: implications for the geodynamic evolution of the central and western Alps. Contributions to Mineralogy and Petrology 132, 269–287.

Rubatto, D. & Gebauer, D. 1999: Eocene/Oligocene (35 Ma) high-pressure metamorphism in the Gornergrat Zone (Monte Rosa, Western Alps):

Implications for paleogeography. Schweizerische Mineralogische und Petrographische Mitteilungen 79, 353–362.

Rubatto, D., Gebauer, D. & Compagnoni, R. 1999: Dating of eclogite-facies zircons: The age of Alpine metamorphism in the Sesia-Lanzo zone (western Alps). Earth and Planetary Science Letters 167, 141–158.

Rubatto, D. & Hermann, J. 2001: Exhumation as fast as subduction? Geology 29, 3–6.

Rubatto, D. & Hermann, J. 2003: Zircon formation during fluid circulation in eclogites (Monviso, Western Alps): implications for Zr and Hf budget in subduction zones. Geochimica et Cosmochimica Acta 67, 2173–2187.

Salters, V.J.M. & White, W.M. 1998: Hf isotope constraints on mantle evolution, Chemical Geology 145, 447–460.

Scherer, E., Cameron, K.L. & Blichert-Toft, J. 2000: Lu–Hf garnet geochronology: Closure temperature relative to the Sm–Nd system and the effects of trace mineral inclusions. Geochimica et Cosmochimica Acta 64 (19), 3413–3432.

Scherer, E., Münker, C. & Mezger, K. 2001: Calibration of the Lutetium-Hafnium Clock, Science 293, 683–687.

Schmid, S.M., Pfiffner, O.A., Froitzheim, N., Schönborn, G. & Kissling, E. 1996: Geophysical-geological transect and tectonic evolution of the Swiss-Italian Alps. Tectonics 15, 1036–1064.

Schmid, S.M., Fügenschuh, B., Kissling, E. & Schuster, R. 2004: Tectonic map and overall architecture of the Alpine orogen. Eclogae gologicae Helvetiae 97(1), 93–117.

Skora, S., Baumgartner, L.P., Mahlen, N.J., Johnson, C.M., Pilet, S. & Hellebrand, E. 2006: Diffusion-limited REE uptake by eclogite garnets and its consequences for Lu-Hf and Sm-Nd geochronology. Contributions to Mineralogy and Petrology 152, 703–720.

Skora, S., Baumgartner, L.P., Mahlen, N.J., Lapen, T.J., & Johnson, C.M. 2006b: Lu/Hf closure temperature estimated from alpine eclogite garnets. 4th Swiss Geoscience Meeting, Bern 2006.

Söderlund, U., Patchett, P.J., Vervoort, J.D., and Isachsen, C.E. 2004: The ^{176}Lu decay constant determined by Lu-Hf and U-Pb isotope systematics of Precambrian mafic intrusions. Earth and Planetary Science Letters 219, 311–324.

Spear, F.S. 1991: On the interpretation of peak metamorphic temperatures in light of garnet diffusion during cooling. Journal of Metamorphic Geology 9(4), 379–388.

Stampfli, G.M. 1993: Le Briançonnais, terrain exotique dans les Alpes? Eclogae geologicae Helvetiae 86, 1–45.

Stampfli, G.M., Mosar, J., Marquer, D., Marchant, R., Baudin, T. & Borel, G. 1998: Subduction and obduction processes in the Swiss Alps. Tectonophysics 296, 159–204.

Stampfli, G.M. & Borel, G.D. 2004: The TRANSMED transects in space and time: Constraints on the paleotectonic evolution of the Mediterranean domain. In: Cavazza, W., Roure, F.M., Spakman, W., Stampfli, G.M. & Ziegler, P.A. (Eds.) The TRANSMED Atlas. Springer, Berlin Heidelberg, 53–80.

Steck, A., Bigioggero, B., Dal Piaz, G.V., Escher, A., Martinotti, G. & Masson, H. 1999: Carte tectonique des Alpes de Suisse occidentale et des régions avoisinantes, 1:100'000, Carte spéc. n. 123 (4 maps). Serv. hydrol. géol. nat., Bern

Thoeni, M. (2002): Sm-Nd isotope systematics in garnet from different lithologies (Eastern Alps); age results, and an evaluation of potential problems for garnet Sm-Nd. Chemical Geology 185 (3–4), 255–281.

Tirone, M., Ganguly, J., Dohmen, R., Langenhorst, F., Hervig, R. & Becker, H.W. 2004: Rare earth diffusion kinetics in garnet: Experimental studies and applications. Geochimica et Cosmochimica Acta 69(9), 2385–2398.

Van Orman, J.A., Grove, T.L., Shimizu, N. & Graham, D.L. 2002a: Rare earth element diffusion in a natural pyrope single crystal at 2.8 GPa. Contributions to Mineralogy and Petrology 142, 416–424.

Vervoort, J.D., Patchett, P.J., Söderlund, U., Baker, M. 2004: The isotopic composition of Yb and the determination of Lu concentrations and Lu/Hf ratios by isotope dilution using MCICPMS. Geochemistry Geophysics Geosystems 5 (11), doi:10.1029/ 2004GC000721.

Manuscript received March 3, 2008
Revision accepted June 6, 2008
Published Online first November 8, 2008
Editorial Handling: Stefan Schmid & Stefan Bucher

1661-8726/08/01S191-16
DOI 10.1007/s00015-008-1281-1
Birkhäuser Verlag, Basel, 2008

Swiss J. Geosci. 101 (2008) Supplement 1, S191–S206

Tracing the exhumation of the Eclogite Zone (Tauern Window, Eastern Alps) by ^{40}Ar/^{39}Ar dating of white mica in eclogites

WALTER KURZ[1], ROBERT HANDLER[2] & CHRISTIAN BERTOLDI[3]

Key words: ^{40}Ar/^{39}Ar dating, white mica, eclogite exhumation, microstructures, Subpenninic nappes, Tauern Window

ABSTRACT

New radiometric ages from the Subpenninic nappes (Eclogite Zone and Rote Wand – Modereck Nappe, Tauern Window) show that phengites formed under eclogite-facies metamorphic conditions retain their initial isotopic signature, even when associated lithologies were overprinted by greenschist- to amphibolite-facies metamorphism. Different stages of the eclogite-facies evolution can be dated provided ^{40}Ar/^{39}Ar dating is combined with micro-structural analyses. An age of 39 Ma from the Rote Wand – Modereck Nappe is interpreted to be close to the burial age of this unit. Eclogite deformation within the Eclogite Zone started at the pressure peak along distinct shear zones, and prevailed along the exhumation path. An age of ca. 38 Ma is only observed for eclogites not affected by subsequent deformation and is interpreted as maximum age due to the possible influence of homogenously distributed excess argon. During exhumation deformation was localised along distinct mylonitic shear zones. This stage is mainly characterised by the formation of dynamically recrystallized omphacite2 and phengite. Deformation resulted in the resetting of the Ar isotopic system within the recrystallized white mica. Flat argon release spectra showing ages of 32 Ma within mylonites record the timing of cooling along the exhumation path, and the emplacement onto the Venediger Nappe. Ar-release patterns and ^{36}Ar/^{40}Ar vs. ^{39}Ar/^{40}Ar isotope correlation analyses indicate no significant ^{40}Ar-loss after initial closure, and only a negligible incorporation of excess argon. From the pressure peak onwards, eclogitic conditions prevailed for almost 8–10 Ma.

---

1 Introduction

Dating of phengitic white mica provides an important tool for understanding the high-pressure evolution within an evolving orogen (e.g., Scaillet 1998), and helps to constrain early decompressional steps within a pressure-temperature-time- (PTt-) path. In contrast to other isotopic systems, e.g. Rb-Sr or Sm-Nd, the use of the ^{40}Ar and ^{39}Ar isotopes only needs one rock forming mineral for dating, despite the fact that the temperature of the Ar-isotopic system in white mica has been reported to be relatively low, and ranges from ca. 350 °C (Lips et al. 1998) to ca. 450 °C (Hames & Bowring 1994; Kirschner et al. 1996) and ca. 500 °C (Hammerschmidt & Frank 1991; Hames & Cheney 1997). Precise closure temperatures depend on grain-size, chemical composition and cooling rate. Phengitic white mica is generally predicted to have a closure temperature of about 500–550 °C, i.e. slightly higher than that of muscovite (Lister & Baldwin 1996; Stuart 2002). Furthermore, several studies (e.g.

Chopin & Maluski 1980; von Blanckenburg et al. 1988, 1989; Dunlap 1997; Bossé et al. 2005) revealed, that temperature is not the only controlling mechanism for setting or re-setting the isotopic system within respective minerals (see also Villa 1998 for a general discussion). Therefore, not only the thermal, but also the tectonic history is to be investigated in detail because of the probably subordinate importance of temperature for re-setting of isotopic systems to deformation (e.g., Chopin & Maluski 1980; Handler et al. 1993; Müller et al. 1999).

In this study we dated phengitic white micas from eclogite-mylonites and their undeformed protoliths from the Eclogite Zone within the Tauern Window (Eastern Alps) (Fig. 1). From this unit a great number of PT- data are available providing a well-constrained PT- history, i.e. from subduction to subsequent exhumation. We have chosen this unit as a testing site for the combination of microstructural and geochronological studies in order to establish a well-based PTt-deformation history of the Eclogite Zone and adjacent units, and in particular the con-

[1] Institut für Erdwissenschaften, Universität Graz, Heinrichstrasse 26, A-8010 Graz, Austria.
[2] Institut für Geologie und Paläontologie, Universität Salzburg, Hellbrunner Strasse 34, A-5020 Salzburg, Austria. present address: forstinger + stadlmann ZT-OEG, Achenpromenade 14, A-5081 Anif, Austria.
[3] Institut für Mineralogie, Universität Salzburg, Hellbrunner Strasse 34, A-5020 Salzburg, Austria.
*Corresponding author: Walter Kurz. E-mail: walter.kurz@tugraz.at, walter.kurz@uni-graz.at

ditions during the exhumation of the eclogite facies rocks. A study on the microstructural evolution of these eclogites has been carried out contemporaneously in order to provide detailed knowledge about the tectono-metamorphic evolution. The results show (1) that detailed knowledge of the tectono-metamorphic history of the dated rocks provides the possibility to date both cooling following the peak of high-pressure metamorphism, and subsequent deformation in eclogites, and (2) that phengitic white micas are useful for the reconstruction of the decompressional paths in high-pressure metamorphic terranes. Inverse $^{36}Ar/^{40}Ar$ vs. $^{39}Ar/^{40}Ar$ isochrone plots prove to be useful for the detection of extraneous (excess) ^{40}Ar-components, which cannot be recognised from $^{40}Ar/^{39}Ar$ spectra, especially when dealing with relatively young rocks with rather low ^{40}Ar-content (Heizler & Harrison 1988). These diagrams are of great convenience for the recognition of atmospheric Ar and excess ^{40}Ar. The age established using an inverse isochron plot, unlike that yielded by a spectrum, is not affected by trapped $^{36}Ar/^{40}Ar$ ratios that are different from the atmospheric Ar ratio (e.g. due to excess ^{40}Ar), and may contribute to a better age interpretation.

2 Geological setting

The Eastern Alps (Fig. 1) are the product of the convergence between Africa and Europe (e.g., Frisch 1979, 1980; Neubauer et al. 2000; TRANSALP working group 2002, 2006; Schmid et al. 2004). Plate tectonic units involved in the Alpine orogen in the area of the Eastern Alps comprise (1) the European continent, represented by the Helvetic Nappes; (2) the European margin, represented by the Subpenninic Nappes (in the Tauern Window these are the Venediger Nappe, Eclogite Zone, and Rote Wand – Modereck Nappe); (3) two partly oceanic basins in the Penninic realm (the Northpenninic Valais and the Southpenninic Piemont-Liguria, represented by the Rhenodanubian Flysch, the Glockner Nappe, the Matrei Zone and the Klammkalk Zone), and (4) the Adriatic (Apulian) microcontinent including the Austroalpine and South Alpine units (Fig. 1a).

Thrusting and nappe stacking within the (Sub-) Penninic units was subsequent to south-directed subduction of the Penninic oceanic units below the Austroalpine Nappe Complex (for summary, see Kurz & Froitzheim 2002). Subsequent continental subduction of the European margin beneath the northern Adriatic continental margin commenced during the Palaeogene and resulted in thrusting of the Austroalpine nappe complex over Penninic units, and internal imbrication of the Penninic and Subpenninic nappes ("late-Alpine" orogeny, see also Kurz et al. 1998b, 2001a, b and Neubauer et al. 2000 for reviews). Thus, the Subpenninic units exposed within the Tauern Window represent the underplated European margin (Kurz et al. 2001a, b; Schmid et al. 2004; Schuster & Kurz 2005) (Fig 1b, c).

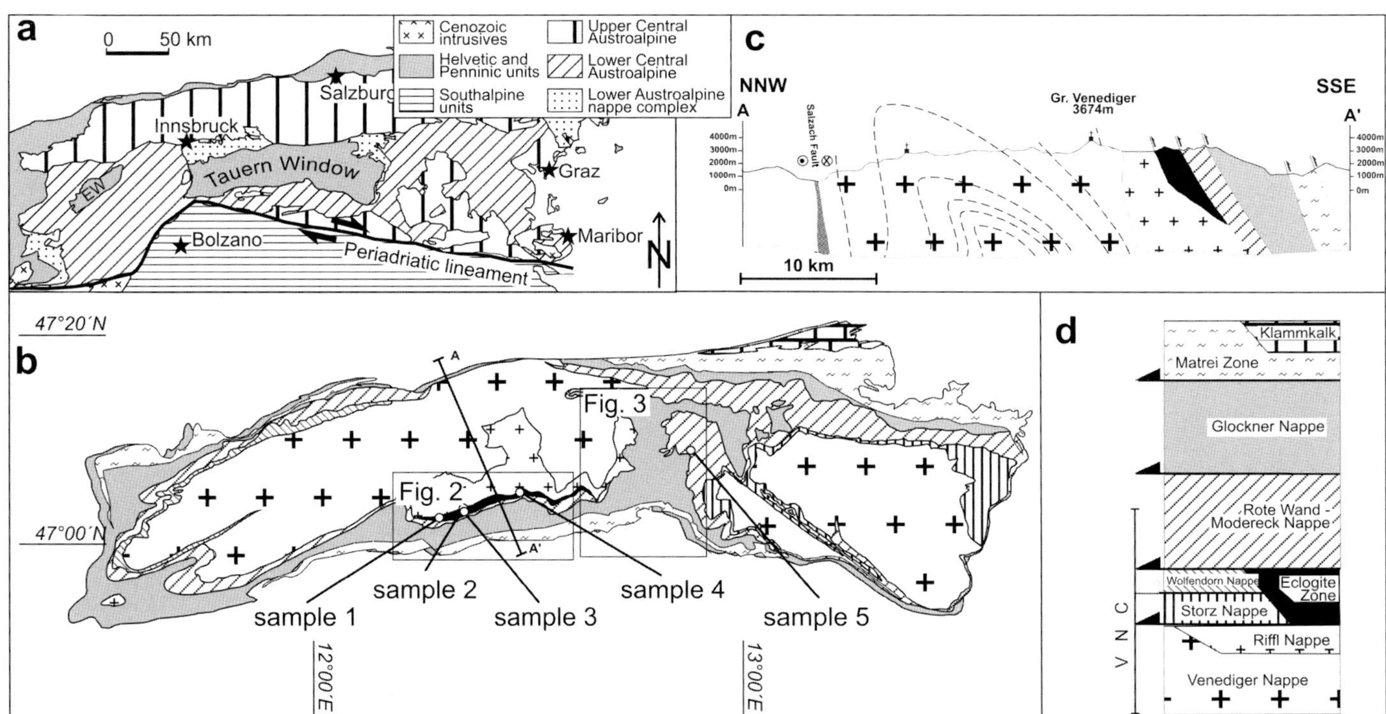

Fig. 1. (a) Simplified tectonic map, showing the major tectonic units of the Eastern Alps (after Kurz et al., 2001a); EW: Lower Engadine Window; (b) Tectonic sketch map of the Tauern Window indicating sample locations (simplified after Kurz et al. 1998a), for legend see Fig. 1d. (c) Structural section across the central Tauern Window showing the structural position of the Eclogite Zone (for location see Fig. 1b) (modified after Kurz et al., 1998a, 2001b). (d) Tectonostratigraphic sketch of the Penninic units in the Tauern Window (modified after Kurz et al. 1998b), thickness of units is not to scale.

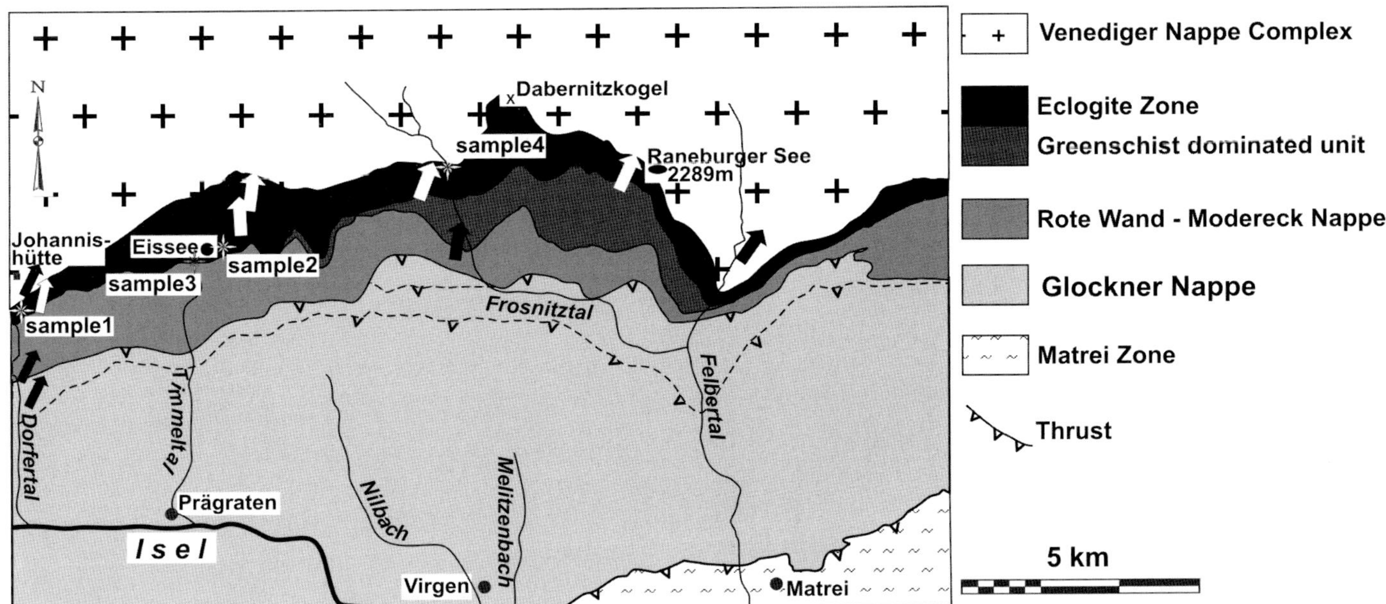

Fig. 2. Tectonic map of the central southern part of the Tauern Window, showing the structural position of the Eclogite Zone (after Kurz et al., 1998a, 2001b); the sampling sites of the eclogites used for microstructural and geochronological analysis are included; arrows indicate the kinematics of nappe emplacement of overlying units contemporaneous to HP metamorphism (D_1 in Fig. 4); the greenschist dominated unit in the upper sections of the Eclogite Zone as defined by Behrmann & Ratschbacher (1989) is characterized by almost complete retrogression of eclogites to garnet amphibolite and/or garnet-bearing greenschists. Kinematic data are available from Behrmann & Ratschbacher (1989), Kurz et al. (1996, 1998a, 2001b).

The units within the Tauern Window can be subdivided into several nappes as described in detail by Kurz et al. (1996, 1998b, 2001a). The general tectonic situation is outlined in Figs. 1b and 1c. In the area of investigation, the Eclogite Zone (Figs. 2, 3) is tectonically imbricated between the Venediger Nappe Complex below, and the Rote Wand – Modereck Nappe above. The Venediger Nappe Complex comprises the Venediger Nappe *sensu stricto*, and the Storz and Wolfendorn Nappes (Fig. 1). Lithologic similarities exist in particular between the Rote Wand – Modereck Nappe and the Eclogite Zone. Both units comprise similar sedimentary sequences of Permian to Triassic quartzites, Triassic metacarbonates and Jurassic breccias, calcareous micaschists and metatuffs. Metamorphic MORBs and gabbros, associated with metasedimentary sequences, particularly occur within the Eclogite Zone (Kurz et al. 1998b, 2001b).

3 Tectonometamorphic evolution of the Eclogite Zone and adjacent areas

The rocks exposed within the Eclogite Zone form a coherent tectonic unit (Figs. 1–3) and were affected by a multiphase tectonometamorphic evolution (for summary, see Miller 1974; Holland 1979; Raith et al. 1980; Dachs 1986, 1990; Stöckhert et al. 1997; Kurz et al. 1998a, 2004; Hoschek 2001; Kurz 2005). The Eclogite Zone is incorporated into a stack of Subpenninic nappes, overlain by the ophiolite-bearing Glockner Nappe (Figs. 1–3) and is the only tectonic unit entirely metamorphosed under eclogite-facies metamorphic conditions. The other units were only partly affected by eclogite facies metamorphism.

Within garnet-micaschists, inclusions in garnet formed by paragonite + zoisite/epidote + quartz, ± phengite, chloritoide, rutile and ore minerals show rectangular to rhombohedral outlines, often interpreted as pseudomorphs after lawsonite (Dachs 1986; Spear and Franz 1986), and document a first stage of prograde metamorphism at ca. 400 °C (e.g., Frank et al. 1987). The eclogite facies rocks were buried to a depth of at least 65 km, indicated by peak pressures of 20–23 kbar at approx. 600 °C (Dachs 1986, 1990; Frank et al. 1987; Zimmermann et al. 1994; Stöckhert et al. 1997; Kurz et al. 1998a; Hoschek 2001) (Ma0 in Fig. 4). Assemblages formed along the exhumation path record approximately 15–16 kbar at 550 °C (Stöckhert et al. 1997; Kurz et al. 1998a) (Fig. 4). The Eclogite Zone was subsequently affected by blueschist facies metamorphism (Ma1 in Fig. 4). Pressures of 7–9 kbar and temperatures of ca. 450 °C are estimated by Raith et al. (1980); 450 °C and 10–15 kbar by Holland (1979) and Zimmermann et al. (1994) (Fig. 4), but the P-T data are not well constrained due to the subsequent strong overprint by amphibolite to greenschist facies metamorphism at approximately 7 kbar and 500 °C (Ma2 in Fig. 4).

For the tectonically underlying Venediger Nappe and Riffl Nappe HP metamorphism at 10–12 kbar has been determined (Selverstone et al. 1984; Cliff et al. 1985; Droop 1985; Selverstone 1993). At this metamorphic stage, corresponding to the blueschist stage of the Eclogite Zone, the Venediger nappe was incorporated into the nappe stack by top-to-the north emplacement, documented by kinematic criteria indicating a top-to-the N to NNE sense of shear both within the Eclogite Zone and in the units below and above (in particular the Venediger Nappe

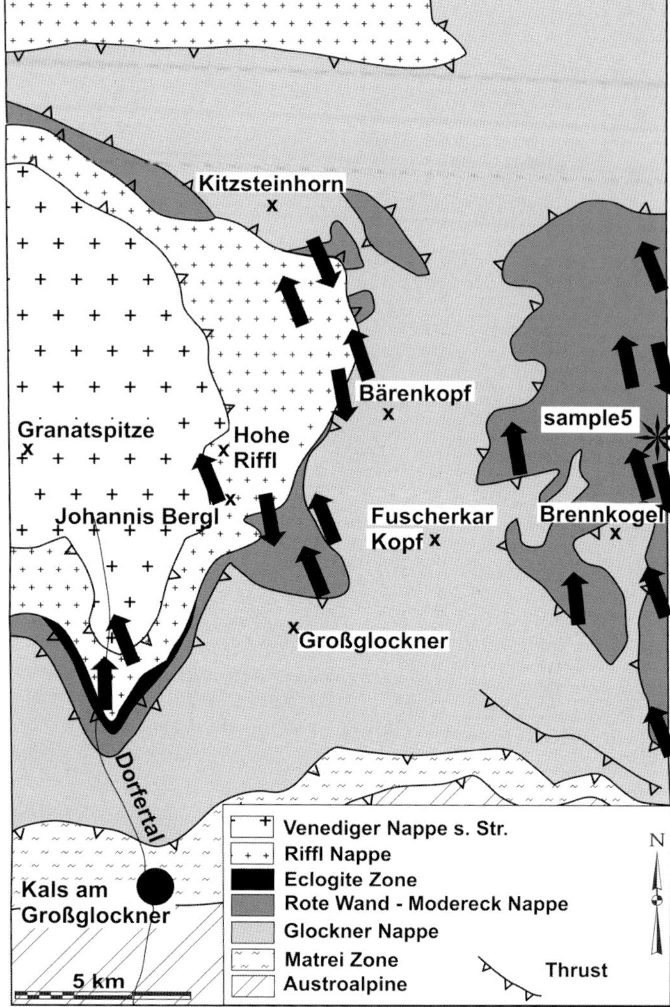

Fig. 3. Tectonic map of the central part of the Tauern Window; the sampling site of micaceous marbles used for geochronological analysis is included; arrows indicate the kinematics of nappe emplacement of overlying units during D_1, contemporaneous to HP metamorphism. Kinematic data are available from Kurz et al. (1996). Top-to-the-south kinematics is related to subsequent folding of the penetrative foliation, resulting in inversion of the shear sense.

and the Rote Wand – Modereck Nappe) (Figs. 2, 3) (D_1 in Fig. 4) (Kurz et al. 1996, 1998a, 2001b). The Venediger Nappe cooled below 300–350° C already at the end of the Oligocene, at about 23 Ma, as indicated by Rb-Sr cooling ages on biotite (e.g., Cliff et al. 1985; Droop 1985; Reddy et al. 1993).

In the units above the Eclogite Zone (i.e. the Rote Wand – Modereck Nappe and Glockner Nappe; Figs. 2, 3) remnants of eclogite facies metamorphism have been locally observed (Proyer et al. 1999; Dachs & Proyer 2001, 2002; Proyer 2003), particularly in internal sections exposed in the southern central part of the Tauern Window. In contrast to the PT- evolution of the Eclogite Zone, the Rote Wand – Modereck Nappe and the Glockner Nappe do not show blueschist facies metamorphism subsequent to the pressure peak (peak conditions of approxi-

mately 16–17 bar at 550 °C) (Dachs & Proyer 2001). After HP metamorphism, these units were affected by Barrovian-type greenschist facies metamorphism (5 kbar, 500 °C), termed "Tauernkristallisation" (e.g., Frank et al. 1987; Selverstone 1993) (Ma2 in Fig. 4).

4 Radiometric ages from the Eclogite Zone and adjacent areas

For a summary of previously published geochronological data from the Eclogite Zone see Thöni (2006). Phengite $^{40}Ar/^{39}Ar$ mineral ages of ca. 36–32 Ma (Zimmermann et al. 1994) from the Eclogite Zone were interpreted as cooling ages postdating eclogite facies metamorphism, and thus the approximate time of emplacement of the Eclogite Zone onto the Venediger Nappe under blueschist facies conditions. Ratschbacher et al. (2004) described $^{40}Ar/^{39}Ar$ ages from high-pressure amphibole, phengite, and phengite + paragonite mixtures. Combined with the thermal evolution showing nearly isothermal decompression from 25 to 15 kbar these were interpreted to document rapid exhumation through ≤15 kbar and >500 °C at ~42 Ma to ~10 kbar and ~400 °C at ~39 Ma. Assuming exhumation rates slower or equal to high-pressure–ultrahigh-pressure terrains in the Western Alps, peak pressures within the Eclogite Zone were reached not long before the high-pressure amphibole age of approx. 42 Ma (Ratschbacher et al. 2004), probably at ≤45 Ma. This is in accordance with dates from the Western Alps (e.g., Droop et al. 1990; for review see Kurz & Froitzheim 2002).

The possibility of a Palaeogene age of HP metamorphism was already discussed by Inger & Cliff (1994). Unpublished Sm-Nd garnet ages of ca. 42 Ma from the Eclogite Zone are cited by Droop et al. (1990), and Inger & Cliff (1994). Significantly younger ages of ca. 26–30 Ma are reported by Inger & Cliff (1994) for meta-sediments associated with the eclogites being penetratively affected by subsequent tectonometamorphic overprint at upper greenschist facies conditions contemporaneous to top-to-the west ductile shearing of the Penninic nappe stack (D_2 in Fig. 4). They interpret their Rb-Sr ages to date cooling of the high-pressure phengites through ca. 550 °C after crystallisation of the rocks of the Eclogite Zone. However they state that the partial resetting of the Rb-Sr isotopic system by subsequent greenschist-facies metamorphic overprint would yield ages, which have to be interpreted as mixtures between post-eclogite cooling and the greenschist overprint.

5 Eclogite microfabrics

The microstructural evolution of eclogites from the Eclogite Zone was described in detail by Kurz et al. (1998a, 2004) and Kurz (2005).

Coarse-grained boudinaged eclogites with a grain-size of up to 1 cm (Fig. 5a) may show a weak foliation (Fig. 5b). Phengite and paragonite usually do not show a preferred orientation. Omphacite shows several features of plastic deformation (D_1

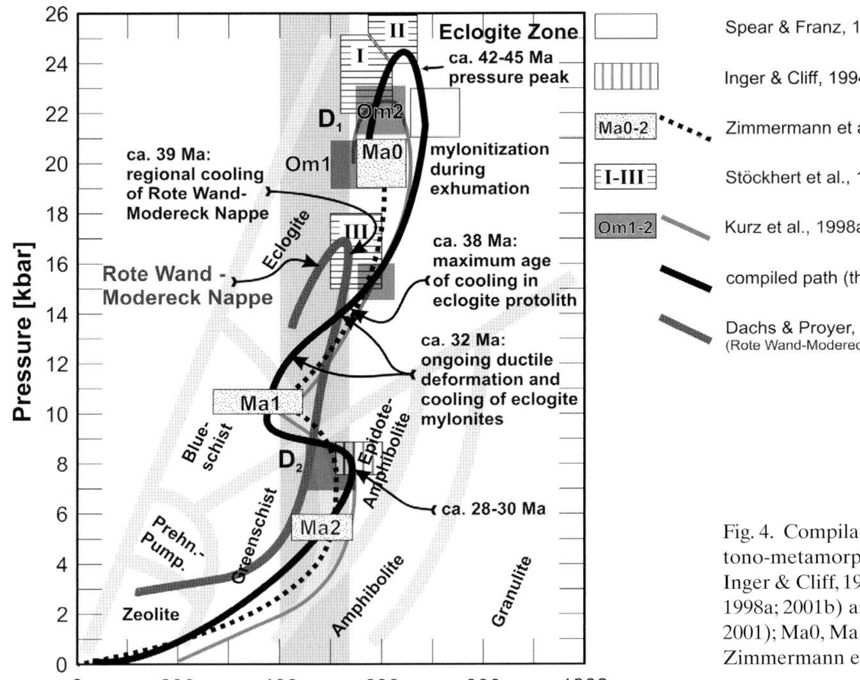

Fig. 4. Compiled Pressure – temperature – time paths, illustrating the tectono-metamorphic evolution of the Eclogite Zone (after Spear & Franz, 1986; Inger & Cliff, 1994; Zimmermann et al., 1994; Stöckhert et al., 1997; Kurz et al., 1998a; 2001b) and the Rote Wand – Modereck Nappe (after Dachs & Proyer 2001); Ma0, Ma1, Ma2 indicate events of Alpine metamorphic overprint after Zimmermann et al. (1994).

in Fig. 4). Coarse grains (omphacite1) are twinned, kinked and bent, and show undulatory extinction and the formation of subgrains (Fig. 5a, b). The subgrain boundaries are generally oriented subparallel to the prism planes. With increasing degree of deformation, the long axes of the subgrains are preferentially oriented subparallel to the trace of the foliation and lineation in XZ- sections (Fig. 5b). Fine grains of recrystallized omphacite (omphacite2) are formed along the grain boundaries of coarse omphacite (omphacite1) (Fig. 5a). Omphacite1 is characterized by jadeite contents of approx. 30 mol%, omphacite2 by jadeite contents of approx. 50 mol% (Kurz et al. 1998a). Omphacite1 is chemically (and optically) zoned (Fig. 5a); the cores and the rims show jadeite contents of approx. 25 mol% and 30–35 mol%, respectively (Kurz et al. 1998a, 2004). Thermobarometric data indicate conditions of 17–20 kbar at approx. 550–580 °C during the formation of omphacite1 (Om1 in Fig. 4), and 20–23 kbar at 600–620 °C at the pressure peak during the formation of omphacite2 (Om2 in Fig. 4) (for details see Kurz et al. 1998a). Therefore, the deformational fabrics document the final section of the prograde evolution up to the pressure peak.

Several stages from coarse-grained eclogites to the formation of fine-grained eclogite mylonites with dynamically recrystallized omphacite2 have been observed. Strongly foliated eclogitic mylonites show a well-developed mylonitic foliation and a stretching lineation defined by elongate garnet, omphacite2, zoisite, kyanite, and glaucophane. Locally, a compositional layering of single-grain-thick quartz layers, garnet, and omphacite is developed. Within the fine-grained mylonites, omphacite2 shows an elongated shape and a shape preferred

orientation subparallel to the penetrative mylonitic foliation (Fig. 5c). Synkinematic phengites are tightly arranged parallel to the penetrative foliation. Glaucophane and barroisitic hornblende, having been formed along the decompressional path, but still under eclogite facies metamorphic conditions (Kurz et al. 1998a), occur in several of the fine-grained, omphacite2 dominated eclogites. Omphacite2 is also partly replaced by these types of amphibole along its rims (Fig. 5d).

The garnet grains are characterized by a round, inclusion-rich core and inclusion-free rims. The grains show straight boundaries and a hypidiomorphic shape (Fig. 5b, c). Garnet is often enriched within monomineralic layers. Within fine-grained mylonites, many garnets show an elongated shape (Fig. 5c), and a shape preferred orientation parallel to the lineation direction, resulting from preferential growth parallel to the kinematic x-axis (Kurz et al. 2004).

From the microstructures and radiometric data described above, the following distinct episodes along the PT- path of the Eclogite Zone can be reconstructed:

1. Deformation at peak pressure conditions indicated by the analysis of microstructures and geothermobarometric data (for details, see Stöckhert et al. 1997; Kurz et al. 1998a, 2004; Kurz 2005) (Fig. 4).
2. Distinct overprints along the decompression path, including a blueschist facies metamorphic overprint; evidence is provided by the analysis of microstructures and geothermobarometric data (for details, see Stöckhert et al. 1997; Kurz et al. 1998a, 2004).

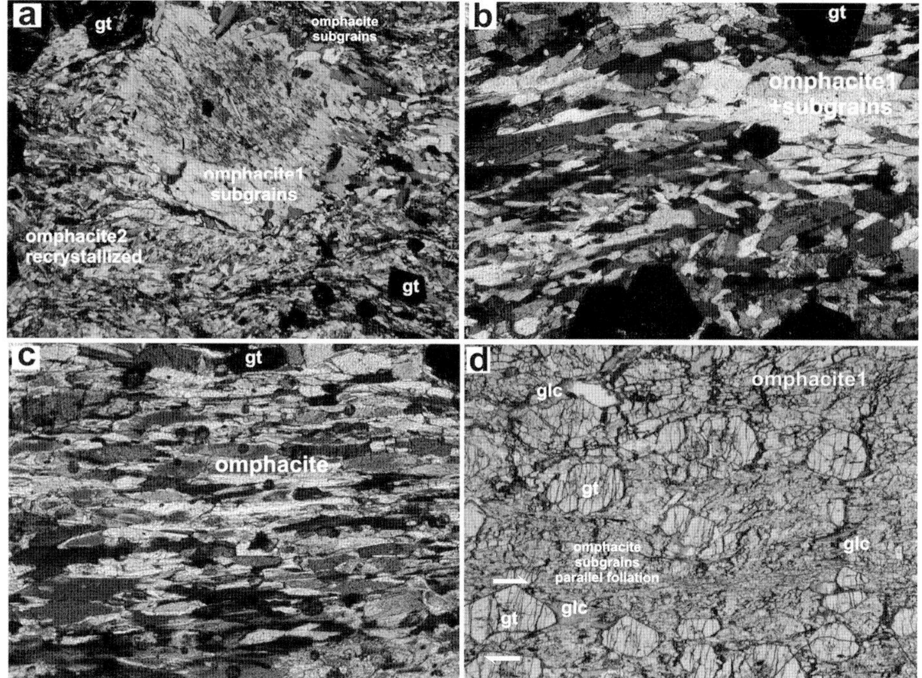

Fig. 5. Microstructures of eclogites from the Eclogite Zone (Tauern Window) (for sample locations see Fig. 2 and the Appendix). a – XZ-section of coarse-grained, un-foliated eclogite, showing omphacite1 with subgrains, and dynamically recrystallized omphacite2 grains along its rim; subgrain boundaries are subparallel to the traces of {010}. b XZ- section of coarse-grained foliated eclogite showing remnants of omphacite1, surrounded by fine-grained dynamically recrystallized omphacite2. c – XZ section of eclogite mylonite showing dynamically recrystallized omphacite2 with well-developed shape preferred orientation in XZ. d – XZ- section of eclogite mylonite showing dynamically recrystallized omphacite2 and glaucophane with well developed shape preferred orientation in XZ; glaucophane is partly replacing omphacite2 and has been formed subsequent to the pressure peak, probably at the transition from eclogite to blueschist facies metamorphism. a, b, c: Crossed polarized Nicols; a–d: long axis of photograph about 4 mm; gt: garnet.

3. Cooling through the closure temperature for the Ar isotopic system of white mica (assumed to lie between 400 °C and 550 °C) (Fig. 4).

The segment between the pressure peak and the cooling through the Ar closure temperature in white mica is less constrained and will be discussed by providing new ^{40}Ar/^{39}Ar mineral ages.

6 Samples analysed by ^{40}Ar/^{39}Ar stepwise heating

The sample sites of analysed eclogites are all located in the northern (internal) part of the Eclogite Zone (see Figs. 1–3 and Appendix), outside the greenschist dominated unit described by Behrmann & Ratschbacher (1989) (Fig. 2). Two types of eclogite samples, not overprinted by subsequent greenschist-amphibolite facies metamorphic assemblages, have been investigated for the presented study: (1) Eclogite-mylonites (samples 1–3) with syn-kinematically grown phengite; the typical mineral assemblage of these samples is shown in Figs. 6a, b, and comprises garnet, omphacite2 (Fig. 4), zoisite, glaucophane, kyanite, phengite, and quartz. Estimates on the PT-conditions for eclogite-facies metamorphism are provided by Kurz et al. (1998a) (Fig. 4) (20–23 kbar, 600–620 °C for the peak assemblage) (eclogite mylonites containing omphacite2, but lacking glaucophane and barroisitic hornblende). (2) Samples of unfoliated fine-grained eclogite (sample 4) (Fig. 6c), not affected by deformational overprint subsequent to the pressure peak, but occurring together with foliated eclogite mylonites, were analysed for comparison. The eclogite facies mineral assemblage comprises garnet, omphacite1 (Fig. 4), kyanite, quartz, parago-

nite and subordinate phengite. Estimates on the PT-conditions for eclogite-facies metamorphism are provided by Kurz et al. (1998a) (Fig. 4) (18–20 kbar at 530–570 °C for unfoliated eclogites containing omphacite1).

Additionally, a calcite-marble (sample 5) (Fig. 6d) from the Rote Wand – Modereck Nappe (for location see Fig. 3) has been analysed to get information on the cooling history of the unit immediately above the Eclogite Zone, because geothermobarometric studies have satisfactorily shown eclogite facies metamorphism in parts of this unit as well (ca. 550 °C, 15 kbar) (Dachs & Proyer 2001). The location indicated in Fig. 3 has been chosen, because this area shows low degree of metamorphic and deformational overprint under greenschist facies conditions (D$_2$ in Fig. 4). This should therefore provide an undisturbed original Ar isotopic composition related to the peak metamorphic conditions.

Samples 1–3 are characterised by similar microstructures, showing a strong penetrative mylonitic foliation with a shape preferred orientation of omphacite2 (with an average grain size of 100–250 μm), zoisite, glaucophane, and phengite (Fig. 6a, b), corresponding to the microstructures of fine-grained eclogite mylonites described above (Figs. 5b, c). Phengites reach a grain size of up to 500 μm. Sample 4 shows a less developed shape preferred orientation compared to previously described samples (Fig. 6c), indicating that penetrative ductile deformation in this sample was less pronounced. Grain-size ranges from 10 to 50 μm, only white mica reaching a grain size of up to 250 μm. The microstructures are indicative for coarse- to medium-grained eclogites described above (Fig. 5a). Although most of the associated lithologies were affected by later retrogression

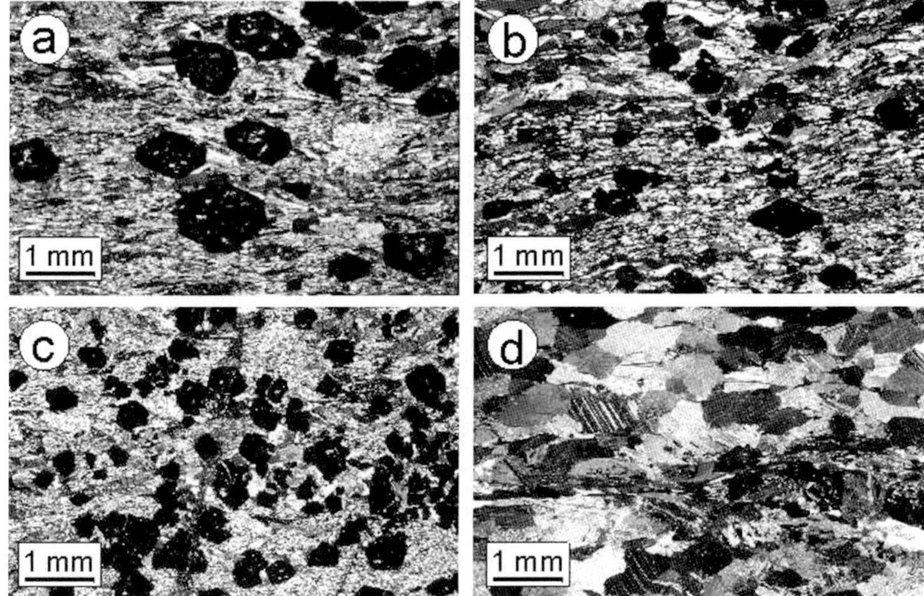

Fig. 6. Examples of typical microstructures of the dated samples. a, b – Eclogite-mylonite samples 1 and 2 indicate strong shape-preferred orientation of phengite, and recrystallized omphacite. Note also the intense elongation of garnet (with an aspect ratio of up to 1.8). c – By contrast, the fine-grained eclogite sample 4 indicates no characteristics for ductile deformation and equidimensional garnet textures. d – Calcite marble sample 5 indicates a penetrative foliation, which is traced by layers of white mica.

at amphibolite to greenschist facies metamorphic conditions, all the eclogite samples described in this study were preserved as meta-stable assemblages, most probably related to the absence of fluids in these rock domains. Sample 5 (Fig. 6d) has been taken from a calcite-marble near the structural base of the Rote Wand – Modereck Nappe. Retrogressed eclogites, intercalated with these marbles, show relics of high-pressure assemblages (Proyer et al. 1999; Dachs & Proyer 2001). Within the marbles the penetrative foliation is traced by layers of white mica sporadically accumulated along isolated foliation (i.e. former bedding) planes.

7 Results

7.1 Electron microprobe analysis

Between 10–20 electron microprobe analyses have been carried out on white mica concentrates of samples 1–5. The chemical formulae for white mica were calculated on the basis of 56 cation charges. Average mica compositions are listed in Table 1 and graphically presented in a muscovite –50% celadonite – paragonite plot (Fig. 7). Analyses are discussed with respect to their muscovite (Ms), celadonite (Cel), paragonite (Pg), and margarite (Mrg) components.

White micas from eclogite-mylonite samples (1A, 1B, 2, and 3) are phengites with similar chemical compositions (average ca. $Ms_{40}Ce_{51}Pg_9$). Both size-fractions of sample 1 (1A: 200–250 μm; 1B: 250–355 μm) yielded similar compositions. White micas from the fine-grained eclogite sample 4 are paragonites with an average composition of $Ms_5Pg_{93}Mrg_2$. White micas from calcite marble (sample 5) have a slightly less phengitic composition than eclogite-mylonite samples 1–3, with an average of ca. $Ms_{64}Cel_{32}Pg_4$.

7.2 $^{40}Ar/^{39}Ar$ dating

$^{40}Ar/^{39}Ar$ analyses have been carried out on six different white mica multi-grain concentrates (ca. 15 grains per sample) of five samples. A detailed description of the analytical techniques and procedures can be found in Handler et al. (2004).

Analytical results are portrayed as age spectra in Fig. 8 together with $^{36}Ar/^{40}Ar$ vs. $^{39}Ar/^{40}Ar$ isotope correlation plots. The detailed analytical results can be requested from the publisher and are provided in the electronic version.

From sample 1 two size fractions (1A: 200–250 μm, and 1B: 250–355 μm) have been analysed. The Ar-release spectrum of sample 1A displays a nearly flat pattern, indicating homogeneous Ar-isotopic composition released through the experiment, and pointing to an undisturbed Ar-isotopic composition of the white mica. Age calculation over all increments yielded

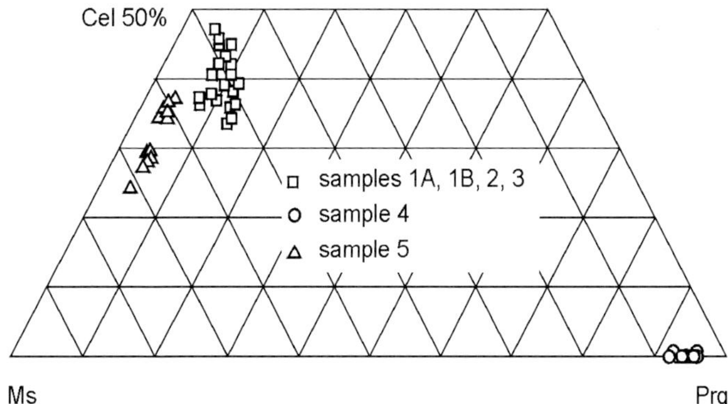

Fig. 7. Muscovite (Ms) –50% Celadonite (Cel) – Paragonite (Prg) triangle-plot of chemical composition of white-mica samples 1–5 analysed by electron microprobe. Each data point represents one individual grain mounted on carbon glass slides.

Table 1. Microprobe analyses of white mica from eclogite samples 1–4, and calcite-marble sample 5 of the Tauern Window, Eastern Alps (Austria). Ms = Muscovite, Pg = Paragonite, Mrg = Margarite (following suggestions of Kretz, 1983), Cel = Celadonite; n = number of analysis on the sample; <3σ = below detection limit of 3σ; values in parentheses are 1σ standard deviation.

	Sample 1A	Sample 1B	Sample 1A + 1B	Sample 2	Sample 3	Sample 4	Sample 5
n	5	16	21	18	12	15	14
SiO2	49.88 (86)	50.81 (99)	50.53 (99)	50.10 (99)	50.69 (90)	46.28 (52)	49.40 (99)
Al2O3	27.15 (26)	27.45 (57)	27.41 (50)	26.70 (70)	27.13 (50)	39.29 (51)	29.50 (60)
TiO2	0.29 (02)	0.28 (04)	0.28 (04)	0.28 (07)	0.26 (03)	0.07 (02)	0.30 (16)
FeO	1.59 (07)	1.55 (07)	1.56 (08)	2.20 (30)	1.90 (15)	0.47 (04)	0.80 (11)
MgO	3.61 (08)	3.65 (18)	3.60 (16)	3.40 (30)	3.43 (20)	0.15 (05)	2.90 (20)
MnO	<3σ	<3σ	<3σ	<3σ	<3σ	<3σ	<3σ
K2O	9.87 (36)	10.13 (21)	10.10 (30)	10.40 (20)	10.20 (13)	0.64 (15)	10.80 (11)
Na2O	0.81 (13)	0.73 (13)	0.70 (13)	0.60 (14)	0.67 (06)	7.36 (26)	0.32 (04)
CaO	<3 σ	<3 σ	<3 σ	<3 σ	<3 σ	0.32 (04)	<3 σ
Total	93.19	94.59	94.18	93.68	94.28	94.57	94.02
Si	6.77 (05)	6.79 (06)	6.79 (05)	6.81 (07)	6.82 (04)	5.97 (02)	6.64 (08)
Al(tot)	4.35 (04)	4.33 (07)	4.34 (07)	4.27 (06)	4.30 (06)	5.97 (04)	4.68 (11)
Ti	0.03 (00)	0.03 (00)	0.03 (00)	0.03 (01)	0.03 (00)	0.01 (00)	0.03 (02)
Mg	0.73 (02)	0.73 (04)	0.72 (04)	0.69 (05)	0.69 (04)	0.03 (01)	0.58 (04)
Fe	0.18 (01)	0.17 (01)	0.18 (01)	0.25 (03)	0.21 (01)	0.05 (00)	0.09 (01)
Mn	<3σ	<3σ	<3σ	<3σ	<3σ	<3σ	<3σ
K	1.71 (07)	1.73 (05)	1.73 (05)	1.80 (05)	1.75 (04)	0.11 (02)	1.85 (03)
Na	0.21 (04)	0.19 (04)	0.18 (04)	0.16 (04)	0.17 (02)	1.84 (07)	0.08 (01)
Ca	<3σ	<3σ	<3σ	<3σ	<3σ	0.04 (01)	<3σ
Cel	38.58 (2.30)	39.70 (2.83)	39.48 (2.74)	40.30 (3.55)	40.85 (1.99)	–	32.16 (3.89)
Ms	50.45 (1.39)	50.50 (1.32)	50.99 (1.34)	51.64 (2.55)	50.08 (1.49)	5.29 (1.20)	63.53 (3.58)
Pg	10.96 (1.46)	9.80 (1.79)	9.53 (1.78)	8.06 (1.87)	9.08 (0.81)	92.49 (1.02)	4.31 (0.54)
Mrg	–	–	–	–	–	2.22 (0.31)	–

an age of 32.5 ± 0.15 Ma. The $^{36}Ar/^{40}Ar$ vs. $^{39}Ar/^{40}Ar$ isotope correlation plot yields a y-axis intercept of $^{36}Ar/^{40}Ar = 0.00310$ (MSWD = 9.6). Because this value is comparable to the present day atmospheric composition of Ar ($^{36}Ar/^{40}Ar_{atm} = 0.00338$), we conclude that no excess Ar has been incorporated at or after the time of initial closure of the isotopic system in these micas.

The coarser grained phengite (sample 1B) indicates slightly higher ages in the first three low-temperature gas release steps, pointing either to optically undetectable intergrowths with higher Ar-retention, or minor influx of extraneous ^{40}Ar-components. Similar ages as for sample 1A are obtained from further steps of the Ar-release pattern (steps 4–13, together comprising 90.3% of the total ^{39}Ar released) (33.3 ± 0.15 Ma). Minor incorporation of excess ^{40}Ar-components is also indicated by the $^{36}Ar/^{40}Ar$ vs. $^{39}Ar/^{40}Ar$ isotope correlation plot, where the ratios of the first three increments define a trend-line, which clearly points to a $^{36}Ar/^{40}Ar$ intercept, being much lower than atmospheric composition. Therefore, a regression line has been calculated only for steps 4–13, which define a plateau in the Ar-release spectrum. The isotope correlation analysis of these steps yielded a y-axis intercept of $^{36}Ar/^{40}Ar = 0.00347$ (MSWD = 0.66), close to the atmospheric Ar-composition.

White mica of sample 2 has the same grain-size (250–355 µm) as sample 1B described above. The Ar-release plot displays fairly consistent ages, except for the first increment, which again displays a slightly older age. Calculation over steps 2–13, together comprising 99.1% of the total ^{39}Ar released, yielded an age of 32.5 ± 0.15 Ma. Regression analyses over these steps within the $^{36}Ar/^{40}Ar$ vs. $^{39}Ar/^{49}Ar$ isotope correlation plot yields a y-axis intercept of $^{36}Ar/^{40}Ar = 0.00309$ (MSWD = 2.6).

For sample 3 coarse-grained (250–355 µm) white micas have been analysed as well. The Ar-release plot displays a flat age spectrum, indicating no disturbance after initial closure of the Ar-isotopic system. Age calculation over all increments yielded 31.8 ± 0.15 Ma, the $^{36}Ar/^{40}Ar$ vs. $^{39}Ar/^{40}Ar$ isotope correlation plot yields a y-axis intercept of $^{36}Ar/^{40}Ar = 0.00289$ (MSWD = 11.7).

White mica sample 4 (200–250 µm) has been separated from a fine-grained eclogite. The Ar-release plot again displays a flat spectrum; only the first step yields a slightly older age. The 1σ error range for this analysis is rather wide related to the low K- content of the analysed paragonite, resulting in low Ar- contents (see Table 1). The total-gas age of this sample is 38.0 ± 0.55 Ma. The regression line of the $^{36}Ar/^{40}Ar$ vs. $^{39}Ar/^{40}Ar$ isotope correlation plot yields a y-axis intercept of $^{36}Ar/^{40}Ar = 0.00329$ (MSWD = 0.53).

White mica sample 5 (250–355 µm) has been separated from a calcite-marble of the Rote Wand – Modereck Nappe. The Ar-release plot displays a slightly disturbed age spectrum. However, the total-gas age of 39.00 ± 0.15 Ma is comparable to the results of sample 4. The regression line calculated

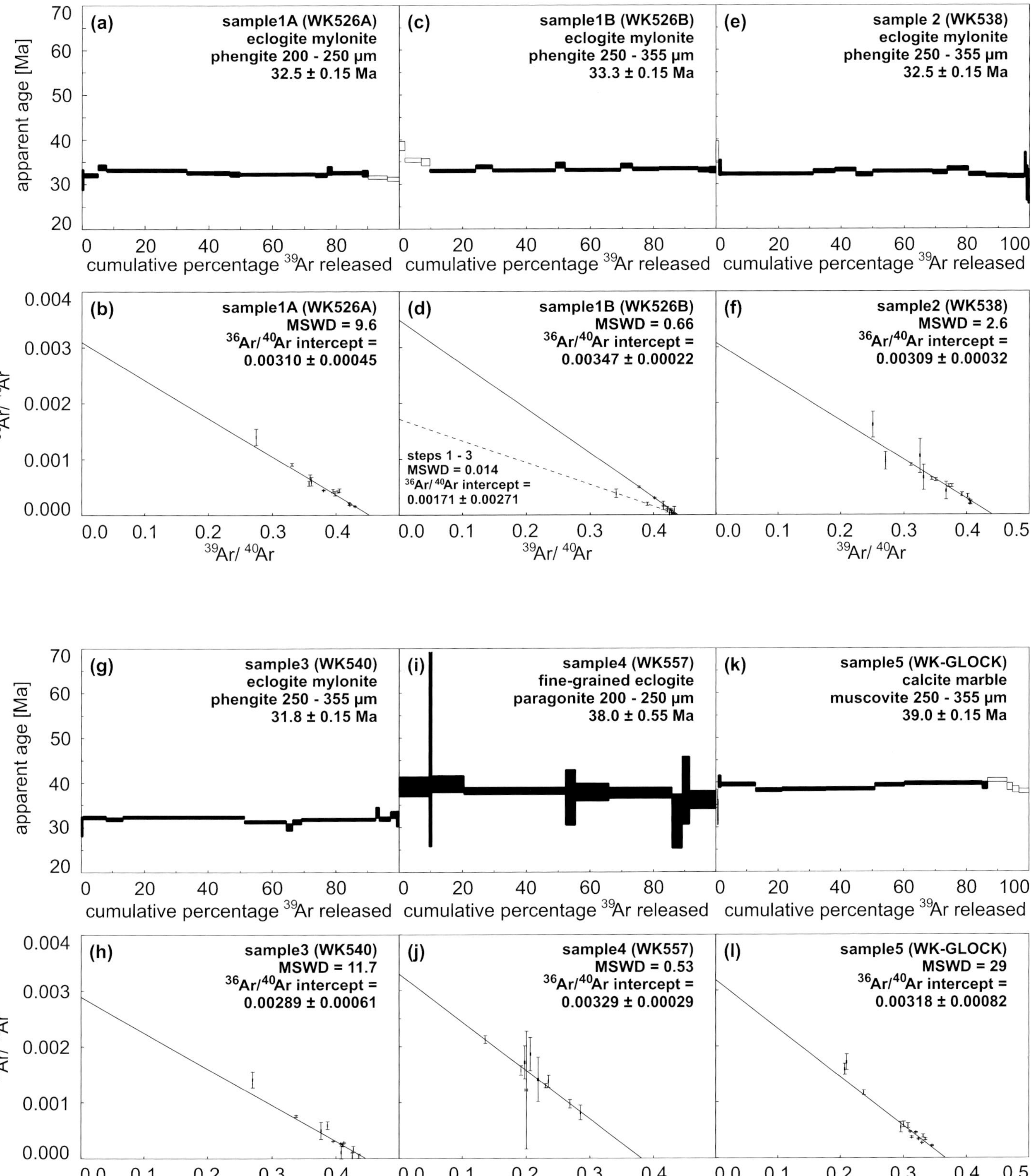

Fig. 8. $^{40}Ar/^{39}Ar$ age spectra and related $^{36}Ar/^{40}Ar$ vs. $^{39}Ar/^{40}Ar$ isotope correlation plots of white mica multi-grain analyses from the Eclogite Zone and the Rote Wand – Modereck Nappe of the Tauern Window; intensity of the laser increases from left to right; increments used for age calculation are indicated; analytical errors (1-sigma) are given by vertical width of bars; half length of error bars in isotope correlation plots indicates 1 sigma error.

over all increments in the ^{36}Ar/^{40}Ar vs. ^{39}Ar/^{40}Ar isotope correlation plot yields a y-axis intercept of ^{36}Ar/^{40}Ar = 0.00318 (MSWD – 29).

7.3 Data quality

Analyses of white micas from eclogite mylonites (samples 1–3) display flat Ar-release spectra with ages ranging between 31.8 ± 0.15 Ma and 33.3 ± 0.15 Ma. Only two of four Ar-release plots display slightly older ages in the first increments (samples 1B, and 2), suggesting minor influx of excess ^{40}Ar-components. When these increments are not taken into account, regression analyses of the ^{36}Ar/^{40}Ar vs. ^{39}Ar/^{40}Ar isotope correlation plots yield y-axis intercepts ranging between ^{36}Ar/^{40}Ar = 0.00289 and ^{36}Ar/^{40}Ar = 0.00347 for all four analyses, being fairly close to the atmospheric isotopic composition of ^{36}Ar/^{40}Ar = 0.00338. There is no correlation between the age of the samples and the extrapolated ^{36}Ar/^{40}Ar ratio. Furthermore, analysis of sample 1A indicates a perfectly flat Ar-release pattern and an extrapolated ^{36}Ar/^{40}Ar ratio nearly identical to the atmospheric value. We therefore conclude that the different ages reported in samples 1–3 do not indicate incorporation of significant amounts of excess ^{40}Ar-components in the Ar-system.

Sample 4, separated from an un-foliated eclogite, and sample 5, separated from a calcite-marble of the Rote Wand – Modereck Nappe display significantly older ages compared to the mylonitic eclogites (samples 1–3). Integrated ages are 38.0 ± 0.55 Ma and 39.00 ± 0.15 Ma, respectively. The relatively large age-error for sample 4 results from the low K- and, therefore, Ar-content of the analysed paragonites. We do not observe any correlation between ages, grain-size within eclogite mylonites, and extrapolated ^{36}Ar/^{40}Ar ratios. The ^{36}Ar/^{40}Ar vs. ^{39}Ar/^{40}Ar isotope correlation plot of sample 4 yields a y-axis intercept of fairly atmospheric composition. In the ^{36}Ar/^{40}Ar vs. ^{39}Ar/^{40}Ar isotope correlation plot of sample 5 the y-axis intercept yields at 0.00318, which may indicate minor incorporation of an excess ^{40}Ar-component. However, the first two steps of the Ar-release spectrum of this sample indicate Ar-loss rather than incorporation of ^{40}Ar. Therefore, the integrated age is interpreted to be geologically meaningful, and the age difference of ca. 7 Ma between white micas separated from eclogite-mylonites (samples 1–3) and white micas separated from calcite marbles (sample 5) to be significant.

8 Discussion and geological implications

The new radiometric ages presented in this study show that phengites formed under eclogite-facies metamorphic conditions retain their isotopic signature, even when associated lithologies were affected by greenschist- to lower amphibolite-facies metamorphic overprint. Despite a few domains showing distinct retrogression to (garnet-) amphibolites, most of the eclogites within the Eclogite Zone were preserved, showing their eclogite facies mineral assemblages. This may be explained in several ways:

1. The Eclogite Zone was exhumed at very high exhumation rates (for discussion, see Glodny et al. 2005) and, therefore, subsequent metamorphic overprint by conductive heat transfer was either prevented or delayed (e.g. Ernst 2006).
2. Subsequent metamorphism proceeded at conditions of heterogeneous fluid distribution, documented by the adjoining occurrence of both well-preserved and completely retrogressed eclogites, and the occurrence of hydrothermal veins.
3. Ongoing deformation under eclogite facies conditions.

Resetting of the Ar-isotopic system during exhumation was highly influenced by concomitant ductile deformation in terms of mylonitization, and heterogeneous regional fluid flow. This is supported by observations and theoretical considerations of previous workers (e.g. Chopin & Maluski 1980; von Blanckenburg et al. 1989; 1993; Villa 1998) that in certain cases ductile deformation is of greater importance for the resetting of isotopic systems than temperature. Furthermore our study shows that different stages of an eclogite-facies metamorphic event can be dated when ^{40}Ar/^{39}Ar dating is combined with microstructural investigations.

The ages presented in this study are in accordance with previously published phengite ^{40}Ar/^{39}Ar mineral ages from the Eclogite Zone of ca. 36–32 Ma (Zimmermann et al. 1994). These were interpreted as cooling ages postdating eclogite facies metamorphism, and thus taken to date the approximate time of emplacement of the Eclogite Zone onto the Venediger Nappe. Similarly, Ratschbacher et al. (2004) described ^{40}Ar/^{39}Ar ages from high-pressure amphibole (ca. 42 Ma), phengite, and phengite+paragonite mixtures (ca. 39 Ma). Based on our new data, we can draw some additional constraints on the available PTt-paths of the Eclogite Zone and adjacent units (Fig. 4). Related to the observed microstructures, deformation at eclogite facies conditions started close to the pressure peak, documented by the formation of dynamically recrystallized omphacite2 (Fig. 5) (Kurz et al. 1998a, 2004), and the synkinematic crystallisation of phengite. Omphacite2 and phengite are present in equilibrium and formed syndeformatively. Inevitably, the pressure peak marks the transition from subduction-related burial to exhumation (Kurz 2005). Deformation was localised mainly along distinct shear zones, and prevailed along the exhumation path (Fig. 4). During decompression and cooling the Ar-isotopic system was initially closed at ca. 38 Ma. These ages are only observed for rocks not affected by subsequent deformational and metamorphic overprint. The Ar release spectra show that the Ar-isotopic system in these white micas was slightly affected by incorporation of excess ^{40}Ar-components or Ar-loss after initial closure. The dated paragonite within these rocks, generally showing a lower closure temperature compared to phengite, is very sensitive for excess Ar. Due to the low K content and the young age only a low amount of excess Ar may have a large influence on the age. Therefore, this age is interpreted as maximum cooling age of the previously established eclogite-facies metamorphic assemblage. Although the isotope correlation plots do not show any influence of excess Ar from an external source (Kelley 1995), and therefore show an atmospheric composition, it is not possible to determine Ar which was set free in the same rock volume above the closure temperature and caught to an undefined amount later on, when the rock passes through the closure window. This may happen when there is only very restricted fluid flow, which may also be responsible for the metastable preservation of peak metamorphic assemblages. Especially a homogeneous distribution of radiogenic excess Ar cannot be detected. Accordingly, the ages provided by paragonite from undeformed eclogites (38 Ma) are approximately 5–6 Ma older than the ages within the mylonite.

From the pressure peak onwards deformation in the eclogite-mylonites (samples 1–3) continued under eclogite-facies conditions along the exhumation path; this is indicated by the synkinematic replacement of omphacite by glaucophane as well. Shear localisation, causing strain softening, resulted in the dynamic recrystallization of white mica within the mylonites. The phengite analyses display flat Ar-release spectra, irrespective of the white mica grain size. Only two of four Ar-release plots display slightly older ages in the first increments (samples 1B, and 2), suggesting minor influx of excess ^{40}Ar-components. This indicates a quite homogenous Ar-retention.

Both types of eclogites analysed (deformed and undeformed), do not show any indication of subsequent overprint by greenschist to amphibolite facies metamorphic mineral assemblages, but show similar sizes of the major rock-forming minerals (garnet, omphacite). This additionally suggests that the different ages may not be attributed to the average grain size of the host rock in terms of intergranular diffusion rates, and not to different grade of metamorphic overprint, but either to deformation enhanced resetting of the Ar-isotopic system, or to homogenously distributed small amounts of excess Ar within paragonite in undeformed samples.

The close relationship between microstructural observations and geochronological ages documents that in samples free of predeformative relics an assemblage in isotopic equilibrium has been frozen during syn-deformational cooling (Glodny et al., 2008). Therefore, the flat Ar release spectra showing ages of 32 Ma within omphacite2- mylonites record the timing of exhumation, with an upper closure temperature limit of up to 550 °C (Hammerschmidt & Frank 1991; Hames & Bowring 1994; Kirschner et al. 1996; Hames & Cheney 1997) (550 °C particularly for phengite). The Rb-Sr ages of approximately 32 Ma presented by Glodny et al. (2005) can therefore be interpreted on this note as well. Referring to exhumation-related assemblages, indicating conditions of 15–16 kbar at approx. 550 °C (Fig. 4) (Stöckhert et al. 1997; Kurz et al. 1998a), the ages of samples 1–3 indicate decompression through 15 kbar at 32 Ma.

Later these rocks experienced a greenschist to amphibolite facies metamorphic overprint at 525 °C and 7,5 kbar (Dachs 1990). The metamorphic temperatures are quite close to the upper closure temperature limit of the Ar-system in phengite, but may have been too low to cause complete Ar resetting. Ages in

the range of 28–30 Ma published by Inger & Cliff (1994) can be interpreted in terms of subsequent resetting; however, these ages are from associated meta-sedimentary sequences, mainly situated within the greenschist dominated unit (Fig. 2). Rb-Sr ages of pseudomorphs of lawsonite from this area (Gleissner et al. 2007), situated within the Glockner Nappe, show that the decomposition of lawsonite occurred at 30 Ma and are interpreted to reflect the onset of greenschist facies metamorphic overprint, just following the final exhumation of the Eclogite Zone. This might indicate that the age of 32 Ma is a cooling age very close to the beginning of thermal overprint.

Sm-Nd garnet ages of ca. 42 Ma from the Eclogite Zone (see Droop et al. 1990, Inger & Cliff 1994), and recently published ^{40}Ar/^{39}Ar ages of ca. 39–42 Ma from high-pressure amphibole, phengite, and phengite+paragonite mixtures (Ratschbacher et al. 2004) may be used to constrain the timing of peak conditions during eclogite facies metamorphism and subsequent cooling, and therefore the formation of omphacite1-bearing eclogites. These ages are quite similar to the Sm-Nd garnet ages and U-Pb SHRIMP ages reported for units of a comparable tectonic position in the Central and Western Alps (e.g. Dora Maira, Adula Nappe) (for summary, see Kurz & Froitzheim 2002). By combining our new data with the previously published ages, we suggest the following tectonic evolution of the Eclogite Zone and associated tectonic units within the Tauern Window (Fig. 9), based on the PTt evolution shown in Fig. 4:

The oceanic lithosphere preserved in the Glockner Nappe was subducted during the Late Cretaceous to Eocene (Fig. 9a), with parts of it reaching eclogite facies metamorphism. Subsequently, the European margin descended into the subduction zone, resulting in eclogite facies metamorphism in the Eclogite Zone at about 40–42 Ma (Fig. 9b). The Eclogite Zone ascended towards the surface within the subduction channel (Kurz et al. 1998a; Kurz & Froitzheim 2002; Kurz 2005), while subduction was still active and, hence, heating was prevented (Kurz et al. 1998a, b; 2001b; Kurz & Froitzheim 2002). This is evidenced by cooling during decompression and the subsequent blueschist facies metamorphic overprint (Fig. 4). An age of ca. 38 Ma from undeformed eclogites is assumed to be a maximum age of this cooling event. Cooling of the Rote Wand – Modereck Nappe occurred approximately at the same time (39 Ma). As the peak temperatures during HP metamorphism within the Rote Wand – Modereck Nappe (ca. 550 °C at 15–16 kbar; Dachs & Proyer 2001) are near to the closure temperature for the Ar system in phengite, this age (39 Ma) is interpreted to be close to the burial age of this unit, and therefore of the European margin. Accordingly, the pressure peak within the Eclogite Zone, originally situated south of the Rote Wand – Modereck Nappe (Kurz et al. 1998b) (Fig. 9b) was reached contemporaneously or slightly before, as indicated by the ages (39–42 Ma) published by Ratschbacher et al. (2004). Syndeformational cooling, related to the emplacement of the Eclogite Zone onto the Venediger Nappe, and the subsequent emplacement of the Rote Wand – Modereck Nappe along a major out-of-sequence detachment above at mid- to lower crustal levels (Fig. 9c, d), may be dated at

Ernst, W.G. 2006. Preservation/exhumation of ultrahigh-pressure subduction complexes. Lithos 92, 321–335.

Frank, W., Höck, V. & Miller, C. 1987. Metamorphic and tectonic history of the central Tauern Window. In: Flügel, H.W. & Faupl, P. (Eds.): Geodynamics of the Eastern Alps. Deuticke/Vienna, 34–54.

Frisch, W. 1979. Tectonic progradation and plate tectonic evolution of the Alps. Tectonophysics 60, 121–139.

Frisch, W. 1980. Plate motions in the Alpine region and their correlation to the opening of the Atlantic Ocean. Mitteilungen der Österreichischen Geologischen Gesellschaft 71/72, 45–48.

Gleissner, P., Glodny, J. & Franz, G. 2007. Rb-Sr isotopic dating of pseudomorphs after lawsonite in metabasalts from the Glockner nappe, Tauern Window, Eastern Alps. European Journal of Mineralogy 19, 723–734.

Glodny, J., Ring, U., Kühn, A., Gleissner, P. & Franz, G. 2005. Crystallization and very rapid exhumation of the youngest Alpine eclogites (Tauern Window, Eastern Alps) from Rb/Sr mineral assemblage analysis. Contributions to Mineralogy and Petrology 149, 699–712.

Glodny, J., Ring, U. & Kühn, A. 2008. Coeval high-pressure metamorphism, thrusting, strike-slip and extensional shearing in the Tauern Window, Eastern Alps. Tectonics 27, TC 4004, doi: 10.1029/2007TC002193.

Hames, W.E. & Bowring, S.A. 1994. An empirical evaluation of the argon diffusion geometry in muscovite. – Earth and Planetary Scientific Letters 124, 161–167.

Hames, W.E. & Cheney, J.T. 1997. On the loss of ^{40}Ar from muscovite during polymetamorphism. Geochimica and Cosmochimica Acta 61, 3863–3872.

Hammerschmidt, K. & Frank, E. 1991. Relics of high-pressure metamorphism in the Lepontine Alps (Switzerland) – ^{40}Ar/^{39}Ar and microprobe analyses on white micas. Schweizerische Mineralogische und Petrographische Mitteilungen 71, 261–274.

Handler, R., Dallmeyer, R.D., Neubauer, F. & Frank, W. 1993. Deformation and isotopic resetting in low-grade metamorphic rocks. Abstracts supplement No. 2 to Terra Nova 5, p. 13.

Handler, R., Neubauer, F., Velichkova, S.H. & Ivanov, Z. 2004: ^{40}Ar/^{39}Ar age constraints on the timing of magmatism and post-magmatic cooling in the Panagyurishte region, Bulgaria. – Schweizerische Mineralalogischen und Petrographische Mitteilungen, 84 (1–2), 119–132.

Heizler, M.T. & Harrison, T.M. 1988. Multiple trapped argon isotopic components revealed by ^{40}Ar/^{39}Ar isochron analysis. Geochimica et Cosmochimica Acta 52, 1295–1303.

Holland, T.J.B. 1979. High water activities in the generation of high pressure kyanite eclogites in the Tauern Window, Austria. Journal of Geology 87, 1–27.

Hoschek, G. 2001. Thermobarometry of metasediments and metabasites from the Eclogite zone of the Hohe Tauern, Eastern Alps, Austria. Lithos 59, 127–150.

Inger, S. & Cliff, R.A. 1994. Timing of metamorphism in the Tauern Window, Eastern Alps: Rb-Sr ages and fabric formation. Journal of Metamorphic Geology 12, 695–707.

Kelley, S. 1995. Ar-Ar dating by laser microprobe. In: Potts, P.J., Bowles, J.F.W., Reed, S.J. B. & Cave. M. R. (Eds.): Microprobe Techniques in the Earth Sciences. The mineralogical Society Series 6, 327–358.

Kirschner, D.L., Cosca, M.A., Masson, H. & Hunziker, J.C. 1996. Staircase ^{40}Ar/^{39}Ar spectra of fine-grained white mica: Timing and duration of deformation and empirical constraints on argon diffusion. Geology 24, 747–751.

Kretz, R. 1983. Symbols for rock-forming minerals. American Mineralogist 68, 277–279.

Kurz, W. 2005. Constriction during exhumation: Evidence from eclogite microstructures. Geology 33, 37–40.

Kurz, W. & Froitzheim, N. 2002. The exhumation of eclogite-facies metamorphic rocks – a review of models confronted with examples from the Alps. International Geology Review 44, 702–743.

Kurz, W., Neubauer, F. & Genser, J. 1996. Kinematics of Penninic nappes (Glockner Nappe and basement-cover nappes) in the Tauern Window (Eastern Alps, Austria) during subduction and Penninic-Austroalpine collision. Eclogae Geologicae Helvetiae 89, 573–605.

Kurz, W., Neubauer, F. & Dachs, E. 1998a. Eclogite meso- and microfabrics: implications for the burial and exhumation history of eclogites in the Tauern Window (Eastern Alps) from P-T-d paths. Tectonophysics 285, 183–209.

Kurz, W., Neubauer, F., Genser, J. & Dachs, E. 1998b. Alpine geodynamic evolution of passive and active continental margin sequences in the Tauern Window (eastern Alps, Austria, Italy): a review. International Journal of Earth Sciences (Geologische Rundschau) 87, 225–242.

Kurz, W., Fritz, H., Piller, W.E., Neubauer, F. & Genser, J. 2001a. Overview of the Palaeogene of the Eastern Alps. In: Piller, W.E. & Rasser, M. (Eds.), Palaeogene of the Eastern Alps: Österreichische Akademisch wissenschaftliche, Schriftenreihe der erdwissenschaftlichen Kommission 14, 11–56.

Kurz, W., Neubauer, F., Genser, J., Unzog, W. & Dachs, E. 2001b. Tectonic Evolution of Penninic Units in the Tauern Window during the Palaeogene: Constraints from Structural and Metamorphic Geology. In: Piller, W.E. & Rasser, M. (Eds.): Paleogene of the Eastern Alps: Österreichische Akademisch wissenschaftliche, Schriftenreihe der erdwissenschaftlichen Kommission 14, 11–56.

Kurz, W., Jansen, E., Hundenborn, R., Pleuger, J., Schäfer, W. & Unzog, W. 2004. Microstructures and Crystallographic Preferred Orientations of omphacite in Alpine eclogites: implications for the exhumation of (ultra-)high-pressure units. Journal of Geodynamics 37, 1–55.

Lips, A.L.W., White, S.H. & Wijbrans, J.R. 1998. ^{40}Ar/^{39}Ar laserprobe direct dating of discrete deformational events: a continuous record of early Alpine tectonics in the Pelagonian Zone, NE Aegean area, Greece. Tectonophysics 298, 133–153.

Lister, G.S. & Baldwin, S.L. 1996. Modelling the effect of arbitrary P-T-t histories on argon diffusion in minerals using the MacArgon program for the Apple Macintosh. Tectonophysics 253, 83–109.

Miller, C. 1974. On the metamorphism of the eclogites and high-grade blueschists from the Penninic Terrane of the Tauern Window, Austria. Schweizerische Mineralogische und Petrographische Mitteilungen 54, 371–384.

Müller, W., Dallmeyer, R.D., Neubauer, F. & Thöni, M. 1999. Deformation-induced resetting of Rb/Sr and 40Ar/39Ar mineral systems in a low-grade, polymetamorphic terrane (Eastern Alps, Austria). Journal of the geological Society of London 156, 261–278.

Neubauer, F., Genser, J. & Handler, R. 2000. The Eastern Alps: Result of a two-stage collision process. Mitteilungen der Österreichischen Geologischen Gesellschaft 92, 117–134.

Ortner, H. & Sachsenhofer, R. 1996: Evolution of the Lower Inntal Tertiary and Constraints on the Development of the Source Area. In: Liebl, W. & Wessely, G. (Eds): Oil and Gas in Alpidic Thrust Belts and Basins of Central and Eastern Europe. EAEG Special Publication 5, 237–247.

Ortner, H., Reiter, F. & Brandner, R. 2006. Kinematics of the Inntal shear zone-sub- Tauern ramp fault system and the interpretation of the TRANSALP seismic section, Eastern Alps, Austria. Tectonophysics 414, 241–258.

Proyer, A. 2003. Metamorphism of pelites in NKFMASH – a new petrogenetic grid with implications for the preservation of high-pressure mineral assemblages during exhumation. Journal of metamorphic Geology 21, 493–509.

Proyer, A., Dachs, E. & Kurz, W. 1999. Relics of high-pressure metamorphism in the Glockner region, Hohe Tauern, Austria: Textures and mineral chemistry of retrogressed eclogites. Mitteilungen der Österreichischen Geologischen Gesellschaft 90, 43–55.

Raith, M., Mehrens, C. & Thöle, W. 1980. Gliederung, tektonischer Bau und metamorphe Entwicklung der penninischen Serien im südlichen Venediger-Gebiet, Osttirol. Jahrbuch der Geologischen Bundesanstalt 123, 1–37.

Ratschbacher, L., Dingeldey, C., Miller, C., Hacker, B.R. & McWilliams, M.O. 2004. Formation, subduction, and exhumation of Penninic oceanic crust in the Eastern Alps: time constraints from ^{40}Ar/^{39}Ar geochronology. Tectonophysics 394, 155–170.

Reddy, S.M., Cliff, R.A. & East, R. 1993. Thermal history of the Sonnblick Dome, south-east Tauern Window, Austria: Implications for Heterogenous uplift within the Pennine basement. International Journal of Earth Sciences (Geologische Rundschau) 82, 667–675.

Scaillet, S. 1998. K-Ar (^{40}Ar/^{39}Ar) Geochronology of Ultrahigh Pressure Rocks. In: Hacker, B.R. & Liou, J. G: (eds.): When Continents Collide: Geodynamics of Ultrahigh Pressure Rocks. Kluwer Academic Publishers/Dordrecht – Boston – London, 161–201.

Schmid, S.M., Fügenschuh, B., Kissling, E. & Schuster, R. 2004. Tectonic map and overall architecture of the Alpine orogen. Eclogae geologicae Helvetiae 97, 93–117.

Schuster, R. & Kurz, W. 2005. Eclogites in the Eastern Alps: High-pressure metamorphism in the context of Alpine orogeny. Mitteilungen der Österreichischen Mineralogischen Gesellschaft 150, 183–198.

Selverstone, J. 1993. Micro- to macroscale interactions between deformational and metamorphic processes, Tauern Window, Eastern Alps. Schweizerische Mineralogische und Petrographische Mitteilungen 73, 229–239.

Selverstone, J., Spear, F.S., Franz, G. & Morteani, G. 1984. High- Pressure Metamorphism in the SW Tauern Window, Austria: P-T Paths from Hornblende- Kyanite- Staurolite Schists. Journal of Petrology 25, 501–531.

Spear, F.S. & Franz, G. 1986. P-T evolution of metasediments from the Eclogite Zone, south central Tauern Window, Austria. Lithos 25, 219–234.

Stöckhert, B., Massonne, H-J. & Nowlan, E.U. 1997. Low differential stress during high-pressure metamorphism: The microstructural record of a metapelite from the Eclogite Zone, Tauern Window, Eastern Alps. Lithos 41, 103–118.

Stuart, F.M. 2002. The exhumation history of orogenic belts from ^{40}Ar/^{39}Ar ages of detrital micas. Mineralogical Magazine 66, 121–135.

Thöni, M. 2006. Dating eclogite-facies metamorphism in the Eastern Alps – approaches, results, interpretations: a review. Mineralogy and Petrology 88, 123–148.

TRANSALP Working Group 2002. First deep seismic reflection images of the Eastern Alps reveal giant crustal wedges and transcrustal ramps. Geophysical Research Letters 29, 92–2–92–4.

TRANSALP working group 2006. TRANSALP – A transect through a young collisional orogen: Introduction. Tectonophysics 414, 1–7.

Villa, I.M. 1998. Isotopic closure. Terra Nova 10, 42–47.

von Blanckenburg, F. & Davies, H.J. 1995. Slab breakoff: A model for syncollisional magmatism and tectonics in the Alps. Tectonics 14, 120–131.

von Blanckenburg, F.& Villa, I.M. 1988. Argon retentivity and argon excess in amphiboles from the garbenschists of the Western Tauern Window, Eastern Alps. Contributions to Mineralogy and Petrology 100, 1–11.

von Blanckenburg, F., Villa, I.M., Baur, H., Morteani, G. & Steiger, R.H. 1989. Time Calibration of a PT-path from the Western Tauern Window, Eastern Alps: the problem of closure temperatures. Contributions to Mineralogy and Petrology 101, 1–11.

Zimmermann, R., Hammerschmidt, K. & Franz, G. 1994. Eocene high pressure metamorphism in the Penninic units of the Tauern Window (Eastern Alps):evidence from ^{40}Ar-^{39}Ar dating and petrological investigations. Contributions to Mineralogy and Petrology 117, 175–186.

Manuscript received 11 October, 2007
Revision accepted 5 February, 2008
Published Online first October 22, 2008
Editorial Handling: Stefan Schmid, Stefan Bucher

Appendix I. Location of samples used for microprobe analyses and ^{40}Ar/^{39}Ar dating.

For samples 1–4 PT-data have been published by KURZ et al. (1998a). To allow easy correlation of the samples, the sample numbers used by KURZ et al. (1998a) are given in parentheses. Numbers and names of the Austrian topographic map ÖK50 are given.

Sample 1 (WK526): Eclogite-mylonite of the Eclogite Zone; Dorfertal 750 m SE of the hut "Johanneshütte"; ÖK50, sheet 152 Matrei; 12° 20' 21" E, 47° 03' 34" N.

Sample 2 (WK538): Eclogite-mylonite of the Eclogite Zone; Timmeltal: north-west-shore of lake "Eissee"; ÖK50, sheet 152 Matrei; 12° 23' 10" E, 47° 03' 56" N.

Sample 3 (WK540): Eclogite-mylonite of the Eclogite Zone; Timmeltal: west-shore of lake "Eissee"; ÖK50, sheet 152 Matrei; 12° 23' 01" E, 47° 03' 48" N.

Sample 4 (WK557): Fine-grained eclogite of the Eclogite Zone; Frosnitz-tal, 250 m north of location "Steinsteig"; ÖK50, sheet 152 Matrei; 12° 27' 01" E, 47° 04' 31" N.

Sample 5: Calcite marble of the Rote Wand – Modereck nappe; location "Hochtor" at the road "Glocknerstraße"; ÖK50, sheet 154 Rauris; 12° 50' 34" E, 47° 05' 04" N.

1661-8726/08/01S207-17
DOI 10.1007/s00015-008-1294-9
Birkhäuser Verlag, Basel, 2008

Swiss J. Geosci. 101 (2008) Supplement 1, S207–S223

Mapping of the post-collisional cooling history of the Eastern Alps

STEFAN W. LUTH[1,*] & ERNST WILLINGSHOFER[1]

Key words: Eastern Alps, Tauern Window, geochronology, cooling, mapping, exhumation

ABSTRACT

We present a database of geochronological data documenting the post-collisional cooling history of the Eastern Alps. This data is presented as (a) geo-referenced *isochrone maps based on Rb/Sr, K/Ar (biotite)* and *fission track (apatite, zircon)* dating portraying cooling from upper greenschist/amphibolite facies metamorphism (500–600 °C) to 110 °C, and (b) as *temperature maps* documenting key times (25, 20, 15, 10 Ma) in the cooling history of the Eastern Alps. These cooling maps facilitate detecting of cooling patterns and cooling rates which give insight into the underlying processes governing rock exhumation and cooling on a regional scale.

The compilation of available cooling-age data shows that the bulk of the Austroalpine units already cooled below 230 °C before the Paleocene. The onset of cooling of the Tauern Window (TW) was in the Oligocene-Early Miocene and was confined to the Penninic units, while in the Middle- to Late Miocene the surrounding Austroalpine units cooled together with the TW towards near surface conditions.

High cooling rates (50 °C/Ma) within the TW are recorded for the temperature interval of 375–230 °C and occurred from Early Miocene in the east to Middle Miocene in the west. Fast cooling post-dates rapid, isothermal exhumation of the TW but was coeval with the climax of lateral extrusion tectonics. The cooling maps also portray the diachronous character of cooling within the TW (earlier in the east by ca. 5 Ma), which is recognized within all isotope systems considered in this study.

Cooling in the western TW was controlled by activity along the Brenner normal fault as shown by gradually decreasing ages towards the Brenner Line. Cooling ages also decrease towards the E–W striking structural axis of the TW, indicating a thermal dome geometry. Both cooling trends and the timing of the highest cooling rates reveal a strong interplay between E–W extension and N–S orientated shortening during exhumation of the TW.

Introduction

The European Alps are one of the most intensively investigated orogens on earth. However, many questions remain, particularly in the Eastern Alps, such as the exact timing of deformation, metamorphism and exhumation of rocks and their related feedback loops with deep (mantle) as well as surface processes.

Many conclusions in the context of mountain building have been derived in the past with the help of geobarometry, geothermometry, and geochronology. The latter is useful to constrain the timing of metamorphic and deformation events under certain temperature and pressure conditions within evolving orogens. Geochronological data from the Eastern Alps is abundant and underpins many findings throughout the last decades. For example, the Eastern Alps bear evidence for two independent collisional events in the Cretaceous and the Paleogene, respectively (e.g. Neubauer et al. 2000 and references therein)

or the temporal variations of peak metamorphic conditions (Inger & Cliff 1994; Hoinkes G. 1999; Neubauer et al. 2000). The wide spread in recorded ages, both, in time and space, in combination with the complex Alpine history has given rise to different interpretations of their meaning (e.g. crystallization, metamorphism, exhumation). A review of geochronological data in a regional framework can be valuable in deciphering orogen-scale trends, which help to better constrain processes at work and to separate local from regional effects.

Following up on the reviews of e.g. Frank et al. (1987a), Thöni, (1999) and Hoinkes et al. (1999), we present updated and geo-referenced compilations of isotope data for the Eastern Alps.

The presented work aims to map the post-collisional cooling history of the entire Eastern Alps by compiling available cooling ages in one single database. From this database we can extract thematic maps regarding the cooling history by using GIS mapping tools. These cooling maps can be used to gain

[1] Vrije Universiteit van Amsterdam, De Boelelaan 1085, 1081 HV Amsterdam.
*Corresponding author: Stefan Luth. E-mail: stefan.luth@falw.vu.nl

insight into regional cooling trends, cooling rates, and their relation with the tectonics, such as the formation of the Tauern Window (TW), the distribution of north-south shortening due to indentation, or the onset of lateral extrusion. Furthermore, the presented database also highlights regions of poor data coverage and, hence, may be of importance for the planning of future dating projects. The database will be updated regularly and can be accessed via the internet (http://www.geo.vu.nl/~wile/).

Geological Setting

The structure of the Eastern Alps is that of a collisional orogenic belt with the continental Austroalpine (AA) unit as highest tectonic unit overlying the Penninic ("Alpine Tethys") suture and the continental units, including their sedimentary cover of European affinity (i.e. Sub-Penninic sensu Schmid et al. 2004) in lowest structural position (Figs. 1a–b). Multiply deformed Austroalpine basement and cover units of Apulian origin are most abundant in the Eastern Alps and their internal structure and last metamorphic overprint is related to Cretaceous stacking of tectonic units following the consumption of the Triassic Meliata ocean farther east (Neubauer et al. 2000 and references therein). The Penninic suture and the Sub-Penninic units are exposed along the central axis of the Eastern Alps within tectonic windows such as the Engadin-, Tauern- and Rechnitz Windows.

They contain remnants of the Penninc oceanic crust, (para) autochthonous, highly metamorphosed cover units of the European continental shelf (Lower Schieferhülle), and allochthonous metasediments, derived from the Penninic Ocean (Upper Schieferhülle). The ophiolites belong to the Jurassic, Alpine Tethys (Penninic Ocean), which separated European from Apulian paleogeographic domains (Frisch 1979; Oberhauser 1995; Schmid et al. 2003). Subduction of this ocean commenced during the late Cretaceous and lasted until the Eocene (Frisch, 1979).

The core of the TW exposes pre-Variscan metamorphic basement and Variscan granitoids, the Zentralgneiss, in a series of domes (Zimmermann 1994; Oberhänsli & Goffé 2004). The domes form the core of a large anticline with an axis parallel to the window's strike as a result of syn- and post collisional shortening coeval with orogen-parallel extension (Lammerer and Weger, 1998).

The Periadriatic Fault separates the AA units from the Southern Alps, a Miocene, south vergent fold and thrust belt of Apulian origin (Castellarin et al. 1992) (Fig. 1). Slip along this fault was right-lateral with a minor (few kilometers) north-side up component during the late Oligocene and Miocene (Ratschbacher et al. 1991; Mancktelow 1995).

Several Tertiary plutons are located along or in the vicinity of the Periadriatic line, such as the Rensen and Rieserferner plutons. Rb/Sr whole rock dating on these granodiorites and tonalites reveals Oligocene ages and geochemical analyses demonstrate a source at the base of thickened crust (Borsi et al. 1978b).

Metamorphism in the Eastern Alps

Austroalpine Units

The AA units are characterized by widespread Cretaceous (Eo-Alpine) metamorphism and a weak post-Cretaceous thermal overprint, which is largely restricted to the vicinity of the TW (e.g. Thöni 1999; Hoinkes et al. 1999).

In general, the grade of the eo-Alpine metamorphism increases from north to south from sub-greenschist to ultra-high pressure conditions (Hoinkes et al. 1999; Oberhänsli & Goffé 2004; Janak et al. 2004). The southern limit of the Alpine greenschist to lower amphibolite metamorphic overprint, as indicated by incomplete resetting of the Rb-Sr white mica system (Borsi 1978) within AA units, coincides with the Defereggen-Antholz-Vals (DAV) Line (Fig. 2b) (Hoinkes et al. 1999 and references therein; Most, 2003). North of the DAV age data within the AA units are only slightly older than those from the TW. In contrast, zircon Fission Track (FT) data to the south of the DAV exclusively record pre-Miocene cooling (Stöckhert et al. 1999; Steenken et al. 2002).

The Paleogene thermal overprint within AA units close to the northeastern corner of the TW is related to thrusting within the lowermost AA units (Liu et al. 2001).

Penninic Units

The metamorphic grade within the Penninic units ranges from eclogite- to greenschist facies. In general, PT-loops derived from different areas in the Eastern Alps show a retrograde path from eclogite facies metamorphic conditions followed by blueschist and amphibolite/greenschist facies metamorphism (Hoinkes et al. 1999).

Eclogite facies metamorphic rocks are restricted to a relatively small strip in the central southern TW (Eclogite Zone) (Fig. 2). According to Frank (1987) the eclogites first cooled to blueschist facies conditions followed by reheating again to 500–600 °C and pressures between 5 and 7 kbar. Further cooling to 375–400 °C and 2–4 kbar took place along similar paths for all Penninic units (Holland 1979).

Timing of eclogitization is speculative, but if dating of the subsequent blueschist event is correct, then the eclogites are of pre-Oligocene age. Eclogite formation occurred under peak pressures of 20–25 kbar and temperatures between 580–650 °C (Holland 1979; Frank 1987; Kurz et al. 1998). These conditions are equivalent to depths of about 60–90 km and were characterized by a very low geothermal gradient of 7–9 °C/km typical for subduction zones (Fig. 2a).

K/Ar white mica ages from the southern border of the central TW range between 34 and 30 Ma and are interpreted as crystallization ages related to blueschist metamorphism (Lambert 1970; Cliff et al. 1985; Zimmermann 1994). Data compiled by Frank et al. (1987) suggest that blueschist formation occurred at temperatures between 400 and 500 °C and pressures around 9 kbar in the TW. The blueschist metamorphic event

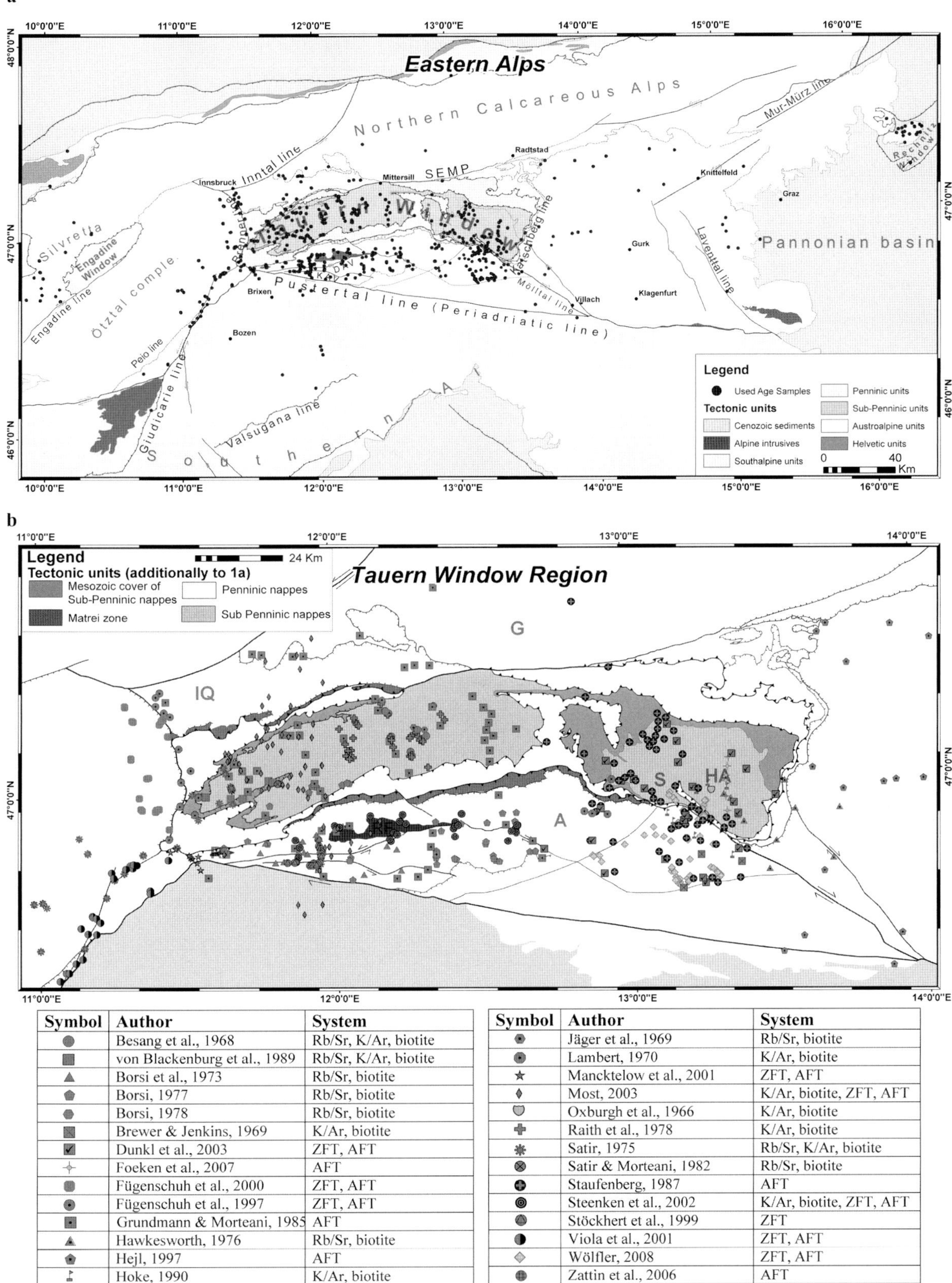

Symbol	Author	System
	Besang et al., 1968	Rb/Sr, K/Ar, biotite
	von Blackenburg et al., 1989	Rb/Sr, K/Ar, biotite
	Borsi et al., 1973	Rb/Sr, biotite
	Borsi, 1977	Rb/Sr, biotite
	Borsi, 1978	Rb/Sr, biotite
	Brewer & Jenkins, 1969	K/Ar, biotite
	Dunkl et al., 2003	ZFT, AFT
	Foeken et al., 2007	AFT
	Fügenschuh et al., 2000	ZFT, AFT
	Fügenschuh et al., 1997	ZFT, AFT
	Grundmann & Morteani, 1985	AFT
	Hawkesworth, 1976	Rb/Sr, biotite
	Hejl, 1997	AFT
	Hoke, 1990	K/Ar, biotite

Symbol	Author	System
	Jäger et al., 1969	Rb/Sr, biotite
	Lambert, 1970	K/Ar, biotite
	Mancktelow et al., 2001	ZFT, AFT
	Most, 2003	K/Ar, biotite, ZFT, AFT
	Oxburgh et al., 1966	K/Ar, biotite
	Raith et al., 1978	K/Ar, biotite
	Satir, 1975	Rb/Sr, K/Ar, biotite
	Satir & Morteani, 1982	Rb/Sr, biotite
	Staufenberg, 1987	AFT
	Steenken et al., 2002	K/Ar, biotite, ZFT, AFT
	Stöckhert et al., 1999	ZFT
	Viola et al., 2001	ZFT, AFT
	Wölfler, 2008	ZFT, AFT
	Zattin et al., 2006	AFT

Fig. 1. a) Simplified tectonic map of the Eastern Alps based on Bigi et al. (1990–92) and Egger et al. (1999). The dots represent the sample locations used in this study. Abbreviations, DAV: Deffereggen-Antholz-Vals Line, KAV: Kalkstein-Vallarga Line, SEMP: Salzach–Ennstal–Mariazell–Puchberg Line. b) Overview of the distribution of sample locations in the Tauern Window region of which geochronological data have been used in this study. Different symbols refer to previous studies as shown in the table. Abbreviations, R: Rensen pluton, RF: Rieserferner pluton, IQ: Innsbruck Quartzphyllite, G: Greywacke zone, A: Altkristallin, S: Sonnblick dome, HA: Hochalm dome.

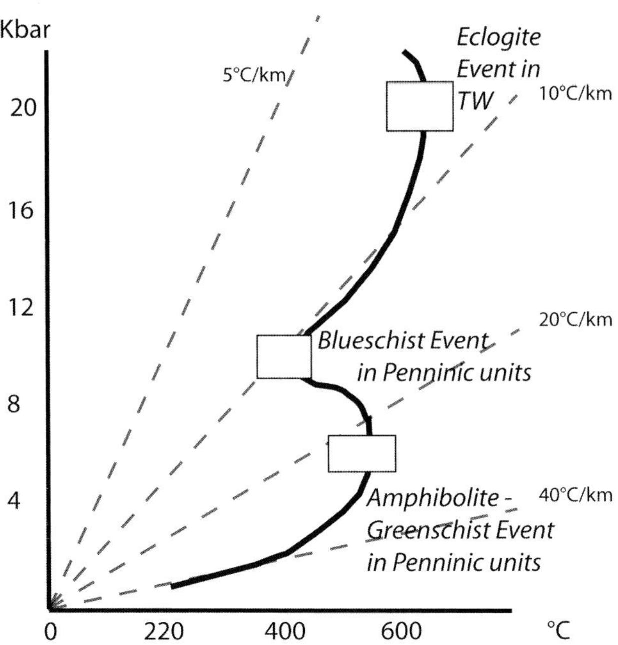

represents a stage of exhumation from ~80 km to ~35 km (e.g. Kurz et al. 1998).

Within the Rechnitz Window remnants of a high-pressure/low-temperature event are found within ophiolitic sequences and yielded metamorphic conditions of 330–370 °C and minimum pressures of 6–8 kbar (Koller 1985). There is no correlation of the blueschist facies metamorphism throughout the Penninic realm due to a lack of reliable age dating.

Within most of the Penninic units, Oligocene greenschist to amphibolite facies metamorphism led to penetrative deformation of rocks and the (re)crystallization of mineral assemblages of which white micas from the eastern TW have been

Fig. 2. a) Generalized pressure-temperature loops of the Penninic units at different locations within the Eastern Alps after Kurz et al. (1998) and Hoinkes et al. (1999). TW: Tauern Window. Dashed lines show different isotherms. b) Map showing Paleogene peak temperatures in the Eastern Alps. After Oberhänsli & Goffé (2004).

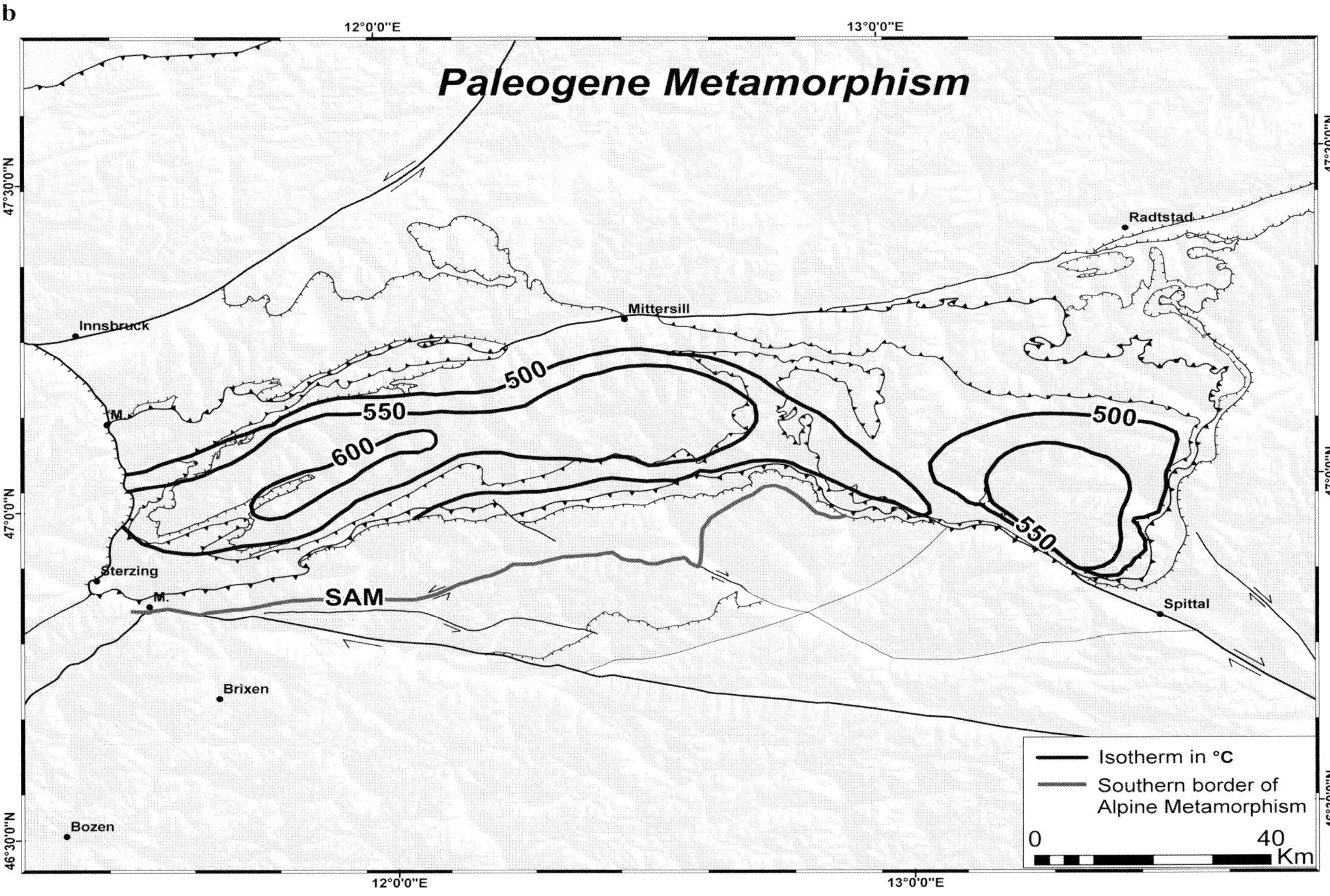

dated between 30 and 28 Ma (Inger & Cliff 1994; Thöni 1999). Peak temperatures of 550 to 600 °C under pressure conditions of 5–7 kbar reflect a geothermal gradient of 20–35 °C/km (Inger & Cliff 1994; Kurz et al. 1998; Hoinkes et al. 1999; Thöni 1999). The timing of subsequent cooling is not well constrained since Rb/Sr white mica data from the western TW yielded ages as young as 16 Ma (Von Blanckenburg et al. 1989). Whether these young ages reflect crystallization, cooling, or partial isotope resetting due to ongoing deformation, is still debated (Cliff et al. 1985; Von Blanckenburg et al. 1989; Hoinkes et al. 1999; Thöni 1999). However, lower temperature isotope systems reveal a clear westward younging of cooling ages, which possibly existed at higher temperatures as well. Post-metamorphic deformation resulted in folding of the Oligocene isograds along the central axis in the western TW and the Sonnblick and Hochalm domes in the eastern TW as well as the formation of normal faults bounding the TW in the west (Brenner fault) and the east (Katschberg fault). A distinct break in metamorphic grade across both faults affirm their normal fault kinematics (Behrmann 1988; Selverstone 1988; Genser & Neubauer 1989; Fügenschuh et al. 1997).

Data Compilation

The presented cooling maps are based on *c.* 600 published age measurements derived from Rb/Sr, K/Ar (biotite) and fission track (zircon, apatite) dating, portraying a temperature range of 375–110 °C (Table 1). The obtained data were published between 1968 and 2008 and are interpreted as cooling ages. The data covers the entire Eastern Alps from the Silvretta/Engadine region in the west to the Rechnitz Window in the east, but are concentrated in several clusters mainly distributed within the TW region (Fig. 1a & b). For a complete reference list for the database, see Figure 1b.

Construction of the Cooling Maps

All dating locations have been plotted on a georeferenced tectonic base-map with a Europe Lambert Conformal Conic projection, which is a combination of the "Structural model of Italy" 1:500.000 (Bigi et al. 1990–92) and the "Geologische Übersichtskarte der Republik Österreich", scale 1:1.500.000 (Egger et al. 1999). Georeferencing and plotting of data has been done with ArcGIS 9® software.

The construction of the cooling maps through time is performed in two steps:

1) The data were categorized after closure temperature for producing three different age contoured *"isochrone maps"* (Fig. 3a–c). The pertinent temperatures to these maps are: 375 °C (K/Ar and Rb/Sr biotite), 230 °C (zircon-FT), and 110 °C (apatite-FT) (Tab. 1). Contouring of the ages was carried out after interpolation of separate data clusters by the nearest neighbour technique. The result was compared with and partly adjusted by manual contouring to minimize

or remove uncertainties owing to data scarcity and to derive at robust interpretations.

2) In order to visualize temperature at certain time periods, *"temperature maps"* were constructed (Fig. 4a–d). The time periods cover the cooling history between 25 and 10 Ma and are identical to the isochrone age intervals on the time maps. Therefore, temperature contours (isotherms) can be deduced from isochrones as well. For example, the *15 Ma temperature map* contains 110 °C isotherms, which are identical to the 15 Ma isochrones on the *110 °C isochrone map*. The combination of isotherms and the closure temperatures of samples belonging to the same age were used as input for a nearest neighbour interpolation of the temperature. For the 15 and 10 Ma temperature maps, younger ages were also included to allow distinction between samples that had already cooled to 110 °C and those that were still hotter as they show ages younger than 10 Ma. In order to visualize this difference, the younger samples were assigned a few tens of degrees above the high temperature limit of the partial annealing zone for the apatite-FT system on these particular maps.

Assumptions, Uncertainties and Simplifications

Since the Rb/Sr, Ar/Ar and K/Ar systems in biotite record lower than peak metamorphic temperatures, which range between 500–650 °C in the Tauern Window, we apply the closure temperature concept and assume that the bulk of the data reflect post-metamorphic cooling. What is considered as closure temperatures are actually averaged temperature ranges. Especially for fission track analysis there is no specific closure temperature but a partial annealing zone defined by a temperature range in which early formed tracks can (partially) anneal and the track lengths can be reduced. Hence, only fission tracks with long track lengths, indicating rapid cooling, are consistent with

Table 1. Closure and partial annealing zone temperatures for the used thermochronometers.

Isotope system	Mineral	Used closure temperature (°C)	Closure temperature range (°C)	References
^{40}Ar/^{39}Ar ^{40}K/^{39}Ar	Biotite	375	300–400	(Grove & Harrison 1996) (Villa 1998)
	Muscovite		375–430	(Kirschner et al. 1996) (Hames & Bowring 1994)
^{87}Rb/^{86}Sr	Biotite	375	250–350	(Jäger et al. 1969)
			400	(Del Moro et al. 1982)
	White mica		450–550	(Jäger et al. 1969) (Purdy & Jäger 1976)
Fission track	Zircon	230	300–180	(Hurford & Green 1983; Zaun & Wagner 1985)
	Apatite	110	90–120	(Green et al. 1986; Gallagher et al. 1998)

the closure temperature concept. However, with respect to the rapid exhumation history of the Alps from the Oligocene onward, average temperatures seem a reasonable approximation since the variation of ages within the partial annealing zone might be of minor importance (Tab.1). The temperature range of the partial annealing zone for Apatite FT is in the order of 120–60 °C, while that for Zircon FT is 300–180 °C (see Tab. 1 for references).

The interpretation of muscovite and phengite ages as cooling ages is still controversial since (re)crystallization of these minerals can occur even below the closure temperature (e.g. (Glasmacher et al. 2003). Some K/Ar muscovite ages in the central TW are younger than nearby K/Ar biotite ages, although the Tc of the former is generally regarded as higher (Raith et al. 1978). Therefore, white mica ages are not shown on a separate time map, but can be used to support interpretations of the *Biotite 375 °C isochrone map* since their closure temperature ranges have some overlap (Tab. 1). Additionally, Rb/Sr white mica ages are regarded as crystallization ages and are, therefore, not considered in this compilation.

In this study we applied a simple filter excluding apatite fission track data from altitudes higher than 2000 m and lower than 1000 m in order to avoid topography-induced complexities of the cooling history, which are not representative for the regional cooling of the Eastern Alps (compare Figs. 3c and d). The chosen elevation range is based on the large amount of available data within this bracket (see histogram in Fig. 3d). Ideally, since positive age vs. altitude relationships are only observed in some sub-regions of the Eastern Alps, corrections for topography should be applied for those sub-areas separately.

Many publications lack accurate coordinates of sample locations. Hence, we applied simple scanning and georeferencing of maps and sketches and adjusted the geographical projections to our base map. Although some maps have untraceable projections and/or oversimplified sketches have been used, the maximum deviation of the plotted locations does not exceed 100 meters and is therefore a minor source of uncertainty.

Cooling Maps

In the following sections the compiled cooling maps will be described starting with the time maps for different isotope systems followed by the temperature maps, which portray regional cooling through time. Isolines with the same time are referred to as "time lines" or "isochrones". References to the underlying data are summarized in Fig. 1b and will not be repeated in the context of the description of the cooling maps.

375 °C Isochrone Map

AustroalpineUnits

The majority of the Austroalpine units cooled below 375 °C before Cenozoic times. However, the unit between the TW and the Periadriatic Line behaved differently and can be divided into a southern unit containing pre-Cenozoic ages and a northern unit, which shows a remarkable decrease of ages toward the TW on the 375 °C map (Fig. 3a) as well as on the 230 °C map (Fig. 3b). Both units are separated by the DAV (for location see Fig. 1), which coincides with the 65 and 35 Ma isochrones. In the area between the south-western corner of the TW and the tip of the Southalpine indenter, isochrone-lines trend parallel to the TW and the Brenner fault, respectively. Of particular note is the considerable north-directed age drop from 65 to 20 Ma within ~10 km distance in this zone.

Penninic Units

The youngest ages within the TW are found in its south-western corner, ranging between 15 and 12 Ma (Fig. 3a). From this location, ages gradually increase towards the northeast. The 20 Ma isochrone surrounds the western TW tracing the Brenner Line in the west and gradually closes along the central axis towards the east. Contouring along the TW's northern boundary is difficult due to lack of data. The higher temperatures there are indirectly deduced, based on relatively old zircon FT ages (Fig. 3b). Two trends appear clearly from contouring in the western TW: (1) A younging trend toward the central ENE–WSW trending structural axis of the window, which starts in the Austroalpine units with the DAV as southern limit. (2) A gradual WSW directed younging trend towards the Brenner Line, across which a distinct age break is observed.

The data from the eastern TW, though limited in amount, suggest earlier cooling than in the western TW, which is supported by K/Ar and Ar/Ar muscovite ages ranging between 30 and 20 Ma in the north-eastern TW (Liu et al. 2001), and between 27–18 Ma in the south-eastern TW (Cliff et al. 1985). Additionally, considering the confinement of Oligocene peak metamorphic conditions within the TW it is expected that the isochrones will largely follow the outline of the TW.

230 °C Isochrone Map

Austroalpine units

Parts of the Austroalpine unit between the TW and the Rechnitz Window referred to as "cold spots" by (Hejl 1997), cooled below 230 °C before Cenozoic times. More cold spots have been found in the Ötztal-Stubai basement complex, the north-eastern part of the Innsbruck Quartzphyllite, the Greywacke zone, and AA basement units north of the Periadriatic Line.

Similar to the 375 °C isochrone map, a high gradient of northward younging from 80 Ma just a few kilometers south of the Periadriatic line to 13 Ma adjacent to the TW is observed north of the Southalpine indentor. Further east, the isochrones are parallel to the DAV.

Penninic units

The south-western corner of the TW cooled through 230 °C between 15 and 12 Ma, similar to the 375 °C map indicating high cooling rates (50 °C/Ma).

In contrast, a gradual westward younging within the western TW is not well expressed, which could be due to the lack of data in the central TW. The eastern Tauern Window is characterized by 19–16 Ma zircon FT ages. Due to the lack of zircon FT ages in the central TW it is not possible to reconcile whether cooling was progressive from east to west or a separation into domains with different cooling ages. A trend towards younger ages from the window's northern border (20 Ma) towards the central axis (12 Ma) is observed within the western TW. North of the DAV ages drop from 25 to 12 Ma in northerly direction.

Cooling below 230 °C in the Rechnitz Window took place between 19 and 14 Ma. The Rechnitz thermal event also led to rejuvenation of the surrounding AA unit as indicated by a 22 Ma age measurement by Dunkl & Demeny (1997).

Southern Alps

Within the Southalpine Basement units zircon fission track ages range between 225 and 213 Ma. Only ages between 81 and 24 Ma from the Permian Brixen pluton close to the Periadriatic Line record Alpine resetting (Mancktelow et al. 2001; Viola et al. 2001).

110 °C Isochrone Map

Apatite fission track (AFT) data are displayed in figure 3c utilizing all of the compiled data. In figure 3d, only data within a selected altitude range (1000–2000 m) have been used in order to avoid dubious interpretations due to topographic affects.

Austroalpine Units

Within the Silvretta basement units west of the Engadine Window, AFT-ages vary from 31 Ma in the east and south to 14 Ma towards the southwest.

In the Ötztal-Stubai area AFT-ages get successively younger approaching the Brenner Line in a SE to E direction. The northwestern part of the Innsbruck Quartzphyllite region cooled below 110 °C not before 15 Ma and, therefore, differs from the bulk of the TW's northern surroundings; in the Greywacke zone AFT ages increase northwards to 60 Ma.

Most ages taken from the Austroalpine units between the Periadriatic line and the TW fluctuate between 20 and 10 Ma and increase slightly towards the southeast, similar to the 375 °C and the 230 °C isochrone maps. The irregular curved shape of isochrones resulting in crossing major tectonic boundaries, such as the DAV (compare Fig. 3c and Fig. 3d), is mainly due to topographic effects. Notice the remarkable difference in timing of cooling at the Brenner- and Katschberg Lines.

Penninic Units

Within the TW, cooling below 110 °C started in the east as early as 20 Ma ago and propagated westwards from that time on. No AFT data are available from the north-eastern TW, which has been contoured by extrapolation of ages from farther south. Somewhat younger ages (15–10 Ma) in the south-eastern corner of the TW may be disturbed by high thermal anomalies associated with the Mölltal fault (Dunkl et al. 2003; Wölfler 2008).

Towards the western part of the eastern TW the age variation becomes larger (16–6 Ma) and practically all AFT-ages within the central and western TW fall between 10 and 5 Ma. Here, westward younging is only limited to the Brenner Line region where ages are as young as 9–4 Ma. A younging trend towards the central axis of the TW is not well constrained. The disturbance of the isochrons may partly be due to the age/elevation correlation.

Only two AFT ages have been found in the Rechnitz Window and are dated at 10 and 7 Ma, indicating coeval cooling within the TW and Rechnitz Window below 110 °C at that time.

Southern Alps

A few ages from the Southern Alps can be interpreted only very broadly and isochrones are restricted to the Brixen pluton, which cooled below 110 °C in the Early Miocene, while further south cooling of the Southalpine Basement through 110 °C was not before 15–11 Ma. The general trend, though poorly constrained, is a north-south younging to 10 Ma near the south-vergent Valsugana thrust (Fig. 1a).

Temperature Maps

The four temperature maps (Figs. 4a–d) are interpolations derived from the isochron maps and contain the same underlying information, albeit visualizing temperature at a certain time. On a single temperature map, information is combined from all the used dating systems and used as additional control on the internal consistency of the isochron maps. This improved the contouring particularly in areas with scarce data coverage and thus, both map types are complimentary.

In the following section, the temperature maps for the time interval from 25 to 10 Ma will be briefly discussed. At 25 Ma (Fig. 4a) temperatures in the entire western and marginal parts of the eastern TW exceeded 375 °C. The "hot cells" of the western and eastern TW, exhibiting the geometry of "elongate thermal domes" are separated by a cooler region in the central TW. The significance of this temperature distribution, however, remains uncertain since few cooling age data are available from the central TW.

At 20 Ma (Fig. 4b) the eastern TW had cooled almost entirely below 200 °C with the exception of the Sonnblick and Hochhalm domes (for location see Fig. 1b).

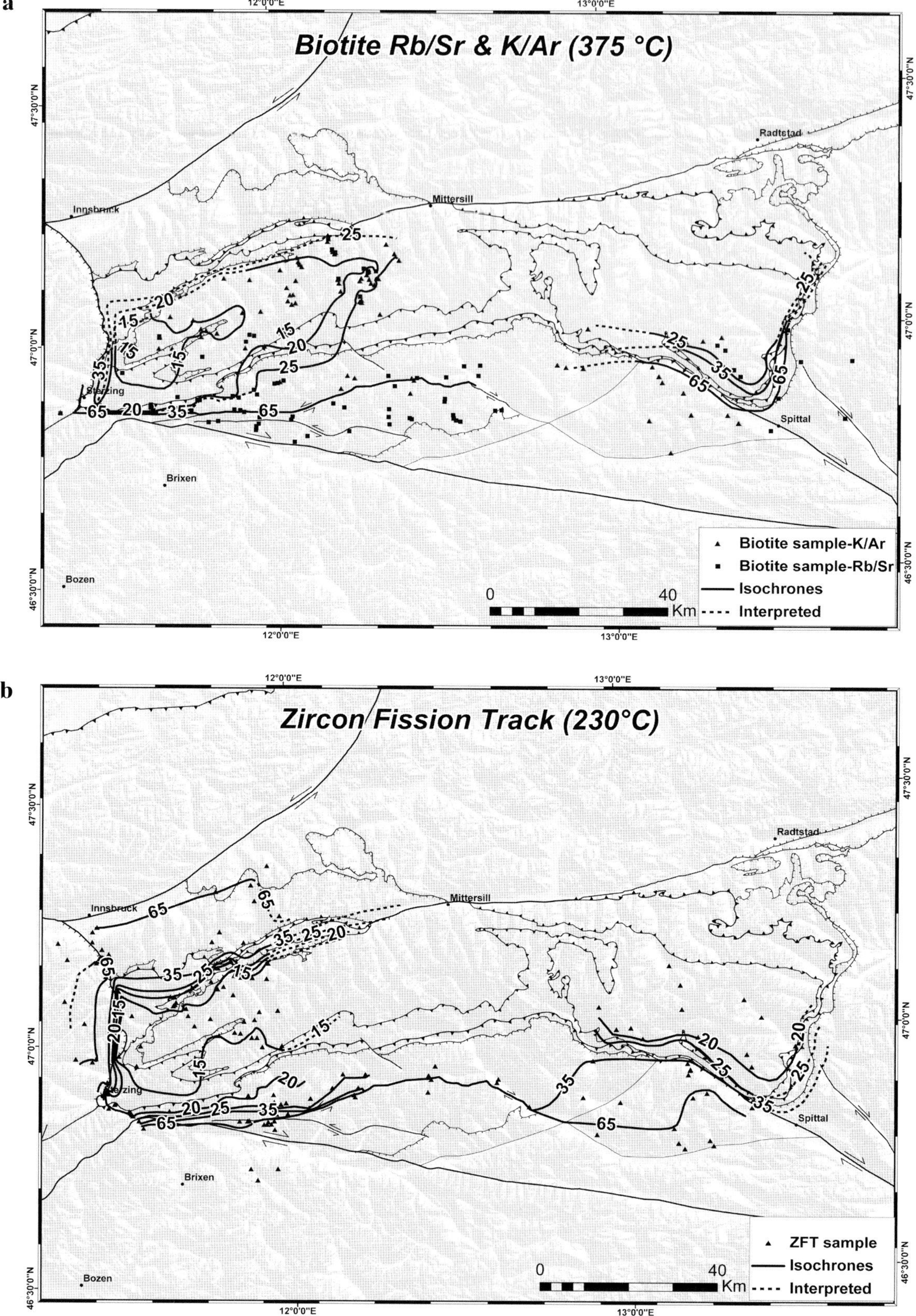

Fig. 3. Isochrone maps for temperatures of 375 (a), 230 (b) and 110 °C (c), respectively. (d) Isochrone map making only use of AFT ages with an altitude range of 1000–2000 m (dark grey in histogram) in order to reduce topographic effects. See text for further explanations. Lower right inset shows the frequency of AFT data with respect to altitude. Upper left inset: Engadine Window. Upper right inset: Rechnitz Window. The solid lines represent isochrones and are based on interpolation of the plotted ages. The dashed lines are extended interpretations. Geological boundaries are based on the Structural model of Italy (Bigi, et al. 1990–92).

c
Apatite Fission Track (110 °C)
12°0'0"E
13°0'0"E
47°30'0"N
47°0'0"N
46°30'0"N
Innsbruck
Mittersill
Radtstad
Sterzing
Brixen
Bozen
Spittal
45
35
25
10
15
10
15
10
20
35
45
15
20
25
35
10
10
10
15
20
10
15
20
25
25
20
35
20
15
15
15
20
0
40
Km
AFT Sample
Isochrones

d
Apatite Fission Track (110 °C)
1000-2000m
Engadin Window
10°0'0"E
12°0'0"E
13°0'0"E
47°30'0"N
47°0'0"N
46°30'0"N
25
25
20
0
30
km
Innsbruck
Mittersill
Radtstad
Sterzing
Brixen
Bozen
Spittal
20
15
20
25
8
8
8
8
15
10
25
20
25
20
15
15
25
10
20
15
35
10
15
0
40
Km
AFT Sample
Isochrones
Histogram Sample Elevation
Sample Count
100
80
60
40
20
1000
2000
ELEVATION

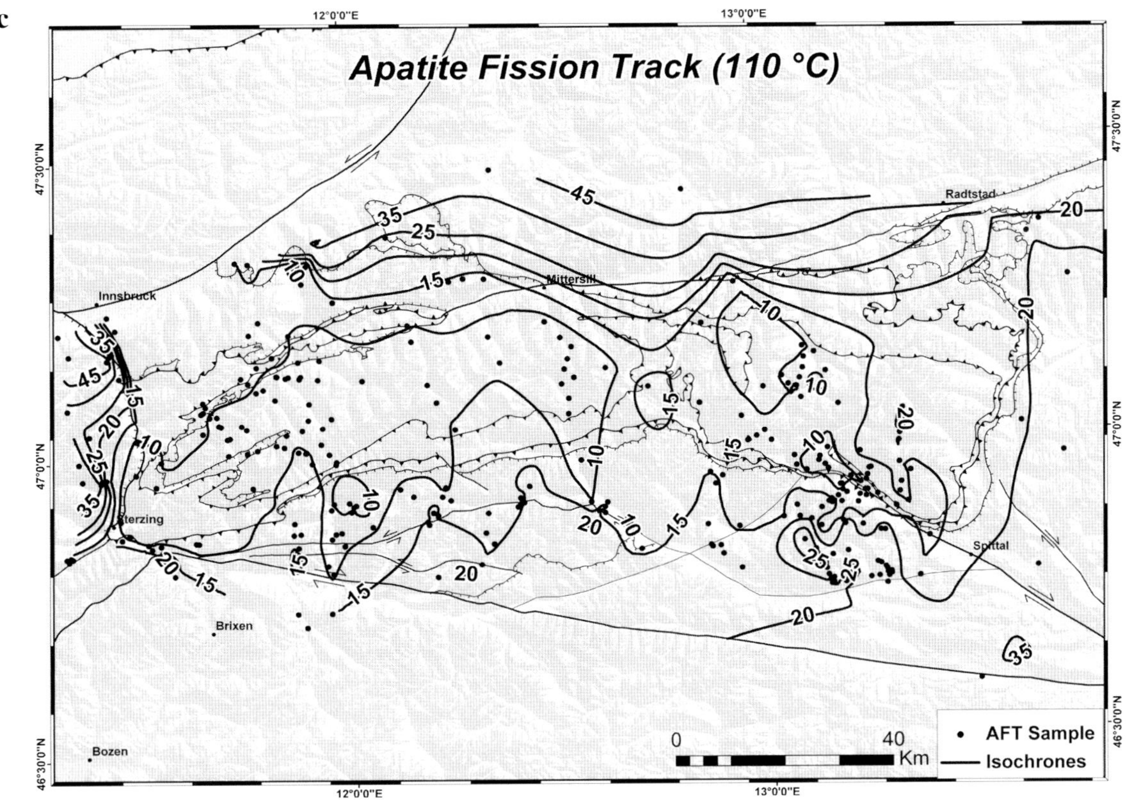

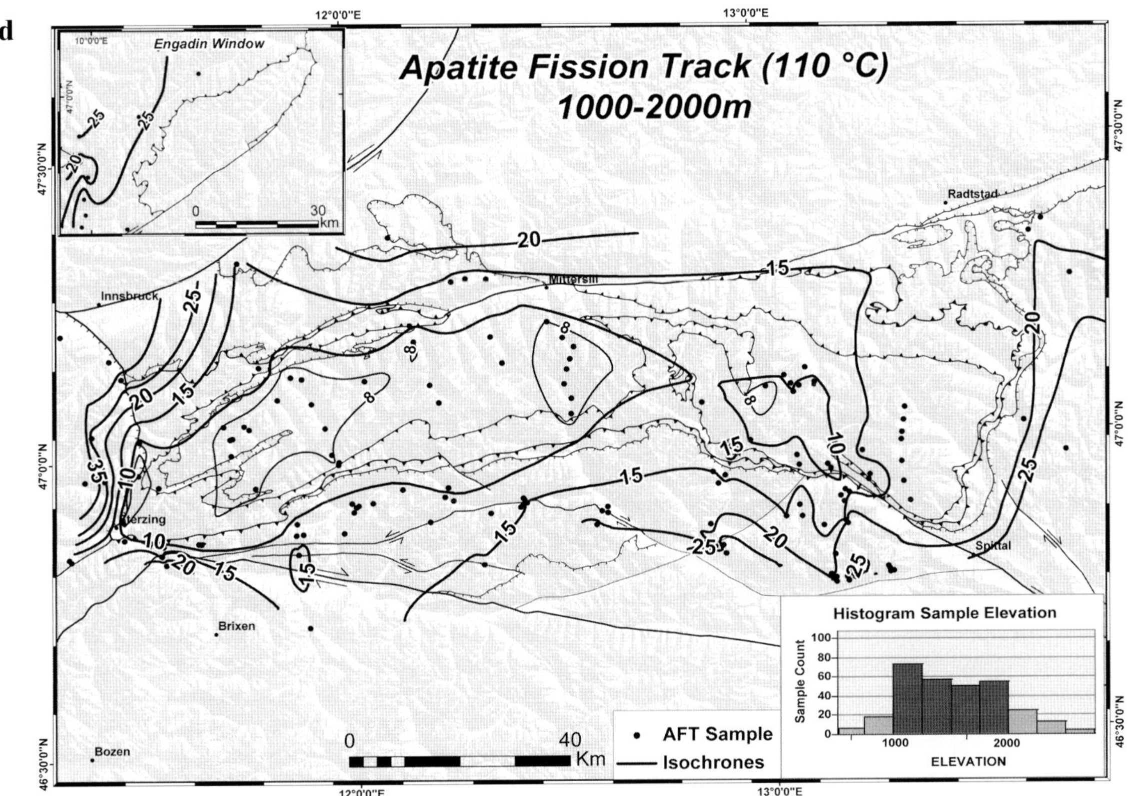

Fig. 3. c, d.

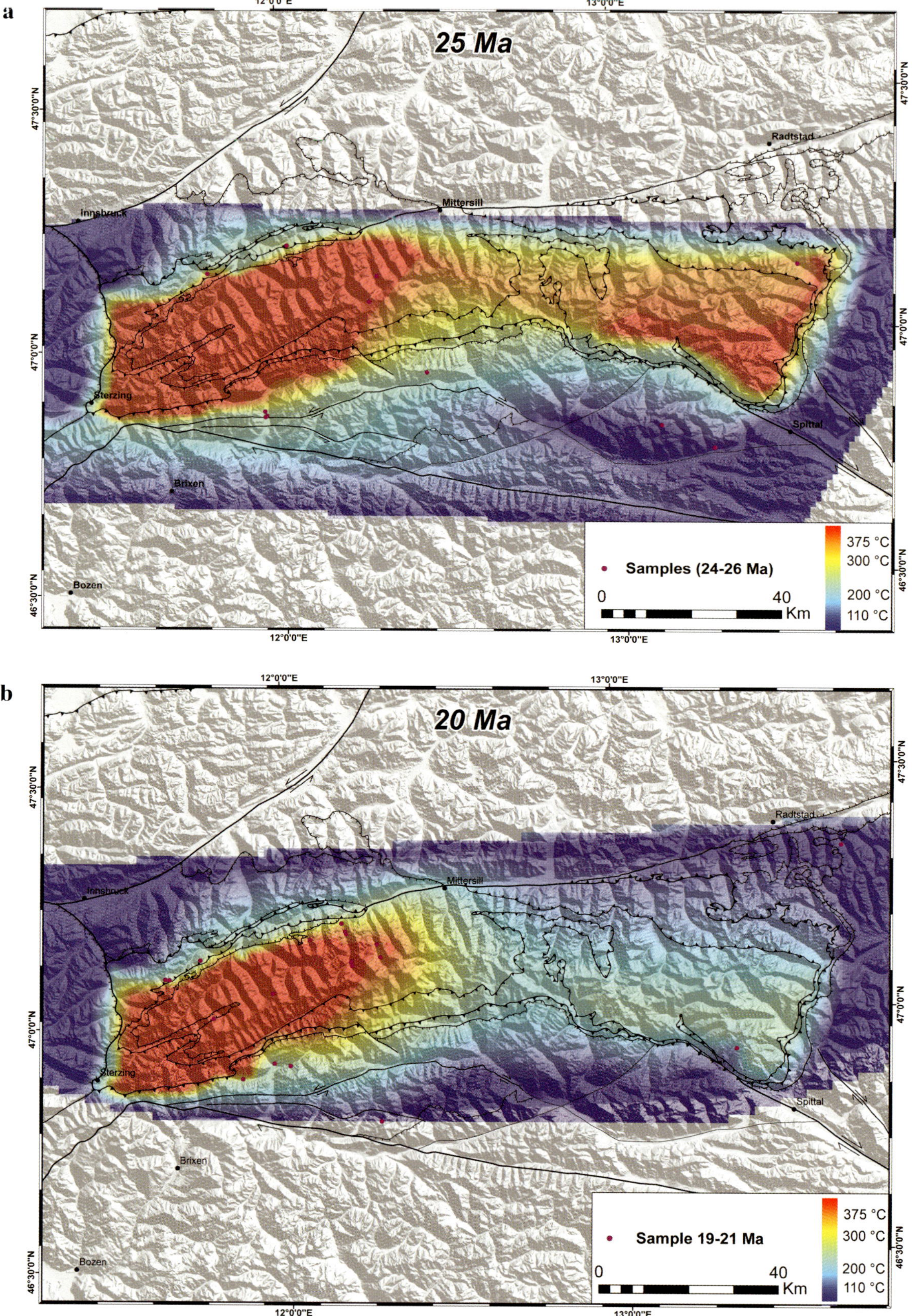

Fig. 4. Temperature maps at 25 (a), 20 (b), 15 (c) and 10 Ma (d), respectively, portraying temperatures, based on the interpolation of cooling age data. The plotted ages together with the time-lines from figure 3 were used as input. Notice the different temperature scale in figure c and d to optimize visualization of low temperatures.

c
15 Ma
47°30'0"N
47°0'0"N
46°30'0"N
46°0'0"N
12°0'0"E
13°0'0"E
Innsbruck
Sterzing
Brixen
Bozen
Mittersill
Radtstad
Spittal
Sample < 11 Ma
375 °C
300 °C
130 °C
110 °C
0
40
Km

d
10 Ma
47°30'0"N
47°0'0"N
46°30'0"N
12°0'0"E
13°0'0"E
Innsbruck
M.
Sterzing
M.
Brixen
Bozen
Mittersill
Radtstad
Spittal
Sample < 16 Ma
375 °C
300 °C
130 °C
110 °C
0
40
Km

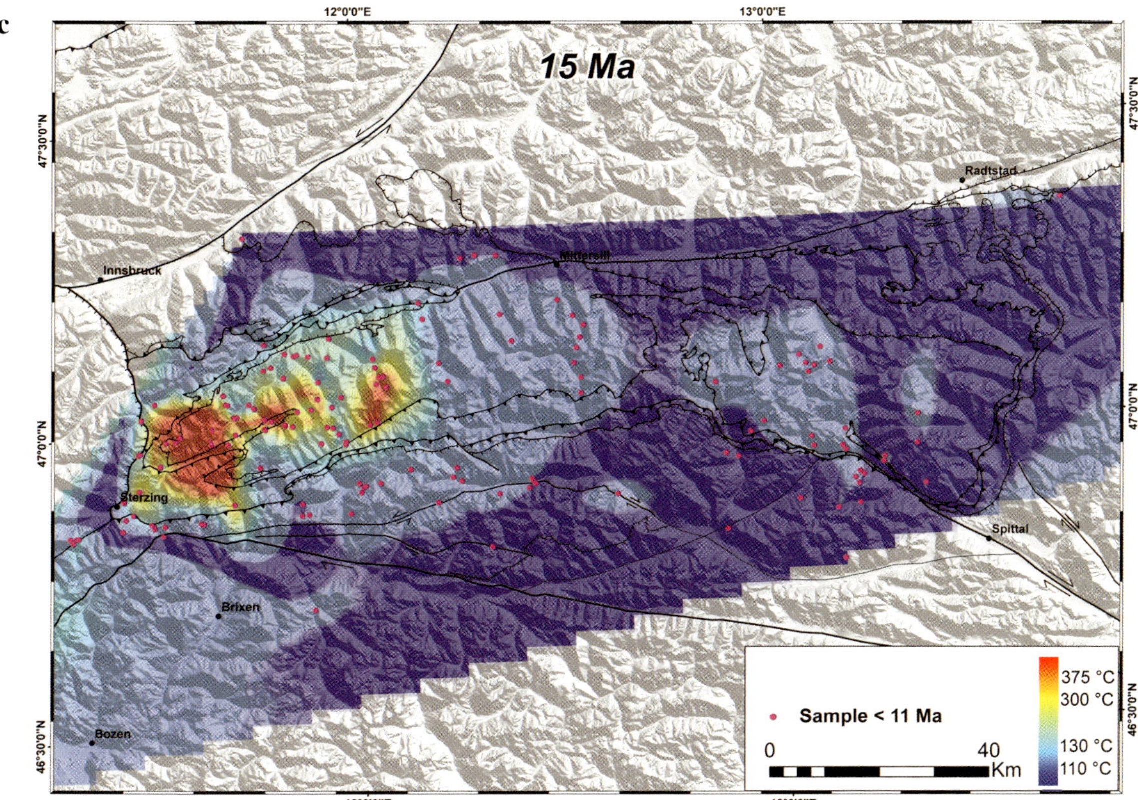

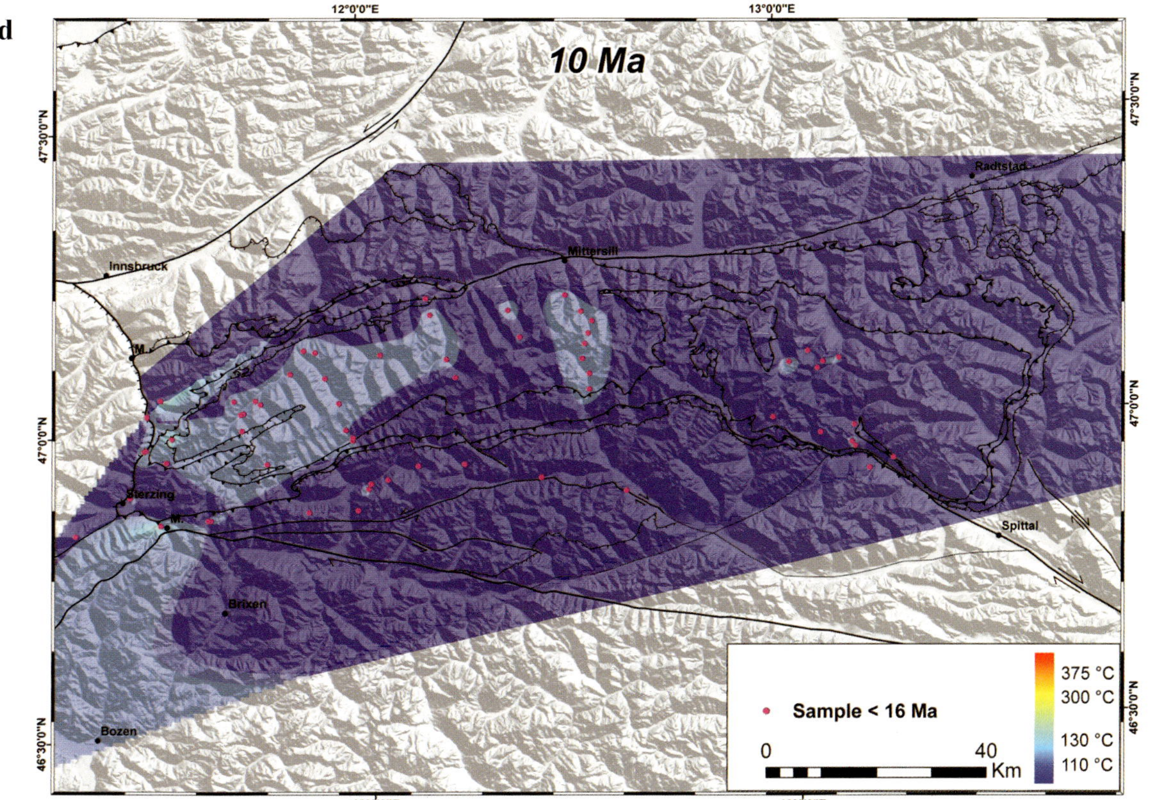

Fig. 4. c, d.

introducing a westerly dipping, normal shear zone within the central TW, which led to the separation of the Zentralgneiss-cored domes of the eastern TW from those of the western TW. Hence, in their model, deformation and subsequent cooling propagated from east to west in response to eastward escape of upper plate brittle crustal wedges and lower plate ductile flow. According to our temperature maps (Fig. 4b), the rocks of the eastern TW were already in a cool environment with temperatures around 230 °C at 20 Ma when lateral extrusion commenced. Sachsenhofer (2001) predicted surface heat flows of up to 200 mWm⁻² for the eastern TW and easterly adjacent AA units in the main phase of the extrusion process (20–15 Ma) mimicking the geometry of an extrusion corridor (see Fig. 5b of Sachsenhofer (2001)). Heat flow of that magnitude would imply temperatures of 350 °C at depths of about 5 km and the onset of partial melting at depths of ca. 10 km. Such high temperatures in shallow crustal levels should have reset the zircon fission track system to younger ages and this has yet to be documented with data. This suggests that the predicted high heat flow is rather a local, fault-controlled phenomenon than a regional-scale feature.

Diachronous cooling may have been partly conditioned by the crustal configuration prior to orogen-parallel extension in the Eastern Alps since crustal thickness and topographic gradients must have existed at that time not only in N–S but also in E–W direction (see Frisch et al. (1998) for a paleo-topography reconstruction). Consequently heat can be efficiently transferred from the thick and hot regions of the eastern TW to the less thick and cool surrounding AA units. Westerly directed heat transfer in the region of the western TW was probably inefficient due to the lack of significant crustal thickness and topographic variations (high mountainous relief from the Ötztal and Silvretta to the Swiss Alps at ca. 29–22 Ma (Frisch et al.1998).

Cooling Rates

Cooling rates in the TW's interior from 375 °C to 230 °C were high (50 °C/Ma) during the Early Miocene, while cooling towards surface conditions slowed down to 25 °C/Ma in the east at around 15–9 Ma and in the west between 12 and 8 Ma.

In the areas close to the major bounding normal faults, near-surface cooling (from 230 °C to 110 °C) was rapid (~50 °C/Ma) and happened in the eastern TW between 18 and 14 Ma and in the Brenner pass region between 14 and 10 Ma (Fügenschuh et al. 1997).

Fügenschuh et al. (1997) calculated exhumation rates in the western TW with the use of isotope data and a time-dependant thermal model for erosion. Their results suggest that Oligocene to early Miocene rapid exhumation predates late Early-Middle Miocene rapid cooling. Hence, lateral extrusion in the Eastern Alps coincides with a phase of rapid cooling. Exhumation related to lateral extrusion, therefore, was not able to maintain the high temperatures of the isothermal exhumation phase suggesting that vertical motions were already slowing down at that stage of TW formation, as also suggested by the geochronologic data of von Blanckenburg et al. (1989).

Dome Forming Mechanisms for the Tauern Window

Different exhumation mechanisms operate at different rates and time-scales and may be diachrounous in space potentially leaving behind different cooling records (Hames & Bowring 1994). Several mechanisms have been suggested for the exhumation of the TW with the primary distinction between the models in the relative importance of the N–S shortening tectonics (Lammerer & Weger, 1998) with respect to the E–W extensional, extrusion related tectonics (Selverstone & Spear 1985; Ratschbacher et al. 1991; Frisch et al. 1998, 2000).

Lammerer and Weger (1998) argue, based on strain measurements, that north-south shortening continued during east-west extension and they conclude that shortening was compensated by extension and, therefore, would not have caused substantial uplift of the TW. In their view the Tauern Window is merely a deeply eroded antiformal stack, which experienced crustal-scale folding. A reduction of crustal strength (Genser et al. 1996) as a consequence of thermal relaxation and early phases of exhumation could have favoured crustal-scale buckling and uplift.

On several accounts it has been suggested that the TW is a metamorphic core-complex (e.g. Frisch et al. 1998, 2000). With reference to classical metamorphic core complexes as described in the Basin and Range province (Lister & Davis 1989) observations in favour of metamorphic core complex interpretation include: (1) the presence of detachment faults separating brittle upper plate from ductile lower plate rocks; (2) a considerable amount of extension within the core of the TW (E–W), and (3) a lateral decrease of cooling ages towards the main detachment faults (Brenner and Katschberg Lines). However, in Basin and Range-type core complexes the kinematics of the main detachments is uni-directional, which is also documented by uni-directional cooling of the exhumed rocks (e.g. (Foster & John 1999). The early deformation history of both bounding normal faults in the Eastern Alps is still poorly understood and additional research with the focus on the relation between the Brenner Line and the Katschberg Line is needed. Furthermore, Cordilleran-type core complexes are characterised by detachments striking parallel to the mountain range (orthogonal in the Eastern Alps) and their formation post-dates shortening (coeval in the Eastern Alps). Consequently, the TW exhibits a large amount of overprinting relations between extensional and contractional structures.

The elliptical cooling pattern in the TW and close surroundings does not support a Basin and Range-type core-complex origin of the TW but argues for its formation as a syn-orogenic metamorphic dome, which was exhumed by a combination of extension and erosion. Uncertainties remain as to whether the entire TW is a single metamorphic dome or a series of domes.

From Regional to Orogen-scale Exhumation and Cooling

Higher temperature isotope systems (Fig. 3a), as well as most of the zircon fission track data (Fig. 3b), document focused cooling and exhumation confined to the Penninic tectonic windows, whereas the main body of the surrounding AA units already cooled to near surface conditions prior to ca. 20 Ma (e.g. Hejl 1997, Fügenschuh et al. 1997, 2000). Cooling through the AFT annealing zone was nearly coeval along the Brenner fault (Fügenschuh et al. 1997), the central axes of the TW (Grundmann & Morteani1985; Staufenberg 1987; Most 2003), the Jaufen, Passeier and Giudicarie faults (Viola et al. 2001), and the Valsugana thrust system (Zattin ct al. 2006), arguing for orogen-scale cooling and exhumation since ca. 12 Ma. Late middle Miocene thrusting and uplift within the Southern Alps, also documented by structural and stratigraphic data (Dunkl & Demeny 1997; Castellarin & Cantelli 2000), were coeval with the termination of extrusion tectonics in the Eastern Alps. This mutual relationship indicates that the coupling between the Southern Alps and the internal part of the Alps (the Alps north of the Periadriatic Line) increased, leading to orogen-scale uplift and cooling to near-surface conditions starting at around 12–10 Ma. Large-scale uplift led to an increase in the catchment area and hence to enhanced sediment discharge compared to the main phase of lateral extrusion (Kuhlemann et al. 2001). Other processes, which possibly contributed to the final widespread uplift stage most likely involve a combination of removal of the lithospheric root either by delamination or convection and surface erosion (Genser et al. 2007).

Conclusion

The compilation of radiometric data and their presentation in "time" and "temperature" maps allow for interpretation of temporal and spatial variations of post-metamorphic cooling within the Eastern Alps.

Based on available isotope data, we argue that the "thermal" differentiation between the western and eastern TW was well established by the end of the rapid, nearly isothermal exhumation phase during the Oligocene (ca. 30–23 Ma). This phase of deformation probably led to the separation of the Zentralgneiss cores, now exposed in the western and eastern TW, in response to orogen parallel extension (e.g. Frisch et al. 2000) and is the main reason for diachronous cooling of the TW. As a result, two thermal domes developed within the TW as documented by an increase of cooling ages away from the centre of the domes. Subsequent decay of the thermal domes was synchronous with the main phase of lateral extrusion in the Eastern Alps, which lasted from ca. 20 to 14 Ma. Hence, cooling by heat conduction and possibly also by convecting fluids dominated over exhumation related upward advection of heat, arguing for a decrease of exhumation rates during that period as suggested by von Blanckenburg et al. (1989).

While N–S cooling trends possibly reflect folding and up-doming within a N–S convergent setting, E–W cooling trends appear to be fault controlled, reflecting top-to-the-W and top-to-the-E normal displacement along the Brenner and Katschberg faults respectively. This cooling pattern is consistent with cooling of detachment-related gneiss domes e.g. (Yin 2004), but differs from uni-directional cooling of Basin and Range-type core complexes.

Orogen-scale uplift involving the Southern Alps post-dates the main phase of lateral extrusion and is documented by late Miocene apatite fission track ages from the Southern Alps, the AA unit and the TW. These data possibly reflect increased coupling between the Eastern and Southern Alps across the Periadriatic Line leading to orogen-scale uplift and erosion in the Alps.

Our synthesis of available isotope data also highlighted that more data are needed from the central and north-eastern TW and across the Katschberg normal fault. In particular, a denser network of (U-Th)/He data is required to be able to link Late Miocene to more recent tectonics (Willingshofer & Cloetingh 2003) of the Eastern Alps. Furthermore, a thorough understanding of the cooling history of the Alps and associated vertical movements should invoke a stronger coupling to the dynamics of the mantle lithosphere.

Acknowledgments

We thank A. Steenken and B. Lammerer for their thorough reviews and suggestions, which helped to improve the manuscript considerably. J. Chadwick is thanked for correcting the English. Furthermore, financial support by NWO-ALW is gratefully acknowledged. This is NSG publication number: 20080601.

REFERENCES

Barth, S., Oberli, F. & Meier, M. 1989. U-Th-Pb systematics of morphologically chracterised zircon and allanite: a high-resolution isotopic study of the Alpine rensen Pluton (northern Italy). Earth and Planetary Science Letters 95(3–4), 235–254.

Behrmann, J. 1988. Crustal-scale extension in a convergent orogen: the Sterzing-Steinach mylonite zone. Geodinamica Acta 2, 63–73.

Besang, C., Harre, W., Karl, F., Kreuzer, P., Lenz, P., Müller, P. & Wendt, I. 1968. Radiometrische Alterbestimmungen (Rb/Sr und K/Ar) an Gesteinen des Venediger-Gebietes (Hohe Tauern, Österreich. Geologisch Jahrbuch 86, 835–844.

Bigi, G., Castellarin, A., Coli, M., Del Piaz, G., Sartori, R., Scandone, P., Vai, G., Cosentino, D. & Parotto, M. 1990–92. Structural Model of Italy, scale 1:500000 sheet 1–3. Consiglio Nazionale delle Ricerche, Progetto Finalizzato Geodinamica, SELCA Firenze.

Borsi, S., Del Moro, A., Sassi, F. & Zirpoli, G. 1973. Metamorphic evolution of the austridic rocks to the south of the Tauern Window (Eastern Alps) Radiometric and Geo-petrologic data. Memoire della Società Geologica Italiana 12, 549–571.

Borsi, S., Del Moza, A., Sassi, F., Zanferrari, A. & Zirpoli, G. 1978a. New geopetrologic and radiometeric data on the alpine history of the Austridic continental margin South of the Tauern window (Eastern Alps). Memorie dell'Istituto della Regia Universitá di Padova 32, 1–20.

Borsi, S., Del Moro, A., Sassi, F. & Zirpoli, G. 1978b. On the age of the periadriatic Rensen massif (Eastern Alps). Neues Jahrbuch fur Geologie und Palaontologie 5, 267–272.

Brewer, M. & Jenkins, H. 1969. Excess radiogenic argon in metamorphic micas from the Eastern Alps, Austria. Earth and Planetary Science Letters 6, 321–331.

Castellarin, A., Cantelli, L., Fesce, J., Mercier, V., Picotti, G., Pini, G., Prosser, G. & Selli, L. 1992. Alpine compressional tectonics in the Southern Alps. Relationships with the N–S Apennines. Annales Tectonicae 6, 62–94.

Castellarin, A. & Cantelli, L. 2000. Neo-Alpine evolution of the southern Eastern Alps. Journal of Geodynamics 30, 251–274.

Christensen, J., Selverstone, J., Rosenfeld, J. & De Paolo, D. 1994. Correlation by Rb-Sr geochronology of garnet growth histories from different structural levels within the Tauern window. Contributions to Mineralogy and Petrology 118, 1–12.

Cliff, R., Droop, G. & Rex D., 1985. Alpine metamorphism in south-east Tauern Window, Austria, 2. Rates of heating, cooling and uplift. Journal of Metamorphic Geology 3, 403–415.

Coney, P. & Harms, T. 1984. Cordilleran metamorphic core complexes: Cenozoic extensional relics of Mesozoic compression. Geology 12, 550–554.

Del Moro, A., Puxeddu, M., Radicatie De Brozolo, F. & Villa, I. 1982. Rb-Sr and K-Ar ages on minerals at temperatures of 300–400 °C from deep wells in de Larderello geothermal field (Italy). Contributions to Mineralogy and Petrology 81, 340–349.

Droop, G. 1985. Alpine Metamorphism in the south-east Tauern Window, Austria. Schweizerische Mineralogische und Petrographische Mitteilungen 61, 237–273.

Dunkl, I. & Demeny, A. 1997. Exhumation of the Rechnitz window at the border of the Eastern Alps and the Pannonian Basin during Neogene extension. Tectonophysics 272, 197–211.

Dunkl, I., Frisch, W. & Grundmann, G. 2003. Zircon fission track thermochronology of the southeastern part of the Tauern Window and the adjacent Austroalpine margin, Eastern Alps. Eclogae Geologicae Helvetiae 96, 209–217.

Egger, H., Krenmayer, H., Mandl, G., Matura, A., Nowotny, A., Pascher, G., Pestal, G., Pistotnik, J., Rockenschaub, M. & Schnabel, W. 1999. Geologische Übersichtskarte der Republik Österreich 1:1500000. Geologische Bundesanstalt.

Foeken, J., Persano, C., Stuart, F. & Ter Voorde, M. 2007. Role of topography in isotherm pertubation: Apatite (U-Th)/He and fission track results from the Malta tunner, Tauern Window, Austria. Tectonics 26(3), TC3016.

Foster, D. & John, B. 1999. Quantifying tectonic exhumation in an extensional orogen with thermochronology: examples from the southern Basin and Range Province. In: Ring, U., Brandon, M., Lister, G. & Willett, S. (Eds): Exhumation processes: normal faulting, ductile flow and erosion,Geological Society London, Special Publications, 343–364.

Frank, W. 1987. Evolution of the Austroalpine elements in the Cretaceous. In: Frank, W., Flügel, H. & Faupl, P. (Eds): Geodynamics of the Eastern Alps, Deuticke, Vienna, 379–406.

Frisch, W. 1979. Tectonic progratation and plate tectonic evolution of the Alps. Tectonophysics 60, 121–139.

Frisch, W., Kuhlemann, J., Dunkl, I. & Brügel, A. 1998. Palinspastic reconstruction and topographic evolution of the Eastern Alps during late Tertiary tectonic extrusion. Tectonophysics 297, 1–15.

Frisch, W., Dunkl, I. & Kuhlemann, J. 2000. Post-collisional orogen-parallel large-scale extension in the Eastern Alps. Tectonophysics 327, 239–265.

Fügenschuh, B., Seward, D. & Mancktelow, N. 1997. Exhumation in a convergent orogen: the western Tauern window. Terra Nova 9, 213–217.

Fügenschuh, B., Mancktelow, N. & Seward, D. 2000. Cretaceous to Neogene cooling and exhumation history of the Oetztal-Stubai basement complex, Eastern Alps: A structural and fission track study. Tectonics 19, 905–918.

Gallagher, K., Brown, R. & Johnson, C. 1998. Fission track analysis and its application to geological problems. Annual Reviews in Earth and Planetary Sciences 26, 519–572.

Genser, J. & Neubauer, F. 1989. Low angle normal faults at the eastern margin of the Tauern Window (Eastern Alps). Mitteilungen der Österreichischen Mineralogischen Gesellschaft 81, 233–243.

Genser, J., Van Wees, J., Cloetingh, S. & Neubauer, F. 1996. Eastern Alpine tectono-metamorphic evolution; constraints from two-dimensional P-T-t modeling. Tectonics 15(3), 584–604.

Genser, J., Cloetingh, S. & Neubauer, F. 2007. Late orogenic rebound and oblique Alpine convergence: New constraints from subsidence analysis

of the Austrian Molasse basin. Global and Planetary Change 58(1–4), 214–223.

Glasmacher, U., Tremblay, A. & Clauer, N. 2003. K–Ar dating constraints on the tectonothermal evolution of the external Humber zone, southern Quebec Appalachians. Canadian Journal of Geosciences 40, 285–300.

Green, P., Duddy, I., Gleadow, A., Tingate, P. & Laslett, G. 1986. Thermal annealing of fission tracks in apatite: 1. A quantitative description. Chemical geology 59, 237–253.

Grove, M. & Harrison, T. 1996. Ar diffusion in Fe-rich biotite. American Mineralogist 81, 940–951.

Grundmann, G. & Morteani, G. 1985. The young uplift and thermal history of the central Eastern Alps (Austria/Italy), evidence from apatite fission track ages. Geologisches Jahrbuch 128, 197–216.

Hames, W. & Bowring, S. 1994. An emperical evaluation of the argon difussion geometry in muscovite. Earth and Planetary Science Letters 124, 161–167.

Hawkesworth, C. 1976. Rb/Sr Geochronology in the Eastern Alps. Contributions to Mineralogy and Petrology, 54, 225–244.

Hejl, E. 1997. "Cold Spots" during the Cenocoic ecolution of the Eastern Alps: thermochronological interpretation of apatite fission-track data. Tectonophysics 272, 159–174.

Hoinkes, G., Koller, F., Rantitsch, G., Dachs, E., Hock, V., Neubauer, F. & Schuster, R. 1999. Alpine metamorphism of the Eastern Alps. Schweizerische Mineralogische und Petrographische Mitteilungen 79, 155–181.

Hoke, L. 1990. The Altkristallin of the Kreuzeck Mountains, SE Tauern Window, Eastern Alps, basement crust in a convergent plate boundary zone. Jahrbuch der geologischen Bundesanstalt 133(1), 5–87.

Holland, T. 1979. High water activities in the generation of high pressure kyanite eclogites of the Tauern Window. Journal of Geology 87, 127.

Hoschek, G. 2007. Metamorphic peak conditions of eclogites in the Tauern Window, Eastern Alps, Austria: Thermobarometry of the assemblage garnet plus omphacite plus phengite plus kyanite plus quartz. Lithos 93(1–2), 1–16.

Hurford, A. & Green, P. 1983. The zeta age calibration of fission-track dating. Isotope Geoscience 1, 285–317.

Inger, S. & Cliff, R. 1994. Timing of metamorphism in the Tauern Window, Eastern Alps; Rb-Sr ages and fabric formation. Journal of Metamorphic Geology 12(5), 695–707.

Jäger, E., Karl, F. & Schmidegg, O. 1969. Rubidium-Strontium-Altersbestimmungen an Biotit-Muskowit-Granitgneisen (Typus Augen- und Flasergneise) aus dem nördlichen Großvenedigerbereich (Hohe Tauern). Mineralogy and Petrology, 13(2–3), 251–272.

Janák, M., Froitzheim, N., Lupták, B., Vrabec, M. & Ravna, E. J. K. 2004. First evidence for ultrahigh-pressure metamorphism of eclogites in Pohorje, Slovenia: Tracing deep continental subduction in the Eastern Alps. Tectonics 23(5).

Johnson, C., Harbury, N. & Hurford, A. 1997. The role of extension in the Miocene denudation of the Nevado-Filabride Complex, Betic Cordillera (SE Spain). Tectonics 16, 189–204.

Kirschner, D., Cosca, M., Masson, H. & Hunziker, J. 1996. Staircase 40 Ar/39 Ar spectra of fine-grained white mica; timing and duration of deformation and empirical constraints on argon diffusion. Geology 24(8), 747–750.

Koller, F. 1985. Petrologie und Geochemie der Ophiolite des Penninikums am Alpenostrand Jahrbuch Geologische Bundesanstalt 128, 83–150.

Krenn, K., Fritz, H., Biermeier, C. & Scholger, R. 2003. The Oligocene Rensen Pluton (Eastern Alps, South Tyrol): magma emplacment and structures during plate convergence. Mitteilungen der Österreichischen Mineralogischen Gesellschaft 94, 9–26.

Kuhlemann, J., Frisch, W., Dunkl, I. & Szekely, B. 2001. Quantifying tectonic versus erosive denudation by the sediment budget: the Miocene core complexes of the Alps. Tectonophysics 339(1–2), 1–23.

Kurz, W., Neubauer, F., Genser, J. & Dachs, E. 1998. Alpine geodynamic evolution of passive and active continental margin sequences in the Tauern Window (Eastern Alps, Austria, Italy): a review. Geologische Rundschau 87, 225–242.

Lambert, S. 1970. A Potassium-Argon Study of the Margin of the Tauernfenster at Dollach, Austria. Eclogae Geologicae Helvetiae 63/1, 197–205.

Lammerer, B. & Weger, M. 1998. Footwall uplift in an orogenic wedge: the Tauern Window in the Eastern Alps of Europe. Tectonophysics 285, 213–230.

Lister, G. & Davis, G. 1989. The origin of metamorphic core complexes and detachment faults formed during Tertiary continental extension in the northern Colorado River region, U.S.A. Journal of Structural Geology 11, 65–95.

Liu, Y., Genser, J., Handler, R., Friedl, G. & Neubauer, F. 2001. ^{40}Ar/^{39}Ar muscovite ages from the Penninic-Austroalpine plate boundary, Eastern Alps. Tectonics 20, 526–547.

Mancktelow, N., Stokli, D., Balz, G., Müller, W., Fügenschuh, B., Viola, G., Seward, D. & Villa, I. 2001. The DAV and Periadriatic fault systems in the eastern Alps south of the Tauern window. International Journal of Earth Sciences 90, 593–622.

Most, P. 2003. Late Alpine cooling histories of tectonic blocks along the central part of the TRANSALP-traverse (Inntal-Gadertal): Constraints from geochronology. Ph.D. Thesis, Tübinger Geowissenschaftliche Arbeiten Reihe A 67, 97 pp.

Neubauer, F., Genser, J. & Handler, R. 2000. The Eastern Alps: result of a two-stage collision process. Mitteilungen der Österreichischen Mineralogischen Gesellschaft 92, 117–134.

Oberhänsli, R. & Goffé, B. 2004. Metamorphic structure of the Alps. Mitteilungen der Österreichischen Mineralogischen Gesellschaft 149, 115–199.

Oberhauser, R. 1995. Zur Kenntnis der Tektonik und der Paläogeographie des Ostalpenraumes zur Kreide-, Paleozän- und Eozänzeit. Jahrbuch der geologischen Bundesanstalt 138, 369–432.

Oxburgh, E., Lambert, S., Baardsgaard, H. & Simons, J. 1966. Potassium argon age studies across the south-east margin of the Tauern Window, the Eastern Alps. Verhandlungen der Geologischen Bundesanstalt, 17–33.

Purdy, J. & Jäger, E. 1976. K-Ar ages on rock-forming minerals from the Central Alps. Memorie dell'Istituto della Regia Università di Padova 30, 31pp.

Raith, M., Raase, P., Kreuzer, P. & Müller, P. 1978. The age of the Alpidic metamorphism in the Western Tauern Window, Austrian Alps, accordign to radiometric dating In: Cloos, H., Roeder, D. & Schmid, K. (Eds): Alps, Apennines, Hellenides, Inter-Union Commission on Geodynamics, 140–148.

Ratschbacher, L., Frisch, W., Linzer, H. & Merle, O. 1991. Lateral extrusion in the eastern Alps, part2: structural analysis. Tectonics 10, 257–271.

Ring, U., Brandon, M., Lister, G. & Willett, S. 1999. Exhumation processes. In: Ring, U., Brandon, M., Willett, S. & Lister, G. (Eds): Exhumation processes: normal faulting, ductile flow and erosion, Geological Society London, Special Publications, 1–29.

Sachsenhofer, R. 2001. Syn- and post-collisional heat flow in the Cenozoic Eastern Alps. International Journal of Earth Sciences 90, 579–592.

Satir, M. 1975. Die Entwicklungsgeschichte der westlichen Hohe Tauern und der Sudlichen Otztalmasse afgrund radiometrischer Alterbestimmungen. Memorie dell'Istituto della Regia Università di Padova, 30(1–84).

Satir, M. & Morteani, G., 1982. Petrological study and Radiometric age determination of the Migmatites in the Penninic rocks of the Zillertaler Alpen (Tyrol/Austria). Tschermarks Mineralogische und Petrologische Mitteilungen 30, 59–75.

Schmid, S., Fügenschuh, B., Kissling, E. & Schuster, R., 2004. Tectonic map and overall architecture of the Alpine orogen. Swiss journal of Geoscience 97(1), 93–117.

Selverstone, J. & Spear, F., 1985. Metamorphic P-T paths from pelititc schists and greenstones from the south-west Tauern Window, Eastern Alps. Journal of Metamorphic Geology 3, 439–465.

Staufenberg, H. 1987. Apatite fission-track evidence for postmetamorphic uplift and cooling history of the Eastern Tauern window and the surrounding Austroalpine (Central Eastern Alps, Austria). Geologisch Jahrbuch 130, 571–586.

Steenken, A., Siegesmund, S., Heinrichs, T. & Fugenschuh, B. 2002. Cooling and exhumation of the Rieserferner Pluton (Eastern Alps, Italy/Austria). International Journal of Earth Sciences 91, 799–817.

Stöckhert, B., Brix, M., Kleinschrodt, R., Hurford, A. & Wirth, R. 1999. Thermochronometry and microstructures of quartz-a comparison with experimental flow laws and predictions on the temperature of the brittle-plastic transition. Journal of Structural Geology 21, 351–369.

Thöni, M. 1999. A review of geochronological data from the Eastern Alps. Schweizerische Mineralogische und Petrographische Mitteilungen 79, 209–230.

Villa, I. 1998. Isotopic closure. Terra Nova 10, 42–47.

Viola, G., Mancktelow, N. & Seward, D. 2001. Late Oligocene-Neogene evolution of Europe-Adria collision: new structural and geochronological evidence from the Giudicarie fault system (Italian Eastern Alps). Tectonics 20, 999–1020.

Von Blanckenburg, F., Villa, I., Baur, H., Morteani, G. & Steiger, R. 1989. Time calibration of a PT-path from the Western Tauern window, Eastern Alps: the problem of closure temperatures. Contributions to Mineralogy and Petrology.

Willingshofer, E. & Cloetingh, S. 2003. Present-day lithospheric strength of the Eastern Alps and its relationship to neotectonics. Tectonics 22(6), 1075.

Wölfler, A. 2008. Tectonothermal evolution of the southeastern Tauern Window and the adjacent austroalpine basement of the Kreuzeck Massif: Evidence from combined fission track and (U-Th)/He analysis. Ph.D. Thesis, Tübingen, 80 pp.

Yin, A. 2004. Gneiss domes and gneiss dome systems. In: Teyssier, C., Whitney, D. & Siddoway, C. (Eds): Gneiss domes and orogeny, Geological Society of America, 1–14.

Zattin, M., Cuman, A., Fantoni, R., Martin, S., Scotti, P. & Stefani, C. 2006. From Middle-Jurassic heating to Neogen cooling: The thermochronological evolution of the Southern Alps. Tectonophysics 414, 191–202.

Zaun, P. & Wagner, G. 1985. Fission track stability in zircons under geological conditions. Nuclear tracks and radiation measurements 10, 303–307.

Zimmermann, R., Hammerschmidt, K. & Franz, G. 1994. Eocene high pressure metamorphism in the Penninic units of the Tauern window (Eastern Alps): Evidence from ^{40}Ar-^{39}Ar dating and petrological investigations. Contributions to Mineralogy and Petrology 117, 175–186.

Manuscript received February 4, 2008
Revision accepted October 27, 2008
Published Online first November 13, 2008
Editorial Handling: Nikolaus Froitzheim & Stefan Bucher

1661-8726/08/01S225-9
DOI 10.1007/s00015-008-1279-8
Birkhäuser Verlag, Basel, 2008

Swiss J. Geosci. 101 (2008) Supplement 1, S225–S233

Apatite fission track and (U-Th)/He thermochronology of the Rochovce granite (Slovakia) – implications for the thermal evolution of the Western Carpathian-Pannonian region

MARTIN DANIŠÍK[1],*, MILAN KOHÚT[2], ISTVÁN DUNKL[3], ĽUBOMÍR HRAŠKO[2] & WOLFGANG FRISCH[1]

Key words: Western Carpathians, Rochovce granite, fission track analysis, (U-Th)/He dating, thermal modelling, Miocene thermal event

ABSTRACT

The thermal evolution of the only known Alpine (Cretaceous) granite in the Western Carpathians (Rochovce granite) is studied by low-temperature thermochronological methods. Our apatite fission track and apatite (U-Th)/He ages range from 17.5 ± 1.1 to 12.9 ± 0.9 Ma, and 12.9 ± 1.8 to 11.3 ± 0.8 Ma, respectively. The data thus show that the Rochovce granite records a thermal event in the Middle to early Late Miocene, which was likely related to mantle upwelling, volcanic activity, and increased heat flow. During the thermal maximum between ~17 and 8 Ma, the granite was heated to temperatures ≳ 60 °C. Increase of cooling rates at ~12 Ma recorded by the apatic fission track and

(U-Th)/He data is primarily related to the cessation of the heating event and relaxation of the isotherms associated with the termination of the Neogene volcanic activity. This contradicts the accepted concept, which stipulates that the internal parts of the Western Carpathians were not thermally affected during the Cenozoic period. The Miocene thermal event was not restricted to the investigated part of the Western Carpathians, but had regional character and affected several basement areas in the Western Carpathians, the Pannonian basin and the margin of the Eastern Alps.

1. Introduction

The Western Carpathians are bordered to the south by the Neogene Pannonian basin and join the Eastern Alps to the west. They record a complex history related to the Variscan and Alpine orogenies. The internal parts of the Western Carpathians are formed by three orogen-parallel, north-vergent principal structural domains – the Gemeric (south, highest), Veporic (medium) and Tatric (north, lowest) domains (Fig. 1a; Andrusov 1968; Plašienka et al. 1997). These domains comprise Variscan crystalline basement and sedimentary cover, and are defined as thick-skinned crustal sheets, which were tectonically juxtaposed through north-directed thrusting in the early Late Cretaceous (Andrusov 1968; Plašienka et al. 1997).

The Alpine evolution of the Western Carpathians is still a matter of discussion: most published concepts rely on the work of Kováč et al. (1994), who concluded from geochronological, stratigraphic, palaeomagnetic and fission track data that the Gemeric and Veporic domains were exhumed and cooled to near-surface conditions in the course of post-Eoalpine un-

roofing during Late Cretaceous to Early Palaeogene times (90–55 Ma). Since then, these domains acted as one individual block, which was not thermally affected for the rest of the Cenozoic period (Kováč et al. 1994). This, however, may conflict with the geodynamic context of the post-collisional evolution in the adjacent Eastern Alps and western Pannonian basin, which was dominated by large-scalc Miocene extension in the course of lateral tectonic extrusion and subduction roll-back (Royden et al. 1982; Ratschbacher et al. 1991; Csontos 1995; Tari et al. 1992, 1999; Frisch et al. 2000; Wortel and Spakman 2000; Sperner et al. 2002). These processes led to rifting and formation of fault-bounded basins, exhumation and unroofing of basement core complexes, crustal thinning, mantle upwelling and volcanism (e.g. Royden et al. 1983; Horváth et al. 1988; Szabó et al. 1992; Kováč et al. 1993, 1994; Frisch et al. 1998, 2000), and could potentially thermally overprint neighbouring parts of the Western Carpathians.

This paper aims to test the hypothesis of Kováč et al. (1994) with low-temperature thermochronological methods [apatite fission track (AFT) and apatite (U-Th)/He (AHe) analyses]

[1] University of Tübingen, Institute of Geosciences, Sigwartstrasse 10, D-72076 Tübingen, Germany. E-mail: martin.danisik@uni-tuebingen.de; frisch@uni-tuebingen.de
[2] Dionýz Štúr State Institute of Geology, Mlynská dolina 1, 81704 Bratislava, Slovakia. E-mail: milan.kohut@geology.sk; lubomir.hrasko@geology.sk
[3] Geoscience Center Göttingen, Sedimentology and Environmental Geology, Goldschmidtstrasse 3, D-37077 Göttingen, Germany. E-mail: istvan.dunkl@geo.uni-goettingen.de
*Corresponding author: Martin Danišík. E-mail: martin.danisik@uni-tuebingen.de

applied on the rocks from the boundary zone of the Veporic and Gemeric domains. Our data allow to identify a distinct Miocene thermal event, which was not restricted to the study area but possibly had a regional character and affected large parts of the Eastern Alpine–Western Carpathian–Pannonian area. Consequently, the actual concepts of this region demand careful reconsideration and adjustment, at least with respect to the thermal evolution.

2. Geological setting

The study area is located in the central part of Slovakia and covers the major tectonic contact between the Veporic and Gemeric domains that developed during Eoalpine shortening in the Cretaceous (Fig. 1a; Plašienka et al. 1997 and references therein).

The Veporic unit, occupying the footwall position, consists of Variscan crystalline basement together with an Upper Palaeozoic-Triassic sedimentary cover, overprinted by low- to medium-grade regional metamorphism in Eoalpine time (Vrána 1964; Vozárová 1990; Plašienka et al. 1999; Lupták et al. 2000; Janák et al. 2001). The Gemeric unit, occupying the hanging-wall position, consists of Palaeozoic mostly low-grade metamorphic rocks intruded by Permian granites that underwent an Eoalpine overprint under very low- to low-grade conditions (Hovorka & Méres 1997; Korikovsky et al. 1997 and references therein).

Post-orogenic collapse and exhumation of the Veporic and Gemeric domains started in the Late Cretaceous as indicated by geochronological data (K-Ar amphibole, feldspar, muscovite and biotite; Rb-Sr biotite; Ar-Ar muscovite ages around 90–80 Ma; Bibikova et al. 1988; Cambel et al. 1990; Hurai et al. 1991; Dallmeyer et al. 1996). At the same time, the thrust plane between the Veporic and Gemeric domains was intruded by a small granitic body, the Rochovce granite (e.g. Poller et al. 2001), which is the only known example of granitic magmatism in the Eastern Alps and Carpathians related to the Cretaceous orogenic cycle. This granite is the target of this study and will be described in more detail in the next section.

Kováč et al. (1994) argue that final cooling and exhumation occurred between 90 and 55 Ma, based on AFT ages of Kráľ (1977), which were interpreted as cooling ages, although critically important track length data were not reported. The Veporic and Gemeric domains should thus lack any thermal overprint younger than 55 Ma. However, remnants of Palaeogene flysch of the Central Carpathian Palaeogene Basin (Vass et al. 1979; Marko & Vojtko 2006) and Neogene subduction- and back arc extension-related volcanism (~17–8 Ma; Figs. 1a,d; Repčok 1981; Lexa & Konečný 1998; Pécskay et al. 2006 and references therein) may challenge such an interpretation.

2.1. The Rochovce granite

The Rochovce granite is a subsurface body discovered by the borehole KV-3 (Figs. 1b,c; Klinec et al. 1980) in the centre of a magnetic anomaly (Filo et al. 1974). Subsequent geophysical and drilling exploration revealed that it rests a few hundred meters below surface and has a diameter of ~5×10 km. Petrographical, mineralogical and geochemical characteristics were reported by Klinec et al. (1980), Határ et al. (1989) and Hraško et al. (1998). It is an I(-A)-type granite, which formed in two phases. The first phase comprises two varieties: (i) coarse-grained biotite monzogranites with pink K-feldspar phenocrysts, locally with mafic microgranular enclaves; and (ii) granite porphyries. Medium- to fine-grained biotite leucogranites and leucogranitic porphyries represent the second phase. Locally, narrow leucogranite veins randomly penetrate the coarse-grained granites of the first phase. U-Pb zircon dating yielded a Late Cretaceous crystallization age: conventional method 82 ± 1 Ma (Hraško et al. 1999), and cathodoluminescence controlled single-grain method 75.6 ± 1.1 Ma (Poller et al. 2001). These ages were recently confirmed by Re-Os dating of molybdenite from porphyry mineralization associated with the second intrusive phase, yielding ~80 Ma (Stein et al. in review).

3. Samples and methods

For thermochronological investigations, three samples of the Rochovce granite were collected from two boreholes (RO-6 and KV-3) from depths between ~400 and ~1400 meters (Table 1, Fig. 1c). Apatite grains were separated using conventional magnetic and heavy liquid techniques and dated by fission track and (U-Th)/He methods. Both analyses were carried out in the Thermochronological Laboratory, University of Tübingen. For fission track analyses we used the external detector method (Gleadow 1981) with the etching protocol of Donelick et al. (1999). The zeta calibration approach (Hurford & Green 1983) was adopted to determine the age. For (U-Th)/He analyses apatite grains were degassed under vacuum using laser-heating and analysed for He using a Pfeiffer Prisma QMS-200 mass spectrometer. Following He measurements, the grains were analysed by the isotope dilution ICP-MS method for U and Th at the Scottish Universities Environmental Research Centre (SUERC) in East Kilbride (Scotland) on a VG PlasmaQuad 2 ICP-MS. For more details on analytical procedures, see Danišík (2005).

4. Results

The results of the AFT and (U-Th)/He analyses are summarized in Tables 1 and 2 and shown in Figures 2a,b.

4.1. AFT and AHe data

All samples yielded Miocene AFT ages between 17.5 ± 1.1 and 12.9 ± 0.9 Ma. Due to limited amounts of suitable apatite grains, we could measure a statistically robust number of horizontal confined tracks only in two samples (KVP-3, RO-6; Table 1). Track length distributions (Fig. 2b) are unimodal, negatively skewed, with mean track lengths of 13.6 and 13.9 µm and stan-

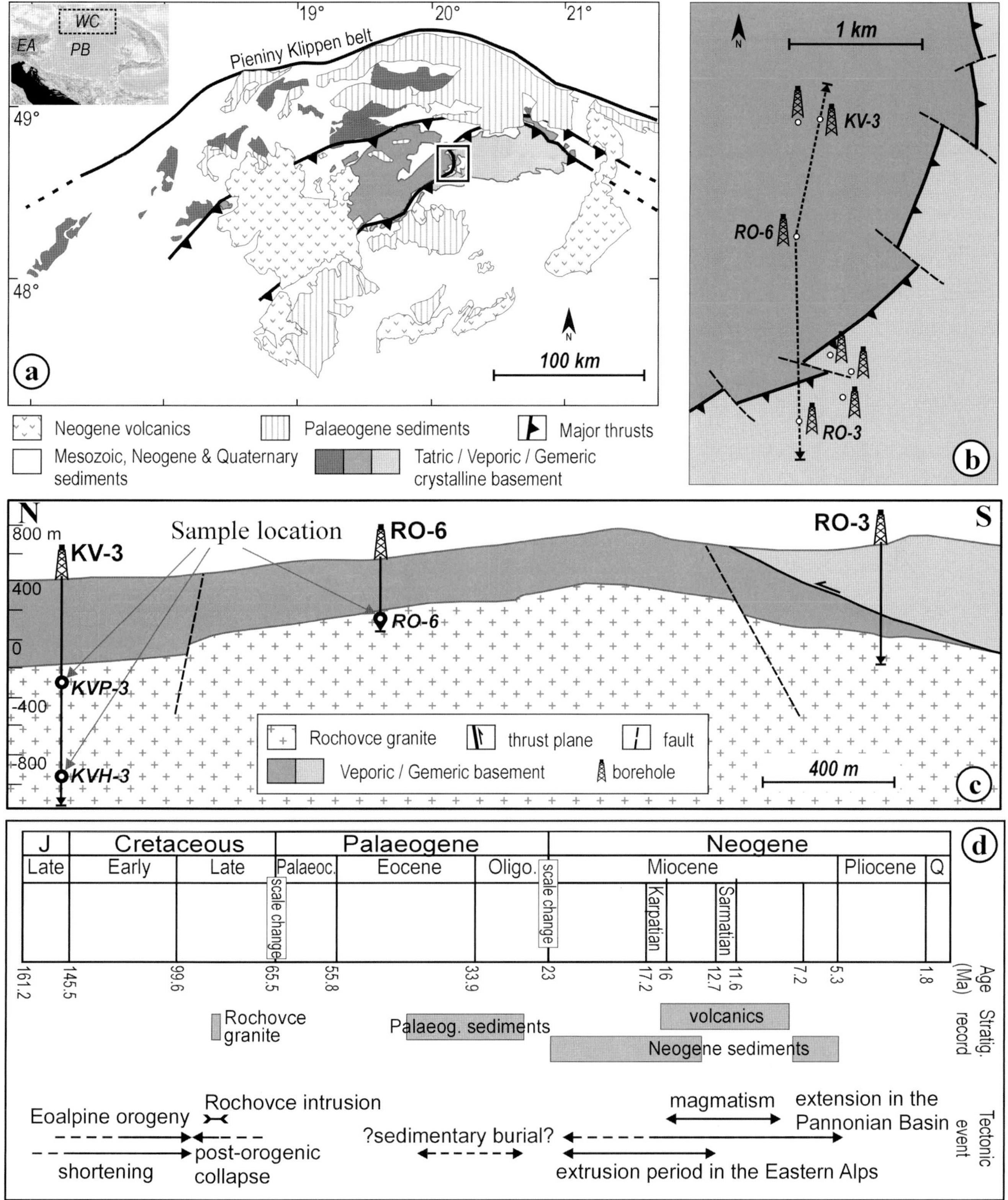

Fig. 1. Inset part of Fig. 1a) Location of the Fig. 1a; WC – Western Carpathians, EA – Eastern Alps, PB – Pannonian Basin. a) Tectonic sketch map of the Western Carpathians with exposures of Variscan crystalline bodies belonging to three principal domains; Tatric, Veporic and Gemeric after Lexa et al. (2000). Location of the study area and map of Fig. 1b are indicated by the rectangle. b) Geological map of the study area with the position of boreholes penetrating the Rochovce granite. c) Schematic profile (for location, see dashed line in Fig. 1b). d) Chronostratigraphic chart of the study area and surrounding regions with relevant geodynamic events (Ratschbacher et al. 1991; Reinecker 2000; Lexa et al. 2000; Poller et al. 2001).

Table 1. AFT data[a].

Sample code	Lat/Lon (WGS84) X	Y	Bore-hole	elevation (m a.s.l.)	Petrography	N	ρ_s	N_s	ρ_i	N_i	ρ_d	N_d	P(χ2) (%)	Age (Ma)	±1σ (Ma)	MTL (μm)	SD (μm)	N (L)
RO-6	48° 41' 41"	20° 17' 31"	RO-6	167	granite	30	1,863	351	11,765	2216	7,214	4959	>95	17,5	1,1	13,6	1,4	63
KVP-3	48° 42' 03"	20° 17' 34"	KV-3	–324	granite	35	2,159	464	13,947	2997	7,230	4959	>95	17,1	0,9	13,9	1,2	60
KVH-3	48° 42' 03"	20° 17' 34"	KV-3	–975	granite	25	1,646	273	14,120	2342	7,262	4959	>95	12,9	0,9	13,9	1,6	9

[a] N – number of dated apatite crystals; ρ_n (ρ_i) – spontaneous (induced) track densities ($\times 10^5$ tracks/cm²); N_s (N_i) – number of counted spontaneous (induced) tracks; ρ_d – dosimeter track density ($\times 10^5$ tracks/cm²); N_d – number of tracks counted on dosimeter; P(χ^2) – probability obtaining Chi-square value (χ^2) for n degree of freedom (where n = No. of crystals – 1); Age ± 1σ – central age ± 1 standard error (Galbraith & Laslett 1993); MTL – mean track length; SD – standard deviation of track length distribution; N(L) – number of horizontal confined tracks measured. Ages were calculated using zeta calibration method (Hurford & Green 1983), glass dosimeter CN-5, and zeta value of 305 ± 4.3 year/cm².

dard deviations of 1.2 and 1.4 µm, respectively, which is typical for rocks with moderate cooling through the partial annealing zone (e.g. Gleadow et al. 1986a,b).

For (U-Th)/He analyses, four to five grains were measured per sample (Table 2), all yielding AHe ages younger than the corresponding AFT ages. Replicates of both samples from the borehole KV-3 reproduce extremely well (all ages are within 1 sigma error). Sample RO-6 revealed a slightly larger spread of single-grain ages. The average AHe ages corrected for alpha ejection range from 12.9 ± 1.8 to 11.3 ± 0.8 Ma.

Both, AFT and AHe ages show positive correlation with elevation that usually allows a direct estimation of long-term exhumation rates, if the closure isotherms of the employed systems remained fairly flat and stationary during cooling (Stüwe et al. 1994). This is, however, not the case with our data, which show decreasing differences between AFT and AHe ages with depth: the AFT ages of the middle and uppermost samples are ~5 Ma older than the AHe ages, whereas the deepest sample (KVH-3) revealed an AHe age overlapping within error with the corresponding AFT age. This pattern shows that the cooling rates changed between ~17 and ~12 Ma (Fig. 2a). The isotherms thus did not remain stationary, and estimation of exhumation rates by the age-elevation relationship is not justified.

4.2. Thermal history

In order to understand the meaning of the data, thermal histories were modelled by the HeFTy modelling program (Ketcham 2005), which allows to compute thermal paths of a sample by combining the fission track annealing and He production-diffusion models. We used an inverse Monte Carlo algorithm (50 000 model searches) with the multikinetic annealing model of Ketcham et al. (1999) and the diffusion parameters of the Durango apatite (Farley 2000).

For modelling, we chose sample KVP-3, whose AHe ages best reproduce (Table 2) and which revealed enough horizontal confined tracks (Table 1). Available information was converted into time-temperature (tT) constraints in the form of

Table 2. (U-Th)/He data[a].

Sample code	Altitude (m a.s.l.)	N_c	Th (ng)	Th error (%)	U (ng)	U error (%)	⁴He (ncc at STP)	⁴He error (%)	TAU (%)	Th/U	Unc. age (Ma)	±1σ (Ma)	F_t	Cor. age (Ma)	±1σ (Ma)	AFT age (Ma)	±1σ (Ma)
RO-6#1	167	1	0,111	2,1	0,074	3,1	0,102	0,9	2,7	1,51	8,4	0,2	0,71	11,8	0,7	17,5	1,1
RO-6#2		1	0,393	1,8	0,178	3,3	0,319	0,9	2,5	2,20	9,7	0,2	0,78	12,5	0,7		
RO-6#3		1	0,346	1,9	0,169	2,5	0,221	0,9	2,3	2,04	7,3	0,2	0,68	10,7	0,6		
RO-6#4		1	0,364	2,2	0,198	1,7	0,404	0,9	2,2	1,84	11,7	0,3	0,80	14,7	0,8		
RO-6#5		1	0,298	2,5	0,171	2,6	0,344	0,9	2,7	1,75	11,8	0,3	0,79	14,9	0,8		
Average age ± Std. dev. (both in Ma)														**12,9 ± 1,8**			
KVP-3#1	–324	1	0,130	1,95	0,052	3,30	0,071	1,0	2,6	2,48	7,1	0,2	0,63	11,4	0,6	17,1	0,9
KVP-3#2		1	0,359	2,73	0,145	2,40	0,238	0,9	2,8	2,48	8,6	0,2	0,73	11,8	0,7		
KVP-3#3		1	0,173	2,14	0,065	3,51	0,101	0,9	2,8	2,67	7,9	0,2	0,66	12,1	0,7		
KVP-3#4		1	0,410	2,16	0,145	2,59	0,240	0,9	2,4	2,82	8,2	0,2	0,73	11,2	0,6		
Average age ± Std. dev. (both in Ma)														**11,6 ± 0,4**			
KVH-3#1	–975	1	0,392	2,3	0,185	2,52	0,274	0,9	2,5	2,12	8,1	0,2	0,78	10,4	0,6	12,9	0,9
KVH-3#2		1	0,178	2,3	0,137	3,27	0,220	0,9	2,9	1,30	10,1	0,3	0,82	12,4	0,7		
KVH-3#3		1	0,350	2,6	0,151	2,56	0,245	0,9	2,8	2,32	8,7	0,2	0,76	11,4	0,7		
KVH-3#4		1	0,498	1,8	0,202	2,41	0,332	0,9	2,2	2,46	8,6	0,2	0,77	11,2	0,6		
Average age ± Std. dev. (both in Ma)														**11,3 ± 0,8**			

[a] N_c – number of dated apatite crystals; TAU – total analytical uncertainty; Unc. age – uncorrected AHe age; F_t – alpha recoil correction factor after Farley et al. (1996); Cor. age – corrected AHe age.

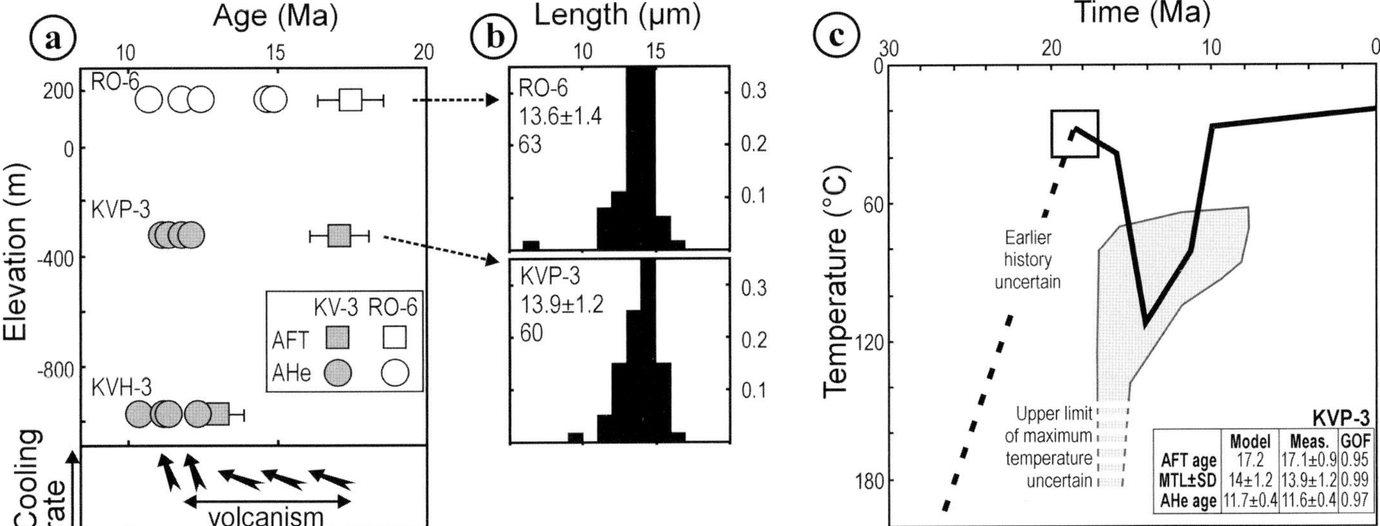

Fig. 2. a) Age-elevation relationship of the Rochovce granite as constrained by AFT (squares) and AHe data (circles; error bars of AHe ages were omitted for readability) and, below, changes of cooling rates with respect to duration of volcanic activity in the study area as estimated from differences of the AFT and AHe thermochronometers. b) Corresponding track length distributions. Explanation in histograms (from top): sample code; mean track length ± standard deviation (both in µm); number of measured tracks. c) Thermal modelling results of AFT and AHe data displayed in a time-temperature diagram modelled with the HeFTy program (Ketcham, 2005). The best fit is shown as a solid black line, shaded polygon shows the values of peak temperature and time at which the cooling began. MTL is mean track length in µm; SD is standard deviation in µm; GOF is goodness of fit (statistical comparison of the measured input data and modelled output data, where a "good" result corresponds to value 0.5 or higher, "the best" result corresponds to value 1). See text for further explanation.

boxes and the modelled tT path was constrained as follows: the beginning of the tT path was set as T = 800 °C at 80–75 Ma, according to the crystallization age of the Rochovce granite (see Section 2.1.). Geological constraints for the time between Late Cretaceous crystallization and Miocene thermal activity recorded by AFT and AHe data are poor. It was proposed that after post-collisional collapse and exhumation, the Veporic-Gemeric domain was exposed at the surface in the latest Cretaceous-Early Palaeogene (e.g. Kováč et al. 1994). However, this indirect conclusion from radiometric ages is not backed by stratigraphic constraints. The evolution during the Palaeogene is also not clear. Evidently the area was buried by flysch of a forearc basin (Central Carpathian Palaeogene Basin; Vass et al., 1979; Kázmér et al. 2003; Marko & Vojtko 2006), however the thickness of the Palaeogene cover remains unclear. To these issues we cannot draw any conclusion from our data and no constraint was set into the model. An important point for the interpretation of the data is the fact that the Veporic-Gemeric domain was at the surface in the Early Miocene prior to Neogene volcanism, which is well constrained by occurrences of Middle Miocene volcanic rocks that overlie the Veporic basement (Marko & Vojtko 2006). Another important evidence is the occurrence of kaolinitic weathering crusts preserved in situ on top of the crystalline basement, which is overlain by Middle Miocene volcanoclastics along the western margin of the Veporic unit (Kraus 1989). This clearly shows that in the Early Miocene, the Veporic-Gemeric domain was at the surface and the Rochovce granite resided at near-surface temperatures. Thus, another constraint was set as T = 20–40 °C at 20–17 Ma.

The end of the tT path was set as T = 20 °C at 0 Ma according to the estimated temperature of the sample in the borehole.

The modelled thermal history (Fig. 2c), which matches both the measured AFT and AHe data, shows that sample KVP-3 experienced reheating between ~17 and 8 Ma, with minimum peak temperatures above 60 °C. Thus, we conclude that the Rochovce granite experienced a thermal event in the Middle to early Late Miocene.

5. Interpretation

The Miocene thermal event can be explained either (i) in terms of changes in the thermal regime, or (ii) in terms of rapid Middle Miocene burial and exhumation of basement units by erosion of a $\gtrsim$1.5 km thick sediment pile, or (iii) a combination of both.

We favour the first option since the age of the thermal event exactly coincides with that of the Miocene volcanism (Fig. 1d) and mantle upwelling, associated with high, extension-related heat flow in the Carpathian-Pannonian arc and back-arc region (Szabó et al. 1992; Tari et al. 1999; Pécskay et al. 2006). The magmatic activity in the studied part of the Western Carpathians occurred between ~17 and 8 Ma and had its climax at ~16–13 Ma when numerous large stratovolcanoes formed (Fig. 1a; Lexa & Konečný 1998; Pécskay et al. 2006). The time interval of 17 to 8 Ma perfectly fits with the thermal overprint revealed by the modelling results and measured AFT and AHe ages (Fig. 2c, Tables 1, 2). After the climax since ~12 Ma, the volcanic activity started to cease as indicated by the decreasing amount of volcanic products in the region (Pécskay et al. 2006). Concurrently, the elevated heat flow started to decrease as documented by the increase of cooling rates at ~12 Ma as recorded by the deepest sample (Fig. 2a). Cooling recorded by the AFT and AHe ages is thus primarily related to the cessation of the heat source and relaxation of the isotherms.

The second option is less likely since there is no direct evidence for a thick cover on the top of the Gemeric-Veporic domain in the Middle Miocene. Neogene sedimentary formations are not preserved in the study area but only in depressions further south. Secondly, the kaolinitic weathering crusts would not survive deep burial. An at least ~1.5 km thick cover that would potentially induce total reset of the AFT system purely by burial is therefore unlikely. However, the possibility that the region was overlain by thin layers of fine-grained sediments with low thermal conductivity that are typical for Neogene successions of the Pannonian basin (Dövényi & Horváth 1988) and might produce thermal blanketing of underlying rocks (Dunkl & Frisch 2002), cannot be ruled out.

6. Discussion

6.1. Late Early to Middle Miocene thermal event and its implications for the Eastern Alpine–Western Carpathian–Pannonian area

Although this study provides the first evidence of Miocene thermal overprint in the Veporic-Gemeric domain, the idea of the existence of a Miocene thermal event is not new. In the Pannonian basin such an event was recognized more than 20 years ago (Royden et al. 1983; Horváth et al. 1988; Ebner & Sachsenhofer 1991; Lankreijer et al. 1995; Lenkey et al. 2002). Since then, Miocene reheating was reported also from crystalline bodies of the Western Carpathians (Danišík et al. 2004, 2005, 2008) and the eastern margin of the Alps (Dunkl & Frisch 2002), which proves its regional character. For instance, Dunkl & Frisch (2002), based on AFT, structural, sedimentological and vitrinite reflectance data, demonstrated that crystalline basement outcrops along the northwestern margin of

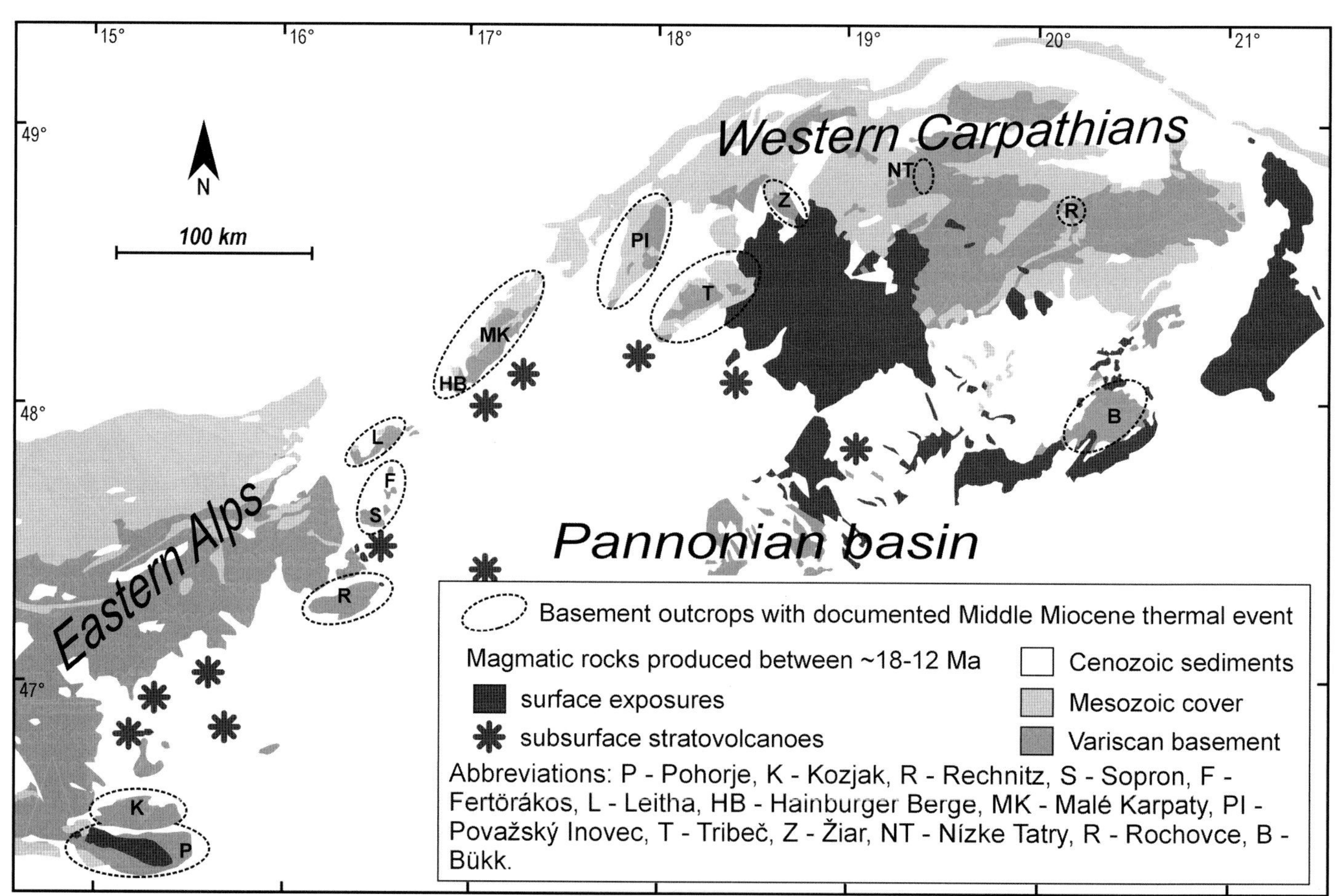

Fig. 3. Spatial relations of Miocene magmatism and crystalline outcrops recording the Miocene thermal event in the Eastern Alpine–Western Carpathian–Pannonian region [distribution of volcanoes modified after Kováč (2000)].

the Pannonian basin, i.e. Kozjak, Pohorje, Rechnitz, Sopron, Fertőrákos and Bükk Mts. (Fig. 3), record a distinct thermal event in the Early/Middle Miocene, when the crystalline bodies were reheated to temperatures >60 °C. The same thermal overprint of several crystalline bodies in the Western Carpathians (i.e., Leitha, Hainburg, Malé Karpaty, Tribeč, Považský Inovec, Žiar and Nízke Tatry Mts., Fig. 3) was reported by Danišík et al. (2004, 2005, 2008) from AFT and thermal modelling data.

Although presence and regional character of the Miocene thermal event are undoubted, opinions regarding its cause are diverse. Dunkl & Frisch (2002) attribute the thermal overprint primarily to burial of the crystalline bodies by Miocene sediments of the Pannonian basin, whereas increased heat flow would play a less important role. Final cooling following the thermal maximum is thus explained by removal of 1–1.5 km of overburden (Dunkl & Frisch 2002).

We favour an alternative explanation and argue that increased heat flow (e.g. Royden et al. 1983; Horváth et al. 1988; Tari et al. 1999) was the main reason for the thermal overprint. Our arguments are the following: The peak of the Miocene thermal event and consequent cooling of the crystalline bodies along the margin of the Pannonian basin occurred at ~18 in the western parts and migrated to the east, where it occurred latest, i.e. at ~8 Ma, as recorded by AFT, vitrinite reflectance and thermal modelling data (this study; Dunkl & Frisch 2002; Danišík et al. 2004, 2008; Fodor et al. 2008). We propose that: (i) this time span conspicuously coincides with late Early to Middle Miocene magmatism of the Carpathian–Pannonian transition area that was constrained to ~20–8 Ma (Lexa & Konečný 1998; Pécskay et al. 2006). (ii) Spatial relation to Miocene magmatism is clear from the distribution of late Early Miocene to Middle Miocene volcanic rocks and basement areas with documented Miocene thermal overprint (Fig. 3). (iii) There are no clear evidences of ~1.5 km deep burial in the Western Carpathians (see Section 4.2.). By contrast, several crystalline bodies were eroded in the Middle Miocene as indicated by stratigraphic record (Kováč et al. 1994; Danišík et al. 2004, 2008). (iv) Heat flow in the Middle Miocene reached extreme values (locally >300 mW/m^2) as demonstrated in the Styrian basin by modelling of vitrinite reflectance data (Fig. 3; Sachsenhofer 1994; Ebner & Sachsenhofer 1995).

These arguments are not meant to disprove the general idea of burial heating in the Middle Miocene (Dunkl & Frisch 2002), but should emphasize the importance of magmatism and elevated heat flow, which relaxes the requirement for deep burial (~1.5 km). One of the implications or our study is that extent and thickness of the Middle Miocene Pannonian basin fill were less than currently believed. Further, there are several examples of extremely high cooling rates in the Miocene reported from crystalline outcrops along the margin of the Pannonian basin, which are readily used as evidence of tectonic denudation (Ratschbacher et al. 1990; Tari et al. 1999). We like to point out that this may merely be an effect of compressed isotherms resulting from elevated heat flow at that time. Thus the observed apparently high cooling rates were in fact moderate and should be interpreted with caution.

7. Conclusion

First AFT and AHe data from the only known Cretaceous Alpine granite in the Western Carpathians are reported. In spite of the fact that our results are based on a limited amount of data, they are of good quality and provide important constraints on the thermal and geodynamic evolution of the entire Eastern Alpine–Western Carpathian–Pannonian region. We conclude:

(i) The Rochovce granite records a distinct thermal event in the Middle to early Late Miocene, likely related to mantle upwelling, magmatic activity, and increased heat flow in the Carpathian-Pannonian region. During the thermal maximum between ~17 and 8 Ma, the granite was heated to temperatures $\gtrsim 60$ °C. Gradual increase of cooling rates recorded by the AFT and AHe data is primarily related to cessation of the heat source and relaxation of the isotherms associated with the termination of the volcanic activity.

(ii) Our thermochronological data disprove the widely accepted concept of thermal stability of the Veporic-Gemeric domain during the Cenozoic period. Instead, we show that this domain was affected by the Miocene thermal event that was not restricted only to the Veporic-Gemeric domain, but had regional character and affected large parts of the basement outcrops in the Western Carpathians, Pannonian basin and easternmost Eastern Alps. We believe that the Miocene thermal event was primarily related to spatially variable increased heat flow and magmatism, and not to burial.

Acknowledgements

This study was funded by the German Science Foundation. Dorothea Mühlbayer-Renner and Dagmar Kost (Tübingen) are thanked for careful sample preparation. An earlier version of the manuscript benefited from constructive reviews by Christoph Glotzbach (Tübingen), Bernhard Fügenschuh (Innsbruck) and László Fodor (Budapest).

REFERENCES

Andrusov, D. 1968: Grundriss der Tektonik der nördlichen Karpaten. Verlag der Slowakischen Akademie der Wissenschaften, Bratislava, 188 pp.

Bibikova, E.V., Cambel, B., Korikovsky, S.P., Broska, I., Gracheva, T.V., Makarov, V.A. & Arakeliants, M.M. 1988: U-Pb and K-Ar isotopic dating of Sinec (Rimavica granites Kohút zone of Veporides). Geologický Zborník – Geologica Carpathica 39, 147–157.

Cambel, B., Kráľ, J. & Burchart, J. 1990: Isotope geochronology of the Western Carpathians crystalline complex (in Czech). Veda, SAV Bratislava, 183 pp.

Csontos, L. 1995: Tertiary tectonic evolution of the Intra-Carpathian area: a review. Acta Vulcanologica 7, 1–13.

Dallmeyer, R.D., Neubauer, F., Handler, R., Fritz, H., Müller, W., Pana, D. & Putiš, M. 1996: Tectonothermal evolution of the Internal Alps and Carpathians: evidence from ^{40}Ar/^{39}Ar mineral and whole-rock data. Eclogae Geologiae Helvetia 89(1), 203–227.

Danišík, M. 2005: Cooling history and relief evolution of Corsica (France) as constrained by fission track and (U-Th)/He thermochronology. Tübin-

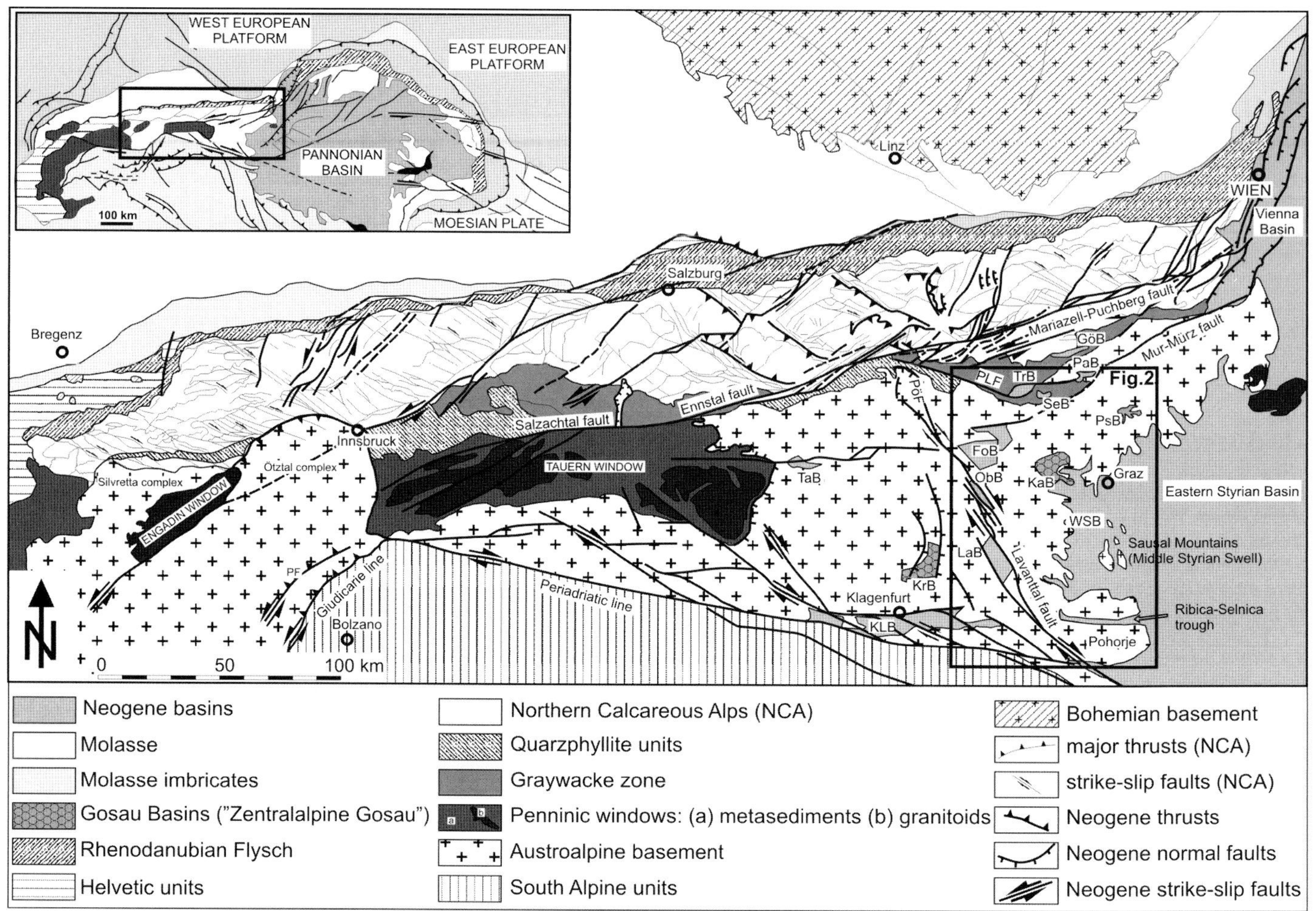

Fig. 1. Tectonic map of the Eastern Alps displaying major and minor Paleogene to Neogene fault systems (after Linzer et al., 2002).
PLF = Palten – Liesing fault; PöF = Pöls fault; GöB = Göriach Basin; PaB = Parschlug Basin; SeB = Seegraben Basin; PSB = Passail Basin; FoB = Fohnsdorf Basin; ObB = Obdach Basin; LaB = Lavanttal Basin; TaB = Tamsweg Basin; TrB = Trofaiach Basin; KLB = Klagenfurt Basin; WSB = Western Styrian Basin; KrB = Krappfeld Gosau Basin; KaB = Kainach Gosau Basin.

insight into the final evolution of the Koralm Complex during the Late Miocene.

2 Geological setting

The major part of the Central Austroalpine nappe pile in the Eastern Alps was already near to the surface during early Cenozoic times, as indicated by zircon and apatite fission track data; these parts were referred to as "Cold Spots" by Hejl (1997). One of these "Cold Spots" is represented by the Koralm Complex (Figs. 1, 2). The pre-Cenozoic evolution of this unit is very well documented by detailed petrological studies (for review, see Habler & Thöni 2001; Kurz & Fritz 2003; Schuster & Kurz 2005). It is characterized by a poly-metamorphic history with signatures of pre-Alpine events and reached amphibolite to eclogite facies conditions during the Cretaceous (Eo-Alpine event). At least three metamorphic events (Variscan, Permian and Cretaceous) are described. The units within the Koralm, Po-

horje, Saualm, and Gleinalm expose high-grade metamorphic units of the Austroalpine Nappe Complex, being incorporated into the Austroalpine nappe stack during the Early Cretaceous (Frank 1987). These units are part of the Lower Central Austroalpine, and in particular the Koralpe – Wölz nappe system (see Janak et al. 2006 and Schmid et al. 2004, respectively) and were formerly referred to as part of the "Middle" Austroalpine unit. These are surrounded by low-grade metamorphic Austroalpine basement units, represented by the Graz Paleozoic in the east, and the Gurktal Nappe in the west, both being part of the Upper Central Austroalpine nappe system and the Drauzug-Gurktal nappe system in particular (see Janák et al. 2006 and Schmid et al. 2004, respectively), formerly referred to as "Upper" Austroalpine. These units are overlain by clastic sediments of Late Cretaceous to Eocene age. Remnants of these clastic sequences are exposed within the Gosau Basins of Kainach and Krappfeld, deposited on top of the Graz Paleozoic and the Gurktal Nappe, respectively (Figs. 1, 2). Alpine cover sequences, building up the main part of the Northern Calcareous Alps, were detached together with their former basement (the Graywacke Zone) from the units below and were thrusted towards north during the Lower Cretaceous. Thus, the units exposed in the eastern central part of the Eastern Alps particularly document the structural evolution of the Austroalpine basement units and the metamorphic evolution related to the Eo-Alpine collision and subsequent exhumation.

Nappe stacking, HP metamorphism and subsequent exhumation of HP units mainly occurred during the Cretaceous and are referred to as Eo-Alpine evolution (Kurz & Fritz 2003).

Petrological and structural studies (for review see Kurz et al. 2002; Kurz & Fritz 2003), including geochronological work, facilitated a detailed reconstruction of the pressure-temperature-time evolution, in particular from the Permian to the Late Cretaceous. At upper crustal levels, the exhumation of the Koralm Complex was accommodated by low-angle normal faults along its southern and north-eastern margins. Extension accommodated by these normal faults triggered the formation of the Gosau sedimentary basins (Fig. 2) during the Late Cretaceous as well (for a review, see Kurz & Fritz 2003). However, the Koralm Complex was not exhumed to the surface at that time as

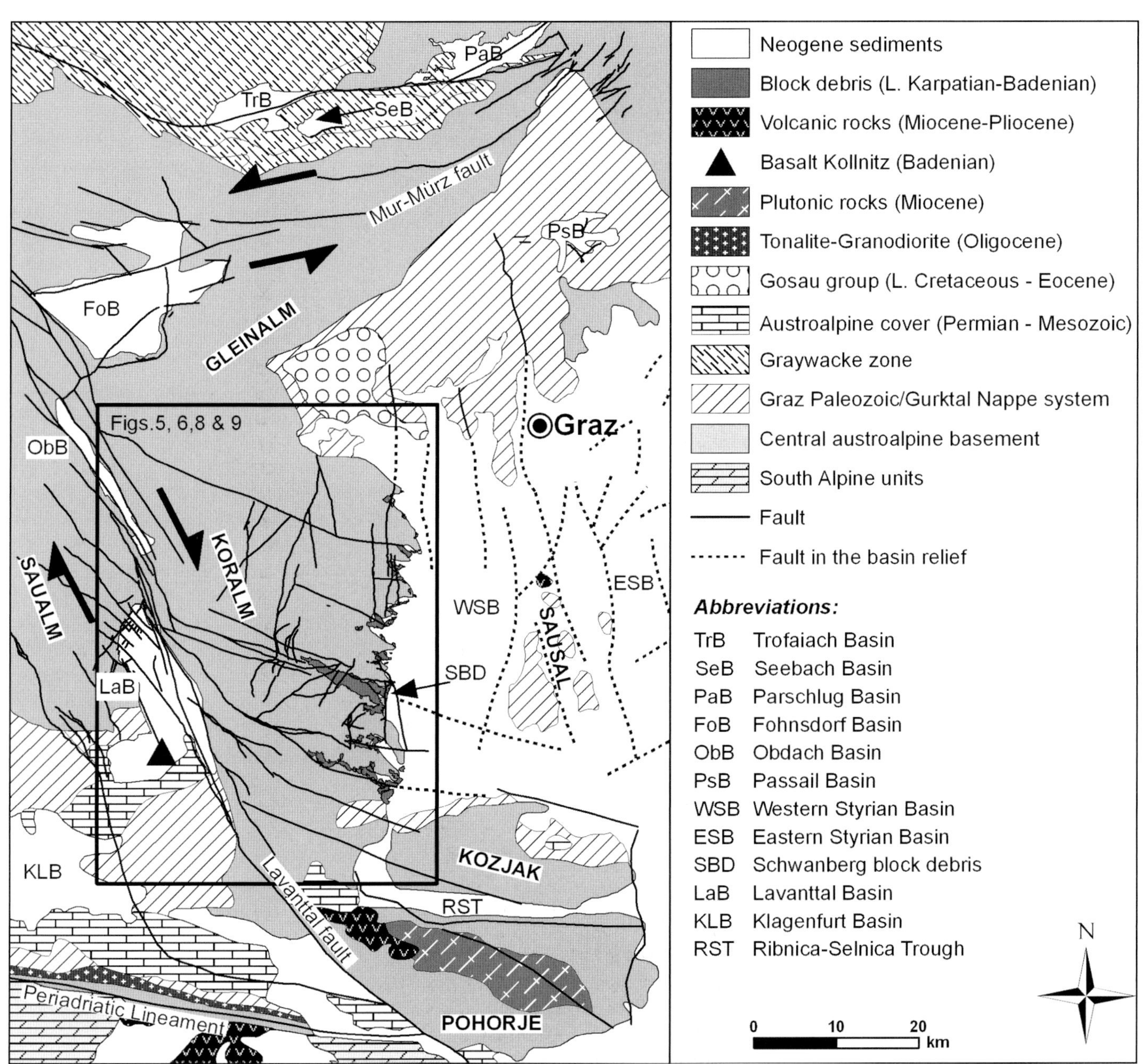

Fig. 2. Geological map of the Koralm Massif and adjacent areas, including the main faults activated during the Miocene.

is indicated by the absence of Koralm-derived pebbles in the Gosau deposits.

The Koralm Complex forms a dome structure with an approximately E–W-trending axis (Kurz et al. 2002). This structure is traced by a penetrative foliation ($s_{1,2}$) dipping to the south in the southern part of the Koralm Complex and the Plankogel Complex, and to the north to northeast in the northern parts; in the central part the penetrative foliation has a subhorizontal orientation. Generally, the foliation is parallel to the lithological and tectonic boundaries, in particular along the southern and northern/northeastern margin of the Koralm Complex.

2.1 Mesozoic/Cenozoic boundary and Cenozoic evolution

The Late Cretaceous to Paleogene tectono-metamorphic evolution of the Koralm Complex and adjacent areas is discussed by Bojar et al. (2001), Fritz et al. (2002), and Kurz & Fritz (2003). The effect of Late Cretaceous to Early Paleogene tectonics and metamorphism is still under debate. Recognition of post-Eoalpine structures and metamorphic assemblages is hampered by the fact that the spatial distribution of Cretaceous/Paleogene structural elements coincide frequently with later, Miocene structures. However, from geochronological and tectono-metamorphic arguments there is strong evidence that the evolution during the latest Cretaceous and Paleogene played a major role in Alpine architecture. This includes: (1) Major tectonic lines, interpreted as Early Cretaceous thrusts are overprinted and sealed by upper greenschist- to amphibolite-facies metamorphism and tectonics. (2) Large rock volumes within eastern sectors of the Eastern Alps cooled down below ca. 250 °C already in Cretaceous times. (3) A large number of geochronological mineral formation ages, previously interpreted to date Eo-Alpine nappe stacking, cluster around *ca.* 80 Ma and may easily be re-interpreted in terms of strike-slip and/or extensional tectonics. In particular, sets of ductile strike slip and normal faults are traced along the southern margin of Austroalpine units (Kurz & Fritz 2003), although frequently obliterated by younger tectonic events along the Periadriatic Lineament (Fig. 1).

Along its margins, the Koralm Complex is surrounded by distinct faults and shear zones. In particular, low-angle normal faults form the northeastern and southern margins of the Koralm Complex. The western margin is formed by a NNW-trending strike slip fault, the Lavanttal fault (Figs. 1, 2). This fault is part of the Pöls-Lavanttal fault system (Frisch et al. 2000a; Reinecker 2000). Along the Lavanttal segment, dextral offset of approximately 10 km was deduced from displaced lithological units. Vertical offset is 4–5 km, whereby the eastern block (Koralm) was up-faulted (Frisch et al. 2000a). Near its southern termination, the Lavanttal fault cuts and offsets the Periadriatic fault by about 20 km. Sedimentary basins (Lavanttal Basin, Obdach Basin) formed along left-handed oversteps. The nature of the Lavanttal Basin is probably an oblique graben structure formed in a transtensional regime (Frisch et al. 2000a); it is assumed to be active since the Early Miocene with peaks in activity between 18–16 Ma and 14–12 Ma (Reinecker 2000). Fault plane solutions display clear dextral strike-slip movements (Reinecker & Lenhardt 1999; Reinecker 2000).

Little is known about the eastern margin of the Koralm Complex, the greatest part of it being hidden below Miocene sediments of the western Styrian Basin (Fig. 1). However, brittle faults and fault-related cataclastic rocks were detected by cored drillings located at the eastern margin of the Koralm Complex (Vanek et al. 2001; Brosch et al. 2001; Pischinger et al. 2005, 2006). Brittle structures that are related to the latest evolution of the Koralm Complex were analysed by Vanek et al. (2001). The few results of tectonic and stress-strain analyses may be correlated with the latest tectonic evolution of the Eastern Alps from the Neogene onwards; this includes sustained N–S-directed extension, being re-oriented and replaced by E–W extension and E–W compression.

Following the descriptions above, the latest clearly documented event within the Koralm Complex is the amphibolite facies metamorphic overprint which occurred at approximately 90 Ma ago. Subsequent cooling is ill-constrained. The final increment of the pressure-temperature-time evolution of the Koralm Complex, i.e. from approximately 90 Ma onwards, is poorly documented, as is the Cenozoic structural evolution of the Austroalpine crystalline complexes in the eastern part of the Eastern Alps. This evolution primarily comprises exhumation, tectonic uplift and surface uplift.

A few data show that crustal stretching, extension and the formation of the Gosau Basins of the Eastern Alps east of the Tauern Window ("Zentralalpine Gosau") coincides with the exhumation of crystalline basement complexes of the Lower Central Austroalpine unit (Fig. 1) (Neubauer et al. 1995). Exhumation resulted in cooling from initial epidote-amphibolite/upper greenschist facies conditions to temperatures below 300 °C at the beginning of the Paleogene. Sphene, zircon and apatite fission track data, for example from the Gleinalm area north of the Koralm Massif, indicate cooling to temperatures below 200–250 °C at 65 Ma (Neubauer et al. 1995). The northern part of the Koralm Complex cooled to temperatures below 200 °C already in the Late Cretaceous (Hejl 1997, 1998). Hence, these regions were already near (approximately 5–8 km) to the surface during the whole Cenozoic. Towards south, the apatite fission track ages within the Koralm Complex gradually become younger. This indicates that the southern parts were exhumed later. In the central part of the Koralm Complex these ages range from approximately 50 to 37 Ma (Hejl 1998; Rabitsch et al. 2007). Approximately 31 Ma are reported from the southern margin of the Koralm Complex, approx. 26 Ma from the western margin (Hejl 1998). Two apatite fission track ages from the central part of the Koralm Complex, close to the Lavanttal fault, show cooling below approximately 120 °C between 28.5 and 18 Ma. West of the Lavanttal fault, apatite fission track ages range from approx. 27 to 12 Ma (Puch 1995). In the Pohorje region early to mid-Miocene cooling of both magmatic and metamorphic rocks is indicated by zircon fission track ages of 26–19 Ma (Fodor et al. 2003). Toward west, in the Gurktal

Nappe Complex (Fig. 1), stronger post-Cretaceous denudation can be observed as compared to the Koralm Complex.

Indirect evidence for the Neogene evolution of the Koralm Complex may be provided by the sedimentary record within the adjacent sedimentary basins (in particular the Styrian and Lavanttal Basins) (Figs. 1–3). The subsidence history of the Styrian Basin as well as that of the Lavanttal Basin, are better constrained due to a well documented stratigraphy (for a summary, see Ebner & Sachsenhofer 1995; Sachsenhofer et al. 1997, 2001). Subsidence started probably at 18 Ma (Ottnangian stage of the Central Paratethys paleogeographic realm) (Fig. 3), followed by a phase of transgression in the Early Karpatian (approx. 17 Ma). In the latest Karpatian (approx. 16.4 Ma) a tectonic event led to the re-organisation of the basin architecture. This event is assumed to be related to block-tilting causing an uplift of the hinterland, represented by the Koralm Massif. This coincided with an eustatic sea level low stand, thus forming a tectonically enhanced sequence boundary. In the southern part of the western Styrian Basin, close to the Pohorje Mountains (Fig. 1), early Miocene sediments lacking a thermal overprint contain apatite grains showing a cooling age of approx. 19 Ma (Eggenburgian), only 1–2 Ma older than the time of deposition (Sachsenhofer et al. 1998). The cooling rate of the mainly Austroalpine source was very fast, pointing to tectonic denudation (Sachsenhofer et al. 1998).

The earliest Badenian (approx. 16 Ma) is characterised by shallow marine conditions; fluvial sedimentation was restricted to the western margin of the basin, i.e. close to the eastern margin of the Koralm Massif. A major sea-level drop at the end of the Badenian (approx. 13 Ma) caused the progradation of (braided-) delta deposits into the western part of the Styrian Basin, followed by a new phase of transgression during the Sarmatian (13–11.5 Ma). This marine influence prevailed up to the Early Pannonian (Sachsenhofer 1996). Limnic and fluviatile sediments replaced this marine period, and from the Late Pannonian onwards, the terrestric sedimentary influence increased due to continuing uplift.

3 Methods

Slickenside and striae data for paleostress orientation analyses were collected following the methods proposed by Angelier & Mechler (1977) and Angelier (1979) both in the field and, due to the restricted occurrence of adequate outcrops, from drill cores. Within the scope of the geological and geotechnical site investigations for the Koralm Tunnel (with a length of 32.8 km to be built under the Koralm Massif) (Steidl et al. 2001) an enormous volume of data and material has been gained and elaborated during the last years. Especially seven deep core drillings (being part of the site investigations for the Koralm Tunnel), reaching depths of up to 1200 m, have extended the access to geological samples into the third dimension.

Criteria used to determine the sense of slip along brittle faults were described by Petit (1987), Angelier (1994) and Doblas (1998). The collected fault-striae data were consecutively used for paleostress analysis. Analysis was performed for each individual outcrop to keep control on possible overprinting relationships and multistage formation of shear fractures and faults. Only rarely data from nearby outcrops were analysed together, and only in cases when the data sampled were too few at a distinct station and provided that the data clearly belonged to the same kinematic set. Orientation distributions of distinct fracture sets and geometrical relationships were analysed by using the program package Tectonics FP 1.6.2, a 32-bit Windows™-Software for Structural Geology (Reiter & Acs 1996–2001; Ortner et al. 2002). The PT-method (P: contraction axis; T: extension axis; Turner 1953) or graphical method (Marret & Allmendinger 1990) provided by this software, was used to calculate the orientation of the kinematic axes from the fault-striae data. Prior to analysis the fault-striae data were separated into homogeneous subgroups (Meschede & Decker 1993). The P-, T- and the B (intermediate) axes were calculated for each fault plane – striae data set by assuming an angle Θ of 30° between the compression axis and the respective fault plane. A Θ of 30° has been shown to be a reasonable value for most cases according to the Mohr-Coulomb failure criterion (Meschede 1994). The mean vectors for the kinematic axes were calculated after Wallbrecher (1986) and represent an approximation of the principal stress axes (Ortner et al., 2002). For highly anisotropic rocks showing reactivation of the foliation planes as frictional shears, however, a different Θ angle was applied that previously was determined by a best-fit analysis (see Tab. 1). Additionally, for each data set the principal stress directions and the stress ratios were calculated with the numerical dynamical analysis (NDA; Spang, 1972).

In contrast to the analysis of field data, the kinematic analysis of discontinuities in drill core samples combines geophysical borehole logging and structural stress/strain analysis (Brosch et al. 2001; Vanek et al. 2001). This procedure consists of two steps. During the first, each discontinuity encountered in the drill cores and identified by an acoustic borehole televiewer is examined with respect to its nature, surface markings, mode and sense of wall displacements, fillings and primary (in-depth) aperture (Brosch et al. 2001). For linear surface markings (striae) the rake angle is recorded with respect to the relative strike line of the discontinuity in the drill core. Secondly, the rake data are transposed into dip and dip direction data and corrected for the deviation of the borehole from the vertical axis. Then the theoretical compression and tension axes are calculated as described above.

4 Brittle Structures and their interpretation

Morphologically the eastern part of the Koralm Massif is characterised by valleys of two main orientations, either trending N–S or WNW–ESE; the widest of the latter contains the Schwanberg block debris (Fig. 2). These morphological features coincide with the two main sets of brittle structures, in particular fault zones and slickensides, recorded in this area. Therefore it can be assumed that the course of most valleys

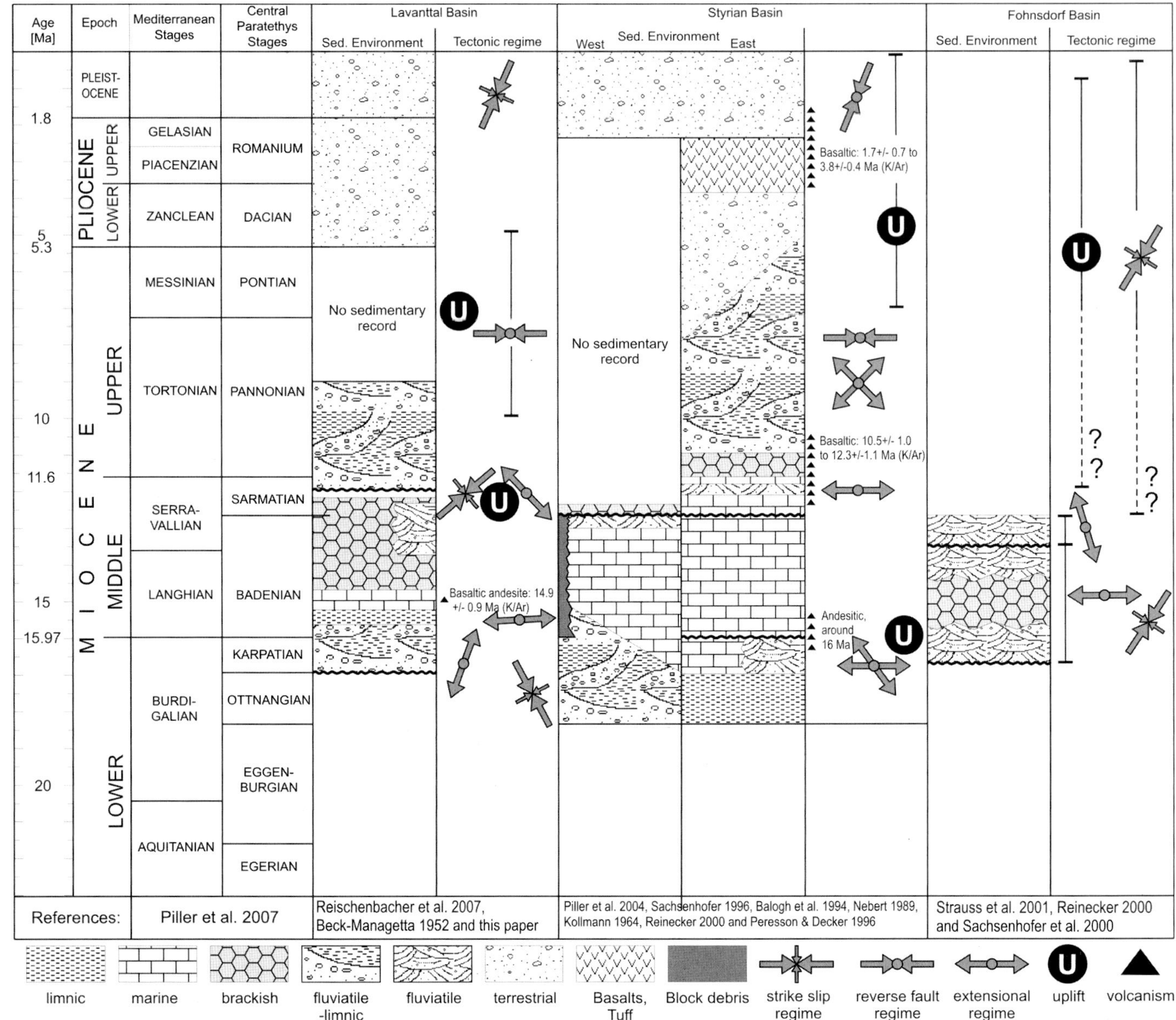

Fig. 3. Comparison of the sedimentary evolution and tectonic regime for the Lavanttal, the Styrian and the Fohnsdorf Basin from the Lower Miocene up to the Pleistocene. Grey filled arrows indicate orientation of paleostress tensors (West is left page margin and East is right page margin).

is fault-controlled, a feature already noted by Stiny (1925) for the northern Koralm. The main faults, together with conjugated secondary fractures, were repeatedly activated during distinct deformational phases.

At map scale mainly two sets of faults can be distinguished; their strike directions range from E to SE and N to NE, respectively (Fig. 2). In general, the E-trending ones are partly covered by block debris deposits; these are crosscut and displaced by NNE-trending faults (Fig. 2). The contact of the Koralm Complex with the Miocene sediments of the western Styrian Basin is badly exposed, as are assumed normal fault zones forming the eastern margin of the Koralm Massif. Both discontinuities were temporally exposed during the excavation of the Koralm pilot tunnel, showing that the eastern margin of the Koralm Complex is formed by a cataclastic shear zone of approximately 1 meter in thickness in this area (Fig. 4). This zone comprises fine-grained cohesive cataclasites (terminology *sensu* Brodie et al. 2002) with a fragment size of approximately 0.5 to 5 cm; the fragments are embedded within a matrix with predominating grain sizes of 0.2 to 0.5 mm (Fig. 4). A highly fractured damage zone with highly variable thickness (several decimetres to several meters), partly grading into a block-in-matrix rock (*sensu* Medley 1994), characterises the footwall. In the hanging-wall the shear zone is covered by slightly com-

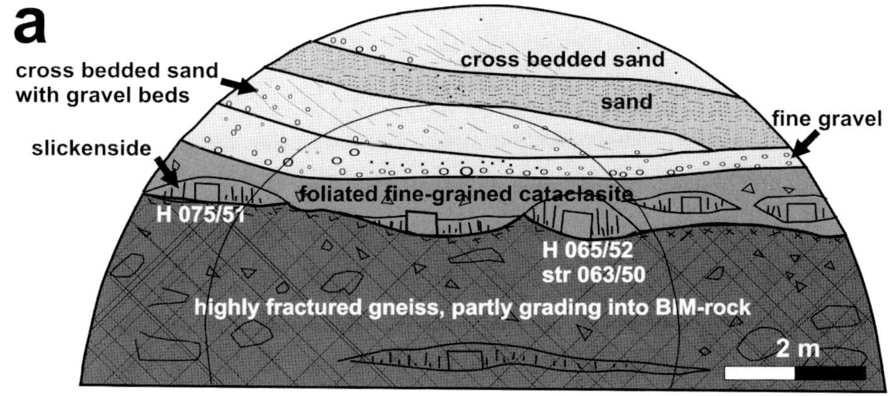

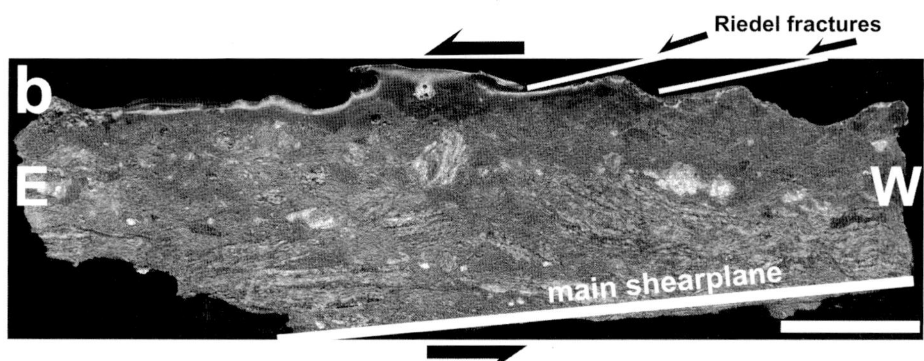

Fig. 4. a) Contact between a cataclastic shear zone and slightly compacted, but undeformed Badenian (?) sand along the eastern margin of the Koralm Complex, as exposed within the pilot tunnel "Leibenfeld" at station 130 m (view is towards east, appr. 12 m wide and 5.5 m high); the cataclastic shear zone rock in the footwall of the sediments is dipping to the east, slickenside striae are plunging subparallel to the fault dip (normal sense of shear); H: slickensided fault plane, str: striation, BIM-rock: Block-in-matrix rock (refer to text for nomenclature). b) polished section of a hand specimen of the cataclasite with gravel-sized, angular to slightly rounded fragments of gneiss in a fine-grained, foliated matrix, Riedel shears support top to east sense of shear. Main shearplane is the same as slickenside in Figure 4a. (Scalebar in the lower right is 3 cm long)

pacted cross-bedded, undeformed sands of probable late Karpatian to early Badenian age (Nebert 1989; Beck-Mannagetta et al. 1991).

At the scale of a few decimetres to meters the sequence of displacements along distinct faults can be derived from overprinting relationships both in outcrops and drill cores. However, these overprinting relationships are restricted to a few key outcrops that provide the basis for the structural analysis at sites with incomplete information about the relative deformation sequence. This sequence comprises four major events of brittle deformation, referred to as D_1 to D_4. The coordinates of outcrop locations with detailed data on the orientation of the evaluated principal stress axes are summarised in Table 1.

D_1 can be subdivided into two sub-phases. D_{1-1}-related structures are restricted to distinct domains. Locally, E to ESE striking sub-vertical fractures were activated as dextral strike-slip faults. These are associated with conjugate N to NW-trending dextral, and NNE-trending sinistral strike-slip faults (Figs. 5a, b). A detailed analysis of the distinct stations shows that either ESE and N-trending, or (N)NW and NNE-trending faults occur as conjugate fracture sets. Locally, NW-trending fractures with dextral displacement occur as single sets. The complete assemblage can be geometrically interpreted to represent ESE-trending Y (main)-, E-trending P-, (N)NW-trending R-(Riedel), and NNE-trending R'-fractures. The results from the analysis of paleostress orientations show a sub-horizontal NNW–SSE orientation of σ_1 and a sub-horizontal orientation of σ_3 in ENE–WSW-direction (Figs. 5a, b). The block debris deposits

of Schwanberg, mainly consisting of components derived from the adjacent basement, are related to major D_{1-1}-related ESE-trending faults, too. According to Nebert (1989), sedimentation of these deposits started during the Late Karpatian/Early Badenian (Fig. 3). In general, the base of the deposits is formed by a zone of highly disintegrated host rock, often accompanied by the development of tectonic breccias and cataclasites. Adjacent to the block debris deposits, the basement protoliths (mainly garnet mica-schist and schistose garnet-bearing gneiss) show severe alteration and deformational overprint of the penetrative fabrics along distinct semi-ductile shear zones. The basement protolith is intensely retrogressed; biotite is mainly replaced by stilpnomelane, plagioclase is mainly replaced by epidote-zoisite and calcite (Egger 2007). Besides the alteration, the development of veins and cracks healed by calcite and subordinate quartz, white mica, and zeolite indicates the presence of hydrothermal fluids during faulting.

The ESE striking dextral strike-slip shears were reactivated during D_{1-2} as conjugate high-angle normal faults. This is indicated by sub-vertical striae associated with top-down displacement criteria overprinting the D_{1-1}-related sub-horizontal striae (Fig. 5d). These high-angle normal faults are far better preserved than the previous strike-slip faults and are locally associated with the development of cm-thick fault gouges. The main valleys and ridges strike parallel to these WNW–ESE oriented structures. As observed in drill cores the D_{1-2}-related normal faults crosscut the lower parts of the block debris deposits of Schwanberg as well (Egger 2007). The

Table 1. Coordinates of outcrop locations used for paleostress analysis with detailed orientations of the principal stress/strain axes determined for each location. Coordinates are in Austrian BMN M34 system. N – total number of data, Nbiv.. number of bivalent data, PT…PT method after Turner (1953), NDA – numerical dynamical analysis after Spang (1972), secfrac – kinematics deduced from secondary fractures e.g. Riedel fractures, Kex – extension fractures (gashes), con – conjugated shear fractures; P – compression axis, B – intermediate axis, T – extension axis, Theta – angle of internal friction, R – shape factor of the paleostress ellipsoid calculated with the numerical dynamical analysis (NDA) after Spang (1972). PT and NDA were calculated with TectonicsFP (Reiter and Acs 1996–2001).

Outcrop-ID	Easting	Northing	Altitude	Method	N	N_{biv}	P [°]		B [°]		T [°]		Theta [°]	R	Event
							dipdir	dip	dipdir	dip	dipdir	dip			
1	663924	196819	531	PT, NDA	4	0	48	72	275	8	181	15	35	0.435	D1-2
1	663924	196819	531	PT, NDA	15	4	138	83	5	5	280	5	35	0.422	D2
5	663787	192855	695	Extension	3	0	270	89	360	1	90	1			D2
7	643581	186454	1554	PT	3	0	227	79	332	1	61	9	30		D2
7	643581	186454	1554	PT	3	0	236	15	123	55	336	31	30		D3-1
7	643581	186454	1554	PT, NDA	4	0	297	80	28	2	121	10	30	0.578	D3-2
8	647102	190757	1384	PT, NDA	15	3	160	76	283	8	11	14	20	0.455	D1-2
9	647176	190763	1364	PT, NDA	8	3	145	69	306	13	40	6	30	0.487	D1-2
10	647070	190665	1384	PT, NDA	6	0	143	88	293	19	27	10	30	0.539	D1-2
22	644596	182403	1351	PT	3	0	173	21	326	69	79	6	35		D1-1
24	661794	193955	803	PT	4	1	212	43	22	51	303	2	30		D3-1
26	662236	194433	772	PT	3	0	84	81	342	2	251	7	30		D2
30	645204	175958	1159	PT, NDA	5	0	283	75	116	14	24	4	30	0.49	D1-2
47	641508	182275	779	PT, NDA	5	1	222	62	353	17	86	20	30	0.508	D2
49	660446	210288	400	PT, NDA	6	1	123	57	347	24	246	19	30	0.471	D2
49	660446	210288	400	conjug	12	0	143	88	53	1	323	2			D3-2
50	655925	184494	957	PT, NDA	12	4	51	43	193	48	308	16	30	0.519	D3-1
52	662334	198465	480	PT	5	4	33	23	273	83	126	7	30		D3-1
53	662576	198465	480	PT, NDA	5	1	34	41	190	47	292	13	30	0.513	D3-1
69	649017	207767	824	PT	3	0	252	74	156	2	66	17	30		D2
70	649543	207315	821	PT, NDA	4	0	332	69	186	17	93	11	30	0.506	D2
71	653276	203732	941	PT	3	0	343	7	128	89	69	5	30		D1-1
72	654254	204047	944	PT, NDA	6	2	69	80	169	2	256	10	30	0.524	D2
72	654254	204047	944	PT, NDA	9	0	144	1	42	78	234	17	30	0.539	D1-1
73	645808	181306	1385	PT, NDA	7	0	242	54	120	22	17	32	30	0.46	D1-2
74	645026	181295	1156	PT, NDA	10	5	140	82	318	9	47	1	30	0.505	D2
75	641028	173255	351	PT, NDA	22	0	186	11	44	77	277	10	30	0.521	D1-1
75	641028	173255	351	PT, NDA	4	0	319	73	112	14	204	7	42	0.427	D1-2
75	641028	173255	351	PT, NDA	12	0	215	6	96	79	306	10	30	0.502	D3-1
75	641028	173255	351	PT, NDA	15	0	240	87	20	4	110	2	30	0.436	D3-2
76	665673	207869	360	PT, NDA	7	0	193	70	349	19	82	7	30	0.525	D2
78	665609	208009	360	Extension	6		260	89	350	1	80	1			D2
78	665609	208009	360	Extension	3		324	89	54	1	144	1			D3-2
79	656552	183317	1003	extension	7		93	79	183	1	273	11			D2
80	654290	182897	1408	PT, NDA	5	0	62	82	273	8	182	3	30	0.483	D1-2
80	654290	182897	1408	PT	3	0	230	50	23	37	124	15	30		D3-1
83	654524	184073	1104	PT	3	0	335	72	85	5	175	16	30		D1-2
83	654524	184073	1104	PT, NDA	4	0	118	76	215	3	306	15	30	0.504	D3-2
83	654524	184073	1104	PT, NDA	6	0	73	5	187	79	335	8	30	0.528	D3-1
84	656922	183178	962	PT, NDA	11	0	102	75	240	11	332	11	30	0.484	D3-2
87	663564	197707	485	PT	3	0	6	24	217	62	100	9	30		D3-1
87	663564	197707	485	PT	3	0	265	79	357	3	88	11	30		D2
88	663482	198540	519	PT, NDA	8	1	190	83	16	6	286	1	30	0.514	D2
91	664252	199059	483	PT, NDA	7	1	79	77	193	3	288	12	30	0.457	D2
92	663094	199115	604	PT	3	0	184	76	17	13	286	3	30		D2
96	665011	197074	434	PT, NDA	7	0	116	79	10	4	281	11	38	0.496	D2
105	653654	203296	946	PT, NDA	6	1	93	58	264	25	350	7	30	0.502	D1-2
111	644448	183782	1512	PT, NDA	12	0	259	59	45	27	142	15	30	0.46	D3-2
113	643169	180605	916	PT, NDA	22	7	49	73	288	9	198	12	30	0.503	D1-2
114	642505	180715	820	PT, NDA	6	0	89	76	241	12	332	4	30	0.419	D3-2
114	642505	180715	820	PT, NDA	9	0	299	76	129	13	39	3	30	0.518	D1-2
114	642505	180715	820	PT, NDA	8	0	91	15	318	71	187	10	30	0.543	D4
116	643499	190857	787	PT, NDA	16	4	210	56	84	16	345	26	30	0.487	D1-2
116	643499	190857	787	PT	3	0	163	52	5	36	267	11	30		D1-1
117	643312	190960	846	PT, NDA	5	0	175	56	3	36	270	3	30	0.452	D2
118	642288	191188	677	PT, NDA	12	0	258	79	65	13	155	5	30	0.416	D1-2
122	641852	191822	835	PT, NDA	9	1	166	67	31	17	295	16	30	0.503	D3-2
125	641585	191533	708	extension	5	0	121	59	31	1	301	31			D3-2
127	641605	191508	670	PT	3	1	126	58	35	9	292	31	30		D3-2

Table 1. (Continued).

Outcrop-ID	Easting	Northing	Altitude	Method	N	N_{biv}	P [°]		B [°]		T [°]		Theta [°]	R	Event
							dipdir	dip	dipdir	dip	dipdir	dip			
129	641008	191033	796	PT, NDA	10	0	153	22	14	62	248	16	30	0.398	D1-1
129	641008	191033	796	PT, NDA	6	0	133	64	12	10	278	22	30	0.443	D2
133	640619	191656	582	PT, NDA	13	2	13	1	172	88	284	2	46	0.412	D3-1
134	643312	190960	846	PT, NDA	19	0	295	72	26	0	114	17	32	0.503	D3-2
134	643312	190960	846	PT, NDA	5	0	206	33	7	56	110	7	42	0.502	D3-1
134	643312	190960	846	PT	3	0	315	40	156	46	53	12	32		D1-1
137	643185	189360	1264	PT, NDA	8	2	152	32	11	47	254	23	30	0.522	D1-1
137	643185	189360	1264	PT, NDA	6	1	231	78	111	6	18	7	30	0.504	D1-2
140	643147	189657	1290	kex, conjug	4	0	192	59	31	29	296	6			D3-2
142	662283	197361	687	PT, NDA	18	0	111	86	10	1	284	4	30	0.412	D2
142	662283	197361	687	PT, NDA	5	0	177	21	328	69	82	15	30	0.587	D1-1
147	646630	164609	420	PT	3	1	342	51	171	49	83	0	30		D1-1
153	643033	184644	1299	PT, NDA	6	0	285	85	109	4	20	1	30	0.524	D1-2
153	643033	184644	1299	PT, NDA	8	0	159	72	344	18	253	2	30	0.503	D2
153	643033	184644	1299	PT	4	3	193	12	25	58	98	22	30		D3-1
155	642137	172272	431	PT	5	3	317	4	209	70	48	6	30		D1-1
158	661345	184999	610	PT	3	0	71	81	184	4	274	9	30		D2
160	661532	184990	627	PT	3	0	268	71	100	19	9	4	30		D1-2
161	655900	187379	950	PT, NDA	8	0	225	76	3	10	94	9	30	0.515	D2
163	660863	185328	593	PT, NDA	9	1	129	79	5	7	276	9	30	0.492	D2
164	658552	185624	728	PT, NDA	7	1	69	1	155	0	221	89	70	0.49	D4
164	658552	185624	728	PT, NDA	5	0	4	4	239	82	92	7	30	0.487	D1-1
166	666079	194998	552	PT, NDA	14	0	355	65	139	21	234	9	30	0.483	D2
166	666079	194998	552	PT, NDA	7	3	289	18	83	70	197	8	30	0.516	D4
167	666222	195122	495	Extension	6	0	250	89	160	1	70	1			D2
172	657335	206228	624	PT	3	0	32	78	169	8	261	11	30		D2
173	662431	209863	393	PT, NDA	7	0	129	53	277	32	17	16	30	0.495	D1-2
175	636706	197628	647	PT, NDA	8	3	179	16	320	69	87	9	30	0.501	D1-1
175	636706	197628	647	PT	5	2	66	74	330	2	240	14	30		D2
177	638417	195115	584	PT	3	0	156	18	13	68	251	14	30		D1-1
177	638417	195115	584	PT	3	0	210	62	90	14	352	22	30		D1-2
178	636596	197512	620	PT	3	1	196	61	329	32	76	17	30		D2
186	660723	191011	515	PT, NDA	9	0	9	11	187	78	278	2	30	0.538	D1-1
187	659822	191359	568	PT	5	2	132	74	12	5	282	19	20		D3-2
187	659822	191359	568	PT, NDA	13	4	0	3	98	77	271	7	20	0.668	D1-1
190	663218	184495	771	PT	7	4	345	3	10	89	75	1	25		D1-1
192	637814	199201	717	PT, NDA	6	0	143	51	8	30	263	22	30	0.507	D2
193	638250	199468	717	PT	3	0	332	28	75	23	195	51	30		D1-1
195	644347	179775	1034	PT, NDA	6	1	350	82	111	8	204	8	30	0.447	D1-2
204	657747	193401	667	PT, NDA	5	1	345	73	153	14	250	2	30	0.474	D2
207	660402	191534	519	PT	3	1	29	13	287	52	126	27	30		D3-1
214	659807	191581	501	PT	6	4	352	12	158	73	262	2	30		D1-1
215	664345	185080		PT, NDA	8	0	88	78	349	2	258	12	30	0.46	D2
216	663811	184927		PT	4	3	176	5	336	71	85	17	30		D1-1
220	650377	182333	1574	PT	3	0	129	23	263	61	30	19	30		D1-1
222	659330	185505	681	PT, NDA	6	1	150	5	21	79	239	11	30	0.511	D1-1
223	660685	184075	630	PT, NDA	7	0	113	82	290	10	20	4	30	0.489	D1-2
224	639826	199759	730	PT,NDA	10	0	320	7	53	45	222	41	30	0.493	D1-1
224	639826	199759	730	PT, NDA	4	0	123	53	310	38	217	2	30	0.483	D1-2
225	633483	185458	472	PT	5	3	73	9	287	69	164	10	30		D3-1
226	663955	184694	563	PT	29	7	108	50	344	27	238	27	45		D2
226	663955	184694	563	PT	6	0	338	78	81	2	172	11	40		D1-2
227	663183	184861	664	PT	27	3	205	55	23	39	111	4	40		D3-2
227	663183	184861	664	PT	3	0	125	60	312	30	218	6	30		D1-2
227	663183	184861	664	PT	4	0	7	14	99	18	240	71	30		D3-1
228	658129	183153	650	PT	19	0	46	52	167	22	271	30	35		D2
228	658128.5	183153	650.455	PT	15	0	121	70	237	8	329	18	32		D3-2
228	658128.5	183153	650.455	PT	17	0	340	77	102	9	188	14	35		D1-2
229	655507.6	185555.5	981.77	PT	12	0	253	71	87	19	352	7	32		D1-2

high-angle faults are additionally associated with sub-vertical, ESE striking extensional veins and open fractures, indicating (N)NE-(S)SW-directed extension. This interpretation is supported by the paleo-principal stress orientations, i.e. σ_1 with a sub-vertical orientation, σ_3 with a (N)NE – (S)SW orientation (Fig. 5c).

Table 1. (Continued).

Outcrop-ID	Easting	Northing	Altitude	Method	N	N$_{biv}$	P [°] dipdir	P dip	B [°] dipdir	B dip	T [°] dipdir	T dip	Theta [°]	R	Event
229	655507.6	185555.5	981.77	PT	8	0	197	18	96	38	305	47	26		D3-1
230	649069.8	184805.2	1465.56	PT	9	0	251	75	70	13	165	3	30		D1-2
230	649069.8	184805.2	1465.56	PT	18	0	72	81	18	1	282	13	40		D2
231	644176.6	183776	1473.04	PT	4	2	180	51	14	29	281	11	35		D2
232	642857.9	182093.9	926.372	PT	10	2	184	79	277	6	16	14	38		D1-2
232	642857.9	182093.9	926.372	PT	9	2	172	67	2	8	271	4	38		D2
233	659741.8	183436.5	930.21	PT	22	0	12	74	113	4	196	15	26		D1-2
233	659741.8	183436.5	930.21	PT	16	0	43	65	216	25	306	7	36		D3-2
233	659741.8	183436.5	930.21	PT	18	0	15	4	105	49	296	45	30		D3-1
233	659741.8	183436.5	930.21	PT	12	0	88	7	171	29	338	56	30		D4
234	651250.7	183919.2	1510.56	PT	60	0	301	86	99	5	189	5	35		D1-2
234	651250.7	183919.2	1510.56	PT	91	0	178	69	3	24	272	3	46		D2
234	651250.7	183919.2	1510.56	PT	13	0	155	17	54	30	265	54	35		D1-1
234	651250.7	183919.2	1510.56	PT	6	0	255	2	354	61	157	29	35		D3-1
235	641981.6	180586.5	700.67	PT	13	0	350	76	100	4	189	14	50	0.507	D1-2
235	641981.6	180586.5	700.67	PT	18	0	355	78	154	12	245	0	52	0.498	D2
235	641981.6	180586.5	700.67	PT	12	0	47	81	229	9	138	2	64	0.542	D3-2
236	656711	181761.2	1233.02	PT	19	0	150	33	13	49	252	23	40	0.511	D1-1
236	656711	181761.2	1233.02	PT	12	0	213	71	91	11	356	13	30	0.564	D1-2
236	656711	181761.2	1233.02	PT	5	0	109	77	330	10	241	8	42	0.53	D2
236	656711	181761.2	1233.02	PT	4	0	52	17	266	72	146	9	30	0.231	D3-1
236	656711	181761.2	1233.02	PT	6	0	211	0	113	2	107	86	40	0.48	D3-1
237	650458.8	181155	1374.76	PT	3	0	162	3	284	87	76	2	30		D1-1
237	650458.8	181155	1374.76	PT	16	0	142	69	355	18	259	11	30	0.505	D2
237	650458.8	181155	1374.76	PT	21	0	158	19	46	50	258	34	30	0.503	D1-1
237	650458.8	181155	1374.76	PT	7	0	50	12	190	72	324	11	20	0.567	D3-1
237	650458.8	181155	1374.76	PT	26	0	197	51	35	38	299	10	30	0.519	D3-2
238	647255.7	180773	1581.38	PT	4	0	217	66	80	17	347	15	30	0.443	D1-2
238	647255.7	180773	1581.38	PT	5	0	68	21	333	11	216	66	30	0.475	D4
238	647255.7	180773	1581.38	PT	3	0	201	1	111	29	292	60	30		D3-1
238	647255.7	180773	1581.38	PT	3	0	164	16	52	56	256	32	30		D1-1
238	647255.7	180773	1581.38	PT	18	0	246	66	99	21	4	13	48	0.464	D1-2
238	647255.7	180773	1581.38	PT	25	0	253	70	29	13	123	13	54	0.452	D3-2
238	647255.7	180773	1581.38	PT	3	0	217	7	128	11	339	79	30		D3-1
239	647255.7	180773	1581.38	PT	9	0	258	73	40	14	132	17	68	0.578	D3-2
239	647255.7	180773	1581.38	PT	4	0	280	67	125	17	38	6	54	0.516	D1-2
240	654045.3	181267	1545.4	PT	55	0	137	72	347	15	255	11	50	0.448	D2
240	654045.3	181267	1545.4	PT	3	0	220	12	99	64	314	22	30		D3-1
240	654045.3	181267	1545.4	PT	13	0	254	60	39	23	133	17	78	0.485	D3-2
240	654045.3	181267	1545.4	PT	6	0	186	75	33	13	302	6	30	0.433	D3-2
240	654045.3	181267	1545.4	PT	7	0	218	61	44	30	313	2	80	0.47	D3-2
240	654045.3	181267	1545.4	PT	5	0	100	15	2	24	212	63	20	0.465	D4

Most of the brittle structures observed in the southern part of the Koralm Massif are related to D$_2$. These are N–S striking slickensides, the major set steeply dipping towards E, and minor conjugate sets dipping to the W (Fig. 6). Slickenside – related striae plunge subparallel or slightly oblique to the slickenside dip direction and show top-down kinematic indicators. In the field these shear fractures are locally associated with sub-vertical extensional veins arranged within an en echelon geometry, mainly filled with quartz or calcite. As seen particularly in drill cores, sub-vertical open en echelon fissures strike in a N–S direction. Altogether, these structures can be interpreted as being related to general E–W-directed extension; they dominate both in surface exposures and in drill cores. The results from the analyses of paleostress orientations show a sub-vertical orientation of σ_1, and a sub-horizontal orientation of σ_3 in E–W to ESE–WNW-direction, locally shifting to a (N)NW-(S)SE orientation (Fig. 6).

These high-angle structures are repeatedly associated with E-dipping low-angle normal faults and shear zones. The latter formed by the reactivation of the penetrative foliation, mainly within smoothly dipping Plattengneis domains. These shear zones are accompanied by cataclastic fault rocks, consisting of very fine grained quartz, white mica and biotite in the matrix with incorporated broken grains of feldspar or cm to decimetre large protolith fragments (Fig. 7a). The damage zone (*sensu* Chester & Logan 1986; Caine et al. 1996) adjacent to the low-angle shear zones is characterised by the formation of closely, millimetre- to centimetre-spaced fractures, 5–20 cm in length, at high angles (70–90°) to the fault zone boundaries, bounding distinct rhombohedral blocks. The fracture-bound blocks show antithetic bookshelf rotation referring to the displacement along the low-angle shear zones, and associated to normal displacement along the high-angle fractures, too (Fig. 7b). Locally

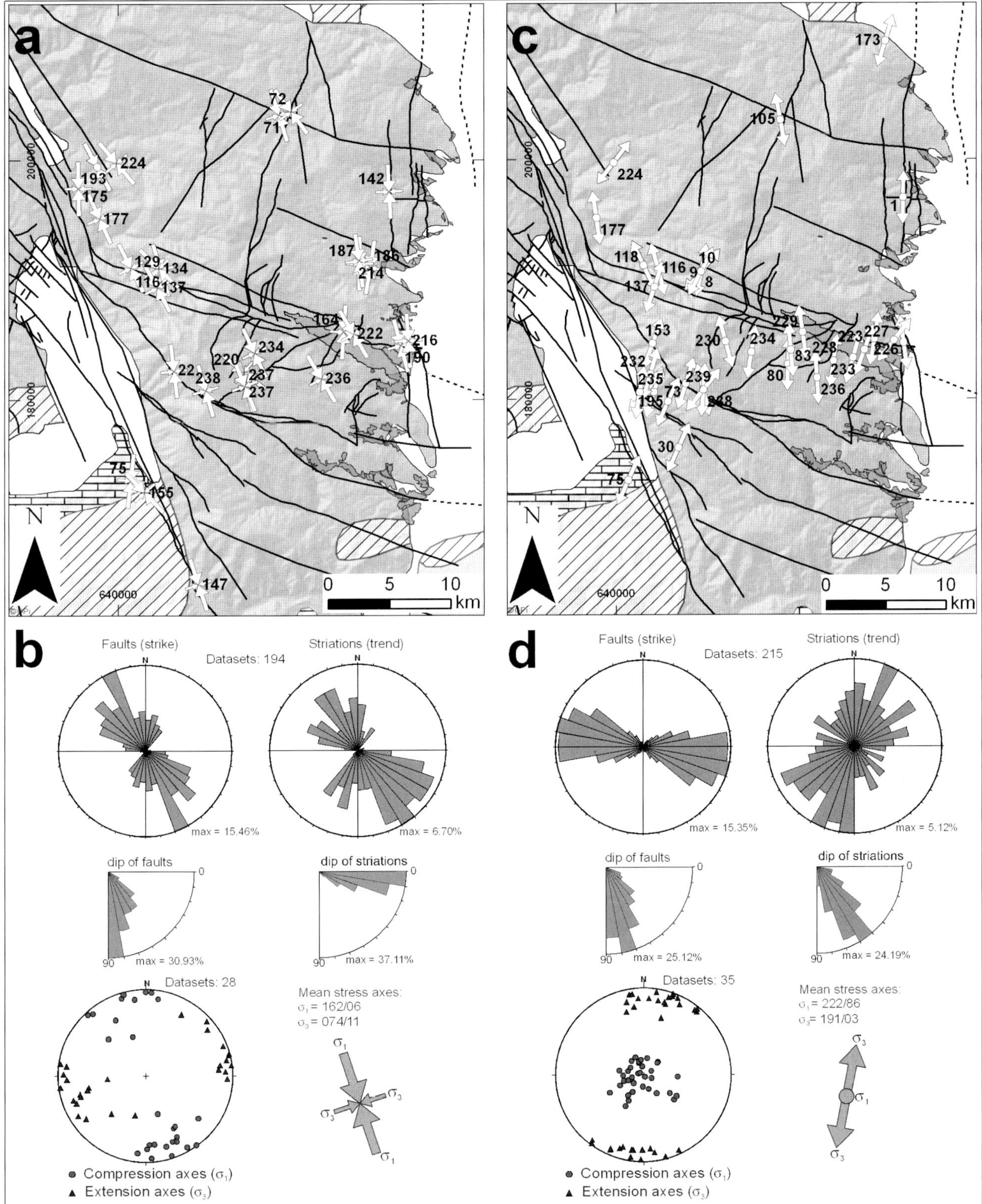

Fig. 5. a) Orientation of σ_1 (large arrows) and σ_3 (small arrows) related to D_{1-1}; labeled numbers refer to Table 1. b) Rose diagrams with strike and dip of fault planes and trend of striations; orientation of σ_1 (filled circles) and σ_3 (triangles) of D_{1-1}-related tensors with mean maximum and minimum principal stress axes. c) Orientation of σ_1 (large arrows) and σ_3 (small arrows) related to D_{1-2}; labeled numbers refer to Table 1. d) Rose diagrams with strike and dip of fault planes and trend of striations; orientation of σ_1 (filled circles) and σ_3 (triangles) of D_{1-2}-related tensors with mean maximum and minimum principal stress axes. Refer to Figure 2 for legend of geological units. Grid is in Austrian BMN M34 system.

these low angle shear zones are associated with antithetic W-dipping conjugate high-angle normal faults (Fig. 7a). Synthetic sets of high-angle fractures continuously curve into the dip direction of the low-angle shear zones and form listric normal faults. Towards the lower tip line the high-angle faults show the development of cataclasites. High disintegration of the protolith may be observed along antithetic high-angle faults, too; in most cases, however, the original structure of the protolith can still be identified, irrespective of the slight alteration of the protolith. This alteration is characterised by the enrichment of feldspar and biotite. Around the upper tip area, the high-angle faults may split up into splays, typically forming horse-tail structures.

N–S striking D_2-related major faults crosscut both previously formed E- to SE trending faults and the block debris deposits of Schwanberg. Locally, the block debris is crosscut by distinct brittle shear zones, a few centimetres wide, as well as by slickensides, indicating a post-sedimentary (re-)activation of distinct faults.

D_3 is characterised by a sub-horizontal orientation of the minimum principal stress σ_3 in SE–NW direction (Fig. 8). Locally, E–W trending fractures are activated as sinistral shears. Additionally ENE- and NNW-striking subvertical fractures are activated as strike slip planes with sinistral and dextral sense of shear, respectively. A detailed analysis of single stations shows that (E)NE- and N- to NW-trending fractures may occur as conjugate shears. The complete assemblage can be geometrically interpreted to represent E-trending Y-, NE-trending R , and (N)NW-trending R'-fractures (Figs. 8a, c). The analysis of paleostress orientations for this deformational event (D_{3-1}) indicates a sub-horizontal NE–SW orientation of the maximum, and a sub-horizontal NW–SE orientation of the minimum principal stress axis (σ_1 and σ_3, respectively) (Figs. 8a, b). Locally, mainly along restraining bends along E–W-striking strike slip faults, σ_3 shifts to a subvertical orientation. This indicates inversion of previously formed E-trending faults and related basins. However, these orientations are poorly constrained because of lack of sufficient data due to subsequent reactivation of fault planes.

SSE to S striking fractures were reactivated as oblique high-angle normal faults with striae dipping toward NW and SE, respectively (D_{3-2}) (Fig. 8d). Foliation planes slightly dipping to

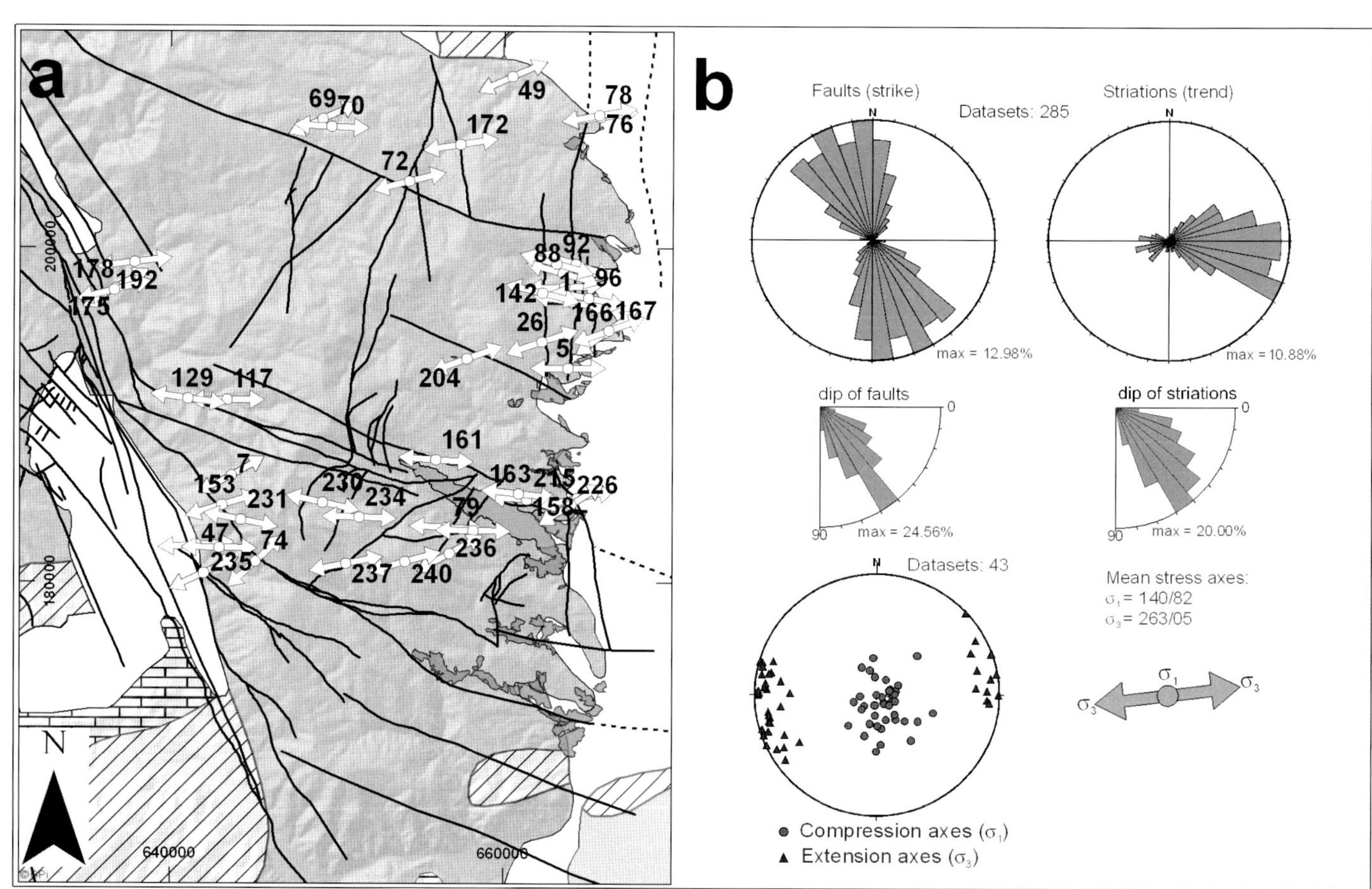

Fig. 6. a) Orientation of σ_1 (large arrows) and σ_3 (small arrows) related to D_2; Labeled numbers refer to Table 1. b) Rose diagrams with strike and dip of fault planes and trend of striations; orientation of σ_1 (filled circles) and σ_3 (triangles) of D_2-related tensors with mean maximum and minimum principal stress axes. Refer to Figure 2 for legend of geological units. Grid is in Austrian BMN M34 system.

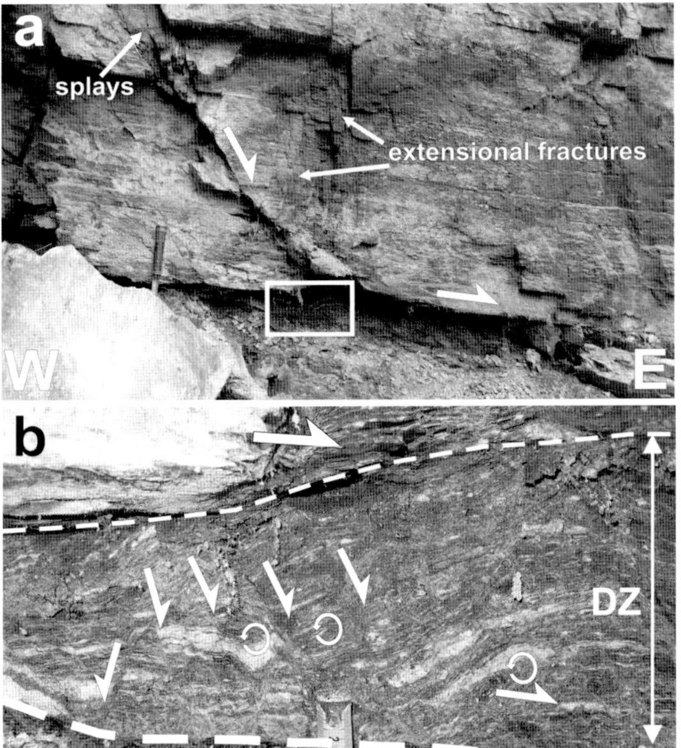

Fig. 7. a) Listric cataclastic shear zone dipping towards E, passing into a shear zone parallel to the foliation of the Plattengneis. White rectangle indicates location of Figure 4b, use hammer handle for scale (approx. 18 cm). b) High-angle fractures associated with foliation parallel cataclastic shear zones with shear-related antithetic rotation of fracture-bound fragments (outcrop-ID: 96, NW of Stainz/Styria).

the E were activated as oblique low-angle normal faults as well. Paleostress orientation analysis of the fault-striae data related to this deformational phase yields a sub-vertical orientation of σ_1 and a sub-horizontal orientation of σ_3 in SE–NW direction (Fig. 8c, d).

Especially sub-horizontal and slightly E- and W-dipping pre-existing foliation planes as well as E- and W-vergent low-angle normal faults were finally re-activated as low-angle reverse faults (thrusts) as can be deduced from associated slickensides and striae indicating reverse to oblique reverse slip (D_4; Fig. 9). The displacements range from of a few centimetres to decimetres. The analysis of paleo-principal stress orientations indicates sub-horizontal σ_1 in E–W direction, and sub-vertical σ_3 (Fig. 9). E–W oriented compression is additionally indicated by the development of kink bands (Fig. 10a) and the normal drag of the pre-existing foliation forming s-type flanking folds in terms of Grasemann et al. (2003) (Fig. 10b).

5 Summary and Discussion

The evolution of the eastern part of the Eastern Alps, in particular the Koralm Massif and the adjacent sedimentary basins,

during the Paleogene is not well known due to the sporadic sedimentary record and the rather low abundance of geochronological data from this period. The Miocene tectonic evolution is better documented for the sedimentary basins, in particular the Styrian and the Lavanttal Basin, located to the east and west of the Koralm Massif, respectively. However, the Paleogene and Neogene evolution of the Koralm Complex still lacks a detailed documentation.

In general, this part was transected by two major sets of faults, coinciding with the general Miocene fault pattern of the Eastern Alps (compare e.g. Ratschbacher et al. 1991; Decker & Peresson 1996; Neubauer et al. 2000). These are:

1. ESE–WNW- to E–W-trending faults, associated with ENE- and NNW-trending conjugate structures;
2. N- to NNE-striking faults, mainly acting as high-angle normal faults, often associated with E-dipping low-angle normal faults along the western margin of the Styrian Basin.

These fault sets were multiply (re-) activated during the Miocene, resulting in a complex pattern of fault interferences. Detailed timing of distinct phases of faulting still remains difficult due to the lack of geochronological data directly dating fault activity, and the lack of exposed interference with sedimentary deposits. Especially along the eastern margin of the Koralm Massif, previously formed E-trending faults and associated structures were covered or sealed by syn- to post-tectonic sediments and may hardly be traced toward east into the Styrian Basin. However, together with the stratigraphic and paleogeographic evolution of the Styrian and Lavanttal Basins and the related subsidence histories (see, for example, Weber & Weiss 1983; Ebner & Sachsenhofer 1995; Sachsenhofer et al. 1997, 1998; Dunkl et al. 2005; Vrabec & Fodor 2005) a rough structural evolution, not provided so far, may be reconstructed for this part of the Eastern Alps.

In general, the Koralm Massif, adjacent basement units north and south of it, and the Styrian Basin are bordered by two major confining fault zones: the ESE-trending Periadriatic fault with dextral sense of displacement in the south, and a system of ENE-trending sinistral fault zones in the north (e.g. Neubauer et al. 2000) (Fig. 11). These are linked by the NNW-trending Lavanttal fault system west of the Koralm Massif (Fig. 11). The evolution of the Styrian Basin can be subdivided into an Early Miocene (Ottnangian to Karpatian: approximately 18–17 Ma) synrift phase and a subsequent postrift phase (Ebner & Sachsenhofer 1995). During the synrift phase, there is a close genetic relation between basin formation and the formation of pull apart structures along predominately E-trending strike slip zones (Ebner & Sachsenhofer 1995; Sachsenhofer et al. 1997, 1998).

During a first phase of deformation (D_{1-1}) WNW–ESE-striking fractures were activated as dextral strike-slip faults in the southern part of the Koralm Complex (Ottnangian to Karpatian: approximately 18–17 Ma) (Fig. 11a). In domains characterised by overlapping fault segments this was associated with

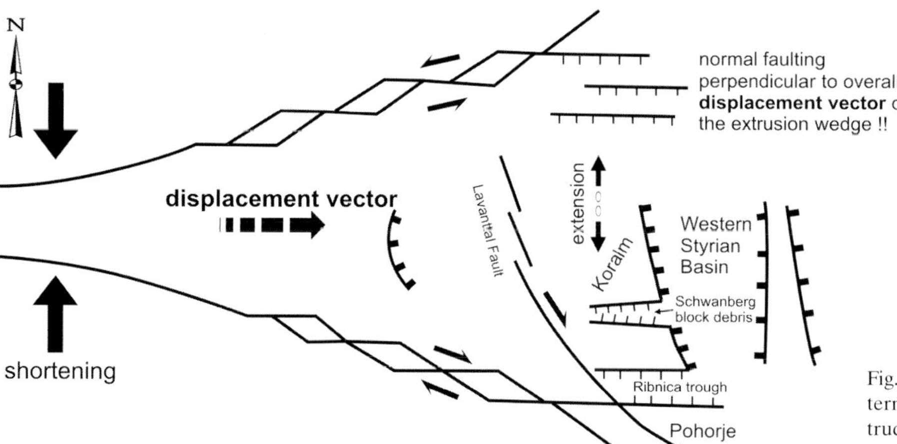

Fig. 12. Scheme showing the development of the fault pattern and related structures within an orogen-parallel extruding wedge, widening toward the direction of displacement. Sketch is not to scale.

ian Basin (Ebner & Sachsenhofer 1995; Reischenbacher et al. 2007).

The formation of the main fault sets in the area of investigation also reflects the structural evolution of an eastward extruding block with increasing width away from the central part of the Eastern Alpine orogen during orogen-parallel escape (e.g. Ratschbacher et al. 1989, 1991; Neubauer et al. 2000) (Fig. 12). This evolution is mainly governed by the northward indentation of a rigid indenter represented by the Southalpine (Fig. 1) accompanied with maximum shortening in the central Eastern Alps, and a continuous decrease of shortening toward east. According to Kuhlemann et al. (2003) this deformation episode occurred between 21 and 12 Ma. In the eastern part of the Eastern Alps, this extruding block is mainly characterised by strike-slip faulting along confining E–W-trending wrench faults associated with the formation of pull-apart basins at oversteps of distinct strike-slip faults, linked by approximately N-trending normal faults accommodating displacement along the mainly E-trending strike-slip fault zones (Fig. 12). These extensional structures form mainly perpendicular to the displacement vector. The eastward increasing width of the extruding wedge implies that progressive lateral displacement causes N–S-directed extension perpendicular to the overall displacement direction; this may be reflected in the re-activation of previously formed E-trending strike-slip faults and by the formation of additional E-trending extensional structures (Fig. 12).

The D_1–D_3 paleostress orientation patterns indicate an apparent clockwise rotation of the minimum principal stress (σ_3) from a N–S to a NW–SE orientation (Figs. 11a-d). However, this may just be related to counter clockwise rotation of crustal blocks, especially north of the Periadriatic fault, and is in accordance with paleomagnetic data indicating counter clockwise block rotation in the eastern part of the Eastern Alps by 30° to 40° in Middle Miocene times (approximately 17–13 Ma) (Fodor et al. 1998; Márton et al. 2000, 2002; Kuhlemann et al. 2003). Consequently, the apparent rotation of the regional stress field just results from passive rotation of the evaluated stress tensors. The subsequent inversion of the regional stress field to E–W oriented compression (Fig. 11e) was described all over the eastern part of the Eastern Alps and the Pannonian Basin (Peresson & Decker 1997) and is interpreted to represent the far-field response of a phase of "soft continental collision" in the Eastern Carpathians. According to Peresson & Decker (1997), deformed Pannonian strata in the eastern Styrian Basin indicate that this paleostress regime may have started at 9 Ma and lasted approximately until 6 Ma.

6 Conclusions

1) The structural evolution of the Koralm Massif during the Neogene is determined by the development of two main fault sets: (a) E- to ESE-trending faults may have formed as strike slip faults with dextral sense of shear, linked by (b) approximately N-trending normal faults. Fault-related E-trending troughs were mainly filled up with block debris ("Schwanberger Blockschutt"); sedimentary deposits up to the Early/Middle Badenian are mainly related to E–W-striking faults.

2) During (N)NW-(S)SE-directed compression the E-trending structures were reactivated as high-angle normal faults or oblique normal faults, indicating a phase of (N)NE-(S)SW extension.

3) Main uplift of the Koralm Massif did not occur before the late Middle Miocene (Sarmatian).

4) Pannonian sedimentation was restricted to the Eastern Styrian Basin; this suggests uplift of the Western Styrian Basin together with the Koralm Complex and the block debris deposits at post-Sarmatian times; this is mainly related to displacement along E-dipping low-angle normal faults during a phase of E–W- to SE–NW-directed extension. The Koralm Complex was elevated by approximately 800 meters during this phase.

5) E–W directed contraction resulted in the reactivation of former low-angle normal faults as W-directed reverse faults, and the re-activation of E-trending structures as related tear faults; this coincides with the inversion of the Styrian Basin.

Acknowledgements

This study has been carried out during a research project (P-17697-N10) granted by the Austrian Science Fund (FWF). We gratefully acknowledge the Österreichische Bundesbahnen (ÖBB) (division Infrastruktur Bau) and the 3G Gruppe Geotechnik Graz ZT GmbH for giving access to pilot tunnel excavations, drill core samples, bore hole logs and geological maps acquired during the Koralm tunnel investigation campaign. Hugo Ortner is thanked for his constructive review comments; also the comments of an anonymous reviewer led to an improvement of the first draft. Stefan Schmid is exceptionally thanked for his thorough editorial comments.

REFERENCES

Angelier, J. 1979. Determination of the mean principal direction of stresses for a given fault population. Tectonophysics 56, T17–T26.

Angelier, J. 1994: Fault slip analysis and paleostress reconstruction. In: Continental Deformation (Ed. by Hancock, P.L.). Pergamon Press, 53–100.

Angelier, J. & Mechler, P. 1977. Sur une méthode graphique de recherche des contraintes principales également utilisable en tectonique et en séismologie: la méthode des dièdres droits. Bulletin de la Société géologique France 19, 1309–1318.

Balogh, K., Ebner, F. and Ravasz, C. 1994. K/Ar-Alter tertiärer Vulkanite der südöstlichen Steiermark und des südlichen Burgenlands. In: Lobitzer, H. et al. (Eds.): Jubiläumsschrift 20 Jahre geologische Zusammenarbeit Österreich-Ungarn, Vol.2. Geologische Bundesanstalt, Wien, 55–72.

Beck-Mannagetta, P. 1952. Zur Geologie und Paläontologie des Tertiärs des unteren Lavanttales. Jahrbuch der Geologischen Bundesanstalt 95, 1–102

Beck-Mannagetta, P., Eisenhut, M., Ertl, V. and Homann, O., 1991. Geologische Karte der Republik Österreich 1:50.000–189 Deutschlandsberg. Geologische Bundesanstalt, Wien.

Bojar, H.-P., Bojar, A.-V., Mogessie, A., Fritz, H. & Thalhammer, O.A.R. 2001. Evolution of veins and sub-economic ore at Strassegg, Paleozoic of Graz, Eastern Alps, Austria: evidence for local fluid transport during metamorphism. Chemical Geology 175, 757–777.

Brodie, K., Fettes, D., Harte, B. & Schmid, R. 2002. Structural terms including fault rock terms. IUGS Subcommission on the Systematics of Metamorphic Rocks, 8 pp.

Brosch, F.J., Pischinger, G., Steidl, A., Vanek, R., & Decker, K. 2001: Improved site investigation results by kinematic discontinuity analysis on drill cores. In: Rock Mechanics – a Challenge for Society (Ed. by Särkkä & Eloranta). Swets & Zeitlinger Lisse, 41–45.

Caine, J.S., Evans, J.P. & Forster, C.B. 1996. Fault zone architecture and permeability structure. Geology 24, 1025–1028.

Chester, F.M. & Logan, J.M. 1986. Implications for mechanical properties of brittle faults from observations of the Punchbowl Fault, California. Pure and Applied Geophysics 124, 79–106.

Decker, K. & Peresson, H. 1996: Tertiary kinematics in the Alpine-Carpathian-Pannonian system: links between thrusting, transform faulting and crustal extension. In: Oil and Gas in Alpidic Thrustbelts and Basins of Central and Eastern Europe (Ed. by Wessely, G. & Liebl, W.). EAGE Special Publication, 69–77.

Doblas, M. 1998. Slickenside kinematic indicators. Tectonophysics 295, 187–197.

Dunkl, I., Kuhlemann, J., Reinecker, J. & Frisch, W. 2005. Cenozoic relief evolution of the Eastern Alps – Constraints from apatite fission track age-provenance of Neogene intramontane sediments. Austrian Journal of Earth Sciences (Mitteilungen der Österreichischen Geologischen Gesellschaft) 98, 92–105.

Ebner, F. & Sachsenhofer, R.F. 1995. Paleogeography, subsidence and thermal history of the Neogene Styrian Basin (Pannonian Basin system, Austria). Tectonophysics 242, 133–150.

Egger, M. 2007. Strukturelle Analyse des Schwanberger Blockschutt Beckens im Hinblick auf die Achse des Koralm-Basistunnels. Magisterarbeit, Institut für Angewandte Geowissenschaften, Technische Universität Graz, 66 pp.

Flügel, H.W. and Neubauer, F. 1984. Steiermark – Erläuterungen zur Geologischen Karte der Steiermark 1:200.000. Geologie der Österreichischen Bundesländer in kurzgefassten Einzeldarstellungen. Geologische Bundesanstalt, Wien, 127 pp.

Fodor, L., Jelen, B., Márton, E., Skaberne, D., Car, J. & Vrabec, M. 1998. Mio-Pliocene tectonic evolution of the Slovenian Periadriatic fault: Implications for Alpine-Carpathian extrusion models. Tectonics 17, 690–607.

Fodor, L. & POSIHU Research Group 2003. Miocene exhumation of the Pohorje-Kozjak Mts., Slovenia (Alpine-Pannonian transition). Geophysical Research Abstracts 5, 11814.

Fodor, L., Gerdes, A., Dunkl, I., Koroknai, B., Pécskay, Z., Trajanova, M., Horváth, P., Vrabec, M., Balogh, K., Jelen, B. & Frisch, W. 2008. Miocene emplacement and rapid cooling of the Pohorje pluton at the Alpine-Pannonian-Dinaric junction: a geochronological and structural study. Swiss Journal of Geosciences, doi: 10.1007/s00015-008-1286-9.

Frank, W. 1987: Evolution of the Austroalpine elements in the Cretaceous. In: Geodynamics of the Eastern Alps (ed. by Flügel, H.W. & Faupl, P.). Deuticke, 379–406.

Frisch, W., Dunkl, I. & Kuhlemann, J. 2000a. Post-collisional orogen-parallel large-scale extension in the Eastern Alps. Tectonophysics 327, 239–265.

Frisch, W., Kuhlemann, J., Dunkl, I. & Brügel, A. 1998. Palinspastic reconstruction and topographic evolution of the Eastern Alps during late Tertiary tectonic extrusion. Tectonophysics 297, 1–16.

Frisch, W., Szekely, B., Kuhlemann, J. & Dunkl, I. 2000b. Geomorphological evolution of the Eastern Alps in response to Miocene tectonics. Zeitschrift für Geomorphologie Neue Folge 44, 103–138.

Fritz, H., Krenn, K., Bojar, A.-V., Kurz, W., Hermann, S., Neubauer, F. & Robl, J. 2002. Late Cretaceous to Paleogene Tectonics and Metamorphism in the Eastern Alps: The "older" extrusion corridor. PANGEO Abstracts, 52–53.

Grasemann, B., Stüwe, K. & Vannay, J.-C. 2003. Sense and non-sense in flanking structures. Journal of Structural Geology 25, 19–34.

Habler, G. & Thöni, M. 2001. Preservation of Permo-Triassic low-pressure assemblages in the Cretaceous high-pressure metamorphic Saualpe crystalline basement (Eastern Alps, Austria). Journal of metamorphic Geology 19, 679–697.

Hejl, E. 1997. ‚Cold spots‘ during the Cenozoic evolution of the Eastern Alps: thermochronological interpretation of apatite fission-track data. Tectonophysics 272, 159–173.

Hejl, E. 1998. Über die känozoische Abkühlung und Denudation der Zentralalpen östlich der Hohen Tauern – eine Apatit-Spaltspurenanalyse. Mitteilungen der Österreichischen Geologischen Gesellschaft 89, 179–200.

Janák, M., Froitzheim, N., Vrabec, M., Krogh Ravna, E.J. & De Hoog, J.C.M. 2006. Ultrahigh-pressure metamorphism and exhumation of garnet peridotite in Pohorje, Eastern Alps. Jounal of metamorphic Geology 24, 19–31.

Kollmann, K., 1964. Jungtertiär im Steirischen Becken. Mitteilungen der Österreichischen Geologischen Gesellschaft, 57(2), 479–632.

Kuhlemann, J., Scholz, T. & Frisch, W. 2003. Postcollisional stress field changes in Eastern Carinthia (Austria). Mitteilungen der Österreichischen Geologischen Gesellschaft 94, 55–61.

Kurz, W. & Fritz, H. 2003. Tectonometamorphic evolution of the Austroalpine Nappe Complex in the central Eastern Alps – consequences for the Eo-Alpine evolution of the Eastern Alps. International Geology Review 45, 1100–1127.

Kurz, W., Fritz, H., Tenczer, V. & Unzog, W. 2002. Tectonometamorphic evolution of the Koralm Complex (Eastern Alps): Constraints from microstructures and textures of the "Plattengneis"-shear zone. Journal of Structural Geology 24, 1957–1970.

Linzer, H.-G., Decker, K., Peresson, H., Dell´Mour, R. & Frisch, W. 2002. Balancing lateral orogenic float of the Eastern Alps. Tectonophysics 354, 211–237.

Márton, E., Fodor, L., Jelen, B., Márton, P., Rifelj, H. & Kevric, R. 2002. Miocene to Quaternary deformation in NE Slovenia: complex paleomagnetic and structural study. Journal of Geodynamics 34, 627–651.

Márton, E., Kuhlemann, J., Frisch, W. & Dunkl, I. 2000. Miocene rotations in the Eastern Alps – paleomagnetic results from intramontane basin sediments. Tectonophysics 323, 163–182.

Marret, R. and Allmendinger, R.W., 1990. Kinematic analysis of fault-slip data. Journal of Structural Geology 12, 973–986.

Medley, E.W. 1994. The Engineering Characterization of Melanges and Similar Block-in-Matrix Rocks (Bimrocks). Ph.D. dissertation, University of California at Berkley, 175 pp.

Meschede, M. 1994. Methoden der Strukturgeologie. Ferdinand Enke Verlag Stuttgart, 169 pp.

Meschede, M. and Decker, K. 1993. Störungsflächenanalyse entlang des Nordrandes der Ostalpen – ein methodischer Vergleich. Zeitschrift der Deutschen geologischen Gesellschaft, 144(2), 419–433.

Nebert, K., 1989. Das Neogen zwischen Sulm und Lassnitz (Suedweststeiermark). Jahrbuch der Geologischen Bundesanstalt 132, 727–743.

Neubauer, F., Dallmeyer, R.D., Dunkl, I. & Schirnik, D. 1995. Late Cretaceous exhumation of the metamorphic Gleinalm dome, Eastern Alps: kinematics, cooling history and sedimentary response in a sinistral wrench corridor. Tectonophysics 242, 79–98.

Neubauer, F., Fritz, H., Genser, J., Kurz, W., Nemes, F., Wallbrecher, E., Wang, X., & Willingshofer, E. 2000: Structural evolution within an extruding wedge: model and application to the Alpine-Pannonian system. In: Aspects of Tectonic Faulting (Ed. by Lehner, F.K. & Urai, J.L.). Springer, 141–153.

Neubauer, F. & Genser, J. 1990. Architektur und Kinematik der östlichen Zentralalpen – eine Übersicht. Mitteilungen des naturwissenschaftlichen Vereins für Steiermark 120, 203–219.

Nievoll, J. 1985. Die brruchhafte Tektonik entlang der Trofaiachlinie (Östliche Zentralalpen, Österreich). Jahrbuch der Geologischen Bundesanstalt 127, 643–671.

Ortner, H., Reiter, F. and Acs, P. 2002. Easy handling of tectonic data: the programs TectonicVB for Mac and TectonicsFP for Windows(TM). Computers & Geosciences, 28(10), 1193–1200.

Peresson, H. & Decker, K. 1997. Far-field effects of Late Miocene subduction in the Eastern Carpathians: E–W compression and inversion of structures in the Alpine-Carpathian-Pannonian region. Tectonics 16, 38–56.

Petit, J.P. 1987. Criteria for the sense of movement on fault surfaces in brittle rocks. Journal of Structural Geology 9, 597–608.

Piller, W.E. & 18 co-authors 2004. Die Stratigraphische Tabelle von Österreich 2004 (sedimentäre Schichtfolgen). Österreichische stratigraphische Kommission und Kommission für die paläontologische und stratigraphische Erforschung Österreichs der Österreichischen Akademie der Wissenschaften.

Piller, W.E., Harzhauser, M. & Mandic, O. 2007. Miocene Central Paratethys stratigraphy – current status and future directions. Stratigraphy 4, 151–168.

Pischinger, G., Brosch, F.J., Fasching, A., Fuchs, R., Goricki, A., Müller, H., Otto, R., Steidl, A. & Vanek, R. 2005. Brittle faulting in the Koralm region – results from the site investigation for the Koralmtunnel. Abstract, Symposium Geologie AlpTransit, Zürich.

Pischinger, G., Brosch, F.-J., Kurz, W. & Rantisch, G. 2006. Normal faulting in a rigid basin boundary block – field evidence from the Koralm Complex (Eastern Alps). Innsbruck University Press Conference Series, PANGEO Austria 2006, 250–251.

Puch, T. 1995: Spaltspurendatierungen an Apatit und Zirkon als Tracer tektonischer Prozesse im Bereich Koralpe-Saualpe-Krappfeld (Kärnten/Österreich), Ostalpen. Unveröff. Diplomarbeit, Naturwissenschaftliche Fakultät, Universität Graz, 124 pp.

Rabitsch, R., Wölfler, A. & Kurz, W. 2007. Fission track dating in fault zones: an example from the Eastern Alps. Geophysical Research Abstracts 9, EGU2007-A-02732.

Ratschbacher, L., Frisch, W., Linzer, H.-G. & Merle, O. 1991. Lateral Extrusion in the Eastern Alps. Part 2: Structural Analyses. Tectonics 10/2, 257–271.

Ratschbacher, L., Frisch, W., Neubauer, F., Schmid, S.M. & Neugebauer, J. 1989. Extension in compressional orogenic belts: The eastern Alps. Geology 17, 404–407.

Reinecker, J. 2000. Stress and deformation: Miocene to present-day tectonics in the Eastern Alps. Tübinger Geowissenschaftliche Arbeiten Serie A 55, 128 pp.

Reinecker, J. & Lenhardt, W.A. 1999. Present-day stress field and deformation in eastern Austria. International Journal of Earth Sciences 88, 532–550.

Reischenbacher, D., Rifelj, H., Sachsenhofer, R., Jelen, B., Coric, S., Gross, M. & Reichenbacher, B. 2007. Early Badenian paleoenvironment in the Lavanttal Basin (Mühldorf Formation; Austria): Evience from geochemistry and paleontology. Austrian Journal of Earth Sciences 100, 202–229.

Reiter, F. & Acs, P. 1996–2001. Tectonics FP 1.6.2, a 32-bit Windows™-Software for Structural Geology.

Sachsenhofer, R.F. 1996. The Neogene Styrian Basin: An Overview. Mitteilungen der Gesellschaft der Geologie und Bergbaustudenten in Österreich 41, 19–32.

Sachsenhofer, R.F., Dunkl, I., Hasenhüttl, C. & Jelen, B. 1998. Miocene thermal history of the southwestern margin of the Styrian Basin: vitrinite reflectance and fission-track data from the Pohorje/Kozjak area (Slovenia). Tectonophysics 297, 17–29.

Sachsenhofer, R.F., Jelen, B., Hasenhüttl, C., Dunkl, I. & Rainer, T. 2001. Thermal history of Tertiary basins in Slovenia (Alpine-Dinaride-Pannonian junction). Tectonophysics 334, 77–99.

Sachsenhofer, R.F., Lankreijer, A., Cloething, S. & Ebner, F. 1997. Subsidence analysis and quantitative basin modelling in the Styrian Basin (Pannonian Basin System, Austria). Tectonophysics 272, 175–196.

Sachsenhofer, R.F., Kogler, A., Polesny, H., Strauss, P. & Wagreich, M. 2000. The Neogene Fohnsdorf Basin: basin formation and basin inversion during lateral extrusion in the Eastern Alps (Austria). International Journal of Earth Sciences 89, 415–430.

Schmid, S.M., Fügenschuh, B., Kissling, E. and Schuster, R., 2004. Tectonic map and overtall architecture of the Alpine orogen. Eclogae geologicae Helvetiae 97, 93–117.

Schuster, R. & Kurz, W. 2005. Eclogites in the Eastern Alps: High-pressure metamorphism in the context of Alpine orogeny. Mitteilungen der Österreichischen Mineralogischen Gesellschaft 150, 183–198.

Spang, J.H. 1972. Numerical Method for Dynamic Analysis of Calcite Twin Lamellae. Geological Society of America Bulletin, 83(2), 467–471.

Steidl, A., Goricki, A., Schubert, W. & Riedmüller, G. 2001. Geological and Geotechnical Ground Characterisation for the Koralm Tunnel Route Selection. Felsbau 19, 14–21.

Stiny, J. 1925. Gesteinsklüftung im Teigitschgebiet. Tschermaks Mineralogische und Petrographische Mitteilungen 38, 464–478.

Strauss, P., Wagreich, M., Decker, K. & Sachsenhofer, R.F. 2001. Tectonics and sedimentation in the Fohnsdorf-Seckau Basin (Miocene, Austria): from a pull-apart basin to a half-graben. International Journal of Earth Sciences 90, 549–559.

Turner, F.J. 1953. Nature and dynamic interpretation of deformation lamellae in calcite of three marbles. American Journal of Science 251, 276–298.

Vanek, R., Pischinger, G. & Brosch, F.J. 2001. Kinematic discontinuity analysis – results and implications of structural geological investigations on drill cores. Felsbau 19, 31–36.

Vrabec, M. & Fodor, L. 2005: Late Cenozoic tectonics of Slovenia: Structural styles at the northeastern corner of the Adriatic microplate. In: The Adriatic Microplate: GPS, Geodesy, Tectonics and Hazards (Ed. by Pinter, N., Grenerczy, J., Weber, D., Medak, D., & Stein, S.). Kluwer Academic Publishers, 1–18.

Wallbrecher, E. (1986): Tektonische und gefügeanalytische Arbeitsweisen. – Enke-Verlag, Stuttgart, 244 pp.

Weber, L. & Weiss, A. 1983. Bergbaugeschichte und Geologie der Österreichischen Braunkohlevorkommen. Archiv für Lagerstättenforschung der Geologischen Bundesanstalt 4, 1–317.

Manuscript received 5 November, 2007
Revision accepted 21 July, 2008
Published Online first October 22, 2008
Editorial Handling: Stefan Schmid, Stefan Bucher

1661-8726/08/01S255-17
DOI 10.1007/s00015-008-1286-9
Birkhäuser Verlag, Basel, 2008

Swiss J. Geosci. 101 (2008) Supplement 1, S255–S271

Miocene emplacement and rapid cooling of the Pohorje pluton at the Alpine-Pannonian-Dinaridic junction, Slovenia

László I. Fodor[1,*], Axel Gerdes[2], István Dunkl[3], Balázs Koroknai[1], Zoltán Pécskay[4], Mirka Trajanova[5], Péter Horváth[6], Marko Vrabec[7], Bogomir Jelen[5], Kadosa Balogh[4] & Wolfgang Frisch[8]

Key words: Alps, Miocene, magmatism, geochronology, thermobarometry, exhumation

ABSTRACT

New laser ablation-inductive coupled plasma-mass spectrometry U-Pb analyses on oscillatory-zoned zircon imply Early Miocene crystallization (18.64 ± 0.11 Ma) of the Pohorje pluton at the southeastern margin of the Eastern Alps (northern Slovenia). Inherited zircon cores indicate two crustal sources: a late Variscan magmatic population (~270–290 Ma), and an early Neoproterozoic one (850–900 Ma) with juvenile Hf isotope composition close to that of depleted mantle. Initial εHf of Miocene zircon points to an additional, more juvenile source component of the Miocene magma, which could be either a juvenile Phanerozoic crust or the Miocene mantle. The new U-Pb isotope age of the Pohorje pluton seriously questions its attribution to the Oligocene age 'Periadriatic' intrusions. The new data imply a temporal coincidence with 19–15 Ma magmatism in the Pannonian Basin system, more specifically in the Styrian Basin. K-Ar mineral- and whole rock ages from the pluton itself and cogenetic shallow intrusive dacitic rocks (~18–16 Ma), as well as zircon fission track data (17.7–15.6 Ma), gave late Early to early Middle Miocene ages, indicating rapid cooling of the pluton within about 3 Million years. Medium-grade Austroalpine metamorphics north and south of the pluton were reheated and subsequently cooled together. Outcrop- and micro scale structures record deformation of the Pohorje pluton and few related mafic and dacitic dykes under greenschist facies conditions. Part of the solid-state fabrics indicate E–W oriented stretching and vertical thinning, while steeply dipping foliation and NW–SE trending lineation are also present. The E–W oriented lineation is parallel to the direction of subsequent brittle extension, which resulted in normal faulting and tilting of the earlier ductile fabric at around the Early / Middle Miocene boundary; normal faulting was combined with strike-slip faulting. Renewed N–S compression may be related to late Miocene to Quaternary dextral faulting in the area. The documented syn-cooling extensional structures and part of the strike-slip faults can be interpreted as being related to lateral extrusion of the Eastern Alps and/or to back-arc rifting in the Pannonian Basin.

Introduction

The Periadriatic Fault Zone (PFZ) represents a major, east-west trending Tertiary shear zone crosscutting the entire Alpine edifice from the Western Alps to the Pannonian Basin (inset of Fig 1a; Schmid et al. 1989). All along its strike the PFZ is associated with Paleogene intrusions (e.g. Laubscher 1983). Salomon (1897) was the first to apply the term "Periadriatic" for these plutons because of their intimate spatial relationship to the PFZ. Since most of these plutons exhibit Oligocene ages (28–34 Ma; for a thorough review see Rosenberg 2004), the term "Periadriatic" is also used, and will be used in this contribution, in a temporal sense (e.g. Laubscher 1983; Elias 1998; Rosenberg 2004).

The Karawanken tonalite (southern Austria, northern Slovenia; Fig. 1b) represents the easternmost outcrop of those Periadriatic intrusions. Its Oligocene age (28–32 Ma) is undisputed and well constrained by U-Pb (zircon), Rb-Sr (biotite), and Ar (amphibole, biotite) analyses (Scharbert 1975, Elias 1998). Further to the east, in the southwestern part of the Pannonian Basin, oil exploration wells reached tonalites below Tertiary sediments (Balogh et al. 1983; Kőrössy 1988). Radiometric ages (29–34 Ma) and detailed geochemical studies of these subsurface rocks clearly indicate that they also belong to the Periadriatic magmatic suite (Benedek 2002).

The Pohorje pluton (northern Slovenia), located 5–10 km north of the PFZ (Fig. 1a), was also regarded as part of the Periadriatic intrusions by many authors (Salomon 1897; Pamić &

[1]*Geological Institute of Hungary, 1143 Budapest Stefánia 14, Hungary. E-mail: fodor@mafi.hu
[2]Institute of Geosciences, Altenhoferallee 1, 60438 Frankfurt am Main, Germany.
[3]Geoscience Center Göttingen, Sedimentology & Environmental Geology, Goldschmidtstrasse 3 D-37077 Göttingen, Germany.
[4]Institute of Nuclear Research, Hungarian Academy of Sciences, H-4026 Debrecen Bem tér 18/c, Hungary.
[5]Geološki Zavod Slovenije, Dimičeva 14, SI-1109 Ljubljana, Slovenia.
[6]Institute for Geochemical Research, Hungarian Academy of Sciences, H-1112 Budapest Budaörsi út 45, Hungary.
[7]University of Ljubljana, Department of Geology, Aškerčeva 12, SI-1000 Ljubljana, Slovenia.
[8]Institute of Geosciences, University of Tübingen, Sigwartstrasse 10, D-72076 Tübingen, Germany.

Pálinkáš 2000; Pálinkáš & Pamić 2001; Rosenberg 2004). Its distant location from the PFZ is not exceptional, since some other Periadriatic intrusions (e.g. Rieserferner pluton, Steenken et al. 2002) are also located along fault zones that are kinematically linked to the PFZ. However, Faninger (1970) concluded that Pohorje cannot be considered as an eastern prolongation of the Oligocene-age Karawanken pluton. Indeed, radiometric data do not support its assignment to the Periadriatic magmatic suite: Rb-Sr and K-Ar data yielded Miocene ages (19–16 Ma) for various rock types of the Pohorje pluton (Deleon 1969; Dolenec 1994). Recently, Fodor et al. (2002b) and Trajanova et al. (2008) presented K-Ar ages of different mineral separates and whole rock samples from the pluton, which all indicated Miocene ages. However, because of the possible thermal effects of the subsequent dacitic volcanism, hydrothermal-metasomatic alteration, and crystal-scale deformation, the interpretation of these data as emplacement ages was not entirely conclusive.

In this study we present new isotopic age data from the Pohorje pluton and spatially related dacites. The data set includes the results of U-Pb Laser Ablation-Inductive Coupled Plasma-Mass Spectrometry (LA-ICP-MS) dating of zircons, K-Ar dating on biotite, amphibole, feldspar and whole rock samples, and zircon fission track dating. These data clearly indicate a late Early Miocene age of the magmatic activity in the Pohorje Mountains. These new geochronological data, in combination with some structural observations, have implications for the structural evolution of the Alpine-Pannonian-Dinaric junction.

Geological setting

The Pohorje (Bachern) and the Kozjak (Possruck) Mountains form the southeastern-most crystalline outcrops of the Eastern Alps and are located at the western margin of the Miocene Pannonian Basin system (Fig. 1a). They consist of basement and cover sequences of the Austroalpine nappe system formed during the Eoalpine orogeny (Frank 1987; Schmid et al. 2004). The deepest tectonic unit in the massif is mainly composed of medium-grade metamorphic rocks (gneiss, micaschist and amphibolite, intercalated with marble and quartzite, and sporadic eclogite lenses; Hinterlechner-Ravnik 1971, 1973; Mioč 1978). High-T retentivity isotopic data (Thöni 2002; Miller et al. 2005; Janák et al. 2007) from these rocks show that they underwent Eoalpine (ca. 90 Ma) high-pressure metamorphism. This rock pile is overlain by low-grade Palaeozoic and non-metamorphosed Permo–Triassic and Senonian sediments (Mioč & Žnidarčič 1977; Fig. 1a), which represent the uppermost tectonic unit in the study area. The original thrust contact between these units was reactivated and/or deformed during Late Cretaceous and Miocene exhumation phases (Fodor et al. 2002a).

The medium-grade metamorphic rocks were intruded by the Pohorje pluton, a 30 km long and 4–8 km wide magmatic body with ESE–WNW orientation (Fig. 1b). Although the boundary of the pluton is locally tectonized, the original magmatic contact, marked by a thin contact metamorphic aureole, was mapped almost continuously along its margin (Mioč & Žnidarčič 1977; Žnidarčič & Mioč 1988). The Pohorje pluton was regarded either as a laccolith (Faninger 1970; Exner 1976), or alternatively, as a batholith (Trajanova et al. 2008).

Two larger occurrences of the host metamorphic rocks are located within the southeastern part of the pluton (Mioč & Žnidarčič 1977). The western one, near the peak of Veliki vrh (1344 m), is topographically and structurally located above the magmatic rocks and has a thermal contact aureole. The eastern one is a narrow, NW trending belt within the narrowing termination of the pluton (Fig. 1a).

The Pohorje pluton predominantly consists of tonalites and granodiorites (Dolar-Mantuani 1935; Faninger 1970; Zupančič 1994a; Altherr et al. 1995; Pamić & Pálinkáš 2000; Trajanova et al. 2008). Both intrusion and country rocks are cut by aplite, mafic and dacite dykes (Kieslinger 1935) and by small shallow subsurface dacite bodies.

Miocene sediments cover the older rock units in the northwestern Pohorje Mountains. They continuously crop out in the Ribnica-Selnica trough, but appear only as scattered occurrences further north, on top of the Kozjak Mountains (Fig. 1). These sediments were tentatively correlated with more than 1 km thick marine successions of the Mura basin, where a late Early Miocene (Karpatian, 17.3–16.5 Ma) age was demonstrated (Márton et al. 2002; Jelen & Rifelj 2003, time scale according to Steininger et al. 1988). In the Ribnica-Selnica trough and on top of the southern Kozjak Mountains, the Karpatian sediments contain dacitic volcanoclastics and small intrusions (Fig. 1.; Winkler 1929; Mioč & Žnidarčič 1977).

Results

Geochronology

U-Pb and Hf isotope data

Zircons of sample 311 from the eastern part of the Pohorje pluton (Fig. 1b) were analysed for U-Pb and Hf isotopes by LA-ICP-MS at Frankfurt University, following the methods described by Gerdes & Zeh (2006, 2008). Spot-selection was guided by internal structures as seen in cathodoluminescence (CL) images of mounted and polished grains. Twenty-five U-Pb and 15 Hf isotope analyses were performed on core and mantle of 16 prismatic zircons (Fig. 2, Table 1a–b). Grains frequently display bright luminescent, rounded cores with relict oscillatory zoning. The oscillatory zoning is often blurred and widened, or fully replaced by CL-homogeneous domains. These cores are enclosed by variably wide CL-dark domains with well-developed oscillatory zoning typical of magmatic growth (zr1, 2, 4, 5, 8, 11 in Fig. 2a). Some CL-dark grains show no cores (zr6 in Fig. 2a), and in other cases the dark overgrowths form only ~20 μm wide rims around large cores (zr4–5 in Fig. 2a). Eleven spot measurements on the U-rich (1500–19800 ppm) CL-dark oscillatory-zoned domains yielded equivalent and concordant results with an age of 18.64 ± 0.11 Ma (2σ). ^{206}Pb/^{238}U ages of

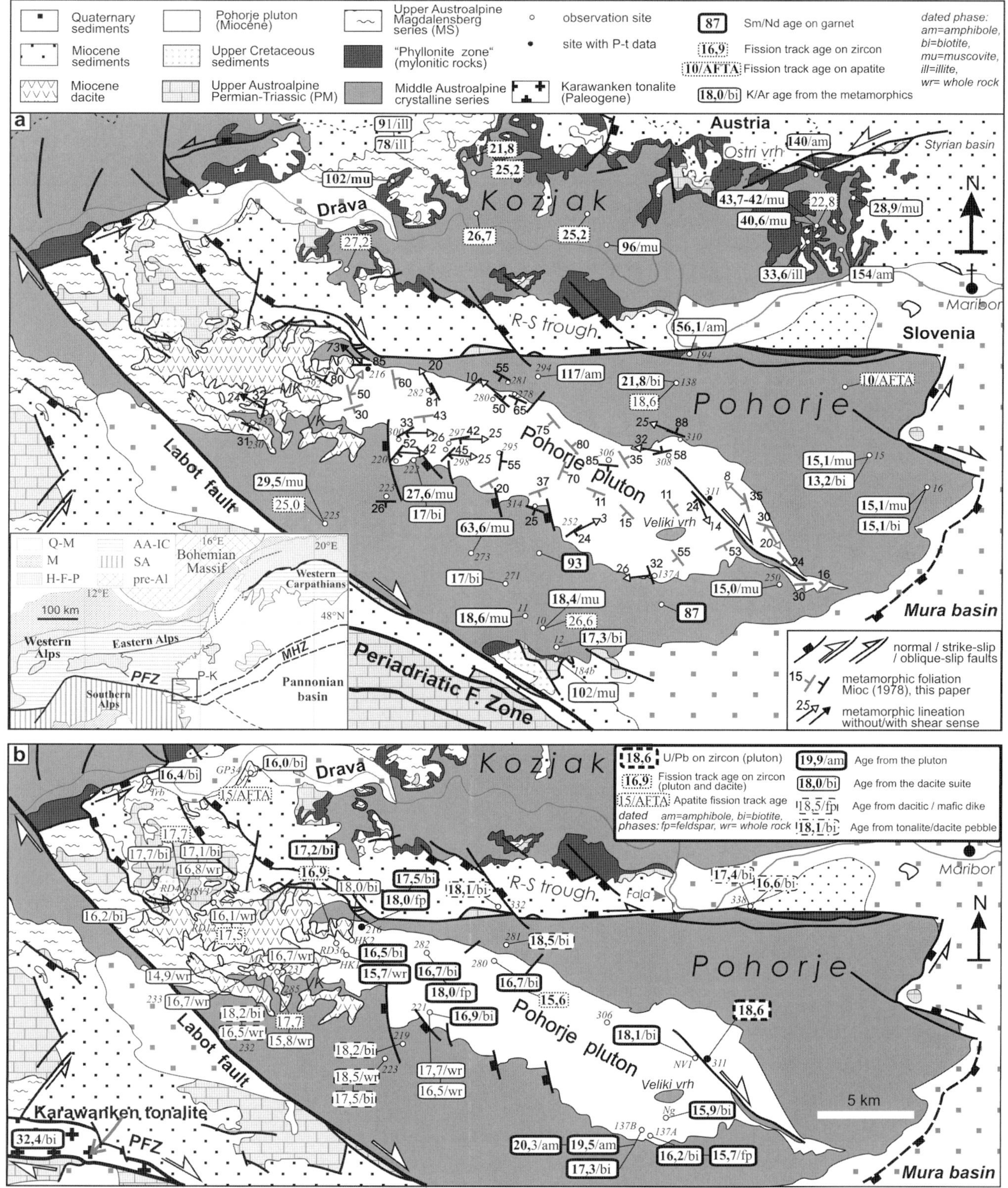

Fig. 1. Geological setting and geochronological data from the Pohorje Mountains 1a: Simplified geological map of the Pohorje Mountains with geochronological data of the metamorphic rocks. Foliation and lineation data of the Pohorje pluton (Mioč, 1978, Žnidarčič & Mioč 1988, and our own measurements) are also shown. R-S trough: Ribnica-Selnica trough; VK: Velka Kopa. Insert shows tectonic sketch map of Eastern and Southern Alps and the western Pannonian Basin showing the location of the Pohorje and Kozjak Mountains MHZ: Mid-Hungarian Zone; Q-M: Quaternary-Miocene; M: Molasse; HFP: Helvetic, Flysch, Penninic units; AAIC: Austroalpine, Inner Carpathian units; SA: Southern Alps; pre-Al: pre-Alpine units. 1b: Geochronological data from the magmatic rocks. In cases of duplicate measurements from the same site (see Table 2), the average age is shown. MK: Mala Kopa.

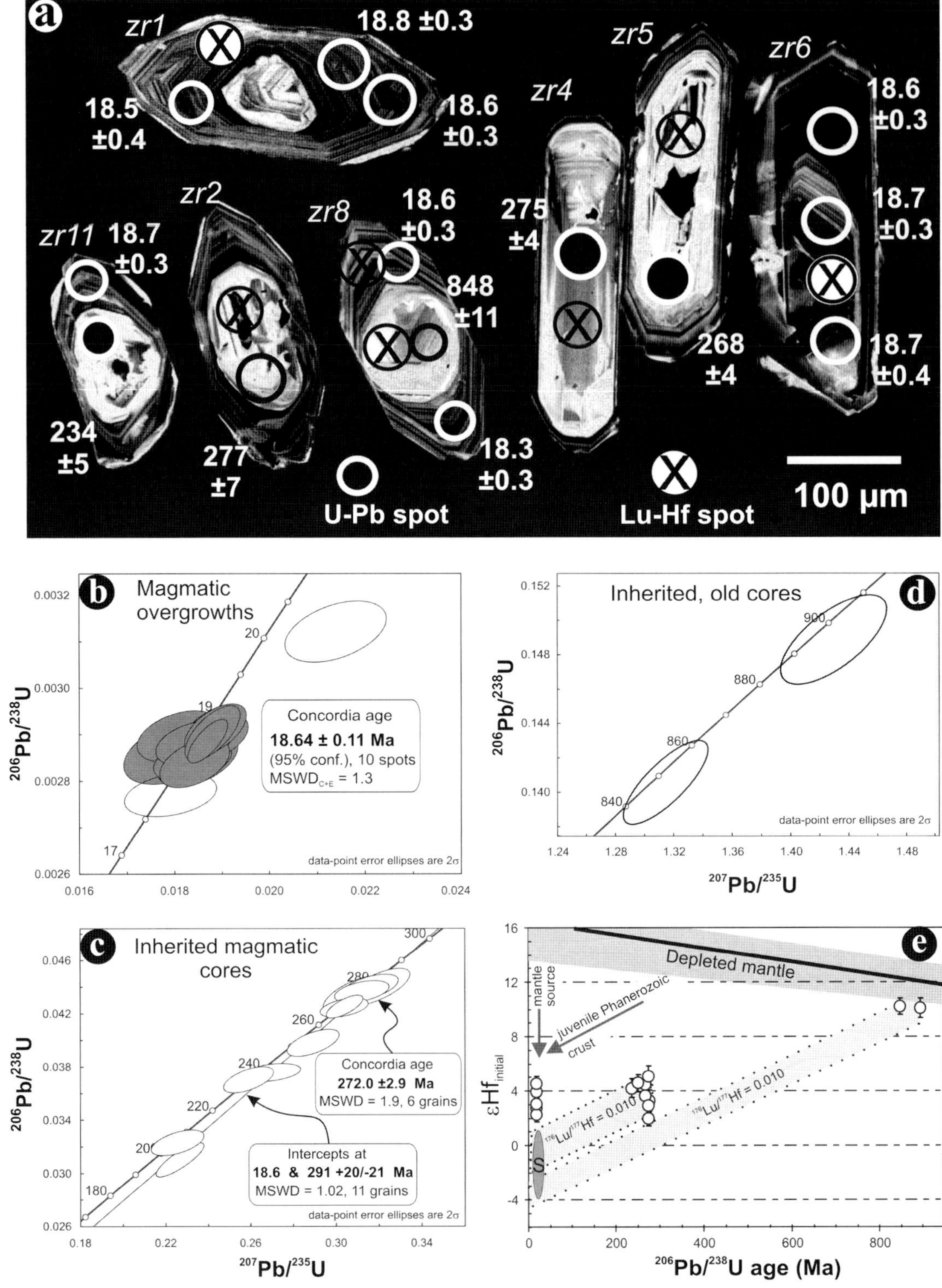

Fig. 2. U-Pb age and Hf isotope data from the eastern part of the Pohorje pluton (site 311). 2a: Representative cathodoluminescence images of zircon with location of U-Pb and Hf isotope spots and the corresponding U-Pb age with 2σ error. Note the uniform age of oscillatory-zoned magmatic rims between 18.3–18.7 Ma. Inherited cores display two age populations of 270–290 Ma and 850 Ma, respectively. 2b, c, d: U-Pb concordia diagrams of the three distinct zircon age populations. 2e: Initial εHf isotope composition versus $^{206}U/^{238}Pb$ age of the different zircon domains. Grey field shows the evolution trend of the crust represented by the Permian and Neoproterozoic cores assuming a mean $^{176}Lu/^{177}Hf$ of 0.010. Also shown is the composition of depleted mantle through time. Note that εHf(t) of Miocene zircon domains are more juvenile as the source component (field with dark S) inferred from the inherited cores. Grey arrows show two possible source components of the Miocene magma to explain the Hf isotope composition.

Table 1. U/Pb age data and Lu/Hf isotopic data from from the Pohorje pluton, sample 311.
1a: Results of U/Pb LA-ICP-MS dating of zircons.

No.	^{207}Pb [a]	U [b]	Pb [b]	$\frac{\text{Th [b]}}{\text{U}}$	$\frac{^{206}\text{Pb [c]}}{^{204}\text{Pb}}$	$\frac{^{206}\text{Pb [d]}}{^{238}\text{U}}$	±2σ	$\frac{^{207}\text{Pb [d]}}{^{235}\text{U}}$	±2σ	$\frac{^{207}\text{Pb [d]}}{^{206}\text{Pb}}$	±2σ	rho [e]	$\frac{^{206}\text{Pb}}{^{238}\text{U}}$	±2σ	$\frac{^{207}\text{Pb}}{^{235}\text{U}}$	±2σ
	(cps)	(ppm)	(ppm)				(%)		(%)		(%)			(Ma)		(Ma)
zr1	1139	1621	4.3	0.09	1430	0.002872	2.0	0.01810	4.7	0.04571	4.3	0.42	18.5	0.4	18.2	0.9
zr1	4210	1539	4.2	0.09	8724	0.002887	1.4	0.01872	3.6	0.04703	3.4	0.38	18.6	0.3	18.8	0.7
zr1	5133	1859	5.1	0.10	10705	0.002921	1.6	0.01899	2.8	0.04715	2.4	0.56	18.8	0.3	19.1	0.5
zr2 c	4386	213	10	0.50	3104	0.04394	2.6	0.3193	3.9	0.05271	2.8	0.68	277	7	281	10
zr3 c	17030	83	15	0.96	24191	0.1489	1.4	1.4301	2.1	0.06967	1.5	0.68	895	12	902	13
zr4 c	2573	103	5.1	0.66	893	0.04360	1.5	0.3119	3.3	0.05188	2.9	0.46	275	4	276	8
zr5 c	12421	276	13	0.61	7641	0.04238	1.4	0.3036	3.0	0.05195	2.7	0.45	268	4	269	7
zr6	11027	6983	19	0.03	4442	0.002906	1.6	0.01820	3.9	0.04542	3.6	0.40	18.7	0.3	18.3	0.7
zr6	24384	19811	53	0.01	33199	0.002894	1.4	0.01882	1.8	0.04716	1.2	0.74	18.6	0.3	18.9	0.3
zr6	11317	7667	21	0.03	5256	0.002907	1.9	0.01884	3.1	0.04701	2.4	0.62	18.7	0.4	19.0	0.6
zr7	3315	2726	7.2	0.03	6814	0.002856	1.7	0.01860	3.8	0.04722	3.3	0.46	18.4	0.3	18.7	0.7
zr8	2499	1776	4.8	0.09	4288	0.002889	1.7	0.01852	3.7	0.04650	3.3	0.45	18.6	0.3	18.6	0.7
zr8	2834	2318	6.2	0.11	3963	0.002846	1.5	0.01853	3.2	0.04723	2.9	0.48	18.3	0.3	18.6	0.6
zr8 c	26231	321	48	0.43	38815	0.1406	1.4	1.3152	1.8	0.06786	1.1	0.79	848	11	852	10
zr9 c	9657	508	22	0.37	18462	0.04246	1.7	0.3038	2.3	0.05190	1.6	0.71	268	4	269	6
zr10 c	2932	263	11.6	0.40	5540	0.04391	1.9	0.3157	3.4	0.05214	2.9	0.54	277	5	279	8
zr11	2485	2012	5.4	0.04	5110	0.002900	1.5	0.01888	3.4	0.04721	3.1	0.44	18.7	0.3	19.0	0.6
zr11 c	10949	227	12	0.63	802	0.03692	2.2	0.2586	3.8	0.05079	3.1	0.59	234	5	234	8
zr12	1775	1418	3.7	0.07	3464	0.002776	1.4	0.01802	4.4	0.04706	4.2	0.32	17.9	0.3	18.1	0.8
zr12	11310	8488	23	0.02	5115	0.002910	1.7	0.01890	2.5	0.04711	1.9	0.67	18.7	0.3	19.0	0.5
zr13 c	2618	163	6.5	0.03	5344	0.04340	2.7	0.3095	4.1	0.05173	3.1	0.65	274	7	274	10
zr14 c	4699	168	8.2	0.95	622	0.03748	1.5	0.2697	4.1	0.05218	3.8	0.36	237	3	242	9
zr15 c	2675	82	3.6	0.47	3104	0.03973	2.1	0.2890	3.5	0.05275	2.8	0.60	251	5	258	8
zr16 c	13095	900	30	0.24	6961	0.03087	3.4	0.2259	4.1	0.05309	2.2	0.84	196	7	207	8
zr16	2327	1508	4.4	0.05	2099	0.003127	1.7	0.02149	4.0	0.04984	3.6	0.43	20.1	0.4	21.6	0.9
Plesov [f]	15255	628	32	0.11	20120	0.05392	1.3	0.3941	1.5	0.05307	0.7	0.72	338.5	4.8	337.4	4.1

c = zircon cores; [a] Within run background-corrected mean ^{207}Pb signal in counts per second. [b] U and Pb content and Th/U ratio were calculated relative to GJ-1 reference and are accurate to about 10% due to heterogeneity of GJ-1. [c] measured ratio corrected for background and ^{204}Hg interference (mean ^{204}Hg = 198 ± 22 cps). [d] corrected for background, mass bias, laser induced U-Pb fractionation and common Pb (Gerdes and Zeh, 2006, 2008); Uncertainties are propagated by quadratic addition of within-run precision (2SE) and the reproducibility of GJ-1 (2SD; n = 12). [e] Rho is the error correlation defined as err206Pb/238U/err207Pb/235U. [f] mean (n = 12) ±2 standard deviation of Plesovice reference zircon (cf. Slama et al. 2008).

1b: Lu/Hf isotopic data of zircons

No.	^{176}Yb/^{177}Hf [a]	±2σ	^{176}Lu/^{177}Hf [a]	±2σ	^{180}Hf/^{177}Hf	^{178}Hf/^{177}Hf	Sig$_{\text{Hf}}$ [b] (V)	^{176}Hf/^{177}Hf	±2σ	^{176}Hf/^{177}Hf$_{(t)}$	εHf(t) [c]	±2σ	T$_{\text{DM2}}$ [d] (Ga)	age [e] (Ma)
zr8	0.0415	13	0.00141	5	1.88667	1.46716	12.1	0.282872	25	0.282871	3.9	0.6	0.70	18.6
zr7	0.0508	19	0.00176	7	1.88666	1.46722	11.7	0.282890	25	0.282889	4.5	0.6	0.67	18.4
zr6	0.0672	26	0.00297	10	1.88663	1.46709	16.6	0.282839	29	0.282838	2.7	0.8	0.77	18.7
zr2 c	0.0429	30	0.00137	12	1.88668	1.46716	11.3	0.282699	29	0.282692	3.3	0.8	0.95	277
zr3 c	0.0706	22	0.00236	7	1.88657	1.46712	7.5	0.282537	28	0.282497	10.1	0.7	1.09	895
zr8 c	0.0504	21	0.00147	7	1.88663	1.46717	11.3	0.282554	26	0.282530	10.2	0.6	1.05	848
zr9 c	0.0445	34	0.00161	15	1.88661	1.46727	10.1	0.282739	26	0.282731	4.5	0.6	0.88	268
zr4 c	0.0531	52	0.00163	15	1.88655	1.46715	8.3	0.282752	30	0.282744	5.0	0.8	0.85	275
zr5 c	0.0257	15	0.00083	4	1.88667	1.46719	11.6	0.282713	25	0.282709	3.6	0.6	0.92	268
zr12	0.0354	9	0.00120	4	1.88665	1.46719	12.1	0.282826	25	0.282825	2.3	0.5	0.79	18.7
zr14 c	0.0786	46	0.00223	12	1.88651	1.46713	10.1	0.282753	27	0.282743	4.2	0.7	0.87	237
zr15 c	0.0564	13	0.00161	5	1.88659	1.46712	9.1	0.282753	26	0.282745	4.6	0.6	0.86	251
zr13 c	0.0178	17	0.00062	6	1.88654	1.46720	12.3	0.282687	24	0.282684	2.9	0.5	0.97	274
zr13 c	0.0191	33	0.00064	11	1.88675	1.46724	8.4	0.282669	25	0.282666	2.3	0.6	1.00	274
zr1	0.0426	14	0.00145	5	1.88673	1.46719	12.0	0.282847	25	0.282847	3.1	0.6	0.75	18.5
GJ-1 [f]	0.0080	3	0.00029	2	1.88672	1.46717	15.4	0.282003	19	0.282000	−13.8	0.7	2.1	600

[a] ^{176}Yb/^{177}Hf = (^{176}Yb/^{173}Yb)$_{\text{true}}$ x ^{173}Yb/^{177}Hf)$_{\text{meas}}$ x (M$_{173(\text{Yb})}$/M$_{177(\text{Hf})}$)$^{\beta(\text{Hf})}$. The ^{176}Lu/^{177}Hf were calculated in a similar way by using the ^{175}Lu/^{177}Hf. Quoted uncertainties (absolute) relate to the last quoted figure. [b] Mean Hf signal in volt. [c] calculated using a decay constant of 1.867×10^{-10}, a CHUR ^{176}Lu/^{177}Lu and ^{176}Hf/^{177}Hf ratio of 0.0332 and 0.282772, and the ages obtained by LA-ICP-MS. [d] two stage model age using the measured ^{176}Lu/^{177}Lu of each spot (first stage = age of zircon), a value of 0.010 for the average granitic crust (second stage; Wedepohl 1995), and a depleted mantle ^{176}Lu/^{177}Lu and ^{176}Hf/^{177}Hf of 0.0384 and 0.28325, respectively. [e] ^{206}Pb/^{238}U LA-ICP-MS ages. [f] mean ±2σ standard deviation of 9 spot analyses of GJ-1 reference zircon.

Table 2. K-Ar ages obtained on the different igneous rocks of Pohorje Mountains. The age calculation is based on the decay constants given by Steiger & Jäger (1977). The analytical errors are quoted for the 68% confidence levels (1σ).

Sample	Rock type	Dated fraction	K (%)	^{40}Ar rad (ccSTP/g)	^{40}Ar rad (%)	K-Ar age (Ma)
		Undeformed dacite				
RD4	dacite	whole rock	2.71	1.781×10^{-6}	61.9	16.8 ± 0.7
RD4	dacite	biotite	5.77	3.870×10^{-6}	69.3	17.2 ± 0.6
RD4	dacite	biotite	5.94	3.926×10^{-6}	93.1	17.0 ± 0.4
MSV1	dacite	biotite	4.67	2.952×10^{-6}	66.6	16.2 ± 0.7
RD12	dacite	whole rock	3.06	1.948×10^{-6}	57.1	16.1 ± 0.6
JV1	dacite	biotite	5.54	3.834×10^{-6}	61.9	17.7 ± 0.7
GP34	dacite	biotite	5.57	3.485×10^{-6}	55.6	16.0 ± 0.6
Trb	dacite	biotite	7.05	4.510×10^{-6}	68.5	16.4 ± 0.5
MK	dacite	whole rock	2.61	1.516×10^{-6}	58.7	14.9 ± 0.6
231	dacite	whole rock	2.74	1.790×10^{-6}	71.1	16.7 ± 0.5
233	dacite	whole rock	2.55	1.668×10^{-6}	73.7	16.7 ± 0.6
285	dacite	whole rock	2.60	1.603×10^{-6}	55.8	15.8 ± 0.7
HK2	dacite	biotite	5.73	4.039×10^{-6}	82.2	18.0 ± 0.7
		Mafic and deformed dacite dykes				
219	mafic dyke	biotite	3.56	2.602×10^{-6}	63.1	18.2 ± 0.7
223	mafic dyke, def.	whole rock	1.17	8.468×10^{-7}	63.8	18.5 ± 0.7
223	mafic dyke, def.	biotite	1.79	1.227×10^{-6}	44.5	17.5 ± 0.7
232	dacite, deformed	whole rock	2.05	1.323×10^{-6}	74.2	16.5 ± 0.5
232	dacite, deformed	biotite	2.55	1.818×10^{-6}	65.9	18.2 ± 0.7
221	mafic dyke	whole rock	1.76	1.214×10^{-6}	78.4	17.7 ± 0.7
221	mafic dyke	whole rock	2.42	1.558×10^{-6}	77.9	16.5 ± 0.6
281	mafic dyke, def.	biotite	4.27	3.090×10^{-6}	64.1	18.5 ± 0.6
		Magmatic pebbles				
332	dacite	biotite	5.67	4.011×10^{-6}	84.6	18.1 ± 0.6
338	granodiorite	biotite	6.91	4.697×10^{-6}	55.4	17.4 ± 0.6
338	granodiorite	biotite	6.36	4.116×10^{-6}	71.3	16.6 ± 0.5
		Pohorje pluton				
HK1	gr-porphyre	whole rock	2.66	1.633×10^{-6}	74.3	15.7 ± 0.6
HK1	gr-porphyre	biotite	5.65	3.647×10^{-6}	62.4	16.5 ± 0.7
RD36	tonalite	biotite	7.65	5.144×10^{-6}	78.7	17.2 ± 0.7
216	gr-porphyre	biotite	4.28	2.931×10^{-6}	60.6	17.5 ± 0.7
216	gr-porphyre	feldspar	2.35	1.656×10^{-6}	65.7	18.0 ± 0.7
221	tonalite	biotite	4.56	3.053×10^{-6}	68.9	17.2 ± 0.6
221	tonalite	biotite	4.52	2.926×10^{-6}	85.4	16.6 ± 0.4
282	granodiorite	biotite	5.87	3.883×10^{-6}	79.0	16.7 ± 0.6
282	granodiorite	biotite	3.32	2.175×10^{-6}	66.2	16.8 ± 0.5
282	granodiorite	feldspar	1.34	9.42×10^{-7}	68.8	18.0 ± 0.6
280	granodiorite	biotite	6.12	3.918×10^{-6}	61.1	16.4 ± 0.7
280	granodiorite	biotite	6.00	3.911×10^{-6}	85.8	16.7 ± 0.4
137A	granodiorite	feldspar	1.36	8.343×10^{-7}	64.1	15.7 ± 0.6
137A	granodiorite	biotite	6.46	4.045×10^{-6}	79.4	16.1 ± 0.6
137A	granodiorite	biotite	6.03	3.835×10^{-6}	85.1	16.3 ± 0.6
NV1	tonalite	biotite	4.29	3.034×10^{-6}	56.9	18.1 ± 0.7
Nag	granodiorite	biotite	5.79	3.583×10^{-6}	62.0	15.9 ± 0.5
137B	cezlakite	biotite	1.63	1.099×10^{-6}	53.9	17.3 ± 0.7
137B	cezlakite	amph	0.73	5.576×10^{-7}	47.9	19.5 ± 0.8
137B	cezlakite	amph	0.60	4.751×10^{-7}	29.3	20.3 ± 1.1
		Karawanken-pluton				
Kw	Tonalite	biotite	4.23	5.376×10^{-6}	76.7	32.4 ± 1.2

CL-light cores define two age clusters at 200–280 and 850–900 Ma, respectively (Fig. 2c–d). Six concordant to slightly discordant analyses of the first group yielded a concordia age of 272 ± 3 Ma, and, together with the four clearly discordant spots, they define a discordia with an upper incept age of 291 ± 21 Ma. We consider the concordia age of 272 ± 3 Ma as a minimum age since it is likely that those grains crystallized at 280–290 Ma but experienced some Pb-loss during Miocene magmatism. Therefore, the upper intercept age represents a more conservative estimate for the crystallization age. A common origin of these cores is supported by a rather uniform Hf isotope composition (εHf$_{(t)}$ = +2.3 to +5.0), which corresponds to the range of late Variscan (~300 Ma), post-collisional granites in the European basement (e.g., Schaltegger & Corfu, 1992). The εHf$_{(t)}$ value points to a juvenile source, which could have been either young crust with an average Neoproterozoic Hf model age, slightly younger than that of the early Neoproterozoic cores (Fig. 2e), or a mixture of a mantle-derived and a crustal component. The εHf$_{(t)}$ values of the two early Neoproterozoic cores (Fig. 2d) is close to the depleted mantle composition at that time (Fig. 2e). Miocene CL-dark domains have εHf$_{(t)}$ values of +2 to +4. Assuming that a crust represented by the Permian and Neoproterozoic cores evolves with an average ^{176}Lu/^{177}Hf ratio of 0.010 (Wedepohl et al. 1995) its Miocene εHf$_{(t)}$ value would plot just below that of the Miocene overgrowth domains (grey field with white S in Fig. 2e). Thus an additional juvenile component, such as a juvenile Phanerozoic crustal component or a Miocene mantle-derived melt (see grey arrows in Fig. 2e), must have been involved as a source of the magma in order to explain its Hf isotope composition.

K-Ar data

K-Ar dating has been performed by the conventional method using high frequency induction heating and getter materials (titanium sponge and SAES St707) for extracting and cleaning the argon. The potassium was determined by flame photometry using a CORNING 480 machine. Details of the analytical techniques and the results of calibrations have been described by Balogh (1985) and Odin et al. (1982). The results of the K-Ar isotopic dating on mono-mineralic fractions and whole rock samples are given in Table 2 and shown in Figures 1a and b. A detailed discussion of K-Ar data is found in Trajanova et al. (2008).

Biotite ages from the pluton range from 18.1 ± 0.7 to 15.9 ± 0.5 Ma (±1σ) (Fig. 1b). They are considered to record cooling below the closure temperature of biotite. Feldspar and whole rock isotope ages scatter between 18.0 ± 0.7 and 15.7 ± 0.6 Ma and fit well with the biotite ages. In two samples (216 and 282) feldspar fractions show somewhat older ages than biotite, but the ages overlap within error limits (Table 2). Two amphibole fractions from a small gabbroic body in the old Cezlak quarry (site 137B) yielded K-Ar ages of 20.3 ± 1.1 Ma and 19.5 ± 0.8. These are basically in agreement with the obtained U-Pb isotope age. The biotite age (17.3 ± 0.7 Ma) obtained from this site corresponds to those determined from tonalitic and granodioritic lithologies of the pluton. Biotite ages from one foliated and one nearly undeformed granodiorite pebble found in the eastern part of the Ribnica-Selnica trough (site 338) are consistent with the ages of intrusion.

Dacitic and mafic dykes exhibiting traces of ductile deformation (referred to as "deformed dykes" in the following) and undeformed mafic dykes show biotite ages between 18.5 ± 0.6 and 17.5 ± 0.7 Ma (Table 2, Fig. 1b). Whole rock ages from these dykes range between 18.5 ± 0.7 and 16.5 ± 0.5 Ma.

Biotite ages from shallow intrusive dacite bodies and undeformed dykes range from 18.0 to 16.0 Ma, whereas whole rock ages yielded slightly younger ages varying from 16.8 to 14.9 Ma (Table 2., Fig. 1b). The oldest biotite age of 18.1 ± 0.6 Ma was obtained from a rhyodacite pebble (site 332, Fig. 1b), which was derived from a surface lava flow.

Fission track data

The results of zircon fission track analyses on five samples from the Pohorje pluton and dacite dykes are listed in Table 3 and shown in Figure 1b. Sample preparation techniques, details of the method and instruments applied are the same as those described by Dunkl et al. (2001). The three FT analyses from the dacites consistently yielded ages around 17.5 Ma. These ages are slightly older than those from the tonalite (16.9 and 15.6 Ma). In the case of tonalite sample 280 the chi-square test failed (Green, 1981). Together with the high dispersion (Galbraith & Laslett, 1993), this indicates that the single grain ages do not follow a Gaussian distribution (Table 3).

Petrology and mineral chemistry

Two samples from the eastern (sample 311) and western (sample 216) part of the Pohorje pluton (Fig. 1) were analyzed with the electron microprobe in order to compare mineral chemistries (Table 4) and to obtain data for thermobarometric calculations. For the instrumental details and correction procedures applied, see Balen et al. (2006).

Both analysed samples contain the same mineral assemblage of amphibole, biotite, plagioclase, K-feldspar, quartz and accessories. No chemical zonation is found within the analysed grains, but some variation exists amongst the individual grains (Table 4). Following the classification of Leake et al. (1997), all amphiboles from sample 216 fall into the magnesiohornblende field, whereas amphiboles from sample 311 are at the boundary between tschermakite and ferro-tschermakite. The investigated tonalites contain the mineral assemblage required for the application of the Al-in-hornblende barometer. We used the calibrations of Hammarstrom & Zen (1986), Hollister et al. (1987) and Schmidt (1992). All three calibrations yielded pressure estimates with a fairly acceptable error range (mostly 0.5 kbar). For sample 216 we calculated 3 to 4 kbar, while sample 311 formed at pressures of about 6–7 kbar. This corresponds to crustal depths of some 8–11 km and 16–19 km, respectively. The higher pressure obtained from sample 311 is in good agreement with previous pressure estimates of 6.8 ± 0.4 kbar by Altherr et al. (1995), who also investigated samples originating from the eastern Pohorje pluton (Cezlak, site 137A, Fig. 1). However, further data are needed to confirm the difference in crystallization depth between the eastern and western segment of the Pohorje pluton. From these pressure values, temperatures were estimated with the method of Blundy & Holland (1990). Sample 216 yielded 750–770 °C, while for sample 311 somewhat higher temperatures of 760–820 °C were calculated.

Deformation of the igneous rocks

Syn-magmatic deformation

Tonalites and granodiorites that were only weakly affected by solid-state deformation often show a foliation defined by the preferred alignment of biotite and the long axes of amphibole grains. Therefore, this foliation seems to have developed during the magmatic stage. Some dacitic dykes show well-developed magmatic flow fabrics at the micro scale.

Two main generations of dykes occur near the Pohorje pluton. A first generation (Fig. 3a) consists of aplites and de-

Table 3. Zircon fission track data obtained on the igneous rocks of Pohorje Mountains.

Sample	Lat. Nord	Long. East	Lithology	Cryst.	Spontaneous		Induced		Dosimeter		P(χ^2) (%)	Disp.	FT age* (Ma ± 1s)
					rs	(Ns)	ri	(Ni)	rd	(Nd)			
RD12	46° 32.2'	15° 10'	dacite	20	53.3	(618)	134.1	(1555)	6.89	(4543)	58	0.02	17.5 ± 0.9
RD4	46° 32.5'	15° 08'	dacite	20	48.6	(731)	121.4	(1826)	6.92	(4543)	22	0.07	17.7 ± 0.9
285	46° 29.2'	15° 11'	dacite	19	57.6	(669)	145.5	(1776)	6.99	(4543)	41	0.02	17.7 ± 0.9
RD36	46° 31.5'	15° 15'	tonalite	20	54.9	(702)	144.7	(1849)	6.95	(4543)	27	0.06	16.9 ± 0.8
280	46° 29.8'	15° 20'	tonalite	20	63.4	(1041)	117.4	(1928)	4.54	(8928)	0	0.19	15.6 ± 0.9

Cryst.: number of dated zircon crystals.
Track densities (ρ) are as measured ($\times 10^5$ tr/cm^2); number of tracks counted (N) shown in brackets.
P(χ^2): probability obtaining Chi-square value for n degree of freedom (where n = no. crystals – 1).
Disp.: Dispersion, according to Galbraith and Laslett (1993).
*Central ages calculated using dosimeter glass: CN 2 with $\zeta = 127.8 \pm 1.6$.

Table 4. Representative mineral chemical data of the Pohorje tonalites, used for thermobarometric calculations from sites 216 and 311.

Amphibole

Sample	Slo-216		Slo-311	
SiO_2	44.79	44.13	40.86	41.81
Al_2O_3	8.26	10.73	12.58	11.19
TiO_2	1.19	1.58	1.07	0.93
FeO	16.30	13.60	19.90	20.68
MnO	0.78	0.71	1.06	0.74
MgO	12.35	12.25	7.78	7.65
CaO	11.42	11.84	11.28	11.67
Na_2O	1.45	1.22	1.59	1.43
K_2O	0.68	0.57	1.50	1.33
Total	97.22	96.63	97.62	97.43
cation numbers on the basis of 23 O				
Si	6.608	6.505	6.203	6.393
Al^{IV}	1.392	1.495	1.797	1.607
Fe^{3+}	0.000	0.000	0.000	0.000
T	8.000	8.000	8.000	8.000
Al^{VI}	0.044	0.369	0.454	0.410
Ti	0.132	0.175	0.122	0.107
Fe^{3+}	0.930	0.580	0.669	0.476
Mg	2.716	2.691	1.761	1.743
Fe^{2+}	1.081	1.096	1.858	2.168
Mn	0.097	0.089	0.136	0.096
C	5.000	5.000	5.000	5.000
Mn	0.000	0.000	0.000	0.000
Ca	1.805	1.870	1.835	1.912
Na	0.195	0.130	0.165	0.088
B	2.000	2.000	2.000	2.000
Na	0.220	0.219	0.303	0.336
K	0.128	0.107	0.290	0.259
A	0.348	0.326	0.593	0.595
Total	15.665	15.521	15.823	15.758

Plagioclase

Sample	Slo-216		Slo-311	
SiO_2	61.01	60.92	60.56	61.54
Al_2O_3	24.70	24.28	24.21	24.50
CaO	6.25	6.24	6.59	6.69
Na_2O	8.21	7.62	7.82	7.49
K_2O	0.33	0.27	0.27	0.27
Total	100.51	99.33	99.45	100.49
cation numbers on the basis of 8 O				
Si	2.703	2.722	2.711	2.720
Al	1.290	1.279	1.277	1.276
Ca	0.297	0.299	0.316	0.317
Na	0.705	0.660	0.679	0.642
K	0.019	0.015	0.015	0.015
An	29.07	30.66	31.29	32.53

K-feldspar

Sample	Slo-216	Slo-311
SiO_2	63.90	63.74
Al_2O_3	17.56	18.64
BaO	0.65	1.99
Na_2O	0.97	0.91
K_2O	15.51	14.93
Total	98.59	100.21
cation numbers on 8 O		
Si	3.009	2.971
Al	0.974	1.024
Ba	0.012	0.036
Na	0.089	0.082
K	0.931	0.888

Biotite

Sample	Slo-216	Slo-311
SiO_2	36.56	36.70
Al_2O_3	14.49	14.55
TiO_2	3.61	2.57
FeO	17.99	20.36
MnO	0.43	0.45
MgO	11.65	10.82
CaO	0.20	0.00
Na_2O	0.17	0.00
K_2O	9.50	9.50
Total	94.61	94.94
cation numbers on 22 O		
Si	5.610	5.663
Al	2.620	2.646
Ti	0.416	0.298
Fe^{2+}	2.308	2.627
Mn	0.056	0.059
Mg	2.664	2.488
Ca	0.033	0.000
Na	0.050	0.000
K	1.859	1.870
Total	15.618	15.651

formed mafic and dacite dykes that trend E–W to NW–SE, with gentle to moderate dips; this generation is oriented sub-parallel to the foliation of the host metamorphic rocks (Fig. 3). Emplacement was probably controlled by the pre-existing foliation. A second generation (Fig. 3b) consists of undeformed mafic and dacite dykes with trend between NNW–SSE and NNE–SSW; only a very few vertical dykes trend WNW–ESE. The dominant set is vertical or steeply dipping and can be interpreted as tensional cracks or conjugate extensional shear fractures. These formed in an ~E–W oriented extensional regime.

Solid-state deformation

The rocks of the Pohorje pluton show clear evidence for solid-state deformation after magmatic crystallization. Where the foliation is well developed, it totally overprints the magmatic fabric. Alignment of biotite and amphibole crystals and elongated lenses of quartz and feldspar locally define a weak stretching lineation. In the northeastern part the foliation is steeply to moderately dipping and is either sub-parallel or at a small angle to the pluton margin (Figs. 1a, 4). The lineation is mostly gently

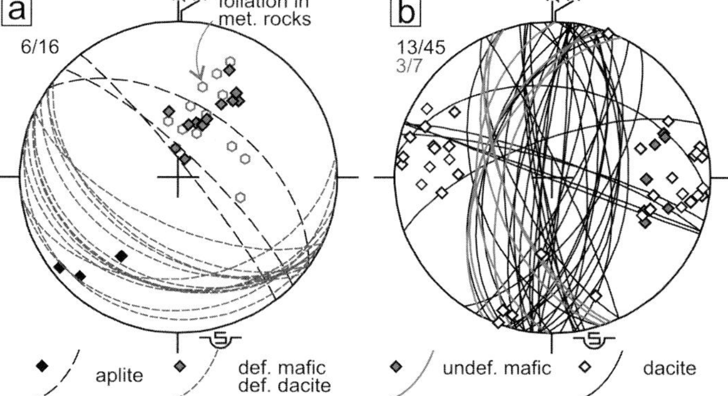

Fig. 3. Stereographs of dykes in the surroundings of the Pohorje pluton; lower hemisphere projection, Schmidt net. Numbers in upper left corner indicate the number of sites and measurements, respectively. 3a: aplite and deformed dykes 3b: undeformed dacite and mafic dykes

dipping. In the southwestern part, the foliation is mainly gently dips to the SE or WSW, partly being sub-parallel to the undulating pluton boundary (Figs. 1a, 4). The lineation mainly trends E–W, being oriented oblique or down dip within the foliation.

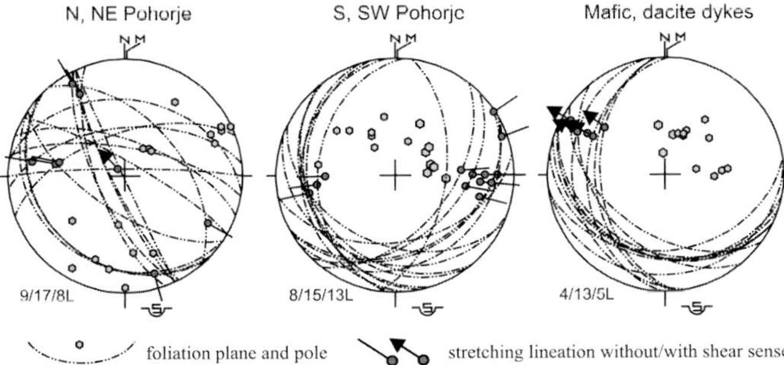

Fig. 4. Stereographs showing the orientations of ductile fabrics found within two domains the Pohorje pluton, and within deformed mafic and dacite dykes. Numbers in lower left corner indicate the number of sites, foliation planes, lineations, respectively.

The scatter in the orientation of the foliations is at least partly due to later deformation.

At the micro scale, solid-state deformation is best evidenced by the complete dynamic recrystallization of quartz into elongated lenses within the most heavily deformed samples (Fig. 5a). Primary biotite was sheared and partly recrystallised into highly elongated, fine-grained tails along the foliation (Fig. 5b). Feldspars mostly show brittle behaviour, but deformation twins, bent twins and incipient core-mantle structures indicate crystal plastic deformation. In the more deformed rocks biotite and quartz tend to form interconnected weak layers, and a typical "augengneiss" structure is developed. In less deformed rocks quartz is partly recrystallised, the relict grains show intensive undulous extinction and formation of subgrains. The microstructural features indicate deformation temperatures characteristic of medium to high-T greenschist facies. This is in accordance with the observation that biotite is generally stable during deformation and transformed into chlorite only to a small extent.

Mafic xenoliths in the tonalite are strongly flattened due to vertical shortening (Fig. 5c). Subvertical aplite dykes suffered folding with gently dipping axial planes. In the Cezlak quarry (site 137A) several moderately E and W dipping ductile shear zones are exposed and form conjugate pairs. They offset aplitic and pegmatitic dykes with a normal sense of shear, and are characterized by grain–size reduction due to incipient mylonitisation (Fig. 5d).

The deformed dacitic or mafic dykes show incipient stages of ductile deformation, mainly along their margins. Their foliation is sub-parallel to the foliation in the host metamorphics (Fig. 4). Site 232 exhibits a penetrative foliation (Fig. 5e), and a weak stretching lineation defined by elongate minerals developed. The shape preferred orientation in dynamically recrystallised quartz aggregates indicates top-to-WNW shear (Fig. 4).

Altogether, our observations demonstrate that the whole Pohorje pluton as well as some related dyke rocks underwent solid state deformation in the upper greenschist facies following the emplacement and crystallisation. This indicates that cooling and uplift was triggered by active tectonics, resulting in vertical flattening associated with ca. E–W stretching mostly, whereas the formation of moderately to steeply dipping foliation with gently dipping lineation in the northeastern part of the pluton is not yet understood.

Brittle deformation

We analyzed brittle deformation by fault-slip analysis and determination of paleostress tensors (Angelier 1984), by observation of cross-cutting relationships in outcrops, and by reinterpretation of map-scale structures. Most of the brittle faults belong to a transtensional deformation phase. Normal faults, trending from NW–SE to NNE–SSW are the most characteristic structural elements both in outcrop- and map scales. The symmetry plane of conjugate normal fault pairs is often not vertical (site 231 on Fig. 6), which indicates a tilting during normal faulting. Normal faults show gradual transition to oblique-normal or even pure conjugate strike-slip faults; sites 137B and 300 show typical examples (Fig. 6).

N–S trending faults displace the southern margin of the pluton at several locations (Fig. 6). NW–SE striking normal or oblique-normal faults bound dacite bodies in the NW Pohorje Mountains, and also the western margin of the Ribnica-Selnica trough. At the eastern margin of the Pohorje Mountains oblique-normal faults are postulated between the metamorphic rocks and the Triassic to Miocene sediments (Fig. 6). Observed synsedimentary normal faults in the areas immediately west and south of Pohorje (Fodor et al. 1998) clearly demonstrate that the E–W tensional phase was associated with Early to Middle Miocene extensional opening of the Pannonian basin and has lasted throughout Middle Miocene.

As typical for transtensional stress field, the calculated stress tensors mostly shows extensional characters, but few sites are marked by strike-slip type stress field with N–S compression. In all cases, the σ_3 stress axes are horizontal, and vary from SW–NE to ESE–WNW, with a prevailing E–W direction. This dispersion of orientations probably reflects spatial variations of the stress field. Extensional and strike-slip type stress tensors show transition, because (1) stress axes σ_2 and σ_1 are similar in value, (sites 137B and 300 on Fig. 6) and (2) strike-slip faults show only minor misfit with respect to the extensional stress tensor.

A few sites in the northern part of the pluton exhibit conjugate strike-slip faults and hence are characterised by a strike-slip type stress field, with a horizontal σ_1 axis trending NE to E (Fig. 6b). These structures represent a separate deformation phase whose importance is minor on map-scale fault pattern. This phase could follow the transtensional deformation (with E–W extension) as a short transient episode at around the Middle to Late Miocene boundary (late Sarmatian to early Pannonian) by using analogies from the Pannonian basin and Eastern Alps (Peresson & Decker 1997; Fodor et al. 1999; Pischinger et al. 2008).

A third deformation phase is characterized by a conjugate set of strike-slip and reverse faults (Fig. 6c). Dextral slip oc-

Fig. 5. Solid-state deformation of the Pohorje pluton and related dykes. 5a: Photomicrograph of a foliated medium-grained tonalite (sample 311). Note the complete dynamic recrystallisation of quartz (upper left) and fine-grained, newly formed feldspars (mirmekite) along the grain boundaries of larger K-feldspars (lower central). +N. 5b: Deformation of biotite: primary magmatic crystals are surrounded by very elongate, anastomosing, sheared and recrystallised tails which form interconnected weak layers defining the foliation (subvertical) with dynamically recrystallised quartz ribbons. Idiomorphic prismatic crystals at the right are amphiboles. Sample 216, 1N. 5c: Hand specimen of a well-foliated granodiorite with an elongate mafic enclave oriented subparallel to the foliation. SE Pohorje, Cezlak quarry (137A). 5d: Steeply dipping, leucocratic dyke crosscutting foliated granodiorite from Southeastern Pohorje, Cezlak quarry, site 137A. The dykes are cut by a ductile shear zone. 5e: Photomicrograph of a strongly deformed dacite dyke from the western Pohorje (site 232). Note well-developed foliation defined by elongated, dynamically recrystallised quartz ribbons and aligned "biotite-fishes" with strongly elongated, recrystallised tails. +N. 5f: Subvertical, leucocratic dyke crosscutting foliated granodiorite and a mafic enclave. The dyke is also flattened vertically, as is indicated by the same, gently dipping foliation observed in the host granodiorite. Note also late, brittle normal faults with decimetre-scale offsets. SE Pohorje, Cezlak quarry (137A).

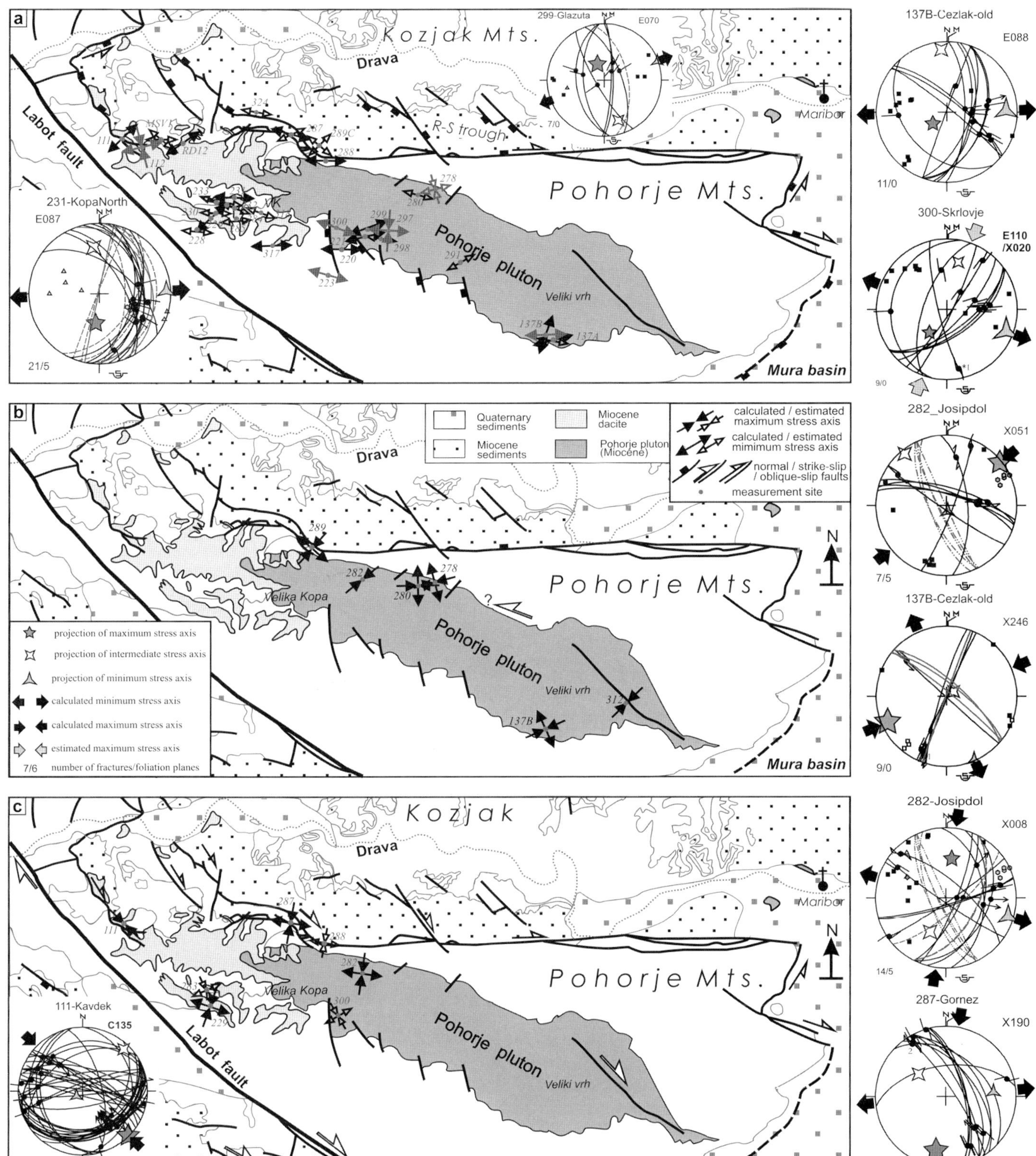

Fig. 6. Fault-slip data, stereographic projections of fractures and paleostress axes calculated with the software of Angelier (1984). For pre-Miocene formations, see Figure 1. Numbers in upper right corner show the stress type: C for compression, X for strike-slip, E for extension with the direction of σ_1 and σ_3 axes (for C, X and E, respectively). 6a: Transtensional phase with ~E–W extension and locally with ~N–S compression. Note tilted fractures at site 231. Grey arrows indicate axes of a stress state being close to strike-slip type stress. 6b: Strike-slip faulting associated with NE–SW to E–W compression. 6c: Strike-slip faulting associated with N–S compression.

curred along the Lavanttal-Labot fault, along other NW–SE trending faults of the NW Pohorje area, and also along the western margin of the Ribnica-Selnica trough. Dextral faulting is suggested in the eastern Pohorje pluton along the fault bounding the narrow metamorphic belt although kinematics is not directly confirmed. The calculated compressional axis is between NNW–SSSE and NNE–SSW, while the tensional axis is perpendicular (Figs. 6c). These strike-slip faults are younger, because (1) they often reactivate normal faults of the E–W extension, proved by relative chronology between dip-slip and strike-slip striae, and (2) because subvertical NW–SE trending dextral faults cut eastward-dipping normal faults north from the Pohorje, in the Kozjak area. The occurrence of strike-slip faulting, combined locally with transpressional deformation, is characteristic for the Late Miocene to Quaternary structural evolution of the wider area (Fodor et al. 1998; Márton et al. 2002; Sölva et al. 2005; Vrabec et al. 2006).

Discussion

Age of emplacement of the Pohorje pluton and related magmatic rocks

The new U-Pb data demonstrate a late Early Miocene age of emplacement (18.64 ± 0.11 Ma) of the Pohorje pluton. This age is markedly different from the Oligocene isotopic ages typically obtained for the Periadriatic plutons (von Blanckenburg and Davies 1995). The age difference is particularly striking in comparison with the nearest Periadriatic intrusion, the Karawanken tonalite, for which we obtained a K-Ar biotite age of 32.4 ± 1.2 Ma (Fig. 1b, Table 2; Trajanova et al. 2008), in agreement with previous data (Scharbert 1975, Elias 1998). The Karawanken and Pohorje intrusions also show considerable differences in terms of their geochemical character and inferred magma sources; the Pohorje pluton is relatively more enriched in large ion litophile elements (LILE) as well as in La, Ce and is hence interpreted as the product of partial melting of amphibolite and eclogite (Zupančič 1994b; Benedek & Zupančič 2002; Márton et al. 2006). Therefore, the Pohorje pluton is not part of the Oligocene-age Periadriatic intrusive suite, as was previously suggested by numerous authors (e.g. Laubscher 1983; Elias 1998; Pamić & Pálinkáš 2000; Pálinkáš & Pamić 2001; Rosenberg 2004). Moreover, the new U-Pb age is in agreement with the Miocene age postulated for this pluton by other authors (Deleon 1979; Dolenec 1994; Altherr et al. 1995; Trajanova et al. 2008).

The oldest K-Ar ages (19.5–20.3 Ma) are obtained on amphiboles from a small gabbroic inclusion within the southeastern Pohorje pluton. These ages are very close (within 1σ error) to the U-Pb age (18.64 Ma) for the Pohorje pluton and in accordance with the assumption that the gabbro crystallized during the early phase of magma evolution.

The average ages obtained from deformed dacite and mafic dykes are slightly younger (18.5–18.2 Ma) than the U-Pb age of the pluton and appear to be older than biotite ages and zircon fission track ages from the dacite. Although the K-Ar ages overlap within the 1σ error range, the paleomagnetic data (Márton et al. 2006) suggest a relatively older formation of deformed dacite and mafic dykes, since they record one additional rotational event in comparison with the dacites.

Regarding the dacite bodies, the biotite K-Ar and the zircon fission track ages are very similar (18–16 and 17.7–15 Ma, respectively). The mean values are close to the assumed age of the Karpatian sediments that also contain pyroclastic intercalations (17.3–16.5 Ma, Steininger et al. 1988). Therefore, we interpret the biotite K-Ar ages and the zircon fission track ages from the undeformed dacites to be close to the emplacement age of the dacites. Dacite magmatism may have already started in the Ottnangian, as is indicated by the oldest biotite ages around 18 Ma, including the age from the rhyodacite pebble. The youngest dacite intrusions occur within Karpatian sediments. The biotite K-Ar age of the Vuzenica body (16 ± 0.6 Ma, sample GP-34 in Fig. 1b) and the slightly younger apatite fission track age (15 Ma; Sachsenhofer et al. 1998) constrain the age of termination of the magmatic activity as early Middle Miocene.

The total time span inferred for the Pohorje magmatism (about 18.5–16 Ma) corresponds to important pulses of volcanism found in the Pannonian Basin system. The oldest Miocene magmatic suite, the so-called "lower rhyolite tuff" has an Ottnangian stratigraphic age (Hámor 1985); the K-Ar mean age was considered to be 19 ± 1.4 Ma for a long time (Hámor et al. 1987; Pécskay et al. 2006). However, new U-Pb and Ar-Ar data from two sites suggest that part of the suite may be as young as 17 Ma (Pálfy et al. 2007). Latest Early to early Middle Miocene igneous activity is well demonstrated in the Styrian basin, located some 20–40 km NE of the Pohorje Mountains in Austria (Ebner & Sachsenhofer 1991). There the age of volcanism (17.3–15 Ma) is constrained by intercalated Karpatian to earliest Middle Miocene marine sediments. Contemporaneous volcanic events produced the so-called "middle rhyolite horizon", a dacitic-rhyodacitic suite in the Pannonian Basin (Harangi et al. 2005). Despite this temporal coincidence, the geodynamic settings of all these magmatic suites may possibly be different.

Cooling history and exhumation of the Pohorje pluton

Magmatic rocks

Our thermo-chronological data permit the reconstruction of the cooling history of the magmatic rocks in the Pohorje Mountains (Fig. 7). The biotite K-Ar ages from the pluton are noticeably younger than its U-Pb age, suggesting that they record cooling and synchronous solid-state deformation of the pluton. Zircon fission track ages are still younger than biotite age from the same sites and thus record a further step along the cooling path. Feldspar ages fit into the general trend of the cooling history although scattered ages may reflect a somewhat disturbed isotope system. Biotite cooling ages from granodioritic pebbles are identical with the postulated Karpatian age of the host sediment (within limits of error), which also indicates rapid exhumation and denudation of the pluton. Considering the youngest biotite and FT ages, as well as the age of the sediments, the entire cooling process below the annealing temperature of fission tracks in zircons must have occurred within 3 Ma (18.6–15.6 Ma).

Metamorphic rocks

Several K-Ar ages were reported from the metamorphic host rocks of the Pohorje and Kozjak Mountains (Fodor et al. 2002b). A detailed discussion of these ages is beyond the scope of this paper and we only briefly consider data that are relevant for inferring the exhumation history of the Pohorje pluton. With one exception, the white mica and biotite K-Ar ages from the host rocks are within the 30–13 Ma range and at least partly reflect the thermal effect from the Pohorje intrusives and possibly that of other buried and unexposed magma bodies (Fig. 1a). This is clearly shown by the existence of thermally "undisturbed" Eoalpine (Cretaceous) mica ages reported from the Kozjak Mountains (except for its eastern portion), as well as by the Cretaceous ages obtained from high-T retentivity isotopic data for the metamorphic rocks in the Pohorje Mountains (Thöni 2002; Miller et al. 2005; Janák et al. 2007). Amphibole K-Ar ages >100 Ma from both the Pohorje and the Kozjak Mountains indicate no resetting (Fig. 1a). One white mica age (102 Ma) from the southernmost Pohorje Mountains, as well as two illite ages from the Kozjak Mountains, which derive from the weakly metamorphosed Upper Austroalpine Paleozoic rocks, are considered as undisturbed Cretaceous cooling ages of the Eoalpine metamorphic event.

The K-Ar white mica ages from the metamorphic rocks reflect a younging trend from west to east (Fig. 1a). In the western part, the muscovite ages, and the zircon fission track ages, are Oligocene and therefore older than the granodiorite intrusion itself; only one biotite age (from close vicinity to the pluton; site 222) is identical with the magmatic biotite ages. These Oligocene ages suggest that the western part of the metamorphic host rocks were cooled below the closure temperature of muscovite (ca. 350 °C) and possibly also below the annealing temperature of the zircon fission tracks (ca. 250 °C) prior to magmatism and were not substantially reheated by the pluton (Fig. 7, cooling path A). Ages from the medium-grade metamorphic rocks in the central southern and northern Pohorje Mountains (sites 271–273, 138) are close to those obtained for the magmatic rocks, and they thus indicate a common cooling history (cooling path B on Fig. 7). While the muscovite and biotite ages in the metamorphic rocks of the easternmost Pohorje Mountains are consistently younger than the biotite age from the pluton, this demonstrates that this part underwent later cooling (Fig. 7, path C). Biotite cooling ages from the pluton itself, however, do not show a similar trend, suggesting that the K-Ar system in biotite was uniformly affected by heat derived from magmatism.

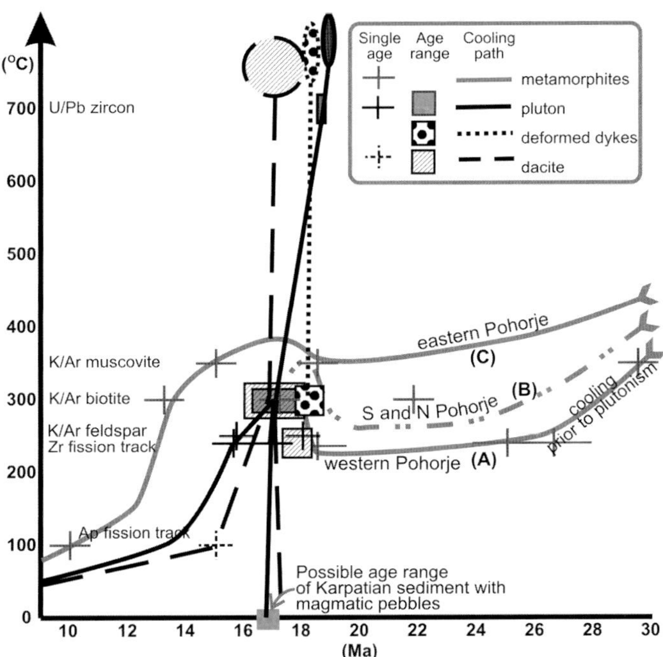

Fig. 7. Cooling history of the Pohorje pluton and surrounding metamorphic rocks. Crosses and boxes indicate the ranges of error concerning age (horizontally) and closure temperature (vertically, and only schematically).

Summary of the emplacement and cooling of the Pohorje pluton

A simplified summary on the evolution of the Pohorje pluton itself, related other magmatic activity and that of the surrounding metamorphic host rocks is shown in Fig. 8. The emplacement of the pluton occurred at 18.6 Ma, as indicated by our U-Pb isotope age data. Inherited zircon cores of Permian and Neoproterozoic age point to melting and/or assimilation of crustal material with Neoproterozoic Hf model ages. The Hf isotope composition of Miocene zircon domains suggests involvement of a juvenile Phanerozoic crust or a Miocene-age mantle-derived melt as a source of the magma. Limited thermobarometric data suggest a relatively larger crystallization depth in the east compared to the west, where the intrusion of the pluton almost reached the weakly metamorphosed Austroalpine unit.

The magma intruded medium-grade metamorphic rocks, which previously underwent Cretaceous (Eoalpine) metamorphism and ductile deformation (Fig. 8a). Fodor et al. (2002a) and Trajanova (2002) demonstrated examples of extensional shear zones, which form an important detachment zone at the top of the medium-grade rocks. By Miocene times, these rocks were already cooled below the closure temperatures of amphibole, and, at least in the western Pohorje Mountains, below the closure temperature of muscovite. On the other hand, advective heat transfer from the magmatic intrusion in the northern, southern and eastern Pohorje area reheated the host metamorphic rocks.

Just after its emplacement, at ~18.5–18.2 Ma, the pluton was cut by aplitic and mafic dykes, which partly intruded into host metamorphic rocks along the pre-existing foliation (Fig. 8b). The pluton shows magmatic flow structures, which, however, are difficult to separate from overprinting solid-state deformation. Solid-state deformation occurred during fast cooling of the pluton, probably under greenschist facies conditions and before cooling below the closure temperature of biotite at around 17.1–16.5 Ma. The subhorizontal foliation planes and E–W mineral lineations in the southwestern part of the pluton indicate vertical flattening and E–W stretching while the moderately to steeply dipping foliation with gently dipping lineation in the northeastern pluton reflect a different type of deformation, probably strike-slip faulting (Fig. 8c).

The emplacement of the dacite stock started at around 18 Ma, and rhyodacitic lava flows also formed at that time (Fig. 8c). Because of the shallow intrusive level, cooling was fast, and hence, no ductile deformation feature was observed. Dykes were emplaced into the westernmost parts of the pluton and both into the medium- and low-grade metamorphic rocks, crosscutting earlier formed detachment zones. N–S trending dacite dykes were presumably controlled by extensional deformation, although the interpretation of WNW trending vertical dykes needs further consideration.

Dacite volcanoclastics were intercalated with sediments during the Karpatian (17.3–16.5 Ma), and dacite magmatism ended during the early Middle Miocene, i.e. around 16 Ma (Fig. 8d). Pebbles were eroded from both the pluton and dacite suites and transported into the Karpatian sedimentary basin. The entire geochronological data set and the presence of magmatic pebbles in Karpatian sediments demonstrate rapid cooling of the pluton after its emplacement through the temperature range typical for ductile deformation and below the annealing temperature of the zircon fission tracks, partly even all the way to the surface within only some 3 Ma.

This process was associated with ductile to brittle deformation of the pluton itself, some of the dykes and the older parts of the dacite bodies (Fig. 8d). Magmatic and solid-state fabrics of the pluton were cut by ductile shear zones and brittle faults, and tilted around horizontal axes. Fault pattern and stress calculations suggest that this deformation phase had a transtensional character, where normal and strike-slip faults were active. In the southwestern part of the pluton, the direction of brittle extension and the orientations of the ductile lineations are parallel to each other, suggesting persistent extensional deformation during ongoing cooling, while strike-slip type deformation could prevail in the NE. Older K-Ar muscovite ages in the host rocks, smaller crystallisation depth, and the close position of weakly metamorphosed uppermost Austroalpine unit with respect to the pluton could all confirm a relatively smaller amount of exhumation in the western part of the Pohorje Mountains.

The activity of brittle E–W extension, locally combined with N–S strike-slip faulting, conforms with lateral tectonic extrusion (a combination of orogenic collapse and continental escape), which affected the whole Eastern Alps in Early to Middle Miocene times (Ratschbacher et al. 1991, Frisch et al. 1998, 2000). Similar data are reported from the northern vicinity of the Pohorje Mountains, in the Koralm region (Pischinger et al. 2008). Frisch et al. (1998, 2000) and Dunkl et al. (2003) showed that an important extensional pulse accelerated E–W extension for a short period of time around 17 Ma. The 18–17 Ma time span also corresponds to the first and most significant lithospheric extensional event of the adjacent Pannonian

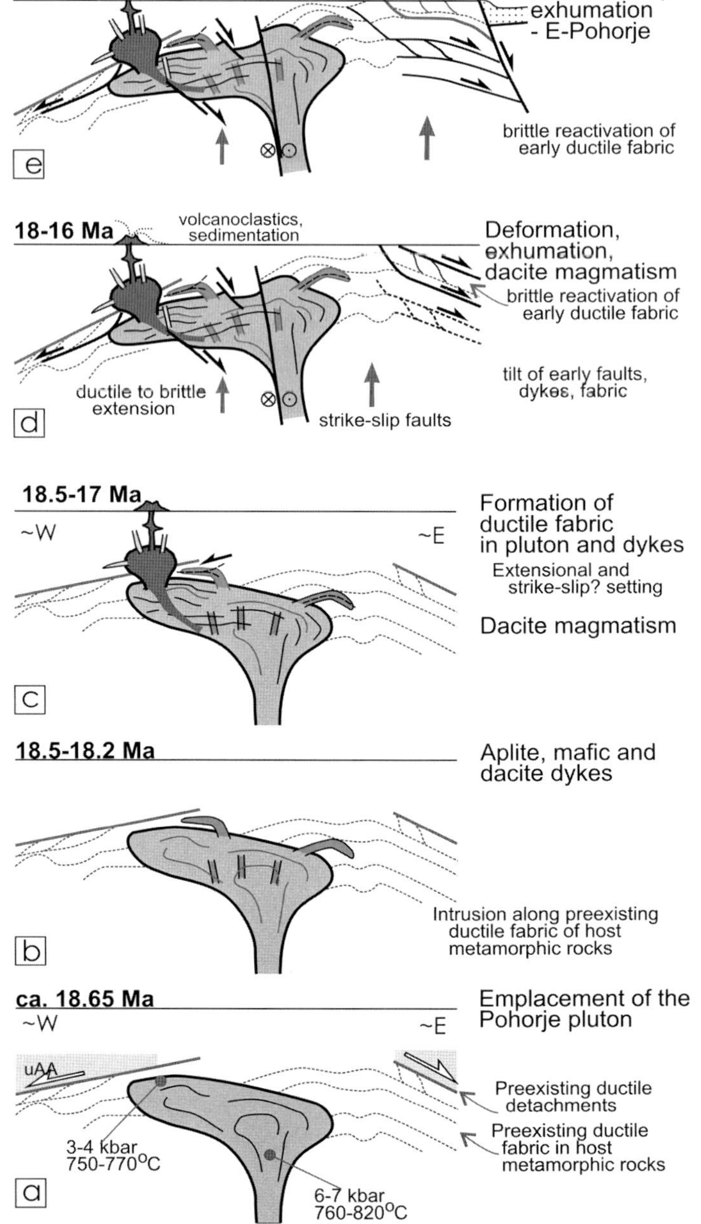

Fig. 8. Sketch for the evolution of the Pohorje pluton and surrounding rocks in terms of a series of ~E–W oriented cross sections. Age ranges are overlapping because some processes were coeval and /or the applied dating methods are not precise enough to separate distinct events. Uppermost weakly metamorphosed units are marked by uAA.

Basin, where high rates of subsidence occurred just east of the Pohorje Mountains, in the Mura-Zala basin (Tari et al. 1992; Sachsenhofer et al. 2001; Márton et al. 2002). In the Pohorje Mountains, this type of deformation was already active from at least 17 Ma onwards, when the pluton and most of the host metamorphics cooled below ca 300–250 C. If the ductile fabric formed by the same type of deformation, this would indicate the onset of this phase about 1 Ma earlier, in agreement with the briefly mentioned regional data.

The next stage in the exhumation occurred at around 15–13 Ma and in the eastern Pohorje Mountains, as is documented in the area east of the pluton (Fig. 8e). Pre-existing ductile shear zones were reactivated and resulted in tectonic exhumation. Formation of the western boundary fault of the Mura-Zala basin and accelerated subsidence in its hanging wall in Karpatian–Badenian time (17,3–13 Ma) may have completed this process.

Strike-slip faults with NE–SW to E–W compression marks the end of major extensional deformation in the area, which is frequently attributed to change in subduction geometry in the Carpathians (Peresson & Decker 1997; Fodor et al. 1999). The final stage of deformation, strike-slip faulting with N–S to NNW–SSE compression is part of a widespread transpressional deformation of the Dinaridic–Alpine–Pannonian junction.

Acknowledgements

Fieldwork campaigns were supported by two bilateral intergovernmental projects between Slovenia and Hungary (Slo5/98, Slo/6 00), donated to B. Jelen, N. Zupančič and L. Fodor (project leaders) and also by research programs in the Geological Institute of Hungary. K-Ar measurements were supported by National Scientific Found of Hungary (OTKA, project T043344, and M041434) donated to K. Balogh and Z. Pécskay, and the mineral chemical and thermobarometric studies by the grant No. F-047322/2004–2007 to P. Horváth. L. Fodor benefited from the Bolyai János scholarship of the Hungarian Academy of Sciences in 2000–2001 and 2006–2007. John Corckery contributed to the fieldwork. The paper benefited from the critical, but highly constructive reviews of C. Rosenberg, U. Schaltegger and S. Schmid. The authors are thankful for the careful and precise work of the staff at the RISØ and Oregon nuclear reactors. Deutsche Forschungsgemeinschaft (DFG) financed fission track analyses in the frame of the Collaborative Research Centre #275 in Tübingen. The authors are grateful for the careful sample preparation made by Dagmar Kost, Gerlinde Höckh and Dorothea Mühlbayer-Renner (Tübingen).

REFERENCES

Altherr, R., Lugović, B., Meyer, H.P. & Majer, V. 1995: Early Miocene post-collisional calc-alkaline magmatism along the easternmost segment of the Periadriatic fault system (Slovenia and Croatia). Mineralogy and Petrology 54, 225–247.

Angelier, J. 1984: Tectonic analysis of fault slip data sets. Journal of Geophysical Research 89, B7, 5835–5848.

Balen, D., Horváth, P., Tomljenović, B., Finger, F., Humer, B., Pamić, J. & Árkai, P. 2006: A record of pre-Variscan Barrovian regional metamorphism in the eastern part of the Slavonian Mountains (NE Croatia). Mineralogy and Petrology 87, 143–162.

Balogh, K. 1985: K-Ar dating of Neogene volcanic activity in Hungary: Experimental technique, experiences and methods of chronologic studies. Manuscript, internal report of the Nuclear Research Institute of Hungarian Academy of Sciences (ATOMKI), D/1, 277–288.

Balogh, K., Árva-Soós, E. & Buda, Gy. 1983: Chronology of granitoid and metamorphic rocks in Transdanubia (Hungary). Annuarul Institului de Geologie şi Geofizică 61, 359–364.

Benedek, K. 2002: Paleogene igneous activity along the easternmost segment of the Periadriatic-Balaton Lineament. Acta Geologica Hungarica 45, 359–371.

Benedek, K. & Zupančič, N. 2002: Periadriatic Intrusive Chain – version 1. In: Dunkl, I., Balintoni, I., Frisch, W., Hoxha, L., Janák, M., Koroknai, B., Milovanović, D., Pamić, J., Székely, B. & Vrabec, M. (Eds.): Metamorphic Map and Database of Carpatho-Balkan-Dinaride Area. http://www.met-map.uni-goettingen.de

von Blanckenburg, F. & Davies, J.H. 1995: Slab breakoff: A model for syncollisional magmatism and tectonics in the Alps. Tectonics 14, 120–131.

Blundy, J.D. & Holland, T.J.B. 1990: Calcic amphibole equilibria and a new amphibole-plagioclase geothermometer. Contributions to Mineralogy and Petrology 104, 208–224.

Deleon, G. 1969: Overview on the results of the radiometric dating of the granitoid rocks in Yugoslavia (in Serbian). Radovi Instituta Geološko-Rudarska Istraživanja Ispitivanja Nuklearnih Drugih Mineralnih Sirovina 6, 165–182.

Dolar-Mantuani, L. 1935: Das Verhältnis der Aplite zu den Tonaliten im Massiv des Pohorje (Bächergebirges). Geoloski Anali Balkanskog Poluostrova 12, 2, 1–165.

Dolenec, T. 1994: New isotopic and radiometric data related to igneous rocks of the Pohorje Mountains. Rudarsko-metalurški zbornik 41, 147–152.

Dunkl, I., Di Giulio, A. & Kuhlemann, J. 2001: Combination of single-grain fission-track chronology and morphological analysis of detrital zircon crystals in provenance studies – sources of the Macigno formation (Apennines, Italy). Journal of Sedimentary Research 71, 516–525.

Dunkl, I., Frisch, W. & Grundmann, G. 2003: Zircon fission track thermochronology of the southeastern part of the Tauern Window and the adjacent Austroalpine margin, Eastern Alps. Eclogae Geologica Helvetica 96, 209–217.

Ebner, F. & Sachsenhofer, R. 1991: Die Entwicklungsgeschichte des Steirischen Tertiärbeckens. Mitteilungen der Abteilung für Geologie und Paläontologie am Landesmuseum Johanneum Graz 49, 96 pp.

Elias, J. 1998: The thermal history of the Ötztal-Stubai complex (Tyrol, Austria/Italy) in the light of the lateral extrusion model. Tübinger Geowissenschaftliche Arbeiten, Reihe A, 36, 172 pp.

Exner, C. 1976: Die geologische Position der Magmatite des periadriatischen Lineaments. Verhandlungen der Geologischen Bundesanstalt 3–64.

Faninger, E. 1970: Pohorski tonalit in njegovi diferenciati. Geologija 13, 35–104.

Fodor, L., Jelen, B., Márton, E., Skaberne, D., Čar, J. & Vrabec, M. 1998: Miocene–Pliocene tectonic evolution of the Slovenian Periadriatic Line and surrounding area – implication for Alpine-Carpathian extrusion models. Tectonics 17, 690–709.

Fodor, L., Csontos, L., Bada, G., Györfi, I. & Benkovics, L. 1999: Tertiary tectonic evolution of the Pannonian basin system and neighbouring orogens: a new synthesis of paleostress data. In: Durand, B., Jolivet, L., Horváth, F. & Séranne, M. (Eds.): The Mediterranean Basins: Tertiary extension within the Alpine Orogen. Geological Society, London, Special Publications, 156, 295–334.

Fodor, L., Jelen, B., Márton, E., Rifelj, H., Kraljić, M., Kevrić, R., Márton, P., Koroknai, B. & Báldi-Beke, M. 2002a: Miocene to Quaternary deformation, stratigraphy and paleogeography in Northeastern Slovenia and Southwestern Hungary. Geologija 45, 103–114.

Fodor, L. Jelen, B., Márton, E. Zupančič, N., Trajanova, M., Rifelj, H., Pécskay, Z., Balogh, K., Koroknai, B., Dunkl, I., Horváth, P., Horvat, A., Vrabec, M., Kraljić, M. & Kevrić, R. 2002b: Connection of Neogene basin formation, magmatism and cooling of metamorphic rocks in NE Slovenia. Geologica Carpathica 53, special issue, 199–201.

Frank, W. 1987: Evolution of the Austroalpine elements in the Cretaceous. In: Flügel, H.W. & Faupl, P. (Eds.): Geodynamics of the Eastern Alps. Deuticke, Vienna, 379–406.

Frisch, W., Kuhlemann, J., Dunkl, I. & Brügel, A. 1998: Palinspastic reconstruction and topographic evolution of the Eastern Alps during late Tertiary extrusion. Tectonophysics 297, 1–15.

Frisch, W., Dunkl, I. & Kuhlemann, J. 2000: Postcollisional orogen-parallel large-scale extension in the Eastern Alps. Tectonophysics 327, 239–265.

Galbraith, R.F. & Laslett, G.M. 1993: Statistical models for mixed fission track ages. Nuclear Tracks and Radiatation Measurements 21, 459–470.

Gerdes, A. & Zeh, A. 2006: Combined U-Pb and Hf isotope LA-(MC)-ICP-MS analyses of detrital zircons: Comparison with SHRIMP and new constraints for the provenance and age of an Armorican metasediment in Central Germany. Earth and Planetary Sciences Letters 249, 47–61.

Gerdes, A. & Zeh, A. 2008: Zircon formation versus zircon alteration – new insights from combined U-Pb and Lu-Hf in situ LA-ICP-MS analyses, and consequences for the interpretation of Archean zircon from the Central Zone of the Limpopo Belt. Chemical Geology, doi 10.1016/j.chemgeo.2008.03.005.

Green, P.F. 1981: A new look at statistics in fission track dating. Nuclear Tracks 5, 77–86.

Hammarstrom, J.M. & Zen, E. 1986: Aluminium in hornblende: an empirical igneous geobarometer. American Mineralogist 71, 1297–1313.

Hámor, G. 1985: Geology of the Nógrád-Cserhát area. Geologica Hungarica series Geologica 22, 234 pp.

Hámor, G., Baranyai-Ravasz, L., Halmai, J., Balogh K. & Árva-Soós, E. 1987: Dating of Miocene acid and intermediate volcanic activity in Hungary. Annals of the Hungarian Geological Institute 70, 149–154.

Harangi, Sz., Mason, P.R.D. & Lukács, R. 2005: Correlation and petrogenesis of silicic pyroclastic rocks in the northern Pannonian Basin, Eastern central Europe: in situ trace element data of glass shards and mineral chemical constraints. Journal of Volcanology and Geothermal Research 143, 237–257.

Hinterlechner-Ravnik, A. 1971: The Pohorje Mountains metamorphic rocks I. Geologija 14, 187–226.

Hinterlechner-Ravnik, A. 1973: The metamorphic rocks of the Pohorje Mountains II. Geologija 16, 245–269.

Hollister, L.S., Grissom, G.C., Peters, E.K., Stowell, H.H. & Sisson, V.B. 1987: Confirmation of the empirical correlation of Al in hornblende with pressure of solidification of calc-alkaline plutons. American Mineralogist 72, 231–239.

Janák, M., Vrabec, M., Froitzheim, N. & de Hoog, C-J. 2007: Kyanite eclogites and garnet peridotites from the Pohorje Slovenia: petrological constraints on the Cretaceous subduction and exhumation in the HP/UHP terrane of the Eastern Alps. Abstract volume of the 8th Workshop on Alpine Geological Studies, Davos, Switzerland, 10.–12. October 2007, 29–30.

Jelen, B. & Rifelj, H. 2003: The Karpatian in Slovenia. In: Brzobohatý, R., Cicha, M., Kováč, M. & Rögl, F. (Eds.): The Karpatian: A Lower Miocene Stage of the Central Paratethys. Masaryk University, Brno, 133–139.

Kieslinger, A. 1935: Geologie und Petrologie des Bachern. Verhandlungen der Geologischen Bundesanstalt 7, 101–110.

Kőrössy, L. 1988: Hydrocarbon geology of the Zala Basin in Hungary. Általános Földtani Szemle 23, 3–162.

Laubscher, H. 1983: The late Alpine (Periadriatic) intrusions and the Insubric Line. Memorie della Societa Geologica Italiana 26, 21–30.

Leake, B.E., Woolley, A.R., Arps, C.E.S., Birch, W.D., Gilbert, M.C., Grice, J.D., Hawthorne, F.C., Kato, A., Kisch, H.J., Krivovichev, V.G., Linthout, K., Laird, J., Mandarino, J., Maresch, W.V., Nickel, E.H., Rock, N.S.M., Schumacher, J.C., Smith, D.C., Stephenson, N.C.N., Ungaretti, L., Whittaker, E.J.W., Youzhi, G. (1997) Nomenclature of amphiboles: Report of the Subcomittee on Amphiboles of the International Mineralogical Association Commission on New Minerals and Mineral names. Mineralogical Magazine 61, 295–321.

Márton, E., Fodor, L., Jelen, B., Márton, P., Rifelj, H. & Kevrić, R. 2002: Miocene to Quaternary deformation in NE Slovenia: complex paleomagnetic and structural study. Journal of Geodynamics 34, 627–651.

Márton, E., Trajanova, M., Zupančič, N. & Jelen, B. 2006: Formation, uplift and tectonic integration of a Periadriatic intrusive complex (Pohorje, Slovenia) as reflected in magnetic parameters and paleomagnetic directions. Geophysical Journal International 167, 1148–1159.

Miller, C., Mundil, R., Thöni, M. & Konzett, J. 2005: Refining the timing of eclogite metamorphism: a geochemical, petrological, Sm-Nd and U-Pb case study from the Pohorje Mountains, Slovenia (Eastern Alps). Contributions to Mineralogy and Petrology 150, 70–84.

Mioč, P. 1978: Geology of the sheet Slovenj Gradec L 33–55 1:100000. Federal Geological Survey of Yugoslavia, Belgrade, 74 pp.

Mioč, P. & Žnidarčič, M. 1977: Geological map of the sheet Slovenj Gradec, L 33–55, 1:100000. Federal Geological Survey of Yugoslavia, Belgrade.

Odin, G.S. et al. 1982: Interlaboratory standards for dating purposes. In: Odin, G.S (Ed.): Numerical Dating in Stratigraphy. Wiley & Sons, Chichester, New York, Brisbane, 123–149.

Pálfy J., Mundil R., Renne P.R., Bernor R.L., Kordos, L. & Gasparik, M. 2007: U-Pb and Ar-40/Ar-39 dating of the Miocene fossil track site at Ipolytarnoc (Hungary) and its implications. Earth and Planetary Science Letters 258, 160–174.

Pálinkaš, L., & Pamić, J. 2001: Geochemical evolution of Oligocene and Miocene magmatism across the Easternmost Periadriatic Lineament. Acta Vulcanologica 13, 41–56.

Pamić, J. & Pálinkaš, L. 2000: Petrology and geochemistry of Paleogene tonalites from the easternmost parts of the Periadriatic Zone. Mineralogy and Petrology 70, 121–141.

Pécskay, Z. et al. 2006: Geochronology of Neogene magmatism in the Carpathian arc and intra-Carpathian area. Geologica Carpathica 57, 511–530.

Peresson, H. & Decker, K. 1997: Far-field effect of late Miocene subduction in the eastern Carpathians: E–W compression and inversion of structures in the Alpine-Carpathian-Pannonian region. Tectonics 16, 38–56.

Pischinger, G., Kurz, W. Übleis, M., Egger, M., Fritz, H., Brosch F.J. & Stingl, K. 2008: Fault slip analysis in the Koralm Massif (Eastern Alps) and consequences for the final uplift of "cold spots" in Miocene times. Swiss Journal of Earth Sciences, this volume.

Ratschbacher, L., Frisch, W., Linzer, H.G. & Merle, O. 1991: Lateral extrusion in the Eastern Alps, part 2: structural analysis. Tectonics 10, 257–271.

Rosenberg C. L. 2004: Shear zones and magma ascent: a model based on a review of the Tertiary magmatism in the Alps. Tectonics 23, TC3002, doi:10.1029/2003TC001526.

Sachsenhofer, R. F., Dunkl, I., Hasenhüttl, Ch. & Jelen, B. 1998: Miocene thermal history of the southwestern margin of the Styrian Basin: coalification and fission track data from the Pohorje/Kozjak area (Slovenia). Tectonophysics 297, 17–29.

Sachsenhofer, R.F., Jelen, B., Hasenhüttl, Ch., Dunkl, I. & Rainer, T. 2001: Thermal history of Tertiary basins in Slovenia (Alpine-Dinaride-Pannonian junction). Tectonophysics 334, 77–99.

Salomon, W. 1897: Über des Alter, Lagerungsform, und Entstehungsart der Periadriatischen granitischkörnigen Massen. Tschermaks Mineralogische und petrographische Mitteilungen Neue Folge 17, 109–283.

Schaltegger, U. & Corfu, F. 1992: The age and source of Late Hercynian magmatism in the Central Alps – evidence from precise U-Pb ages and initial Hf isotopes. Contributions to Mineralogy and Petrology, 111, 329–344.

Scharbert, S. 1975: Radiometrische Altersdaten von Intrusivgesteinen im Raum Eisenkappel (Karawanken, Kärnten). Verhandlungen der Geologischen Bundesanstalt 4, 301–304.

Schmid, S.M., Aebli, H.R., Heiler, F. & Zingg, A. 1989: The role of the Periadriatic line in the tectonic evolution of the Alps. In: Coward, M.P., Dietrich, D. & Park, R.G. (Eds.): Alpine tectonics. Geological Society, London, Special Publications 45, 153–171.

Schmid, S.M., Fügenschuh, B., Kissling, E. & Schuster, R. 2004: Tectonic map and overall architecture of the Alpine orogen. Eclogae Geologicae Helvetia 97, 93–117.

Schmidt, M.W. 1992: Amphibole composition in tonalite as a function of pressure: an experimental calibration of the Al-in-hornblende barometer. Contributions to Mineralogy and Petrology 110, 304–310.

Slama, J., Košler, J., Condon, D.J., Crowley, J.L, Gerdes, A., Hanchar, J.M., Horstwood, M.S.A., Morris, G.A., Nasdala, L., Norberg, N., Schaltegger, U., Schoene, B., Tubrett, M.N. & Whitehouse, M.J. 2008: Plešovice zircon – a new natural reference material for U-Pb and Hf isotopic microanalysis. Chemical Geology 249, 1–35.

Sölva, H., Stüwe, K. & Strauss, P. 2005: The Drava River and the Pohorje Mountain Range (Slovenia): geomorphological interactions. Mitteilungen des Naturwissenschaftlichen Vereines für Steiermark 134, 45–55.

Steenken, A., Siegesmund, S., Heinrichs, T. Fügenschuh, B. 2002: Cooling and exhumation of the Rieserferner Pluton (Eastern Alps, Italy/Austria). International Journal of Earth Sciences 91, 799–817.

Steiger, R.H. & Jäger, E. 1977: Subcommission on geochronology – convention on use of decay constants in geochronology and cosmochronology. Earth and Planetary Science Letters 36, 359–362.

Steininger, F., Müller, C. & Rögl, F. 1988: Correlation of Central Paratethys, Eastern Paratethys, and Mediterranean Neogene stages. In: Royden, L.H. & Horváth, F. (Eds.): The Pannonian Basin. American Association of Petroleum Geologists Memoir 45, 79–87.

Tari, G., Horváth, F. & Rumpler, J. 1992: Styles of extension in the Pannonian Basin. Tectonophysics 208, 203–219.

Thöni, M. 2002: Sm–Nd isotope systematics in garnet from different lithologics (Eastern Alps): Age results and an evaluation of potential problems for garnet Sm–Nd chronometry. Chemical Geology 185, 255–281.

Trajanova, M. 2002: Significance of mylonites and phyllonites in the Pohorje and Kobansko area. Geologija 45, 149–161.

Trajanova, M., Pécskay, Z. & Itaya, T. 2008: K-Ar geochronology and petrography of the Miocene Pohorje Mountains batholith (Slovenia). Geologica Carpathica 59, 247–260.

Vrabec, M., Pavlovčič-Prešeren, P. & Stopar, B. 2006: GPS study (1996–2002) of active deformation along the Periadriatic fault system in northeastern Slovenia: tectonic model. Geologica Carpathica 57, 57 65.

Wedepohl, K.H. 1995: The composition of the continental crust. Geochimica et Cosmochemica Acta 597, 1217–1232.

Winkler, A. 1929: Über das Alter der Dacite Gebiet des Dradurchbruchs. Verhandlungen der Geologischen Bundesanstalt 169–181.

Žnidarčič, M. & Mioč, P. 1988: Geological map of the sheets Maribor and Leibnitz, L 33–56 and L 33–44, 1:100 000. Federal Geological Survey of Yugoslavia, Belgrade.

Zupančič, N. 1994a: Petrografske značilnosti in klasifikacija pohorskih magmatskih kamnin. Rudarsko-metalurški zbornik 41, 101–112.

Zupančič, N. 1994b: Geokemične značilnosti in nastanek pohorskih magmatskih kamnin. Rudarsko-metalurški zbornik 41, 113–128.

Manuscript received 18 January, 2008
Revision accepted 8 July 2008
Published Online first November 1, 2008
Editorial Handling: Stefan Schmidt & Stefan Bucher

1661-8726/08/01S273-22
DOI 10.1007/s00015-008-1288-7
Birkhäuser Verlag, Basel, 2008

Swiss J. Geosci. 101 (2008) Supplement 1, S273–S294

A map-view restoration of the Alpine-Carpathian-Dinaridic system for the Early Miocene

KAMIL USTASZEWSKI [1,*], STEFAN M. SCHMID [1], BERNHARD FÜGENSCHUH [1,2], MATTHIAS TISCHLER [1,**], EDUARD KISSLING [3] & WIM SPAKMAN [4]

Key words: Tectonics, kinematics, palinspastic restoration, Alps, Carpathians, Dinarides, Adriatic plate

ABSTRACT

A map-view palinspastic restoration of tectonic units in the Alps, Carpathians and Dinarides reveals the plate tectonic configuration before the onset of Miocene to recent deformations. Estimates of shortening and extension from the entire orogenic system allow for a semi-quantitative restoration of translations and rotations of tectonic units during the last 20 Ma. Our restoration yielded the following results: (1) The Balaton Fault and its eastern extension along the northern margin of the Mid-Hungarian Fault Zone align with the Periadriatic Fault, a geometry that allows for the eastward lateral extrusion of the Alpine-Carpathian-Pannonian (ALCAPA) Mega-Unit. The Mid-Hungarian Fault Zone accommodated simultaneous strike-perpendicular shortening and strike-slip movements, concomitant with strike-parallel extension. (2) The Mid-Hungarian Fault Zone is also the locus of a former plate boundary transforming opposed subduction polarities between Alps (including Western Carpathians) and Dinarides. (3) The ALCAPA Mega-Unit was affected by 290 km extension and fits into an area W of present-day Budapest in its restored position, while the Tisza-Dacia Mega-Unit was affected by up to 180 km extension during its emplacement into the Carpathian embayment.

(4) The external Dinarides experienced Neogene shortening of over 200 km in the south, contemporaneous with dextral wrench movements in the internal Dinarides and the easterly adjacent Carpatho-Balkan orogen. (5) N–S convergence between the European and Adriatic plates amounts to some 200 km at a longitude of 14° E, in line with post-20 Ma subduction of Adriatic lithosphere underneath the Eastern Alps, corroborating the discussion of results based on high-resolution teleseismic tomography.

The displacement of the Adriatic Plate indenter led to a change in subduction polarity along a transect through the easternmost Alps and to substantial Neogene shortening in the eastern Southern Alps and external Dinarides. While we confirm that slab-pull and rollback of oceanic lithosphere subducted beneath the Carpathians triggered back-arc extension in the Pannonian Basin and much of the concomitant folding and thrusting in the Carpathians, we propose that the rotational displacement of this indenter provided a second important driving force for the severe Neogene modifications of the Alpine-Carpathian-Dinaridic orogenic system.

1. Introduction

1.1 Plate tectonic setting

The Alps, Carpathians and Dinarides form a topographically continuous, yet highly curved orogenic belt, which bifurcates and encircles the Pannonian Basin. They are part of the much larger system of Circum-Mediterranean orogens (Fig. 1). Despite such a continuous topographic expression, this orogenic system is characterised by dramatically diachronous deformation stages along-strike. Its various parts comprise different paleogeographic domains, and major thrusts have opposing polarity (Schmid et al. 2008): in the Western and Eastern Alps as well as in the Carpathians thrusts face the European foreland, whereas in the Southern Alps and the Dinarides thrusts face the Adriatic foreland. In a pioneering article, Laubscher (1971) suggested that the Alps and the Dinarides owe their different structural facing to opposing subduction polarities. Deep reflection seismic profiling and seismic tomography have since shown that the western and central segments of the Alps are underlain by a south-dipping lithospheric slab, attributed to subducted European lower lithosphere (Schmid et al. 1996, 2004a; Schmid & Kissling 2000), whereas the Dinarides and Hellenides are underlain by a northeast-dipping lithospheric

[1] Institute of Geology and Palaeontology, Bernoullistrasse 32, University of Basel, CH-4056 Basel.
[2] Institute of Geology and Palaeontology, University of Innsbruck, A-6020 Innsbruck.
[3] Institute of Geophysics, ETH Hönggerberg, CH-8093 Zürich, Switzerland.
[4] Faculty of Geosciences, Utrecht University, NL-3508 Utrecht, Netherlands.
*Corresponding author, now at: Department of Geosciences, National Taiwan University, No. 1, Sec. 4, Roosevelt Road, Taipei, 10617 Taiwan.
E-mail: kamilu@ntu.edu.tw, Tel: +886-2-2363 6450 #205
**Now at: StatoilHydro ASA, 4035 Stavanger, Norway.

slab (e.g. Wortel & Spakman 2000; Piromallo & Morelli 2003) south of 44° N.

Recent studies suggest that a northeast-dipping "Adriatic" lithospheric slab of about 200 km length also underlies the Eastern Alps in the area east of the Giudicarie and Brenner Fault (Lippitsch et al. 2003; Kissling et al. 2006). Schmid et al. (2004b) and Kissling et al. (2006) tentatively proposed that this slab is a remnant of Palaeogene orogeny in the Dinarides, which impinged onto the Alps during the Neogene by a combination of northward motion of the Adriatic plate indenter and dextral wrench movement along the Periadriatic Fault. This was a speculative assignment, however, since there is no geophysical evidence for a continuous, northeast-dipping Adriatic lithospheric slab between the Eastern Alps and the Dinarides north of 44° N (Piromallo & Morelli 2003; this study).

In this contribution we present in map-view a palinspastic restoration of the substantial displacements and rotations the various tectonic mega-units of the Alpine-Carpathian-Dinaridic orogenic system underwent during the last 20 Ma (e.g. Balla 1987), providing better insight into their early Neogene configuration. We propose that the Mid-Hungarian Fault Zone (Fig. 2) originated from a transform fault, across which the subduction polarity changed from the Alpine to the Dinaridic polarity. Furthermore, our restoration implies c. 200 km of N–S shortening between the Adriatic and European plates, in line with the estimated length of the northeast-dipping slab (Lippitsch et al. 2003). This suggests that this slab represents Adriatic lithosphere that was subducted since the early Neogene. Based on this interpretation, we discuss possible reasons for the current discontinuity of the northeast-dipping slab between the Eastern Alps and the Dinarides.

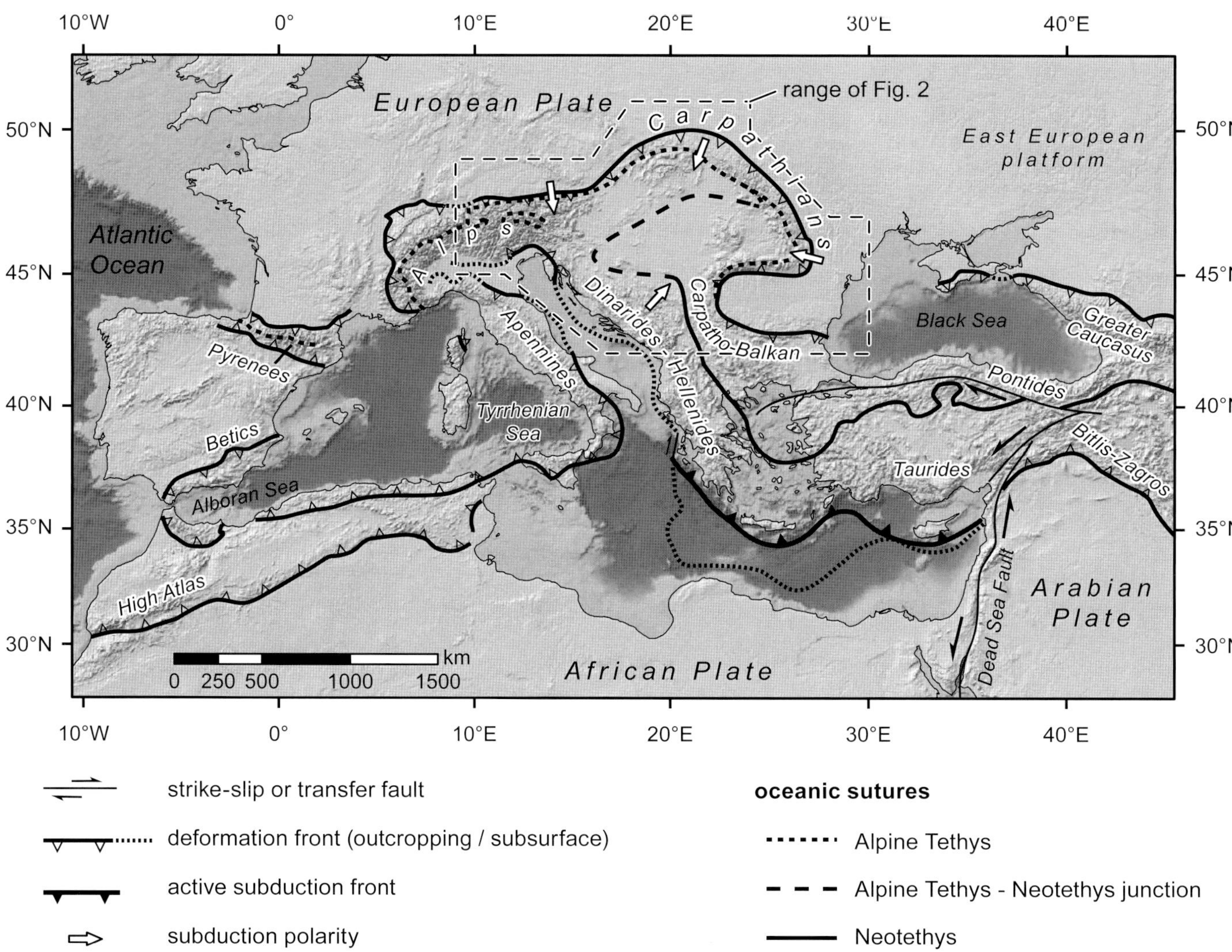

Fig. 1. Overview of the circum-Mediterranean orogenic belt with the positions of present-day deformation fronts and subduction zones (modified after Cavazza et al. 2004). Traces of oceanic sutures in the Eastern Mediterranean are modified after Stampfli & Borel (2004), in the Alps, Carpathians and Dinarides after Schmid et al. (2004b, 2008), in the Pyrenees after Dewey et al. (1973). Topography and bathymetry are from the ETOPO5 dataset (NOAA, 1988). Continental shelf areas are light grey (less than 2000 m water depth); areas of thinned continental or oceanic crust are dark grey (greater than 2000 m water depth). Subduction polarities are only shown for the range of Fig. 2.

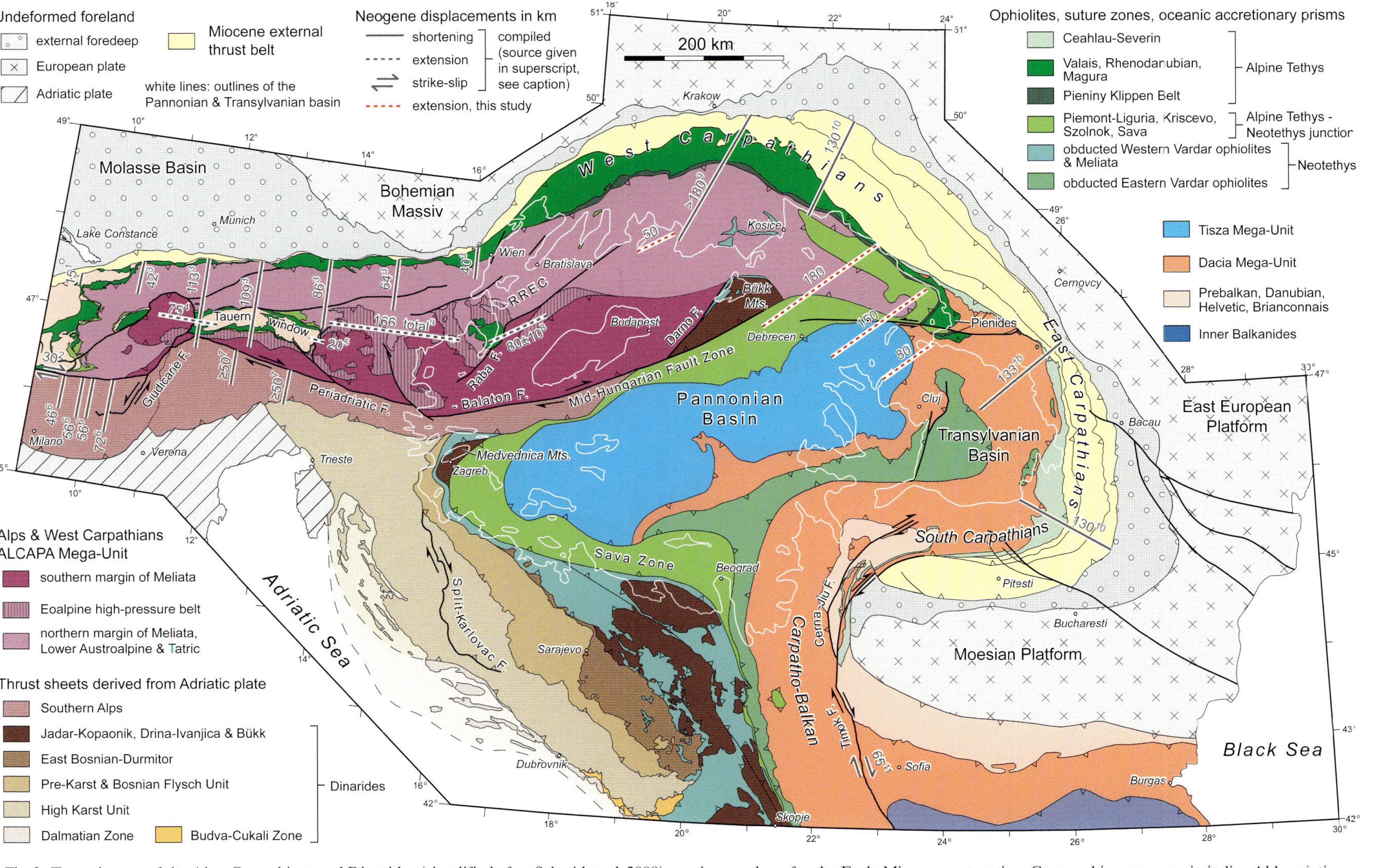

Fig. 2. Tectonic map of the Alps, Carpathians and Dinarides (simplified after Schmid et al. 2008), serving as a base for the Early Miocene restoration. Geographic names are in italics. Abbreviations: F. = fault, Mts. = mountains, RREC = Raba River extensional corridor. The estimates on the magnitude of Neogene (post-20 Ma) displacements (in km) were derived from the following sources given in superscripts: 1) Schmid et al. (1996), 2) Grasemann & Mancktelow (1993), 3) Frisch et al. (1998), 4) Fügenschuh et al. (1997), 5) Genser & Neubauer (1989), 6) Schönborn (1992, 1999), 7) Nussbaum (2000), 8) Tari (1996), 9) Roca et al. (1995), 10) Roure et al. (1993), 11) Fügenschuh & Schmid (2005). See Fig. 4 regarding extension estimates in the Pannonian Basin.

The eye-catching oroclinal architecture of the Alps, Carpathians and Dinarides has since long inspired map-view retro-deformations that improved our understanding of the pre-collisional plate tectonic assemblage (e.g. Channell & Horváth 1976). Burchfiel (1980) demonstrated that many of the continental fragments involved underwent significant shortening during the Mesozoic to Cenozoic convergence of Europe and the Apulia/Adria promontory. However, the substantial extension, which the ALCAPA (Csontos & Vörös 2004; Schmid et al. 2008) and Tisza-Dacia mega-units underwent in the Late Cretaceous (Froitzheim et al. 1997) and particularly in the Neogene during their emplacement into the Carpathian embayment, remained unaccounted for. As a mega-unit we define an assembly of tectonic units, which as a whole has a common paleogeographic origin that is distinct from adjacent tectonic units or other mega-units, from which it is separated by faults. In contrast to a microplate, its original lithospheric underpinnings are either lacking or unknown. Hence, this definition is a somewhat more loose definition of a terrane, which implies that a crustal allochthon was separated from its own lithospheric mantle underpinnings via accretion to a larger plate. Also, the term mega-unit complies with its frequent usage in the regional geological literature (see e.g. Haas 2001; Haas & Pero 2004).

Based on paleomagnetic data, Balla (1987) showed that the ALCAPA and the Tisza-Dacia mega-units north and south of the Mid-Hungarian Fault Zone, respectively, floor the area that is presently occupied by the Pannonian Basin (e.g. Csontos & Nagymarosy 1998; Tischler et al. 2007). Starting in the late Early Miocene, these fragments experienced opposed sense rotations during their emplacement into the Carpathian embayment (see Fig. 7 in Balla 1987). Subsequent palinspastic restorations (e.g. Royden & Baldi 1988; Csontos et al. 1992; Csontos 1995; Fodor et al. 1999; Csontos & Vörös 2004) generally built on Balla (1987). Many of these studies took paleomagnetic constraints on the rotations of tectonic units rather quantitatively into account, whilst the translation and concomitant deformation of these units was only qualitatively treated.

Another set of palinspastic restorations aimed at restoring tectonic units based on the quantification of shortening amounts derived from balanced cross sections or from restoring offsets along major strike-slip faults. Two such examples are Frisch et al. (1998) and Linzer et al. (2002), who provided a palinspastic restoration of the Eastern Alps for Late Oligocene to Miocene times. Roure et al. (1993), Ellouz & Roca (1994), Roca et al. (1995), Morley (1996) and Behrmann et al. (2000) provided estimates of amounts of shortening in the Carpathian thrust belt based on balanced cross-sections. Except for Morley (1996), however, most previous studies did not consider the system of the Alps-Dinarides-Carpathians and the Pannonian Basin in its entirety and hence failed to link its evolution to the role played by the rigid Adriatic indenter during Miocene to recent times.

We aim at a semi-quantitative palinspastic restoration of the Early Miocene tectonic configuration, based on a recently compiled new tectonic map (Fig. 2) of the Alpine-Carpathian-Dinaridic orogenic system (Schmid et al. 2008). We chose the 20 Ma timeframe for our restoration, as this time roughly corresponds to the onset of numerous interrelated tectonic events (Fig. 3), which are briefly reviewed in the following.

1) The E-directed lateral extrusion (escape and concomitant extension) of the Austroalpine units into the Pannonian Basin (Ratschbacher et al. 1991) lasted from about 23 to 13 Ma according to Frisch et al. (1998). A younger (20 Ma) age for the time of onset of this event was inferred from mineral cooling and fission track ages obtained across low angle normal faults that exhumed Lower Austroalpine and Penninic series during this extrusion (e.g. the age of the Brenner normal fault; von Blanckenburg et al. 1989; see also Fügenschuh et al. 1997 and references therein). This is roughly in line with the 18–15 Ma age range of Ottnangian to Badenian sediments that accumulated in intra-montane pull-apart basins associated with conjugate strike-slip faults, which evolved during the eastward escape of the Eastern Alps (Strauss et al. 2001; Linzer et al. 2002 and references therein).

2) Shortening in the Southern Alps started between 20 and 17 Ma, i.e. contemporaneously with the sinistral offset of the Periadriatic Fault by the Giudicarie Fault system (Schönborn 1992, 1999; Schmid et al. 1996; Stipp et al. 2004) and lasted until the Messinian. The major shortening in the external Southern Alps occurred during Serravallian-Tortonian times (14–7 Ma). By contrast, thrusting of the Austroalpine nappes onto the North Alpine Foreland Molasse Basin had practically ended at around 17 Ma (Linzer et al. 2002 and references therein).

3) Back-arc-extension and related sedimentation in the rift- and wrench-related troughs of the Pannonian Basin complex started in Eggenburgian to Ottnangian times (Bérczi et al. 1988; Nagymarosy & Müller 1988; Horváth & Tari 1999), i.e. between 20.5 and 17.5 Ma in the time scale of Steininger & Wessely (1999).

4) Neogene thrusting in the external flysch belt of the West Carpathians (Royden 1988) reached the Dukla and Silesian units of the West Carpathians by Early Burdigalian (Eggenburgian) times (Kováč et al. 1998; Oszczypko 2006), i.e. at about 20 Ma. In the East Carpathians, thrusting also initiated at around 20 Ma in the most internal units of the Moldavides (Convolute flysch and Audia-Macla nappes) but reached more external flysch units only by Late Burdigalian times, i.e. at around 17 Ma (Săndulescu 1988; Matenco & Bertotti 2000). Some authors invoke along-strike younging of orogenic activity (e.g. Jiříček 1979) that supposedly correlates with a general decrease of the ages of calc-alkaline magmatism in the internal Carpathians and the Pannonian

Basin from west towards southeast (e.g. Seghedi et al. 1998; Harangi et al. 2006). In any case, thrusting continued longer in the East Carpathians (Matenco & Bertotti 2000). The formation of this most external thrust belt made up of flysch sediments was associated with subduction of a retreating lithospheric slab underneath an upper plate formed by the ALCAPA and Tisza-Dacia mega-units, including the previously accreted Magura flysch, that started at around 20 Ma. Note that earlier thrusting affected more internal units only; we discuss here the onset of Neogene thrusting that started to propagate into the external flysch belt and which is widely accepted as being contemporaneous with back-arc extension in the Pannonian Basin (e.g. Royden 1988; Kováč et al. 1998).

5) In the Early Burdigalian (20.5–18.5 Ma) SE-directed thrusting also initiated in more internal units of the Carpathians and led to the formation of the Pienide nappes (northern Romania) due to the collision of the ALCAPA and the Tisza-Dacia mega-units (Săndulescu 1988; Tischler et al. 2007). This thrusting occurred during the initial stages of emplacement of the ALCAPA and Tisza-Dacia mega-units into the Carpathian embayment and was contemporaneous with their soft collision with the European foreland across the external flysch belt (Márton et al. 2007).

2.1 Premises for Early Miocene palinspastic restoration

The Early Miocene restoration was performed with respect to a fixed European foreland under the following premises:

1) Changes to the geometric configuration of tectonic units were applied only to those that underwent significant extension or bending around vertical axes during the timeframe under consideration (20–0 Ma), namely the ALCAPA and Tisza-Dacia mega-units.

2) In areas of shortening the outlines of tectonic units, which are separated by major thrusts were left unchanged. Consequently, gaps opening between the restored and actual position of such individual units permit to assess the amount of shortening those areas underwent during the last 20 Ma.

3) The foreland was considered as non-deformed, thus retaining a fixed geometric configuration. The undeformed European foreland (including the external foredeep) served as a fixed reference frame, the position of which remained unchanged. Based on geological arguments it was assumed that the Adriatic Plate behaved as a rigid block that was subjected to translation and rotation only.

4) Magnitudes of shortening, extension and strike-slip displacements were compiled from published sources (Figs. 2 and 4). These data underlay our basic palinspastic restoration of the area of interest. Such data are strictly distinguished from deformations that had to be applied for pure 2-D compatibility reasons in areas that are not or ill-constrained by data. To a large extent the latter are a result of our restoration.

5) Rotations were applied to tectonic units only as required from the following map-view geometric and kinematic arguments: (i) along-strike shortening gradients, (ii) restoration of displacements along curved strike-slip faults and (iii) geometric adjustments to the regional strike. We emphasise that paleomagnetically constrained rotations were not considered as input to our restoration, except for a few small blocks. This was done deliberately in order to test whether a purely kinematic approach yields results that are qualitatively compatible with the paleomagnetic data (e.g. Márton et al. 2003, 2007).

2.2 Database

Our palinspastic restoration of the Alpine-Carpathian-Dinarides domain (Figs. 2 and 6) integrated the results of a number of earlier published studies to which we refer below. For the Alpine part of the ALCAPA Mega-Unit we used the restoration by Frisch et al. (1998), which we combined with data from Burkhard & Sommaruga (1998), Fügenschuh et al. (1997), Genser & Neubauer (1989), Grasemann & Mancktelow (1993), Gratier et al. (1989), Lickorish & Ford (1998), Philippe et al. (1996), Ratschbacher et al. (1991), Schmid et al. (1996), Schönborn (1992, 1999), Nussbaum (2000) and Stipp et al. (2004). Estimates on the magnitude and direction of extension in the eastern (Pannonian Basin) part of ALCAPA were mainly adopted from Tari (1996), and also from Fodor et al. (1998) and Szafián et al. (1999). In addition, we used the crustal thinning factor map for the Pannonian Basin by Lenkey (1999) to estimate the magnitude of extension affecting the ALCAPA and Tisza parts of the Pannonian Basin. For the West Carpathians, we took the restorations by Roure et al. (1993), Roca et al. (1995), Morley (1996) and Behrmann et al. (2000) into consideration for estimating the total shortening that was achieved in the Miocene external thrust belt. These data permitted us to constrain the position of the most external edge of ALCAPA. Furthermore, we used the restorations by Tischler (2005) and Márton et al. (2007) for restoring the boundary area between the ALCAPA and Tisza-Dacia mega-units.

Concerning the Tisza-Dacia Mega-Unit we incorporated a restoration of the South Carpathians bend zone that was proposed by Fügenschuh & Schmid (2005) and that is essentially based on data by Balla (1987), Berza & Draganescu (1988), Berza & Iancu (1994), Kounov et al. (2004), Kräutner & Krstić (2006), Matenco et al. (1997, 2003), Moser (2001), Patrascu et al. 1994, Schmid et al. (1998) and Săndulescu et al. (1978). Fügenschuh & Schmid (2005) retro-deformed the tectonic unroofing of the Danubian window as well as strike-slip movements along the curved Cerna-Jiu and Timok faults, which delimit the "mobile" Tisza-Dacia Mega-Unit against the stable Moesian promontory. However, this part of our restoration does not fully constrain the Early Miocene location of the frontal tip of the Tisza-Dacia Mega-Unit, since the westernmost part of this unit underwent contemporaneous stretching during its displacement into the Carpathian embayment. Therefore we used

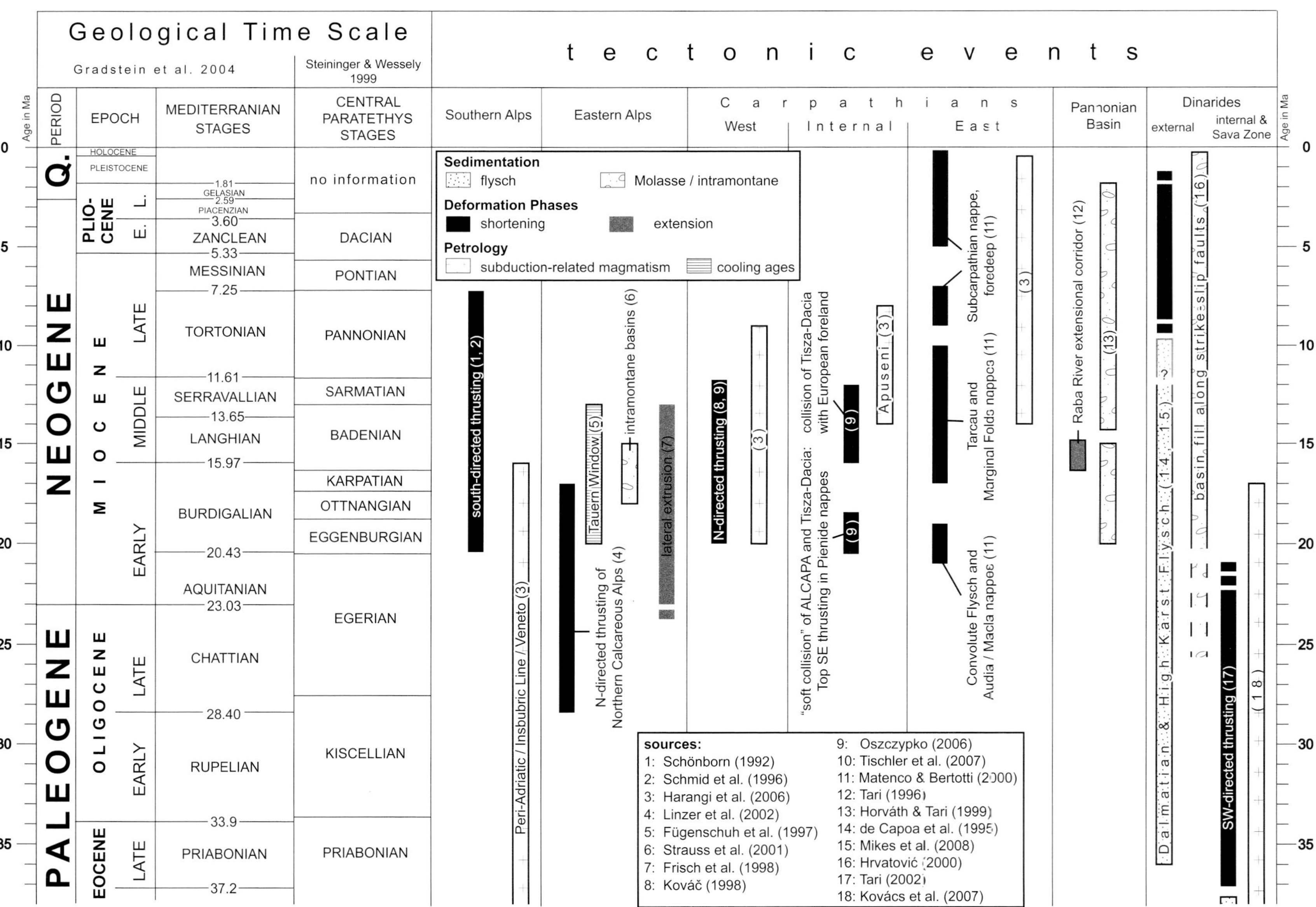

Fig. 3. Timetable illustrating major Neogene tectonic events in the Alpine-Carpathian-Pannonian-Dinaridic domain.

the restoration by Morley (1996) in order to anchor the frontal tip of the extending Tisza-Dacia Mega-Unit.

Note that no pre-existing reconstructions or published balanced cross sections could be used as input for the restoration of the Dinarides. The position of the different Dinaridic units was indirectly derived, based on the restoration of the other parts of the orogenic system (see below).

2.3 Estimating the magnitude of extension in the Pannonian Basin

In order to restore the tectonic units underlying the Pannonian Basin, estimates of the magnitude and direction of Neogene extension are needed. To this end we adopted the contour map of the crustal thinning factor by Lenkey (1999), which covers the entire Pannonian Basin. This map was calculated by taking into account the present-day depth to the pre-Neogene basement and the heat flow, assuming an initial crustal thickness of 35 km. Lenkey (1999) derived the depth to the pre-Neogene basement from the isopach map of the Neogene sediments in the Pannonian Basin that was published by Horváth and Royden (1981). We also implemented the fault pattern in the Pannonian Basin as given by Horváth (1993) and Horváth et al. (2006).

For our reconstruction we chose five transects that cover the entire Pannonian Basin (A–A' to E–E', Fig. 4). According to Csontos & Vörös (2004), transects A–A' and B–B' are underlain by the ALCAPA Mega-Unit and trend parallel to the dominant extension direction (ca. N060° E), in which the Rechnitz Window core-complex (termed "Raba River extensional corridor" by Tari 1996, "RREC" in Fig. 2) opened, as deduced from ductile transport lineations observed in outcrops from the lower plate of the core complex. Transect A–A' exactly coincides with the cross section of Tari (1996), along which he estimated the amount of extension across the Raba River extensional corridor to amount to 80 ± 10 km by correlating footwall (in outcrop) and hangingwall cut-offs (in subsurface, see Figs. 4b and 7 in Tari 1996). Transect B–B' was chosen parallel to A–A', but is offset to the north by some 15 km. It covers the largest Neogene depocenter of the Little Hungarian Plain. Thus the combined transects A–A' and B–B' traverse the entire northwestern part of the Pannonian Basin along the known extension direction.

Transects C–C' to E–E', each of them traversing the entire Pannonian Basin, are underlain by the Tisza-Dacia Mega-Unit (e.g. Csontos & Vörös 2004). Also these transects were constructed parallel to the dominant extension direction during the Karpatian-Badenian synrift phase, as derived from basin-wide kinematic analyses of fault-slip data (ca. N050° E; Fodor et al. 1999).

While along transect A–A' the magnitude of extension was adopted from Tari (1996), we need to discuss below how we estimated extension along transects B–B' to E–E'. These four transects traverse numerous normal faults. We subdivided these transects at their intersection points with normal faults into several segments i, which are characterised by variable crustal thinning factors. For each transect segment we estimated the magnitude of extension as follows:

$$l_0 = \frac{l_1}{\delta} \tag{1}$$

where l_0 is the initial length of a segment prior to extension, l_1 its present-day length after extension and δ the crustal thinning factor as given by Lenkey (1999). Once l_0 was obtained for each transect segment, the magnitude of extension dl (length change) along each segment could be estimated:

$$dl = l_1 - l_0 \tag{2}$$

Along a given transect, the total amount of extension dl_{tot} was calculated by summing up the incremental length changes dl along its constituent segments i:

$$dl_{tot} = \Sigma \, dl_i \tag{3}$$

This approach yielded highly variable magnitude of extension for the four transects under consideration (combined transects A–A', B–B' and transects C–C', D–D' and E–E'), ranging from about 78 km for transect E–E' to 180 km for transect C–C' (Fig. 4). Results of this analysis suggest that the magnitude of extension is largest in the central parts of the Pannonian Basin that are closely associated with the Mid-Hungarian Fault Zone. In the following, we will implement these estimates of the amount and direction of extension in our Early Miocene restoration of the geometry and position of the ALCAPA and Tisza-Dacia mega-units.

2.4 Restoring the ALCAPA Mega-Unit

Combining the extension values of transects A–A' and B–B' (Fig. 4) yields a net extension of 130 ± 10 km for the northwestern part of the Pannonian Basin, which is floored by the eastern parts of the ALCAPA Mega-Unit. Combined with 160 km of extension in the Eastern Alps (Frisch et al. 1998), this totals to 290 ± 10 km of overall extension that affected ALCAPA during the Neogene. This value was used to restore the Early Miocene outlines of ALCAPA (Fig. 5a). The Raba River extensional corridor, bounded to the west by the Raba River Fault, is mainly responsible for the Neogene exhumation of the Lower Austroalpine, Tatric and Valais units below higher tectonic units (Fig. 2; for more details see Plate 1 in Schmid et al. 2008). This extensional corridor is located exactly at the longitude, where the E–W-strike of the Eastern Alps turns into the NE-strike of the West Carpathians. The available time constraints (Fig. 3) show that shortening in the Miocene thrust belt of the West Carpathians started earlier and lasted longer (Eggenburgian to Sarmatian) than core complex formation in the Raba River extensional corridor (Karpatian to Early Badenian). Kinematically this implies that the eastern part of the ALCAPA Mega-Unit underwent a counter-clockwise rotation prior, during and after its extensional deformation.

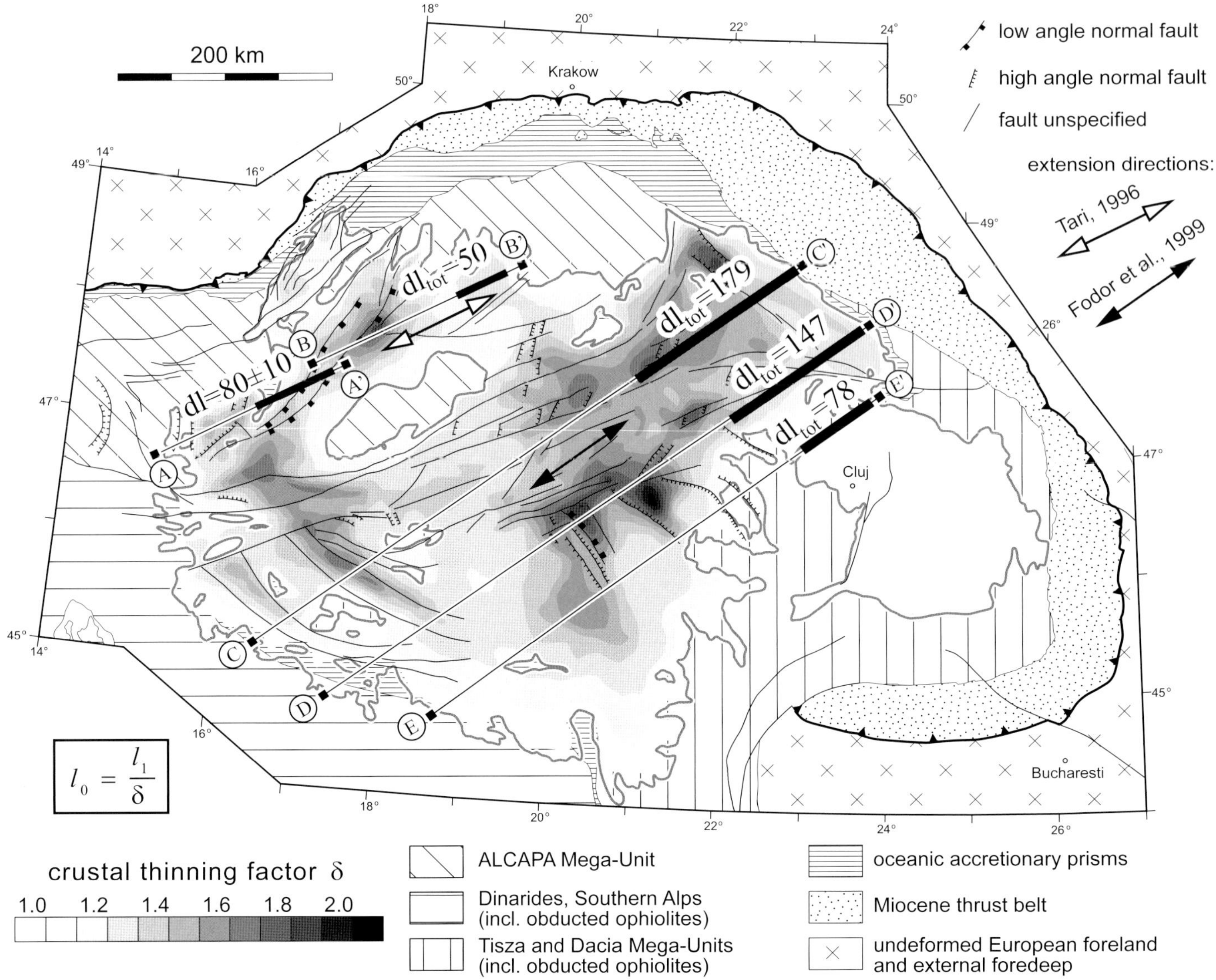

Fig. 4. Estimates on the extension amount in the Pannonian Basin. Fault pattern in the Pannonian Basin was adopted from Horváth (1993). The contour map of the crustal stretching factor is taken from Lenkey (1999). All thrusts except the most external one are omitted for reasons of legibility.

In order to restore geometry and location of the eastern parts of ALCAPA prior to rotation, a 23° clockwise rotation around a pivot point located at the SW tip of the Raba River Fault was applied to ALCAPA. Such a rotation leads to a linear striking Rhenodanubian – Magura Flysch Belt and results in a position of the northern edge of ALCAPA that is in agreement with previous studies (Morley 1996; Behrmann et al. 2000; Fig. 5a). Moreover, this rotation also straightens out the Periadriatic Fault and its extension into Hungary (Fodor et al. 1998). We consider the Darno Fault, which separates ALCAPA from the Bükk Mountains, to represent part of the Periadriatic Fault (Schmid et al. 2008). We applied an additional 30° clockwise rotation (i.e. 23 + 30 = 53°) to the Bükk Mountains in order to align the Darno Fault along-strike with the main trend of the Periadriatic Fault.

In a next step, the eastern margin of ALCAPA was translated westward along a roughly E–W-trending trajectory by c. 280–300 km, corresponding to the extension value arrived at by the above described estimates. The available time constraints suggest that at 20 Ma the easternmost tip of ALCAPA was just getting in contact with the Tisza-Dacia Mega-Unit, as evidenced by Burdigalian SE-directed thrusting in the Pienides (Tischler et al. 2007; Márton et al. 2007).

2.5 Restoring the Tisza-Dacia Mega-Unit

For those parts of the Pannonian Basin, which are floored by the Tisza-Dacia Mega-Unit, extension values appear to decrease systematically SE-ward away from the Mid-Hungarian Fault Zone towards the un-extended Transylvanian

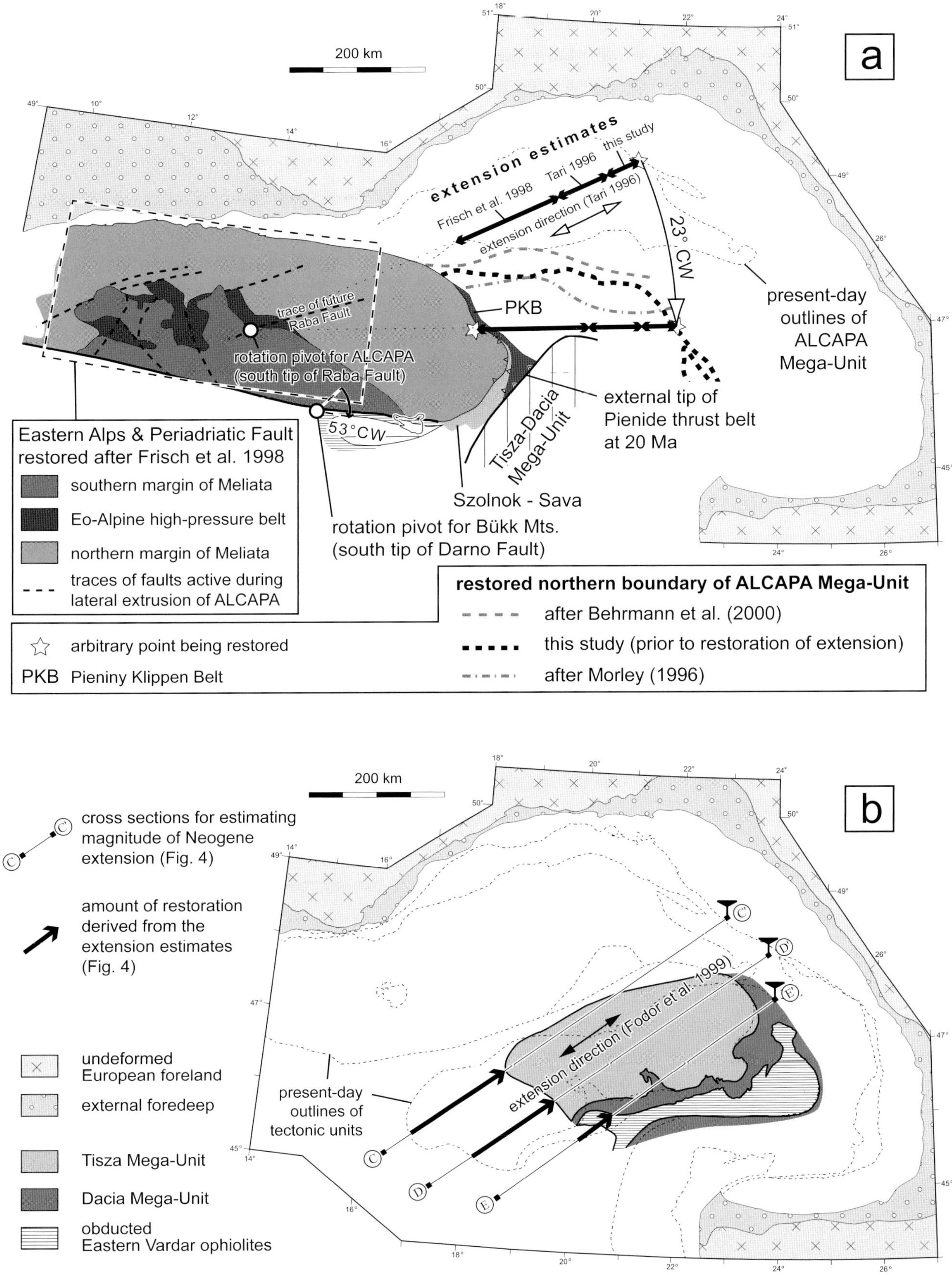

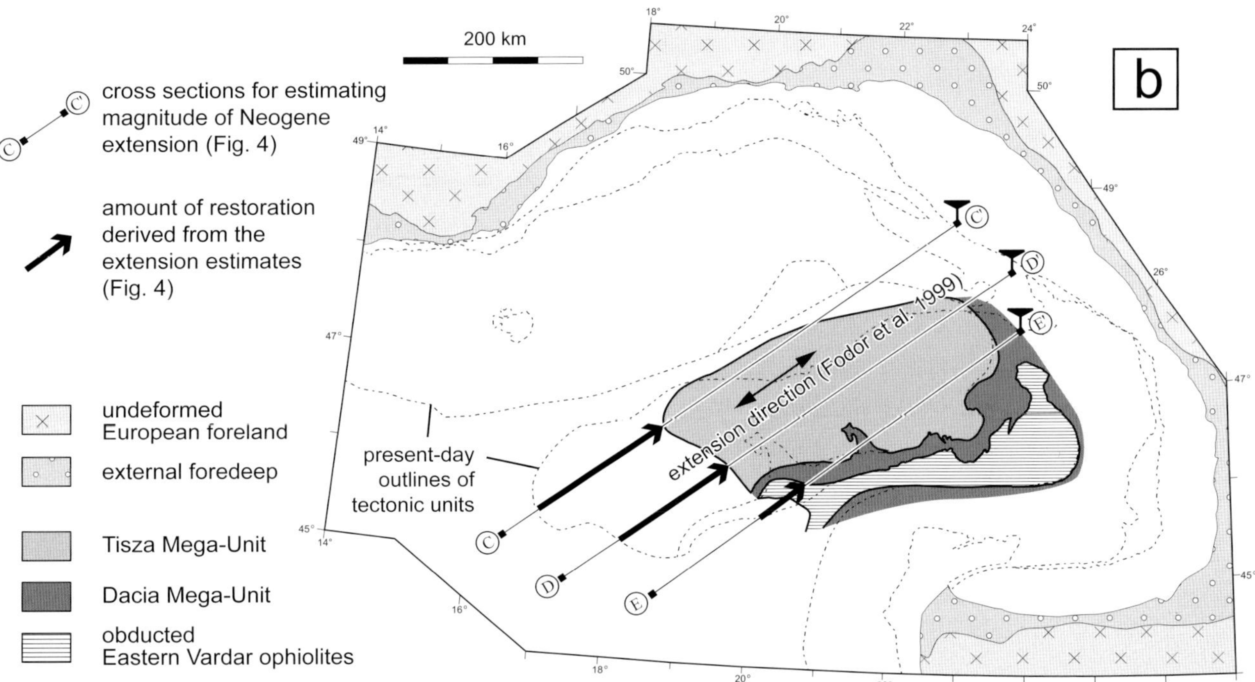

Fig. 5. Reconstruction of (a) the ALCAPA and (b) Tisza-Dacia mega-units according to the constraints described in the text.

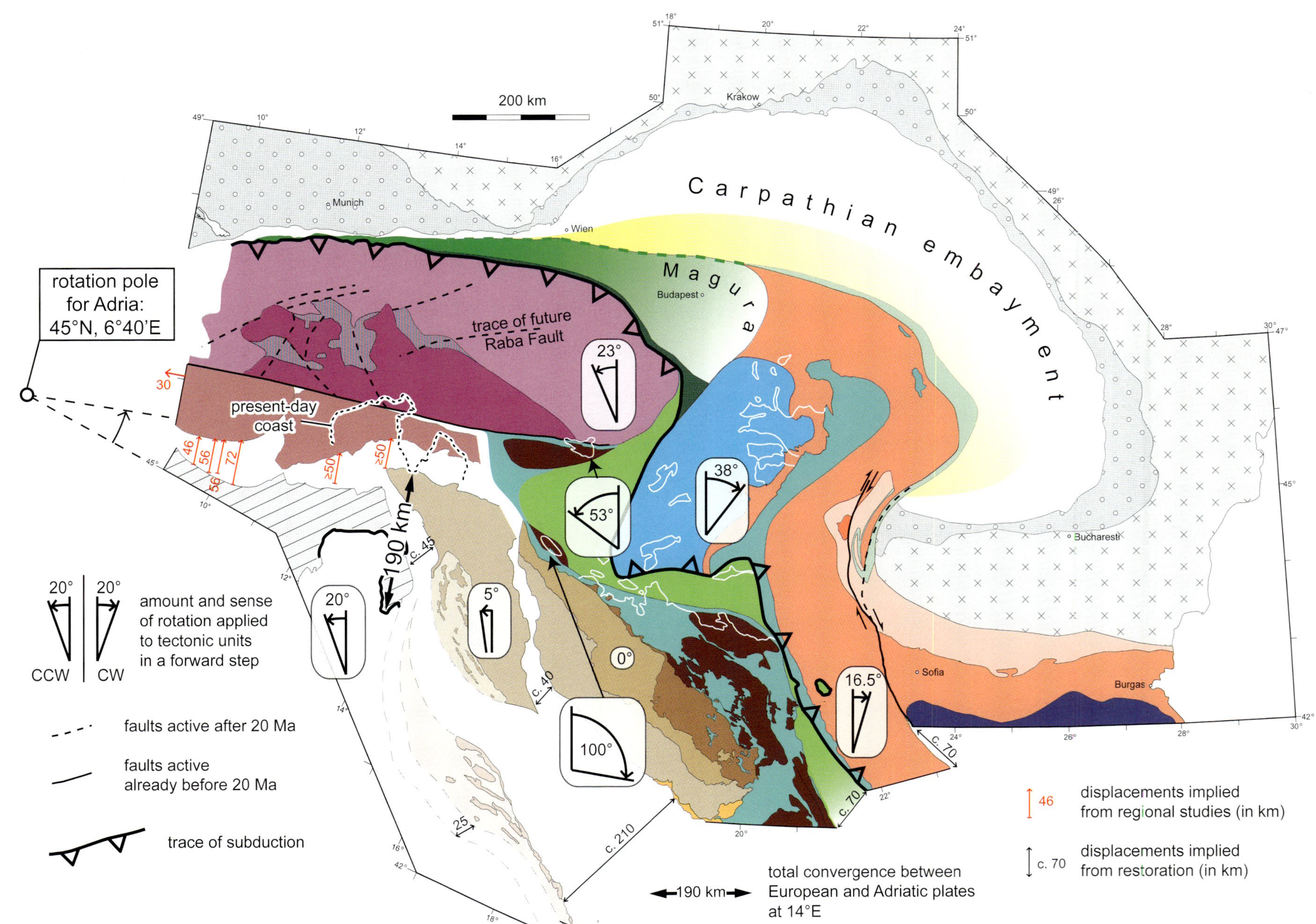

Fig. 6. Restoration of tectonic units in the Alps, Carpathians and Dinarides domain for the Early Miocene. Colours and patterns of tectonic units corresponds to those in Fig. 2.

Basin (Krézsek & Bally 2006) and the Moesian Foreland (Fig. 4). This is also compatible with the fact that the Neogene fault-bounded basins that affect the northwestern part of the Apuseni Mountains shallow out towards southeast (Săndulescu et al. 1978); this directly indicates decreasing amounts of extension towards SE. The restored configuration of Tisza-Dacia was arrived at by shortening the entire mega-unit along transects C–C' to E–E' by the corresponding extension values (Fig. 5b).

For reconstructing the position of the external, leading edge of Tisza-Dacia (external with respect to the Carpathian thrust facing) during the Early Miocene, we adopted the restoration of Fügenschuh & Schmid (2005). This restoration is mainly constrained by retro-deforming 65 km of dextral strike-slip displacement of the Tisza-Dacia Mega-Unit relative to the Moesian foreland along the curved Timok Fault (see Figure 9c in Fügenschuh & Schmid 2005; Moser 2001).

2.6 Restoring the location of the Adriatic Plate

The Early Miocene position of the Adriatic Plate was reconstructed by retro-deforming 1) the Neogene shortening recorded in the Southern Alps, 2) the Neogene dextral strike slip displacements along the Periadriatic and Giudicarie Faults, and 3) the post-20 Ma shortening across the Western Alps.

In the Southern Alps west of the Giudicarie Fault the magnitude of Neogene shortening systematically decreases from east to west (Schönborn 1992, 1999) and approaches zero along the ECORS-CROP profile near Torino (Schmid & Kissling 2000, outside of Fig. 2). The available estimates on the magnitude of Neogene shortening in the Southern Alps east of the Giudicarie Fault are minimum estimates only (Nussbaum 2000) and hence it is unclear whether the systematic increase of shortening continues further towards east. We consider the Giudicarie Fault kinematically and temporally related to post-20 Ma shortening within the Southern Alps (Stipp et al. 2004) and therefore eliminated the later offset of the Periadriatic Fault (Schmid et al. 1999) by the Giudicarie Fault. The restoration of the Southern Alps east of the Giudicarie Fault is hence constrained by assuming an initially straight Periadriatic Fault and satisfying the minimum shortening estimates of Nussbaum (2000). Note, however, that there is no general agreement about an initially straight Periadriatic Fault (see e.g. Viola et al. 2001 for a contrasting opinion). Our palinspastic restoration (Fig. 6) reveals that the Periadriatic Fault aligns with the Balaton Fault and its eastern extension along the northern margin of the Mid-Hungarian Fault Zone (Fig. 2), a geometry that allows for the eastward lateral extrusion of the ALCAPA Mega-Unit.

The magnitude of Neogene dextral strike slip movements along the Periadriatic Fault and its eastern extension varies along strike. This is an effect of the E–W extension of the ALCAPA Mega-Unit during its lateral extrusion and displacement with respect to the Southern Alps, which were not affected by this E–W extension (Ratschbacher et al. 1991). Hence, the

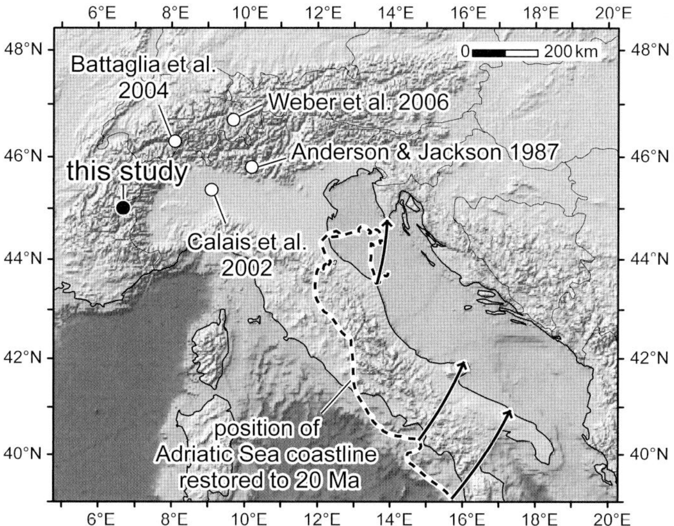

Fig. 7. Comparison of the rotation pole of the Adriatic plate indenter derived in this study (black) with previously derived poles (from the inversion of seismotectonic and/or geodetic data). The assumption of a largely rigid, rotating Adriatic plate indenter implies an increase of displacement away from the rotation pole (shown by arrows that are concentric around the rotation pole). The coastline of the Italian part of the Adriatic Sea (lying on the stable Adriatic plate) would thus be restored back to a position shown by the dashed line.

Periadriatic-Balaton Fault system represents a stretching fault in the sense of Means (1989).

In restoring the Adriatic Plate to its Early Miocene position, we took into account an extra 30 km of dextral displacement that is not related to extrusion of the eastern part of the Alps as a constituent of ALCAPA. This extra displacement is the result of E–W extension across the Lepontine dome in the Central Alps, which is largely accommodated along the Simplon normal fault (just outside the western margin of Fig. 2; Grasemann & Mancktelow 1993).

Post-20 Ma shortening across the Western Alps occurred beyond the area covered by Figs. 2 and 6, but was taken into account since it had an effect on the relative movements between the Adriatic Plate and the ALCAPA Mega-Unit. Neogene shortening across the Western Alps is relatively small, however, and varies from about 35 km near Geneva (e.g. Gratier et al. 1989) to less than 10 km further to the south in the area of the Argentera Massiv (e.g. Lickorish & Ford 1998).

The three categories of interrelated constraints on the Neogene motion of the Adriatic Plate mentioned above jointly indicate that it underwent a combination of translation and simultaneous rotation. Once the north-directed displacement of the Adriatic Plate related to shortening in the Southern Alps has been restored to the pre-20 Ma situation, a 20° counterclockwise rotation had to be applied to the Adriatic Plate that is described by a pivot pole located in the vicinity of Torino (at 45° N and 6°40'E, Figs. 6 and 7). We emphasise that this rotation is a product of the above-mentioned constraints obtained from outside the Adriatic Plate. Independent estimates however, based either on paleomagnetic data (see below), the inversion of seismotectonic (Anderson & Jackson 1987) or GPS data (Calais et al. 2001; Battaglia et al. 2004; Weber et al. 2006) also consistently indicate a counter-clockwise rotation of the Adriatic Plate (Fig. 7).

2.7 Restoring the tectonic units of the Dinarides

The restoration of the major units of the Dinarides is not based on quantitative estimates of post-20 Ma shortening but is indirectly constrained by the retro-deformations applied to the Adriatic Plate and the Tisza-Dacia Mega-Unit. Within the Dinaridic-Southern Alpine domain we allowed for displacements along the following faults that were active during the past 20 Ma: (1) South-directed thrusting in the eastward continuation of the Miocene Alpine retro-wedge (Southern Alps and their continuation into Slovenia and Hungary; Schmid et al. 2008 and references therein), (2) ongoing shortening along the frontal thrust of the Dinarides offshore Dalmatia (e.g. Bennett et al. 2008) and in line with evidence for thrusts affecting Pliocene to Quaternary sediments in the Albanian foredeep, offshore Montenegro and Albania (Picha 2002), (3) transpression along the Split-Karlovac Fault (Chorowicz 1970, 1975 and Geological Map of former Yugoslavia) and (4) thrusting along the frontal thrust of the High-Karst Unit and Budva-Cukali Zone, respectively, which override the Dalmatian Zone and its southward continuation, the Kruja Zone of Albania (Aubouin & Ndojaj 1964) and the Gavrovo-Tripolitza Zone of Greece (Jacobshagen 1986). This thrusting affects also Mid-Miocene sediments according to new data (de Capoa et al. 1995; Mikes et al. 2008). The partitioning of displacements across the faults chosen in Fig. 6 is not constrained by hard data and will be discussed below.

The internal edge of the Tisza-Dacia Mega-Unit (internal with respect to the external one facing the Carpathian thrusts) is located adjacent to the Sava Zone, which is considered as forming the suture zone between Dinarides and Tisza-Dacia Mega-Unit (Schmid et al. 2008; Ustaszewski et al. submitted). Final closure of the Neotethyan oceanic basins along the Sava Zone had occurred in Maastrichtian to Early Paleogene times (Pamić 1993, 2002). This is constrained by the observation that Early Miocene syn-rift sediments of the Pannonian Basin system seal Mid-Eocene siliciclastics, which are affected by the last stages of post-collisional thrusting (Tari 2002; Ustaszewski et al. submitted). Consequently, the internal Dinarides are assumed to have remained attached to the Tisza-Dacia Mega-Unit during the last 20 Ma. Hence the position of the internal Dinarides 20 Ma is essentially controlled by the restoration of the Tisza-Dacia Mega-Unit. Nevertheless, some minor geometrical adjustments were required across the Sava Zone (Fig. 6). These adjustments, which are relatively minor and uncontrolled by data, are thought to relate to extensional and subsequent Pliocene to recent inversion tectonics along the southern margin of the Pannonian Basin in Croatia and Serbia (Tomljenović & Csontos 2001; Saftić et al. 2003). Dextral strike slip movements along the Sava Zone in Southern Serbia and Macedonia are a consequence of the restoration (see below).

3. Discussion

3.1 Implications for the Southern Alps – Dinarides realm

The Early Miocene restoration presented in Fig. 6 implies that the Neogene rotational northward motion of the Adriatic plate with respect to the stable foreland of the European plate resulted in some 190 km total N–S shortening in the Trieste – Eastern Alps area, at a longitude of 13° E to 15° E. This shortening is partitioned between south-directed thrusting in the Southern Alps, shortening within the ALCAPA Mega-Unit and very minor N-directed thrusting onto the northern Alpine foreland. Due to the counter-clockwise rotation of the Adriatic plate shortening in the Southern Alps decreases westward towards the rotation pivot point (Fig. 6). This decrease in shortening is compatible with estimates of Miocene N–S shortening across the Alps of Eastern Switzerland amounting to 61 km during the last 19 Ma (Schmid et al. 1996).

Conversely, the counter-clockwise rotation of the Adriatic plate indenter implies that with increasing distance E-ward away from the rotation pivot point the magnitude of shortening progressively increases (Fig. 7). In keeping with this relationship, small changes in the assumed rotation angle lead to drastic variations in the magnitude of shortening in the SE-most parts of the Dinarides, for which the deduced total shortening of 235 km is subject to large uncertainties. We interpret this shortening (25 + 210 km in Fig. 6) to have been predominantly taken up in the external Dinarides (offshore Dalmatia, along the Split-Karlovac Fault, along the frontal thrust of the High-Karst Unit and, further south, in front of and within the Budva-Cukali-Pindos Zone). It is very difficult to assess how this total post-20 Ma shortening in the external Dinarides is partitioned between the thrusts at the front and at the rear of the Dalmatian Zone (and its southward continuation, the Gavrovo-Tripolitza Zone). In Fig. 6 we gave preference to large amounts of shortening at the rear of the Dalmatian Zone since the amount of shortening in front of the Kruja Zone, i.e. offshore Albania appears to be relatively small (Picha 2002) while by far more substantial shortening occurred across the more internal Budva-Cukali or Krasta-Pindos Zone during the Neogene (Kilias et al. 2001). Note that Neogene to recent shortening in the Hellenides, however, occurred in a more external domain, in front of the Gavrovo-Tripolitza Zone (van Hinsbergen et al. 2005).

The deduced differential northward displacement of the Adriatic plate implies that the thrusts of the Dinarides also accommodated dextral strike-slip displacements. Substantial strike-slip faulting apparently overprinted the thrust faults of the internal Dinarides and the Sava-Vardar suture of the Sava Zone (see also Morley 1996). Dextral transpression is directly evidenced by the presently observed steep dip of these thrusts (see profile 5 of Plate 2 in Schmid et al. 2008).

3.2 Extension in the Pannonian Basin and contemporaneous thrusting in the external Carpathians

Our estimate of some 290 km SW–NE extension in the ALCAPA Mega-Unit roughly corresponds to the minimum estimate of some 260 km shortening that was accommodated in the NE Carpathian thrust belt (Behrmann et al. 2000). This illustrates that Miocene thrusting and extension in these two domains are not only coeval, as proposed by Royden et al. (1983), but also of similar magnitude as shown by Behrmann et al. (2000). This supports the concept of a retreating Carpathian subduction zone, causing coeval back-arc extension. Thereby the rate of subduction is mostly or exclusively taken up by extension while the plate convergence rate, as defined by Royden & Burchfiel (1989), approaches zero.

Comparing Figures 2 and 6 reveals that the invasion of the ALCAPA and Tisza-Dacia mega-units into the Carpathian embayment and ultimately their docking to the European foreland across the Miocene thrust belt was largely accommodated by their contemporaneous extension. Most of the significant and complex motions between these two extending mega-units took place across the Mid-Hungarian Fault Zone (Csontos & Nagymarosy 1998). These motions were variably of a transpressional, strike slip and/or transtensional nature and were accompanied by block rotations that will be discussed in the light of paleomagnetic data below (see Tischler 2005; Tischler et al. 2007; Márton et al. 2007 for an analysis of the NE tip of the Mid-Hungarian Fault Zone).

Our restoration brings the Magura flysch belt and the Sava Zone, both forming part of the Cenozoic Alpine-Dinaridic suture zone, closely together. It also leads to a good alignment of the internal Dinarides and their two fragments, which are preserved in the Bükk and Medvednica Mountains (Tomljenović et al. 2000; Tomljenović 2000, 2002; Dimitrijević et al. 2003; Schmid et al. 2008; Tomljenović et al. 2008). The opposed subduction polarity between Alps-Western Carpathians and Dinarides demands, however, substantial post-collisional geometrical modifications of this suture (Laubscher 1971). Our model, which in many respects is similar to that of Royden & Baldi (1988), proposes that the precursor of the present-day Mid-Hungarian Fault Zone acted as a transform fault between the opposed Alpine and Dinaridic subduction zones in Paleogene times. Since the wide area of the Carpathian embayment, depicted in Fig. 6, must have largely disappeared as a result of subduction zone retreat of the European Plate rather than by plate convergence, we speculate that this area was at least partly underlain by old, i.e. dense oceanic lithosphere rather than by thick and buoyant continental lithosphere.

3.3 Inferences from block rotations and comparison with paleomagnetic data

Our restoration predicts counter-clockwise and clockwise rotations of the ALCAPA and Tisza-Dacia mega-units, respec-

tively, during their advance into the Carpathian embayment. This is in qualitative, albeit not quantitative, agreement with the results of paleomagnetic studies on these two mega-units (e.g. Patrascu et al. 1994; Márton & Fodor 1995, 2003; Márton & Márton 1996; Panaiotu 1998, 1999; Márton 2000). In general our restoration predicts smaller rotations than those inferred from paleomagnetic data. This can partly be explained by the fact that the two extending mega-units do not represent rigid blocks, and that smaller blocks forming perhaps part of larger scale fault zones may be subjected to more intense rotations. On the other hand, the timing of magnetization, and hence the timing of these rotations cannot always be determined accurately. In the case of the Tisza-Dacia Mega-Unit Fügenschuh & Schmid (2005) argued, based on purely geological-tectonic arguments, that much of a total of 90° clockwise rotation, generally reported to be entirely of Miocene age (e.g. Patrascu et al. 1994), pre-dates the Miocene.

In the case of the counter-clockwise rotation of the ALCAPA Mega-Unit we interpret this rotation to result from the combined effect of slab retreat and its collision with the simultaneously advancing Tisza-Dacia Mega-Unit (e.g. Tischler et al. 2007). However, we regard the clockwise rotation of the Tisza-Dacia Mega-Unit to be driven by slab retreat only; it was accommodated by dextral displacements along the Timok and Cerna-Jiu Faults (Fügenschuh & Schmid 2005).

Concerning the counter-clockwise rotation of the Adriatic Plate and the external Dinarides we again obtain a smaller amount of rotation (20°) from our reconstruction compared to the 30° inferred from paleomagnetic data for latest Miocene to recent times (Márton et al. 2003). It is evident that rotations larger than those predicted by our restoration would not only be inconsistent with the data we used as input for our restoration but also would lead to massive overestimates of shortening in the southern Dinarides, located far from the pole of rotation.

The clockwise rotations of up to 100° recorded on fragments of the internal Dinarides exposed in the Medvednica Mountains are interpreted as being restricted to small blocks (Tomljenović 2002; Tomljenović et al. 2008) rather than applying to the Dinarides as a whole. The 53° counter-clockwise rotation, which we applied in our reconstruction to another fragment of the internal Dinarides, the Bükk Mountains, was arrived at by leaving this fragment attached to the ALCAPA Mega-Unit and by straightening out the Periadriatic-Balaton-Darno Faults. This qualitatively agrees also well with paleomagnetic data from the Bükk Mountains reporting counter-clockwise rotations between 30 and 80° (Márton & Fodor 1995), as well as tectonic and paleogeographic data (Csontos 1999, 2000). The 16.5° clockwise rotation indicated in Fig. 6 for the southernmost part of the Tisza-Dacia Mega-Unit is, on the other hand, a mere consequence of our restoration and is not supported by paleomagnetic data.

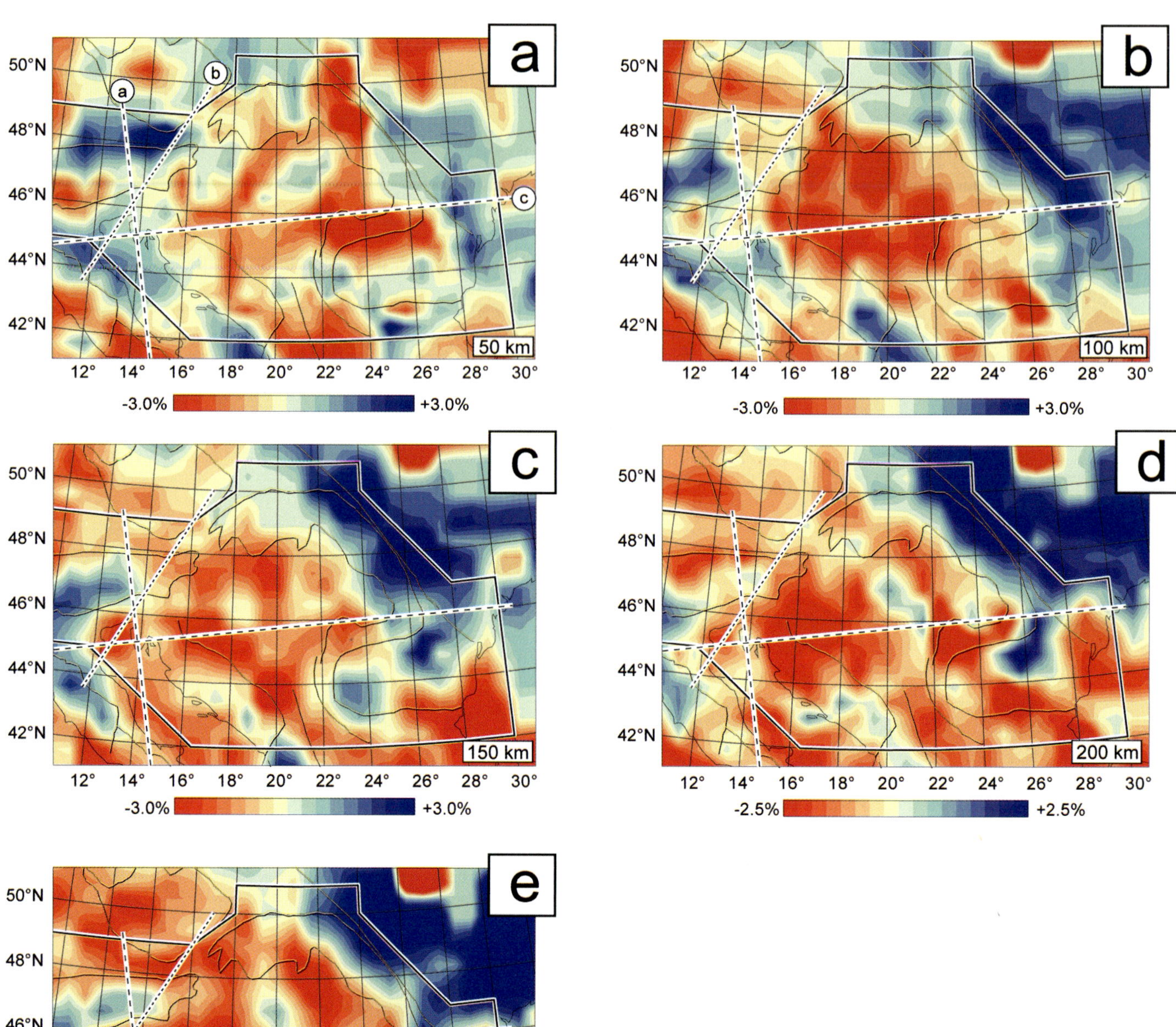

Fig. 8. Horizontal tomographic sections through the study area ranging from 50 km (a) to 250 km depth (e). The black polygon shows the range of Fig. 2; dashed lines show the traces of the vertical sections (Fig. 9). The sections are derived from the seismic tomography model of Bijwaard & Spakman (2000). See text for further details.

3.4 Implications regarding the present-day lithosphere-scale configuration

Our reconstruction has bearings on the interpretation of the present-day lithospheric configuration of the Alpine-Carpathian-Dinaridic orogenic system (Figs. 8 & 9). In particular, it potentially provides an explanation for the change in the lithosphere-scale configuration that is observed in the Eastern Alps according to recent results of high-resolution teleseismic mantle tomography (Lippitsch et al. 2003; Schmid et al. 2004a,b; Kissling et al. 2006). This work demonstrated the existence of a NNE-dipping lithospheric slab ("Adriatic slab") underneath the Eastern Alps in the area east of the Giudicarie Fault (Fig. 9b; Lippitsch et al. 2003), while another slab ("European slab") preserved west of this fault dips to the SE (Kissling et al. 2006), as is expected for the European lithosphere which represents the southwards subducted lower plate during Alpine collision. The consistent overall architecture of the crustal configuration during Alpine collision, unchanged all the way into the Western Carpathians (Schmid et al. 2008), implies that the European lithosphere formerly represented the lower plate also in the area east of the Giudicarie Fault; this in turn suggests that a late-stage post-collisional modification of the lithosphere-scale geometry must have occurred in the easternmost Alps. In a first step we discuss the time constraints for the onset of this change in subduction polarity, and in a second step we suggest possible mechanisms, which could have led to this severe modification.

The length of the subducted "Adriatic slab" east of the Giudicarie Fault was estimated to about 210 km along a NE–SW section (Fig. 9b; Lippitsch et al. 2003) across the Eastern Alps. Within error this value is perfectly compatible with our estimate of a total of 190 km N–S-shortening (Fig. 6) inferred for a N–S transect through the Trieste – Eastern Alps area, which intersects the tomography section C–C' of Lippitsch et al. (2003; see Fig. 9b). This strongly suggests that the NE-directed subduction of the Adriatic lithosphere was associated with post-20 Ma crustal shortening. The average plate convergence rate would be in the order of 1 cm yr⁻¹ for the last 20 Ma across

the Eastern and Southern Alps along the transect depicted in Fig. 9b. According to our palinspastic restoration (Fig. 6) the post-20 Ma plate convergence rate decreases westwards, which agrees with the plate convergence rate estimates between 0.3 and 0.5 cm yr⁻¹ derived from an Alpine transect through Eastern Switzerland and based on geological estimates (Schmid et al. 1996). Note that present-day convergence rates between the western parts of the Adriatic plate are also smaller, i.e. in the order of 0.5 cm yr⁻¹ (Battaglia et al. 2004).

Based on literature data and the results of our restoration we propose that a dramatic change has occurred at about 20 Ma, when the S to SE-wards subducting European Plate gave way to the NE-wards subducting Adriatic Plate below the Alps east of the Giudicarie Fault. Comparison of Figs. 2 and 6 reveals that the change in subduction polarity in the easternmost Alps was associated with substantial strike slip displacements and hence cannot be understood without considering the entire Alpine-Carpathian-Dinaridic system in three dimensions.

In order to analyse the geometry of the mantle lithospheric configuration east of the Giudicarie Fault in more detail, we present five horizontal tomographic slices (Fig. 8) that were derived from the seismic tomography model of Bijwaard & Spakman (2000). Inconsistencies between these sections and those presented in the area of the Eastern Alps are primarily due

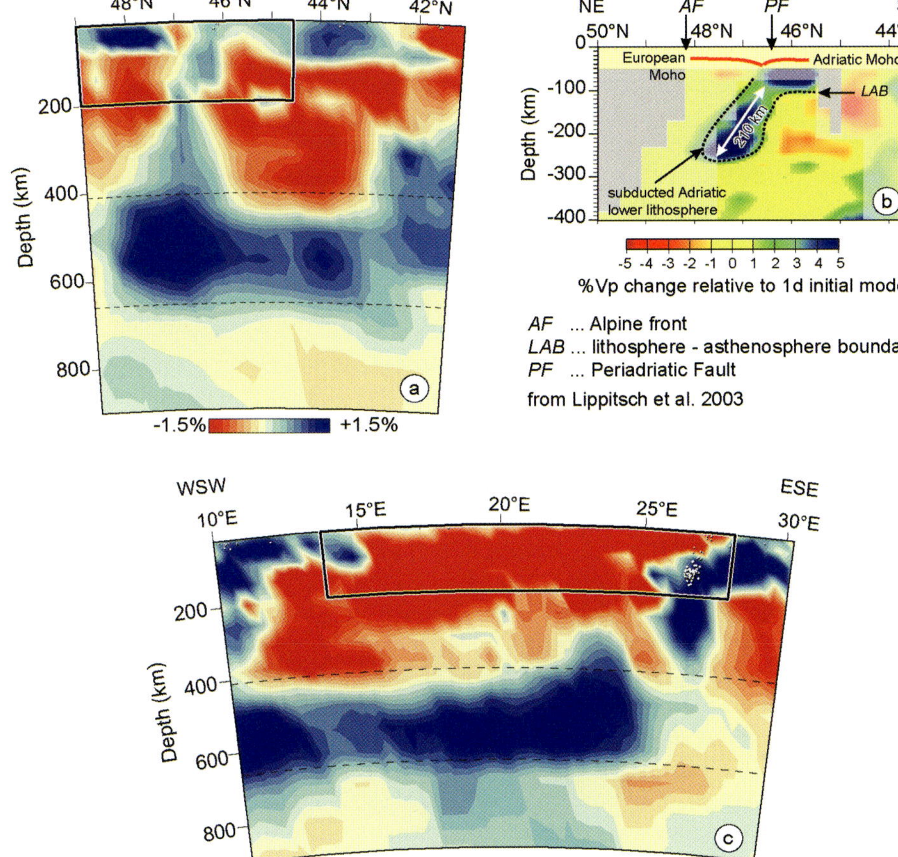

Fig. 9. Vertical tomographic sections through the study area. (a) and (c) are sections derived from the seismic tomography model of Bijwaard & Spakman (2000), (b) represents section C–C' of Lippitsch et al. (2003). The bold black rectangles in (a) and (c) show the extent of the vertical walls of the block diagram of Fig. 11.

to the higher resolution achieved by Lippitsch et al. (2003) in the Alpine region, who applied corrections for lateral velocity changes in the crust. The sections show the following first-order characteristics that can also be seen in the horizontal sections of Piromallo & Morelli (2003, their Fig. 8): (1) High V_p anomalies in the sections at 50 and 100 km depth (Figs. 8a & b) suggest a continuous Adriatic lithosphere between the Eastern Alps and the Dinarides. At depths greater than 100 km, high V_p anomalies underneath the Dinarides are confined to latitudes south of 44° N, but are absent further north (Fig. 8c-e). This suggests that an Adriatic lithosphere slab dipping underneath the Dinarides is only present south of 44° N and that a connection to the slab underneath the Eastern Alps identified by Lippitsch et al. (2003) cannot be inferred from the data available. (2) The area of high V_p below the northernmost Eastern Alps (north of latitude 47° in Fig. 8a) is part of the European lithosphere, which also surrounds the Eastern Carpathians based on geological arguments. Note that the immediate connection of high velocity mantle underneath the Eastern Alps to the circum-Carpathian European lithosphere is not obvious from Fig. 8. Further east, however, the European lithosphere is again well depicted by fast velocities in Figs. 8 b to e, and connects southeastwards with the sub-vertically oriented Vrancea slab north of Bucharest that is a part of Moesia (Matenco et al. 2007). (3) Low velocities characterise the area of the Pannonian Basin. The area of high velocities underneath the Eastern Alps depicted in the horizontal sections of Fig. 8 cannot unequivocally be attributed to either the Adriatic slab identified by Lippitsch et al. (2003) nor, alternatively, to a remnant of the former S-ward subducting European slab that must have retreated or broken off at 20 Ma, giving way to the Adriatic slab.

In order to further investigate this matter we looked at three vertical tomographic sections (Fig. 9). Fig. 9a clearly shows a horizontal high velocity body located at the northernmost rim of the Alps north of about 47° latitude and extending into the Bohemian massif (see also Fig. 8a). The lower interface of this body extends down to a depth of less than 100 km. Because of its location, this high velocity body clearly has to be correlated with the European lithosphere. The downward tapering low velocity anomaly at 47° N coincides with the topographically highest part of the Eastern Alps along this transect and very likely represents a deep crustal root underneath the Eastern Alps (see Kissling & Spakman 1996). The high velocities south of 47° N represent Adriatic lithosphere that dips steeply northward underneath the European lithosphere. The exact length of this slab cannot be estimated from Fig. 9a due to "blurring" (upward conical high velocity anomalies that are likely caused by the incident angles of the ray paths used in the tomography).

The section in Fig. 9b represents the high-resolution tomographic section of Lippitsch et al. (2003) and is shown at exactly the same scale as the section in Fig. 9a. The Adriatic slab is clearly shown to be dipping underneath the Eastern Alps reaching a depth of c. 270 km. In combination, the sections Fig. 9a and 9b clearly suggest that Adriatic lithosphere is dipping underneath the European lithosphere. The section in Fig. 9c shows that the high velocity body representing the Adriatic lithosphere at the western end of the section does not continue further to the east beyond 15° E (longitude of Zagreb), neither horizontally nor as a dipping slab. The Pannonian Basin is characterised by low velocities throughout, suggesting a very reduced lithosphere thickness there. However, assuming that an average stretching factor δ around 1.6 (see section 2.3) also affected the mantle lithosphere, initial total lithosphere thickness being c. 100 km, one would still expect a lithosphere thickness of roughly 60 km from tectonic stretching alone, i.e. much more than observed in Fig. 9c. This discrepancy confirms the widely accepted view first proposed by Sclater et al. (1980) that the lithosphere underneath the Pannonian Basin has been, in addition to tectonic stretching, substantially thermally attenuated. This is also in line with the widespread occurrence of Late Miocene to Pleistocene alkalic and mafic magmatics with an inferred asthenospheric source, postdating subduction-related magmatism in the area (e.g. Wilson & Downes 2006).

The horizontal depth slices at 100 km depth by Lippitsch et al. (2003, their Fig. 12a), Piromallo & Morelli (2003, their Fig. 8) and our Fig. 8b were used for mapping the location of the upper mantle-lithosphere boundary of the NE-ward subducted Adriatic slab underneath the Eastern Alps and internal Dinarides, or the eastern limit of the Adriatic lithosphere where no slab can be discerned dipping underneath the Dinarides. We did this by contouring the transition from positive to negative V_p variations along the presumed NE edge of the Adriatic lithosphere as it intersects these three horizontal sections. We are fully aware that the 0% V_p variation contours in the horizontal sections are not comparable, since they strongly depend on the V_p mantle velocity models used by the different authors as well as on the accuracy of these models. Nevertheless we plotted these contours onto the tectonic map of Fig. 2 for a qualitative assessment (Fig. 10). In combination with Figs. 8 and 9, the contours in Fig. 10 suggest the following: (1) The Adriatic lithosphere can be traced continuously from the Eastern Alps into the Dinarides. (2) However, a lithospheric slab can only be confidently identified in the high-resolution area of Lippitsch et al. (2003) and south of 44° N (roughly the latitude of Sarajevo) in the Dinarides, but is absent according to the tomographic data available in the area between. Given the currently available tomographic models, the southeastern limit of the Adriatic slab identified by Lippitsch et al. (2003) is speculative. Horváth et al. (2006) suggested its south-eastern termination along the Zagreb Line (corresponding to the western part of the Mid-Hungarian Fault Zone, Fig. 2). This problem should be addressed by future geophysical studies. Interestingly, the area where no Adriatic slab can be seen in the tomographic models coincides with the location of the western rim of the Pannonian Basin in parts of the Dinarides, a basin that has experienced substantial tectonic stretching and asthenospheric uprising. It is therefore conceivable that asthenospheric upwelling underneath the Pannonian Basin led to a severe modification of a formerly continuous Adriatic lithospheric slab dipping underneath the Dinarides,

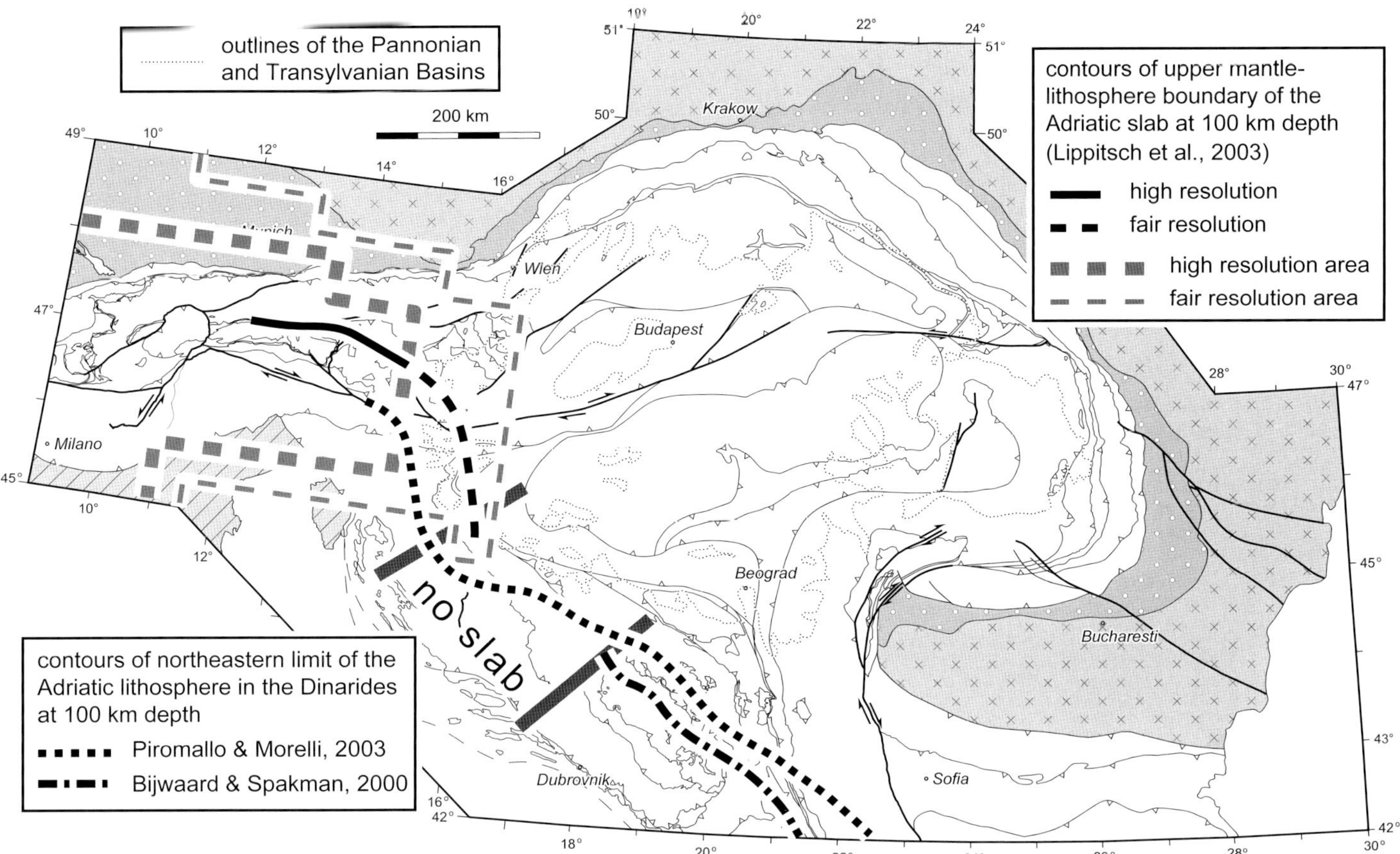

Fig. 10. Contours of the upper mantle-lithosphere boundary of the Adriatic slab at 100 km depth as inferred from the 3D V_p-model of Lippitsch et al. (2003, their Fig. 12) in comparison with the contours of the eastern limit of the Adriatic lithosphere underneath the Dinarides using the horizontal depth slices at 100 km depth of Piromallo & Morelli (2003, their Fig. 8) and Bijwaard & Spakman (2000; our Fig. 8b). The contours were derived by tracing the transition from positive to negative V_p variations of each model along the presumed NE edge of the Adriatic lithosphere. The contours are superimposed onto the outlines of the tectonic units of Fig. 2. See text for further details.

just as it caused strong thermal attenuation of the lithosphere underneath the Pannonian Basin itself, creating pathways for asthenosphere-derived melts.

Fig. 11 presents a 3D sketch, which integrates the present-day crustal structure of the Alpine-Carpathian-Dinaridic system (Schmid et al. 2008) with the present-day lithosphere-scale configuration inferred from the results of mantle tomography discussed above. This sketch allows for a brief discussion of the mechanisms, which could have led to changes of the original lithospheric configuration that induced the severe post-20 Ma crustal displacements, rotations and deformations that are evident from comparing Figures 2 and 6.

Clearly, the post-20 Ma emplacement of a once continuous Adriatic slab underneath the Eastern Alps was only possible once the European slab underneath the Eastern Alps gave way by slab break-off, a process that initiated earlier, i.e. between 40 and 35 Ma, in the transition area between Western and Eastern Alps (von Blanckenburg & Davis 1995) but took place at around 20 Ma in the Eastern Alps and the area of the Carpathian embayment. Whether or not this break-off was a continuous process, systematically migrating eastward (Spakman & Wortel 2000), is not yet fully understood. The Carpathian embayment was formerly underlain, according to most authors (e.g. Balla 1982; Mason et al. 1998), at least partly by oceanic lithosphere. Subduction and slab retreat (e.g. Royden 1988; Wortel & Spakman 2000; Sperner et al. 2002, 2005), starting at around 20 Ma in the area of the Carpathian embayment, created the necessary space that allowed for the invasion of the ALCAPA, Tisza and Dacia mega-units and the formation of the highly arcuate Alpine-Carpathian orogenic system. However this retreat, associated with severe crustal thinning and upwelling of the asthenosphere underneath the Pannonian Basin, cannot be the only driving force for this substantial post-20 Ma reorganization.

Rotation and north-directed translation of the Adriatic Plate indenter, also including the underpinnings of the adjacent Dinaridic orogen, provided a second and probably equally important driving force for these modifications. The displacement of the Adriatic Plate, of course in combination with slab break off or retreat of the European Plate, led to the change in subduction polarity along the transect through the easternmost Alps depicted in Fig. 11. Due to the simultaneous counter-clockwise rotation of the Adriatic Plate indenter, its effects in

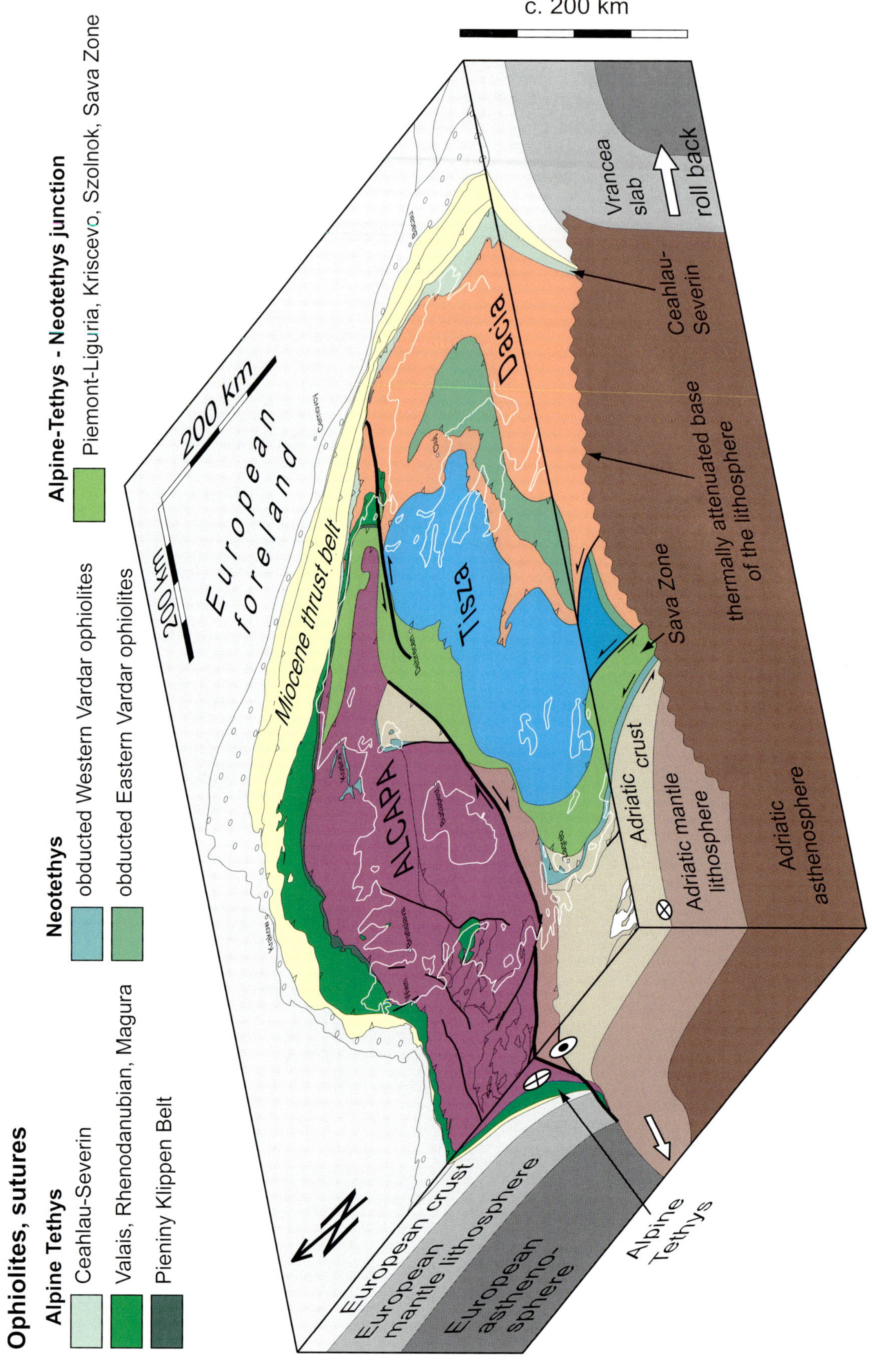

Fig. 11. Interpretative block diagram showing present-day lithospheric structures in the Eastern Alps, Carpathians and northern Dinarides. The horizontal lid of the block represents a simplified version of the tectonic map presented in Fig. 2; the vertical block walls are interpretative representations of the vertical tomographic sections (Fig. 9a and c). The MOHO depth in the N–S-trending, western edge of the block is roughly after Brückl et al. (2007). In the E–W-trending edge of the block, no seismic MOHO is shown, but the fossil crust-mantle boundary in order to better document the lithosphere tectonics. In addition to tectonic stretching, the lithosphere underneath the Pannonian Basin is also strongly thermally attenuated, giving a possible explanation for the absence of the Adriatic lithospheric slab underneath the Dinarides.

terms of post-20 Ma shortening are relatively minor in the Western Alps, but become increasingly important eastwards (N–S transect corresponding to the western edge of the block model in Fig. 11). After the Latest Miocene to Pliocene, i.e. the onset of intra-plate folding and overall Quaternary inversion (e.g. Matenco et al. 2006; Cloetingh et al. 2007) compression due to the Adria push also largely governed the easternmost parts of the Alpine-Carpathian-Dinaridic system of orogens.

4. Conclusions

We conclude that:

1) The Mid-Hungarian Fault Zone was the locus of a Paleogene plate boundary transforming opposed subduction polarities between Alps (including Western Carpathians) and Dinarides before 20 Ma ago.
2) The ALCAPA Mega-Unit fitted into an area W of present-day Budapest, while the Tisza-Dacia Mega-Unit was already on its way invading the Carpathian embayment.
3) That slab retreat of the partly oceanic European lithosphere provided an important driving force for extension and compression in the Pannonian-Carpathian realm considered as a closed system associated with negligible plate convergence. However, we conclude that the rotating Adriatic Plate indenter provides an additional important driving force, interfering with slab retreat but primarily responsible for:
4) Neogene shortening amounts to over 200 km across the southern external Dinarides, contemporaneous with dextral wrench movements in the Dinarides and the adjacent Carpatho-Balkan orogen, and:
5) N–S convergence between the European and Adriatic plates and amounts to some 200 km at a longitude of 14° E is in line with post-20 Ma emplacement of a formerly possibly continuous Adriatic slab underneath the Eastern Alps, corroborating the interpretation of results based on high-resolution teleseismic tomography.

Acknowledgements

We acknowledge financial support through the Swiss National Science Foundation (projects "Tisza" Nrs. 200021-101883/1 and 200020-109278/1). We thank Liviu Matenco (Amsterdam) and Peter Ziegler (Basel) for critical remarks on an earlier version of the manuscript. We further acknowledge the very thorough and constructive reviews of Ewald Brueckl (Vienna), Niko Froitzheim (Bonn) and Philippe Agard (Paris).

REFERENCES

Anderson, H. & Jackson, J. 1987: Active tectonics of the Adriatic Region. Geophysical Journal of the Royal Astronomical Society 91, 937–983.

Aubouin, J. & Ndojaj, I. 1964: Regards sur la géologie de l'Albanie et sa place dans la géologie des Dinarides. Bulletin de la Société Géologique de France 7/6, 593–625.

Balla, Z. 1987: Tertiary paleomagnetic data for the Carpatho-Pannonian region in the light of Miocene rotation kinematics. Tectonophysics 139, 67–98.

Battaglia, M., Murray, M.H., Serpelloni, E. & R., B. 2004: The Adriatic region: An independent microplate within the Africa-Eurasia collision zone. Geophysical Research Letters 31, L09605, doi:10.1029/2004GL019723.

Behrmann, J.H., Stiasny, S., Milicka, J. & Pereszlenyi, M. 2000: Quantitative reconstruction of orogenic convergence in the northeast Carpathians. Tectonophysics 319, 111–127.

Bennett, R.A., Hreinsdóttir, S., Buble, G., Bašic, T., Marjanović, M., Casale, G., Gendaszek, A. & Cowan, D., 2008. Eocene to present subduction of southen Adria mantle lithosphere beneath the Dinarides. Geology 36, 3–6.

Bérczi, I., Hámor, G., Jámbor, Á. & Szentgyörgyi, K. 1988: Neogene sedimentation in Hungary. In: Royden, L.H. & Horváth, F. (Eds.): The Pannonian Basin. A Study in Basin evolution, AAPG Memoir 45. The American Association of Petroleum Geologists and the Hungarian Geological Society, Budapest, Hungary, 57–67.

Berza, T. & Drăgănescu, A. 1988: The Cerna-Jiu fault system (South Carpathians, Romania), a major Tertiary transcurrent lineament. Dări de Seamă Institutul de Geologie si Geofizică 72–73, 43–57.

Berza, T. & Iancu, V. 1994: Variscan events in the basement of the Danubian nappes (South Carpathians). ALCAPA II Field Guidebook. Romanian Journal of Tectonics and Regional Geology 75, 93–104.

Bijwaard, H & Spakman W. 2000: Non-linear global P-wave tomography by iterated linearized inversion. Geophysical Journal International 141, 71–82.

Brückl, E., Bleibinhaus, F., Gosar, A., Grad, M., Guterch, A., Hrubcová, P., Keller, R., Majdański, M., Šumanovac, F., Tiira, T., Yliniemi, J., Hegedűs, E. & Thybo, H. 2007: Crustal structure due to collisional and escape tectonics in the Eastern Alps region based on profiles Alp01 and Alp02 from the ALP 2002 seismic experiment. Journal of Geophysical Research 112, B06308, doi:10.1029/2006JB004687.

Burchfiel, B.C., 1980. Eastern European Alpine system and the Carpathian orocline as an example of collision tectonics. Tectonophysics 63, 31–61.

Burkhard, M. & Sommaruga, A. 1998: Evolution of the western Swiss Molasse basin: Structural relations with the Alps and the Jura belt. In Mascle C. et al. (Eds.): Cenozoic Foreland Basins of Western Europe. Geological Society Special Publications 134, 279–298.

Calais, E., Nocquet, J.-M., Jouanne, F. & Tardy, M. 2001: Current strain regime in the Western Alps from continuous Global Positioning System measurements, 1996–2001. Geology 30, 651–654.

Cavazza, W., Roure, F., Spakman, W., Stampfli, G.M. & Ziegler, P. (Editors), 2004: The TRANSMED Atlas. The Mediterranean Region from Crust to Mantle. Springer, Berlin, Heidelberg, New York, 141 pp. + CD-ROM.

Channel, J.E.T. & Horváth, F. 1976: The African/Adriatic promontory as a paleogeographical premise for Alpine orogeny and plate movements in the Carpatho-Balkan region. Tectonophysics 35, 71–101.

Chorowicz, J. 1970: La transversale de Zrmanja (Yougoslavie). Bulletin de la Société Géologique de France 12/6, 1028–1033.

Chorowicz, J. 1975: Mechanics of Split-Karlovac transversal structure in Yugoslavian Dinarides. Compte-Rendu Académie des Sciences Série D 280/20, 2313–2316.

Cloetingh, S., Bada, G., Matenco, L., Lankreijer, A., Horvath, F. & Dinu, C. 2006: Modes of basin (de)formation, lithospheric strength and vertical motions in the Pannonian-Carpathian system: inferences from thermomechanical modelling. In: Gee, D.G. & Stephenson, R.A. (Eds.): European Lithosphere Dynamics, Geological Society London Memoirs 32, 207–221.

Csontos, L. 1995: Tertiary tectonic evolution of the Intra-Carpathian area: a review. In: Downes, H. & Vaselli, O. (Eds.): Neogene and related magmatism in the Carpatho-Pannonian region). Acta Vulcanologica 7, 1–13.

Csontos, L. 1999: Structural outline of the Bükk Mts. (N Hungary). Text in Hungarian, abstract and figure captions translated. Földtani Kölöny 129 611–651.

Csontos, L. 2000: Stratigraphic re-evaluation of the Bükk Mts (N Hungary). Text in Hungarian, abstract and figure captions translated. Földtani Kölöny 130, 95–131.

Csontos, L., Nagymarosy, A., Horvath, F. and Kováč, M., 1992. Tertiary evolution of the Intra-Carpathian area: A model. Tectonophysics 208(1–3): 221–241.

Csontos, L. & Nagymarosy, A. 1998: The Mid-Hungarian line: a zone of repeated tectonic inversions. Tectonophysics 297, 51–71.

Csontos, L. & Vörös, A. 2004: Mesozoic plate tectonic reconstruction of the Carpathian region. Palaeogeography, Palaeoclimatology, Palaeoecology 210, 1–56.

de Capoa, P., Radoicic, R. & D'Argenio, B., 1995. Late Miocene deformation of the External Dinarides (Montenegro and Dalmatia). New biostratigraphic evidence. Memorie di Scienze Geologiche 47, 157–172.

Dewey, J.F., Pitman III, W.C., Ryan, W.B. F. & Bonnin, J. 1973: Plate tectonics and the evolution of the Alpine system. Geological Society of America Bulletin 84, 3137–3180.

Dimitrijević, M.N., Dimitrijević, M.D., Karamata, S., Sudar, M., Gerzina, N., Kovács, S., Dostály, L., Gulácsi, Z., Less, G. & Pelikán, P. 2003: Olistostrome/mélanges – an overview of the problems and preliminary comparison of such formations in Yugoslavia and NE Hungary. Slovak Geological Magazine 9, 3–21.

Ellouz, N. & Roca, E. 1994: Palinspastic reconstructions of the Carpathians and adjacent areas since the Cretaceous: a quantitative approach. In: Roure, F. (Ed.): Peri-Tethyan Platforms. Édition Technip, Paris, 51–78.

Fodor, L., Jelen, B., Márton, E., Skaberne, D., Car, J. & Vrabec, M. 1998: Miocene-Pliocene tectonic evolution of the Slovenian Periadriatic fault: Implications for Alpine-Carpathian extrusion models. Tectonics 17, 690–709.

Fodor, L., Csontos, L., Bada, G., Györfi, I. & Benkovics, L. 1999: Tertiary tectonic evolution of the Pannonian Basin system and neighbouring orogens: a new synthesis of palaeostress data. In: Durand, B., Jolivet, L., Horvath, F. & Séranne, M. (Eds.): The Mediterranean Basins: Tertiary Extension within the Alpine Orogen. Geological Society London Special Publications 156, 295–334.

Frisch, W., Kuhlemann, J., Dunkl, I. & Brügel, A. 1998: Palinspastic reconstruction and topographic evolution of the Eastern Alps during late Tertiary tectonic extrusion. Tectonophysics 297, 1–15.

Froitzheim, N., Conti, P. & van Daalen, M. 1997: Late Cretaceous, synorogenic, low-angle normal faulting along the Schlinig fault (Switzerland, Italy, Austria) and its significance for the tectonics of the Eastern Alps. Tectonophysics 280, 267–293.

Fügenschuh, B., Seward, D. & Mancktelow, N.S. 1997: Exhumation in a convergent orogen: the western Tauern window. Terra Nova 9, 213–217.

Fügenschuh, B. & Schmid, S.M., 2005. Age and significance of core complex formation in a very curved orogen: Evidence from fission track studies in the South Carpathians (Romania). Tectonophysics 404, 33–53.

Genser, J. & Neubauer, F. 1989: Low angle normal faults at the eastern margin of the Tauern window (Eastern Alps). Mitteilungen der Österreichischen Geologischen Gesellschaft 81, 233–243.

Gradstein, F., Ogg, J. & Smith, A.G. 2004: A Geologic Time Scale. Cambridge University Press, Cambridge, 589 pp.

Grasemann, B. & Mancktelow, N.S. 1993: Two-dimensional thermal modelling of normal faulting: the Simplon Fault Zone, Central Alps, Switzerland. Tectonophysics 225, 155–165.

Gratier, J.-P., Ménard, G. & Arpin, R. 1989. Strain-displacement compatibility and restoration of the Chaînes Subalpines of the Western Alps. In: Coward, M.P., Dietrich, D. & Park, R.G. (Eds.): Alpine Tectonics, Geological Society Special Publication 45, 65–81.

Haas, J. 2001: Geology of Hungary. Eötvös University Press, Budapest, 317 pp.

Haas, J. & Pero, C. 2004: Mesozoic evolution of the Tisza Mega-unit. International Journal of Earth Sciences 93, 297–313.

Harangi, S., Downes, H. & Seghedi, I. 2006: Tertiary-Quaternary subduction processes and related magmatism in the Alpine-Mediterranean region. In: Gee, D.G. & Stephenson, R. (Eds.): European Lithosphere Dynamics. Geological Society of London Memoir 32, 167–190.

Horváth, F., 1993. Towards a mechanical model for the formation of the Pannonian basin. Tectonophysics 226(1–4), 333–357.

Horváth, F. & Royden, L.H. 1981: Mechanism for formation of the intra-Carpathian basins: a review. Earth Evolution Sciences 1, 307–316.

Horváth, F. & Tari, G.C. 1999: IBS Pannonian Basin project; a review of the main results and their bearings on hydrocarbon exploration. In: Durand, B., Jolivet, L., Horváth, F. & Seranne, M. (Eds.): The Mediterranean basins; Tertiary extension within the Alpine Orogen. Geological Society London Special Publications 156, 195–213.

Horváth, F., Bada, G., Szafián, P., Tari, G., Ádám, A. & Cloetingh, S. 2006: Formation and deformation of the Pannonian Basin: constraints from observational data. In: Gee, D.G. & Stephenson, R. (Eds.): European Lithosphere dynamics, Geological Society London Memoir 32, 191–206.

Hrvatović, H. 2000: Post-Orogenic Intramontane Basins. In: Pamić, J. & Tomljenović, B. (Eds.): PANCARDI 2000. Field trip guide book, Dubrovnik. Vijcsti Hrvatskoga Geoloskog Drustva 37, 73–74.

Jacobshagen, V. 1986: Geologie von Griechenland. Beiträge zur regionalen Geologie der Erde, 19. Gebrüder Bornträger, Berlin, 363 pp.

Jiříček, R. 1979: Tektogeneticky vyvoj karpatskeho oblouku behem oligocenu a neogenu (in Czech, translated title: Tectogenetic development of the Carpathian Arc in the Oligocene and Neogene). In: Machel, M. (Ed.): Tectonic profiles through the West Carpathians. Geol. Ustav Dionyza Stura, Bratislava, 203–214.

Krézsek, C., Bally, A.W. 2006: The Transylvanian Basin (Romania) and its relation to the Carpathian fold and thrust belt: insight in gravitational salt tectonics. Marine and Petroleum Geology 23, 405–442.

Kilias, A., Tranos, M., Mountrakis, D., Shallo, M., Marto, A. & Turku, I. 2001: Geometry and kinematics of deformation in the Albanian orogenic belt during the Tertiary. Journal of Geodynamics 31, 169–187.

Kissling, E. & Spakman, W. 1996: Interpretation of tomographic images of uppermost mantle structure: examples from the Western and Central Alps. Journal of Geodynamics 21, 97–111.

Kissling, E., Schmid, S.M., Lippitsch, R., Ansorge, J. & Fügenschuh, B. 2006: Lithosphere structure and tectonic evolution of the Alpine arc: new evidence from high-resolution teleseismic tomography. In: Gee, D.G. & Stephenson, R.A. (Eds.): European Lithosphere Dynamics, Geological Society London Memoirs 32, 129–145.

Kováč, M., Nagyymarosy, A., Oszczypko, N., Csontos, L., Slączka, A., Maruntieanu, M., Matenco, L. & Márton, E. 1998: Palinspastic reconstruction of the Carpathian-Pannonian region during the Miocene. In: Rakus, M. (Ed.): Geodynamic development of the Western Carpathians. Geological Survey of the Slovak Republic, Bratislava, 189–217.

Kovács, I., Csontos, L., Szabó, C., Bali, E., Falus, G., Benedek, K. & Zajacz, Z. 2007: Paleogene-early Miocene igneous rocks and geodynamics of the Alpine-Carpathian-Pannonian-Dinaric region: An integrated approach. In: Beccaluva, L., Bianchini, G. & Wilson, M. (Eds.): Cenozoic volcanism in the Mediterranean Area, Geological Society of America Special Paper 418, 93–112.

Kounov, A., Seward, D., Bernoulli, D., Burg, J.-P. & Ivanov, Z. 2004: Thermotectonic evolution of an extensional dome: the Cenozoic Osogovo-Lisets core complex (Kraishte zone, western Bulgaria). International Journal of Earth Sciences 93, 1008–1024.

Kräutner, H.G. & Krstić, B. 2006: Geological map of the Carpatho-Balkanides between Mehadia, Oravita, Nis and Sofia. CD-version provided at the 18th Congress of the Carpathian-Balkan Geological Association. Geoinstitut, Belgrade.

Laubscher, H. 1971: Das Alpen-Dinariden-Problem und die Palinspastik der südlichen Tethys. Geologische Rundschau 60, 813–833.

Lenkey, L. 1999: Geothermics of the Pannonian Basin and its bearing on the tectonics of basin evolution. PhD Thesis, Vrije Universiteit Amsterdam, Amsterdam, 215 pp.

Lickorish, W.H. & Ford, M. 1998: Sequential restoration of the external Alpine Digne thrust system, SE France, constrained by kinematic data and synorogenic sediments. In: Mascle, A., Puidgefàbregas, C., Luterbacher, H.P. & Fernàndez, M. (Eds.): Cenozoic Foreland Basins of Western Europe, Geological Society London Special Publication 134, 189–211.

Linzer, H.-G., Decker, K., Peresson, H., Dell›Mour, R. & Frisch, W. 2002: Balancing lateral orogenic float of the Eastern Alps. Tectonophysics 354, 211–237.

Lippitsch, R., Kissling, E. & Ansorge, J. 2003: Upper mantle structure beneath the Alpine orogen from high-resolution teleseismic tomography. Journal of Geophysical Research 108(B8), 2376, doi:10.1029/2002JB002016.

Means, W. 1989: Stretching faults. Geology 17, 893–896.

Márton, E. 2000: The Tisza Megatectonic Unit in the light of paleomagnetic data. Acta Geologica Hungarica 43/3, 329– 343.

Márton, E. & Fodor, L. 1995: Combination of palaeomagnetic and stress data – a case study from North Hungary. Tectonophysics 242, 99–114.

Márton, E. & Fodor, L. 2003: Tertiary paleomagnetic results and structural analysis from the Transdanubian Range (Hungary): rotational disintegration of the ALCAPA unit. Tectonophysics 363, 201–224.

Márton, E. & Márton, P. 1996: Large scale rotations in North Hungary during the Neogene as indicated by palaeomagnetic data. In: Morris, A. & Tarling, D.H. (Eds.): Palaeomagnetism and Tectonics of the Mediterranean Region, Geological Society London Special Publication 105, 153–173.

Márton, E., Pavelić, D., Tomljenović, B., Avanić, R., Pamić, J. & Márton, P. 2002: In the wake of a counter-clockwise rotating Adriatic microplate: Neogene paleomagnetic results from northern Croatia. International Journal of Earth Sciences 91, 514–523.

Márton, E., Drobne, K., Ćosović, V. & Moro, A. 2003: Paleomagnetic evidence for Tertiary counterclockwise rotation of Adria. Tectonophysics 377, 143–156.

Márton, E., Tischler, M., Csontos, L., Fügenschuh, B. & Schmid, S.M. 2007: The contact zone between the ALCAPA and Tisza-Dacia mega-tectonic units of Northern Romania in the light of new paleomagnetic data. Swiss Journal of Geosciences 100, 109–124.

Mason, P.R.D., Seghedi, I., Szakacs, A & Downes, H. 1998: Magmatic constraints on geodynamic models of subduction in the Eastern Carpathians, Romania. Tectonophysics 297, 157–176,

Matenco, L. & Bertotti, G. 2000: Tertiary tectonic evolution of the external East Carpathians (Romania). Tectonophysics 316, 255–286.

Matenco, L., Bertotti, G., Dinu, C. & Cloetingh, S. 1997: Tertiary tectonic evolution of the external South Carpathians and the adjacent Moesian platform (Romania). Tectonics 16, 896–911.

Matenco, L., Bertotti, G., Cloetingh, S. & Dinu, C. 2003: Subsidence analysis and tectonic evolution of the external Carpathian–Moesian Platform region during Neogene times. Sedimentary Geology 156, 71–94.

Matenco, L., Bertotti, G., Leever, K., Cloetingh, S., Schmid, S.M., Tarapoanca, M. & Dinu, C. 2007: Large-scale deformation in a locked collisional boundary: Interplay between subsidence and uplift, intraplate stress, and inherited lithospheric structure in the late stage of the SE Carpathians evolution. Tectonics 26/4, TC4011, doi:10.1029/2006TC001951.

Mikes, T., Dunkl, I., von Eynatten, H., Báldi-Beke, M. & Kázmer, M. 2008: Calcareous nannofossil age constraints on Miocene flysch sedimentation in the Outer Dinarides (Slovenia, Croatia, Bosnia-Hercegovina and Montenegro). In: Siegesmund, S., Fügenschuh, B. & Froitzheim, N. (Eds.): Tectonic aspects of the Alpine-Dinaride-Carpathian System, Geological Society London Special Publications 298, 335–363.

Morley, C.K. 1996: Models for relative motion of crustal blocks within the Carpathian region, based on restorations of the outer Carpathian thrust sheets. Tectonics 15, 885–904.

Moser, F. 2001: Tertiäre Deformation in den rumänischen Südkarpathen: Strukturelle Analyse eines Blattverschiebungskorridors am Westrand der mösischen Plattform. Tübinger Geowissenschaftliche Arbeiten, Reihe A, 63, Tübingen, 169 pp.

Nagymarosy, A. & Müller, P. 1988: Some aspects of Neogene biostratigraphy in the Pannonian Basin. In: Royden, L.H. & Horváth, F. (Eds.): The Pannonian Basin. A Study in Basin evolution, AAPG Memoir, 45. The American Association of Petroleum Geologists and the Hungarian Geological Society, Budapest, Hungary, 69–77.

NOAA, 1988. Data Announcement 88-MGG-02, NOAA National Geophysical Data Center, Digital relief of the Surface of the Earth. http://www.ngdc.noaa.gov/mgg/global/etopo5.html, Boulder, Colorado.

Nussbaum, C. 2000: Neogene tectonics and thermal maturity of sediments of the easternmost Southern Alps (Friuli Area, Italy). PhD Thesis, Université de Neuchâtel, Neuchâtel, 172 pp.

Oszczypko, N. 2006: Late Jurassic-Miocene evolution of the Outer Carpathian fold-and-thrust belt and its foredeep basin (Western Carpathians, Poland). Geological Quarterly 50, 169–194.

Pamić, J. 1993: Eoalpine to Neoalpine magmatic and metamorphic processes in the northwestern Vardar Zone, the easternmost Periadriatic Zone and the southwestern Pannonian Basin. Tectonophysics 226, 503–518.

Pamić, J. 2002: The Sava-Vardar Zone of the Dinarides and Hellenides versus the Vardar Ocean. Eclogae Geologicae Helveticae 95, 99–113.

Patrascu, S., Panaiotu, C., Seclaman, M. & Panaiotu, C.E. 1994: Timing of rotational motion of Apuseni Mountains (Romania) – Paleomagnetic data from Tertiary magmatic rocks. Tectonophysics 233, 163–176.

Philippe, Y., B. Colletta, E. Deville, and A. Mascle, The Jura fold-and-thrust belt: a kinematic model based on map-balancing. In Horvath F. (Ed.): Structure and prospects of Alpine Basins and Forelands; Peri-Tethys Memoir 2. Mémoires du Musée National de l'Histoire de la Nature Paris 170, 235–261.

Picha, F.J. 2002: Late orogenic strike-slip faulting and escape tectonics in the frontal Dinarides-Hellenides, Croatia, Yugoslavia, Albania, and Greece. AAPG Bulletin 86, 1659–1671.

Piromallo, C. & Morelli, A. 2003: P wave tomography of the mantle under the Alpine-Mediterranean area. Journal of Geophysical Research 108 (B2), doi:10.1029/2002JB001757.

Ratschbacher, L., Frisch, W., Linzer, H.-G. & Merle, O. 1991: Lateral extrusion in the Eastern Alps; Part 2, Structural analysis. Tectonics 10, 257–271.

Roca, E., Bessereau, G., Jawor, E., Kotarba, M. & Roure, F. 1995: Pre-Neogene evolution of the Western Carpathians: Constraints from the Bochnia-Tatra Mountains section (Polish Western Carpathians). Tectonics 14, 855–873.

Roure, F., Roca, E. & Sassi, W. 1993: The Neogene evolution of the outer Carpathian flysch units (Poland, Ukraine and Romania): kinematics of a foreland/fold-and-thrust belt system. Sedimentary Geology 86, 177–201.

Royden, L.H. 1988: Late Cenozoic tectonics of the Pannonian Basin system. In: Royden, L.H. & Horváth, F. (Eds.): The Pannonian Basin. A Study in Basin evolution, AAPG Memoir, 45. The American Association of Petroleum Geologists and the Hungarian Geological Society, Budapest, Hungary, 27–48.

Royden, L.H., Horváth, F. & Rumpler, J. 1983: Evolution of the Pannonian Basin System 1. Tectonics. Tectonics 2, 63–90.

Royden, L.H. & Báldi, T. 1988: Early Cenozoic tectonics and Paleogeography of the Pannonian and surrounding regions. In: Royden, L.H. & Horváth, F. (Eds.): The Pannonian Basin. A Study in Basin evolution, AAPG Memoir, 45. The American Association of Petroleum Geologists and the Hungarian Geological Society, Budapest, Hungary, 1–16.

Royden, L.H. & Burchfiel, B.C. 1989: Are systematic variations in thrust belt style related to plate boundary processes? (The Western Alps versus the Carpathians). Tectonics 8, 51–61.

Saftić, B., Velić, J., Sztanó, O., Juhász, G. & Ivković, Ž. 2003: Tertiary subsurface facies, source rocks and hydrocarbon reservoirs in the SW part of the Pannonian Basin (Northern Croatia and South-Western Hungary). Geologia Croatica 56, 101–122.

Săndulescu, M., Kräutner, H.G., Borcos, M., Năstăseanu, S., Patrulius, D., Stefănescu, M., Ghenea, C., Lupu, M., Savu, H., Bercia, I. & Marinescu, F. 1978: Geological map of Romania 1:1.000.000. Institut de Géologie Roumain, Bucharest.

Săndulescu, M. 1988: Cenozoic tectonic history of the Carpathians. In: Royden, L.H. & Horváth, F. (Eds.): The Pannonian Basin. A Study in Basin evolution, AAPG Memoir, 45. The American Association of Petroleum Geologists and the Hungarian Geological Society, Budapest, Hungary, 17–25.

Schmid, S.M., Aebli, H.R. & Zingg, A. 1989: The role of the Periadriatic Line in the tectonic evolution of the Alps. In: Coward, M.P., Dietrich, D. & Park, R.G. (Eds.): Alpine Tectonics, Geological Society London, Special Publication 45, 153–171.

Schmid, S.M., Pfiffner, O.A., Froitzheim, N., Schönborn, G. & Kissling, E. 1996: Geophysical-geological transect and tectonic evolution of the Swiss-Italian Alps. Tectonics 15, 1036–1064.

Schmid, S.M., Berza, T., Diaconescu, V., Froitzheim, N. & Fügenschuh, B. 1998: Orogen-parallel extension in the Southern Carpathians. Tectonophysics 297, 209–228.

Schmid, S.M. & Kissling, E. 2000: The arc of the western Alps in the light of geophysical data on deep crustal structure. Tectonics 19, 62–85.

Schmid, S.M., Fügenschuh, B., Kissling, E. & Schuster, R. 2004a: Tectonic map and overall architecture of the Alpine orogen. Eclogae Geologicae Helvetiae 97, 93–117.

Schmid, S.M., Fügenschuh, B., Kissling, E. & Schuster, R. 2004b: TRANSMED Transects IV, V and VI: Three lithospheric transects across the Alps and their forelands. In: Cavazza, W., Roure, F., Stampfli, G.M. & Ziegler, P.A.

(Eds.): The TRANSMED Atlas: The Mediterranean Region from Crust to Mantle. Springer, Berlin, Heidelberg.

Schmid, S.M., Bernoulli, D., Fügenschuh, B., Matenco, L., Schuster, R., Schefer, S., Tischler, M. & Ustaszewski, K. 2008: The Alpine-Carpathian-Dinaridic orogenic system: correlation and evolution of tectonic units. Swiss Journal of Geosciences 101, 139–183.

Schönborn, G. 1992: Alpine tectonics and kinematic models of the central Southern Alps. Memorie di Scienze Geologiche, Istituti di Geologia e Mineralogia dell' Universita di Padova XLIV, 229–393.

Schönborn, G. 1999: Balancing cross sections with kinematic constraints: The Dolomites (northern Italy). Tectonics 18, 527–545.

Sclater, J., Royden, L., Horvath, F., Burchfiel, B., Semken, S. & Stegena, L. 1980: The formation of the intra-Carpathian basins as determined from subsidence data. Earth and Planetary Science Letters 51, 139–162.

Seghedi, I., Balintoni, I. & Szakács, A. 1998: Interplay of tectonics and neogene post-collisional magmatism in the intracarpathian region. Lithos 45, 483–497.

Sperner, B., Ratschbacher, L. & Nemcok, M. 2002: Interplay between subduction retreat and lateral extrusion: tectonics of the Western Carpathians. Tectonics 21(6), 1051, doi:10.1029/2001TC901028.

Sperner, B. & CRC 461 Team 2005: Monitoring of Slab Detatchment in the Carpathians. In: Wenzel F (Ed): Perspectives in modern Seismology. Lecture Notes in Earth Sciences 105, 187–202.

Stampfli, G.M. & Borel, G.D. 2004: The TRANSMED transects in space and time: constraints on the paleotectonic evolution of the Mediterranean domain. In: Cavazza, W., Roure, F.M., Spakman, W., Stampfli, G.M. & Ziegler, P.A. (Eds.): The TRANSMED Atlas: The Mediterranean Region from Crust to Mantle. Springer, Berlin and Heidelberg, 53–80.

Steininger, F.F. & Wessely, G. 1999: From the Tethyan Ocean to the Paratethys Sea: Oligocene to Neogene Stratigraphy, Paleogeography and Paleobiogeography of the Circum-Mediterranean region and the Oligocene to Neogene Basin evolution in Austria. Mitteilungen der Österreichischen Geologischen Gesellschaft 92, 95–116.

Stipp, M., Fügenschuh, B., Gromet, L.P., Stünitz, H. & Schmid, S.M. 2004: Contemporaneous plutonism and strike-slip faulting: A case study from the Tonale fault zone north of the Adamello pluton (Italian Alps). Tectonics 23: TC3004, doi:10.1029/2003TC001515.

Strauss, P., Wagreich, M., Decker, K. & Sachsenhofer, R.F. 2001: Tectonics and sedimentation in the Fohnsdorf-Seckau Basin (Miocene, Austria): from pull-apart basin to a half-graben. International Journal of Earth Sciences 90, 549–559.

Szafián, P., Tari, G., Horváth, F. & Cloetingh, S. 1999: Crustal structure of the Alpine-Pannonian transition zone: a combined seismic and gravity study. International Journal of Earth Sciences 88, 98–110.

Tari, G.C. 1996: Extreme crustal extension in the Rába River extensional corridor (Austria/Hungary). In: Decker, K. (Ed.): PANCARDI workshop 1996. Dynamics of the Pannonian-Carpathian-Dinaride system, Lindabrunn, Austria. Mitteilungen der Gesellschaft der Geologie- und Bergbaustudenten in Österreich 41, 1–17.

Tari, V. 2002: Evolution of the northern and western Dinarides: a tectonostratigraphic approach. EGU Stephan Mueller Special Publication Series 1, 223–236.

Tischler, M. 2005: A combined structural and sedimentological study of the Inner Carpathians at the northern rim of the Transylvanian Basin (North Romania). PhD Thesis, University Basel, Basel, 124 pp.

Tischler, M., Gröger, H., Fügenschuh, B. & Schmid, S.M. 2007: Miocene tectonics of the Maramures area (Northern Romania): implications for the Mid-Hungarian fault zone. International Journal of Earth Sciences 96, 473–496.

Tomljenović, B. 2000: Zagorje – Mid-Transdanubian zone. In: Pamić, J. & Tomljenović, B. (Eds.): PANCARDI 2000 Fieldtrip Guidebook, Dubrovnik. Vijesti Hrvatskog geološkog društva, 37/2, 27–33.

Tomljenović, B. 2002: Strukturne Znacajke Medvednice i Samoborskoj gorja (in Croatian, translated title: Structural characteristics of Medvednica and Samoborsko gorje Mts.). PhD Thesis, Zagreb University, Zagreb, 208 pp.

Tomljenović, B. & Csontos, L. 2001: Neogene-Quaternary structures in the border zone between Alps, Dinarides and Pannonian Basin (Hrvatsko Zagorje and Karlovac Basins, Croatia). International Journal of Earth Sciences 90, 560–578.

Tomljenović, B., Csontos, L., Márton, E. & Márton, P. 2008: Tectonic evolution of the northwestern Internal Dinarides as constrained by structures and rotation of Medvednica Mts., North Croatia. In: Siegesmund, S., Fügenschuh, B. & Froitzheim, N. (Eds.): Tectonic aspects of the Alpine-Dinaride-Carpathian System, Geological Society Special Publication 298, 145–168.

Ustaszewski, K., Schmid, S.M., Lugović, B., Schuster, R., Schaltegger, U., Bernoulli, D., Hottinger, L., Kounov, A., Fügenschuh, B. & Schefer, S. submitted: Late Cretaceous intra-oceanic magmatism in the Internal Dinarides (northern Bosnia and Herzegovina): Implications for the collision of the Adriatic and European plates. Lithos special issue "Balkan Ophiolites", A. Robertson, S. Karamata, K. Resimić-Šarić (Eds.).

van Hinsbergen, D., Hafkenscheid, E., Spakman, W., Meulenkamp, J.E. & Wortel, R. 2005: Nappe stacking resulting from subduction of oceanic and continental lithosphere below Greece. Geology 33, 325–328.

Viola, G., Mancktelow, N.S. & Seward, D. 2001: Late Oligocene-Neogene evolution of Europe-Adria collision: New structural and geochronological evidence from the Giudicarie fault system (Italian Eastern Alps). Tectonics 20, 999–1020.

von Blanckenburg, F. & Davies, J.H. 1995: Slab breakoff: A model for syncollisional magmatism and tectonics in the Alps. Tectonics 14: 120–131

von Blanckenburg, F., Villa, I.M., Baur, H., Morteani, G. & Steiger, R.H. 1989: Time calibration of a PT-path from the Western Tauern Window, Eastern Alps: the problem of closure temperatures. Contributions to Mineralogy and Petrology 101, 1–11.

Weber, J., Vrabec, M., Stopar, B., Pavlovčič-Prešeren, P. & Dixon, T. 2006: The PIVO-2003 Experiment: A GPS study of Istria Peninsula and Adria microplate motion, and active tectonics in Slovenia. In: Pinter, N., Grenerczy, G., Weber, J., Stein, S. & Medak, D. (Eds.): The Adria Microplate: GPS Geodesy, Tectonics and Hazards, NATO Science Series, Series IV: Earth and Environmental Sciences. Springer, 305–320.

Wilson, M. & Downes, H. 2006: Tertiary-Quaternary intra-plate magmatism in Europe and its relationship to mantle dynamics. In: Gee, D.G. & Stephenson, R.A. (Eds.): European Lithosphere Dynamics, Geological Society London Memoirs 32, 147–166.

Wortel, M.J.R. & Spakman, W. 2000: Subduction and slab detachment in the Mediterranean-Carpathian region. Science 290, 1910–1917.

Manuscript received March 3, 2008
Revision accepted June 6, 2008
Published Online first November 8, 2008
Editorial Handling: Nikolaus Froitzheim & Stefan Bucher

1661-8726/08/01S295-16
DOI 10.1007/s00015-008-1287-8
Birkhäuser Verlag, Basel, 2008

Swiss J. Geosci. 101 (2008) Supplement 1, S295–S310

The Rila-Pastra Normal Fault and multi-stage extensional unroofing in the Rila Mountains (SW Bulgaria)

CHRISTIAN TUECKMANTEL[1,2], SILKE SCHMIDT[1,3,*], MARKUS NEISEN[1], NEVEN GEORGIEV[4], THORSTEN J. NAGEL[1] & NIKOLAUS FROITZHEIM[1]

Key words: Extensional tectonics, Alpine orogeny, Rhodope Metamorphic Province, Rila-Pastra Normal Fault (RPNF), Southwest Bulgaria, Mylonite

ABSTRACT

The Rhodope Metamorphic Province represents the core of an Alpine orogen affected by strong syn- and postorogenic extension. We report evidence for multiple phases of extensional unroofing from the western border of the Rila Mountains in the lower Rila valley, SW Bulgaria. The most prominent structure is the Rila-Pastra Normal Fault (RPNF), a major extensional fault and shear zone of Eocene to Early Oligocene age. The fault zone includes, from base to top, mylonites, ultramylonites and cataclasites, indicating deformation under progressively decreasing temperature, from amphibolite-facies to low-temperature brittle deformation. It strikes E–W with a top-to-the-N- to NW-directed sense of shear. Basement rocks in the hanging wall and footwall both display amphibolite-facies conditions. The foliation of the hanging-wall gneisses, however, is discordantly cut by the fault, while the foliation of the footwall gneisses is seen to curve into parallelism with the fault when approaching it. Two ductile splays of the RPNF occur in the footwall, which

are subparallel to the foliation of the surrounding gneisses and merge laterally into the mylonites of the main fault zone. The concordance between the foliation in the footwall and the RPNF suggests that deformation and cooling in the footwall occurred simultaneously with extensional shearing, while the hanging-wall gneisses had already been exhumed previously. The RPNF is associated with thick deposits of an Early Oligocene, syntectonic breccia on top of its hanging wall. Integrating our results with previous studies, we distinguish the following stages of extensional faulting: (1) Late Cretaceous NW–SE extension (Gabrov Dol Detachment), exhumation of the present-day hanging wall of the RPNF; (2) Eocene to Early Oligocene NW–SE to N–S extension (RPNF); (3) Miocene to Pliocene E–W extension (Western Border Fault), formation of the Djerman Graben; (4) Holocene to recent N–S to NW–SE extension (Stob Fault), reactivating the SW part of the Western Border Fault.

Introduction

In the past decades, the geology of the Rhodope Metamorphic Province in southern Bulgaria and northern Greece has undergone a major reinterpretation. Earlier, the province had been regarded as a rigid continental block sandwiched between Alpine chains, the Balkan Mountains to the north and the Hellenides to the south (Kossmat 1924; Kober 1928; Hsü et al. 1977; Burchfiel 1980; Dercourt et al. 1986; Boncev 1988). Numerous structural and geochronological studies have now established the Alpine shaping of the Rhodope Province (Meyer 1966, 1968; Kronberg & Raith 1977; Ivanov 1988; Burg et al. 1996; Ricou et al. 1998; Mposkos & Krohe 2000; Liati 2005). The recognition of Alpine eclogites (Kolcheva et al. 1986; Liati 1988) and ultrahigh-pressure metamorphic rocks (Mposkos & Kostopoulos 2001; Perraki et al. 2006) further increased the interest in the geology of the Rhodope Province. Structural studies

established that crustal shortening and the related metamorphism were followed by extensional tectonics, leading to the formation of detachment faults and metamorphic core complexes (Bonev et al. 1995; Dinter 1998; Bonev et al. 2006). In spite of all this progress, key elements of the tectonic evolution of the area are still poorly understood. This applies to the pre-orogenic arrangement of continents and oceans, the timing and geometry of subduction and collision processes, and the respective roles of crustal shortening and extension in the formation of the major shear zones and tectonic boundaries.

The Rhodope Metamorphic Province comprises not only the Rhodope Mountains in a geographic sense but also the Rila and Pirin Mountains and, according to some authors (e.g. Ricou et al. 1998; Dinter 1998), also the southern part of the Serbomacedonian Massif. The province can be subdivided into two main tectonic units (Fig. 1). The lower unit, referred to as Pangaion-Pirin Complex (Ivanov et al. 2000), contains abun-

[1] Steinmann-Institut, Universität Bonn, Nußallee 8, 53115 Bonn, Germany.
[2] Present address: Rock Deformation Research, School of Earth and Environment, University of Leeds, Leeds, LS2 9JT, UK.
[3] Present address: Geologisch-Paläontologisches Institut, Universität Münster, Corrensstraße 24, 48149 Münster, Germany
[4] Department of Geology and Paleontology Sofia University St. Kliment Ohridski, 15 Tzar Osvoboditel Blvd., 1000 Sofia, Bulgaria.
[*] Corresponding author: S. Schmidt. E-mail: schmidts@uni-bonn.de

dant marble and Tertiary granitoid intrusions and experienced an Alpine regional metamorphic overprint, which in most areas did not exceed greenschist-facies conditions. It comprises the Pangaion Unit (Kronberg et al. 1970) and its extension into the Pirin Mountains in Bulgaria. The overlying units include both continental and oceanic rocks (gneisses, schists, marbles and metaophiolites) and are intruded by Alpine granitoids as well. We summarize these units as the Upper Complex instead of using the more detailed subdivision schemes of other au thors (e.g. Krohe & Mposkos 2002) because we do not know to which of the units defined by these authors the gneisses in our study area belong. The Upper Complex underwent Alpine regional amphibolite-facies metamorphism and contains several sites with eclogites and remnants of ultrahigh-pressure metamorphism (Mposkos & Kostopoulos 2001; Perraki et al. 2006). The timing of the high-pressure and ultrahigh-pressure metamorphic event(s) is controversial; geochronological data, interpreted to date high-pressure metamorphism, range from ca. 180 Ma to 42 Ma (Wawrzenitz & Mposkos 1997; Liati & Gebauer 1999; Mposkos & Krohe 2000; Liati et al. 2002; Liati 2005; Bauer et al. 2007).

The Rhodope Metamorphic Province is structurally overlain and framed by greenschist-facies, partly ophiolitic rocks, referred to as the Circum-Rhodope Belt in the older literature (Kaufmann et al. 1976; Kockel et al. 1977). However, these rocks do not represent a coherent tectonic unit. Greenschists to the southwest of the Rhodope Metamorphic Province belong to the Vardar Zone *sensu lato* (Fig. 1), a complex, Meso- to Cenozoic oceanic suture zone (Ricou et al. 1998). To the east, the Alexandropolis and Mandrica Greenschists represent an island arc and accretionary complex related to southward subduction during the Jurassic, emplaced from the south on top of the Upper Complex (Bonev & Stampfli 2003, 2008). Greenschists overlying the Rhodope Metamorphic Province to the northwest (Frolosh Greenschists), however, were formed from a Cadomian ophiolite complex (Haydoutov et al. 1992; Graf 2001) and do not represent an Alpine suture zone. The Frolosh Greenschists are intruded by the Struma Diorite Formation, which is latest Proterozoic in age (Graf 2001; Kounov 2002).

The formation of the Upper Complex is generally attributed to subduction and accretion since at least the Late Jurassic (Dinter 1998; Ricou et al. 1998; Bauer et al. 2007). The lower-grade units exposed in the Pangaion-Pirin Complex were overthrust by the Upper Complex after the exhumation of the latter from high-pressure conditions but possibly while high temperatures still prevailed (Mposkos & Krohe 2000; Krohe & Mposkos 2002). The present-day architecture of the Rhodope Metamorphic Province was significantly influenced by late and post-orogenic extension. According to Dinter (1998), the exposure of the Pangaion-Pirin Complex is related to a Miocene southwest-directed detachment fault, the Strymon Valley Detachment (Sokoutis et al. 1993; Fig. 1). Within and at the borders of the Upper Complex, prominent extensional structures of Alpine age have been identified as well, such as the detachment faults bounding the Kesebir-Kardamos Dome

in the Eastern Rhodopes (Bonev et al. 2006) or the Gabrov Dol Detachment to the northwest (Bonev et al. 1995; Fig. 1). In the Eastern Rhodopes, the occurrence of Eocene amphibolite-facies metamorphism closely below unmetamorphic Eocene sediments and volcanics documents the existence of significant unroofing faults (Krohe & Mposkos 2002).

As compared to the eastern, central, and southern parts of the Rhodope Metamorphic Province, the northwestern part (Rila Mountains and surroundings) is much less known in terms of structure and tectonic evolution. In order to start filling this gap, the present study focuses on the lower Rila valley located in the northwestern part of the Upper Complex (Figs. 1, 2, 3). This area is largely made up of various amphibolite-facies gneisses intruded by granitic bodies. The gneisses occupy a structurally high position within the Upper Complex according to Burg et al. (1996) and Ricou et al. (1998). The Alpine age of the metamorphism and the igneous bodies is generally accepted and bordering extensional detachment faults have already been proposed to explain unroofing of this complex (Bonev et al. 1995; Shipkova & Ivanov 2000). The Gabrov Dol Detachment dips shallowly northwest and separates the Upper Complex in the footwall from the Frolosh Greenschists and Struma Diorite Formation in the hanging wall (Bonev et al. 1995). The shear sense is top-to-the-northwest. The Gabrov Dol Detachment is crosscut by the ca. 73 Ma old Plana Pluton (Fig. 1; Boyadjiev 1981) and must therefore be Late Cretaceous in age or older (Ricou et al. 1998). Shipkova & Ivanov (2000) described the northwest-dipping Djerman Detachment Fault, an important, mylonitic to cataclastic, moderate- to low-angle normal fault. It forms the northwestern border of the Rila Mountains ca. 15 km north of our study area (Fig. 2).

Here we present evidence for large-scale normal faulting within the Rila Mountains (Figs. 2, 3, 4). We describe the Rila-Pastra Normal Fault (RPNF), a major extensional fault and shear zone which is younger than the Gabrov Dol Detachment, probably Eocene to Early Oligocene, and interferes with the latter in a complex way. The RPNF and the underlying mylonites display progressively decreasing, amphibolite-facies to sub-greenschist conditions. It is associated with thick deposits of an Early Oligocene (Cernjavska 2000), syntectonic breccia on top of the hanging wall, close to the normal fault (Padala Formation, Zagorchev et al. 1999; Fig. 3). The fault strikes E–W and shows a top-to-the-north- to northwest-directed shear sense. Even though basement rocks in the hanging wall and footwall both display amphibolite facies conditions, we will show that cooling in the footwall occurred simultaneously with shearing along the RPNF whereas the hanging-wall gneisses had already been exhumed earlier. Previous to our study, Shipkova & Ivanov (2000) found that the Padala Formation is underlain by a major normal fault. They connected this fault with the Djerman Detachment exposed further to the north (Fig. 2), which is not confirmed by our mapping. Westaway (2006) also describes and interprets the tectonics of the lower Rila Valley. He did not notice the zone of mylonites and cataclasites that forms the RPNF.

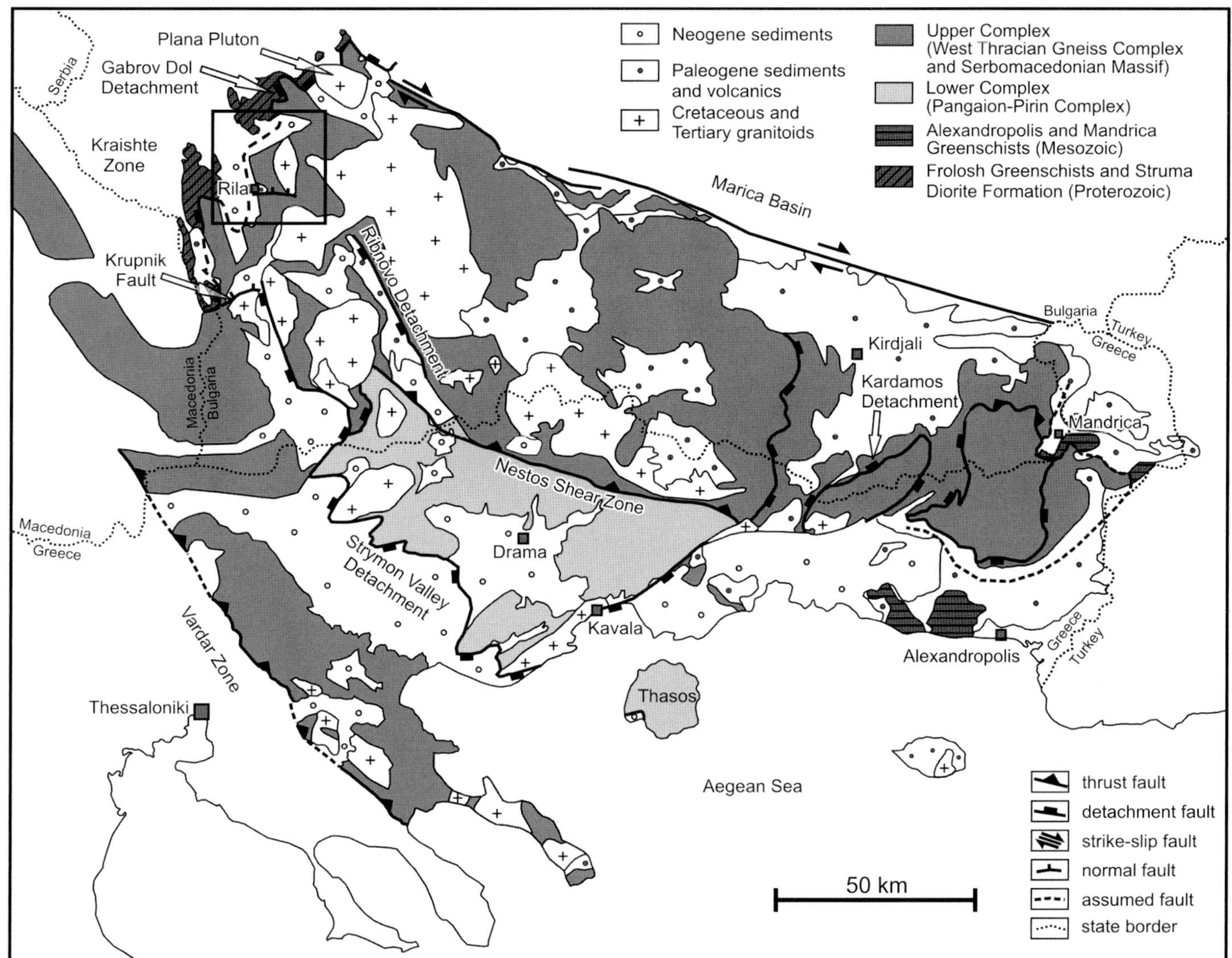

Fig. 1. Tectonic overview of the Rhodope Metamorphic Province, modified after Burg et al. 1996, Ricou et al. 1998 and Bonev et al. 2006. The box indicates the outline of Fig. 2.

In the following section, we describe the geological units in the study area in detail. After that, we present the structural record, especially features associated with the RPNF. Finally, we will reconstruct the extensional evolution of the northwestern-most Rhodope Metamorphic Province.

Geological edifice of the lower Rila valley

The lower Rila valley is dominated by amphibolite-facies gneisses (Gneiss Series in Fig. 3). They are bordered to the west by Neogene clastic sediments of the Djerman Graben. The associated west-dipping high-angle normal fault, termed Western Border Fault in the following, crops out only in few localities. In most places, Pleistocene alluvial-fan deposits cover the fault. The E–W-striking, mylonitic to cataclastic RPNF and two associated ductile splays in its footwall are located in the

southern part of the study area. The main fault crosses the Rila valley 3 km east of Rila town. Immediately north of the fault, an unmetamorphic sedimentary breccia, the Padala Formation (Figs. 4, 5a, b), rests on the hanging-wall basement. Towards the north, the breccia is bounded by the brittle, steeply dipping Padala Fault. In the west, the hanging wall of the RPNF comprises a diorite body that is fault-bounded on all sides and represents an extensional klippe of the Struma Diorite Formation on top of the Upper Complex (see below). Two granite bodies intruded into the hanging-wall Gneiss Series, the Kalin Granite to the northeast and the Badino Granite to the northwest.

Struma Diorite Formation

A diorite body is exposed around the eastern part of Rila town (Fig. 3). It consists of plagioclase and amphibole and is greenish

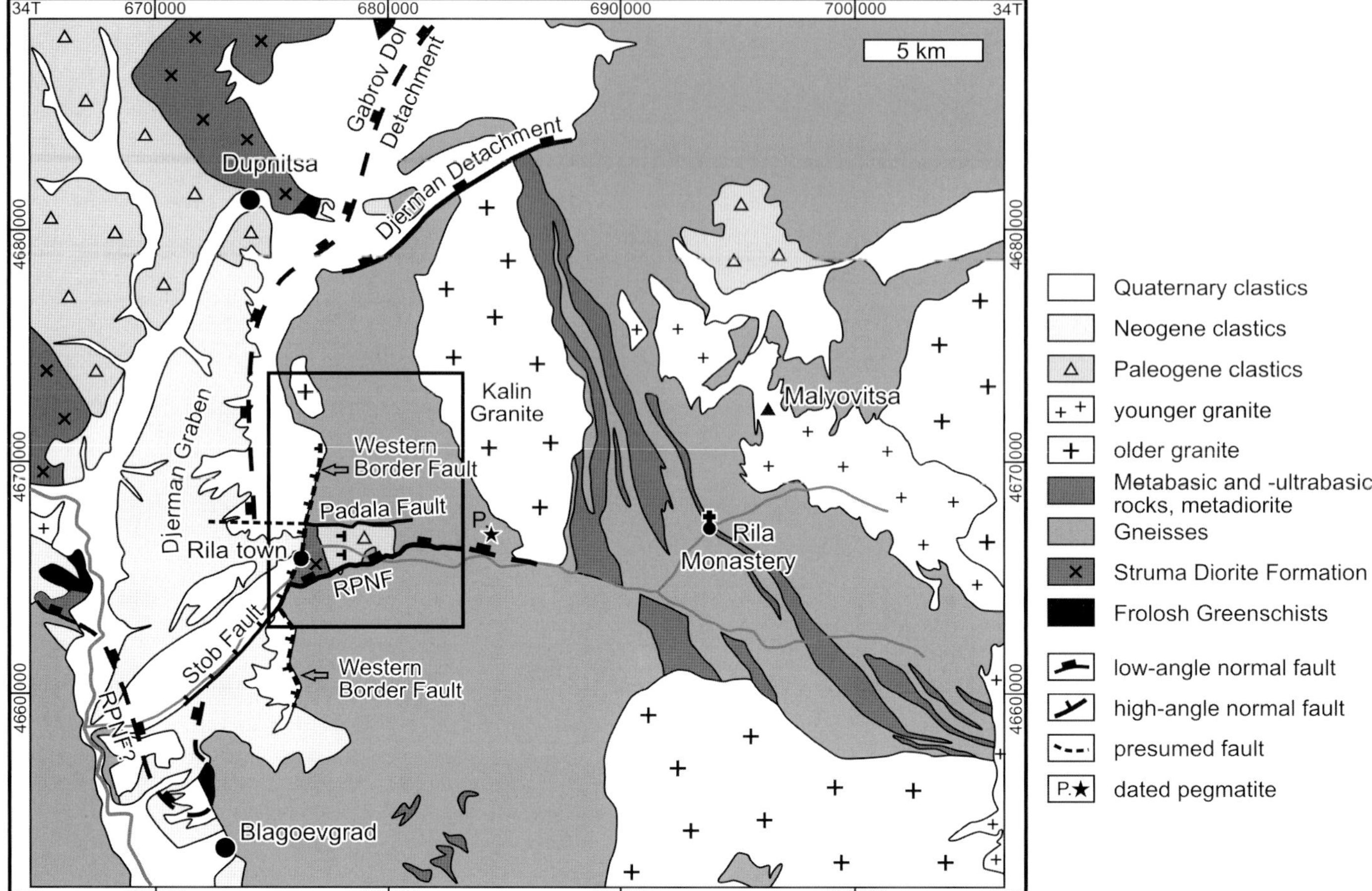

Fig. 2. Geological overview of the Rila Mountains, based on the geological map (1:100'000), sheet Blagoevgrad (Marinova & Zagorchev 1991). The box marks the outline of the study area (Fig. 3). UTM-Coordinates (WGS84) are given in units of meters.

in hand specimens. Additional mineral phases are biotite and clinopyroxene; the latter was partly replaced by amphibole. The diorite is crosscut by basalt dykes (Fig. 6 g) and thin epidote and quartz veins. The basalt has a glassy matrix and phenocrystals of plagioclase, clinopyroxene and some clinozoisite. Again, clinopyroxene was partly replaced by amphibole. Furthermore, a small fraction of the amphibole and clinopyroxene was replaced by prehnite, indicating metamorphism under prehnite-pumpellyite-facies conditions. This very low metamorphic overprint is in contrast to the amphibolite-facies metamorphism of the surrounding Gneiss Series. Towards the east, the diorite is separated from the sedimentary breccia by a subvertical layer of cataclasite and tectonic breccia (Fig. 3: Orlitsa Fault). These fault rocks are partly formed from the diorite, partly from a granite which is found only within these fault rocks. Granite fragments consist of quartz, microcline, plagioclase and chlorite. Biotite, muscovite and zircon occur in small quantities. Most of the biotite was replaced by chlorite.

The Struma Diorite Formation to which the diorite body at Rila belongs, was described by Stephanov & Dimitrov (1936) and Haydoutov et al. (1992). It comprises intrusive and vol-

canic rocks of dioritic and granitic composition, created in an island-arc setting (Haydoutov et al. 1992). The Struma Diorite intruded into the Frolosh Greenschists (Bonev et al. 1995; Zagorchev 2000). U-Pb zircon dating of the Struma Diorite Formation yielded ages of ca. 569 to ca. 544 Ma (Graf 2001; Kounov 2002).

Gneiss Series

The Gneiss Series is the main lithological unit of our working area. It comprises different ortho- and paragneisses with intercalated mica schists, amphibolites, foliated and unfoliated pegmatites, and unfoliated aplites. A distinction between ortho- and paragneisses in the field is difficult, but the occurrence of large feldspar porphyroclasts in most gneisses points to a preponderance of orthogneisses. In large parts of the working area locally developed leucosomes indicate beginning melting (Fig. 5f; see also Shipkova & Ivanov 2000). The gneisses consist of quartz, plagioclase, microcline, biotite, white mica and partly garnet in various proportions. Accessory minerals are apatite, zircon, clinozoisite, sphene and ilmenite. Abundant kyanite was found in three garnet-mica schist samples, two from the hanging wall (Fig. 6 h) and one from the footwall of the RPNF. In one sample from the hanging wall, kyanite is overgrown by staurolite, indicating conditions of the high amphibolite facies. East of the Kalin Granite, Kolcheva & Cherneva (1999) described kyanite- and staurolite-bearing metapelites. Amphibolites display an amphibole and plagioclase assemblage with accessory quartz, garnet, sphene, clinozoisite and epidote. In gneisses, mica schists and amphibolites outside the shear zones, garnet was partly transformed into biotite during the formation of the main foliation (Fig. 6 h). This indicates the breakdown of garnet and white mica during decompression from elevated pressures at amphibolite facies conditions.

Petrographic differences between the high temperature rocks in the hanging wall and footwall of the RPNF are limited. Footwall gneisses contain only little white mica and much biotite, whereas gneisses of the hanging wall contain more white mica. Units in the hanging wall close to the RPNF were altered penetratively by brittle deformation and contain abundant pegmatitic dykes (Fig. 5e). Furthermore, the variety of rock types seems to be larger in the hanging wall, with abundant garnet-mica schists, pegmatites and amphibolites besides gneisses. On the other hand, the footwall basement is dominated by slightly migmatized, biotite-feldspar gneisses. Undeformed pegmatites in the hanging wall of the RPNF (Fig. 2) were dated at ca. 63 Ma using K-Ar on muscovite (Boyadjiev & Lilov 1976).

Although the footwall and hanging-wall gneisses are petrographically similar, they exhibit pronounced structural differences. In the footwall of the RPNF, the foliation mostly dips northwest rather uniformly at angles of 35 to 45° (Fig. 7b). In the hanging wall, in contrast, the Gneiss Series shows very heterogenic dip angles and dip directions (Fig. 7a) and is affected by large-scale folds (Fig. 4). Such folds were not observed in the footwall. Pegmatites in the hanging wall often crosscut the foliation, whereas syn-tectonic, foliation-parallel migmatitic leucosomes and pegmatites characterize the footwall. However, older, foliated pegmatites occur in the hanging wall as well. Outside the mapped shear zones, stretching lineations are predominantly oriented N–S (Fig. 3, 7).

Granite intrusions

The margin of the Kalin Granite body is exposed in the northeastern part of the study area (Fig. 2, 3). The mineral assemblage of this granite is quartz, K-feldspar, plagioclase, biotite and epidote. Zircon and ilmenite are less abundant. Growth zoning and growth twins can be observed in the plagioclase grains, most of which are idiomorphic and contain epidote inclusions. Biotite has a dark brownish colour due to a high Ti content and is often transformed into chlorite. Most of the grains are about 2 mm in diameter. The Kalin Granite shows no evidence of ductile deformation and its contact in the study area crosscuts the foliation of the gneisses. This indicates post-tectonic intrusion with respect to the deformation in the hanging-wall gneisses.

The granite yielded K-Ar biotite ages between 54 and 42 Ma (Boyadjiev & Lilov 1976) and a U-Pb zircon age of ca. 46 Ma (Arnaudov et al. 1989).

The Badino Granite is a small intrusion in the northwestern part of the study area, exposed over an area of ca. 2.5 km². Most of its grains are about 0.5 mm in diameter, significantly smaller than in the Kalin Granite. The mineral assemblage is plagioclase, quartz, K-feldspar, biotite, and ilmenite. Biotite is dark greenish-brown and often contains ilmenite. Plagioclase grains are idiomorphic and have a clear growth zoning as well as growth (Albite and Karlsbader) and deformation twins. Most of the quartz grains show bulging and subgrain-rotation recrystallization. In contrast to the Kalin Granite, the Badino Granite shows a penetrative foliation often parallel to the contacts of the granite body and to the foliation in the adjacent Gneiss Series. Thus, it is not post-tectonic and may be older than the Kalin Granite. However, it is also possible that the ductile deformation lasted longer in this western part of the study area than in the east.

Padala Formation

The Padala Formation (Zagorchev et al. 1999) crops out to the east of Rila town, overlying the Gneiss Series. It is mostly a very poorly sorted and matrix-free breccia with angular to subrounded components ranging from sand size to blocks of several meters in diameter (Fig. 5b). In most places no bedding is visible. The breccia consists mainly of gneiss and mica schist components. Sporadically pegmatite, amphibolite, mylonitic gneiss and, near the boundary to the diorite, diorite fragments can be found. The mapping showed that the breccia was deposited on a non-planar, rugged surface with small valleys and ridges. The breccia is well lithified and forms rock towers and steep escarpments (Fig. 5a, b). In rare outcrops, up to 1 m thick greywacke layers can be found. In contrast to the unbedded breccia, these greywacke layers have bedding planes dipping gently to variable directions. Coal seams are also described from the Padala Formation (Zagorchev et al. 1999), which we did not find. The minimum thickness of the breccia is about 280 m.

The unsorted, matrix-free character of the breccia and the angular, up to several meters large clasts indicate deposition by rock avalanches and rockfalls. In contrast, the greywacke layers represent fluvial sediments. The dating of pollen and spores in the Padala Formation by Cernjavska (2000) established an Early Oligocene deposition age. Whereas the breccia rests on the hanging-wall gneisses with a depositional contact, it is separated from the footwall gneisses by the RPNF. An exception is an area south of the eastern end of the diorite, where the breccia rests on the footwall gneisses with a depositional contact (Fig. 3). This place is at a high level in the Padala Formation and, because of the overall shallowly-oriented layering, represents its youngest part. We therefore assume that the formation was mainly deposited during the activity of the RPNF (see below), its youngest layers postdating the activity.

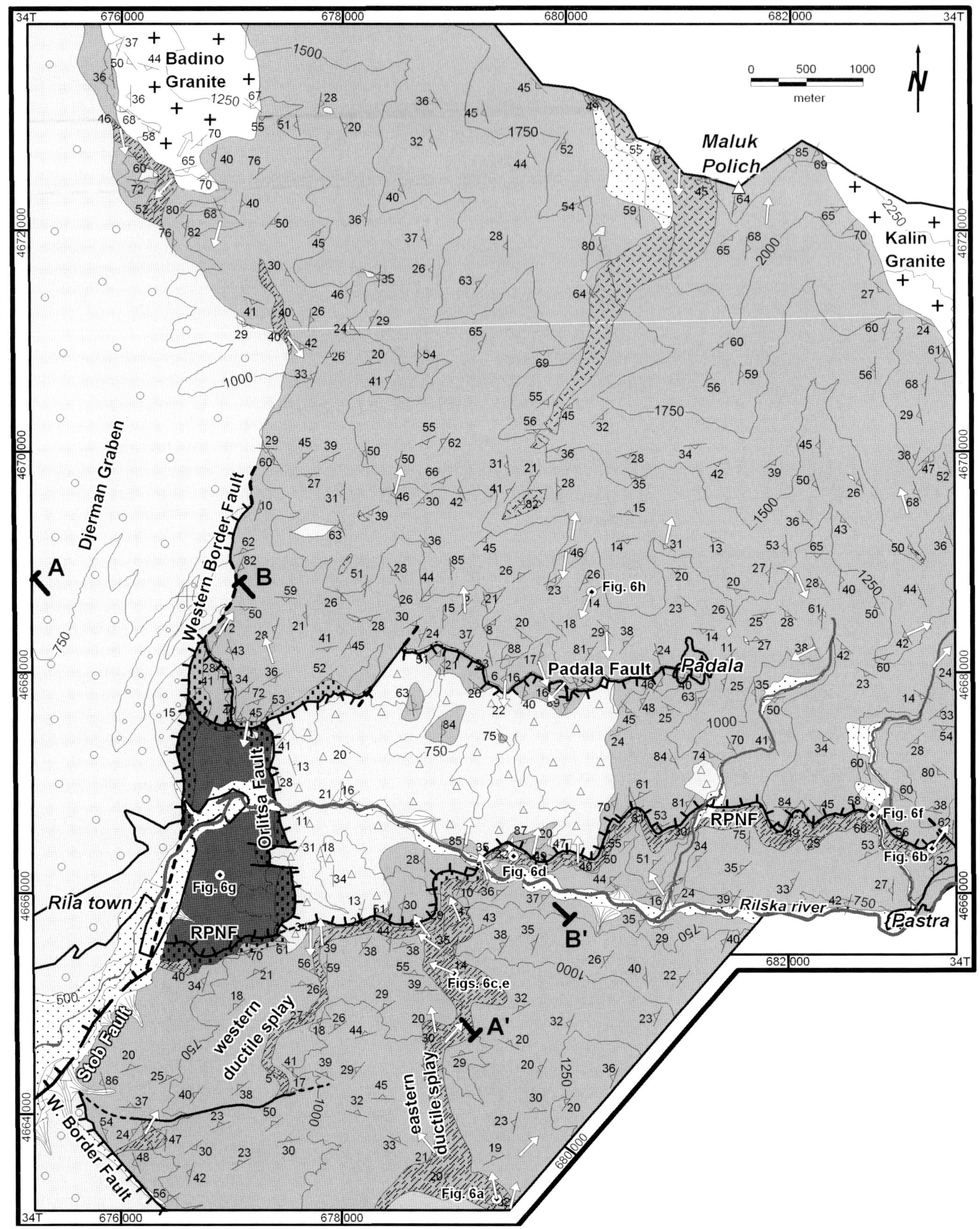

Fig. 3. Geological map of the study area. Senses of shear are indicated. UTM-coordinates (WGS84) are given in units of meters.

Neogene clastic sediments

The western part of the study area is covered by Neogene clastic sediments. Two units can be differentiated. The lower one is the Barakovo Formation of Miocene to Pliocene age (Zagorchev 1992). It is dominated by reddish sand with only few gravel components. The upper unit, the Badino Formation of Pleistocene age (Zagorchev 1992), is dominated by gravel and boulders of up to 2 m diameter. The components in both units are slightly to well rounded and comprise gneisses, mica schists, granites and diorites. Bedding is mostly subhorizontal. Near the western border of the exposed basement, bedding dips gently to the west. Both clastic units represent alluvial fan deposits filling the Djerman Graben.

Structures and tectonic evolution of the lower Rila valley

The study area records a long history of exhumation related to several generations of extensional shear zones and faults. The most prominent structure is the RPNF with its associated footwall mylonites. The older Gabrov Dol Detachment is not exposed in the study area, but most likely underlies the diorite. We suggest that the Padala Formation with a depositional contact covers its trace at the surface. Following the RPNF activity, the poorly exposed brittle Western Border Fault controlled the deposition of Neogene sediments to the west. Present day deformation takes place along the Stob Fault, reactivating a part of the Western Border Fault to the southwest of Rila town.

Legend of Fig. 3

Holocene		
Alluvial fan		
Badino Fm. (Pleistocene clastics)	Neogene	
Barakovo Fm. (Mio- & Pliocene clastics)		
Padala Formation	Early Oligocene	
Cataclastic diorite and granite	Struma Diorite Formation	
Struma Diorite		
Kalin Granite	Granites	
Badino Granite		
Cataclasite		
Mylonite		
Aplite	Gneiss Series (amphibolite facies)	
Amphibolite		
Gneiss and mica schist		

normal fault		cross section	
fault		foliation	
presumed fault		sense of shear	

Pre-Rila-Pastra-Normal-Fault structures;
Gabrov Dol Detachment

The foliation and the predominantly N–S-striking stretching lineation of the Gneiss Series in the hanging wall of the RPNF are probably the oldest preserved tectonic structures in the lower Rila valley. Although the foliation is pervasive, some zones are distinguished by their particularly strong, mylonitic deformation (Fig. 3). Thin sections parallel to the stretching lineation show both top-to-the-north and top-to-the-south shear senses. It appears that quartz was predominantly recrystallized by grain boundary migration in rocks that display a top-to-the-south sense of shear, and that top-to-the-north shearing is mainly associated with subgrain-rotation recrystallization of quartz. Since grain boundary migration occurs under higher temperatures than subgrain rotation (Stipp et al. 2002; Passchier & Trouw 2005), and the overprinting of deformation structures in individual thin sections generally indicates decreasing temperatures, it is likely that top-to-the-south shearing took place before the top-to-the-north shearing. These relations require, however, a more detailed study before they can be interpreted in terms of tectonic evolution. The shearing of the Gneiss Series in the hanging wall took place before the intrusion of the post-tectonic Kalin Granite (ca. 46 Ma, Arnoudov et al. 1989) and also before the intrusion of the undeformed pegmatites dated at ca. 63 Ma (Boyadjiev & Lilov 1976), that is, before the earliest Tertiary. Therefore the exhumation of the hanging-wall Gneiss Series is related to unroofing events prior to the formation of the RPNF. Possible unroofing faults are the Gabrov Dol and Djerman detachment faults described by Bonev et al. (1995) and Shipkova & Ivanov (2000), respectively.

The contrast in metamorphic grade between the Struma Diorite body at Rila town, which was overprinted under prehnite-pumpellyite facies conditions only, and the surrounding and underlying high-amphibolite-facies gneisses, requires important relative displacements along faults separating these units. To the south, this could be the RPNF. To the north, a similar contrast exists between the diorite and the gneisses in the hanging wall of the RPNF. At the surface they are only separated by a steeply dipping, brittle fault zone with a north-side-up sense of movement (Padala Fault, Figs. 3, 4), which cannot explain the contrast in metamorphism. As outlined above, north of the study area the Gabrov Dol Detachment separates the Struma Diorite Formation and Frolosh Greenschists in its hanging wall from the Upper Complex in its footwall (Bonev et al. 1995; Fig. 1). It is a major detachment fault with a northwestward to westward sense of shear that unroofed the Rhodope Metamorphic Province (Bonev et al. 1995). The Gabrov Dol Detachment was formed under greenschist facies conditions and active under progressively decreasing temperatures. The 73 Ma old Plana Pluton (Boyadjiev 1981) cuts the Gabrov Dol Detachment (Ricou et al. 1998; Fig. 1). Consequently, this detachment fault must also be older than the Early Oligocene Padala Formation in our study area.

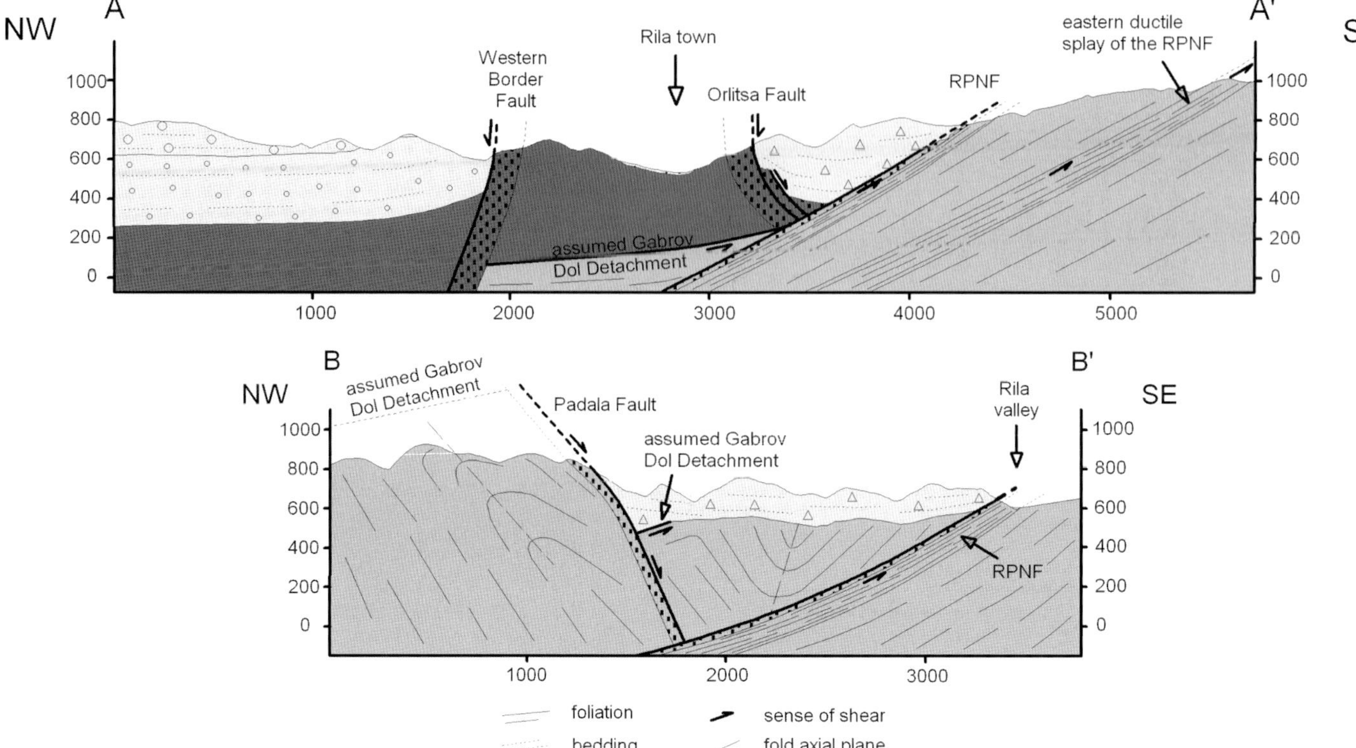

Fig. 4. NW–SE running cross sections AA' and BB' (for location see Fig. 3). Lithological signatures are the same as in Fig. 3.

By combining our observations with the results of Bonev et al. (1995) we suggest that the Gabrov Dol Detachment lies in a gently dipping orientation at the base of the diorite. East of the diorite, it reached the paleo-surface in the area now covered by the Padala Formation (Fig. 4). South of the diorite and the breccia, the Gabrov Dol Detachment was excised by the RPNF. North of the diorite it was cut by the Padala Fault.

Rila-Pastra Normal Fault (RPNF)

The RPNF crops out in the southern part of the study area. It is an E–W-striking brittle fault, underlain by cataclasites and mylonites (Fig. 4), and two associated ductile fault splays within its footwall. Both splays trend N–S to NE–SW and merge into the main fault zone to the north (Fig. 3).

The RPNF forms the boundary between the Struma Diorite Formation and the Gneiss Series in the western part of the study area. To the east, it represents the boundary between the Padala Formation and the Gneiss Series and still further east it lies within the Gneiss Series. The fault is made up of 10 to 30 m thick cataclasites throughout the study area, derived from diorite and gneiss in the west and derived from gneiss and mica schist in the east. In general, the cataclastic rocks show no foliation or lineation.

A mylonitic gneiss and mica schist layer of 30 to 140 m thickness underlies the cataclasites. The footwall splays are made up exclusively of mylonitic gneiss and mica schist. The mylonitic rocks exhibit a pronounced foliation and lineation (Figs. 5c, 5d, 6a-d, 7). The foliation of the mylonites directly below the RPNF strikes E–W and shallowly to moderately dips north, whereas the foliation of the ductile splays strikes more NE–SW and dips northwest. The mylonite lineation strikes NW–SE to NNE-SSW and mostly plunges gently to the NW to NNE (Fig. 7f). Shear-sense criteria at various scales show that shearing was consistently top-to-the-north to -northwest (Figs. 3, 5, 6) with only few exceptions, which can be explained by conjugate shear domains.

The mylonites show signs of deformation under high greenschist- to amphibolite-facies conditions and a variable degree of retrograde overprinting. Several samples show dynamic recrystallization of quartz by grain boundary migration, typical for high greenschist- to amphibolite-facies conditions (above 500 °C, Stipp et al. 2002), and only minor retrograde overprinting (Fig. 6a). In these samples, feldspar is often dynamically recrystallized, leading to core-and-mantle structure of porphyroclasts (Fig. 6b). In other samples, subgrain rotation is the dominant recrystallization mechanism in quartz (between 400 and 500 °C, Stipp et al. 2002). Bulging recrystallization, typical for the temperature range 280 to 400 °C (Stipp et al. 2002) is observed as well (Fig. 6e). Lower-greenschist-facies overprinting (chloritisation) is almost ubiquitous within mylonites of the RPNF itself, close to the cataclasites. The structurally uppermost mylonites are in some places ultramylonites with an extremely small grain size (Fig. 6d). The cataclastic fault rocks

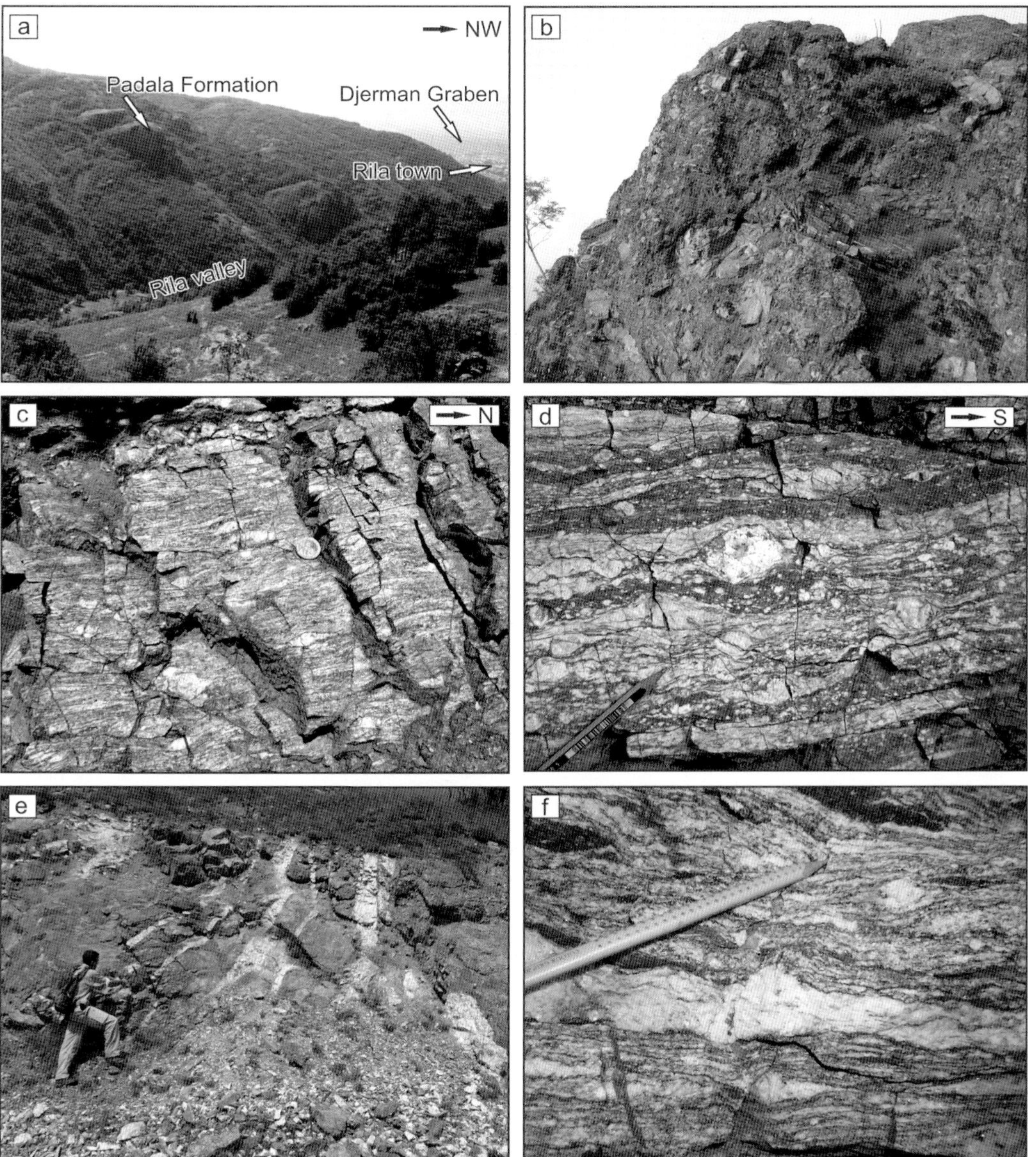

Fig. 5. Field pictures. a) Looking southwest across the lower Rila valley. These escarpments are built by the Padala Formation. Roughly southward-dipping bedding is visible. b) Escarpment of the Padala Formation. The components consist mainly of gneiss and mica schist fragments (34 T, r 677931, h 4666343). c) Top-to-the-north, greenschist-facies mylonite of the RPNF. The coin is 3 cm in diameter (34 T, r 679380, h 4666385). d) Top-to-the-north, greenschist-facies mylonite of the RPNF. Quartz flows around large feldspar porphyroclasts. Shear bands and sigma clasts show sinistral shear-sense (34 T, r 680121, h 4666272). e) Cataclastic gneiss with numerous pegmatite dykes in the hanging wall of the RPNF (34 T, r 680451, h 4667417). f) Migmatitic, biotite-rich orthogneiss in the footwall of the RPNF, 2 km west of Pastra village (34 T, r 680894, h 4665984).

are derived partly from mylonites, partly from hanging-wall rocks (diorite) and have a variable, generally small grain size (Fig. 6f). Thus, the RPNF shows a successive development from amphibolite-facies ductile flow to near-surface brittle deformation. The fault rocks reflect progressively lower temperatures towards the hanging wall, as is typical for extensional faults and shear zones. Hanging wall rocks are partly unmetamorphic to very-low-grade metamorphic (Struma Diorite Formation, Fig. 6 g), partly experienced similar conditions as the footwall rocks but at an earlier time (Gneiss Series, Fig. 6 h).

On a whole, the RPNF is oblique to the pervasive, amphibolite-facies foliation of the footwall (Fig. 3). However, when approaching the RPNF, the foliation of the footwall smoothly curves into parallelism with it. Moreover, the two ductile splays in the footwall are approximately parallel to the footwall foliation and merge upward into the mylonites underlying the RPNF. Therefore, the RPNF nowhere truncates structures of the footwall. We suggest that the pervasive foliation of the Gneiss Series in the footwall is related to the same extensional deformation that produced the RPNF. It de-

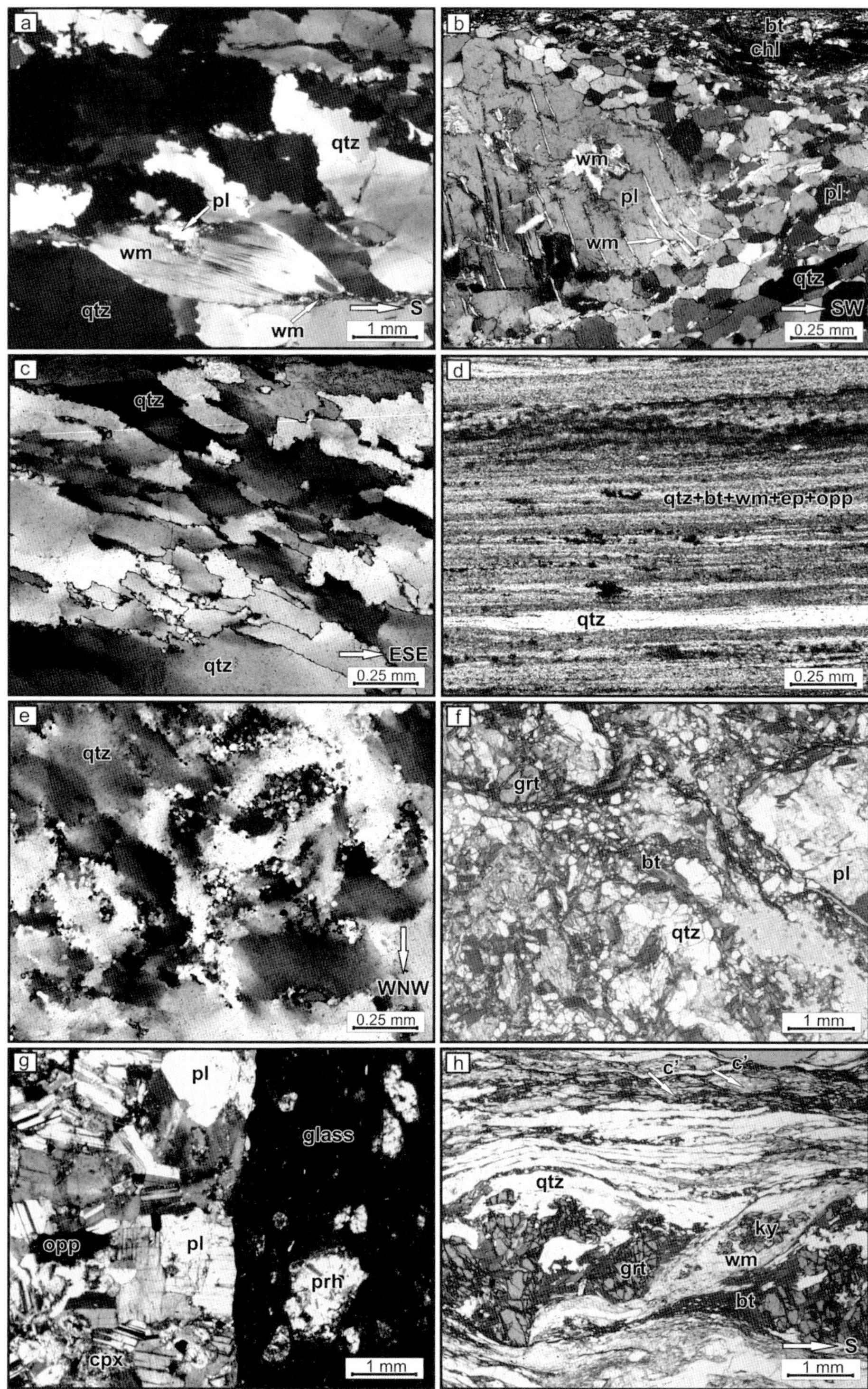

Fig. 6. Thin section micrographs of RPNF fault rocks and two samples from its hanging wall (g, h). From a) to f) the pictures show deformation under progressively colder conditions. For sample localities see Fig. 3. a,b,c,e: crossed polars; d,f,g,h: plane polarised light. Mineral abbreviations after Kretz (1983), wm = white mica, opp = opaque phase. If not mentioned otherwise the thin section shows the X-Z plane of strain parallel to the stretching lineation and normal to the foliation. a) Quartz recrystallized by grain boundary migration. The large mica fish shows a top-to-the-north sense of shear. b) Strongly mylonitized orthogneiss. Large plagioclase crystal surrounded by recrystallized plagioclase grains (core-mantle structure) showing triple junctions. c) Quartz recrystallized by subgrain rotation. Oblique foliation showing a top-to-the-WNW sense of shear. d) Ultramylonitic gneiss, containing few bright quartz-rich layers. e) Bulging recrystallization of quartz in an X-Y thin section. f) Cataclastic gneiss of the RPNF. g) Diorite of the Struma Diorite Formation (left) with basalt dyke (right) in the hanging wall of the RPNF. The magmatic texture is well preserved in both. Note the occurrence of prehnite in the basalt. h) Garnet-mica schist from the hanging wall of the RPNF, showing garnet partly transformed into biotite and a mica fish containing kyanite. Shear bands (parallel to the arrows) and sigma clasts indicate top-to-the-south sense of shear.

veloped before the extensional deformation was localized in relatively narrow shear zones. The different strike directions of the RPNF (E–W) and the footwall shear zones (NE–SW) suggest that during the development of the extensional fault system, the extension direction rotated from NW–SE to N–S.

This may also explain the large scatter of stretching lineations (Fig. 7f).

The Padala Formation was deposited on top of the RPNF hanging wall and consists of rock fragments derived from the exhumed footwall, except for minor diorite clasts near the

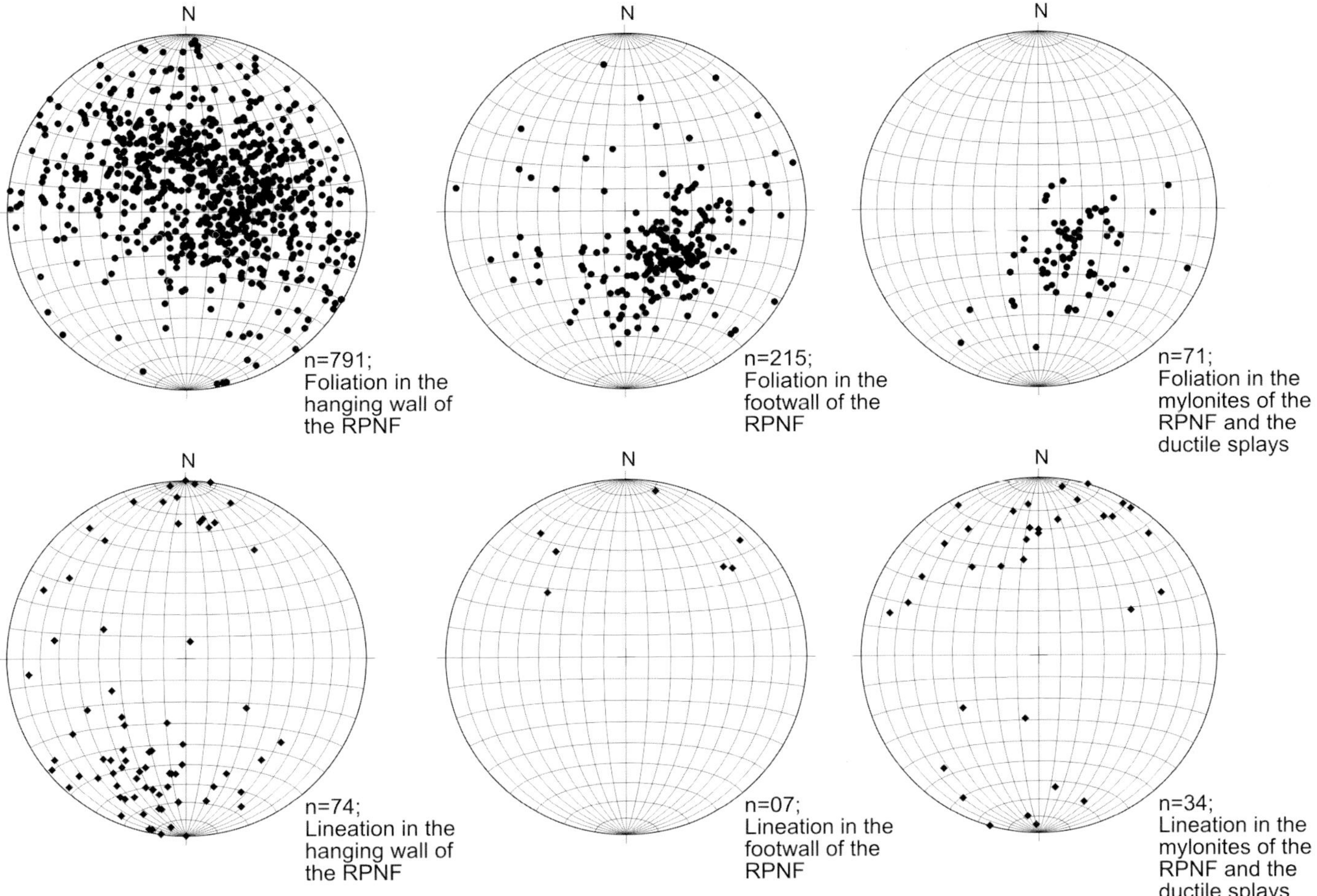

Fig. 7. Stereographic projections of measured foliations and lineations, lower hemisphere.

contact to the diorite. The onlap of the youngest parts of the breccia on the footwall southeast of Rila town indicates that the breccia deposition locally outlasted the activity of the fault. In the remaining area, the breccia is separated from the footwall by the fault. Therefore we assume that the breccia deposition was syn- to post-tectonic with respect to the RPNF. As the breccia was deposited in the Early Oligocene (Cernjavska 2000), we interpret the RPNF as an Eocene to Early Oligocene structure.

The brittle Padala Fault north of Rila town trends E–W and is marked by a zone of 10 to 30 m thick cataclastically deformed gneisses, mica schists and pegmatites. This fault defines the northern limit of the Padala Formation and the Struma Diorite Formation. The outcrop pattern indicates that the orientation of the fault changes significantly along its length. The dip direction varies from north- to southward and the dip angle from ~30 to 90°. This may be partly explained by recent slope instability. However, the map pattern indicates that the Padala Fault developed from several independent fault segments subsequently linked together to form the observed, variably oriented fault. We assume that these faults developed as steeply dipping nor-

mal faults, conjugate to the RPNF (Fig. 8). The Orlitsa Fault, i.e. the eastern, steeply dipping, N–S-striking border fault of the diorite against the Padala Formation (Fig. 3, 4), probably formed as an antithetic, hanging-wall-block-bounding fault of the Gabbrov Dol Detachment. It may have been reactivated during the activity of the RPNF and deposition of the Padala Formation.

Western Border Fault

Extension along the brittle Western Border Fault postdates the activity of the RPNF. This structure crops out in few locations along the border between the Neogene sediments and the Gneiss Series, but is mostly covered by sediments. The outcrop pattern indicates that the Western Border Fault strikes roughly N–S and dips steeply west. Where it is exposed (north of Rila town), the fault truncates the foliation in the gneisses, and is characterized by cohesionless fault breccia and gouge. We did not find any mylonites related to this fault. Sand and gravel of Pleistocene age (Badino Formation, Zagorchev 1992) cover parts of the fault, showing that the activity of the Western Border Fault mostly ended before the Pleistocene. It controlled the sedimentation of the Neogene clastics in the Djerman Graben.

Due to its completely brittle character and steep orientation, the Western Border Fault cannot be the continuation of the Djerman Detachment (Shipkova & Ivanov 2000), a moderately northwest-dipping, greenschist-facies mylonite zone overlain by cataclasites. We assume that the Western Border Fault truncates the Djerman Detachment north of our study area. Since the Djerman Detachment has a similar orientation as the RPNF, these two faults may be of the same age.

Stob Fault and active tectonics

The present-day deformation in southwestern Bulgaria as determined by GPS data and earthquake fault plane solutions (Kotzev et al. 2006) is still extensional, but with a N–S direction of extension, and forms the northernmost part of the Aegean extensional domain. Presently active normal faults mostly trend WNW–ESE and ENE–WSW (Tranos et al. 2006). The earthquake-generating Krupnik Fault (Fig. 1; Zagorchev 1970; Tranos et al. 2006) is located ca. 20 km south of our study area. Within the study area the active Stob Fault (Tranos et al. 2006) is the NE–SW striking part of the Western Border Fault to the southwest of Rila town, reactivated in the Holocene. It forms the northwestern border of the southernmost basement outcrops (Fig. 2, 3). Further southwest, the fault represents the boundary between relatively uplifted Neogene sediments to the southeast and Pleistocene to Holocene alluvial sediments of the Rilska River to the northwest. Tranos et al. (2006) estimated a Holocene vertical offset of ca. 1 m along the Stob Fault.

Discussion

Comparison with earlier tectonic studies of the Rila area

Shipkova & Ivanov (2000) described the northwest-dipping Djerman Detachment Fault that forms the northwestern border of the Rila Mountains ca. 15 km north of our study area. We have visited this area and agree with Shipkova & Ivanov (2000) that the Djerman structure is an important, mylonitic to cataclastic, moderate- to low-angle normal fault. They assumed that this detachment continues south along the western slope of the Rila Mountains, that is, where we mapped the Western Border Fault, turns east at Rila town, follows the northern border of the Struma Diorite body and the Padala Formation (our Padala Fault), curves around the eastern end of the outcrop area of this formation, and continues along the fault which we have described as the RPNF. Thus, they already described the mylonites and cataclasites of the RPNF in the Rila valley as belonging to an extensional detachment fault. This is confirmed by our work. However, we cannot confirm their assumption that the Djerman Detachment Fault

continues into the RPNF. We found no mylonites related to the Western Border Fault but just a steep zone of brittle deformation, neither did we find a mylonite zone along the Padala Fault. Instead, our mapping indicates that the RPNF extends east along the Rila valley to Pastra and probably further on.

Westaway (2006) also reports some observations from the profile in the Rila Valley. He "could see no evidence of mylonitization of the underlying basement" (below the Padala Formation) and therefore assumes that the basement / Padala Formation contact is an unconformity surface. This is true for the Rila valley road itself (Fig. 3), because there the Padala Formation rests on a thin layer of hanging-wall gneiss, but only some hundred meters away, on the small side road to Padala, mylonites and cataclasites directly underlying the Padala Formation are well exposed (e.g., Fig. 6d).

Prolongation of the Rila-Pastra Normal Fault towards the east and west

We followed the RPNF up to Pastra village, at the eastern border of the map area (Fig. 3). Its continuation further east is still speculative. The map pattern (Fig. 2) shows that the southern end of the Kalin Granite lies along the eastward extrapolation of the RPNF trace. If the fault in fact continues like this, the Kalin Granite should be older than at least the late activity stages of the RPNF, which is in line with the ages determined for the Kalin Granite (ca. 46 Ma; Arnaudov et al. 1989) and for the syn-RPNF Padala Formation (Early Oligocene, Cernjavska 2000). The large granite mass exposed in the southeast corner of the map area (Fig. 2) might then even represent a lower part of the Kalin Granite pluton, exhumed by top-to-the-northwest displacement along the RPNF. Southeast of the Kalin granite, the map pattern suggests a sinistral offset of metabasite layers by ca. 5 km. This offset appears quite small for the RPNF, if it continues in this direction. However, in view of the top-to-the-north to -northwest kinematics of the RPNF, this sinistral offset may just be the horizontal component of a much larger displacement. On the other hand, it is possible that the RPNF follows a different trace towards the east, and that the sinistral offset is caused by some other fault. Additional field and laboratory work is necessary to answer these questions. Towards the southwest, a low-angle normal fault contact between Frolosh Greenschists and gneisses near Blagoevgrad (Fig. 2) may represent the continuation of the RPNF.

Age of metamorphism in the hanging-wall gneisses

Since the Gabrov Dol Detachment is, according to Ricou et al. (1998), cut by the ca. 73 Ma Plana Pluton, it is probably Late Cretaceous in age or older. A Late Cretaceous age appears most likely because the detachment is nowhere sealed by strata older than Tertiary (Ricou et al. 1998). Having explained the contrast in metamorphism between the diorite and the Gneiss Series by the unroofing Gabrov Dol Detachment, we then have to assume that the metamorphism of the Gneiss Series north of the Rila Detachment is also Cretaceous in age (or, less likely, older). This is supported by the ca. 63 Ma age of the undeformed, post-metamorphic pegmatite (Boyadjiev & Lilov 1976), although this K-Ar muscovite age may not be very reliable. In contrast, the metamorphism in the footwall of the RPNF is probably Eocene in age, but not younger than Early Oligocene because the footwall rocks were at the surface in the Early Oligocene when they were locally covered by the Padala Formation.

Tectonic evolution of the western Rila Mountains

South of the Rila valley, the Strymon Valley Detachment was active in the Miocene (16 to 3.5 Ma) as a top-southwest, low-angle detachment fault (Dinter 1998). Thus, it is younger than the RPNF and localized at a deeper level, forming the top of the Pangaion-Pirin Complex. From the existing maps and our own observations, we assume that the Strymon Valley Detachment does not continue to the north as far as our study area, but that it looses displacement and dies, which may be explained by a vertical-axis relative rotation of the two fault blocks (Brun & Sokoutis 2007). The northward loss of displacement of the Strymon Valley Detachment is paralleled by the narrowing and final disappearance of the Pangaion-Pirin Complex, which was exhumed by the detachment (Fig. 2). The west-dipping Ribnovo Detachment Fault along the east side of the Mesta Graben (Burchfiel et al. 2003) is Late Eocene to Oligocene in age, similar to the RPNF. Taken together, three generations of important normal faults can be observed in the western part of the Rhodope Metamorphic Province: A Late Cretaceous one at the top of the Upper Complex (Gabrov Dol Detachment), an Eocene-Oligocene one within the Upper Complex (RPNF, Ribnovo Detachment), and a Miocene one at the base of the Upper Complex (Strymon Valley Detachment).

This situation is similar to the Alps where Late Cretaceous extensional detachment faults are found in the uppermost units, the Austroalpine nappes (e.g. Schlinig Fault, Froitzheim et al. 1997), Eocene to Oligocene detachments at an intermediate structural level (e.g. Turba Normal Fault, Nievergelt et al. 1996), and Miocene detachments at the deepest levels (e.g. Simplon Fault, Mancktelow 1985). It should be noted that the extensional faults in the uppermost units (Austroalpine) were active before these units were emplaced by thrusting on the lower structural levels. A similar situation may apply to the Western Rhodopes, where the emplacement of the Upper Complex on the Pangaion-Pirin Complex along the Nestos Thrust (Fig. 1) probably occurred in the Early Tertiary (Dinter 1998), postdating the activity of the Gabrov Dol Detachment. This suggests repeated changes between crustal shortening and extension during the evolution of the Rhodopes, as also observed in the Alps (Froitzheim et al. 1994; Beltrando et al. 2007).

The northwestern Rhodope Province shares several characteristics with the Eastern Rhodopes. The Alpine evolution of both areas was determined by extensional detachment fault systems that cut through the Upper Complex. Exhumation and subsequent cooling of detachment fault footwall units in the Eastern Rhodopes (Kardamos and Kechros domes) occurred between 55 and 35 Ma (Lips et al. 2000; Krohe & Mposkos 2002; Marchev et al. 2003). North of the Kardamos dome a coarse, syn-detachment breccia, comparable to the Padala Formation, crops out. Phytofossils in marl and clayey limestone beds within this breccia point to a Maastrichtian to Paleocene age (Bonev et al. 2006). This region was also affected by intrusions of granitoids, between 70 and 53 Ma in age (Ovtcharova et al. 2003; Marchev et al. 2004). An important difference is that the marine transgression in the Eocene, which took place in the Eastern Rhodopes during the Eocene, did probably not reach the Rila area. A further distinction is the lack of evidence for a high-pressure or ultrahigh-pressure metamorphic history in the northwest. However, this may be due to the fact that this region has been less well examined so far. It is possible that the Gneiss Series we described has been exhumed from deep structural levels, comparable to the metamorphic units in the Eastern Rhodopes.

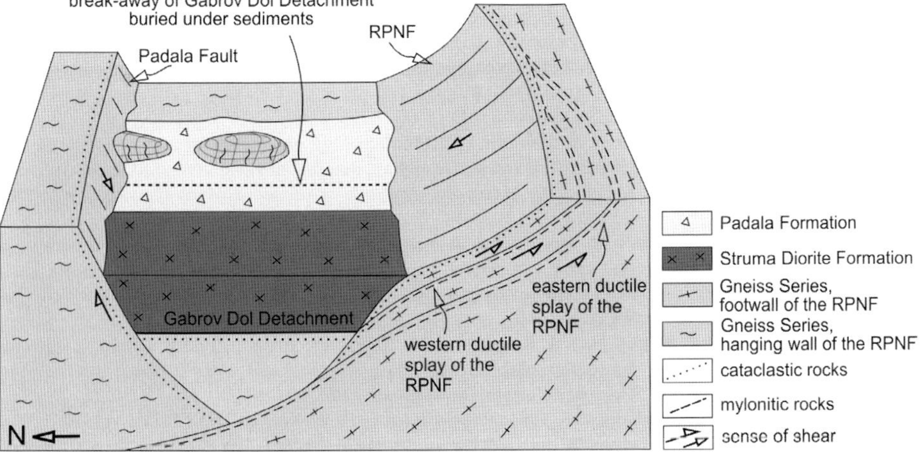

Fig. 8. Block diagram schematically showing the main tectonic features of the study area: The RPNF and its two ductile splays, the buried Gabrov Dol Detachment and the brittle Padala Fault.

Mposkos, E.D. & Kostopoulos, D.K. 2001: Diamond, former coesite and su-persilicic garnet in metasedimentary rocks from the Greek Rhodope: a new ultrahigh-pressure metamorphic province established. Earth and Planetary Science Letters 192(4), 497–506.

Nievergelt, P., Liniger, M., Froitzheim, N. & Ferreiro Mählmann, R. 1996: Early to mid Tertiary crustal extension in the Central Alps: The Turba Mylonite Zone (Eastern Switzerland). Tectonics 15(2), 329–340.

Ovtcharova, M., Quadt, A.V., Heinrich, C.A., Frank, M., Kaiser-Rohmeier, M., Peycheva, I. & Cherneva, Z. 2003: Triggering of hydrothermal ore min-eralization in the central Rhodopean core complex (Bulgaria) – insight from isotope and geochronological studies on tertiary magmatism and migmatisation. In: Eliopoulos, D.G. et al. (Eds.): Mineral exploration and sustainable development. Rotterdam, 367–370.

Passchier, C.W. & Trouw, R.A.J. (2005) Microtectonics. Springer-Verlag, Ber-lin, Heidelberg, 366 pp.

Perraki, M., Proyer, A., Mposkos, E., Kaindl, R. & Hoinkes, G. 2006: Raman micro-spectroscopy on diamond, graphite and other carbon polymorphs from the ultrahigh-pressure metamorphic Kimi Complex of the Rhodope Metamorphic Province, NE Greece. Earth and Planetary Science Letters 241(3–4), 672–685.

Ricou, L.-E., Burg, J.P., Godfriaux, I. & Ivanov, Z. 1998: Rhodope and Vardar: the metamorphic and the olistostromic paired belts related to the Creta-ceous subduction under Europe. Geodinamica Acta 11(6), 285–309.

Shipkova, K. & Ivanov, Z. 2000: The Djerman Detachment fault – An effect of the late Tertiary extension in the north-west part of the Rhodope Massif. Comptes rendus de l'Académie Bulgare des Sciences 53(2), 81–84.

Sokoutis, D., Brun, J.-P., Van den Driessche, J. & Pavlides, S. 1993: A major Oligo-Miocene detachment in southern Rhodope controlling north Ae-gean extension. Journal of the Geological Society (London) 150, 243–246.

Stephanov, A. & Dimitrov, Z. 1936: Recherches géologiques dans la région de Kjustendil (in Bulgarian). Review of the Bulgarian Geological Society 8(3), 1–28.

Stipp, M., Stunitz, H., Heilbronner, R. & Schmid, S.M. 2002: The eastern Tonale fault zone: a 'natural laboratory' for crystal plastic deformation of quartz over a temperature range from 250 to 700 °C. Journal of Structural Geol-ogy 24(12), 1861–1884.

Tranos, M.D., Karakostas, V.G., Papadimitriou, E.E., Kachev, V.N., Rangue-lov, B.K. & Gospodinov, D.K. 2006: Major active faults of SW Bulgaria: implications of their geometry, kinematics and the regional active stress regime. In: Robertson, A. H. F. & Mountrakis, D. (Eds.): Tectonic develop-ment of the Eastern Mediterranean region. Geological Society, London, Special Publication 260, 671–687.

Wawrzenitz, N. & Mposkos, E. 1997: First evidence for Lower Cretaceous HP/HT-Metamorphism in the Eastern Rhodope, North Aegean Re-gion, North-East Greece. European Journal of Mineralogy 9(3), 659–664.

Westaway, R. 2006: Late Cenozoic extension in SW Bulgaria: a synthesis. In: Robertson, A. H. F. & Mountrakis, D. (Eds.): Tectonic development of the Eastern Mediterranean region. Geological Society, London, Special Publication 260, 557–590.

Zagorchev, I.S. 1970: On the neotectonic movements in a part of South-West Bulgaria (in Bulgarian with French summary). Bulletin of the Geological Institute, Bulgarian Academy of Science 19, 141–152.

Zagorchev, I.S. 1992: Neotectonic development of the Strouma (Kraištid) Line-ament, southwest Bulgaria and northern Greece. Geological Magazine 129(2), 197–222.

Zagorchev, I.S. 2000: Rhodope and Vardar: the metamorphic and the olistos-tromic paired belts related to the Cretaceous subduction under Europe. Comment: Rhodope facts and Tethys self-delusions. Geodinamica Acta 1, 55–59.

Zagorchev, I.S., Goranov, A., Vulkov, V. & Boyanov, I. 1999: Palaeogene sedi-ments in the Padala graben, northwestern Rila Mountain, Bulgaria. Geo-logica Balcanica 29(3–4), 57–67.

Manuscript received 29 January, 2008
Revision accepted 9 May, 2008
Published Online first November 1, 2008
Editorial Handling: Stefan Schmid & Stefan Bucher